Understanding Body Shapes of Animals

Holger Preuschoft

Understanding Body Shapes of Animals

Shapes as Mechanical Constructions and Systems Moving on Minimal Energy Level

Springer

Holger Preuschoft
Bochum, Nordrhein-Westfalen, Germany

ISBN 978-3-030-27667-6 ISBN 978-3-030-27668-3 (eBook)
https://doi.org/10.1007/978-3-030-27668-3

This Springer imprint is published by the registered company Springer Nature Switzerland AG
The registered company address is: Gewerbestrasse 11, 6330 Cham, Switzerland

Preface

Why This Book?

It is a commonplace, indeed, to state that animals display divergent shapes. The eye-catching differences are located, aside from external coverage (moist or horny skin with or without osteoderms, feathers, pelt), in the locomotor apparatus and the jaw apparatus but not in those most essential parts like the central and peripheral nervous system; the sensory organs; the autonomous nervous system; the vascular system; the digestive apparatus, excluding the jaw apparatus in a wide sense (as mentioned above); the respiratory system; and the urogenital system.

All these organ systems follow their specific constructional principles but are integrated into the external body shape. The construction of these systems yields important aspects for the relationships between and among the animals, that is, for taxonomy, in particular on higher levels.

The diversity of vertebrates leaves us confronted with the old, big riddles of zoology:

- Why are the various shapes of animals connected with the modes of life?
- By which laws are shapes tied to these particular modes of life?
- What have more or less similar shapes to do with each other?
- How do divergent shapes develop?
- What makes animals at all different?

The last three questions are and have been, since a long time, the subject of intensive discussions. The focus of this book will be the first two questions.

The *locomotor apparatus* is built up by the skeleton and the muscles, both of them together with their respective structures of connective tissue, which are called fascia, ligaments or tendons. It is generally believed to be "*adapted*" to the *mode of life*. The latter is understood in this book as the complete behaviour comprising all acts which the animal performs. The most basic part of all behaviour is *body posture and locomotion*. Both require the input of energy and ultimately lead to the gain of energy. Essential for any sort of adaptation is a positive energy balance. Already, the development of the characteristics during ontogeny requires energy, which must not be neglected. The use of the characters also rests on the input of energy. If total energy required for the ontogenetic development and the use of a trait is more expensive than its advantages, it turns to become a selective disadvantage.

Clearly, the completely developed, "ripe" locomotor apparatus must be strong enough to resist the greatest of the regularly occurring forces, and it must be able to work at relatively low energetic costs. The highest and easily predictable forces come to action during locomotion, because the whole body mass comes into play. For the acquisition of food, high forces are required for catching prey or destruction of its body parts in the case of predators, for mechanical disintegration of resistant plant material and for breaking down hard or tough structural compounds of plants by chewing in the case of herbivores.

An explanation of the essence of such "adaptations" is by no means simple. Bock and v. Wahlert (1965) recommended strongly to discriminate between the "biological role", which a trait plays for the survival of an animal in a particular environment, and its mechanical function. The latter is the preferred subject of this book.

A long string of books exists aside from innumerable articles in scientific journals which promise a "functional" interpretation of morphological traits in animals. Recent examples are Hildebrand's and Goslow's *Analysis of Vertebrate Structure*; Carter and Beaupré, 2001; Hildebrand et al. 1985 or McNeill Alexander's monographs (1967, 1971, 1975, 1985, 2006). Without casting doubt on these publications (which I myself use often and esteem them highly), I would like to remark that they sometimes do not proceed on their way to a functional understanding as far as is possible. Several points seemed not satisfying for me, my coworkers and my students:

First, too often, no clear lines are drawn between "biological role", which a trait plays for improving the animal's chances of survival and to reproduce successfully, and its "mechanical function", both in the sense of Bock and v. Wahlert (1965). Admittedly, plausible assumptions about the value of traits are intermingled with really functional approaches (Biological role is essentially what Mayr, 1993, has termed "ultimate", mechanical role what he has called "proximate" causation.).

Second, a classical case is the common notion that the pelvis of australopithecines documents bipedal posture and locomotion. If, however, the reasons are sought for the general assumption that pelvic shape is an indicator of body posture, nothing can be found aside from the descriptive statement that the fossil pelvis is morphologically closer to that of humans than to that of living apes.

Third, the same line of argument is just being established for the feet of australopithecines. The feet of living apes are rather different in shape (and in the way they are used). Fossil remains show traits of modern humans as well as traits

---

My understanding of "Functional Morphology" is very similar to the definition of "Constructional Morphology", as given in Schmidt-Kittler and Vogel, 1991.

known to exist in apes. As a rough generalisation, it can be stated that they assume an intermediate position. But (nearly) nowhere the question is raised, and seriously discussed, which relationship in fact exists between foot shape and bipedality.

Fourth, quite frequently, elaborate and sophisticated statistical comparisons (often under consideration of allometry) are taken as substitutes instead of searching for the *reasons*. All statistical procedures, including allometry, are by their nature descriptive methods, which prove correlations beyond doubt, but do not explain, why they exist. A really reliable discussion of the physical mechanisms behind morphological traits is not even attempted.

Other books, for example, those of Blickahn, Carrier, Heglund, Hochmuth, Nachtigall, etc., are absolutely perfect with concern to physics – far better than what I can do! They are aimed at a presentation of biomechanical rules, and therefore difficult to understand for biologists, if not deterrent. The major interest of biologists, palaeontologists and anatomists (or those medical students and doctors who are not exclusively focussed on illness and treatment) is on a widely differing level of thinking. Books and presentations of the above type first of all intimidate the readers or the audience by their wealth of physics. As a result, most biologists give up (too early!) and believe that this sort of approach is just too complicated.

WHY, after all, THIS BOOK? Based on the selected examples, it intends to show the existence of really hard arguments (reasons), which can be pursued down to the laws of physics, for the development of particular morphological traits. These lines of thinking are sometimes very simple, so simple that some readers may prematurely think: "this is what we knew since Olim´s times"! (e.g. from Boeker, 1935). In spite of this criticism, I will prefer approaches which are based on unchallenging and well-known laws. Complicated methods will only be brought into play if absolutely necessary. In the sequence of their appearance in the fossil record, frequently occurring combinations of traits will be sketched, and reasons for these morphologies will be searched on the field of mechanics. The most preferred subjects are those which we have analysed ourselves. Most of them are quadrupedal animals; we have never worked on limbless lizards, snakes, amphisbaenids nor caecilians.

The considerations presented here do not aim at exact and detailed descriptions of morphology or behaviour but rather at combinations of basic behavioural acts and often existing morphological characters, which have existed during long periods, sometimes in various evolutionary lines. These morphologies are familiar to every biologist and can be seen in the collections of any museum. I also do not regard necessary to document in detail the modes of locomotion which can be seen every day (or at least every second day) in TV programmes. Definitely, I do not intend to rewrite evolutionary history. I just want to raise the question *why* the well-known and often observed changes have taken place – and to focus on searching nature-dictated answers in at least some cases.

On the basis of solid mechanical arguments, sometimes the *advantage* can be estimated quantitatively, which the possession of a particular trait includes for an animal, and the disadvantages or *expenses* connected with this trait. So our approach opens the possibility to identify the influence exerted by body size (mass or unidimensional measurements) and so the reasons for the increase or decrease of body size (Preuschoft 2010, in Kyoto). In the last decades, a considerable number of papers appeared in which traits are checked for their correlation with size indicators. For this purpose, commonly multivariate statistics and allometric regressions in double logarithmic coordinate systems are used. Their inclination and intercepts with the y-axis are easy to describe and to compare. Studies of this type are extremely worthful as a proof for the existence of evolutionary tendencies, but by no means, they can *identify the reasons* for size differences!

My general, basic hypothesis concerning the shapes of animals is the following: *all traits of the locomotor apparatus, which occur repeatedly in various groups, rest on the same mechanical conditions* – no matter whether they are classified as "analogies" or "homologies". It may be added that *all evolutionary changes must be developed on the basis of the morphological preconditions existing in the forerunner*. The reliability of these hypotheses will be checked in a number of cases. I hope that this book will help to develop better, more precise arguments for the evaluation of "selective advantages/disadvantages" of traits, which make it possible to discard vague and general, often subjective opinions and to replace them by precise calculations. I do not claim completeness of explanations. Instead, a selection of examples – hopefully of the best known locomotor groups – will be presented. The examples were selected on the basis of my personal knowledge and preferably of work done in my former working group "Functional Morphology".

On the background of the generally accepted theory of evolution (or given the correctness of the generally accepted theory), the *major question WHY* can be detailed: *What is the evolutionary advantage (or disadvantage) of a particular trait, and how big is the advantage (or disadvantage) derived from possession of a trait?*

The choice of animals for the following investigation is largely a matter of hazard and depends on what we have been working on.

The mechanical laws will be sketched and formulae will be provided where needed, simply to avoid placing too much emphasis on the physical and technical aspects.

Bochum, Nordrhein-Westfalen, Germany

Holger Preuschoft

Contents

Chapter 1
Introduction: General Considerations, Working Hypotheses, Questions, and Starting Points

Keywords Mechanically relevant tissues · Basic mechanics · Generally accepted results · Inductive approaches · Deductive approaches

The *confusing diversity of shapes* among animals has always been a problem. The first really compatible attempt to classify organisms, made by Linnaeus in the early eighteenth century, was based alone on (morphological) similarity. From the very beginning, not so much the general external shape or size (like in H. Melville's *Moby Dick*, where whales are subdivided in folio (that means sperm and toothless whales), octavo (grampus, orca), duodecimo (porpoises) etc. formates) was taken into consideration but the similarity of internal organisation. This soon became the dominating way to investigate animals. Cuvier (1831) already complaint about researchers who were more interested in sorting superficial characters of organisms for taxonomic purposes than in understanding them "physiologically" (cited from Maier 1984a, b).

In the context of the *theory of evolution* (Darwin 1859), the morphological classification of animals according to their similarity has received a depth of time and an inherent logic – which has immediately been seen (and misunderstood) as the *cause* for diversification. An arrangement of forms like in the famous evolutionary tree of Haeckel (1866, Fig. 1.1a) by no means can be set aside as historical reminiscence. Instead, the classification based on similarity has won new and actual interest with modern *cladistic approaches*. The present state-of-the-art "phylogenetic classification" according to Hennig (1982) also implies chronological depth as a background and as a means to control the traditionally accepted "taxonomy". The procedure to establish an arrangement has been greatly improved in comparison with the traditional mode. This becomes particularly evident if biometrical data (Fig. 1.1b), or the presence/absence of particular traits, are taken as a starting point. Exactly the same procedures can be applied if DNA units are taken as a basis (Fig. 1.1c). The distances between forms (genera, species, subspecies, populations) depend from the degree of similarity, and these values are relative units, which depend from the groups under consideration. If the probability of mutations is assumed to be constant during time, the (morphological or molecular) distance between the compared groups can be taken as proportional to the

H. Preuschoft, *Understanding Body Shapes of Animals*,
https://doi.org/10.1007/978-3-030-27668-3_1

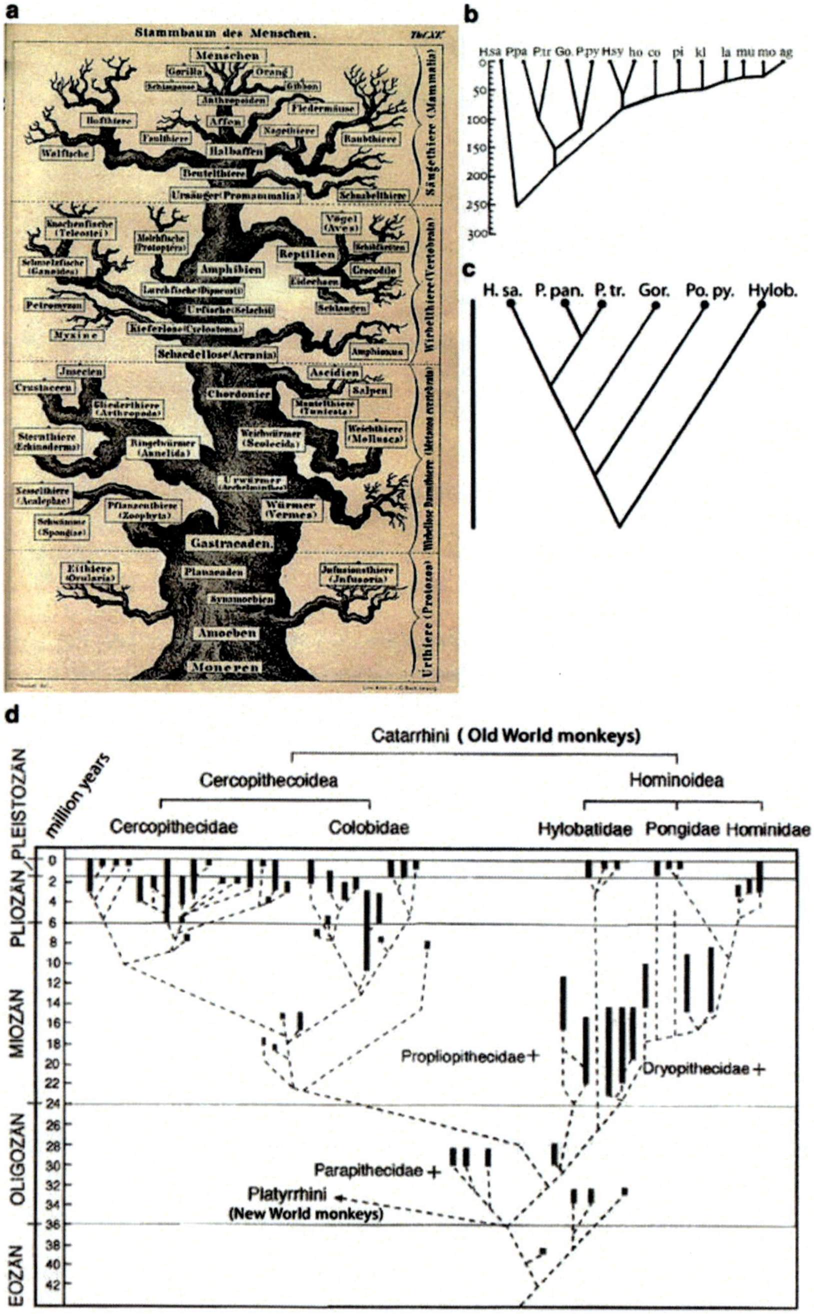

Fig. 1.1 Evolutionary trees and cladograms. (**a**) The animal kingdom with *H. sapiens* as the crown species, according to Haeckel (1866). The classes and orders of animals are arranged according to similarity of their internal structure and the families and genera according to external similarity. The relations between taxa rest on Darwin's theory. (**b**) Cladogram of hominoid primates as an example

chronological distance from the last common ancestor ("*molecular clock*"). When going his way, the "cladogram" wins depth in time and transforms into an evolutionary tree. If applied to morphological data, the weak points of this procedure are the choice and definitions of the "traits". What is a "trait" and, worse, how can the observer know, whether a trait is taken over from ancestors (plesiomorph) or newly acquired (apomorph)? This problem is known since considerable time: without independent information (like, e.g. stratigraphy), the direction to read evolutionary lines is a matter of hazard.

The causes of more or less pronounced similarity of forms and their development are searched by zoologists nearly exclusively with the aid of *phylogeny*. If the origin of a particular character in a particular form is investigated, the first thing to do is looking for the character state in ancestors and which transformations have occurred during time. The approach implies "homology" of the traits. The next step can be the question for which purpose or "function" the trait may be suited. An attempt to answer this question usually aims at the "biological role" in the sense of Bock and von Wahlert (1965), while its "mechanical function" is looked at thereafter.

The answers found on the basis of evolutionary theory seem to be conclusive to such an extent that virtually no space is left for other attempts to find different solutions. Surprisingly, however, such attempts are revived from time to time because of religious or political reasons (e.g. "creationism"). For several decades, the theory of recapitulation (ontogeny recapitulates phylogeny, Haeckel 1866) dominated the current opinions. Comparisons of ontogenetic stages with phylogenetic changes are sometimes drawn (especially in behavioural science, as by Nakano 2002) but rarely defended explicitly (Niemitz 2002, Maier 2014). Still, all these approaches are descriptive by nature and leave open the question WHY a trait has been acquired. Molecular analyses are of little use in solving these problems; they just transfer the problems from shape to genes.

Even if all rational arguments, as well as the (biological and physical) facts, which are known since the nineteenth century, are taken into consideration, it is difficult to comprehend why and how fundamental principles of body construction have diverged. Answers to these questions and generally valid rules are worked out in the "*Frankfurt critical evolutionary theory*", developed by the group working in the

Fig. 1.1 (continued) for a cladistic grouping. The genera, species and subspecies or populations are arranged according to their morphological similarity. In this special case, "similarity" is measured in the form of Cartesian coordinates of defined points at the skulls. The values along the vertical axis are relative, characterising alone the morphological distances between taxa, beginning with the smallest (*H. agilis* and *H. moloch* on the right side). (From Creel and Preuschoft 1984). The same or very similar statistical procedures can be – and are often – used to handle DNA sequences. Along the horizontal axis, forms are arranged according to their similarity with *H. sapiens*, and the distances along the vertical indicate taxonomic distance. In essential, cladogram **c** is similar to **b**, although the position of *Pongo* is different in **b** and **c**, where the hylobatids are treated as one homogenous group. (**d**) Under the assumption that the rate of mutations is statistically the same over millions of years and in all forms, the vertical (similarity-)axis can be seen as proportional to time, as a "molecular clock". So the cladogram is transformed into an evolutionary tree. (Modified from Szalay and Delson 1979)

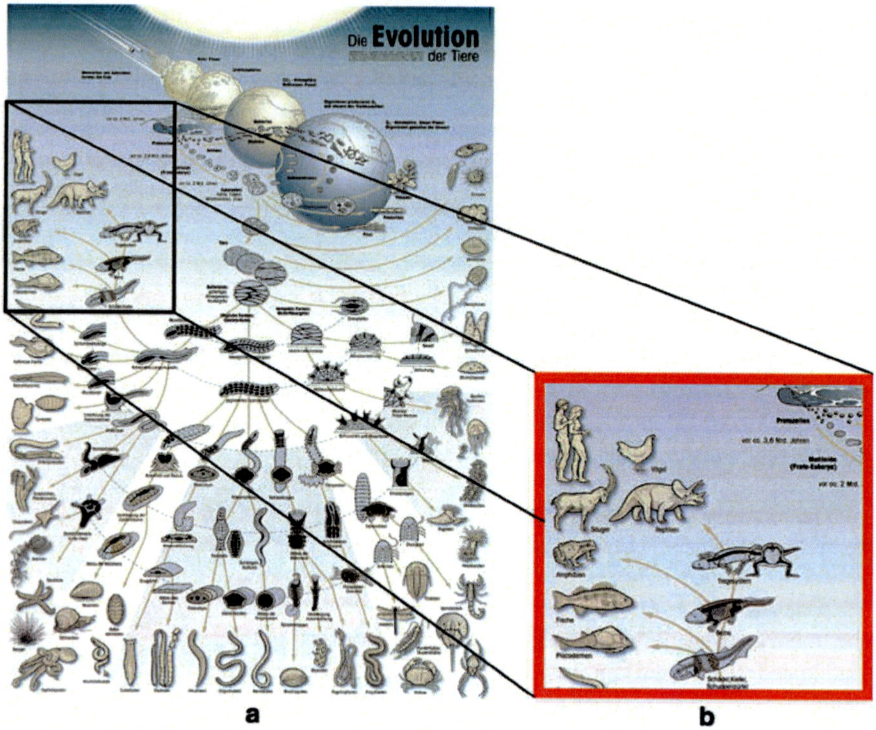

Fig. 1.2 The evolution of animals according to their constructions. The frame surrounds the vertebrates, on which the present book is focussed. (From Grasshoff and Gudo 2001). The basal position of *Trichoplax* and sponges as well as the Ctenophora has been confirmed by Adoutte et al. 1999, and Balavoine & Adoutte, 1998 using molecular biology

1970s through 1990s at the Senckenberg Museum (Fig. 1.2, summarised in Grasshoff and Gudo 2001, Gudo and Grasshoff 2002, Grasshoff 2014, there further literature). Characteristics of all evolutionary lines are principles of body construction. In the course of evolution, many generations of living organisms have developed widely different constructional principles, and this took place without any failure of the construction – which would have resulted in immediate extinction. New constructions can only develop on the basis of existing forms, which have solved the problems of survival. Only these successful constructional principles can be the fundament for new developments. The underlying arguments seem to be superficial (at least for dedicated empiricists), rather hypothetical but are in fact convincing, plausible and logical if seen from the viewpoint of researchers who ask the big question "why". The results lead to a grouping of living organisms which is in agreement with "new animal phylogeny", implying a basal position of the Ctenophora and independent evolution of *Trichoplax*, Cnidaria, sponges and bilateria. These items had been proposed early by the Frankfurt group and were confirmed by means of molecular genetics in several studies (Adoutte et al. 2000; Balavoine and Adoutte 1998, 2003).

A classification of diversity following alone externally visible shapes, however, sometimes camouflages deep-reaching differences of internal organisation or modes of life. For example, the well-known streamline shapes of rapid swimmers, with their dorsal fins, slender peduncles and sickle-shaped flukes, characterise animals which are very different with regard to other important characteristics: gill-breathing fishes have their shape in common with air-breathing reptiles and mammals, exotherm fishes and reptiles are similar to endotherm mammals, and oviparous fishes and reptiles possess the same external shapes as viviparous mammals. Another example for these so-called *analogies* or *convergences* can be found in the head, particularly in the jaws: slender jaws with pointed teeth are generally taken as indicators for predatory habits, while large contacts between upper and lower teeth are considered as reliable indicators for the intake and processing of hard or tough food materials. Analogous or convergent traits meet the same requirements of the animals' environments; therefore they are considered to be "adaptations" – where too often no precise concept exists to what exactly the traits are in fact "adapted". In many cases, the so-called adaptations are part of the locomotor apparatus in wide sense and therefore of interest for the present study.

In contrast to the above-mentioned "analogies", traits of common (evolutionary) origin are termed *homologies*. Among them are characteristics of reproduction (laying eggs versus giving birth to living young); breathing with the aid of gills versus lungs; and the divergent structure of lungs in reptiles, birds on one side and mammals on the other and the ability to maintain constant body temperature (endothermy) or to allow it to change with external, environmental temperature. The construction of most sensory organs is the same among all vertebrates; the general body construction with four extremities is the same; and the development of the skull on the basis of ossification centres or the construction of the jaw apparatus, including its musculature, is the same. The origins of homologies can be explained easily as common evolutionary heritage. The origins of *analogies*, however, deserve more detailed explanations. These will be the preferred topic of this book.

As stated above, the *locomotor apparatus* of vertebrates is taken as "adapted to" the animal's *mode of living*. Attempts to explain in detail the essentials of such an adaptation are often by no means simple but yield insight. It has been noted (Grasshoff, pers. comm.) that the concept of "being adapted" contains an allusion to teleology. Therefore, I prefer the more neutral terms "specialised or optimised for" or just note that a structure is "suited to sustain" the sort of loading connected with a behaviour. *Mode of living* in our context is seen as the sum of all behaviours. The most basic constituents of all behaviour are body posture and locomotion (the "postural behaviour" according to Prost 1967).

All traits entail *advantages* as well as *disadvantages*. The development of morphological traits requires the input of energy. If the energy requirement for the ontogenetic development does not pay, the possession of this trait becomes a selective disadvantage. Once developed, the locomotor apparatus must be suited to sustain the highest of the regularly occurring forces and must be expected to perform its functions on the lowest possible energetic level. The same holds true for the jaw apparatus and the acquisition and intake of food. Locomotion can be opposed to movements on the same spot (ideomotion). High and easy to predict forces, however, occur during locomotion, where the whole body mass comes into action, but the acquisition of food may also lead to very high forces. For each trait, the question must be raised: Which is the advantage connected with the possession of this trait? This question can be approached *qualitatively*, pursuing the problem, which conditions determine the necessity of this trait. Of course, the principles of physics have their relevance for living beings not less than for dead objects. As long as we are looking at the locomotor apparatus, advantages as well as disadvantages are on the field of mechanics. This discipline has identified the relevant facts and their relations clearly and in great precision. Therefore, biologists should learn and make use of the terminology of mechanics (and not try to invent new rules nor to introduce new terms for known facts). After a qualitative relation between a trait and the surrounding external or internal conditions has been found, we sometimes may be able to find out *quantitatively* how many % of energy can be saved by the possession of this trait (in comparison to nonpossession) or by how many % the chances to survive are improved if the trait is present.

Aside from advantages, the *disadvantages* also need attention. For example, the reinforcement of a skeletal element entails increased strength but also increased weight. The methods of mechanics allow to determine the gain of strength and compare it to the increase of weight. It must be expected that the adaptation to increased stresses is limited on one side by the increasing weight and on the other by the forces determined by weight. An "adaptation" to extreme but rarely occurring and not foreseeable stresses as in accidents cannot be expected, because it would make the animal's body heavier than compatible with every day conditions. The consumption of energy for the development of body structures would be greater, as well as for the maintenance of body shape and for locomotion. Therefore, each analysis of advantages of a trait must be supplemented by an analysis of the accompanying disadvantages.

All this raises the hen and egg question: what was first, morphology, mode of living or the environment? If shape was first, this can be called a "preadaptation", although this term implies the idea that evolution is an aimed process. The relation between factors such as shape, behaviour and environment has been depicted by M.E. Morbeck (1979) as the "Pretzel of Life" (Fig. 1.3). You can start wherever you

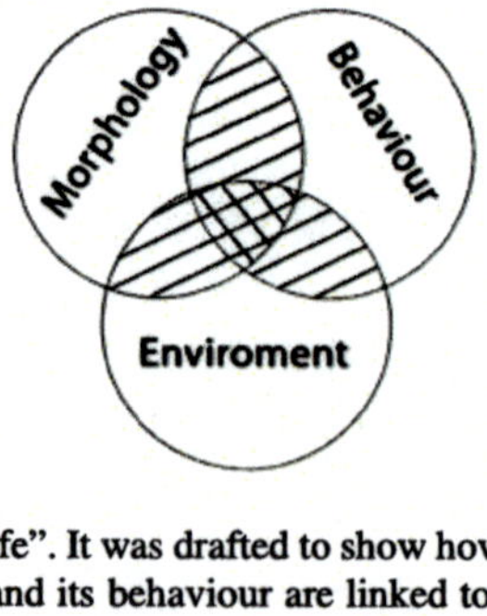

Fig. 1.3 Morbeck's "Pretzel of Life". It was drafted to show how morphology of an animal's body and its surrounding environment and its behaviour are linked to each other. No matter where you begin, you have to take into consideration the other two. The question raised in this book is: In which way are the complexes tied to each other? (Modified from Gomberg et al. 1979)

want, with the segment of environment used by a special animal, with its behaviour within this environment or with its shape, in any case you will arrive at a connection to other complexes.

In the present book, the starting point is always the visible *shape of an animal*, or a group of animals, and a short look at its (their) evolutionary history. This leads to the questions:

1. *What does the animal do with its body?* In other words, what is his behaviour in the widest possible sense?
2. *How exactly does the animal use its limbs, its jaws and its body stem?* This implies detailed knowledge of sequence of movements ("kinematics") during locomotion in its particular environment and as well during acquisition and processing of food. It also includes the extremely informative force plate and EMG measurements. This point also touches the "proximate" function of the body. The next step is analytical and based on mechanical theory.
3. *Which stresses are evoked in the segments?* (Entrance into biomechanics: analysis of segments, especially the skeletal elements, under static conditions; finally analysis under the conditions of kinetics.)
4. *How do the parts fit to the above determined stresses?* This is the core of "functional morphology". First the static and later the more complex kinetic loads are compared with the strength properties, found empirically or estimated theoretically from the shapes of the segments.
5. *Which advantages obtains the animal from possession of a trait and the use made of it for survival?* (How does an organism profit from possession of the trait under consideration? (Preuschoft et al. 1991) This is the ultimate question for the biological role of a trait in a special environment. In optimal cases, the answers found may be quantitative, not alone qualitative.
6. *Which process leads to "adapted" or "optimised" shapes?*
7. *Does the general hypothesis hold true,* stating that all traits of the locomotor apparatus which animals have in common are based on biomechanical conditions?

Questions 1 through 3 can be answered alone by an *empirical, inductive approach*, which is so highly esteemed in biology. Freely moving animals must be observed, and details must be recorded (sometimes under experimental conditions) to find out exactly what the animal does and in which way it moves and exposes its parts to loads, as well as which muscles are active.[1] Point 4 can also be determined

[1]Reliable and detailed knowledge about locomotion can be obtained alone from living and essentially unrestricted animals. Locomotor habits may vary considerably within related taxa, and there is no convincing relationship between the locomotor modes used by a species and its taxonomic position. This observation is not in conflict with the fact that in each evolutionary line only those traits can be "adapted", "optimised" or "specialised" which exist in the representatives of this line. Any evolutionary change can take place alone on the basis of the existing structures. Especially marked examples will be investigated in the following. This investigation will include the factors body size, body shape and locomotor mode.

empirically – BUT measuring stresses is not possible without invasive methods. There are many cases, in which these methods can by no means be applied, in others a more or less massive drawback of movements cannot be avoided, so that the results obtained do not represent the naturalistic motion. Finally, the well-being, or even the survival, of the experimental animal is on risk. In all cases the expenses for experiments are immense. Experience shows that "theoretical" calculations of internal forces (stresses) without experimental overkill in the majority of cases lead to fair results. The methods developed by engineering sciences provide the framework for this approach and the necessary formulae. The strength properties of biological materials have been investigated experimentally; for our purposes, values like those mentioned below under the headline "biological materials" can be used as constants – as long as no better information is available. Points 5 and 6 require the not highly ranked and therefore rarely used *deductive approach*. As a justification, we have to keep in mind that the starting points of our deductions have the reliability of physical laws – what cannot be claimed for most other biological rules, or "laws". For question 6, there exist two well-known possible answers:

First occurrence of undirected mutations, out of which natural selection eliminates those, which are less fit and therefore produce fewer offspring.
Second "functional morphogenesis" in the sense of F. Pauwels 1965, often termed "Wolff's law".

The 1st option is what every biologist thinks first, it represents the most often proposed explanation. The second option is considered rarely among biologists. In medicine, however, the rules of "functional adaptation intravitam" are fundamental for any therapeutic intervention in surgery and orthopaedics. Each healed fracture, each correction of a malformation, proves the correctness and validity of the rules. In the following text and in Chap. 10 "Open Questions", these points will be touched repeatedly, but are not the focus of interest.

This book does neither pretend to deal with all riddles of animal morphology, nor does it claim to present the only true, nor complete, solutions. Therfore, no attempt is made to answer question 7 directly, but the present examples speak for themselves.

The choice of animals for the following investigation is largely a matter of hazard and depends on what we have been working on. The mechanical laws will be sketched, and formulae will be provided where needed, simply to avoid placing too much emphasis on the physical aspects.

In order to direct emphasis away from the individual case, I prefer the use of vernacular names of animals. Only if necessary, the scientific names will be used. Dispute over taxonomic items, especially its most recent fluctuations, is by no means intended. In contrast, when talking about anatomical details, I prefer to use the correct anatomical denominations rather than colloquial English names. The Nomina Anatomica presents an elaborate list of precise terms. Likewise, the correct denominations of mechanical facts may sound unusual and repetitive, but again, they precisely pin down what is mentioned and so minimise misunderstanding.

1.1 Basic Facts and Accepted Results

1.1.1 Tissues

In the *tissues* which constitute the locomotor apparatus, the intercellular matrix often plays an essential role. Its total volume sometimes is bigger than that of the cells, and the mechanical properties of tissues are determined by the composition of this matrix. In all cases, great amounts of collagenous fibres are present.

Bone material, which forms most of the skeleton, can sustain compression. It is present in the form of compact bone (like in the walls of long bones) but also as spongy substance and divided into slender trabeculae, like in the vertebral bodies, in the tarsal and carpal bones or in long bones near the joints. This substantia spongiosa is also present between thin external plates in the scapula or in the pelvis. Bone is a classical compound material. Its matrix is heavily mineralised, and its content of about 45–60% mineral substance (calcium-hydroxyapatite) gives it the characteristic rigidity. It also contains lots of collagenous fibres, primarily of type I. The cells are located in lacunae left open by the matrix and are connected by a meshwork of thin processes. Depending from the arrangement of fibres in the matrix, we can discriminate between woven bone, which is produced early in ontogeny and during healing processes, but replaced in most vertebrates by the stronger fibro-lamellar bone. In the latter, the fibres are arranged in thin lamellae parallel to the surface of the bony element, or in the Haversian systems, or osteons. In these secondary structures, the fibres form concentric layers wrapped around central capillaries. The fibres are spirally wound around the centre. The slope of the contra-rotating fibres seems to depend on the dominating stresses: low slope at places where compression is particularly high and very steep slope at places where tensile forces may occur.

Long bones usually have tubular, hollow cross sections. The central cavity is filled with fatty marrow and blood vessels. The filling material seems to contribute some 10% to the strength of the whole bones, but this influence is here not taken into consideration. More attention deserves the tubular shape of these skeletal elements: If the amount of bone material necessary for sustaining the usual compressive loads is arranged along the walls of a tube, it can sustain forces which deviate from the bone's long axis. Because bending strength depends on

$$\sigma_{\max} = \Delta A * y^2, \tag{1.1}$$

where ΔA is a small unit of the cross-sectional area and y the distance from the neutral fibre, the arrangement of bone material along the walls of a tube is an efficient measure to increase the strength against excentric loads, which seem to occur under normal conditions and provide additional safety against accidental loads. The strength of compact bone material varies within one element (as emphasised by Amtmann 1971) around 2000 N/cm^2 against compression and about the same size order against tension. Other authors have reported different (mostly lower) values; Wainwright et al. (1976) list 1700–220 N/cm^2 under compressive and 1840–2000 N/m^2

under tensile loads. For spongy bone substance, values of about 600 N/cm^2 have been found, but without consideration of the direction of trabeculae in the tested sample. Bone is a rather stiff material, with *Young's modulus* of about 180.5 N/cm^2 under compression and 17.2 GN/m^2 under tension.

Cartilage is also a compression-resistant material which contributes to the formation of the skeleton: in mammals less, in reptiles more and in sharks, skates and rays, as well as chimaeras, the whole skeleton is built of this material. In young embryos, the skeleton is mostly preformed by hyalinous cartilage, which is replaced during growth gradually until not more than the joint condyles (in reptiles and in subadult birds and mammals) or only the contact areas between the articulating bones (in adult birds and mammals) persist in cartilaginous form. An exception are the most ventral parts of the ribs, which persist even in mammals as cartilaginous parts of the skeleton.

Cells are more or less round and spun into a dense meshwork of fibrous scleroproteins. The amorphous intercellular matrix consists of 60–70% water with proteoglycans, which are responsible for the marked tendency to expand by inclusion of water. Wainwright et al. (1976) compare cartilage with a hydrostatic system, "in which the fluid element is provided by the hydration water of the proteoglycan gel and the container provided by a collagen meshwork which immobilises the molecules of this gel. Thus, the rigidity or turgor of this hydrostatic system arises from the osmotic swelling of the proteoglycan gel against the constraints imposed by the collagen fibre swelling". If loaded constantly over periods of several minutes, cartilage slowly gives way, because of a slow loss of water, which is squeezed out of the matrix (Wainwright et al. 1976). The swelling after removal of the load is limited alone by the fibres (mostly of type II), which are ubiquitous in the matrix, especially surrounding the rounded cells. Hyalinous cartilage under realistic conditions can sustain nothing than compressive forces, if distributed over large areas, preferably hydrostatic pressure. Under loads which are concentrated on small areas, or which exert bending moments, cartilage will fail – unless it is reinforced by other materials, like collagenous or elastic fibres, as, for example, in a fibrous perichondrium. If the amount of fibres is high, so that you have to search for isolated cartilage cells, the tissue is called fibrous or elastic cartilage. In amphiarthroses, in intervertebral discs or in pseudarthroses, the content of collagenous fibres is very high, and in the deformable nasal, or external ear skeletons, elastic fibres are the most obvious parts. Hyalinous cartilage can sustain about 300 N/cm^2. Its Young's modulus varies around 160 N/cm^2.

We all (all biologists) are accustomed to assume that the decisive information about the shape of an animal is contained in its skeleton. In fact the great amounts of tension-producing (in large animals hundreds of kg) material (= muscles, tendons and ligaments) require obviously considerable amounts of compression-resistant material, just to maintain body shape. Likewise, the mass exposed to gravity needs the existence of compression-resistant props of constant length and shape to sustain weight force. This has the very practical reason that the bony skeleton is not really palatable for humans and can be preserved in dried state over long times and can be mounted in positions similar to those of the living animal. This habituation is useful for interpretations of fossil documents, which consist of nearly exclusively on the hard substances. All these facts make a precise knowledge of the skeleton very helpful – but for understanding their shapes, we need urgently to know the relations between shape of the skeleton and mode of living of a particular being.

Vertebrate bodies contain considerable amounts of *connective tissue* in a variety of variants: loose, cellular, tense, parallel-fibred and elastic. Characteristic are its cells which are connected to form a three-dimensional meshwork by long and very slender extensions. The intercellular matrix contains collagenous fibres, primarily of type I, intermingled with some elastic fibres of types III and V. These fibres can be arranged at random and be relatively isolated (pliant connective tissue). This material as a whole has no own characteristic shape and fills the empty interspaces between organs and parts thereof. It is often located between more rigid elements, like muscles or tendons, where it permits easy sliding of the organs against each other. It is rich in cells and ready to regenerate after lesions. The number of fibres can be so great that the densely packed bundles of fibres replace the more or less liquid matrix completely and that the few cells are jammed in and seem to be compressed between the bundles of fibres. The dermis is a good example for this type of very *tough connective tissue*. The share of elastic fibres is greater in younger and less in aged individuals. The numerous and thick collagenous fibres of type I can also be oriented parallel. In these cases, the fibres form *ligaments*, *tendons* and *aponeuroses* and cover the entire skeleton as a sort of hide, which is called periosteum. The same hide crosses the interval between the neighbouring bones as the completely closed joint capsula and continues on the next skeletal element. From this description follows that the joint cartilage has no perichondrium. At particular places, the joint capsula is reinforced by so-called joint ligaments. The most characteristic property of these structures is strength against tensile forces without much lengthening. The observed values are between 1500 and 2000 N/cm^2. Under load, tendons of ungulates have been found to stretch by some 4% of their length, (Ker 1993; Wainwright et al. 1976; Young's modulus about 10^5 N/cm^2), whereby the first 3% just stretch out the normally crimped arrangement of tendon fibres. That means tendons and ligaments are rather stiff. Their properties limit movements between skeletal elements, and the forces produced by muscles are transmitted without delay and without loss to the respective bones – the essential precondition for exact neuronal control of movements.

If loaded continuously over longer periods (hours), even within physiological limits, fibres become longer; they undergo plastic deformation. This property sets limits to their function: permanent, continuously acting forces cannot be sustained by ligaments. On the other hand, lengthening by moderate forces is used in rehabilitation after lesions or in the training of athletes. Under the influence of intermittent mechanical stimuli (see below), fibrous tissues become stronger, and this gives them the potential to heal after lesions.

Among the unelastic, tense fibres can always be found small amounts of *elastic fibres*. Chemically, tense and elastic fibres are closely related, but while the former are nearly straight, unbranched and densely packed, the latter appear under the microscope as branching fibres or undulated bands. In some structures, these elastic elements can dominate. In the human and other primate's bodies, they play a minor role, but in hooved animals, the nuchal ligament contributes much to keeping the horizontal neck in balance and to carry the head, without inhibiting a very deep head position completely. In birds, the tendons of the mighty pectoral muscles are elastic. *Elastic ligaments* stretch under load by factors of 40–50% of their shortest length (measured by Kienapfel and Preuschoft (2011) in the nuchal ligaments of horses). They exert per cm^2 of cross-sectional area from 200 to 300 N (Bucher and Wartenberg 1989). According to Gellmann and Bertram (2002), the force produced by stretching the equine nuchal ligament (one of the most impressive elastic elements found among animals) varies between a compliant zone (modulus of elasticity $8*10^5$ N/m^2, a value close to that found by Wainwright et al. (1976) for isolated elastic fibres) and a stiff zone (modulus of elasticity $1*10^8$ N/m^2), the latter being determined by collagenous fibres.

Musculature, which represents as much as 40% of total body mass, consists, in contrast to the above described tissues of cells ("smooth" muscles of the vessels and the digestive apparatus, myocardial muscle), or big, cell-like "fibres" which contain many nuclei below their sarcolemma ("skeletal" musculature). In our context alone the latter is of importance.

The *ability to contract* is due to the myofibrils inside the "cells". These are combinations of actin and myosin filaments, which are irregularly distributed inside "smooth" muscle cells but parallel and on the same height in myocardium and skeletal muscles. This arrangement gives them the well-known "striation" visible under the microscope. While in muscle cells of the myocardium only one nucleus is located in the centre, the cell-like fibres of skeletal muscles contain hundreds of nuclei, located near the sarcolemma, giving way to the masses of myofibrils. Each cell-like muscle fibre is wrapped in a fine cover of loose connective tissue (perimysium internum). Groups of fibres are so combined to fibres of higher order. At both ends of the high-order fibre bundles, these fibres (perimysium internum) unite to a "perimysium externum" and the muscle's fascia. The fibres of the tendons are deeply inserted into the sarcolemma at both ends of the muscle fibres. All these structures of connective tissue pass on to finally reach their insertions. This can either be a strong aponeurosis or fascia or the periosteum of a skeletal element.

The muscle fibres can either be arranged parallel to each other (e.g. the large muscles of the ventrolateral trunk wall) or converge to a tendon in a form called pinnation, either from one side only or from both sides (like in most limb muscles). In some muscles several systems of pinnation are combined to form very complex arrangements (e.g. m. deltoideus of humans). The force, which a muscle can produce, depends on the number of muscle fibres. This can be approached after cutting at right angles to the long axis of the fibres. Clearly, this "physiological cross section" deviates much from the simple morphological cross section through a muscle belly. The distance by which a contraction shortens the muscle is proportional to fibre length. Therefore the pinnated muscles of distal limb segments produce great forces, but cannot cause wide excursion angles.

The characteristic of musculature is their ability to assume or maintain variable lengths: Muscles can shorten after being stimulated neurally (= isotonic contraction), or they can maintain a given length against a load (= isometric contraction). The forces measured in active contractions are extremely variable, between 300 and 1000 N/cm^2 of their "physiological cross section". The strength of muscles is much less than that of their tendons (see above: 1500–2000 N/cm^2), and therefore the muscle bellies are much thicker than their often surprisingly slender tendons. Within the force-transmitting chains of the locomotor apparatus, muscles represent the weakest link. To produce or just to sustain the required force, the cross sections of muscles (= their weight) must be great in comparison to other mechanically loaded parts. According to Grand (1977, 1978) the share of musculature – and that of the skeleton – varies considerably among mammals and reptiles. In the agile land-living mammals, it amounts to 40–50% of total body weight, about two times the mass of the skeleton with 9–19%. The experience of butchers regarding pigs and cattle agrees with these values.

Working against a load consumes energy but only in the stimulated muscle. If not stimulated, muscles behave similar to *rubber bands*; that means they exert an elastic resistance against being stretched. The forces produced under these conditions are poorly known, aside from the simple fact that they grow faster than linear (Fig. 1.4c). An attempt to determine these forces was made by Witte et al. (1995a, b). This elastic behaviour seems to be due to the considerable amounts of titin fibres present in muscles parallel to the actomyosin fibrils (Rao, 1998). Well known and easy to observe is the fact that muscles produce higher forces in pre-stretched than in pre-shortened state. The force, which a muscle can produce, grows with pre-stretching of the fibres again much faster than linearly but seems to level out, or even to become smaller, at greater lengths. While the principles of production of muscle force have been described early, since the 1960s, it has been pointed out that the investigated muscle lengths do not at all occur in vivo, because the length of the muscle-tendon units are limited by the incorporation in the body and restricted by the joint movements (Kostuj, 2002, Kostuj et al., in prep.). Kostuj has measured the length differences of muscle-tendon units in human forearm muscles. Observed shortening of course takes place alone in muscle, not in the tendon. She has found considerable variation among the individual forearm muscles. The average "operating range" (defined as divergence in both directions from the "neutral", extended position) between extreme positions of the joints was −12% to +25,6%. In other words, most muscles are stretched under in vivo conditions by about 35–40% of their shortest lengths.

Fatty tissue is not, in a strict sense, part of the locomotor apparatus. However, it fulfils in some cases mechanical functions. The amount of fat in a body varies considerably, and it is present at various places (subcutaneous, filling of empty interspaces between muscles and in noncongruent joints). Typically, volar surfaces of the hands and the plantar surfaces of feet are covered by fatty cushions. Cells have one huge vacuole ("white" fatty tissue) or several, smaller vacuolae ("brown" fatty tissue). These are filled with lipids that can be metabolised if necessary. These cells are surrounded by a meshwork of fibres, which connects them to groups located in loose connective tissue (if empty spaces are filled, which change their shapes during movements). For our purposes of particular importance are the fatty cushions

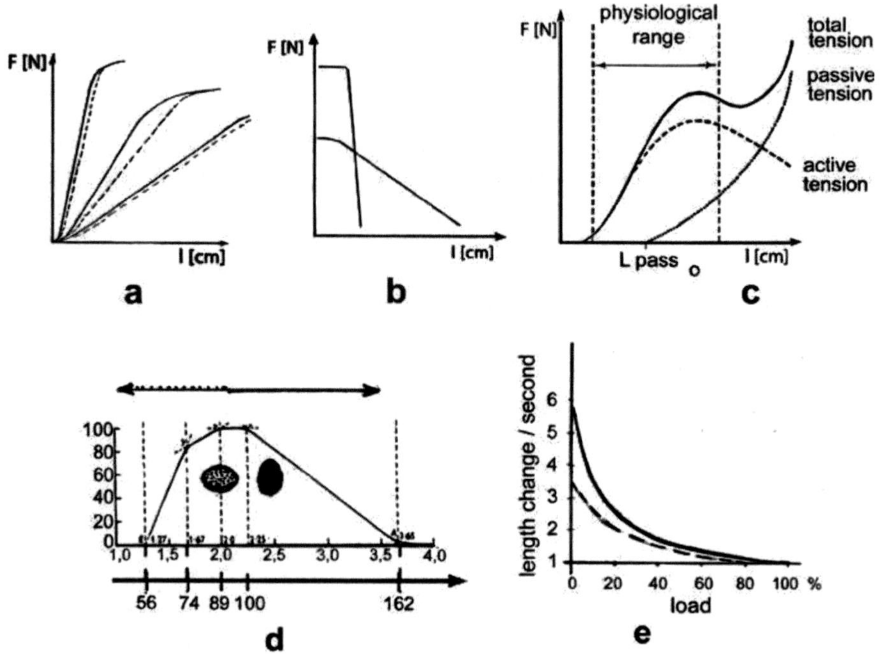

Fig. 1.4 Stress/strain plots, indicating changes of dimensions under load. (**a**) Stretching under tensile forces. "Stiff" or "rigid" material (most tendons and ligaments) with high Young's modulus on the left, compliant material (elastic ligaments, inactive muscle) with low Young's modulus on the right. Continuous lines show increasing, broken lines decreasing loads. The reduction of length after relieving the load lags behind former lengthening. This "hysteresis" can be more (graph in the middle) or less marked (graphs on the left and right). The slightly curved "toe region" of the graphs, indicating very low resistance against initial deformation, is typical for mammalian tendons. In other materials, this part can be absent. The longest part of each graph indicates linear deformation proportional to the loads. This part of the graphs is characterised by the calculated "Young's modulus". At very high loads, the length increase becomes more and more rapid, while the loads grow more slowly. This part of the graph can be short, indicating a complete rupture, or long in the case of a gradual giving way. Deformations in this region are usually not elastic, but lead to permanent plastic changes of length. (**b**) Schematic diagram of shortening under compressive loads. Rigid material (compact bone) undergoes less deformation than more compliant substances (cartilage) and sustains much higher loads. If ultimate strength is approached, the total breaking down of resistance in brittle material leads to a steep decrease, while softer material yields gradually. Usually, diagrams of this type are drawn like in (**a**), with the abscissa showing the relative change of dimension (= strain, extension/length) and the ordinate measured in units of stress (*F*/*A*). (**c**) Skeletal muscles show, if not actively contracted, elastic behaviour, and stiffness increases with length. If actively contracted, muscles produce more force than in elastic stretching. In "isotonical" contraction, they show the same behaviour as in elastic stretching; in "isometrical" action, muscle force passes a maximum before decreasing again. It should be noted that the length differences shown here as "physiological range" have been found in isolated frog muscles and are by no means realistic for muscles, the lengths of which are determined by the angular excursions of the joints they are crossing. From Kandel et al. when? (**d**) Force produced under realistic conditions (scatter ellipses) plotted against sarcomere length in human forearm muscles. Palmar flexion of the wrist joint to the left, dorsiflexion on the right. Sarcomere lengths vary between 1, 1.5, 2.0, 2.5, and 3.0 μm; forces are given in 20%, 40%, 60%, 80%, and 100% of maximum values. (From Kostuj 2002). (**e**) Under increasing loads (horizontal axis, measured in % of the isometric maximum force), muscle contractions become slower (vertical axis, measured in muscle length per second). (Modified from Brandes et al. 2019)

covering the soles of the forelimbs and hindlimbs. In these structures ("Baufett" in German), which are never resorbed even in cases near starvation, the groups of fat cells are surrounded tightly by tough connective tissue. The former function as "pneus", which means tension-resistant wrappings filled with a liquid that can be deformed, but their volume cannot be reduced mechanically. The wrapping keeps the liquid-filled elements in their place. Because the pneus are small and tightly connected, they form a cushion of constant form. The whole system serves to distribute concentrated external forces evenly over a greater area, thus reducing the danger of overloading or hurting the structures protected by them.

The teeth are consisting of "*dentine*", a material very similar to bone, but its matrix has a greater percentage of *mineral/anorganic* substance (70%, bone 60–45%). The major difference is that the dentine-producing cells are not on all sides wrapped in by the hard substance but remain at the inner surface of the tooth wall. While bone cells are connected to each other by tiny channels which form together a three-dimensional meshwork, such channels in dentine run from the cells mainly parallel to each other toward the periphery. In teeth the potential to regenerate is very restricted. The external surface of tooth crowns in most craniota is covered by the hardest substance of the body with a mineral content of 95%: *enamel.* It is produced early during ontogeny by cells which afterward disappear completely, so that absolutely no regeneration is possible. Enamel in tetrapods consists of lengthy crystallites ("prisms") which extend between dentine and external surface of the tooth. The patterns in which these crystallites are arranged may vary considerably and obviously has to do with mechanical loading of the respective tooth.

The body of a vertebrate is composed of *segments* that can be moved against each other. Each segment consists of a skeletal element between two neighbouring joints and is wrapped in soft part: periosteum, ligaments, muscles and tendons. In particular the muscles are rather far away from the joint axes, so that they possess long lever arms. Because of the bony elements, the segments are not flexible, but the soft parts can be moved (within limits) against the skeleton "Schwabbelmassen" ("wobbling masses?").

1.1.2 Basic Mechanics

Some *basic mechanical concepts* are necessary for understanding "function" and must be defined, and discriminated. I have used Schreyer et al., Praktische Baustatik from 1964 and a new edition of Dubbel´s Taschenbuch (Grote and Feldhusen, 2007).

Forces (*F*) can act as "external" forces, like weight, or ground reaction force, or mass inertia from outside on a body segment. "Internal" forces, like the pull of muscles or their tendons, joint ligaments, or joint forces, or bending moments, are produced (by muscles) or occur as a reaction to external forces inside the body. Forces have different qualities:

Compressive force (Fig. 1.5a) acts mainly on the skeleton. It distributes evenly across the whole cross-sectional area of the loaded element.

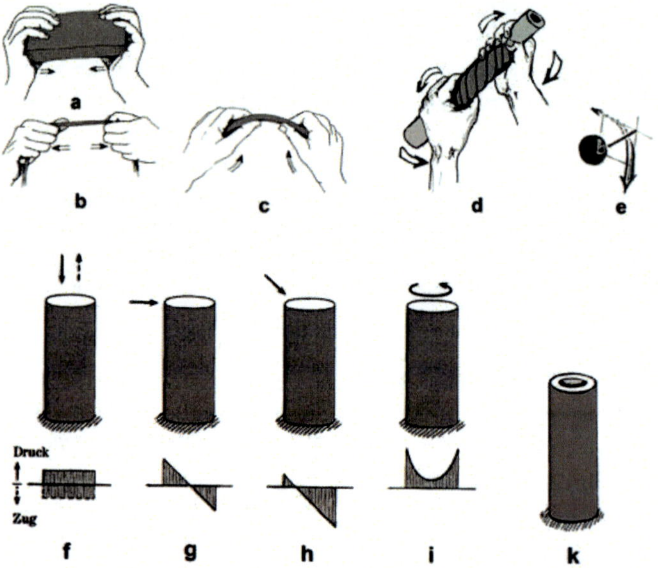

Fig. 1.5 Qualities of stressing, produced by external forces (arrows). (**a**) Compression; (**b**) Tension; (**c**) Bending; (**d**) Torsion, with emphasis on shortening of the external layer. (**e**) Torsion, with emphasis on the pressure exerted from the external layer against the central parts. (**f–k**) Columns anchored to the ground and loaded by **f** compression/tension, **g** bending, **h** compression superimposed by bending, **i** and **k** torsion. Below the columns: distribution of compressive/tensile stresses on the cross section. In **f**, stresses distribute evenly over the available cross section. In **g**, one side is extended; the other is compressed. Near the centre, the stresses pass through zero. In **h**, stresses increase at one side and must be subtracted on the other. This sort of stress distribution occurs often in models of long bones. (**i**) Torsion leads to high stress near the periphery and low or even zero stresses near the centre. The stress concentrations favour tube-shaped profiles and allow a reduction of material near the centres. (**k**) A tube instead of the column offers higher bending and rotating strength, if the same amount of material is available as in **i**, and its resistance against "normal" forces (= cross-sectional area) remains the same. (**d** and **e** are redrawn from Preuschoft and Christian 1999)

Tensile force (Fig. 1.5a) occurs seldom by the action of an external force; instead it is mostly produced by muscles and transmitted through tendons, aponeuroses or ligaments. Tension spreads also across the entire cross-sectional area.

Bending (Fig. 1.5b) occurs in parts (mostly segments) of the body, if forces in non-axial directions act against a body segment which cannot move. Aside from the acting force (F), the length of the lever arm (l) is a decisive factor. Multiplication of $F * l$ yields the *bending moments*. Bending leads to compressive stresses on one side of the loaded element and tensile stresses on the other. In bones, the bending moments usually are superimposed by compression (Fig. 1.5c), so that the compressive stresses are increased and the tensile stresses reduced, either to small values, to zero or to low compressive stresses. That means the acting force is excentric to the longitudinal axis of the bone.

Torsion (Fig. 1.5d) is evoked by force components acting at right angles to, but distant from, the longitudinal axis of a long bone or a jaw. As in bending, the lever arm length (i.e. the distance from the force to the long axis) is a likewise essential factor to yield the *torsional moment*. The stresses evoked by torsion grow rapidly near the periphery of the loaded element, while the centre is not much stressed.

Moments (*M*) are forces multiplied with their lever arms (*l*). They occur as bending moments over a body part, as rotating moments about the axis of a joint or as torsional moments along a body segment, e.g. the trunk. In many cases, bending moments change along the length of a body part. These changes can be illustrated as areas showing for each unit of length the bending moment prevailing at this point. Long bones often show curvatures of their long axis. These curvatures approach the bony element to the resultant of all forces acting along the respective section and so reduce the bending moments. According to new calculations, it has become doubtful that pure bending stresses come into action within bones (Sverdlowa and Witzel 2010). Instead, the resulting compressive forces seem just to assume excentric positions relative to the bone's long axis. Therefore, the considerations below will be focussed on the respective body segments, instead of the skeletal elements. The bending stresses are distributed on different structures: Compressive on bone or cartilage and tensile on ligaments, tendons or muscles.

Forces evoke inside an element "internal forces" or *stresses*:

$$(\sigma = F/A), \tag{1.2}$$

which are distributed in different ways on the cross section. Compression and tension (often combined as "normal stresses") spread evenly over the total cross-sectional area (Fig. 1.5a); bending moments lead to compressive stresses on one side of the loaded element and tensile stresses on the other. By superimposing compression on bending, the compressive stresses are increased and the tensile stresses reduced, either to small values, to zero or to low compressive stresses (Fig. 1.5b). The stresses evoked by torsion grow rapidly toward the periphery of the loaded element, while they remain on a low level near the centre (Fig. 1.5c). This is comparable to the stress distribution in bending. In fact all load cases shown in Fig. 1.5b–d have in common the presence of the highest stresses near the periphery and low stresses in the centre. Deposition of bone material at places, where stresses are high, and resorption, where stresses are low, lead to the often-observed tubular shape of long bones. This arrangement of material is advantageous, because it combines the lowest possible weight with the greatest strength against various stresses. Under the condition that the cross-sectional area is the same as in Fig. 1.5a, the same "normal" stresses, but also much greater bending and torsional moments, can be sustained. Under the condition that the diameter of the marrow cavity is 0.57 of the external diameter (as observed in human metacarpals), an elongation of the external diameter of a column by not more than 21% leads to the same cross-sectional area, combined with about 30% greater bending strength.

Strength of segments, which is their ability to sustain forces or moments, depends upon the quality of loads and can differ between compression, tension, torsion and bending. It is dictated by the *material properties*, as well as by *the shapes of the loaded elements*. Approximate values for material strength are given under "biological materials". Strength is easy to determine in the case of parallel-fibred tendons or ligaments, or of compact bone, or even of cartilage. Very difficult is the determination of strength in the case of muscles (see above). In the rigid elements of the skeleton, strength is correlated with shape; strength against "normal" forces, compression and tension depends on the cross-sectional area: $\sigma = F/A$. Bending and torsion are opposed by the "moment of resistance" (W), defined as area moment of inertia I/maximal distance of cross section from neutral fibre ($\Sigma\ \Delta a * y^2/y_{max}$). Ultimate strength, or strength at failure (σ_{max}), shows the extreme limits of a given structure. It is usually a multiple of the "normally" occurring loads, the divergence between them being the "safety factor". According to measurements of the Alexander Group (e.g. Ker 1993), this safety factor varies considerably between individual tendons and groups of animals.

Any material undergoes *deformation* under the action of a force. This deformation varies and is often made visible by plotting stress (F/A) over length changes (Fig. 1.4; stress-strain diagram). The ability to deform under load and reassume the original form after relieving is called *elasticity*. The amount of deformation is proportional to the strength of the acting force. Elasticity (E) is measured by the relation between stress (σ that means F/A), divided by the change of length (ε, provoked by stretching or compressing).

$$E = \sigma/\varepsilon, \tag{1.3}$$

is called *Young's modulus of elasticity*. The higher its value, the stiffer or more rigid is the material, and the smaller the value, the more compliant is the material. Bones, tendons and ligaments are rather stiff materials, which permit only small deformation and possess high Young's modulus, while elastic ligaments or muscles behave like rubber bands, possessing low Young's modulus.

As can be seen from Fig. 1.4, important biological materials (tendons, tight ligaments) are, under low loads, rather compliant, and therefore loads are transmitted slowly. This "toe region" can be absent, like in bone. If loads are increased, the elongation is linearly proportional to the acting force. The slope of this part is measured as Young's modulus, and it varies among tissues (see above). Under decreasing loads, the return to the former dimensions is often delayed (hysteresis). If loads increase further, the deformation may take place faster than the increase of force, and the former dimensions are not restored. That means the deformation becomes permanent "plastic" and is no longer elastic. In other, more brittle materials (bone), the ultimate strength is clearly marked by a sudden and complete reduction of the supported load.

Force closure, closed linkage and circle of forces and balance of acting forces. While compressive forces can be transmitted across gaps between skeletal elements – as in joints – transmission of tensile forces is only possible if continuity between the elements exists. This continuity is accomplished in joint capsules, ligaments as well as tendons by the insertions into bone. As a general rule, "joints" in the sense of interruptions of continuity between bony skeletal elements can transmit forces, but not moments. In spite of this, very dense fibrous cartilage can permit a limited transmission of torsional or bending moments (like in the iliosacral connection, the pubic symphysis or the mandibular symphysis). All forces act only between the bony insertions of the involved structures and so form a "circle of forces". All forces occurring within such a circle must sum up to zero, and all moments must compensate each other, so that each part remains in equilibrium at its place. No external forces come into action within such a circle of forces.

For analysing the forces, and the flow of stresses within a body segment, it has turned out most appropriate to start off from the external reaction forces, which are evoked by the activities of the animal on its environment (in the sense of "actio efficit reactionem" after Newton 1687). In water-living animals, it is mainly the friction of water or the inertia of the body; in land-living animals, the weight or the inertia of the animal evokes a reaction force in the supporting substrate. These *external forces are concentrated* on rather small areas. This is in contrast to swimming in water, where the external reaction *forces are distributed* over large parts of the animal's surface. Ground or substrate reaction forces (GRF or SRF) may assume variable directions in relation to the supporting limb. Very frequent is such a change during the rolling off: While a foot is on the ground, that part of body weight carried by this foot moves forward. The leg usually rotates about the joint next to the ground, while the foot remains in contact. In some cases, the whole foot may roll off on the supporting surface. The GRF swings with stylopodium and zeugopodium, so that a regular shift of the load can be observed: first the posterior part of the foot, then its middle and finally the most anterior part of the foot carry most of the load. At touchdown, not only weight is taken over (admittedly a rather small part) but also the movement of the forward swinging foot is braked, by a horizontal force component directed against movement of the body. At about midstance, the vertical weight component is greatest, while neither braking nor accelerating force components come into action. In the second half of the stance phase, the vertical weight component becomes smaller, but a horizontal force component, pointing in the direction of movement, accelerates the body and the limb.

1.2 Animals in Focus: Vertebrates or Craniota

Most considerations in this book are restricted to *vertebrates* (or more exactly *craniota*[2]), which are emphasised in Fig. 1.2. Neither the attribution of forms to higher taxa nor their derivation from acranial chordates is subject to reasonable doubt. Although the representatives of the classes appear rather different, all have in common as the most important characteristic of their construction an axial skeleton.

[2]Westheide and Rieger, Schultze (both 2004), as well as Maier (2014) have pointed out that "vertebrata" is not a perfectly fitting term to characterise these animals. First, because the vertebral column can be replaced during the whole lifetime by its embryological ancestor, the notochord (e.g. *Latimeria*), second, because vertebrae seem to have evolved several times independently since the Devonic (in e.g. lepospondyli, rhachitomi, stereospondyli, embolomeri). The following considerations refer in a strict sense to the craniota, the skull-posessing animals, this the more so, since we have nothing new to contribute concerning acranial forms.

This can either be a notochord or a column of vertebrae. Both are compression-resistant rods, which keep the length of the animal constant when the muscles of the trunk wall or tail are contracting and producing tensile forces in longitudinal direction (Gutmann 1994 and elsewhere). The existence or the missing of a skull does not matter in this context. In any case, the approach chosen here must investigate what exactly the axial skeleton must sustain during locomotion:

- How muscles and skeleton are stressed under defined and comparable conditions
- Whether the vertebral column in land animals has the same mechanical function as in the more primitive aquatic animals
- Whether limbs are exposed to the same qualities of load as the axial skeleton
- Whether the musculature is arranged according to the same principles in aquatic and in land-living forms
- Whether skull and jaws follow the same rules as does the locomotor apparatus

Investigation of mammals challenges our understanding of mechanical necessities but also promises rich results: Mammals are those animals in which morphology and behaviour are known best. The very diverse locomotor habits of mammals require very diverse activities of muscles and shapes of skeletal elements. Because the results obtained can easily be transferred to members of other classes, their investigation can unveil general rules.

Reliable and detailed knowledge about locomotion can be obtained alone from living and unrestricted animals. Therefore, many attempts to compel an experimental animal to show "reproducible" locomotor patterns are loaded with a bias. Normal locomotor patterns may vary within closely related taxa, and there is no convincing relationship between the species-specific locomotor pattern and its taxonomic position. This statement is not in conflict with the fact that in each evolutionary line, only those traits can be developed ("optimised" or "specialised", often called "adapted", a word, which, however, implies an aimed process!), which are existing in this evolutionary line. Evolution can take place alone on the basis of those elements which are present in the respective ancestor as was emphasised repeatedly by the Frankfurt School around Gutmann Peters, Grasshoff. Some of the most marked examples shall be treated here, and the influence exerted by body size and locomotor mode on body shape shall be the first focus of interest.

1.3 Some not so Common Aspects of Commonly Known Facts: What About Growing Big?

That *body size* has a relation to *morphology* and is principally known and discussed frequently. Numerous studies on "allometry" pursue the statistical correlations between "size" and morphological characters. In the first of these studies, Snellius (1891) has investigated brain size.

The relation was formulated as

$$H = c * K^{r}, \tag{1.4}$$

where H is brain size, K is body size, the exponent r characterises the dependence of H on K and can be below zero ("scales with 0.33 body weight") and c stands for all factors which have, aside from body weight, influence on brain weight (as, e.g. the degree of cerebralisation).

The great number of these studies camouflages the fact that they do not yield answers to questions aiming at causes. This holds true, in particular, if "explanations" are sought on the level of biological roles – which themselves are vaguely defined. Of course it is completely pointless to spend efforts on analysing relations between traits which are not statistically correlated. But too often the limited value of correlation statistics is lost out of sight: it is a descriptive method, which leaves all questions regarding causes open.

After identifying the facts, the problems can be considered, which are raised by increased body size. Not before you know the traits which are influenced by size, and in which way, it makes sense to search for consequences which arise from size and for causes.

"Size" in the contexts of this book is *principally defined as* "mass" and measured in kg. Under the conditions existing on earth, mass is perceived as weight, that is, mass (m) times earth acceleration or gravity (g, assumed here to be 9.81 m/s^2). Weight is a force (F), measured in N. In movements, the acceleration (a) given to the mass of the whole body, or parts of it, may become as, or even more important than, gravity. The mass of an animal (m) is proportional to its volume (V, which sometimes is easier to determine). The proportion between both is the specific weight, or "density", and varies from one body segment to the next and possibly among groups of animals. This is on one side due to gas-filled organs, like the lungs, or guts, or pneumatised bones, and on the other side due to the share of heavy, compact bone, like in the skull. A common experience is that the bodies of most animals float, if immersed into water. That means their specific density is about the same as that of water. Well-known exceptions are birds (which pop up if immersed) and parts of the locomotor apparatus (which do only float, as long as they are connected with lungs or guts). Anyway, in the context of this book, density or specific weight of body segments is assumed to be near 1, the density of water.

Cope (1887 and 1892) has already observed that representatives of many evolutionary lines tend to become bigger with time. This observation has been generalised as "Cope's rule". The validity of Cope's rule is the subject of numerous discussions (e.g. Sander et al. 2010; Klein et al. 2011). In spite of this widely spread interest, the question WHY the evolutionary younger forms are often bigger than the older ones is asked rarely. If at all an explanation is sought, and after the facts are documented, it should be found out which advantages an individual obtain from an increase in body size. In other words, if we do not contend ourselves with the idea that Cope's rule is an inherent, genetically fixed (but aimed!) law, not simply the name for an observation, we have to answer the questions: where the advantages of big size are to be found and what keeps size increase within limits.

As discussed in many studies, the increase of body mass entails a favourable ratio between heat-producing mass and heat-dissipating surface. *Mass is proportional to volume*, a third power of linear dimensions, and *surface area*, that is skin, is a second power of linear dimensions. This relation has been formulated in the late nineteenth century as "Bergmann's rule" and has been taken as the reason for evolutionary or geographical increase of size. Neither their validity for all observed size differences between species and genera is per se obvious nor whether the size differences lead to the same advantages or disadvantages in all animals.

An example has been worked out among *primates*. Reichard and Preuschoft (2016) started from the lesser apes, a group which is closely related but exists, sometimes sympatrically, in two size classes: most of them weighing not more than 5, perhaps 6 kg (gibbons s.s.), and one species weighing about 12 kg (siamang). In the same geographical region, that is, in a very similar biotope, exist much smaller forms, the slender and the slow loris (*Loris tardigradus*, about 0.2 kg, and *Nycticebus coucang* of about 1.5–2 kg) and the much larger orangutan (females weighing 30–40 kg, males 80 kg). The values found empirically (weight) and those calculated (surface area) are set out in Table 1.1 and Fig. 1.7. The general rule is confirmed by these results, and they are in accordance with the hypothesis that the larger body size allows inhabiting environments with colder and more rainy climate. Unfortunately, this argument does not hold true for the enormous dimorphism in female and male orangutans!

In several taxa, the *size difference between the sexes* ("sexual dimorphism") is quite impressive. In primates, for example, the females of some species weigh not

Table 1.1 Weights, calculated volumes, and calculated surface areas of various primates living in roughly the same geographical region

				Proportions	
	Weight[a] (kg)	Body volume (cm^3)	Body surface (cm^2)	V/A	A/V
Slender loris	0.231	194	393	0.49	2.03
Slow loris	0.823	1040	970	1.07	0.93
Gibbon[b]	5.690	5757	3880	1.48	0.67
Siamang	10.700	10,913	5644	1.93	0.52
Orangutan	35.700	46,223	13,354	3.46	0.29
Orangutan (M)	78.200	83,891	18,825	4.46	0.22

From Reichard and Preuschoft (2016)
Note the parallel between empirical weights and calculated volumes
M male
[a]Average female mass after Smith and Jungers (1997)
[b]Excluding Hoolock and Crested Gibbons

more than 40% or 50% of the males (orangutan, gorilla, baboons, *proboscis monkey*). This will be detailed below at the end of Chap. 8 "Primates".

What regards keeping *body temperature* continuously on a high level, Röhrs (1968) has argued, that more effective than an increase of body size is a thicker fur with longer hair. The density of pelage, which may play a similar role, varies, for example, between >2000 hairs/cm^2 in gibbons and siamangs and about 900 hairs/cm^2 in orangutan and other large apes (Schultz 1933).

For a representative group of giants of the past, the *sauropod dinosaurs*, the reasons for size increase and the problems connected with huge body size have been investigated by an interdisciplinary group of scientists (Klein et al. 2011; also Sander et al. 2010). In giant ectotherm reptiles (like sauropods), big size implies indeed the possibility to maintain *body temperature* on a favourably high level. According to brand-new investigations (study in review), the problem in the case of sauropods is not so much how to maintain body heat but to get rid of it ("waving sail" according to Warsinger, 2018).

Size also implies a very long *digestive tract* and therefore extended periods for digestion. This allows herbivores to exploit fibre-rich, difficult to digest food, which is commonly available (Hummel and Clauss 2011).

Mechanical arguments for large body size do also exist: *Travelling speed* on a low energetic level is greater in large than in small animals (Preuschoft et al. 1991, 2011; see also below, Chapters 6 and 7); *reaching distance* for obtaining food becomes longer. Increased mass improves *fighting power*, required in intraspecific conflicts (where it decides about success in reproduction!) and as a means against predation pressure on the side of prey animals (Hone and Benton 2005; Sander et al. 2010). Body mass is also a means to provide the superiority of the predator over its victim. Not alone mass but also a large body volume provides advantages as a defence against being bitten by smaller predators. All these points have been discussed on the level of biological roles; their mechanical background has hardly been investigated in the past.

The advantages of large body size are opposed by *disadvantages*. Obvious is the fact that absolutely more food is required by a larger animal. In the case of falling, the danger is greater in a large, than in a small animal, because the impact on the ground is greater. This is partly due to the greater falling height of body segments, partly due to the greater masses which are stopped at contact with the ground. The falling body (or parts of it) contains an impulse (*P*) of

$$P = m * v = m * (2gh)^{1/2} \tag{1.5}$$

At the end of the fall, the body is stopped by a braking force (*F*), which is determined by mass (*m*) and the height of falling (*h*, see also Chapter 8.2):

$$F = P/t = m * (2 * g * h)^{1/2} \tag{1.6}$$

A further disadvantage is that all movements become slower with size, because the mass moment of inertia (J) grows with mass and with the square of linear dimensions:

$$J = m_i * \text{length}\, {l_i}^2, \tag{1.7}$$

while *muscle force lags behind body size*. This is due to the above-mentioned, sufficiently known relationship: Weight is a third power function of lengths or diameters. The strength of, for example, the weight-bearing limb bones, depends upon the cross-sectional areas, which are second power functions of the diameters. As much as bones need strength, the equilibrium of joints, as well as movements, requires *strength of muscles*. This depends upon the "physiological cross section", strictly speaking the cross sections through all muscle fibres. In a first and general approximation, this can be assumed to be proportional to the cross-sectional area of the muscles (Fig. 1.7). Clearly these grow with body size, but like bone strength, they follow a second power function of diameters (Figs. 1.6a and 1.7a), which grows slower than mass. The force reserves of animals of increasing *size grow with mass*$^{2/3}$

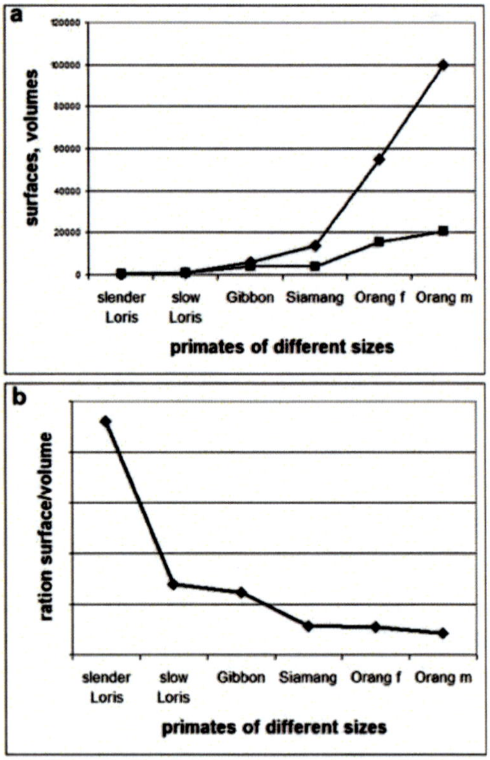

Fig. 1.6 (**a**) Volume and calculated body surface in slender loris, slow loris, small gibbon, siamang, a subadult orangutan (only a bit smaller than an adult female) and a fully grown male orangutan. Graph is essentially the same as in Fig. 1.7, but based on empirical data. (**b**) Ratio between surface and volume, calculated for primates of varying body mass

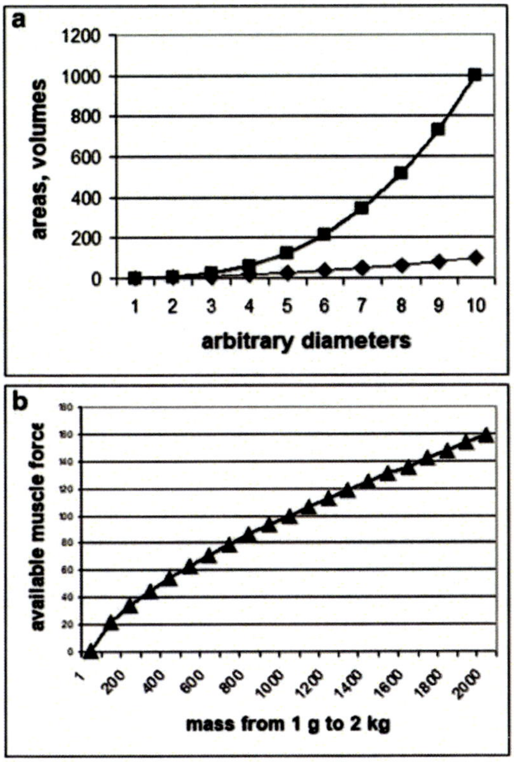

Fig. 1.7 (**a**) Body mass is proportional to its volume, a 3rd power of the linear dimensions forming the abscissa (squares). Muscle force is proportional to the "physiological cross section" of the muscles (diamonds), a 2nd power of linear dimensions. Because a 3rd power function yields greater values than a second power function, small animals always can dispose of greater force reserves than larger ones. (**b**) The total available force grows with mass$^{2/3}$. Therefore, growing mass implies an absolute increase of available force, but the rate of increase becomes less with growing body mass (triangles)

(Fig. 1.7b). Empirically, the share of musculature of mammals varies with the specific mode of locomotion between 25% and more than 50% of total body weight (Grand 1977, 1984, 1990).

Often has been noted that large animals tend to keep the big joints of the limbs in *more extended positions* than smaller ones, in order to compensate this loss of force reserves. The extension of limb joints reduces the load arm lengths of GRF and so saves (under static conditions) energy. The same effect can be reached by an elongation of the force arms of muscles, so that less muscle force is required to control the respective joint (see Chapter 8.8). Long force arms, however, entail slower movements and therefore are not compatible with all lifestyles (see below Chapters 6 and 8.10). Under the (kinetic) conditions of movement, rapid excursions require not only the force to balance the joints, but the muscle forces must overcome the mass moment of inertia (J). The plot of resistance over size follows a fourth power, if lever arms are taken into consideration, even a fifth power (Fig. 1.8c). The speed of muscle contraction can be handled like a given factor, although it is influenced by the load to be lifted (see Fig. 1.4e). A heavily loaded muscle cannot contract as fast as if under a minor load, but there is yet no evidence which allows to

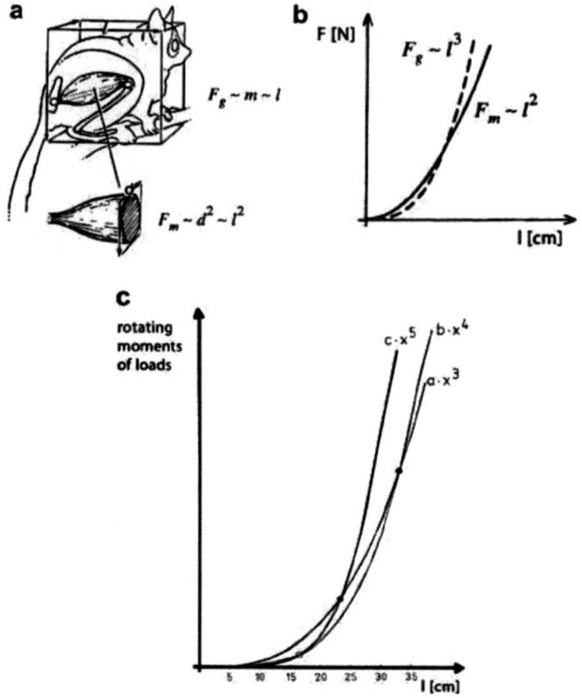

Fig. 1.8 Relation between force and mass. (**a**) Bushbaby in crouched posture fills nearly the volume of a cube (3rd power function of linear dimension, *l*), which is approximately proportional to mass (*m*). Multiplied by earth acceleration, mass exerts a force (F_g) on its substrate. The muscle force (F_m) which is available to an animal depends upon the cross-sectional area of the muscles (2nd power function of the linear dimension, *l*). This is shown, as an example, for the ***m.*** vastus, an important extensor of the knee. (**b**) As expected, the graph proportional to weight, or mass (broken line), is steeper than that of available force (solid line). At low values of *l* (= small-sized animals), the forces, however, exceed those of the load! (**c**) Similar relationships come into play if the rotational moments about joint axes are calculated. Under static conditions, multiplication with lever arm length leads to 4th power functions, and under the kinetic conditions of movement, mass moment of inertia must be overcome, the resistance following a 5th power function. The lever arms are assumed to grow isometrically, but the proportionality factors a, b, and c are different. The surprising point is that at smaller values for length (= small animals), the functions of high powers grow slower than those of lower power functions. Intersections are marked by dots (•). The exact positions of these dots depend upon factors (mass/force ratio) not considered in the present study. While the increase of the graphs in animals of large size is as expected, there are greater force reserves available at smaller absolute sizes below the dots. (Taken from Günther et al. 1989, (**a**, **b**) and from Preuschoft et al. 1998 (**c**))

estimate the degree to which contraction speed is lost in particular cases. Velocity of muscle contraction depends from the load to be moved and increases from 0.1 to 4 m/s, if the load decreases from 65 to 5 units.

The graphs of second, third and higher powers may deviate, depending on proportionality factors, from each other and show quite unexpected relations (as,

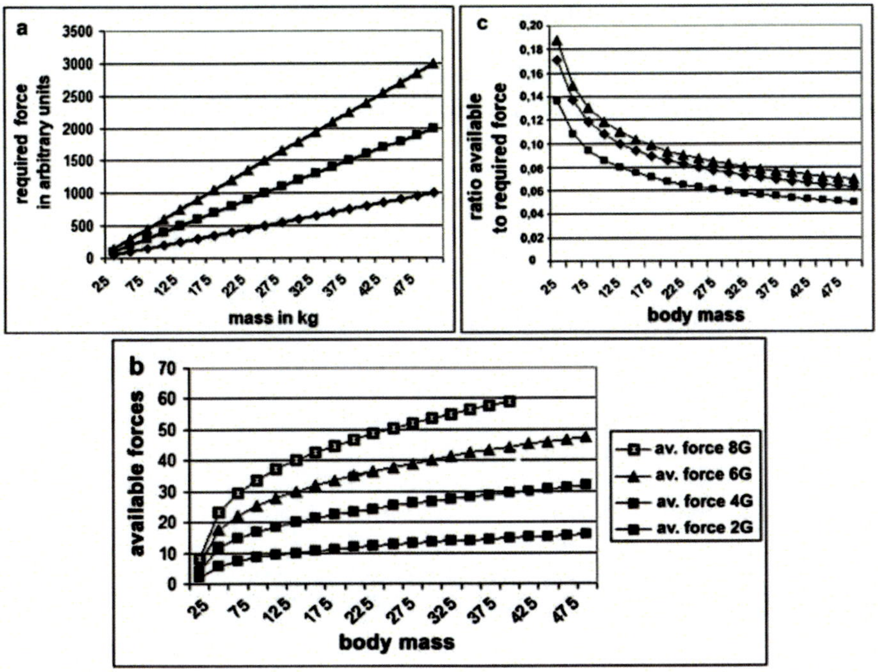

Fig. 1.9 (**a**) Forces required for a particular acceleration grow linearly with mass. (**b**) Because mass is proportional to a 3rd power of diameters, and muscle force from the 2nd power of the muscle's cross section, the total muscle force available in the body of an animal is proportional to an exponent of 2/3. Therefore, the available force reserves become less with growing mass (horizontal axis), no matter which acceleration is chosen. If body weights are below 25 kg, the total available forces grow rapidly. Between 25 and 125 kg, the increase is smaller, and beyond 125 kg, the increase is slow. The greater the acceleration, the more force is required. (**c**) The ratio available force/required force is plotted against body masses between 1 and 125 kg in steps of 25 kg. It becomes smaller with growing body mass and of course is smaller at low accelerations. During low accelerations of 2 G, the ratio is more favourable than during high accelerations of 6 G. In this inversion, the decrease becomes slower with growing mass. The missing forces in large animals can be compensated by long lever arms, but with longer lever arms, the angular speed of the resulting movement also becomes smaller (see also Chap. 7, Fig. 7.10)

e.g. Figs. 1.8b, c). Therefore it makes sense to take a close look at the situation under consideration of the really existing dimensions. Examples can be found below, in the Chap. 6.4.2 and Chap. 7.3.2 (Fig. 1.9).

If mass, and lever arm lengths, and the speed of muscle contraction remain constant, the acceleration of any joint excursion depends alone from the available force because

$$a = F_{\mathrm{a}}/m, \quad \text{or} \quad F_{\mathrm{a}} = a * m, \tag{1.8}$$

If muscle force is limited, the acceleration is greater in small mass animals than in those with a bulkier body. Since the limits of muscle force are less restrictive in smaller animals, the habit of leaping and hopping in most cases is confined to smaller body sizes, without regard to which class or order they belong (See Chaps. 6 and 7). By contrast, larger animals have less force available for accelerations than smaller forms. For example, the force (mass times acceleration) acting on a 500 kg horse to accelerate it to two times earth acceleration (2 g) would give an antelope of 250 kg an acceleration of 4 g and a lion weighing 125 kg an acceleration of 8 g. The other way round, the force, which accelerates an antelope or a deer by 6 g, would push a horse to not more than 3 g. The "relative" force, of which a 500 kg horse can dispose to accelerate to 2 g, would be sufficient for accelerating a 75 kg leopard to 4 g or a 25 kg duiker to 6 g.

Mass is proportional to a third power of diameters and muscle force from the second power of the muscle's cross section; the total muscle force available in the body of an animal depends from/is proportional to its mass with an exponent of 2/3 (see Figs. 1.7b and 1.9b). If body weights are below 25 kg, the total available forces grow rapidly. Between 25 and 125 kg, the increase is less, and beyond 125 kg is rather slow. The greater the acceleration, the more force is required.

Because all graphs showing relations between mechanical necessities and size follow exponential functions, a sharp borderline cannot be expected. The respective trait just requires a growing exploitation of its inherent/the existing possibilities and so becomes more demanding for larger animals.

1.4 Essentials of Chapter 1

The aims of the book are detailed. After my working hypotheses, basal information about the tissues of the locomotor apparatus are presented, as well as those relatively simple, but fundamental concepts of mechanics which are absolutely necessary for understanding the following text. The group of animals (craniota) which will be in the focus, is defined, and some often neglected aspects of statistical/allometrical approaches are emphasised.

Chapter 2
Head: What Is Making First Contact with Environment? Teeth and Jaws

Keywords Fang teeth · Crushing teeth · Folded dentine · Fixation of teeth · Classes of mammalian teeth · Lower jaw · Upper jaw · Fusion with brain capsule · Head as a link · Jugal arches · Lateral snapping · Suction feeding · FESA · FESS · Post-natal development · Marsupials

All vertebrates, even the most primitive among them, have in common the general arrangement of parts: *head, gill apparatus or neck and trunk*. The latter is often elongate and gradually passes over into the *tail* (Fig. 2.1). Characteristic for the great majority of vertebrates, the "*Craniota*", is the head as the most anterior part of the body. It is the place of sensory organs, olfaction, vision, electrical sense (as far as present) and hearing (as far as present), as well as the sense of equilibrium. The stimuli collected are processed in the brain, near the place of perception, and responses to the stimuli are prepared. In addition, the head contains the organ system by which animals get in touch with their environment, check its consistence and – perhaps – break off parts which are digested. Traditional anatomy discriminates between the locomotor apparatus at one side and the head and its parts on the other. This convention will also be followed in this book, because all approaches to the head have much in common and must take into account conditions, which are quite different from those acting on the locomotor apparatus. The development of the head during evolution has been in the focus of many attempts (a recent review: W. Maier 2014, 2017). There are fish-like animals which have many characteristics in common with Craniota, but do not possess jaws, but have reinforced skeletal elements (and that means a structure comparable to a "cranium") at their most anterior end (*Agnatha*, comprising the living Cyclostomata, slime eels Myxinidae and Petromyzontidae; and among fossils the Silurian ostracoderms (Pteraspidomorphi and Cephalapidomorphi). In the fossils, the anterior end is covered by a heavy bony armour. A concentration of sensory organs and stimuli-processing neurons has recently been observed in a very early placoderm from the Silurian (Dupret et al. 2014), and most primitive elements of a jaw apparatus have been found in another Silurian placoderm from China (419 MY, Zhu et al. 2013). According to Bliek (2017), there should exist "cephalized, unossified species" in the early Cambrian of

H. Preuschoft, *Understanding Body Shapes of Animals*,
https://doi.org/10.1007/978-3-030-27668-3_2

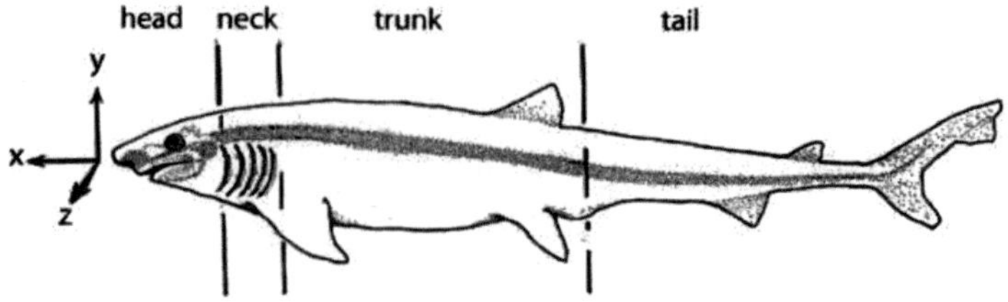

Fig. 2.1 General scheme of body parts in primitive aquatic vertebrates. "Head" comprises the skeletal skull (jaws grey), muscles, mouth cavity, sense organs (dark grey) and brain (stippled outline). "Neck" can be used for the gill apparatus and the structures connecting the head and trunk. In land-living forms, the newly acquired gas-exchange organ (lungs) originates from this region, and its construction enables mobility of the head against the trunk. "Trunk" expands from the anterior pair of limbs to the posterior pair, ending with genital organs and gut exit. It includes the great body cavity, surrounded mainly by the muscular trunk wall and containing the digestive and urogenital organs. Their openings mark the end of the trunk and the begin of the tail. While all other body parts are mirror-symmetric (therefore the term "bilateria"), the digestive track is not, thus deviating from the otherwise maintained arrangement. The axes x (= horizontal and longitudinal, direction of locomotion), y (vertical at right angles to the former) and z (horizontal and bilateral, at right angles to x) are used throughout the text

China. "Gnathostomes and particularly placoderms" became "dominant in the Devonian".

It has never been shown conclusively that transition from water to land, or mode of locomotion (sprawling or extended limbs), or the length of the neck possesses any direct influence on the morphology of the head and jaws. In spite of this, interrelations of this type repeatedly have been the aim of new attempts (like that of Porro et al., SVP in Berlin, 2014).

Because the biological role of *sensory organs* is to inform the locomoting organism about the newly reached environment, it seems immediately plausible that they occupy a place at the anterior end of the moving body. Indeed, in the most primitive of the living Craniota (sturgeons, sharks, batoids), the olfactory, optical and electrical and gustatory sensors, as well as "acoustical" sensors, are parts of the head, most of them enclosed in skeletal capsules of the skull. The same holds true for the organ of equilibrium which is closely connected with the acoustic organ. In bony fishes, amphibians, reptiles, birds and mammals, these capsulae are ossified (at least in part) and firmly fused with the brain capsule. Immediate active contact between the organism and its environment is made by *teeth* and *jaws*. Therefore, their position at the anterior part of the body seems also plausible without closer examination, especially in water-living animals. The upper jaw is tightly connected with the above complex (maybe because the adducting muscles distribute their origins on the bones of the skull) and merges with the above described complex (in mammals, and amphibians, partly in birds and reptiles). The lower jaw, however, retains its independence from the upper jaw and the whole skull. The processes leading to this configuration can be observed in ontogeny as well as in phylogeny. The shape and function of the head are closely related to shape, length and mobility of the neck – which will be discussed below.

For all attempts to understand the shapes of jaws, especially the arrangement and the classes of teeth within a dental arcade, it is tempting to search for explanations on the level of biological roles, following the scheme: For acquisition of grass, a transversely orientated cutting edge (blade, array of cutting teeth) is useful; for diminution of the cell walls at the beginning of the digestive tract, batteries of molarised teeth are essential; or for catching swift and slippery prey, numerous, pointed teeth in rather long and slender jaws promise the best results. Explanations on the level of mechanical function are more difficult to identify, but possible: As a first step, the forces occurring from outside the organism at intake of food must be identified. Biting, chewing and sucking lead to forces acting against teeth or directly against jaws. The next step is determining the equilibrium in the joints between lower and upper jaw, and the third the investigation of the resultant forces and the flow of stresses inside the skeletal elements.

Right from the beginning, the question must be raised, if connective and supporting tissues in the *skull are also optimised for precisely defined mechanical functions* as in the locomotor apparatus. Only if this is the case, the rules of "Wolff's law" or of Pauwels' causal morphogenesis will lead to forms, which are suited with a minimum of material to sustain the loads which occur during daily life. The first approach to an answer is searching the pattern of stresses within the structures of the skull under the loads which occur during "normal" function. The next step is finding out, whether the existing shapes fit to the stresses, so that they can sustain them with a minimum of material (=weight). These questions will be the focus of the following considerations. The existing literature on skull shape does by no means favour or correspond to this idea. Instead, attempts are quite common to understand the head and in particular the skull as an assembly evolved in order to "protect" the brain or sensory organs. Such an idea, by the way, implies teleological expectations![1]

2.1 Teeth

The forces relevant for biting and chewing in most Craniota (exceptions are some fishes, amphibians, turtles and birds) act against teeth, from where they are transmitted to the jaws. Teeth can be expected to be suited for sustaining the mechanical forces acting on them every day. The same holds true for the connections between tooth and jaws, technically the "bearings" of the teeth.

[1] A recently discussed example is the browridges in macaques, which have been presented by Ross in Greifswald (2004) and in Florida (2005) as structures which do not fulfil a mechanical function. The underlying approach is weak in two respects: first, the starting off from the existing shape by no means can clarify which structures are necessary for sustaining the occurring stresses, and second, the spreading of stresses in the skull showed in fact that fairly high stresses flow through the browridges in some phases of the chewing cycle.

Seen from an active animal's viewpoint, food objects can be classified either as "sedentary" or as "mobile". Both may be small and individualised, very small and ingested in great numbers (herbivores while harvesting plants, baleen whales) or large. In the latter case, either the prey must be captured and swallowed as a whole (e.g. snakes), or the large bodies of prey animals must be disconnected – which brings crocs in trouble.

Teeth play different biological roles: catching, holding, disconnecting and comminution. A survey of the mechanical requirements which teeth have to fulfil for performing their biological roles can be obtained Lucas 1979, and especially from Lucas and Luke (1984) and Maier (1984b). These roles have nothing to do with taxonomic position of the animal under consideration. Nobody doubts that teeth, in particular tooth crowns, must have shapes which are suited to play these roles. The acting forces vary with mode of living, but in each case, teeth must be strong enough to withstand them. Whether teeth in fact possess shapes which in a strict sense are optimised for sustaining stresses to which they are exposed has been investigated by Preuschoft et al. (1974) in the case of shark teeth.

2.1.1 Selachian Teeth

Among selachians exist roughly two variants (Figs. 2.2 and 2.3): fangs (catching) and cutting teeth on one side (in most sharks) and crushing teeth on the other (in rays and skates), where both shapes may well occur in the same dentition (as in *Heterodontus*, Fig. 2.2).

The *fang teeth* (Fig. 2.3a–c) mentioned first either hook behind spines or into the joints between parts of the rigid external armour of invertebrates or are inserted into the relatively soft tissue of the prey. The reaction forces acting against catching teeth are either concentrated on the tips or distributed over the external surface or along their margins. Figure 2.4 shows examples of all three variants. Shark teeth can be

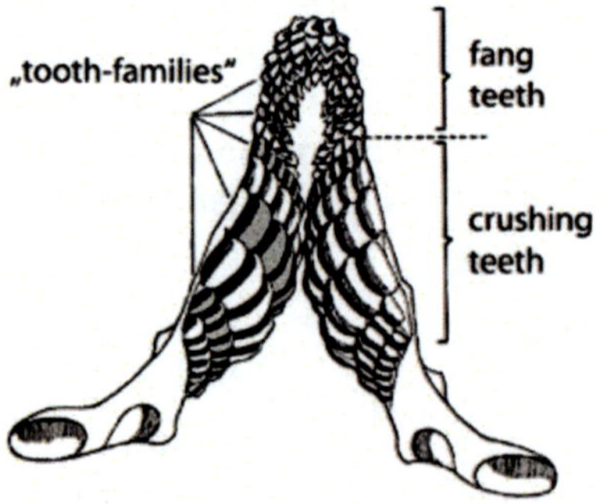

Fig. 2.2 Lower jaw of a bull-headed shark (*Heterodontus portusjacksoni*) showing the diverging shapes of teeth – which is rare among sharks. The closeness of the right and left halves of the mandible allows full muscle activity on both sides (see also below). The joint condyles are nearly at right angles to the tooth rows. (Modified from Preuschoft et al. 1974)

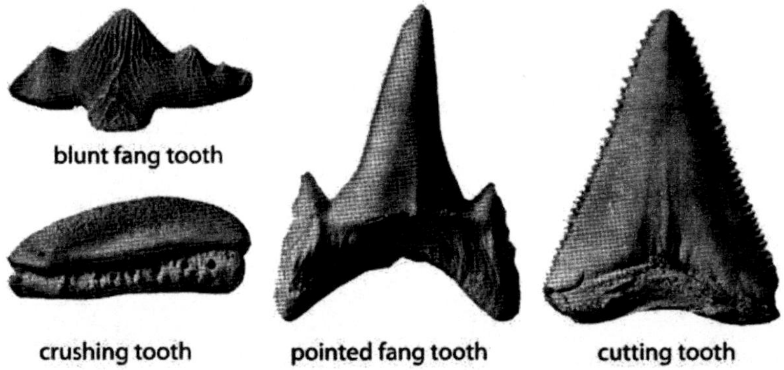

Fig. 2.3 Divergent teeth in Chondrichthyes. From right to left: cutting tooth of *Carcharodon*, pointed catching tooth of *Carcharias*, blunt fang tooth from the anterior part of a yaw in *Heterodontus*, crushing tooth from the posterior part of the jaw in the same species. (From Preuschoft et al. 1974)

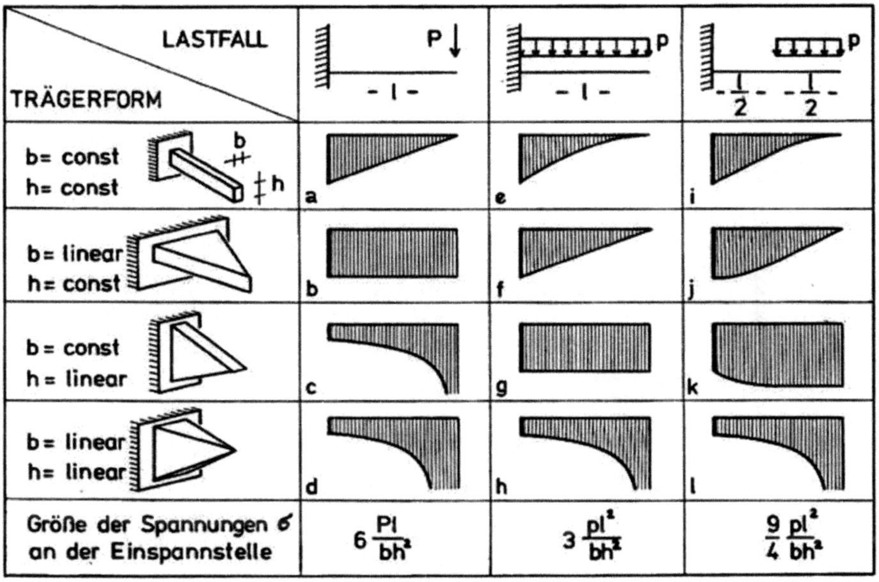

Fig. 2.4 Mechanical stresses in catching and cutting teeth. Three load cases are shown on the horizontal: external force concentrated on the tip, distributed along the whole margin, and distributed over the distal half of the margin. In vertical direction, four characteristic profiles of cantilevers (= teeth) are shown. For each combination, the bending moments are presented in graphical and in mathematical form (bottom line). *P* stands for the external load (in other illus called F_e). (From Preuschoft et al. 1974)

simplified as plates, supported alone at their bases (=cantilevers). Loads are acting either against the, in most cases, triangular labial or lingual surfaces (catching) or against the margins in the plane of the plate (cutting). Depending upon the direction and distribution of the external forces, bending moments are evoked (Fig. 2.4a–l). These can be quite high, particularly near the tips, which are in danger to break off.

It is possible to calculate those dimensions of a cantilever, at which σ = const, that means the dimensions at which the loaded body possesses equal strength at all points: The bending moment (σ) under concentrated load depends upon

$$M = F * \text{lever arm} \tag{2.1}$$

which is a second power.

$$\text{If} \quad \sigma = M/W \tag{2.2}$$

is assumed to be constant, the stresses in the loaded cantilever follow a root function, and the same must be expected by the outer contours. If the load is distributed evenly along the outline of the cantilever, a concave contour leads to equal stress at all places. Figure 2.5 shows shapes of cantilevers (left) or teeth (right), which are suited

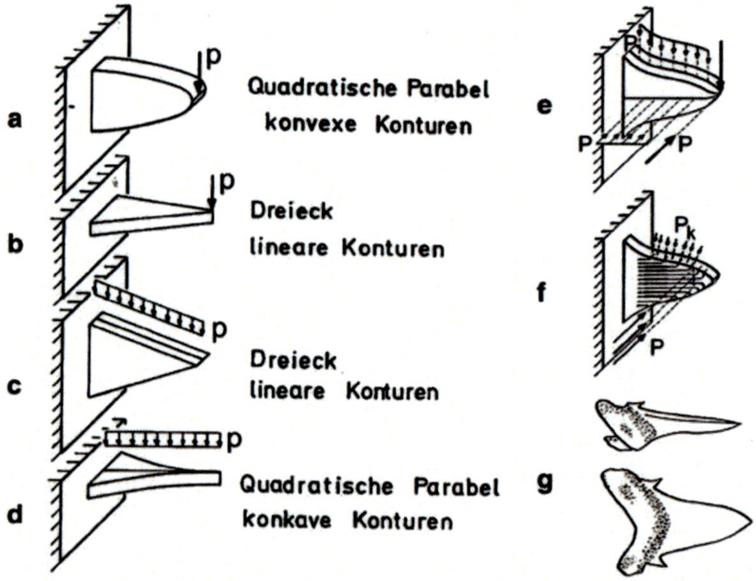

Fig. 2.5 Catching and cutting teeth are bodies of equal strength. **(a–d)** Cantilevers exposed to different external loads (*P*), the upper two concentrated, the lower two distributed. In all cases, those outlines of a cantilever are shown which offer the best fitting strength properties. (From Preuschoft et al. 1974). **(e, f)** Combinations of load cases require combined profiles. External forces acting against a tooth in labio-lingual direction **(f)** are countered not by shape, but by high-resistance fibres covering the plain surface. **(g)**: examples of shark teeth which are similar to the bodies of equal strength against combined loads

to sustain particular stress patterns, because the stresses (σ) assume the same values at all points, so representing bodies of equal strength.

The bending moment M is proportional to bite force F_b, applied to an arbitrarily selected point of the tooth, multiplied by the distance between the investigated point and the basis of the tooth. Bending moment is maximal, if F_b is concentrated on the tip of a tooth with height l.

The moment of resistance (W) is

$W = I/y_{max}$ and shape-dictated strength

$$I = \Sigma \Delta A * y^2 \tag{2.3}$$

with y being the distance between ΔA and the neutral plane of bending.

Many sharks are shaking their heads in the occlusal plane after having inserted their teeth into large prey. Upper and lower jaws are moved in the direction of tooth rows, so that the teeth act like a saw. The individual tooth is exposed to forces, distributed along the mesial or distal margins. Because of its long diameters in mesiodistal direction, teeth have very high moments of resistance in this plane.

An attacking shark can swim faster or slower than its prey. In both cases, the teeth are, once inserted into the prey, exposed to forces distributed over their labial or lingual surfaces. Attempts of a seized prey animal to escape lead to the force application mentioned last. In labio-lingual direction, the fangs of many sharks are flat and therefore offer but low resistance against being bent. Figure 2.6 shows the

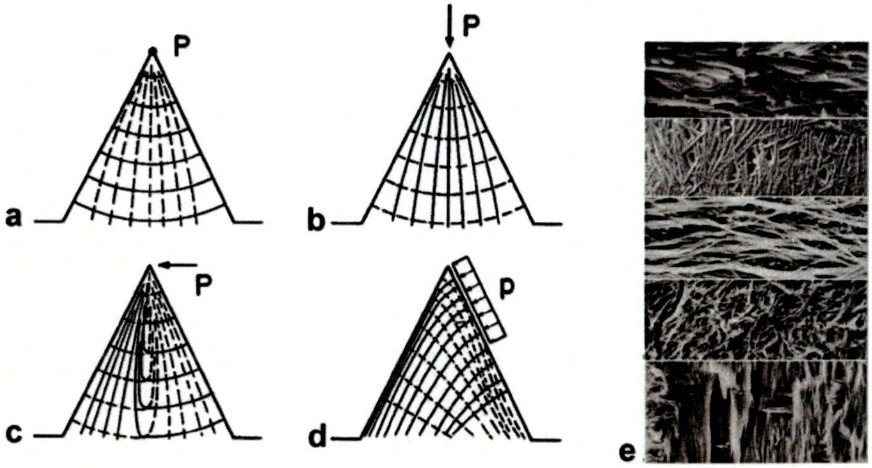

Fig. 2.6 **(a–d)** Trajectories of stresses in a triangular plate as found by application of the photoelastic method. **(a–c)** External loads P (equivalent to the otherwise used symbol F_e) concentrated on the tip, **(a)** perpendicular to the drawing plane, **(b)** directed toward the basis, **(c)** in the plane of the drawing from right to left, **(d)** load distributed over the apical half of the tooth, acting in the plane of drawing. **(e)** Fibres of the "enamel-like substance" found in selachians. (From Preuschoft et al. 1974)

trajectories of stresses found experimentally with the photoelastic method. a, b and c just show the general occurrence of stresses near the surface in longitudinal direction, while being not realistic in their concentration on the tips, Fig. 2.6d indicates a fanning out of the trajectories toward the margins of the tooth under distributed loads. Both surfaces of shark teeth are covered by fibres of enamel-like substance, most of them running in parallel from the basis toward the tip, or the edges of the teeth (Fig. 2.5f and 2.6). This substance offers about six times the strength of shark dentine, if loaded in the direction of the crystallites (Preuschoft et al. 1974). The longitudinal orientation of trajectories and fanning out near the margins are similar to the arrangement of enamel-like fibres. They obviously reinforce the teeth in the plane which must be flat (that is weak) in order to fulfil their biological role. Slender shapes or at least narrow diameters in one direction favour deep insertion into the tissues of the prey. The same effect can be improved by specialised cutting edges, either in the form of hollow ground or by fine serrations. The cutting edges of the teeth consist of enamel-like substance rather than dentine, as shown in Fig. 2.6[2] and 2.7 which is again arranged to fit to the trajectories of stresses.

The *crushing teeth* in rays and the bull-headed shark (Nobiling 1977) possess a very resistant form and a surface, which concentrates the forces exerted between upper and lower jaws on small areas of the prey (Figs. 2.7a and 2.8). By means of the so-created enormous bite pressure, they can destroy, or break down, or crash the hard shells of bivalves or crustaceans. On the side of the teeth, the high and concentrated forces in unpredictable, varying directions are best sustained by dome-like constructions. The enamel-like substance in these crushing teeth does not consist of a thin layer of parallel fibres, but forms sturdy caps which can sustain three or four times the breaking stress of dentine (Preuschoft et al. 1974) by which they are underlain and supported. The individual teeth represent a three-dimensional structure, curved in two planes. The low vault of these “domes” prevents bending stresses. The close-packed arrangement of the teeth (resembling a paved road) allows a bracing of the hard enamel caps against each other. So a continuous hard cover of the jaws is formed. A part of the force applied from outside is sustained by the hard enamel-like caps, and a part is passed on to the underlying dentine. The external surface is often crenulated, whereby the wrinkles reinforce the caps, at the same time concentrating bite pressure on tiny areas and excluding a slipping of the prey out of the jaws.

A comparison of the “optimised” forms in Figs. 2.5 and 2.7 with the teeth of various sharks yields good coincidence of fang teeth with mechanical necessities (Fig. 2.7) and biological roles. If the first are in conflict with the latter, the observed shapes have the character of a compromise. In some cases, it seems plausible that the behaviour of the animal is adapted to the properties of the teeth, that is, teeth are used in such a way that undue stresses are avoided.

The outline of a second power parabola (Fig. 2.5a), as can be seen in fang teeth, provides optimal resistance against loads which are concentrated on the margins near the tip, but is not optimal for easy and deep insertion into a prey. Against the same

---

[2]Why cutting edges facilitate insertion is discussed below.

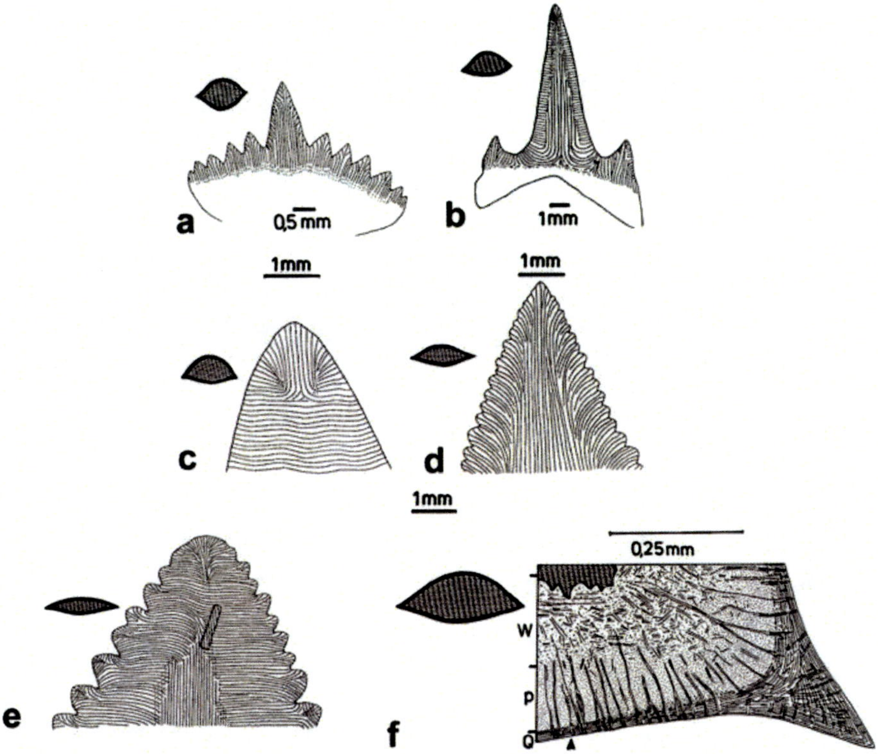

Fig. 2.7 (**a–e**) Arrangement of enamel-like fibres on the surfaces of shark teeth. In all teeth are cross sections added. (**a–c**) The cutting edges at the margins are smooth, and in (**d** and **e**), the margins are denticulated. On the surface of these denticles, the fibres also fan out toward the highly stressed cutting edges. (**f**) Cross section showing the arrangement of crystallites on one margin of a tooth with smooth cutting edge. The fibres form a pattern similar to what is technically required, as shown in Fig. 2.6d. (**a**) *Ginglymostoma*, (**b**) *Odontaspis* (*Carcharias*), (**c**) *Isurus*, (**d**) *Carcharhinus*, (**e**) *Carcharodon*, (**f**) *Odontaspis*, cross section. (From Preuschoft et al. 1974)

load, when concentrated on the labial or lingual surface, a triangular, pointed outline offers the required strength. If the apical part of the tooth is inserted into the tissue of the prey, the external force is distributed over a part of the margin or of the surface. In this case, the stresses grow less than linearly, and a sharp, pointed tip is not in danger to break off.

2.1.2 *Fixation of Teeth*

The teeth of all Chondrichthyes sit on top of the jaw cartilage on their broad basis (Figs. 2.2 and 2.9), by which bite pressure is distributed on a large surface. Shark teeth are used for only a short period and are soon replaced by new ones, which

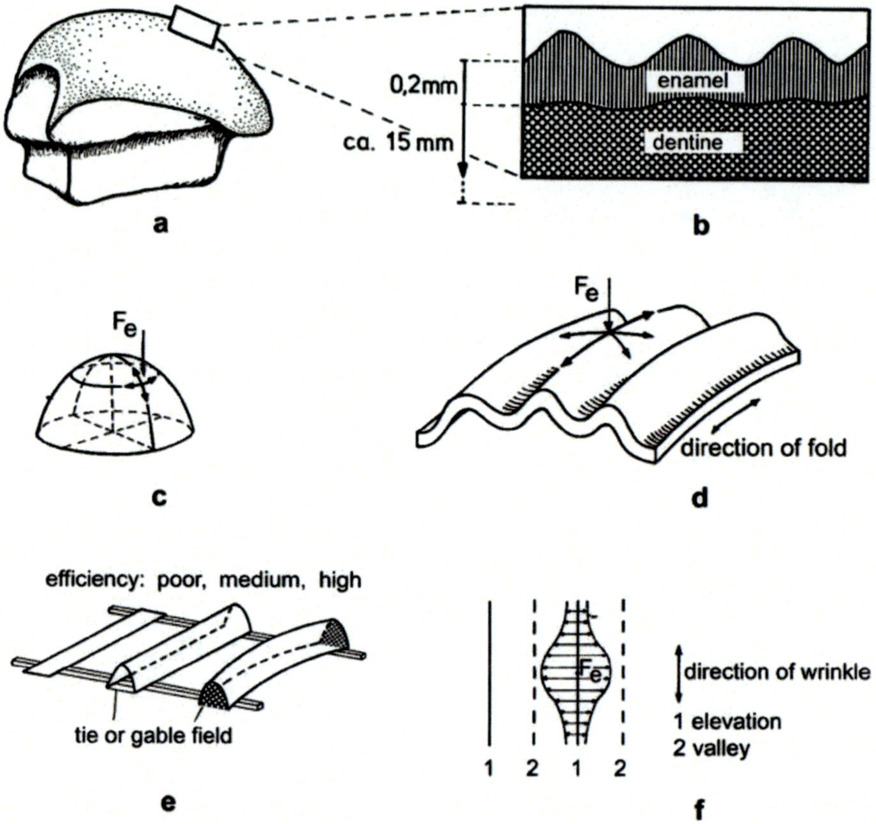

Fig. 2.8 **(a–b)** Isolated crushing tooth of *Heterodontus*. The enamel-like substance covering its surface is underlain and supported by the more compliant dentine. **(c)** A load acting from outside (F_e) on a dome-shaped structure evokes stress flows in two directions. **(d)** Schematic illustration of the wrinkled hard cap of enamel-like substance. The directions in which stresses distribute are indicated. **(e)** A thin layer of strong material offers little resistance against being bent if flat, much more resistance if curved in one direction and still stronger if curved in two directions. Strength of the folds can be increased by a tie between the sloping walls (Zugband) or by a gable wall (Giebelfeld). **(f)** Distribution of the concentrated load F_e on a wrinkle curved in two directions. (From Preuschoft et al. 1974)

follow on the lingual side (Fig. 2.9). The pneu-like (hydraulic) construction of the cartilage (see below, Fig. 2.21) does not permit an imbedding of the teeth similar to that observed in other animals. Comparative anatomy discriminates between acrodont, pleurodont and thecodont anchoring of teeth in the jaws.

The teeth of reptiles and of mammals, in contrast to shark-like fishes, are imbedded in niches of the jaws, the so-called alveolae. Transmission of bite force takes place from the roots of teeth through the strong fibres of the periodontium to the bony walls of alveolae. Especially in humans this sort of fixation by oblique

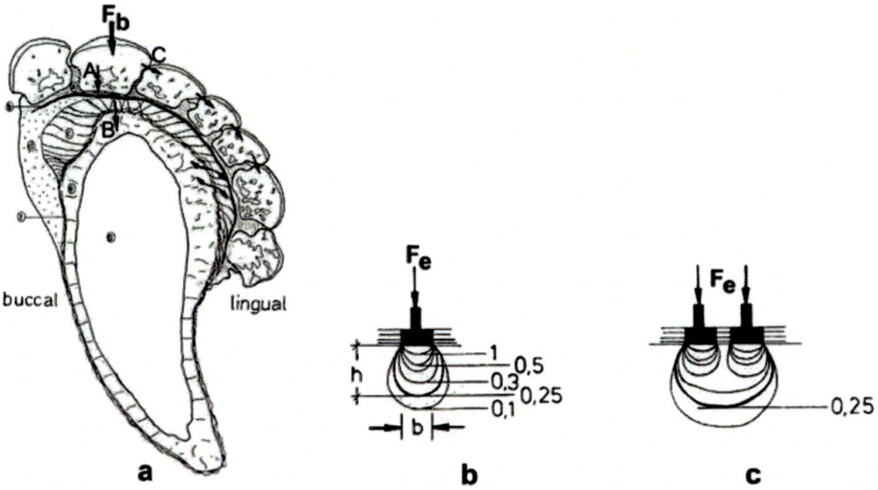

Fig. 2.9 Anchoring of teeth and transmission of bite force from a tooth basis on the underlying material. **(a)** Cross section through the posterior part of the lower jaw of *Heterodontus*. Only a part of bite force (F_b) is transmitted directly to the basis of the tooth; another part is transmitted to the neighbouring teeth (arrows). Material immediately below the tooth basis is compressed; on the lingual side of the jaw, tensile forces grow more, if farther away from the bite point. **(b)** Stresses are high below the basis and decrease gradually in the form of onion shells. If several force applications are close to each other (**c**), the zones of stress merge with depth below the surface. (From Nobiling 1974)

Sharpey's fibres is well-known (Fig. 2.10a). Witzel et al. (1998) have documented by two-dimensional FE models the mechanical relevance of the arrangement and direction of fibres, as well as of splitting the roots into several pillars. The apical openings, used by capillaries as entrance into and exit from the dentine cavity, are mechanically not stressed.

An outstanding example of the fixation of highly loaded mammal teeth has been investigated by Stefen et al. (2015). According to an earlier study (Stefen et al. 2011), beavers can (as probably all rodents) separate chewing completely from gnawing. The latter is mainly effected by the lower incisors which are a perfect tool for cutting and planing, while the upper incisors form a prop, fixed to the gnawed tree or branch, against which the lower teeth are working. The tips of the gnawing teeth are moved forward-upward during the power stroke. The resultant of adduction of the lower jaw and forward movement has exactly the direction of the lower incisor, but because of the curvature of the tooth, this resultant deviates in a narrow angle from the tooth axis (Fig. 2.10c). The very rigid tooth so is exposed to bending. At its compressed side, the thick enamel layer is the strongest material of the entire body. Tensile forces between the lingual (occipital) margin of the tooth and its alveola are sustained by the densely packed Sharpey's fibres of the periodontium. The pull of these collagenous fibres redirects the resultant force gradually in such a way that the longitudinal axis of the tooth coincides everywhere with the acting force

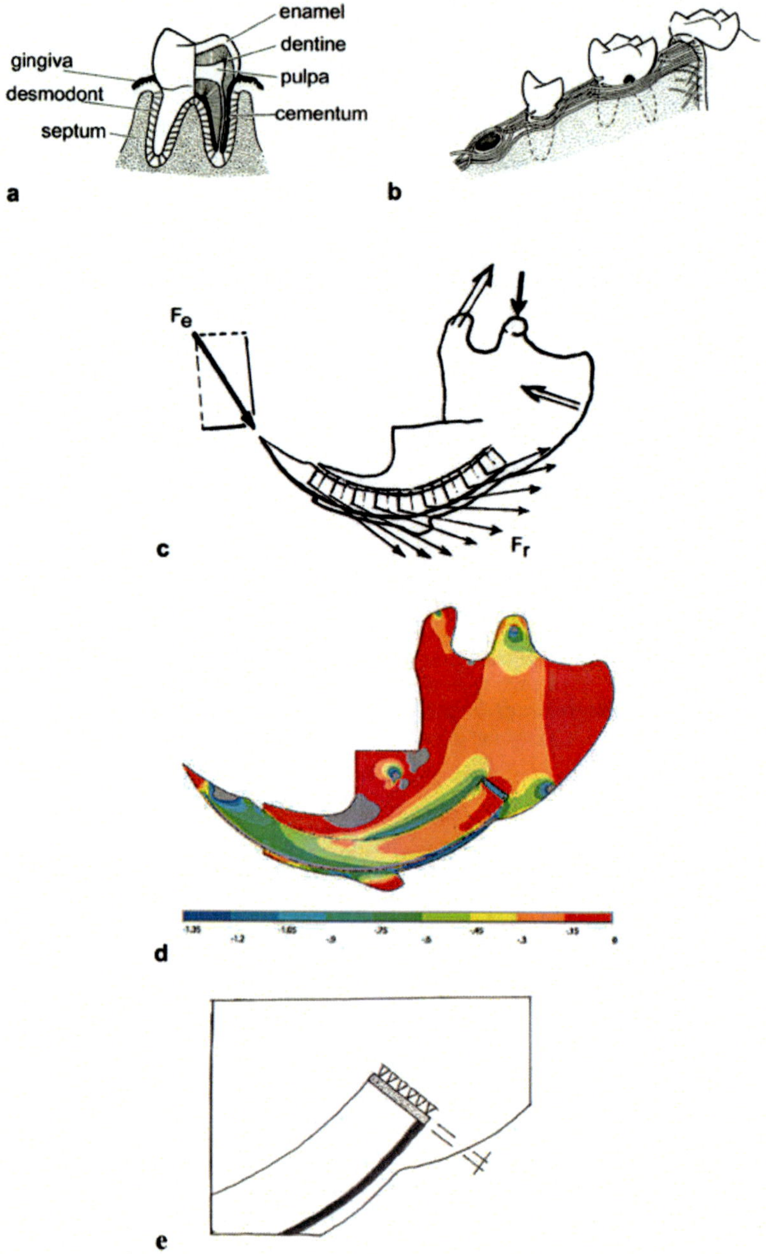

Fig. 2.10 Fixation of teeth in mammals. (**a**) Human molar, right half cut open to show the dentine (hatched) and enamel cap (white). The oblique Sharpey's fibres are concentrated in the middle of the root, in full accordance with the distribution of tensile stresses found by Witzel et al. (1998) in FE models. (**b**) The collagenous periosteum, immediately below the gingiva is remarkably thick, especially along the tooth rows. For the sake of overview, the distances between the teeth are enlarged and the anterior premolar is removed. The major directions of fibres are emphasised. (**c–e**)

(**b**). At the rostral side of the tooth, however, the periodontium forms nothing but a thin layer which transmits the remaining compressive components to the alveolar wall and allows minimal sliding of the tooth. The apical part of the permanently growing tooth is exposed to minor displacements, but not to relevant forces.

2.1.3 Teleost and Reptilian Teeth

In many bony fishes, amphibians, reptiles (in both groups fossil and recent) and the canines of mammals, fang teeth are more or less *conical in shape*. This shape is observed so frequently that it is taken as a reliable indicator of predatory habits. If the most frequent two-dimensional load cases to which teeth are exposed (as shown schematically in Fig. 2.11) are rotated about the long axis of the tooth (vertical in Fig. 2.11), the optimised two-dimensional outlines must occur in all directions and so characterise a conical shape with outwardly curved contours. The three load cases require slightly different optimal shapes. The tooth shapes that can be observed in existing animals seem to represent compromises, which offer sufficient strength to the most probable loads. While sharks with their frequent replacement of teeth can afford extremely sharp cutting edges and sharply pointed fang teeth, although they are prone to failure, the tips in mammals (where they have to endure for a lifetime) are not so sharp, but offer more strength against breaking off. If an animal uses its teeth exclusively for, e.g. catching fish, high external forces concentrated on the tooth tip are not probable and occur rarely. Therefore, the tips can be more pointed. If teeth are primarily used for perforating prey in order to allow the secretes of digestive glands better access, they do not need very sharp points (which make them fragile) but can assume a stronger, highly resistant form. Indeed both variants can be observed, for example, in the jaws of *Tomistoma* or *Crocodylus porosus*, where the slender, spiky anterior teeth are optimised for catching small, not armoured and swift prey and the stout, blunt posterior teeth optimised for perforating

Fig. 2.10 (continued) FES Analysis of the lower jaw plus incisor of a beaver (*Castor fiber*). (**c**) The lower jaw is hung up by mm. temporalis and masseter/pterygoideus lateralis; braced at the joint; the external force (F_e) acts concentrated on the cutting edge of the gnawing tooth. Within the tooth, the same external force deviates from the long axis of the tooth. Because the tooth is tied by numerous, radially orientated Sharpey's fibres to the lingual side of the alveola (the relevant components are indicated), the resultant forces (F_r) are gradually redirected (and reduced), so that the remaining resultant forces fall in line with the tooth axis. (**d**) The FESA model of the lower jaw shows high compressive forces at the labial side of the tooth (green), where a thick layer of enamel gives additional strength. Compressive stresses (green) also occur at the lingual side of the tooth root, where they reduce the tensile forces between the alveola and tooth, so giving reason for the observed concentration of tension-resistant fibres in the middle of the gnawing tooth. Stresses are low in the pulpa and near the apex of the root. (**e**) Minimal movements of the apex during the power stroke of gnawing. (Modified from Stefen et al. 2015)

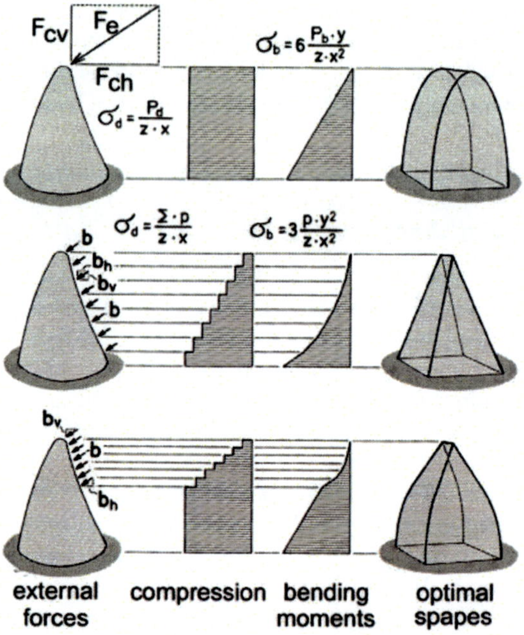

Fig. 2.11 The stress patterns of typical fang teeth, as present in many bony fishes, amphibians (fossil and recent) and reptiles, as well as in canines of mammals, depend on how deep they are inserted. Teeth have conical shapes and more or less rounded cross sections (left side). Three load cases are shown schematically: On top, the tooth receives a load concentrated on its tip (F_e). In the middle, the deeply inserted tooth is exposed to the same load, which is, however, distributed on its height. The partial loads are called *b*. On bottom, only one half of the tooth is inserted into the prey, and therefore the partial loads (*b*) are confined to the area close to the tip. The loads exert compressive components (left diagram) and bending moments (diagrams on the right). On the far right, bodies of equal strength against the same loads are shown. If the loads are rotated about the long axis of the teeth, bodies of equal strength need a cone-like shape, instead of a 2D outline. In the formulae, *P* stands for total force, *p* for a part of it. *y* is the height of the tooth, or the distance of the investigated plane from the tooth tip, *x* is the "height of the tooth in the plane of the drawing, and *z* stands for breadth perpendicular to the drawing plane. The teeth can be understood as bodies of equal strength against bending loads that act in arbitrary planes. (From Preuschoft et al. 1991. See also Fig. 9.35)

the tissues of the prey animals with high bite forces. Similar divergence among teeth can be seen in some theropods.

In a majority of cases, fang teeth are not straight, but possess an *occipitally concave curvature* (Figs. 2.11, 2.12, and 2.13). This follows more or less the periphery of a circle around the mandibular joint. These teeth follow the trajectory of movement in closing the jaws. This seems fine, because in high and slender teeth, the crown follows the canal opened by the sharp tip. Completely straight teeth would be bent more strongly and require more force to being inserted, because they must cut open an elongate opening, rather than a simple puncture. The curvature of the

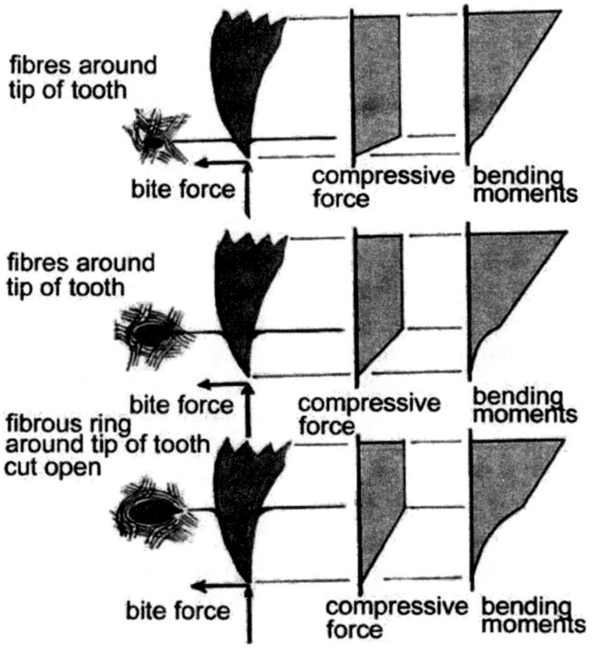

Fig. 2.12 Three stages of inserting a fang tooth into the dermis of a victim. The external force is distributed over an increasing part of tooth height, but at the tooth basis, the same bending moments are reached in all cases – as long as the bite force is kept constant. The first column shows the effect exerted on the fibres of the subcutaneous fibres of the corium. The tip of the tooth pushes the fibres aside. If the tooth is inserted deeper, its increasing diameters squeeze the fibres into a circle surrounding the tooth. When the fibres are tensed, no deeper insertion is possible, unless tension in the fibrous ring is released. One or two cutting edges cut the fibrous ring open, permitting deeper and less force-demanding insertion of the tooth. Columns 3 and 4 show the compression and the bending moments along the tooth. (From Preuschoft et al. 2011)

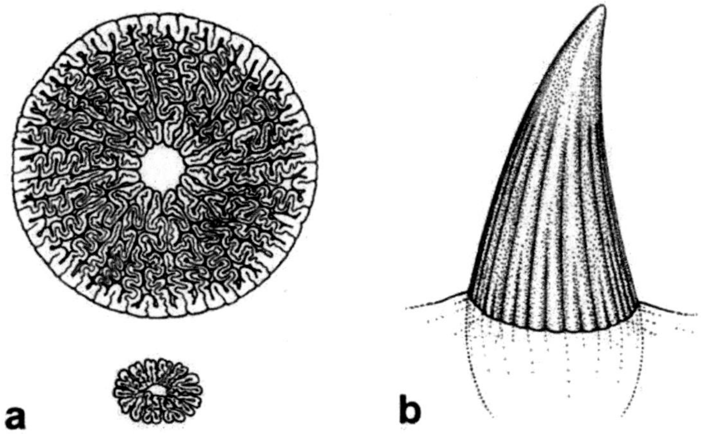

Fig. 2.13 Fully developed labyrinthodont tooth. (**a**) Cross section near the root (top) and near the tip (bottom), (**b**) lateral view on a "folded" tooth

long axes of fang teeth may as well be a means to counteract the attempts of the prey to escape. By such attempts, a force component occurs in rostral direction (Figs. 2.5, 2.11 and 2.12), so that the total bite force is somewhat inclined and greater than without this component. In a curved tooth, the bite force does not deviate so much as in a straight tooth, and this means the bending moments acting along the tooth are smaller. In addition, the danger of slipping out of the mouth is reduced.

Figure 2.12 shows the insertion of a pointed tooth into the skin and the subdermal connective tissue of a victim. The tooth tip first pushes aside the fibres of the corium. These become stretched and tensed, until they together form a tight, close-fitting ring of fibres around the tooth, which eventually prevents it from being inserted deeper, because the fibres cannot be further elongated. The more slender the tooth and the more acute its tip, the lower will be the resistance against deeper insertion. The inclusive disadvantage of slender teeth, however, is the danger of breaking. A sharp cutting edge along the tooth slits open this fibrous ring and so allows deeper insertion. Cutting edges are very common among predatory animals and may exist on two sides of a tooth or on only one side, which may be hollow ground or serrated (sharks, theropods, lizards, basal synapsids, some carnivorous mammals, pigs and primates). The claws of cats also possess such cutting edges (Homberger et al. 2007). Humans, producing spearheads or arrowheads, knew and made use of this effect since early prehistoric times: Everywhere in the world, the points of thrusting weapons were equipped with cutting edges.

A number of animals of diverging taxonomic position possess so-called labyrinthodont teeth (Preuschoft et al. 1991; Figs. 2.13 and 2.14). The characteristic is the longitudinally folded dentine, covered by a thin layer of enamel. The characteristic "plicidentine" and tooth shape were formerly used as a taxonomic marker, applied to the earliest Devonian terrestrial "amphibians" and their forerunners (Crossopterygians), but these teeth are also found in Jurassic and Cretaceous ichthyosaurs, fossil and recent Varanidae and recent *Lepisosteus*, forms which are taxonomically not related. Instead, the folding of the dentine represents an interesting variant of "adaptation": Dentine in general, and plicidentine as well, is deposited from the periphery inward, the dentine-building cells (odontoblasts) being part of the tooth pulp. The folding gives the tooth considerable strength, even if the dentine layer is thin. With increasing age, more and more dentine is deposited on the internal side of the tooth wall, until the whole cross section is filled up with dentine (Figs. 2.13 and 2.14a). The mature labyrinthine tooth reaches the strength of a tooth with compact walls. As shown in Fig. 2.14b, c, based on empirical data from an *Ichthyosaurus* specimen, the folding gives the immature tooth considerable strength during its rapid growth. This allows it to play its biological role in an early stage, before it has developed its permanent shape. Still its strength lags behind the situation in mature teeth (Fig. 2.14c). In other forms (*Benthosuchus*), the strength values of immature teeth are much closer to the mature state. The knack is that these labyrinthine teeth occur in animals, which possess but shallow jaws (the shapes of which will be discussed below in Chapter 2.2). In these mandibles and maxillae, big teeth do not find the space required for developing their definite shape, but the folding allows

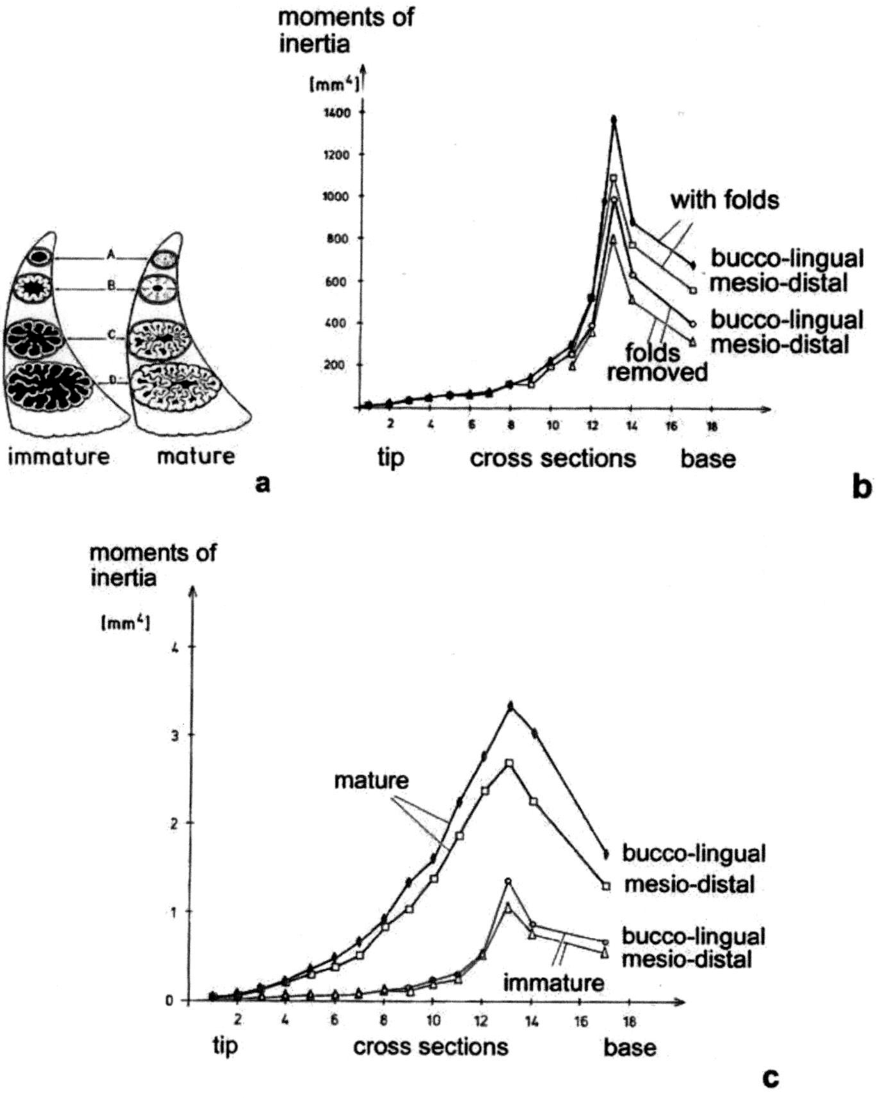

Fig. 2.14 (**a**) Immature (left) and fully developed (right) labyrinthodont teeth, the former with a thin layer of dentine at the external surface and the latter with thick dentine layer, which fills the space between the external surface and marrow cavity. (**b**) Area moments of inertia as indicators of strength against being bent on 17 cross sections (numbers 1–18). Values are calculated in mesiodistal as well as in labio-lingual direction; they increase from the tip (left) to the root (right). The high values are from young teeth with thin dentine layers; the two lower values are calculated after removing the folds on the side of the pulpa. Obviously, the folding yields increased bending resistance. (**c**) Comparison of area moments of resistance in immature and mature teeth (with thick dentine layer, forming a compact wall). Both series are from *Ichthyosaurus*. (From Preuschoft et al. 1991)

them to grow early beyond the level of the jaws, so surmounting the alveolar margins, and to sustain their biological loads already in an immature state.

2.1.4 Mammalian Teeth

In mammals, different *classes of teeth* are distinguished: incisors, canines, premolars and molars (Fig. 2.15a–d. The names of these classes already anticipate ideas about

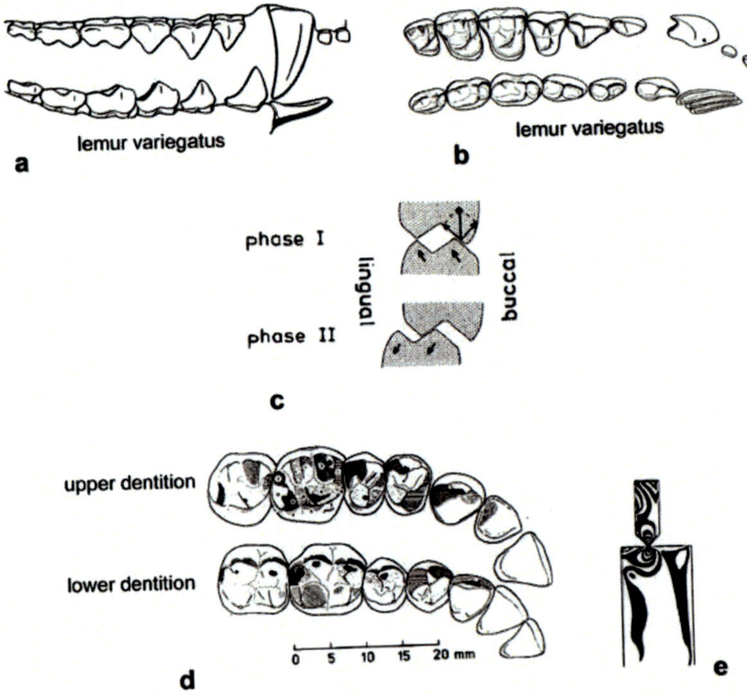

Fig. 2.15 (**a, b**) Tooth rows of a strepsirrhine primate (*Lemur variegatus*) in side view (**a**) and occlusal view (**b**) to show the different classes of teeth (from left to right: 3 molars, 3 premolars, canine, 2 incisors in the upper, canine plus two incisors combined to form a tooth comb in the mandibular tooth row). Deviating from most other mammals, the anterior premolar looks and functions in many strepsirrhines like a lower canine. (**c**) Two cusps of a molar during power stroke, phase I on top and phase II on bottom. The pair of short arrows indicate the movement of the lower teeth. At the contacts to the upper teeth are forces evoked, which can be broken down into components parallel and "normal" to the contact surfaces. Sliding along the flanks of the cusps leads to scratches. (**d**) Dental arcades of *Homo* as example for "higher" primates. The upper tooth row is mirrored. Wear facets resulting from contacts between the cusps of molars and movements of lower against upper teeth during the phases I and II of the power stroke (**c** and Fig. 2.17c) are marked by the same hatching. (From Maier 1980). (**e**) According to a photoelastic model, the concentration of force entails intolerably high stresses in sharp tips as well as immediately below plane surfaces. This leads to breaking away of lens-shaped chips, which leave pits behind

the biological roles played by the teeth. At least in humans, the four upper and lower incisors form, thanks to their chisel-like shapes, something like forceps (in prehistoric and early historic times) or scissors (in most living humans). In our closest biological relatives, the apes, the incisors assume after a short time of abrasion very large contact areas with their opponents, which provide but little possibilities of "incision". While human canines fit into the row of incisors (from which they are separated by their longer roots and somewhat more pointed shape), they stick out in other primates, in carnivores, insectivores and some hooved animals. This means that they come into contact with the bitten object or victim first, before other teeth, and their pointed shape favours piercing the surface of the bitten. Several Old World monkeys, as well as pigs (Suidae) and tragulids, sharpen the margins of their (upper) canines permanently by sliding contacts with their opponents (lower canines and first premolars). The so produced cutting edges favour deep insertion into soft parts and indeed are used to inflict deep and extended wounds to the soft parts.

The exact role of the *premolars* is little known and may be quite variable in view of their variable shapes, which span from canine-like fangs to molar-like devices for coping with extremely resistant objects.

The role and function of *molars*, however, have been investigated in much detail.

Lucas and Luke (1984) took their starting point not from the teeth and their movements, but from the food material consumed and its properties (Table 2.1). They discriminated not more than three categories, among which the first is characterised by the ability to resist compression and bending and the last by its resistance against tensile forces. The juicy, fluid contents of the third category are wrapped in as a whole or subdivided in compartments by sometimes quite resistant covers (cuticula of arthropods, shells of mussels or snails). The functions, to which teeth are optimised, are destroying the shells of seeds or nuts (category 1), breaking

Table 2.1 Food categories

	1 Hard, brittle	2 Juicy	3 Tough
Deformability	Low	Low	Pliant
Strength	High	Low	Moderate
Toughness	Low	Low	High
Strain at failure	Low	Low	High
Examples	Unripe fruits	Ripe fruits	Leaves
	Seeds	Grasses	
	Nuts	Animal tissues	
	Arthropods	Arthropods	Arthropods

Taken from Lucas and Luke (1984)

Insects, which are preferred by small primates (<200–300 g), are comparable to juicy fruits, but may be armoured by a hard, brittle and tough cuticula. Specialised fruit eaters weigh around 1000 g, while species above >2000 g take in a growing share of leaves, belonging to column 3 (Maier 1984a, b)

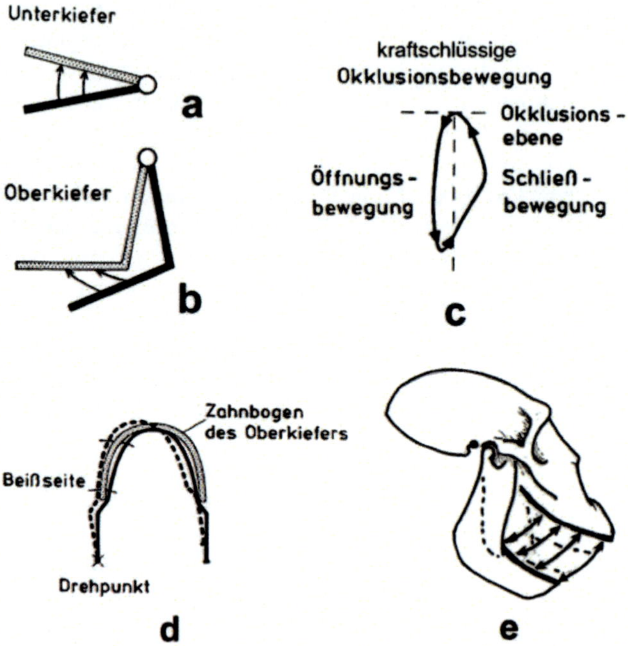

Fig. 2.16 Movements of the mandible against the upper jaw in mammals. (**a**) Mandibular joint in the height of the occlusal plane. Bite forces are nearly perpendicular to the upper margins of the mandible. (**b**) Joint elevated above the occlusal plane. The lower teeth have a marked component of movement along the upper tooth row in rostral direction. To avoid a discrepancy of contacts between anterior and posterior teeth, and to secure equal contacts between lower (= mandibular, black) and upper (= maxillary, grey) teeth, the tooth rows must be slightly curved (as in **e**) rather than straight. (**c**) Movements of the most rostral point of the mandible in anterior view during each chewing cycle. It begins with the downward-directed "opening stroke", changes its direction for the closing, or "approaching stroke", and exerts force against the bolus in the "power stroke", which follows the profiles of the cusps in phases I and II in Fig. 2.15c. (**d**) View from above on the occlusal plane of primate tooth rows. Upper dental arcade grey, lower jaw at the beginning of the power stroke a broken line, at the end of the power stroke, a continuous heavy line. The circles of forces between the lower and upper jaw are closed through the bolus on only one, the "bite" side. On this side, the mandibular condyle rests in its socket. (From Wolff 1984). (**e**) Fancy mammal with joint elevated above the tooth rows. To secure equal contacts between upper and lower teeth, both tooth rows are slightly curved

down the resistance of compartment walls or the external body wall of category 2 and squeezing out the contents and finally cutting the fibres of tough materials (category 3). The fibres contained in the connective tissues of animals are easily digested once swallowed to reach the digestive tract. Simple cutting blades are adequate means to produce seizable portions (Fig. 2.16a). The fibres contained in plant cell walls, however, cannot be attacked by digestive juices of vertebrates. They lack the enzymes needed to metabolise structural polysaccharides of the cell wall.

So, in order to get hold of the rich resources contained in the cell, most, if not all, herbivorous mammals harbour a gut microflora capable of cellulose digestion (Waterman 1984). These microbial cultures thrive best on large surfaces and therefore must be provided with mechanically comminuted material. This mechanical breakdown is the proper function of mammalian molars.

Nearly at the same time, Hiiemae (1976, 1978, 1984), Kay and Hiiemae (1974), Crompton and Hiiemae (1970), Kay (1975, 1978), Maier (1980, 1984a, b) and Maier and Schneck (1981) have found out the different ways how comminution of food material is accomplished between the *multi-cusped molars*. With the aid of x-ray films, Hiiemae showed that many (probably all) mammals chew alternatingly on only one, the "working (or chewing, or bite) side". The mandibular condyle of this side is in contact with its socket. The other, "balancing side", is not loaded, and its condyle not in contact with its socket (Fig. 2.16d). After few (about three) cycles, the sides are changed, so that the former balancing side takes over the bite forces. During each cycle, three phases occur regularly: lowering the mandible in the "recovery", or "opening stroke"; approaching the lower teeth to those in the upper jaw in an upward laterally directed movement, until the substance between the teeth ("bolus") is fixed ("approach" or "closing stroke"); and finally the squeezing of the bolus between the upper and lower teeth. This phase (the so-called "power stroke") is executed with great force, as documented by EMG. This leads first to a further approaching of the tooth rows in phase I (Figs. 2.16c and 2.15c) and second to a minimal moving downward of the mandible (phase II). In both phases upper and lower teeth touch each other, the flanks of their cusps sliding along each other, while muscle activity is not interrupted. This sliding along the slopes of the cusps leads to characteristic wear facets first in the enamel (if this is worn down in the dentine), which are visible under the microscope (Kay 1975; Maier 1980). Grine (1981) has differentiated between scratches and pits (("microwear"). *See also in* Chap. 9). The former are consequences of friction (Figs. 2.15c and 2.17k); the latter are caused by concentration of high forces on very small surfaces (Fig. 2.15e). This is typical for contacts between hard, brittle objects (e.g. seeds) and the teeth. As can be seen from Fig. 2.15e in a photoelastic model, the highest concentration of forces occurs slightly below the (lower) flat surface. Breaking away of the lens-shaped flake leaves behind a pit. By the way, the same model also shows the stress concentration in the tip, which has been mentioned above. Kay and independently Maier have investigated the wear facets of primates (Fig. 2.15d) and related the lower teeth and their wear facets to those of the upper teeth. This allowed reconstruction of phases I and II of the power stroke (see below).

While the above-mentioned authors were focussed on primates, v. Koenigswald (1980, 1982, 1988, 1990) has looked first at the enamel structure in at the teeth of rodents, later in other mammals. Janis (1979) and Fortelius (1985) investigated primarily ungulates (Fig. 2.18). The same principles seem to be realised in these animals on diverging pathways. Different from primates, they are consumers of hard, tough, fibre-rich plant material, especially grass. They possess very high cusps, the

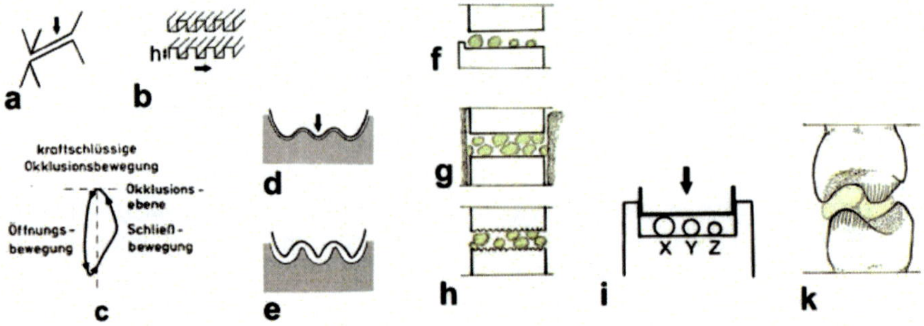

Fig. 2.17 Teeth suited for breaking down food materials as listed in Table 2.1. (**a**) The simplest way of clipping long fibres (no matter whether of plant or animal origin) into pieces is cutting blades. Each bite yields just two fragments. A notch in one of the blades (e.g. carnassials in carnivores, **c**) prevents slipping of the bolus along the tooth rows in rostral direction. It may (but not necessarily) be opposed by a cusp on the opposing side. The cutting system works well if few, big lumps of flesh, connective tissue, or bone, must be separated. If, however, great masses of grass fibres must be broken down into particles of few millimetres in diameter, the number of cutting edges must be multiplied (**b**), while their lengths can be reduced. Such an arrangement, which produces not only two but numerous fragments by each stroke, is realised in many folivores and graminivores (Fig. 2.18a–g) Juicy, more or less liquid content can be pressed out of its envelop by surfaces parallel to each other (**c–i**), no matter whether they are plane or varying in height. Hard and brittle objects of small size (seeds) are exposed to extremely high force concentrations between smooth plates (**f–h**). Escaping of food items from the occlusal surface is inhibited in (**f**) by a high margin on one side of the lower tooth and in (**d, e, g,** and **i**) on both sides. In mammals, the food particles are usually fixed between the teeth by the deformable, slippery cheeks (**g**, left) and the tongue (**g**, right.). Another efficient means to keep objects between the teeth and to exclude slipping away are wrinkles on the occlusal surfaces (**d, e**). Larger objects between the teeth will be broken down repeatedly during one approach of the molar teeth (**i**). The larger the crown surfaces, the more hard objects can be crushed by the same power stroke (provided that enough muscle force is available!). The flatter the crown surfaces, the smaller and the more numerous the fragments. In other words, the efficiency of chewing grows with the size of the occlusal surfaces and is limited by the force, which is concentrated on each object (e.g. seed). (**k**) The mortar (stippled) and pestle (hatched) system, as common among primates. Juicy fruits or arthropods are squeezed between the cusps as "pestle" and "mortar". The system also works well to comminute hard and brittle material. (**a–c**, **i** taken from Lucas and Luke 1984)

interspaces of which being filled with cementum.[3] In several lineages, these teeth grow permanently. On the chewing surfaces of the teeth, food material subsequently comes into contact with lamellae of three different properties, dentine, enamel and cementum, among which enamel resists abrasion longest. The diverging speed of abrasion leads to formation of sharp cutting ridges on top of the enamel lamellae, which are able to destroy even the highly resistant plant fibres, as noted above the

[3]Cementum is a bone-like hard substance, which is produced on the external side of a developing tooth. It can be found in primates on the external surface of neck and roots of teeth. The fibres of the periodontium (or desmodont) find their insertions in this layer. Near the apex of the roots, cementum is rich in cells and thick; in the region of the tooth neck, cells are rare, and the cementum layer is thin. In the extremely high cusps of rodents and hooved animals, cementum fills the narrow gaps between the cusps, so that the chewing surface after initial abrasion consists of three substances, enamel (extremely hard, nearly exclusively mineral substance) and dentine and cementum, both containing <70% mineral substance and dentine being harder than cementum.

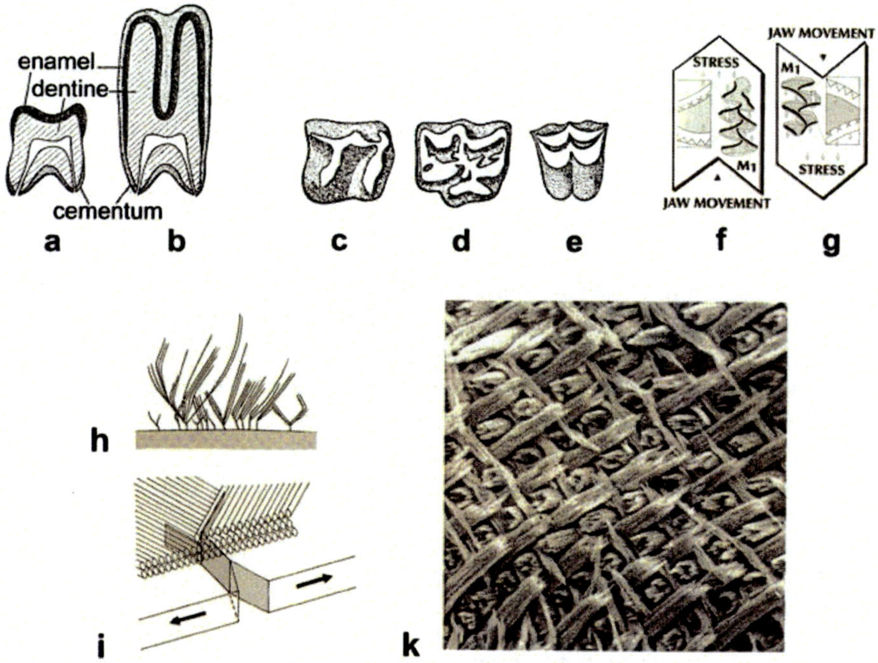

Fig. 2.18 (**a**) Low-crowned molar of, for example, a primate. (**b**) Permanently growing molar of a hooved animal. By chewing, the enamel caps of cusps are worn down (black) and the less resistant dentine or hatched (white) or cementum (stippled) is exposed. Wear leads to the characteristic patterns visible on the occlusal surfaces of (**c**) rhino, (**d**) horse and (**e**) cattle (from Romer and Frick). The three materials form grinding surfaces which provide series of cutting edges in different directions. (**f, g**) Chewing surfaces of lower (**f**) and upper (**g**) molars in a mouse. The tooth rows slide on each other in longitudinal direction. The enamel ridges stick out from the dentine surface and are exposed to high stresses. These may lead to mini-cracks in the enamel (**h, i**), but the cracks are distributed under loss of depth (**i**) because of the arrangement of prisms. (**k**) Example for the arrangement of enamel prisms in a rodent incisor. (**h, i, k** from von Koenigswald and Pfretzschmer 1991)

structural polysaccharides of the cell walls (Waterman 1984). The enamel ridges form species-specific patterns, produced by wearing down the original high cusps. Tooth rows are moved against one another in such a way that many of these cutting ridges are crossed under a narrow angle. It seems that rodents move their jaws more in longitudinal direction than primates, although Hiiemae found exactly the same strokes in both.[4]

[4]Enamel consists of "prisms", which are secreted by the *adamantoblasts* which during development at the external side of the tooth caps. These prisms are in humans orientated more or less perpendicular to the external surface of the tooth. The electron microscope shows on etched surfaces that each prism consists of bundles of calcium hydroxyapatite crystallites. The so-formed systems are interwoven in three dimensions, whereby the patterns seem to be species-specific (Boyde 1976; von Koenigswald 1980, 1982, 1990; von Koenigswald and Sander 1997). A connection between species-, food- and tooth-specific stressing of the enamel seems highly probable, but has rarely (e.g. by Pfretschmer, 1988, von Koenigswald and Pfretzschmer 1991) and not at all completely been investigated.

The masticatory apparatus of mammals functions only if devices exist which keep the chewed *bolus of food between the teeth*. These devices are the cheeks on the outside and the tongue on the inside of the tooth rows. Both consist of musculature, covered by a permanently moist and slippery mucous membrane. The cheeks in, for example, rodents and hooved mammals cover the entire post-canine row of molars and premolars, leaving open not more than a restricted opening between the mobile lips in front of the incisors. In other forms (e.g. carnivores, cetaceans, insectivores), the closure on the outside is less complete, but still exists in the form of mobile lips. These permit, however, the exposure and the use of the entire tooth rows, including the post-canine teeth. The tongue does not alone keep a bolus between the teeth but also is responsible for food transport through the oral cavity to the pharynx (Hiiemae 1984). The tongue is also used for additional comminution of soft materials between its dorsum and the ridges of the hard palate. Finally, it controls the space available in the oral cavity (e.g. in baleen whales or in sucking).

2.2 Jaws

Teeth are positioned on or in jaws (Figs. 2.2, 2.9, 2.10, and 2.20). To fulfil their primary biological role, at least upper and lower jaws must be connected by muscles. This is the standard in selachians, the jaw apparatus of which is loosely attached to and in a mobile connection to the rest of the skull. As an example of such mobile, independent jaws, we have investigated the bullhead shark *Heterodontus portusjacksoni* (Nobiling 1977). Like in all selachians, the upper and lower jaws are so similar that the following statements are not restricted to the lower jaw but hold true for both. Regarding this item, selachians diverge from other Craniota (see below)

The shapes of *both jaws* in the shark are quite similar, the upper being slightly longer than the lower. The adducting, strongly developed musculature is split into more separate muscle individuals than in other sharks, and its insertions are distributed from the rostral fang teeth to the occipital crushing teeth. The simpler arrangement in most sharks enables the jaw adductors to exert relatively great force even in wide opening of the gape (Wroe et al. 2008). When biting on the posterior crushing teeth, the load arms are short.

The closeness of left and right jaws (Fig. 2.2) also permits making full use of the muscle forces produced on both sides. Obviously, bullhead sharks are able to exert unusually high bite forces on (hard) objects between their teeth.

A preliminary analysis should start off from the concentrated joint and bite forces, while muscles are taken as “snow loads”. The joint is at about the same height as the tooth rows, and the longest diameters of the condyles are at right angles to the median sagittal plane, so permitting nothing than “hinge movements” about one axis. Under the assumption of the above loading, and of homogeneity of the material “cartilage”, the outlines of the jaws are the same as those of bodies of equal strength, if their diameters perpendicular to the drawing plane are assumed to be constant. Figure 2.20e–g shows the moments of resistance of cross sections through the jaws.

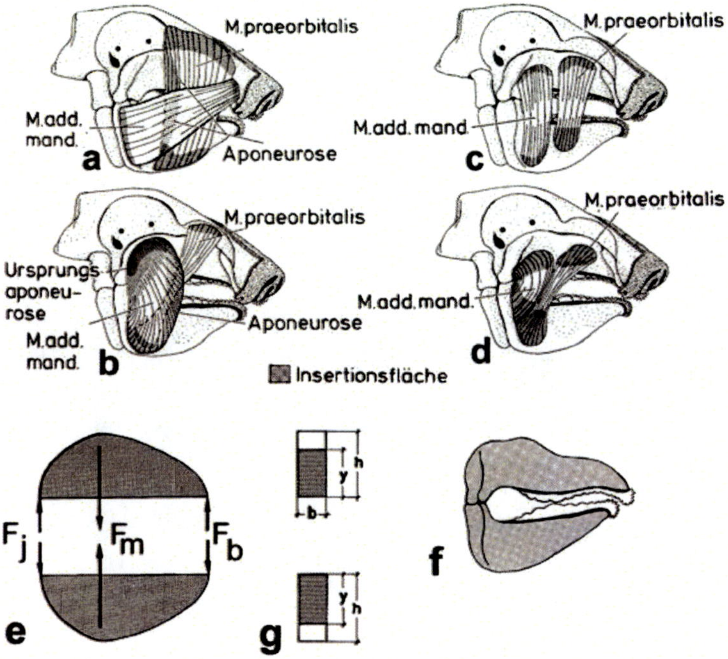

Fig. 2.19 **(a–d)** The adductors of the mandible are highly differentiated in *Heterodontus*, and their insertions are stained in grey. **(f)** Upper and lower jaw of a bullhead shark. **(e)** Muscles are taken as "snow load", distributed over some length and their resultant simplified in **(e)** as the arrows F_m. The joint and bite forces are concentrated as F_j and F_b. For the sake of simplicity, both jaws are modelled as beams with quadrangular cross sections **(g)**. Both jaws **(f)** are bodies of equal strength against concentrated bite forces (F_b) and joint forces (F_j). (From Nobiling 1977)

The increase from rostral in occipital direction is faster than linear (compare below). The irregularities in the sloping graphs coincide with the insertions of muscles.

The skeletal material, however, in reality is not homogenous. As in all Chondrostei, the jaws consist of a matrix, containing but few cells, and much water. The matrix has a marked tendency to soak with water and to swell. It is enveloped by a very tough fibrous perichondrium, encrusted with calcified plaques. These constituents form a unit resembling a hydraulic "pneu". Deformation is limited by fibres of the envelope. From this layer originate fibre systems connecting both sides of the jaw. Three systems of fibres exist, two of them crossing the matrix, as shown schematically in drawing a. These fibres form a pattern which agrees exactly with the directions of tensile stresses inside the skeletal elements under the loads shown schematically in Fig. 2.21. The teeth are attached to the perichondrium by sigmoid fibres which sustain the compression derived from bite force (Figs. 2.9 and 2.10).

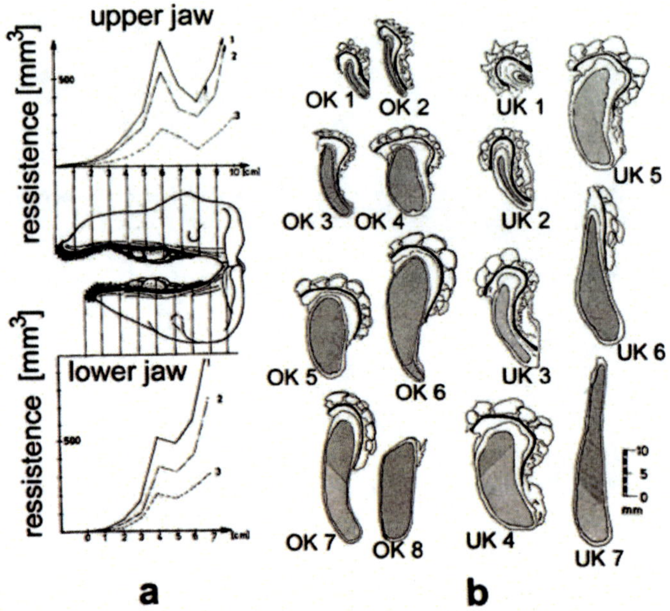

Fig. 2.20 Upper and lower jaw (**a**) of a bull-headed shark with selected cross sections through jaws with teeth (**b**). The moments of resistance (y-axes) calculated for the cross sections are shown in the diagrams. The increase is faster than linear. The interruptions in the sloping are due to the insertions of muscles. (From Nobiling 1977. See also Fig. 2.9)

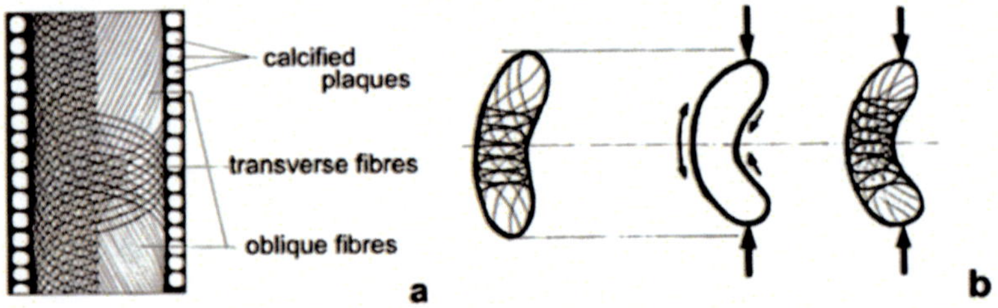

Fig. 2.21 (**a**) Jaws of selachians consist, like the whole skeleton, of cartilage, the matrix of which contains few cells and much water, which makes the system incompressible, provided that the matrix is enveloped by a very tough external coat. Tension-resistant fibres within the matrix are shown schematically. The fibres in transverse direction (see also **b**) exclude an increase of the diameter under compressive load; fibres in oblique direction (**a**) resist deformation by bending on the convex and give way on the concave side. (From Nobiling 1977)

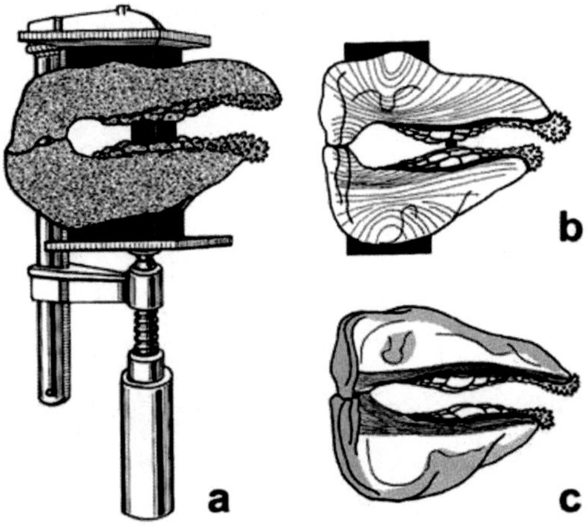

Fig. 2.22 (**a**) Cast of the jaws of a bull-headed shark. Instead of the muscles, the branches of a vice-pin press the jaws against an object between the teeth. Applications of forces are drawn in black. (**b**) Tensile forces are indicated by cracking of the lacquer coat under a bite force, concentrated on the crushing teeth. The highest tensile stresses occur at places, where the perichondrium is thick and parallel-fibred (**c**, dark grey). After removal of the load, resilience of the jaw model leads to cracks in the lacquer indicating compressive stresses (not shown). Compression occurs at places where the cartilage is particularly thick (**c**, light grey). (From Nobiling 1977. See also Fig. 2.23)

Casts of the jaws have been investigated by the "stress-coat technique" (Figs. 2.22 and 2.23).[5] The results obtained show the greatest tensile stresses at those places, where the perichondrium is thickest and obviously orientated in the direction of the stresses. Concentrations of compressive stresses are at places, where the cartilage is particularly thick.

2.2.1 *Lower Jaw*

Among all Craniota, the lower jaw remains a morphologically independent element of the skeleton. Its biological role is carrying the teeth and getting hold of objects and/or comminution of food particles by representing one branch of the forceps formed by both jaws. Catching as well as comminution of food can be investigated with the methods of mechanics, and the lower jaw seems at a first glance to raise few problems against analysis. Like the locomotor apparatus, the lower jaw is mirror-

[5]Under the influence of minimal stretching or compression, cracks occur in a thin layer of lacquer, sprayed on the surfaces of objects. If this is loaded, the dried lacquer will show cracks perpendicular to the direction of stresses, so indicating the directions of tensile forces. If the lacquer is put on the loaded object, cracks will appear after unloading perpendicular to the direction of compressive forces. The highest forces are indicated by the first appearance and greatest density of cracks.

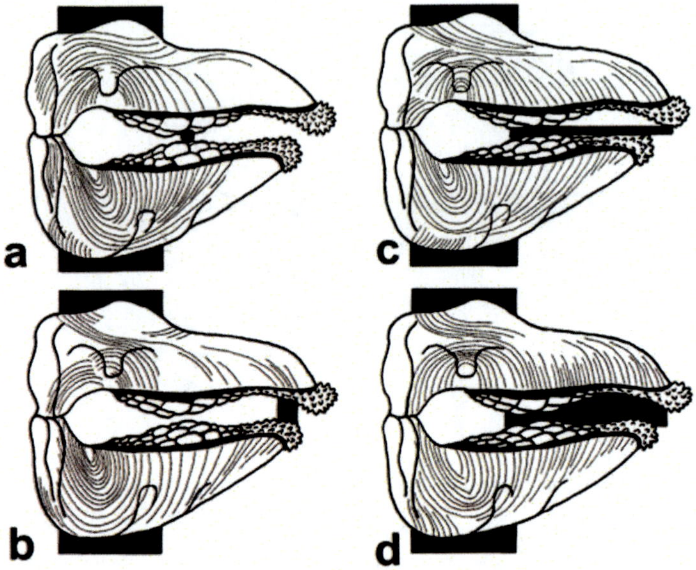

Fig. 2.23 Compressive stresses as documented by the stress-coat technique if biting is done on various places, showing several "load cases". **(a–b)** Bite forces (black) concentrated on points, **(c–d)** bite forces distributed on several teeth. Tensile stresses are not shown; they run perpendicular to compressive stresses. See also Fig. 2.22. (From Nobiling 1977)

symmetric, and therefore both halves can be treated separately. If attempts are made to go beyond the simplified side view, the analysis has to discriminate between the balancing side (without tooth forces) and the biting or chewing side (on which tooth forces are applied). In many representatives of various evolutionary lines (*Amia*, many teleosts, Lepidosauria, opossum and relatives, smaller carnivores, Strepsirrhini; fossil in *Eusthenopteron*, *Acanthostega*, *Greererpeton*, *Crassigyrinus*, *Megalocephalus* and *Lydekkerina*, all listed by Porro et al. 2014), the halves of the mandible are connected by a mobile symphysis. Joints like this are formally called "amphiarthrosis" and consist mainly of so-called fibrous cartilage, in fact massive bundles of fibres, interspersed with few round cartilage cells. Joints, by definition, can transmit forces, but no moments. The mobile symphysis leads to a number of consequences:

First, if muscles are placed lateral to the straight line between symphysis and joint, this leads to a rotation of the mandibular body inward. If muscles are acting medial to the line connecting the joint with the symphysis, the mandible is rotated teeth outward. Both rotating moments cannot be countered by the (mobile!) symphysis. To maintain balance between all forces applied to a mandible, its symphysis, muscles and mandibular joint, as well as the tooth rows, where external forces are acting, must be arranged along or at least close to one straight line (Fig. 2.24, taken from Wolff 1984, 1985). This straightness is not in conflict with the pear-shaped outline of the upper tooth rows (see below, *Varecia*, *Canis*),

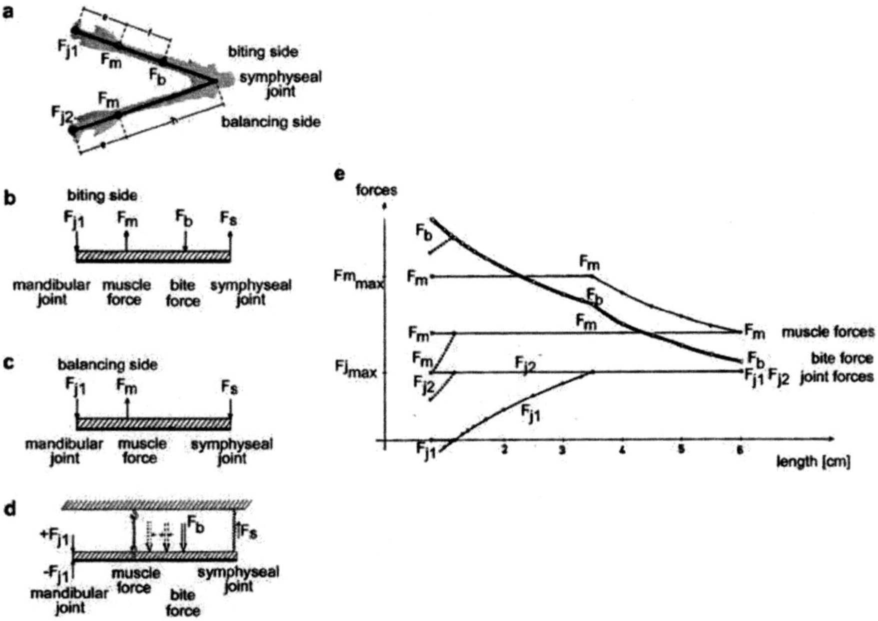

Fig. 2.24 (**a**) Occlusal view on the mandible of a strepsirrhine with mobile symphysis. Bite force (F_b) is applied to the left side. F_{j1} or F_{j2}, joint force; F_m, forces of the adducting muscles; *f*, distance between muscle and bite point; *h*, lever arm of symphyseal force. (**b**) Lateral view on the biting side, (**c**) on the balancing side. At the symphysis, forces (F_S) alone, but not moments, can be transferred to the other side. (**d**) The mandible is compared in side view to a scale, the beam of which is hung up by a chain (= combined muscle forces). One or more bite forces are acting against teeth and pushing the right side downward. On the left side, the scale tends to go upward and exert compressive force against the joint socket (F_j). In support of the muscle force, the symphyseal force (F_s) tends to move the rostral part upward. This increases the bite forces and may contribute, if bite forces are high enough, to a swing of the occipital part of the mandible downward. (**e**) Bite forces exerted at various distances from the (idealised) muscle, calculated for the above model, between joint (left) and symphysis (right). Continuous line F_b: bite forces. Muscle forces on active side (F_m) and on balancing side (F_m), joint forces on the biting side (F_{j1}) and on the balancing side (F_{j2}). While muscles and joint on the balancing side are exposed to constant loads, no matter where the bite point is located, the joint forces on the biting side (F_{j1}) become smaller with rearward moving bite points, until they finally become tensile. (From Wolff 1984, 1985)

because occlusion is established at a time between one or two teeth of only one side. The necessary lateral movements of the lower jaw (in order to reach occlusion) are facilitated by the symphyseal joint.

Second, adducting muscles insert at the medial (lingual) side of the mandible as well as on its lateral (buccal) side. Therefore the mandibular ramus can be kept in equilibrium by exerting more force by one and less force by the other adducting muscle or by a shift of a muscle insertion. Such a shift around the lower margin of the mandible can be seen in various/diverse animals (rodents, carnivores, crocodiles). It represents a permanent, built-in arrangement (rather than a means to stabilise the actual imbalance) and requires more energy than an arrangement along a straight line.

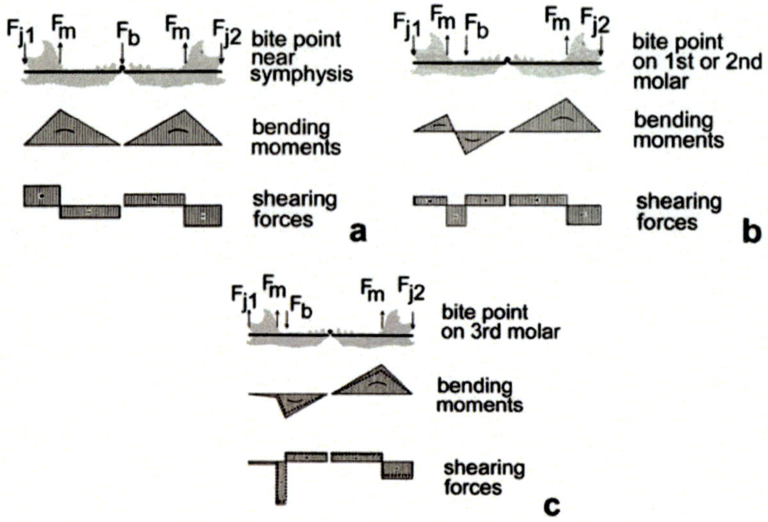

Fig. 2.25 Maximum bending moments and shearing forces calculated for models of a mandible with mobile symphysis. Conventions as in Fig. 2.24. Both halves of the mandible are drawn in the same plane, and the muscles are assumed to exert the same forces on both sides. (**a**) The bite force applied to the most anterior teeth evokes the same stress patterns in both halves. (**b**) Bite force acting on a premolar (left side of the reader) leads to bending of the active side in two directions, while the balancing side is loaded alone by the symphyseal force. (**c**) Bite force acting on the posterior part of the last molar. The bending moments on the chewing side (left side of the reader) again change their sign, the joint is relieved or under tension, and the balancing half is exposed to the same bending moments as in the other cases. Stippled are the bending moments if muscle force is reduced – as concluded above from Fig. 2.24e. (From Wolff 1985)

Third, the muscle forces of both sides can be exploited only, if the force on the balancing side is transmitted to the biting or chewing side. Transmission of force (but not moments!) is possible alone at the symphysis. In fact, forces are transmitted from the active side to the side on which muscles are quiet. As can be seen from Fig. 2.25c, about 2/3 of the bite force exerted on the active side occurs at the inactive side of the anaesthetised animal.

Fourth, transmission of force to the contralateral side at the symphysis leads to maximal lengths of lever arms and to maximal stresses in the mandible on the balancing side (Figs. 2.25 and 2.26). Shifting the bite point rearward in occipital direction leads to greater bite forces, as in the simple models shown in Fig. 2.24.

Figure 2.25b, c shows an effect which has yet found little attention. It occurs, regardless if the mandibles possess a mobile or a rigid symphysis: If a bite force is applied to posterior molars, the joint force decreases, until it finally changes its sign and becomes tensile, instead of compressive. This is because the rearmost parts of the dental row are behind, not in front of the resultant of the muscle force. Tensile stresses luxate the mandibular joint condyle from its socket if posterior bite points are chosen. To avoid luxation of the condyle, tooth force obviously is reduced as long as nerval control is maintained (stippled line in Fig. 2.24e). Under such normal

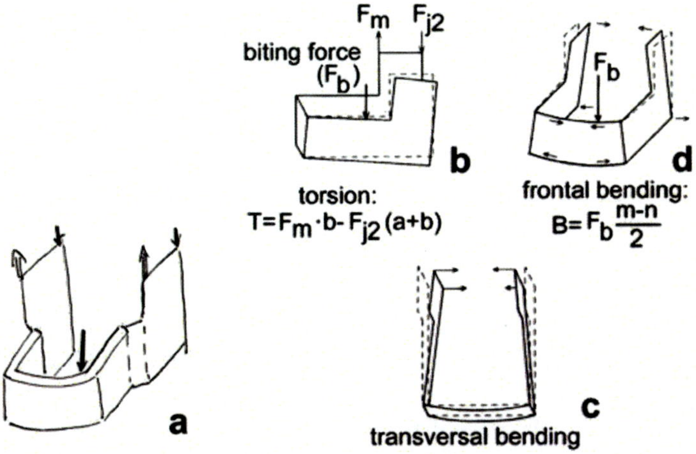

Fig. 2.26 Deformations of a mandible with rigid symphyseal region. From Wolff (1984). (**a**) Simplified model with points of application of "external" forces, seen from the top left. (**b**) The same model in side view after deformation by a load applied to the left molars. The asymmetrical application of bite force leads to torsion of the symphyseal region. (**c**) The same model in top view under the influence of the mm. pterygoideus mediales, which lead to bending in the plane of occlusion ("wishboning"). (**d**) The same model seen from the top left deformed by a load applied to the left second incisor. Biting on the lateral incisor or canine causes bending of the symphyseal region, plus inward swinging of the rami. In all load cases, the lateral parts of the mandible are exposed to changing bending moments in side view (as shown in Fig. 2.24 for a mandible with mobile symphysis). The symphyseal region, however, is exposed to torsional and (downward convex) bending moments. Bite forces on the anterior premolars evoke a teeth-outward rotating moment. This is countered by a teeth-inward rotating moment of muscle insertions lateral to the tooth rows

conditions, the muscle forces are reduced by nervous reflexes before a dangerous situation is reached.[6] Reduced bite force on the most posterior molars, however, entails reduced bite pressure exerted on food material. A decrease of the crown surface on the last molar (in humans and many chimpanzees), or a narrowing of the posterior part of the last molar (in baboons), allows to exert the same bite pressure on objects between the teeth in spite of the reduced bite force.

In the present context, it is absolutely irrelevant, whether a mandible consists of only one element, the dental, or of a combination of several bones (dentale, articulare, angulare, coronoid, etc.; see current textbooks). In all cases, its overall shape must be suited to withstand the occurring stresses. Indeed, the shape of the mandible in all yet investigated animals is optimised to the "normal" forces and moments, being determined by muscle activity and bite point (Porro et al. 2014, Berlin). The composition of the lower jaw out of several bony cores has been

[6]This nerval control can be excluded by anaesthetising the experimental animal, while a constant muscle force is produced by not varied electrical stimulation. If bite forces are applied to molars M2 or M3, tensile forces lift the mandibular condyle out of its socket (own observations).

emphasised by descriptive and evolutionary morphology. Aside from the largely theoretical possibility to move one element against the other (kinetic skull –or kinetic mandible? Witzel, oral communication), there is yet no evidence for a biological role nor a functional necessity, which favours the combining of several bony elements to a lower jaw (Wolff 1985, in spite of the attempts of the group in London – as presented at the SVP-meeting in Berlin 2014).

Aside from lower jaws with a joint between the two halves, there also exist fused *lower jaws*, forming just *one, rigid unit*[7] (Fig. 2.26). There are two variants: in one of them both halves closely parallel each other, so that a seemingly unpaired element is formed (as in long-snouted crocs, many birds, sperm whales, some dolphins). This variant seems to be common among water-living and swift-prey-catching animals. It obviously allows rapid closure of the jaws, because water resistance is kept at a minimum (see also below, Chapter 2.2.2). Most hooved herbivores have mandibles not too far away from the described form. Both halves of the mandible are close to each other, diverging slightly in rostral or occipital direction, while the symphysis is partly, but not completely ossified. No mechanically satisfying analysis of such shapes seems to exist.

The other variant of mandibular fusion, namely, a complete unification of both sides, is realised in monkeys, apes and humans. It transforms the two bony elements into a single, curved beam, which is hung up at both sides and loaded most often asymmetrically. This sort of symphyseal region has far-reaching consequences for the mechanical situation. It will be detailed below in Chap. 8 and especially 9 "Human Evolution".

It is exposed to the deformations shown in Fig. 2.26. Fusion allows the transmission of moments from one side to the other; it implies, however, high stresses, especially in the symphyseal region. A biological role as reason for this particular development of the symphysis is not known. As an example for a "functional" analysis of mandibles with rigid symphysis, we have investigated the lower jaws of primates of the superfamily Hominoidea (see Chapter 9.7). The results are not restricted to the investigated taxa, but hold true for all forms. The mandibular body has roughly the shape of an arcade, overtopped by the ascending rami (Fig. 2.26). Both, the arcade-like shape and the height of the rami, give bending as well as torsional moments long lever arms and therefore require considerable strength, greater than in mandibles with symphyseal joint.

In animals with rigid mandibular symphysis, the force of both sides can be exploited best, if the tooth rows of both sides are as close to each other (see Chaps. 8 and 9). In many individuals, especially among apes, the *molars are approaching the midline* and so allow greater bite forces than otherwise (Fig. 2.30). This, however, is only possible and therefore connected with the rigid one-element mandible.

One of the variable traits of the mandible is the *height of the joint* above the occlusal surface. It occurs in mandibles with joints, as well as in rigid mandibles.

[7]There is no sharp borderline between "mobile" and "rigid" symphyses: in bears, and in other carnivores, and in many hooved animals, the ventral part is united by connective tissue, while the dorsal part is completely ossified.

(See also above.) Figure 2.16a shows the classical hinge effect of a mandibular joint at the height of the tooth rows. It can be observed in most living reptiles, insectivores, whales and carnivores – pinnipeds as well as fissipeds. Bite forces are nearly perpendicular to the upper margins of the mandible. When the jaws are closed, tooth tips move along the periphery of circles about the joint as a centre. This makes the common curved long axes of fang teeth advantageous (see above). The most posterior teeth will come closer to each other than the anterior teeth, and a force component along the tooth row in rostral direction will occur. This sliding of the bolus along the tooth rows is impeded, if the teeth are curved in occipital direction, or if notches are present, like in the carnassials of carnivores. In Fig. 2.16b, the joint is elevated above the occlusal plane, like in many herbivore mammals (elephants, rhino, Suidae, hippos, even-toed as well as odd-toed hooved animals and primates). Rodents vary in this trait; I could not find any general rule among them. In several (probably herbivorous) dinosaurs, the joint was located not above, but below the tooth row. The effect of both dislocations is basically the same, so that alone the situation in mammals will be investigated further.

The movement of the lower teeth during the power stroke has a marked component along the upper tooth row in rostral direction. This makes sense, because it implies a sliding of the lower on the upper teeth in longitudinal direction, which leads to the crossing of many sharp ridges as dealt with in Fig. 2.18. Occlusion takes place in posterior teeth earlier than in anterior teeth. That means posterior teeth come into action earlier than the more anterior teeth and exert more force on the food particles. To avoid this discrepancy, and to secure equal contacts between several lower and upper teeth at the same time, the upper tooth row is commonly somewhat convex (as illustrated schematically in Fig. 2.16e).

2.2.2 Upper Jaw, Nasal Cavity and Neural Cranium

In most Craniota, the dorsal opponent of the lower jaw is an unpaired, complex element, including, aside from the *upper jaw*, the "neurocranium", that is, the capsules of the sense organs and the brain. While the jaws in selachians can be interpreted as the mirror-symmetric pair of branches of a forceps, the majority of reptilian, avian and mammalian skulls are discriminated by this melting of several elements from the above-described situation in selachians (where the upper jaws are separated and independent elements without force-bearing connection to the neurocranium). The more or less tightly connected parts are of different origins: some are preformed during development by cartilage; others are dermal bones.[8] All constituents of this complex have their own requirements regarding space and

[8]Comparative anatomy discriminates between various types of fixation of the upper jaw to the neurocranium: "autostylic condition" without connection of jaws and neurocranium in placoderms and several primitive fishes, holocephalians and dipnoans; bracing alone the jaw joint by the dorsal part of the second brachial element, the hyomandibular; "hyostylic condition" in most selachians; contact of the anterior (rostral) part of the upper jaw and bracing the joint by the hyomandibular, the "amphistylic condition" in sarcopterygians and primitive fishes.

strength. Taken together, they represent just one, mechanically coherent, robust structure. The combination of bones was investigated in much detail by Herring and her group (e.g. Herring et al., 2001).

This "upper" skull is loaded by forces which require much strength: First, the muscles which adduct the lower jaw take their origins from the neurocranium (see, e.g. Figs. 2.19, 2.22 and 2.23), so favouring the posterior bracing of the upper jaw. Second, the transmission of all forces between the head and the trunk (head as a link; see sub-chapter 2.2.3) takes place through the neck muscles and the joint at the occipital wall of the brain capsule (Figs. 2.27 and 2.28).

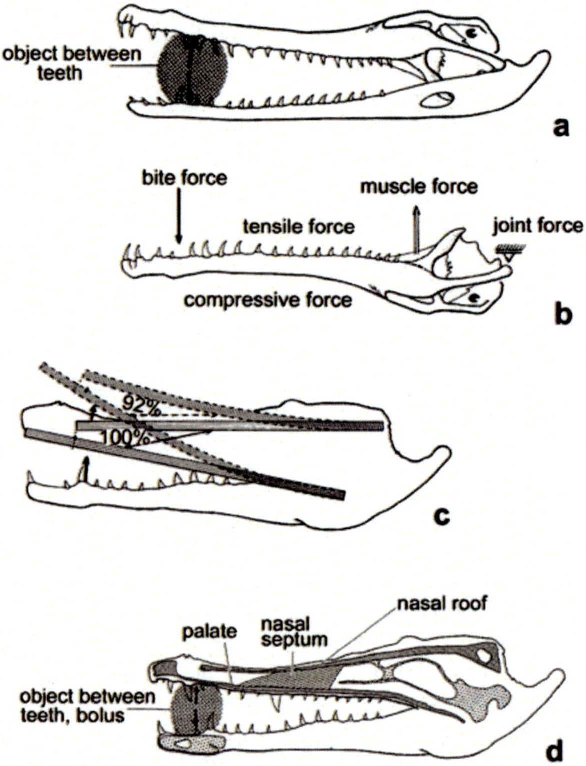

Fig. 2.27 Upper jaws in crocodiles. The long jaws of crocodiles are pressed against each other by the jaw muscles. (**a**) The skull of a slender-snouted Sunda-Gavial (*Tomistoma*) in side view is exposed primarily to bending moments between the bite point and the mandibular joint, with a maximum at the muscles (see also Fig. 2.28c, f) – exactly as the lower jaw discussed above. (**b**) The same upper jaw turned upside down to better illustrate the mode of loading, which is comparable to a balance scale, suspended by the muscles, like in Fig. 2.24d. These are represented by their common resultant force (white arrow). (**c** and **d**) saltwater crocodile (*Crocodylus porosus*). The object between the teeth offers resistance as reaction to the bite force. (**c**) Under the influence of the bite force, not alone the lower jaw is bent (as discussed above), but also the upper jaw, including the palate, plus the nasal roof. The distance between arbitrarily selected points on the nasal roof and the palate (double-headed arrow) is shortened. (**d**) Such a deformation is prevented by the bony nasal septum, above all the vomer, which maintains the distance between the nasal roof and palate. From Preuschoft et al., 1986a

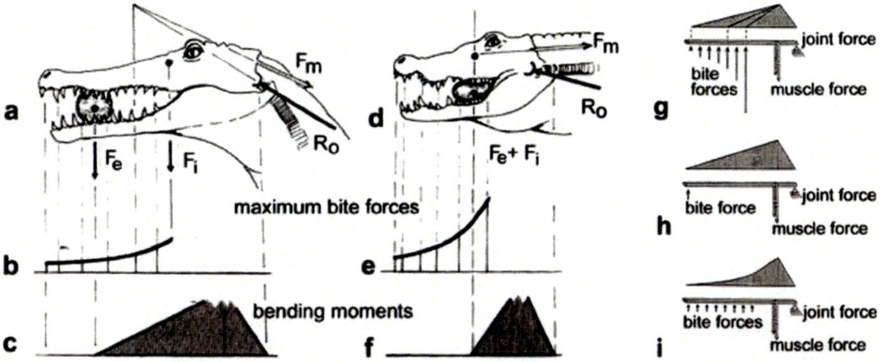

Fig. 2.28 Bite forces in crocodiles are determined by the available muscle force, multiplied by the force arm. **(a)** Long-snouted saltwater crocodile (*C. porosus*), compared to **(d)**, the short-snouted *Osteolaemus cataphractus*. **(b, e)** The bite force (F_e) can be applied to more anterior **(a)** or more posterior teeth **(d)**, where in the latter case it assumes greater values **(e)**. **(c and f)** bending moments which occur on selected bite points. In addition to bite forces, the head exerts a weight force (F_i) concentrated in the head's CoM; the latter may be added to the weight of the seized object (F_e). Both must be balanced by the force of neck muscles (F_m) which keep the head in its position, regardless of neck posture. R_o is the resultant compressive force, consisting of F_i, F_e and F_m, which act against the occipital condyle. **(g)** Depending upon the length of the load arms, bite forces (vertical arrows) become greater the closer the bite point to the joint. Greater bite force leads to steeper increase of bending moments, like in **(e)**. If bite force is concentrated on the most anterior teeth **(h)**, overall bending moments reach their maximum values, even though bite force is limited (as in **g**). **(i)** If bite forces are distributed over several teeth, the increase of bending moments grows slower and non-linear. Because muscle force is assumed to be the same in all drawings, the peak value and the decrease of bending moments toward the joint are identical. (Modified from Preuschoft and Witzel 2002)

Because the skin plus subdermal layers of connective tissue are rather thin and closely tied to the underlying periosteum, the shape of the skull resembles closely the shape of the head, except at the origins of jaw muscles. Up to now hardly observed structures of connective tissue need to be given more attention, since they take over a considerable part of the highest stresses (e.g. thickened periosteum, nasal septum). Since the adducting muscles occupy parts of the braincase, the mechanical situation becomes more complicate than in the anterior part (i.e. the upper jaw in the strict sense), behind the anterior root of the zygomatic arch, or the eyes, respectively. This is particularly obvious in crocodiles, which therefore offer themselves as models. Crocs are discriminated from other reptiles by their possession of a completely closed bony palate. Their long and slender jaws represent the longest part of the head (Figs. 2.27 and 2.28).

The absolute bite forces depend on muscle recruitment and on the position of the bite point. They have been determined for, e.g. sharks (Snodgrass and Gilbert 1967 cited in Nobiling 1977 and Preuschoft et al. 1974), dogs (Lindner et al. 1995), crocodiles and other reptiles (Sinclair and McNeil, 1987), and estimated for fossil theropods. These values, however, are worthless without information on bite points and mass of the respective animal, and they are of limited value for understanding shape. Of general importance are the following facts:

First, bite forces reach very high values. We measured in an agonistically snapping opossum (see below, Figs. 2.35a–d and 2.37b, d) between the anterior teeth more than three times body weight and in biting a maximum of nine times body weight between the premolars. In this case, bite force increased slowly, so indicating its growth to be due to increased recruitment of more and more muscle fibres – not braking a rapid closure of the jaws so that deceleration exerts much less influence than active biting. In an anaesthetised opossum, the force during electrical stimulation was on the stimulated side 191,3 N and on the inactive side 174,6 N (see also above unfused mandibles). These values fit roughly to those measured in awake opossums.

Second, bite forces in the lower and upper jaw are the same, but act in opposite directions.

Figure 2.27 shows how the *upper jaw* is deformed by the bite force: The roof of the nose is compressed and the palate stretched; both skeletal parts are bent dorsally concave. The coherent bone plate of the palate forms a tension-resistant structure, the strength of which is increased by the thick, parallel-fibred periosteum. In addition, the hollow upper as well as lower jaws enclose long extensions of the adductor muscles. When actively contracted during biting, these muscles form efficient ties which take over the tensile stresses caused by bending. That implies that bony elements are not necessarily exposed to bending, especially in those sections where bending moments are particularly high. An approximation of the nasal roof and the palate is blocked by the nasal septum, of which the lamina perpendicularis ethmoidei and the vomer exhibit visible striations in the oblique direction of the compressive force. The neural part of the skull is not affected by this bending. The lower jaw in side view is bent by the same forces dorsally convex, so that its lower margin is compressed and its upper margin, near the tooth rows, stretched. In this region, the periosteum is particularly thick and parallel-fibred, and the above-mentioned extension of the adducting muscles within the hollow mandible also acts as an efficient tie.

The force resulting from all jaw adductors (F_m) is assumed to be the same in all cases. The skull is kept in equilibrium by the joint force (F_g) on one side and the bite force (F_b) on the other. Lever arms l_g and l_b determine the size of the forces, because

$$F_b * l_b = F_g * l_g \tag{2.4}$$

Bite forces on the occipital teeth are higher than those on the rostral teeth, and their respective bending moments increase steeper or slower (Fig. 2.28), while the maximal bending moment is always determined by the opposing moment of the muscles.

$$F_m = F_b + F_g \tag{2.5}$$

The increase of the bite forces from anterior to occipital teeth follows an exponential function, as shown in Fig. 2.28b, e.

Alone the stresses caused by biting favour generally observed similarities of skull shape: In many “primitive” Craniota (amphibians and reptiles), the *outline of the*

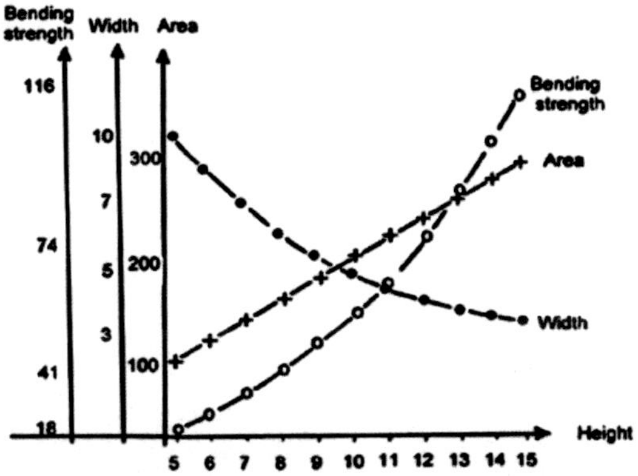

Fig. 2.29 Mechanical conditions which determine the cross sections through simplified, primitive skulls. With increasing height, the bending strength grows exponentially, the cross-sectional area linearly. Skull width can be reduced without loss of strength, as long as height increases. As detailed in the text, the static strength obviously is not the only determinant for skull shape, which instead depends to a large extent on its biological role. (From Hohn-Schulte et al. 2013)

skull in side view resembles a body of equal strength against the stresses evoked during biting on the anterior teeth (Fig. 2.29 and see above, Chapter 2.1). A very similar result has been obtained by application of FESA and published by Jenkins et al. (2002). Because bending stresses (σ_B) are greatest near the external surface of the bent object, the outer parts of the cross section provide most of the strength. Stresses seem to exceed a threshold value, above which tissues are stimulated to build additional bone. The parts in the interior of the bent object, however, receive little stress, which is below the threshold, and therefore can be reduced. This allows saving of weight and explains why jaws (and also long bones, see above, Subchapter 1.1.1 , and below, Chapter 5) are normally hollow, tube-like structures. The hollow space in the upper jaw encloses the nasal cavity; its walls house the olfactory organ. Likewise, torsional stresses concentrate near the periphery. Long diameters permit the walls to be very thin (and that means light!), since, according to the formulae

$$\sigma = M/W, \quad \text{or} \quad \sigma = F_b * l_b/\Sigma\Delta A * y^2/y_{max} \tag{2.6}$$

the stresses σ can remain constant, when the area (ΔA) decreases, but the diameter y increases. With the height of the cross section grows its area linearly – what is equivalent to growing weight – and its bending resistance exponentially, while its breadth may decrease, as do the jaws of many mammals (Preuschoft et al. 1989). So simple this approach seems to be, it helps to understand, why in the evolution of predatory forms narrow and rather high upper jaws have been developed so often (Preuschoft et al. 1988), especially in reptiles (carnosaurs!) and in mammals. It seems plausible without further explanation that long, slender jaws are well suited to seize swift-moving, small prey individuals, while broader and stouter jaws are better suited to overcome large prey and require less precise coordination of biting.

The flat and broad skull shape in some primitive amphibians and in crocodilans, however, becomes even more enigmatic (see below).

An additional reason for *narrowness of the jaws* in animals catching their prey in water is the resistance offered by water against closing the jaws (see also below, subchapter 2.2.3). The faster the peripheral speed of the jaws, the greater is the resistance of water. A reduction of resistance is possible by reducing the width of one (lower jaw of sperm whale) or both jaws (dolphins, narrow-snouted crocodiles).

A similar effect on water resistance is accomplished by a soft and very compliant floor of the mouth cavity, which lags behind the movement of the jaws. In animals with slender mandibles and very wide mouth (as in baleen whales), this leads to reductions by factor 1/5 or more. Huge masses of water stuck with tiny organisms can be taken in by just one mouthful gulp. Even if the mandible itself is rather thick and the floor of the mouth limited, we have found a reduction of the resisting area by factor 1/2 (e.g. in *Tiktaalik*; see Hohn et al. 2013)

Especially in rapid snapping, it is a matter of hazard on *which side* of the tooth row the bite will find resistance in the catched prey or victim. Asymmetry of bite force leads to *torsional moments* about the long axis of the snout (see Fig. 2.31, also in Metzger et al. 2005). Narrow tooth rows keep the torsional moments in one-sided biting low. This is a good – though not the only – reason for the tooth rows to be close to each other, especially in protruding, long jaws. This indeed can be observed, for example, in *Sus*, *Canis*, *Varecia*, *Cercopithecus*, *Pliopithecus* and *Aegyptopithecus* (Fig. 2.30). For the molars, this condition is unimportant, although chewing also takes place on one side only and although bite forces are very high (see above). But the molars are located at the level of the anterior roots of the zygomatic arches, where the adducting muscles take their origins and rotate in the direction opposite to bite forces. These diverging conditions make the pear-shaped outline of the tooth row advantageous, which can be observed in many mammals.[9]

The *length of the jaws* (l) obviously increases the reaching distance of the snap. In addition, it leads to a rapid movement of the upper and lower anterior teeth. If the insertions of the adducting muscles in relation to the joint, as well as the speed of muscle contraction (together determining the angular velocity ω), are taken as constant, the peripheral speed of the jaws grows with length.

$$v_p = \omega * l \tag{2.7}$$

This advantage of long jaws, however, is opposed by two disadvantages. First, the bite force becomes rather low, because

$$F_b * h_b = F_m * h_m, \quad \text{and} \quad F_b = F_m * h_m / h_b. \tag{2.8}$$

[9]All this does not hold true for Hominoidea, in which the anterior root of the zygomatic arch is much higher than the tooth rows (see Chapter 9).

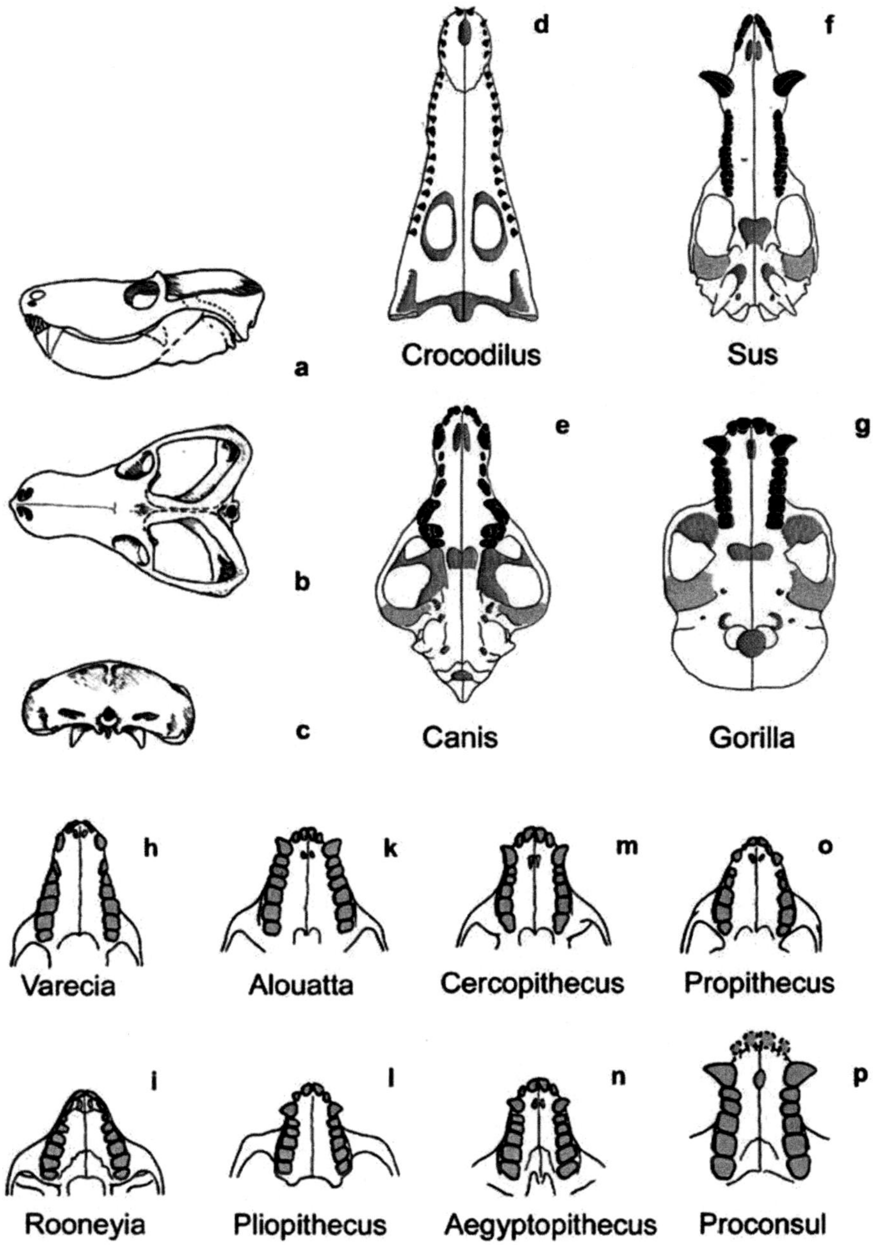

Fig. 2.30 Tooth rows and breadth of upper jaw. (**a–c**) Three views of the synapsid (therocephalian) *Theriognathus* from early Triassic as an example of a "primitive" skull in a land-living animal. (**d**) Dental arcades of *Crocodylus*, (**e**) dog, (**f**) domestic pig, (**g**) *Gorilla*. (**h, k, m, o**), extant, and (**i, l, n, p**), extinct primates. The narrowness of the anterior part of the dentitions, rostral to the muscle origins on the zygomatic roots, keeps the torsional moments limited. The widening of the tooth rows in occipital direction, however, favours the exploitation of the muscle force alone on the biting side. See text above. (**a–c** from Preuschoft and Witzel 2005; **d–g** from Preuschoft et al. 1986a, b; **h–p** from Preuschoft et al. 1986a, b)

Table 2.2 Torques of weight and mass moments of inertia of head segments in three mammals

	Opossum			Dog	Cat
	Upper jaw	Lower jaw	Total	Total	Total
Static torque, kgm					
Segment 1			$0.679*10^{-3}$	0.0493	$0.757*10^{-3}$
Segment 2			$1.023*10^{-3}$	0.0566	$1.002*10^{-3}$
Segment 3			$1.469*10^{-3}$	0.1089	$2.073*10^{-3}$
Segment 4			$1.569*10^{-3}$	0.0481	$2.000*10^{-3}$
Total static torque			$4.740*10^{-3}$	0.2628	$5.832*10^{-3}$
Transformation into Nm			$46.499*10^{-3}$	2.5783	$57.212*10^{-3}$
Mass moment of inertia about condyles, kgm^2					
Segment 1			$5.84*10^{-5}$		$5.901*10^{-5}$
Segment 2			$6.75*10^{-5}$		$6.311*10^{-5}$
Segment 3			$7.05*10^{-5}$		$10.156*10^{-5}$
Segment 4			$4.55*10^{-5}$		$5.000*10^{-5}$
Prey					
Total mass moment of inertia			$24.20*10^{-5}$		$29.155*10^{-5}$
Transformation into Nm^2			$237.408*10^{-5}$		$286.012*10^{-5}$
Mass moment of inertia about mandibular joint					
Segment 1	$2.597*10^{-5}$	$1.274*10^{-5}$			
Segment 2	$2.275*10^{-5}$	$1.600*10^{-5}$			
Segment 3	$1.975*10^{-5}$	$1.157*10^{-5}$			
Segment 4	$0.494*10^{-5}$	$0.421*10^{-5}$			
Total mass moment of inertia	$7.342*10^{-5}$	$4.452*10^{-5}$			
Transformed into Nm^2	$72.023*10^{-5}$	$43.674*10^{-5}$			

Taken from Preuschoft and Witzel (2005)
Calculated in kgm and kgm^2, respectively; sums transformed into Nm

Second, the mass moments of inertia, which resist the movements of the jaws (Table 2.2), grow by the square of length and therefore outgrow the increase of speed (which is a linear function of length l):

$$M_{\mathrm{i}} = \Delta m * l^2 \tag{2.9}$$

As an example, Table 2.2 shows the increase of mass moments of inertia in the case of an American opossum. As can be seen from Figs. 2.28a, d and 2.35a, d, e, these loads are balanced at the occipital joint by the neck muscles, requiring sufficient strength to permit rapid snapping movements. The lever arms of jaw adductors are in most animals short in relation to jaw length. Because the speed of muscle contraction does not vary much, really high peripheral speeds can only be reached, if the ratio between lever length of bite force and lever length of muscles is shifted in favour of the former. The low bite force as a consequence of jaw length permits nothing but very slender and pointed anterior teeth, as in *Gavialis* or *Thomistoma*. The question has been raised, whether the disadvantage of low bite force in long jaws can perhaps be avoided.

The opposite of slender anterior teeth, an impressively massive anterior dentition, which combines several large fang teeth with a massive upper jaw and a similarly massive symphysis of the mandible can be observed in gorgonopsians from later Permian times and in moschorhinid therocephalians (Jenkins et al. 2002). The authors interpreted the accumulation of heavy bone material as a means to produce in biting a hammer-like effect ("*malleognathism*"). Preuschoft and Witzel (2005) have investigated the question whether high peripheral velocity and mass distribution in the jaws may entail a reinforcement of the bite force in the sense of a "hammer-like effect". As shown in Fig. 2.35, the partial masses (m_i) of sections of the upper and lower jaw in opossum, dog and cat have been determined (Table 2.2, from Preuschoft and Witzel 2005). Multiplication with the squares of the respective lever arms (l^2) yields for each section the mass moment of inertia (M_i). Total mass moment of inertia of each jaw (M_t) then is

$$M_t = \Sigma m_i * l^2 \tag{2.10}$$

In the living animals, the partial masses of rostral jaw sections are very low, and their mass moments of inertia therefore are small. This permits very precise positioning of the jaws before contact is made between teeth and the bitten object without expenditure of energy. The rapid snapping in agonistic contexts takes place under clearly observable muscle activity (Fig. 2.37b, c), rather than a passive braking of an actively induced movement. The relatively slow increase of bite force clearly follows the growing muscle activity. Even in anaesthetised experimental animals, the bite force was regulated down, if the cushions on the measuring device were not soft enough. At least the investigated animals found it more advantageous to coordinate the movements precisely with low-energy input, than to increase the bite force by great mass moments of inertia. Our reserve against "malleognathism" becomes even greater, since the accumulation of bone material in the fossil synapsids can be explained as well by concentrations of stresses at the roots of the mighty anterior teeth. Jenkins et al. (2002) themselves have presented results about bone reinforcements at places where stresses are particularly high (in their Figs. 6 and 7) but have not evaluated these results.

Whereas the above-shown mechanical approaches are fairly rough approximations, FESA allows the investigation of three-dimensional structures quasi-experimentally. Figure 2.31 shows the distribution of stresses in a tube with cylindrical profile, which has been used as a model for a narrow jaw. The stresses made visible in these analyses must be sustained by the lengthy, slender upper jaws of, e.g. long-snouted crocodiles as *Thomistoma* or *Gavialis*.

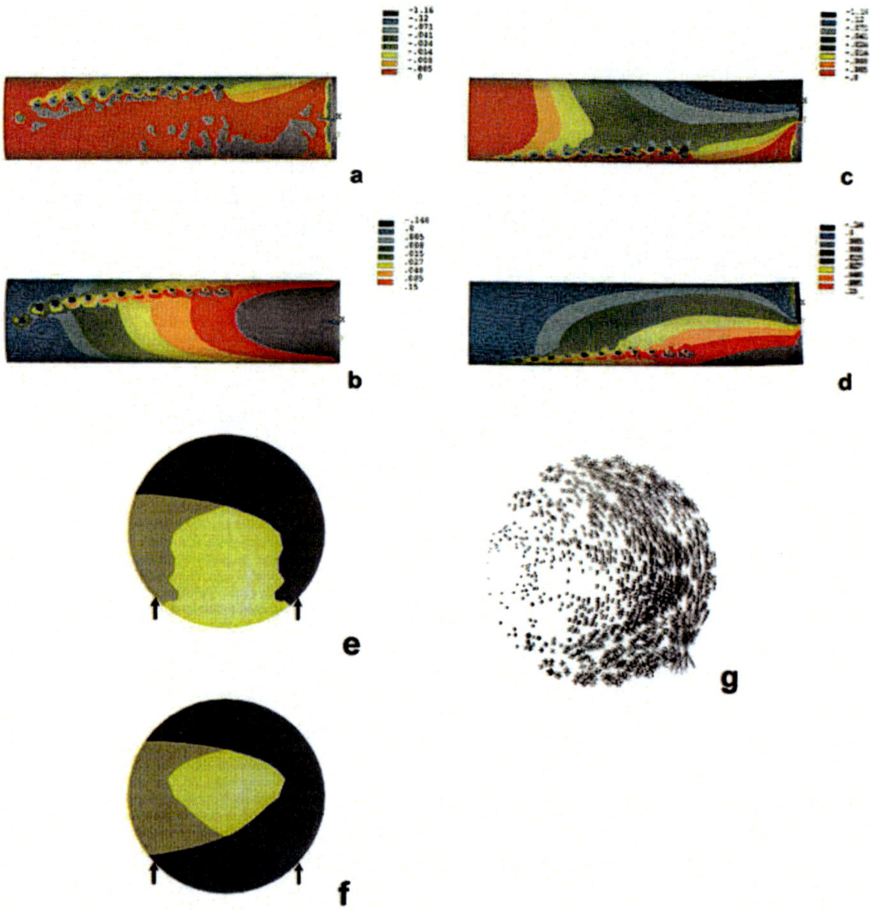

Fig. 2.31 FESA Analysis of a lengthy upper jaw, modelled as a cylinder. Bite forces are applied along the lateral periphery. (**a, b**) Occlusal and (**c, d**) lateral views show the occipital increase of bending stresses, compressive components on top and tensile components below. The deformation in side view (where the nose is deflected upward) is not illustrated in (**c** and **d**). The red colour in (**a**) indicates the absence of compressive forces, while the change from dark blue to red in the image below (**b**) shows the increase of tensile stresses. Compressive stresses (dark blue) are greatest near the dorsal bearings on the right side (**c**) along the nasal roof. (**e, f**) Cross sections through the model in the occipital part of the upper jaw show high von Mises stresses (dark blue) in loading by bite forces applied to the right side (**e**), or green/light grey under a bite force applied to the left side. To withstand these stresses, a U-shaped profile is optimal. Under bilateral loading by bite forces on both sides at the same time, the stressed parts of the cross section merge along the palate (**f**), so that a circular profile is required. (**g**) Trajectories of the von Mises stresses in the rostral part of the upper jaw during biting on the right side alone. The stresses resemble the arrangement in (**f**), and concentrate near the periphery, leaving the interior of the rod (= the nasal cavity) largely unstressed. (Taken from Preuschoft and Witzel 2002)

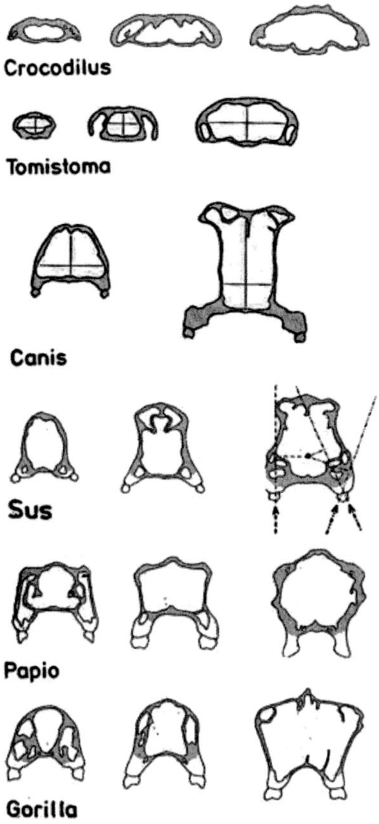

Fig. 2.32 Cross sections through long upper jaws of different animals (horizontal rows) are arranged from rostral (left) to occipital (right). Only the closed (grey) profiles are considered, and the alveolar processes are neglected. The profiles vary considerably, as does existence or absence of maxillary cavities, the walls of which are often very thin. In spite of this, the shape-dictated strength properties grow from the most rostral to cross sections just in front of the orbits. The arrows in the drawing *Sus* indicate potential bite forces with their lever arms. (From Preuschoft et al. 1983, 1986a, b)

As morphological expressions of the strength properties of upper jaws, Preuschoft et al. (1983, 1986a, b) have calculated the areas (*A*), area moments of inertia (I) and polar area moments of inertia along the upper jaws in various long-snouted animals. The calculations were confined to the parts rostral to the anterior root of the zygomatic arch or the orbit, respectively, and cross sections perpendicular to the occlusal plane (Fig. 2.32) were considered to estimate bending and torsional strength against the stresses caused by biting on anterior teeth (which evoke, as seen above, the greatest stresses).

In saltwater crocodiles (*Crocodylus porosus*), Sunda-Gavial (*Thomistoma*), dog, pig and several long-snouted primates (*Gorilla*, *Papio*), the cross-sectional areas possess different profiles (Fig. 2.32), and the walls of the nasal cavity are surprisingly thin. This clearly has to do with the long diameters (see also above), because

$$I = \Sigma \Delta A * y^2 \tag{2.11}$$

where y is one half of the respective diameter.

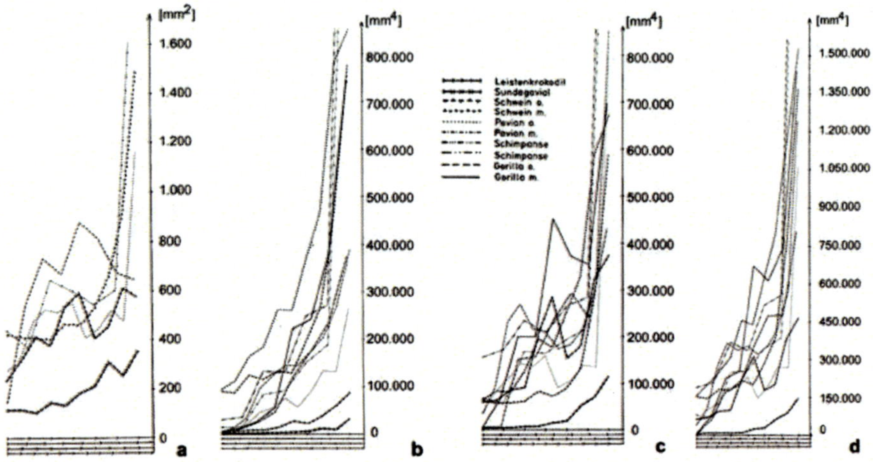

Fig. 2.33 Shape-dictated mechanical strength of cross sections through upper jaws (as in Fig. 2.32), with (*m*) and without (*o*) the alveolar processes. On the left side, the most rostral, and on right side, the most occipital values. All upper jaws (= distance from alveolar margin of first incisor to anterior root of jugal arch or posterior margin of palate) are reduced or elongated to the same length. The relative positions of the sections are given on the abscissa. (**a**) Areas as indicators for shearing strength; (**b**) Area moments of inertia as indicators of bending strength in side view; (**c**) Area moments of inertia as indicators of strength against bending in the occlusal plane; (**d**) Polar area moments of inertia as indicators for torsional strength. Note the general predominance of concave slopes, in spite of the variation of individual values. (Re-combined from Preuschoft et al., 1983, 1986, a, b)

In all investigated animals, the cross-sectional areas as expression of strength against shearing forces grow from the tip of the snouts in occipital direction. The increase is by no means continuous; instead it takes place in unpredictable steps. Taken as a whole, the increase is considerable. The area moments of inertia (I) in both planes, as well as the polar moments of inertia (Fig. 2.33) also increase in steps, but as a whole, the slope follows a parabola, in other words, the shape-dictated strength properties increase faster than linearly. An increase was to be expected, because the bending moments grow linearly (Fig. 2.28). The requirements caused by the mechanical function of the jaws are not only fulfilled; the strength properties obviously surpass the linearly growing bending moments. A detailed FESA (see Chapter 7) suggests that this is a by-product of the stress flow around the orbits and forehead. If the parts of a baboon skull are loaded separately and according to their teeth like cantilevers (Fig. 2.34a), their respective profiles form bodies of equal strength (Fig. 2.34 b, c, d).

2.2.3 *Head as a Link, Jugal Arches and Lateral Snapping*

Aside from biting and chewing, an additional sort of loading the head is too often overlooked: predators as well as herbivorous animals often use their *head as a link*,

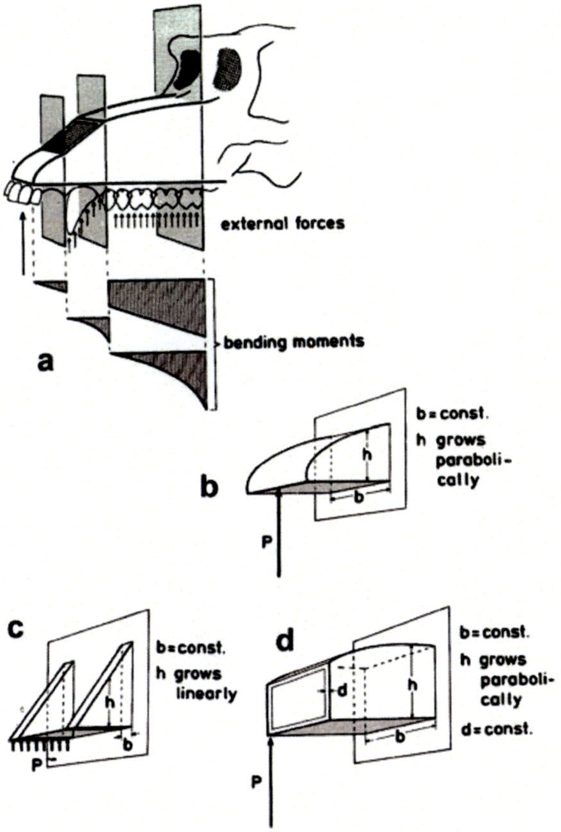

Fig. 2.34 Segments of a baboon skull (**a**) are compared with bodies of equal strength against bending stresses which probably are exerted by tooth forces in the respective anatomical structures. The relevant bending moments are indicated by the hatched diagrams. From the tip of the snout rearward: premaxilla (**b**) as a body of equal strength against loads acting concentrated against the medial incisors; margins of the nasal aperture (**c**) as body of equal strength against external forces distributed along the canine; upper jaw plus nasal cavity (**d**) as quadrangular box offering equal strength against bending moments. (From Preuschoft et al. 1986a, b)

connecting external objects with the animal. There exist two basic variants of loading the head as a link between the environment and the major part of the animal's body: first, application of external forces directly on the head, as in pecora and rhinos, and second, in combination with the use made of jaws (Figs. 2.35f and 2.36). The mechanical implications of the first variant have not received much attention and can be dealt with in short form. In rhinos, several reptiles (Chamaeleonidae) and the fossil Ceratopsia exist horns at the most anterior extension of the head, sometimes accompanied by additional horns which are attached farther rearward, sometimes behind the eyes. Observations have shown that living forms indeed use their horns

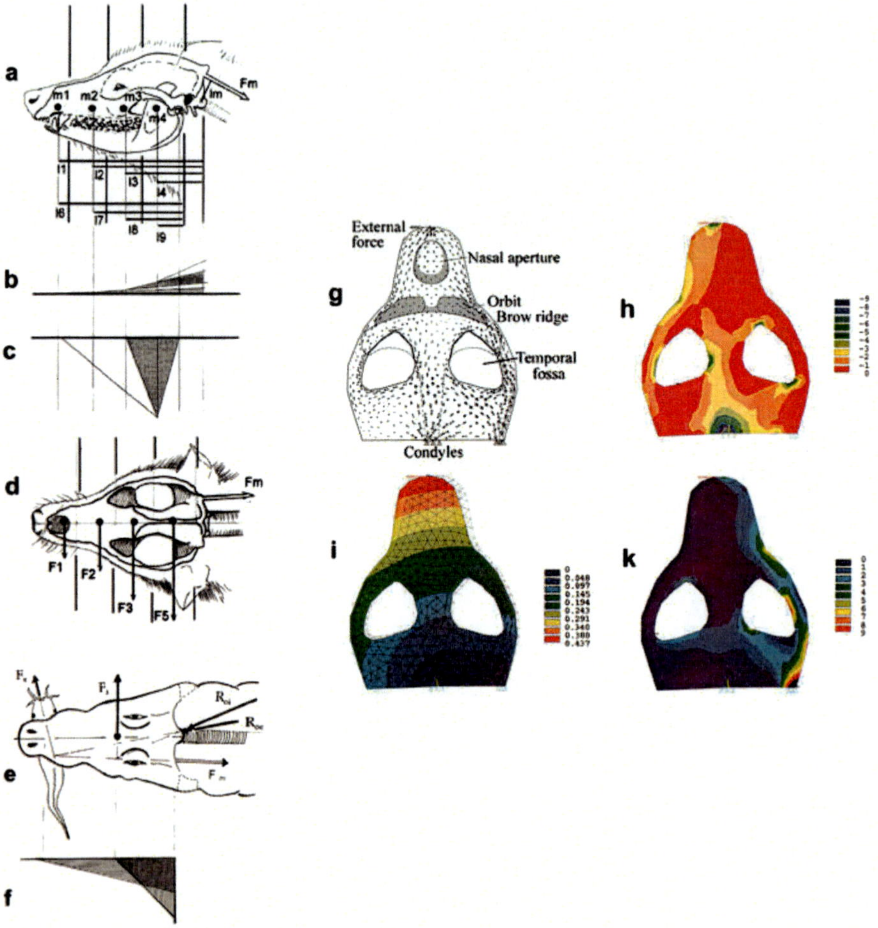

Fig. 2.35 Influence of own weight and external forces on the head. **(a–d)** Opossum in side view **(a)** and in top view **(d)**. The head is divided into four sections, the weights of which (F_1–F_4) are concentrated in their respective CoM (m_1–m_4) in **(a)**. The values are proportional to the lengths of the fat horizontal lines l_1–l_9 in **(a)** and to the fat horizontal arrows in **(d)** (see also Table 2.2). At the occipital joint, the sectional weights exert moments, which depend upon the sum of weight sections and load arm lengths (l_1–l_4, shown in **b**). Equilibrium is established by the dorsal neck muscles F_m. The illustrated "static" bending moments can be superimposed by the dorsally concave bending moments which result from biting (**c**, white diagram, relatively low bite force on canines; grey diagram, high bite force on molars). The effect of rapidly accelerated movements can be seen best in top view, where no weight forces are acting (**d** and **e**). Mass moments of inertia come into action instead of static moments. The former grow by the square of the load arms (analogous to $l_1 - l_4$ in **a**) and by segment masses, which evoke forces F_1–F_4. (From Preuschoft and Witzel 2005). **(e)** Head of a crocodile in top view. Instead of partial masses, the whole mass of the head is assumed to be concentrated in the CoM. Two variants of loading the head are shown: first a catfish seized by the anterior teeth and struggling for escape, while exerting the external force F_e, and second, the mass moment of inertia acting against being moved rapidly to the left side of the crocodile. The force exerted by inertia at the CoM is F_i. For both forces, F_i and F_e, and for the tensile force exerted by the neck muscles, the resultants R_i and R_e are shown. **(f)** Bending moments caused by the forces F_i and

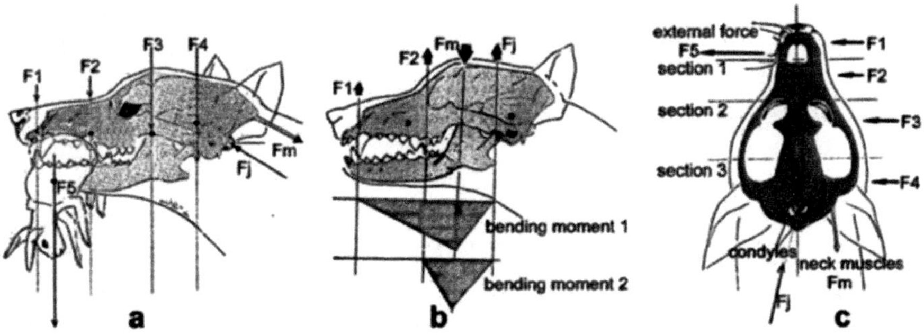

Fig. 2.36 Dog as another example for the influence of external forces on the head. In (**a**) the head is again cut into four segments. The weights of these segments (F_1–F_4, magnitudes indicated by the lengths of the arrows) after multiplication with load arms (not shown) cause ventrally flexing moments at the occipital joint. The carried prey has a long lever arm; its weight is bigger than that of even the heaviest segment (arrow). The sum of all moments must be balanced at the occipital joint by a contraction of the dorsal neck muscles (F_m) and leads to a compression of the atlanto-occipital joint. Because the load of the prey acts to open the mandible, a contraction of the mandibular adductors (*Fm* in b) is necessary. The posterior part of the m. temporalis acts (like in many other mammals) as a tie, which reduces the dorsally convex bending moments in the posterior skull. (**b**) The skull is stressed by the weight forces described in (**a**) (acting dorsally convex) plus the opposing bite forces (acting dorsally concave). Two variants are shown, biting on the canines and on the carnassials. (**c**) In shaking prey, force components are exerted which act in the occlusal plane. The inertial forces of head segments F_1–F_4 and the prey F_5 are represented by arrows, like in (**a**). Again the segments of the head and the seized prey or victim possess load arms of different lengths, the greatest forces being exerted by the prey-like object held between the anterior teeth (canines). (From Preuschoft and Witzel 2005)

in agonistic contexts; in the case of extinct ceratopsians, sound evidence suggests that the same was true in these giants.

Within the skull, application of an external force on the tip of the nose leads to high bending moments, which grow linearly toward its most posterior part. In the most rostral section, the external forces act on the nasal bone and then on the lateral

---

Fig. 2.35 (continued) F_e. In this case, the influence of the inertial force is greater than that of the prey. (From Preuschoft and Witzel 2002). (**g–k**) FES Analysis of the skull of a chimpanzee, loaded by a force in the occlusal plane. (**g**) Skull outline with the trajectories of stresses evoked by a force acting from the right to the left side. The skull is supported at the occipital condyles and balanced by the pull of the neck muscles. Note the density of trajectories along the right zygomatic arch. In this experiment, the influence of the m. temporalis is not considered, which in many cases takes over a considerable part of the tensile forces. (**h**) The right half of the skull is not under heavy compression, because the circle of forces is closed at the occipital joint in the median plane. (**i**) The outline of the meshed upper jaw shows (exaggerated) the deformation caused by the load and indicates the linear increase of moments from low (red), through orange, yellow, light green and dark green, to high, light blue to dark blue. (**i**) Compressive stresses (orange, yellow, green) primarily on the left side (while red indicates zero stress!). (**k**) Tensile stresses (light blue, green yellow, red) prevail on the right side (dark blue means zero stress). (From Witzel et al. 2004) walls of the nasal cavity, formed by the maxillae. At the rearmost part of the head, moments are counteracted by the dorsal neck muscles, which establish equilibrium at the occipital condyles. Several authors have observed that the nasal bone in these animals is strongly developed and that the voluminous neck muscles possess long lever arms.

Among Cervidae, Giraffidae, Antilocapridae and Bovidae (the "*pecora*" of Romer 1945), at least the males, often both sexes, possess bony outgrows of the head, which can be and actually are used as weapons. This use has several times been observed in cases of defence against predators, but by far the most frequent observations of weapon use have been made during intraspecific contests. These contests take place essentially in the same widespread method of fighting between males: pressing the foreheads against each other and trying to shove the enemy backward or disturbing his balance (See Chapter 7.6). The agonistic contacts are enforced by a run-off, especially in sheep. In goat, however, a rearing on the hindlimbs and crushing down with the proximal parts of the horns making first contact leads to the same effect. The "weapons" are just the instrument to transmit force on the enemy or to guide the foreheads into contact. Regardless of the shapes (horn-covered and unbranched "horns" in bovids, branched antlers in cervids, skin-covered pegs in giraffes, horn-covered and branched in pronghorn, Antilocapridae), these structures have their solid bases behind and above the orbits on the nasal, frontal or parietal bones. Because of this location, the lever arms of the external force exerted by the partner are short in relation to head length, so that balance at the occipital joint can be established by the force developed by the neck muscles (see Chapter 7). In view of the spectacular forces exerted by the rivals, it is surprising that no deep-reaching studies have been made on the stresses occurring in the skulls of pecora, nor ideas have been developed, how the animals cope with these loads. It should be noted that horses (as well as camels ?), which do not employ this characteristic method of fighting, and do not possess frontal (or nasal, like rhinos) weapons, do possess neither long lever arms at the AOJ nor obviously strong nuchal muscles (Kienapfel & Preuschoft, *Not yet published*).

As examples for the combination of bite forces with *external loads* may be mentioned predators, catching, killing or bringing down large prey, or carrying prey (as in Figs. 2.35 and 2.36), or herbivores digging out roots (as pigs often do), or tearing branches from a stem or twigs from branches (as domestic hooved animals do in various situations, see Schlunk 2010). Grass stalks and grass blades have considerable tensile strength, which must be surpassed again and again by grazers. Because seized objects may be quite heavy, or firmly connected with their environment, the external forces to which animals expose their heads can be considerable. This the more so, since great masses may be accelerated, so that inertia plays an important role. The external forces, including those derived from inertia, and the accelerations are poorly known. To make a reasonable estimate of the required forces, we have determined in the case of dog, cat and opossum the mass moments of inertia of head segments (see above and Table 2.2) and the forces exerted on bitten objects (Figs. 2.35 and 2.36). Against all these forces, the head must be balanced in the occipital joint by muscle contractions. The muscles passing the joint on its lateral, dorsal, or (rarely) ventral side insert primarily into the occiput, so that the part of the head between its rearmost part and the point of force application has to sustain additional stresses. A number of examples are given, details of loading are explained and stress distributions are analysed in Figs. 2.35 and 2.36. The examples show the size and potential variability of these values. Opossums (Fig. 2.37b) often shake their heads after biting. The first snap is exerted with more than 60 N, that is, about three times body weight (of about 2 kg). Then follow several jerks.

Chris Beard observed that the canines of felids have more or less round cross sections. These teeth sustain the same bending moments in all directions. Canids, however, have canines with longer rostro-occipital than labio-lingual diameters. This means that the strength of these teeth against bending in the direction of the jaws is considerably higher. Indeed forces of this direction are typically exerted by dogs, as we have measured: after gripping hold of a victim (or of any soft element when playing), dogs try the strength of this object once or two times, before they really decide to put in efforts. In intervals of half a second they exert pulling forces of about their own body weight (Fig. 2.37a). The acceleration is about two times earth gravity, and therefore the acting forces are also in the size order of two times body weight. The long snout in these cases is directed from the canines to the occiput and the neck in line with the force, and therefore exposed mainly to tensile stresses. The same holds true, when the dog is lifted from the ground, hang up alone by its teeth. Bending moments become not very great, neither in the head nor in the neck. In felids such a behaviour is not at all common. In addition, jaws in cats are short, so that moments of external forces, exerted by the prey cannot be very great. A reinforcement of the canines in the direction of the jaws is superfluous.

While in biting the highest compressive stresses occur perpendicularly to the occlusal plane along the nasal roof and the braincase (Figs. 2.27 and 2.28), the lifting of a load (prey) leads to tensile stresses along the upper contour, which reach their maximum at the occiput, where they are passed on to the dorsal neck muscles. Compression and bending acts along the lower jaw (Fig. 2.36a). The two stress patterns are in essential opposite to each other. In dorsal view (Figs. 2.35d, e and 2.36c) bending occurs as well, but the direction of bending may change. Equilibrium is established in all cases by the neck muscles. As an example for bending in the occlusal plane, the stress flow in a chimpanzee's skull is shown in Fig. 2.35g–k. In which way these various stresses combine is a difficult problem to solve. Its solution depends mainly from the relation between the acting forces: bite force, size of external force which evokes bending in the form of added load or acceleration of the head plus load. Alone the bite forces have (sometimes) been investigated. In order to get a first impression, we have measured in a couple of experiments the bite forces and the loads that are taken over by the animal, using its skull like a link. Results are shown in Fig. 2.37. The forces exerted from outside are often in the size order of two times body weight. Loading by bite forces repeatedly has been dealt with above. At a first glance, the wedge-like shapes of most skulls (Figs. 2.35 and 2.36) seem to fit roughly to the bending strength required in top view.

The width of the skulls is roughly determined by the *zygomatic arches*, which are connected to the posterior skull by the massive otic capsulae and rostrally by the

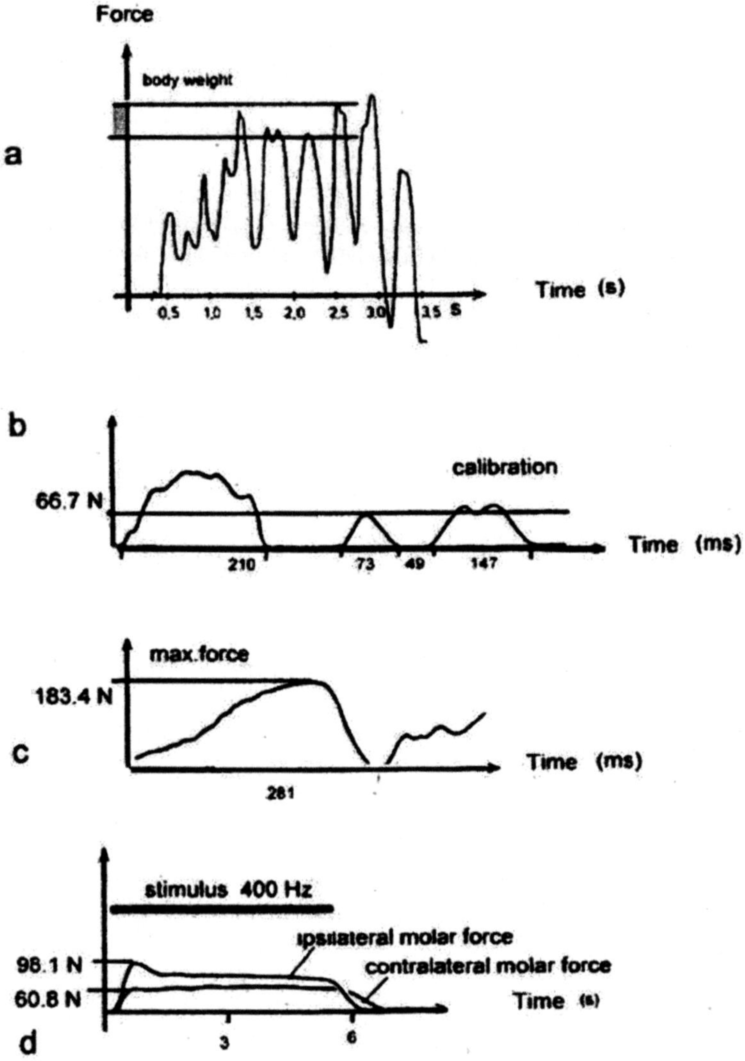

Fig. 2.37 External forces, documented by measuring instruments. **(a)** One out of 12 dogs, pulling after anchoring the teeth in a bite meter, leads to the transmission of forces close to the acting dog's body weight. Acceleration is about 2 G. **(b)** Adult, wake opossum, snapping in defence to preclude approaching of the measuring device, followed by shaking the head. The highest peak is at about 130 N, much more than the body weight (straight line! see also Thomasson et al., 1990). **(c)** Same opossum during biting. Note the rather slow increase of bite force, which indicates increasing muscle activity. **(d)** Bite forces measured in an anaesthesised opossum at defined teeth after electrical stimulation of mandibular adductor muscles. Bite force is greater on active than on contralateral side. (From Preuschoft and Witzel 2005)

anterior zygomatic roots. It seems plausible to interpret the arches as tie on the stretched side or a brace on the compressed side (see, e.g. Preuschoft and Witzel 2005). This the more so, since the anterior parts of the braincases are often narrow, e.g. in pinnipeds and several fissiped carnivores, many insectivores, and in primates the charactristic postorbital incision. The distance of the arches from the midline makes them efficient (because strength is determined by "area moment of inertia", which follows the square of this distance (formulae 2.6). If the neck muscles expand their insertions to the ear region, the former ideas seem to be supported, because muscle force is directly transmitted to the longitudinally orientated element. Nevertheless, the curvature, slenderness and persistence of a suture in the middle of the arch, between the processes of the squama temporalis and of the maxillare, keep the question open, how such a fragile-looking structure could endure over long periods of evolution, in spite of considerable changes of skull shape. Indeed the jugal arches especially of mammals have received much attention (see in particular Herring and Rafferty 2000).

Witzel et al. (2004) have observed on a coronal/transversal section through a primate (chimpanzee) head that the tough fascia temporalis is expanded between the linea temporalis and the jugal arch (Fig. 2.38), so that it transmits the pulling force of the m. masseter to the skull roof. By the well-known thickening of the m. temporalis during contraction, it seems to be pre-stretched. The angle between the lowest part of the facia temporalis and the m. masseter is wide in the anterior part of the jugal arch, but more acute in its posterior. Because the resultant of the two forces follows the cosine of the angle between them, the redirectional forces are much smaller anteriorly and become longer posteriorly. This requires minor strength of the arch in its anterior and considerable strength against being bent in its posterior part. This way of stressing has been confirmed, detailed and quantified by Curtis et al. (2011). These authors have shown by means of FESA with and without temporal fascia that the vaulting of the fascia by contraction of the m. temporalis leads to pre-stretching with forces high enough to compensate the pull of the m. masseter and so to keep the arch during biting free from bending moments in side view. The zygomatic arches, if seen from the top, are comparable to a beam, which is firmly fixed on both sides. A well-known construction of this type is the Tower Bridge in London. The necessary joint is in the middle between the bearings[10] in the jugal arch; it has the form of the always persisting suture. In its Z shape and in its histological construction, this suture permits exactly what a constructing engineer would have to provide in the middle of a beam, immovably fixed at both sides, in order to keep stresses down and to exclude moments.

Above and in Fig. 2.29, has been mentioned that height of the skull includes mechanical advantages, concerning strength and mobility. On this background, we may and should wonder why the heads of many, if not most, "primitive" vertebrates

---

[10]Beams on two supports are in most cases resting on both sides, where bearings can sustain vertical loads, but at least one of them has no horizontal force components. An important example for this structure is the horizontal trunk of tetrapods, as described in Chapter 4.

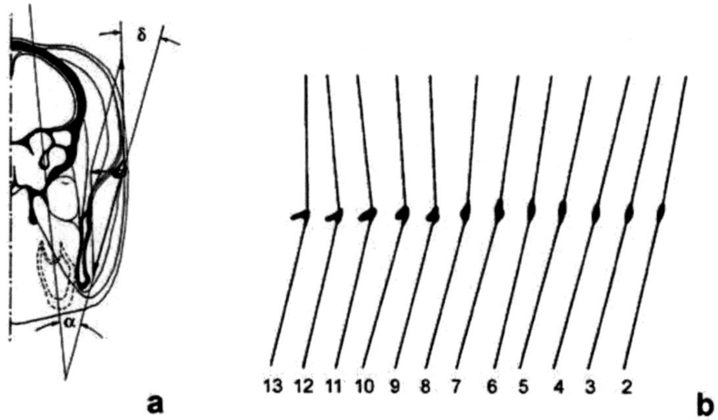

Fig. 2.38 Stressing of zygomatic arch. (**a**) Frontal section perpendicular to the occlusal plane through the fossa temporalis of a chimpanzee. Skull in the plane of the section is black; the mandible is cut through the flat ascending ramus. A more rostral cross section through the mandible is shown by the stippled outline. Double lines on top of the jugal arch: fascia temporalis; below the jugal arch: superficial fibres of m. masseter. Simple lines show arbitrarily selected fibres of mm. temporalis, masseter and pterygoideus lateralis. (**b**) Cross sections through the jugal arch, beginning on the right side close to its anterior root, ending on the left side close to its posterior root at the temporal bone. In the planes of the sections, the forces exerted by the fascia temporalis (upward) combine with the forces exerted by the m. masseter (downward) to resultants, which obviously are small in the more anterior part, while becoming greater toward the posterior root. Therefore, the jugal arch does not need, and does not provide, much bending resistance near the anterior root, in contrast to its posterior parts. (Modified from Witzel et al. 2004)

are flat and often allow extremely wide opening of the jaws. In fact there exist conditions, which make a flat head advantageous.

Well-known representatives of *flat-headed animals* are crocodiles. The flatness of croc skulls is commonly attributed on the level of "biological role" to their way of capturing large mammals: their whole body suspended in water, with only the eyes and nostrils above the surface, they are nearly invisible when approaching their victims slowly. The advantage of flatness of their heads is seen in the far-reaching, and nearly complete immersion in water. Once in reaching distance, the crocodile rushes forward or moves the head to the side for the often-described "lateral strike". The same movement (lateral striking) is used in intraspecific conflicts, often without opening the mouth and exposing the teeth. Hohn et al. (2011) have kept in mind crocodilian behaviour as a model for interpretation of the Late Devonian tetrapod *Tiktaalik*.

In the latter fossil, the conditions of embedding, as well as the lifestyle of many modern newts and crocodiles, suggest that the animals spend a large part of their time in shallow water. The weight of the replaced water makes it unnecessary to waste energy for maintenance of body shape. In accordance with Archimedes' principle ("a body, immersed in water, loses so much weight, as the replaced volume

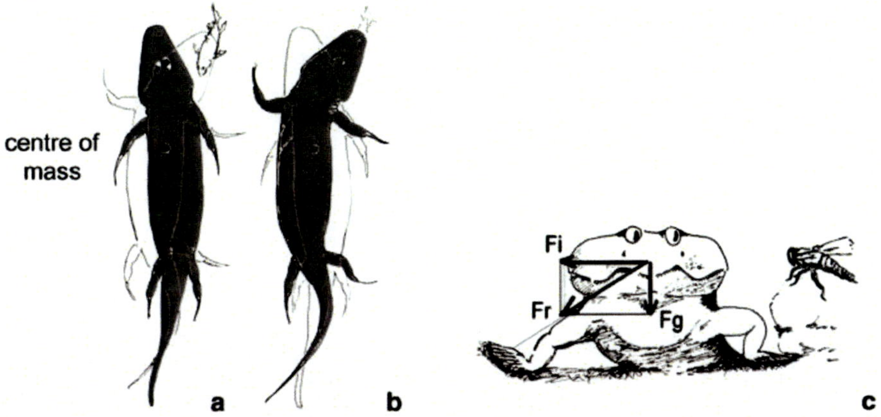

Fig. 2.39 (**a**) Fossil *Tiktaalik*, swimming slowly, or lying in wait for potential prey, which is spotted on its right side. (**b**) A rapid lateral strike of the head (grey, the thin, unshaded outlines show the situation after and before the strike) is the least energy-consuming method for capturing a prey. The inevitable recoil movement of the body about the CoM is counterbalanced passively by the mass moment of inertia of the elongate body and/or by actively coordinated movements of the forelimbs. The other possibilities: rushing forward in longitudinal direction or rotating the whole body about a vertical axis require more energy. (After Hohn-Schulte et al. 2013). (**c**) Lateral snapping can also be performed on land. Overcoming of mass inertia requires a force F_i in horizontal direction. This force plus weight force of the head, neck and shoulder region (F_g) combines to a resultant F_r, which passes along the abducted forelimb to reach the ground within the outline of the manus. This force has short lever arms at the limb joints

of water weighs"), buoyancy is used best if the whole animal is submerged. To immerse the whole body completely even in shallow water, it needs a dorsoventrally depressed outline of its cross-sections. This general flatness includes the head.

A more mechanical hypothesis for explanation of the shallow head shape is shown in Fig. 2.39a. A crocodile, or a primitive tetrapod like *Tiktaalik*, may lie in wait in shallow water while a prey is approaching. The victim can be reached by a rapid lateral flexion of the head alone, or by rushing forward in longitudinal direction, or by a turning of the whole body about a vertical axis through its CoM. The lateral movement of the head requires contraction of the lateral muscles of the head-neck-shoulder region, for overcoming the mass moments of inertia of and water resistance against the head (see, Figs. 2.39 and 2.40). Water resistance is (as noted above) proportional to the area A (grey), multiplied by the angular velocity ω. Because of the swiftness of potential prey, ω must be rather high. The active muscles take their origins from the trunk and therefore also exert moments on the trunk. A movement, however, is opposed by the (very large) mass moment of inertia of the lengthy trunk. Likewise, a backward stroke of the left forelimb (as in (b)), or pushing against the ground, may evoke the counterforce. Rushing forward is opposed by the inertia of the whole body mass. Inertia must be overcome by forces, exerted by the limbs and/or the propulsive tail against the surrounding water, or by pushing the limbs against objects connected firmly to the ground. A turning of the whole body can be initiated by the limbs, or by the unpaired fins of the body axis. All

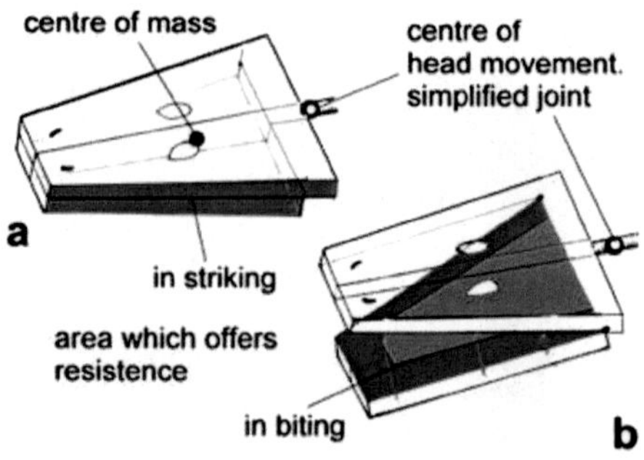

Fig. 2.40 Flat head of a primitive tetrapod. (**a**) Water resistance against lateral striking depends on the exposed area (grey) multiplied by the square of angular speed (ω, Eq. 3.7). The pivot for lateral striking is assumed to be the base of the (short) neck. (**b**) Water resistance against closing the jaws. The mandibular joints close to the posterior margin of the head are marked as dots. They define the axis of the angular movement in which the anterior teeth pass a longer distance than the posterior teeth. The speed of the closing movement and the exposed area (grey) determine resistance. A compliant floor of the oral cavity (light grey) reduces the exposed area. (From Hohn-Schulte et al. (2013))

methods of attacking require the input of force, that is, energy. A calculation of the minimal force requirements has been performed using the measurements of *Tiktaalik*. Movement alone of the head needs 59,25 N; rushing forward of the whole body needs 163,9 N, rotation of the whole body 592,5 N. Lateral snapping obviously is the least energy-demanding method. To arrive at this result, the low, flat shape of the head, with its rather small area exposed to water resistance (Fig. 2.40), provides a contribution. In fact the "lateral strike" is in crocodiles the most frequently observed method of capturing prey, and therefore a good reason to develop a shallow head. Whether opening of the mouth includes a suction effect has never been investigated.

2.2.4 Suction Feeding

Very common among water-living animals is *suction feeding* (Schwenk, 2000). While lateral snapping can be performed as well under the conditions of living on land, suction is much less efficient if the animal is not immersed in a dense and viscous medium like water. The movements of the whole animal or at least its head in water lead to increased pressure against the most anterior parts, commonly visible in the well-known bow wave of vessels. This stow compression pushes prey objects away from the mouth. The magnitude of water resistance depends on the speed of movement and the mass of displaced water, and therefore the long and slender jaws, as present in, e.g. long-snouted crocodilians (discussed above), keep it on a low level. An active reduction of the pressure against the moved head, by sucking of a volume of water into the explosively expanded mouth cavity, permits the intake of small or large food particles regardless of the speeds of prey and predator. The elements existing in teleosts have been described by Hofer, 1945 and also by Westneat. In suction feeding, these elements are moved suction feeding takes place very rapidly (<100 ms) by a precisely coordinated opening and protrusion of the jaws, lowering the hyoid-complex, lateral expansion of the gill cover (operculum; Muller and Osse 1984; see also Day et al. 2015). In front of the circular opening of the mouth, a funnel of low pressure is created, the diameter of which is close to that of the fish (Muller et al. 1982; van Leuwen, 1985; Osse et al. 1985). The water streaming into this low-pressure volume carries small food particles into the mouth cavity, where they are kept, while the water volume leaves it through the gill openings ("unidirectional flow"). The acceleration of the volume requires energy, and the whole process entails a side effect to the acceleration of the animal's body forward: the fish so to say "sucks itself forward". Suction has also be observed in predatory tadpoles (Debanand Olson, 2002).

A number of studies on suction feeding have been performed on aquatic and semiaquatic amphibians (Heiss et al. 2013b, 2018) and turtles, which perform suction in a way different from teleosts. The absence of a gill apparatus requires divergent techniques, which have been studied by Lauder (1985), van Leeuwen and Muller (1984, 1985), van Damme and Aerts (1997), Aerts et al. (2001) and Lemell et al. (2002). In the matamata and in *Heosemys grandis*, after slowly approaching the head to the prey, the head is pushed rapidly forward, and the jaws are opened, while the hyoid is lowered. So water pressure in the oropharyngeal space is reduced. This pressure reduction excludes at least the formation of a bow wave in front of the forward moved jaws. Whether indeed water is aspirated into the mouth is not completely beyond doubt. In favour of this assumption, it may be noted that the same process can be repeated to transport the prey from the mouth into the throat (Lintner 2010, Lintner et al. 2012). The sucked-in water leaves the mouth cavity ("bidirectional flow") through the slightly opened margins of the mouth. This technique is used to capture prey of medium size, limited by breadth and length of the oral cavity. Accelerating the volume of water requires the input of considerable energy.

Suction feeding also was observed in marine mammals: toothed cetaceans and pinnipeds. Some suction occurs aside from "ram feeding" in dolphins (Marshall et al. 2008) but becomes really vital in beluga whales, walruses and bearded seals (Kane and Marshal 2009). In all these forms, the lips are pursed for lateral closure of the gape, and the tongue is retracted, with the hyoid apparatus lowered to enlarge the oral volume synchronised with opening the jaws. The mammals mentioned last can even reverse the process and produce a forward-directed jet of water to dig out invertebrates from the sediment.

A particular technique has been described in detail by Heissig et al. (2013a) in the Chinese giant salamander *Andrias davidianus* (Fortuny et al. 2015). In this amphibian of about 1 m length, the rapid opening (70 ms!) of the plate-like upper and lower jaws (Fig. 2.41) leads to a considerable reduction of water pressure in the mouth cavity with the consequence of a rapid influx of about 1.2 l water with a speed of

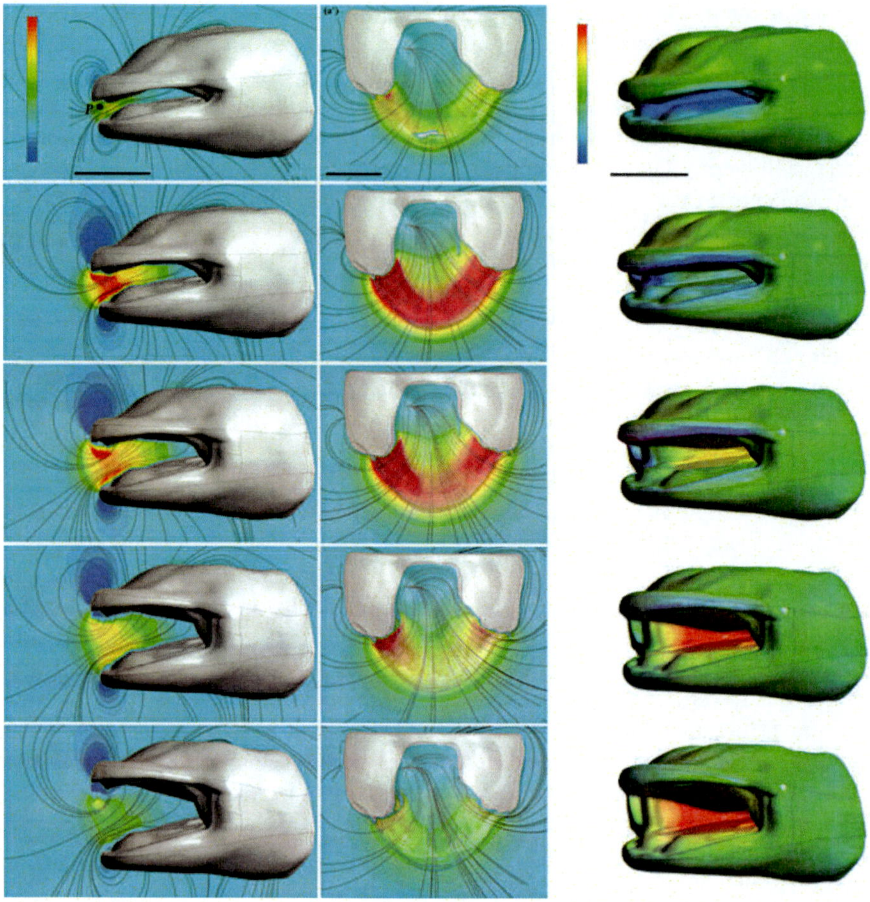

Fig. 2.41 FES Analysis of water flow at the head of a giant salamander during suction feeding. *Left* column: lateral view on the left side. During opening, high compressive forces act against the margins of upper and lower jaws (dark blue). Jaw opening leads to a rapid flow of water (trajectories shown as curved lines) into the oral cavity. The velocity of instreaming water is indicated yellow and red. *Middle* column: top view, where the upper jaw is transparent. Streamlines are +/− perpendicular to the margins of the mouth opening, and the highest velocity is reached along the lower jaw. *Right* column: water pressure in the oral cavity as seen from front left. Dark blue and light blue, low pressure in early stages; yellow and red, high pressure, while water is filling up the cavity. In this stage, the jaws must be closed to keep the prey inside the mouth. In the living animal, this increase of intra-oral pressure is reduced or delayed by lowering of the mouth floor. The surplus water slowly flows out between the jaws (bidirectional water flow). (From Heissig et al. 2013a, b)

0.95 m/s. This accelerates the prey by about 40–50 m/s^2. Not more than 51 ms later, the oropharyngeal space is widened to engulf the water (containing the prey animal) before it leaves the mouth cavity slowly through the slightly opened jaws (bidirectional flow). According to the observations of the authors, this enlargement of the

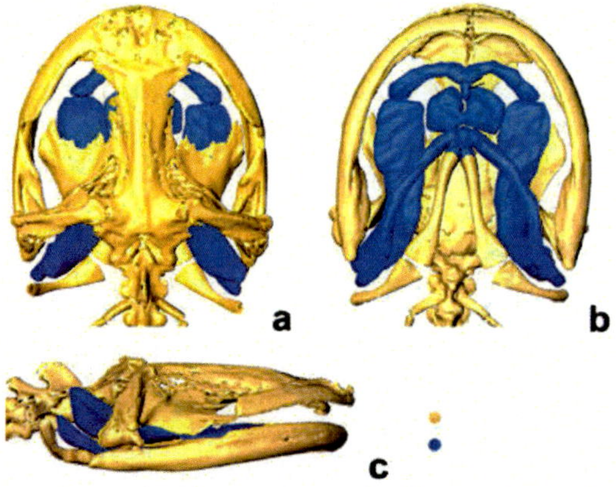

Fig. 2.42 Skull (yellow) plus cartilaginous hyoid complex (blue) in Chinese giant salamanders *Andrias davidianus*. Viewed from dorsal (**a**), ventral (**b**) and right side (**c**). It should be noted that the arcades formed by bony upper and lower jaws coincide with the places where the highest forces are applied to the skull. (From Heissig et al. 2013a, b)

oral cavity is induced passively by the inward streaming water. Whether enlargement is supported actively by muscles is left open. Anyway, it resembles the lagging of the compliant oral floor behind the closing movement of the jaws, as can be observed, for example, in baleen whales (or is detailed above (Fig. 2.40b) for *Tiktaalik*.

The lifting of the upper jaw is obviously induced by contraction of the autochthonous, epaxial dorsal musculature and the lowering of the lower jaw by the straight hypaxial ventral muscle (in Heissig et al. termed "m. rectus cervicis"), which continues as m. geniohyoideus to the mandible. These muscles evoke high compressive stresses in the occipital joint – a fact in contrast to jaw muscles (as described above) that are without any influence on the neck region. The jaws of *Andrias* are exposed in the rapid opening to exactly the opposite forces as in biting. This needs appropriate strength of the skull. The upper jaw receives the necessary strength in longitudinal direction by its melting with the braincase (Fig. 2.42); the cartilaginous hyoid complex seems to be in essential a means for stiffening and enforcement of the oral floor. Alone fairly rigid, wide and long plates yield, when separated rapidly, the high velocity of instreaming water which is necessary for capturing swiftly moving fishes. The authors document clearly by FESA that opening of the jaws is performed against high water resistance, especially near the (therefore ossified!) margins of the jaws, and requires the input of much energy. According to the published calculations, water flows on all sides perpendicular to the jaws into the oral cavity. Therefore, the action must not be aimed precisely – in contrast to the above-described suction feeding in bony fishes. Provided that the speed of jaw opening can be maintained, the longer and wider the arcades of the jaws are, the stronger will be the stream of water. This fact may well be the reason for the often mentioned and indeed obvious size of the mouth opening in many fossil and recent amphibians. The high input of energy for this sort of suction feeding can be expected to be balanced by a reward in the form of fairly large prey animals.

The series of pictures in Fig. 2.41 make visible a fact which is of importance for the development of shape: The highest forces are acting against the margins of the mouth opening. This is in accordance with the well-known morphological trait, namely, the arcade-shaped upper and lower jaw skeletons (Fig. 2.42), which are rather typical for amphibians. In the upper jaw, these arcades are braced along the midline by nasal capsula and braincase, at both sides by the otic capsulae. Between the branches of the lower jaw, the hyoid complex is located, which may either give way to increased water resistance or resist being deformed because of its skeletal elements.

In mammals, the nasal capsula and the braincase are usually wider than in reptiles and therefore exhibit greater bending strength. The diameters of the nasal capsula can be so long that they together represent a tube-like shell construction (in shrews of the genus *Sorex*), which does not need any additional reinforcement by jugal arches. These can be reduced completely (Preuschoft and Witzel 2002). This situation is in contrast to that in carnivores, especially pinnipeds, which possess an extremely narrow postorbital or interorbital incision and therefore seem to offer but little bending strength parallel to the occlusal plane. In these animals, a bracing by the jugal arches seems to be a necessity. This interpretation fits to the situation in diapsid reptiles, in which the zygomatic arches are continued in rostral direction by the lateral margins of the antorbital openings.

Aside from bending in at least two planes, there may also occur torsion of the entire head.

This is best exemplified in the "death role" of crocodiles. After seizing a big prey, the animal twists about its longitudinal axis, bringing down its prey from the legs, suffocating it or tearing it into pieces. The death role is used in water as well as on land. The lengthy and slender body of the predator does not contain much resistance against being twisted; the rotational moments are created by movements of the limbs or the tail against water and transmitted through the neck on the head with its jaws. On land or in very shallow water, crocs know to lift their body highly on the limbs, so that body weight works on a long lever arm when swinging about the feet, so creating a considerable tilting moment. The structures involved in force transmission between head and trunk, especially the neck vertebrae, are strongly developed (Preuschoft and Klein 2013).

According to the above-described analysis, the nasal cavity must be seen as a stress-free region of the upper jaw, which does not take part in sustaining the internal forces (Witzel and Preuschoft 2005; Preuschoft and Witzel 2002, 2005). This fact is visible in the FESA models of Fig. 2.31 and will again be dealt with in Chap. 9, "Human Evolution". Figure 2.31e shows symmetrical biting on both sides at the same time, favouring a U-shaped outline without palatal plate, while Fig. 2.31f shows the stresses in asymmetrical biting on one of the sides alone, which requires a tube closed on all sides by the nasal roof and the palate.

All the above approaches are confined to the parts of the skull anterior to the orbits. The part behind the orbits seemed to all authors too complex for proposing a mechanical model.

2.2.5 *Finite Element Systems Analysis and Finite Element Systems Synthesis*

The most important progress for understanding the relation between mechanical loading of the skull and its shape was made through applying of the FE method by U. Witzel. For biological problems, the FE method is extremely valuable, because the stress flow can be demonstrated in detail in highly complex, three-dimensional structures (e.g. in Fasnacht et al. 2002; Jenkins et al. 2002; Ross et al. 2005; Richmond et al. 2005; Metzger et al. 2005; see also E. Rayfield, 2004, 2005, 2007; Porro et al. 2014). *Witzel* emphasised the discrimination between *inductive* Finite Element Systems Analysis (*FESA*) and *deductive* Finite Element Systems Synthesis (*FESS*; Witzel and Preuschoft 2005; Preuschoft and Witzel 2005; Witzel et al. 2011; Curtis et al. 2011). The former (FESA) takes its starting point from the existing shapes and investigates the stress flow within the existing elements. This procedure is generally accepted and highly esteemed, in particular if the existing shape is described in all existing details, by using, for example, CT or micro-CT technique, and if the muscle forces are known precisely from experiments. However, an answer to the question, how, or because of which (mechanical?) selective pressures the shapes have been evolved, can never be found on this way, simply because the evolutionary result of the analysis (i.e. the shape of a structure) is taken as a starting point for the investigation, which means all conclusions rest on circular reasoning. A reasonable answer to the question why a morphology has evolved can only be found if conditions are identified, which have led to the existing shape. In the FES Synthesis, an ample volume ("Bauraum") is provided, within which the stresses can spread without restrictions. Our procedure is exemplified by the skull of *Diplodocus* in the Senckenberg Museum, Frankfurt. This volume is filled by a three-dimensional meshwork of tetrahedral finite elements (Fig. 2.43a, b). To this meshwork the forces are applied (Fig. 2.43c–f). The programme ANSYS 7.0 calculates for each of the finite elements the stresses and shows their relative sizes, coded in colours (Fig. 2.44). In addition, the directions of stresses (trajectories) can also be shown in the form of arrows, the lengths of which are proportional to stress magnitudes. Biologically meaningful loads are applied, and the highest of the stresses that occur in various load cases are stored ("CAD"). Only those parts of the "Bauraum" which contribute to sustaining the loads are used in a next step to create a new, much more detailed and complex, "reduced" Bauraum, in which all unstressed elements do no longer exist. The process is accomplished by cutting the loaded "reduced" Bauraum into slices, of which alone the stress-bearing parts are combined. The new three-dimensional solid or "reduced" Bauraum is exposed to the same loads as before. During this procedure, again parts of the model turn out to be not stressed, while others receive a greater share of the acting forces. In most cases so far investigated, the three-dimensional presentation of the loaded parts in the "reduced" model is quite similar to that of the skull of the animal under investigation (Figs. 2.46 and 2.48, after reiteration). Additional precision can be provided by repeated reiteration of the described procedure. With this procedure, we synthesised

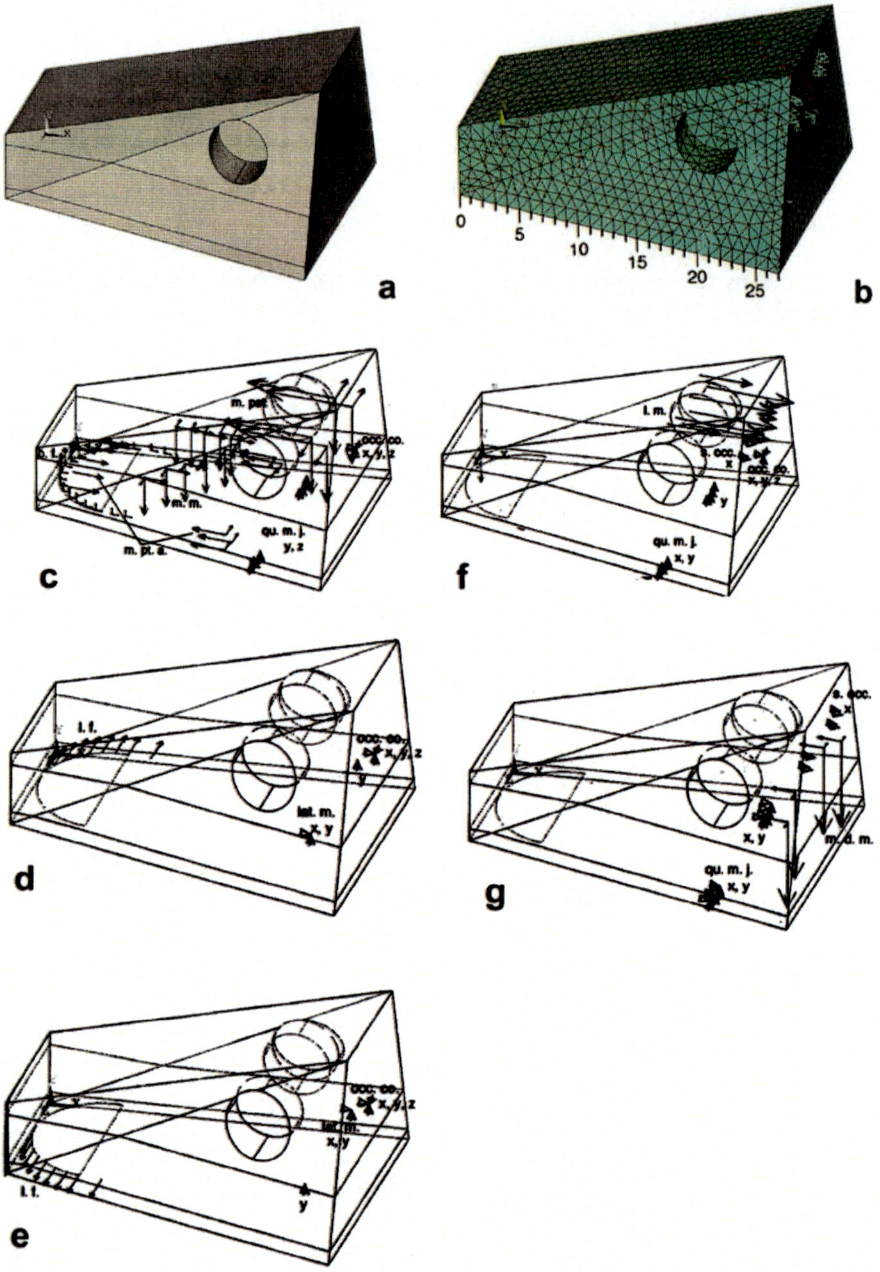

Fig. 2.43 (**a**) "Bauraum" provided for the synthesis of a *Diplodocus* – skull. (**b**) 3D meshwork filling the bauraum shown in (**a**) and position of cross sections for creating the reduced model at lower border; (**c–g**) Five load cases: (**c**) directions of resulting forces exerted by jaw muscles and evoked bite forces, (**d**) tooth forces on the right half of the upper jaw, with strong, laterally directed components, comparable to tearing off leaves from branches. (**e**) Same as (**d**), but lateral component directed to the left. (**f**) Weight of the head, concentrated in the CoM and balanced at the occipital joint by dorsal neck muscles. (**g**) Opening of the jaws by the muscles available for this function. (From Witzel and Preuschoft 2005)

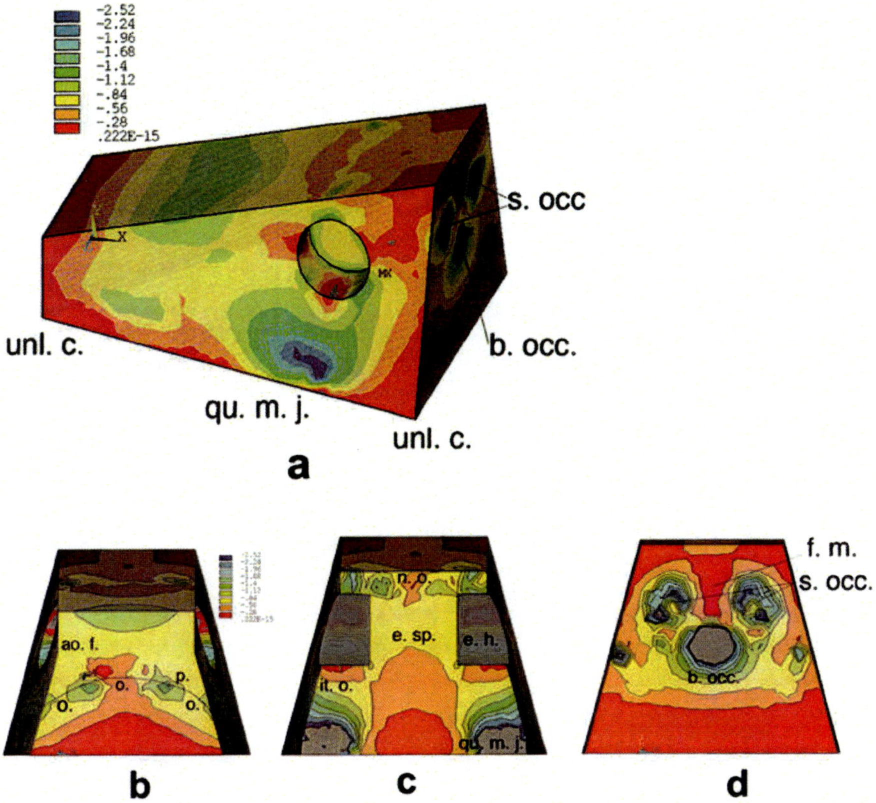

Fig. 2.44 (**a**) FES Analysis of compressive stresses within the "Bauraum", seen from the top left. Red are unstressed parts; yellow, green, light blue or dark blue are parts under stress. (**b–d**) Cross sections through the loaded bauraum, seen from in front. (**b**) At the middle part of the snout, note the high stresses on the dorsal side (green) and the lack of stress adjacent to the oral cavity (red). (**c**) Section through the orbits (grey) and through the mandibular joints (grey). (**d**) Section close to the most occipital margin, showing the occipital condyle of the basioccipital (grey) and the supraoccipital bones (green, dark blue), as well as the foramen magnum (red: free of stress). (From Witzel and Preuschoft 2005)

the human midface (see Chap. 9), the facial skeleton of hominoids (see Chap. 9) and the highly complex skull of *Diplodocus* (Witzel et al. 2011). In the majority of cases, the reduced model matches the really existing skull of the genus under investigation (Fig. 2.46 and 2.48), a 2nd or 3rd reiteration of the same process yields additional precision, and that means increased similarity to the real skulls.

Crucial for the success of this approach is the a priori consideration of traits which seem to be determined by their respective biological roles. According to our experience, only few factors exert a decisive influence: size of the brain (see above), location and size of eyes and their surroundings, arrangement of muscles and outline

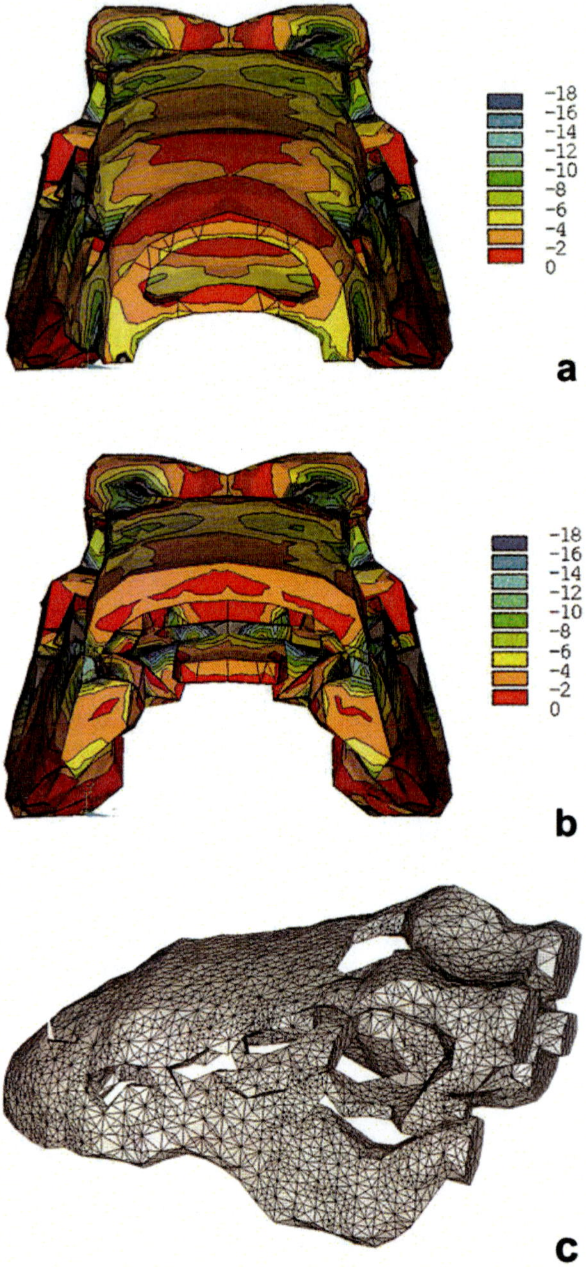

Fig. 2.45 In slices through the Bauraum (**a**), like those shown in Fig. 2.44b–d, all parts are eliminated, which do not receive compressive stress above a given threshold. The stresses become again visible on cross sections (more brilliant colours): (**a**) section through the anterior part of the snout. The more posterior parts are slightly shaded and therefore darker. (**b**) Section at the middle of skull length. (**c**) The remaining parts of the sections are recombined to a "reduced" model (**c**). This is meshed and again exposed to the same loads as in Fig. 2.43

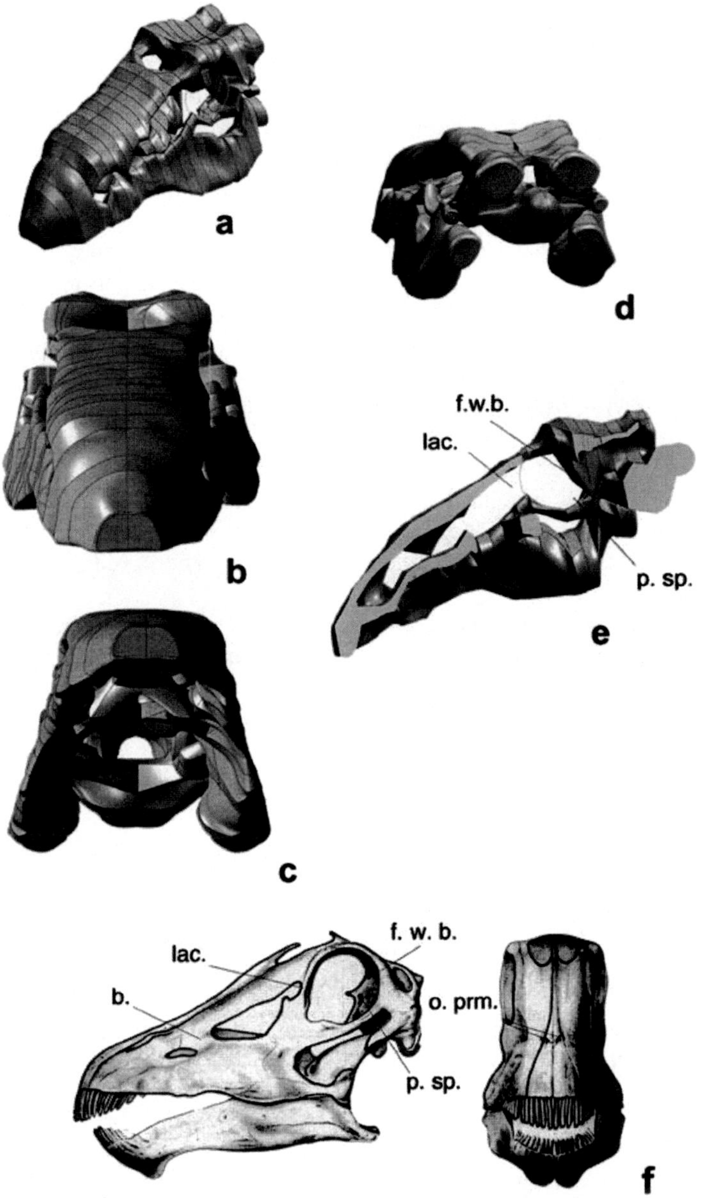

Fig. 2.46 (**c**) After repeated/second elimination of the stress-free areas of the sections, the remaining parts are combined to a three-dimensional model and smoothed. This model has sufficient strength to withstand all load cases shown in Fig. 2.43. Four views of the resulting shape are shown: (**a**) from top left, (**b**) from in front and above, (**c**) from behind, (**d**) from top left behind. (**e**) Longitudinal section in the medial sagittal plane. The space between the frontal and nasal bone and the palate is without stress and therefore leaves room for the nasal cavity. Likewise, the medial aperture between the orbits and the antorbital opening are free from stress. The existence of an occipital condyle, as well as the brain cavity and the foramen magnum, should also be noted. For comparison, the fossil skull is shown in side view and anterior view (**f**). The remarkable similarity between the reduced model and the real skull shows that the enigmatic skull shape of this dinosaur is perfectly suited for providing the necessary strength of the construction. (From Witzel and Preuschoft 2005)

and length of the dental arcade. The relationship between the size of the head and the size of the brain is the most relevant factor. The position and size of the eyes are obviously determined by their biological role: They depend on the use made by the organs in the natural environment. They occupy a part of the Bauraum and force flows must be directed around them. The dental arcades and their spatial relationship to the braincase (e.g. in front of, below, inclination against the skull base (the basisphenoid)) also depend obviously on the use made of the jaws, as well as the shapes and sizes of the teeth. Again the arrangement existing in the investigated animal must be taken as precondition. The arrangement of jaw muscles is different among reptiles on one side and mammals and their ancestors on the other. Since it does not seem to be prescribed directly by the function of the jaw apparatus, independent knowledge of muscle arrangement (e.g. from insertions visible in fossils) is also one of the preconditions for applying FESS.

In case that different load cases can be expected and therefore must be investigated (Fig. 2.43), the calculated stress pattern can be stored and superimposed on stress patterns found in other load cases (CAD). So we get the morphology, which is suited to withstand all load cases considered or, in other words, to which it is optimised.

Following the FESS of *Diplodocus* (Witzel and Preuschoft 2005; Witzel et al. 2011), R. Gössling (2010) has synthesised the skull of the prosauropod *Plateosaurus*. The first problem he had to overcome was restoring the completely preserved but severely distorted fossil skull. The basis for the reconstruction were x-ray sections through the skull in the Tübingen collection (Fig. 2.47). In each section, the better preserved side was taken for granted, and the less preserved side was symmetrically reconstructed. In cases of doubt, the external aspect was taken into consideration. The results of this procedure are shown in red in Fig. 2.47. The sections were combined to a three-dimensional image of the skull (Fig. 2.48). Independent from this reconstruction, the Bauraum for synthesis was defined according to the preserved rough outline of the head, as above in the case of *Diplodocus*. Gössling's study goes beyond the former approach by including the lower jaw. For this purpose, the author first synthesised a lower jaw separately (Fig. 2.49). The synthesised "reduced" model of the skull matches nicely the reconstruction of skull shape in Fig. 2.48.

2.2.6 *Does Skull Shape Obey the Same Rules as the Locomotor Apparatus? Development of Shape in Ontogeny in a Marsupial* (Monodelphis)

The search for arguments proving the dependence of shape from mechanical function (like in the postcranial skeleton) leads to investigating the ontogenetic development. The majority of experiments aimed at the changes of shape during embryonic development have been made on birds (chicken). Egg shells can be

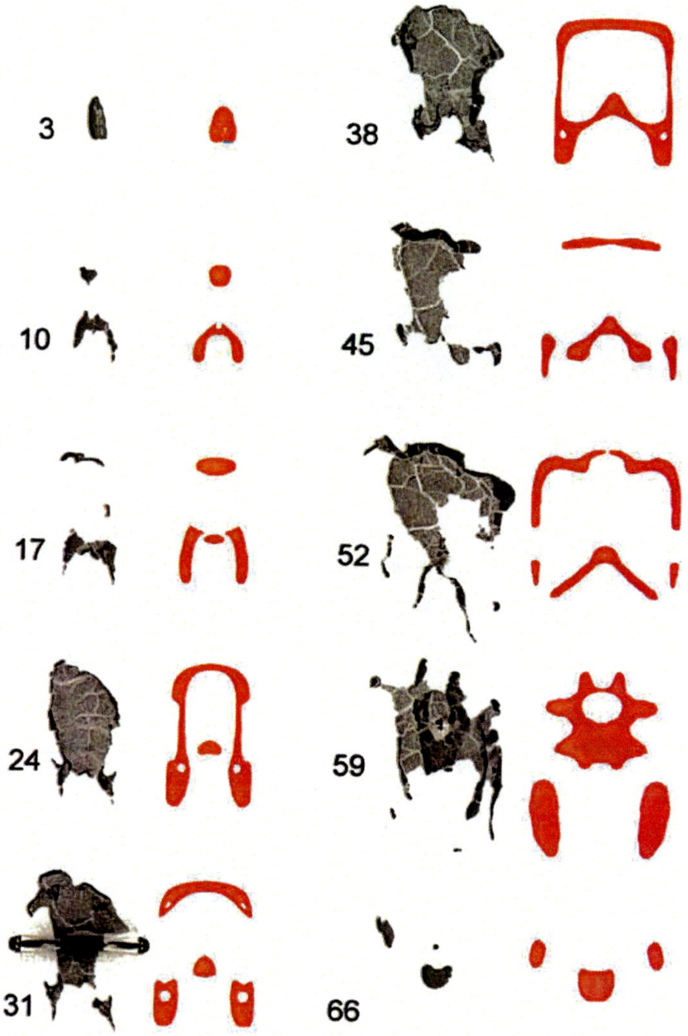

Fig. 2.47 Reconstruction of cross sections (red) through a partly imbedded and severely deformed skull of *Plateosaurus* (grey). The reconstructed (red) skull resembles closely the "reduced model" in Fig. 2.48. (From Gössling, 2010)

opened and the mechanical situation in the growing embryo can be changed (e.g. Christ et al. 1993; Hall, 2005, 2009), so that deviations from "normal" can be observed. Similar experiments are much more complicated in mammals, because of the development inside the uterus. In the case of the skull and the jaws, it is not known whether at all, and if yes, in which way mechanical forces act on the eutherian skull during intrauterine development. After birth, the requirements on the skull seem to diverge widely in sucklings and in the older stages which acquire

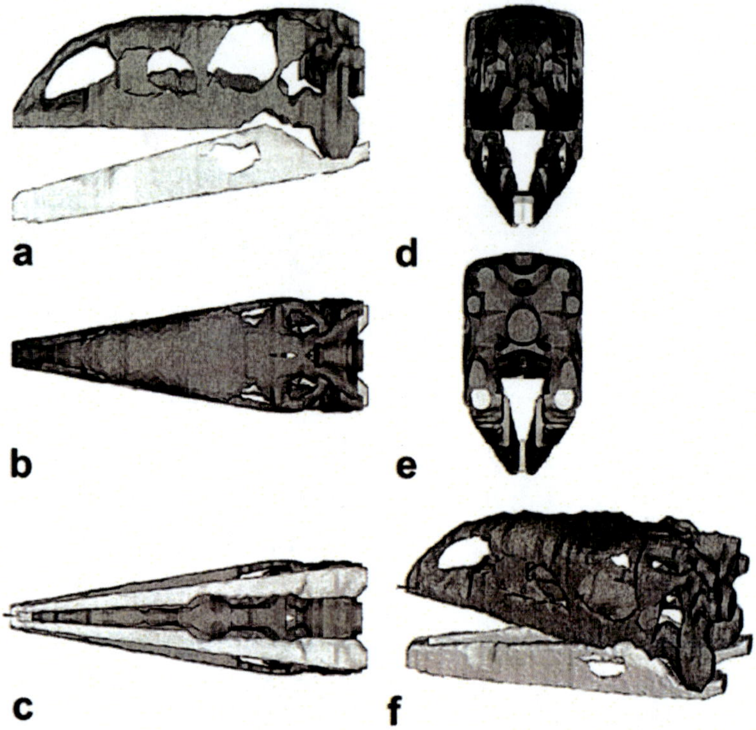

Fig. 2.48 (**a–c**) On the basis of the cross sections reconstructed views of the formerly deformed *Plateosaurus* – skull. (**a**) left side, (**b**) top, (**c**) ventral view, all with mandible (light grey, skull dark grey). (**d**, **e**) Examples of cross sections. (**f**) Combination of the cross sections to a synthesised model of the skull, which consists exclusively of compression-sustaining finite elements (as 2.47a–c, from Gössling, 2010)

and take in formed nutrition. Although the morphological changes are described scrupulously in many studies ("heterochronic changes"), no comparisons seem to exist of the morphological changes under consideration of the respective functional needs. There is only one group of mammals, in which the skull is exposed to clearly defined, relatively high external loads in early ontogenetic stages: the marsupials (Maier, 1993). While kangaroos show a number of specialisations, in most marsupials the chondrocranium and the ossification centres are rather similar in most genera of this group. The newborn young crawls on the shortest way to the maternal tit, regardless of its location in a pouch (Australian forms) or not (as in NW marsupials). The jaws and lips are closed around the tit, while the pouch young dangles downward from the head (Fig. 2.50). In a mechanical sense, the tit forms a joint, through which (tensile) forces, but no moments can be transmitted. Stresses are restricted to the head and (though limited) to the neck plus trunk, the latter giving way to all external forces (Strotdrees 1998). A critical point is how the lower jaw is adducted. Our first assumption was that the adductors of the mandible are

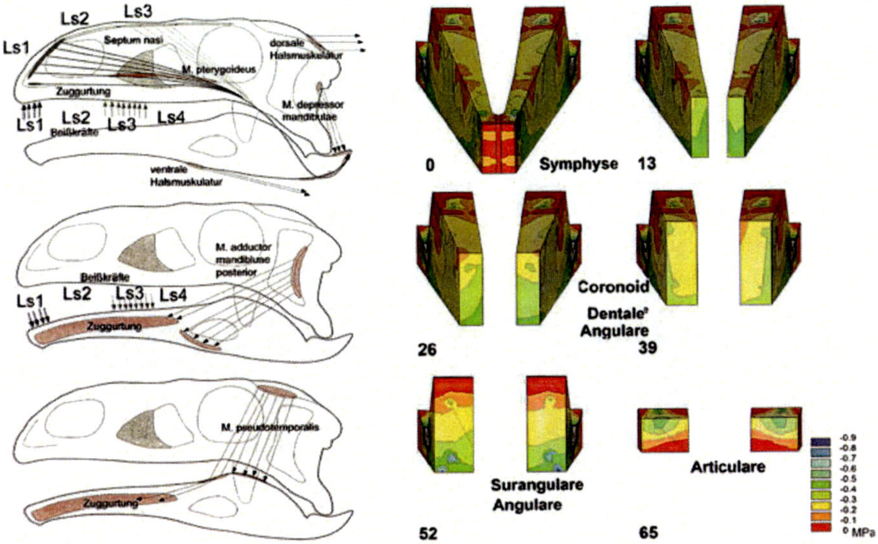

Fig. 2.49 Finite elements systems synthesis of the mandible of *Plateosaurus. Left side:* possible bite forces and three layers of the (reconstructed) muscles with their origins in the periosteum (pink areas). The muscles and their insertions form tensile ties for the bony elements of the skull. *Right side*: distribution of compressive stresses on various cross sections through the Bauraum provided for the mandible, all frontal view. Section 0 is the symphysis, Section 39 is the quadrato-mandibular joint, and Sections 52 and 65 are the post-articular processes. The reduced model, which resulted from the synthesis, is illustrated in Fig. 2.48f and from Gössling, 2010

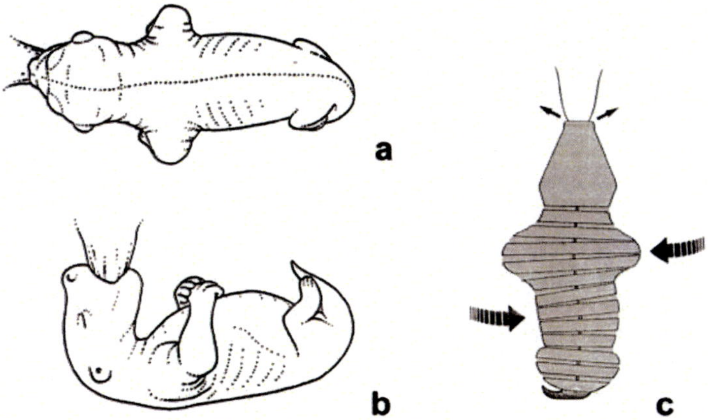

Fig. 2.50 *Monodelphis.* Pouch young seen from dorsal (**a**) and right side (**b**). It is fixed to the mother's tits by sucking. (**c**) The external forces are applied to the head of the young through the jaws. No resistance exists against movement of the animal's body in relation to the tit, and only little resistance exists inside its body. The longest link is the head, and inside the head occur relevant stresses. (From Strotdrees 1998)

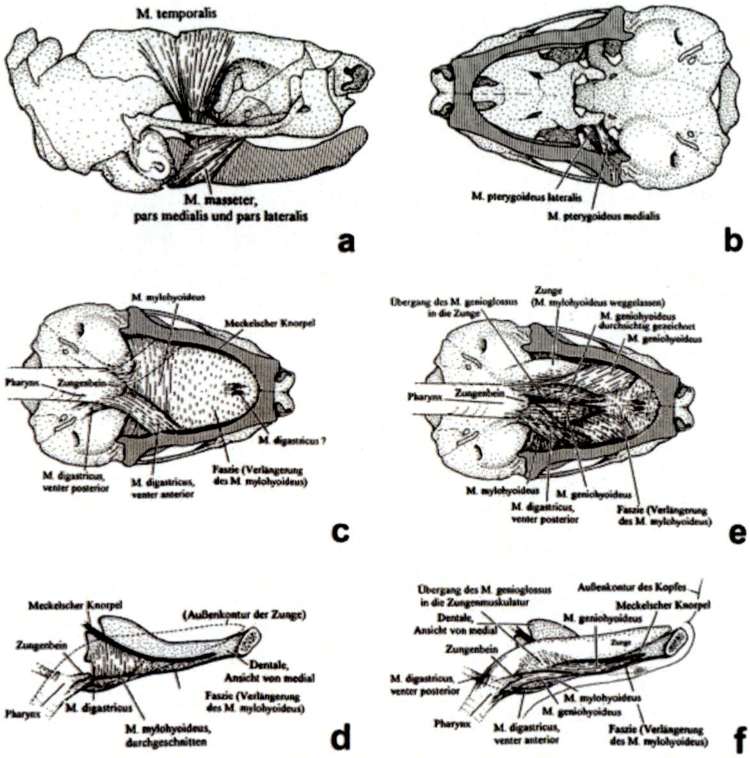

Fig. 2.51 Chondrocranium with tension-resistant structures of *Monodelphis*, as an example of a metatherian skull. (**a**) Viewed from right side, with mandibular adductors, (**b**) ventral view of the cranium, mandible in dark grey, (**c**) ventral view, including parts of the oral floor (*m.* mylohyoideus and the fascia which prolongs it to the symphysis; see also in **d**), (**d**) medial view of the left half of the mandible (dentary) with transverse muscles of the oral floor. The heavy black line shows the cartilago meckeli. (**e**) Ventral view of oral floor with longitudinal muscles; (**f**) median sagittal section through the lower jaw, showing the left half of the mandible with longitudinal muscles of the oral floor redrawn from Strotdrees, 1998

responsible. Simple surface EMG recordings of human infants (living pouch young *Monodelphis* were not available for a direct control), however, show no activity in the m. temporalis nor the m. masseter during sucking, but strong and rhythmic activity of the m. mylohyoideus and other muscles of the oral floor (Trappmann 1990). In this case, a vacuum is produced, and the external air pressure approaches the upper and lower jaws, while they are kept apart by the resistance offered by the tit, comparable to a bite force distributed over a large area. In very young *Monodelphis* (Fig. 2.51), both muscle groups are well-developed. No matter from where the adducting force comes, the jaws are pushed apart at their margins, and both jaws are exposed to bending moments. These bending moments increase in occipital direction and reach a maximum rostral to the mandibular joint (Fig. 2.52).

Fig. 2.52 Metatherian skull and physiological loads acting on it. Simplified, schematic views of the chondrocranium illustrated in Fig. 2.51. (**a**) From the right side, mandible isolated, outline of dura mater stippled, cranial base dark grey, the primary centres of ossification shown schematically, in light grey. (**b**) From the top, outline of dura mater stippled, (**c**) lateral view on the chondrocranium as in (**a**), with the cross sections investigated below, (**d**) lower jaw isolated. In both elements, the earliest ossifications are indicated by stippled outlines. (**e**) Forces acting on the skull, approximate sizes indicated by thickness of black arrows. The upward-directed force is distributed over the length of the upper jaw, while balance is maintained in part by the jaw adductors, which take their origins from the skull, as shown in Fig. 2.51. The mandibular joint is under compression. The forces evoke the bending moments shown in (**e**). (**f**) The skull is propped on the occipital condyles, and its position controlled by the neck muscles, as shown in (**a** and **b**) as white arrows. If it is fixed in a position against external forces (so forming a link), it is subject to bending as shown in the respective diagram. (Modified from Strotdrees, 1998)

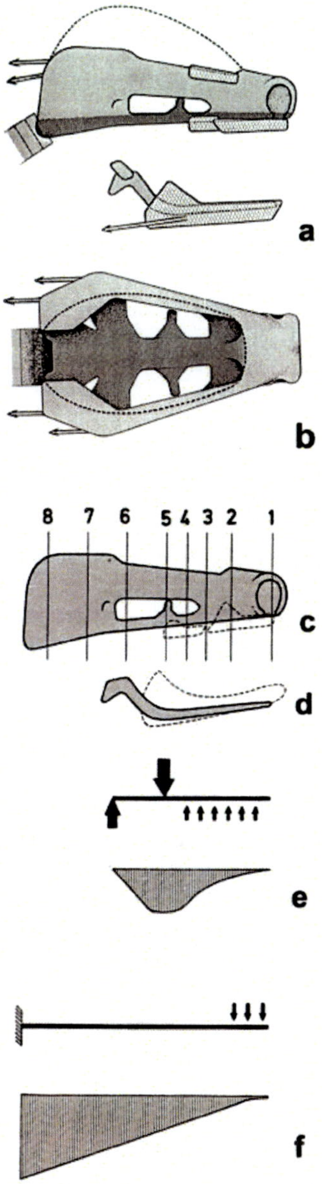

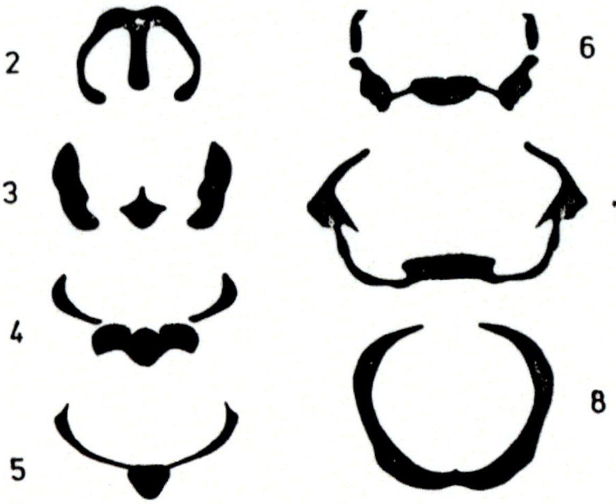

Fig. 2.53 Cross sections through the chondrocranium of *Monodelphis* as an example of a developing, but externally loaded, skull. Seven sections are used for calculating the areas, area moments of resistance as indicators of bending strength in the dorsoventral and in the medio-lateral (= occlusal) planes. (From Strotdrees, 1998)

The T-shaped and early developed nasal cartilage transmits the compressive force acting against the palate to the roof of the nasal capsula. The U shape of skull base and side walls of the skull (Fig. 2.52a) also offers considerable bending strength in the dorsoventral as well as in the occlusal plane. The areas of the cross sections, as well as the moments of resistance, increase in occipital direction (Fig. 2.53b). The first ossification of dermal bones (palate, nasal, frontal, jugal, dental) occurs at places which are exposed to high mechanical stresses. The brain does not possess a cartilaginous nor bony cover but instead is wrapped in a tough fascia, resembling the dura mater. From this fascia, fibres of the m. temporalis take their origins (Fig. 2.51a; see also above, under fascia temporalis and jugal arch). Compressive stresses from bending the upper jaw in dorsal direction, as well as contractions of the temporal muscle, lead to an increase of pressure in the liquor cerebrospinalis in which the brain is (weightlessly!) suspended. This rhythmic increase of intradural pressure has indeed been shown experimentally by controlling the anterior fontanelle in sucking babies by Trappmann (1990). The system of liquor and surrounding tough fascia acts like a "pneu" (in the sense of Frei Otto) or "hydraulic system" (in the sense of the Frankfurt theory). Equilibrium is established, if the sum of bending moments is zero. On the one side, the counter-pressure of the tit against the jaws (F_b, applied to the palate on dorsal, to the dentary on ventral side) is multiplied by its lever arm (h_b); on the other side, the compression of the pneu (F_p) is multiplied by its lever length (h_p). The latter is determined by the distance between the skull base and the centre of the cross-sectional area through the pneu (Figs. 2.52, 2.53, and 2.54).

$$F_p * \text{lever arm}\, h_p - F_b * \text{lever arm}\, h_b = 0 \tag{2.12}$$

Since the force

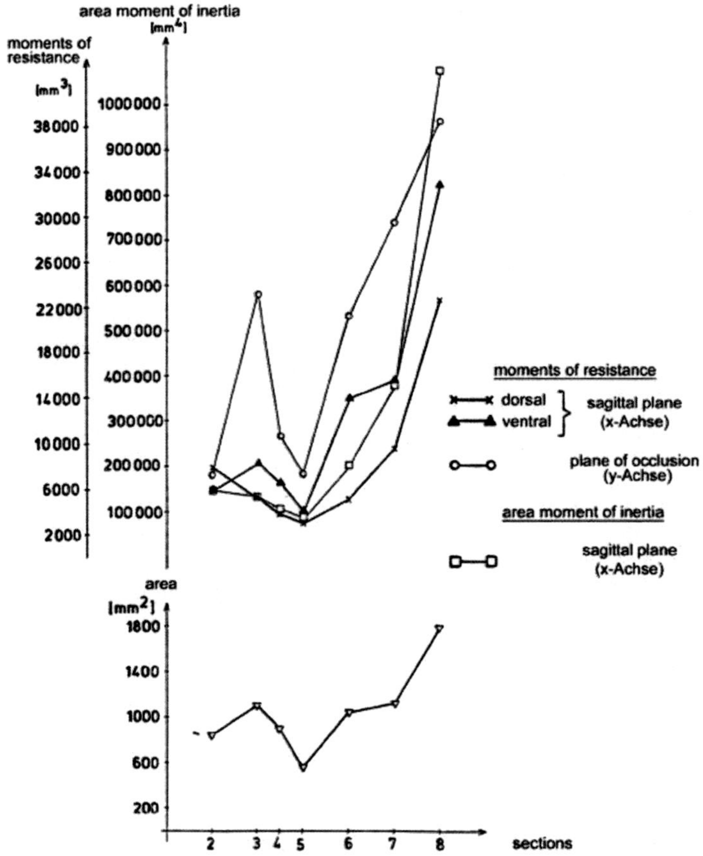

Fig. 2.54 *Monodelphis* as an example of an externally loaded, embryonic skull. The shape-dictated strength increases from the most rostral (2) to the most occipital section (8). In Sections 4 and 5, the jugal arch is ignored, and the strength probably underestimated. Numbers of sections like in Fig. 2.52c. (From Strotdrees, 1998)

$$F_p = \text{pressure within the liquor-filled space/cross sectional area of liquor-filled space}, \quad (2.13)$$

a very limited increase of pressure of few percent is sufficient to produce substantial forces for resisting against dorsally bending moments (compare the action of the compressed belly in leaping tarsiers, Chap. 8.10).

2.3 General Conclusions Concerning Tooth and Head Morphology

The head comprises, aside from the tooth-bearing jaws, the capsulae of sense organs (olfactory, optic, statoacoustic) and the case for the brain, which collects all these stimuli before transforming them into action of the whole animal. These parts do hardly exert mechanical forces but, since they play important biological roles, must be located at appropriate places and require space and supporting structures. The head also is closely related to form, length and mobility of the neck, although the most primitive state is the complete absence of a "neck" in the usual sense. The organs or parts of the body which make primary contacts of the organism with parts of its environment and really transmit forces from the environment to the animal's body are the teeth and jaws. The big question is whether the same rules concerning the development of shapes (Wolff's law, Pauwels' causal morphogenesis) which hold true for the locomotor apparatus also hold true for the head.

The answers we found are diverse: *Teeth* in very primitive vertebrates (selachians), in more advanced forms (labyrinthodonts, reptiles) and in mammals possess shapes, which are suited to sustain the forces acting on them and, at the same time, to fulfil special biological roles. Their shapes are complete before they are exposed to mechanical loads.

By contrast, the jaws (and the *whole skull*) are exposed to external loads (bite and muscle forces) during growth, and because of Wolff's law we can expect shapes which fit perfectly to the stress patterns. Indeed they do so, at least in the various forms investigated: primitive sharks, crocodiles, synapsids and *Monodelphis*.

The results obtained by U. Witzel by using the FE method show that even the complicated framework forming the skulls of sauropod and prosauropod dinosaurs exactly fits to the stress pattern which occurs under reasonable conditions. The spatial relation of tooth rows to braincase, capsules of sense organs, the arrangement of muscles (reptile-like or mammal-like) and the mandibular joints must be given in advance. These factors are evidently related to biological roles and determine the flow of forces which is analysed by FE.

The procedure used in FESS is essentially similar to and in fact repeats the process of evolution: selecting the fitting shapes out of an uncontrolled multitude. Evolution is commonly understood as a similar process: Undirected offers are made in the form of mutations, and out of this stock, the optimised variants are selected (Schmidt-Kittler 1991).

The skulls of Eutheria already possess their general shape at the time of birth, when they are exposed to physiological loads. Therefore, it is difficult to observe the influence exerted by mechanical forces on the skull. In marsupials, by contrast, the skulls are exposed to mechanical loads at an early stage, and the pouch young are better to investigate than the intrauterine embryos of Eutheria. So the influence of mechanical loads can be estimated. In the case of a juvenile *Monodelphis*, an overall coincidence has been observed between the pattern of stresses evoked by normal behaviour and shape.

All these results do not contradict, but are in agreement with the conclusion that the formation of the skull (with jaws) obeys the same rules as the postcranium.

Chapter 3
Axial Skeleton in Aquatic Animals

Keywords Undulatory propulsion · Tail fin · Rapid swimmers · Gills · Pectoral fins · Segmentation of trunk wall · Strength of axial skeleton · Whales · Porpoises · Sirenia · Pinnipeds · Amphibious mammals

3.1 Definitions, Existence of Vertebrates (Including Agnatha) in Phylogeny

Even the most primitive among the "vertebrates" have in common the axial structure: the whole body is arranged around a flexible rod, characterised by its not compressible, constant length. This element may be a notochord (= *chorda dorsalis*), or a *vertebral column*, the name-giving skeletal organ of the group (Fig. 3.1b). This is regardless of the existence of a complete head (by which "craniota" are separated from "acrania": the adult urochordates, *Branchiostoma*, and cyclostomids). In the majority of these water-living forms, the slender body is equipped with a coherent fin seam along the posterior trunk and tail, or unpaired fins along the trunk and tail. The spinal cord with its nerval cells is dorsal to the notochord, and the digestive tract, however, on its ventral side. The anterior part of the latter permits intake of food as well as gas exchange by means of a gill apparatus. Farther behind, a large digestive gland ("liver") opens into the digestive tract. Close to the paired excretory organs (kidneys), gonades are arranged in the posterior part of the trunk cavity and have their openings near the anal opening of the gut. The segmented musculature of the trunk fills dorsally and on both sides of the notochord the interspace between the axial skeleton and skin. Because they show exactly this arrangement of essential organs, the acranial larvae of *Branchiostoma* are often taken as the simplest model of a "vertebrate".

Forms without the skull and jaws but with the axial skeleton show up in the fossil record earlier than the oldest skull-possessing craniota (Chap. 2). About 100 individuals of fossil *Metaspriggina* are known from the Canadian Burgess Shales, 521–509 MY old (Morris and Caron 2014). They possess an elongate body with notochord and segmented trunk wall musculature, as well as a gill apparatus

H. Preuschoft, *Understanding Body Shapes of Animals*,
https://doi.org/10.1007/978-3-030-27668-3_3

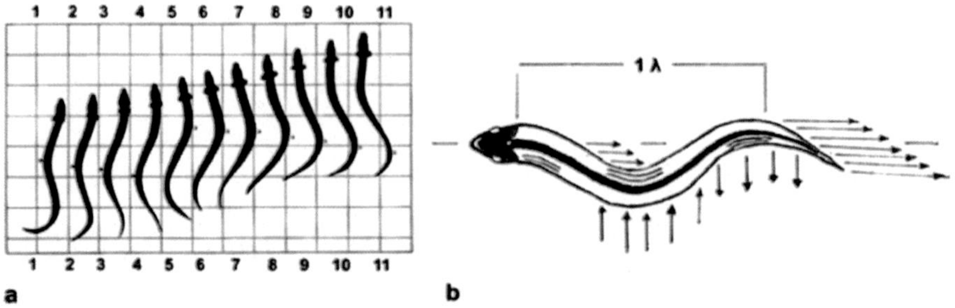

Fig. 3.1 (**a**) The undulating, wave-like movement in one plane of an elongate body (here an eel-like teleost) yields propulsion with a minimum energy. The body wave travels faster from the head to the tail than the swimming speed of the animal, and its amplitude increases rearward. (From Gray 1968). (**b**) If the propulsive movements are performed by the entire body, muscles are activated alternatively on both sides (double lines). The force components (arrows) in transverse *z*-direction, which the surrounding water exerts against the movements of the body, are shown on the left side of the animal. Forces are not proportional to the lengths of these arrows. On the animal's right side, the speed of the rearward accelerated water is symbolised by lengths of the longitudinal arrows. To exclude irregular deformations of the body under the influence of muscle activity, an incompressible rod of constant length must be arranged between the muscles of both sides: the notochord and vertebral column (black). In many slender forms, a complete wavelength (1*λ*, as indicated) is performed by the body, so that the force components exerted in *z*-direction (=perpendicularly to the long axis) on both sides compensate each other, summing up to zero

supported by six pairs of gill arches. Most primitive elements of a jaw apparatus have been found in fossil placoderms from the Silurian (419 MY old, Dupret et al. 2014; Zhu et al. 2013; see also Chapter 2).

Among living animals, the notochord persists into the adult stage in *Branchiostoma*, jawless hagfishes (myxinids) and lampreys (petromyzontids), sturgeons, other chondrostei and the coelacanth *Latimeria* (Westheide and Rieger 2004). In the latter animal (as well as in the big, several metres long sturgeons), the notochord is a tube of tough connective tissue filled with a highly viscous liquid. The outermost layers of the connective tissue may consist of elastic fibres and so allow bending of the element in all directions despite its rather long diameters. The liquid filling is not compressible, and the tough fibres of the "pneu", or "hydraulic system", exclude any increase of diameter, so that the length remains constant. In more highly developed vertebrates, the notochord is partly surrounded by bony arches, which allow limited torsion, but are not in direct contact with each other (Parrington 1967; Witzmann and Schoch 2012). The exact arrangement of these bony arches is taken as the name-giving trait for discriminating the most primitive, sometimes land-living, animals: embolomeres, rachitomes, temnospondyli and lepospondyli (Parrington 1967, pp. 271–275; Starck 1979; Witzmann and Schoch 2012). Of the Devonian *Tiktaalik*, bony limb bones, girdle skeletons, ribs and skulls are preserved, but no complete vertebral column (Daeschler et al. 2006; Shubin et al. 2006). This suggests the existence of a (not fossilised) notochord, which was surrounded only in parts by bony elements.

In most other living animals, the chorda dorsalis exists alone in early embryonic stages. According to Vassilieva et al. (2013), the newly hatched larvae of Asian clawed salamanders (Hynobiidae) possess a complete vertebral column. In mammals, the notochord is replaced during prenatal growth by bony vertebral bodies, which originate from enchondral bone cores. By means of the intravertebral discs, they make contact in longitudinal direction with each other (Starck 1975) and so can transmit compressive forces. The vertebral arches and their processes develop later than vertebral bodies, though also in prenatal stages.

3.2 Why an Axial Skeleton?

The observed morphological facts leave us alone with the *question why*, after all, is the *notochord, or the vertebral column*, present in so many animals and so widely spread among animals belonging to a variety of evolutionary lines? (Portmann, 1950).

Aside from the "internal forces" provoked by the trunk muscles and described in the caption of Fig. 3.1, and in Sub-Chap. 3.2.1, there exist "external" forces, which must be countered by a stiff rod. Water resists against moving an immersed body because of its density and its viscosity. The former evokes at the head a resisting compressive force in x-direction F_w ("ram"- or "bow"-"resistance"), directed against the movement, its size proportional, according to Nachtigall (1977) and Gersten (2003), to

$$F_w = \text{Area of cross section} * \text{square of speed} \tag{3.1}$$

In addition, a sort of "friction resistance" (F_r) acts widely distributed along the whole length of the body. It is proportional to

$$F_r = \text{Surface area} * \text{square of speed} \tag{3.2}$$

3.2.1 Continuous Propulsion by Undulatory Movements Required in a Viscous Medium

If a body is suspended in water, it becomes weightless, but to overcome the permanently acting water resistance, aquatic animals have to *propel their bodies continuously*. A widely spread technique of propulsion, which can be observed among various stems of the animal kingdom, is undulation (Alexander, 1967; Bone and Marshall, 1985; See also the review in Morphisto, 2008). Undulatory

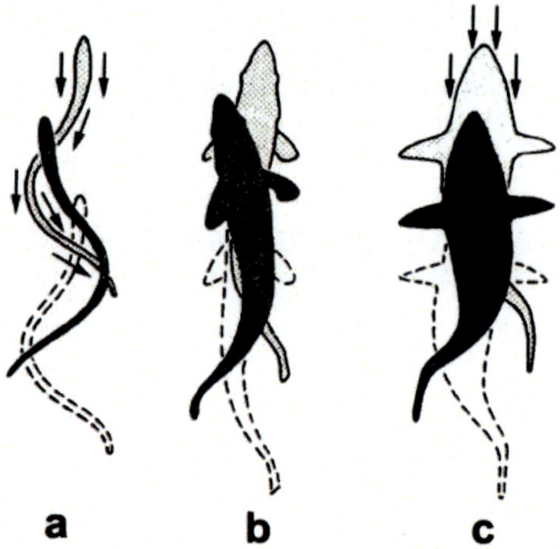

Fig. 3.2 Transition of relative wavelengths from undulating eel-like (anguilliform, (**a**)) through perciform (**b**) movement to oscillation of the tail alone (thunniform, (**c**)). Extreme excursions to the left are shown in black and to the right by broken outlines. In (**a**), the "frictional resistance" of water is illustrated (arrows) which acts along the entire surface of the animal, while in (**c**) the "ram resistance" of water against the head is shown. (Modified from Reif 1981b, who followed Lindsey 1978)

movements can be seen in trypanosoms, plathelmints, annelids, free-swimming, marine snails (Grasshoff and Gudo 2001), as well as vertebrates. In the latter animals, undulatory propulsion is initiated by muscles, which are arranged along the whole length of the trunk and tail. Transmission of the propulsive force to the head and body surface requires a compression-resisting element between the head and body surface on one side and the propelling muscles on the other. The least energy-consuming structure suited to sustain these compressive forces is a notochord or a vertebral column: the flexible axial skeleton allows bending in all directions, but no shortening. Torsion of the elongate body is impeded by a spiral-shaped arrangement of fibres, collagenous in the notochord and muscular in the trunk wall (Fig. 3.1b).[1] Flexion of the trunk is produced by the coordinated contraction of segmented musculature (Sub-Chapter 3.3.2) which fills the space between the axial skeleton and skin. The muscular activity alone evokes compressive forces between the anterior and posterior fixations of the pulling muscles, which are also sustained by the axial skeleton. A segmentation of these body wall muscles allows confining muscle activity on parts of the length. Alternating flexions of body parts in both

[1]As was emphasised by Gutmann (1989) in annelids, the segmented, liquid-filled coelom forms a series of compression-resistant hydraulic chambers, or "pneus", which can be exposed to compression by a contraction of the annular muscles. The pressure in their interior excludes, because liquids are not compressible, any shortening of the chambers under the pulling force of the longitudinal muscles, which cause flexion of the trunk.

directions are used for propelling the animal, because flexions progress from the head rearward with growing amplitudes (Figs. 3.1 and 3.2), in the case of eels up to 60% of body length. The body so performs wave-like ("undulating") movements. Formation of these waves, that is, deviation of body parts from the straight line, is performed against water resistance (Fig. 3.1b). It seems plausible that elastic recoil of the axial skeleton assists undulatory movements, but no studies seem to have pursued this point in detail. Deviations cause force components at right angles to the direction in which progress is made. While being of no direct use for propulsion, they require energy and corrections by steering. If these force components add up to zero during each swinging cycle, they annihilate each other, and no efforts are necessary to keep the animal on course. This seems to be a plausible reason for the frequent use made of relative wavelengths around 100% of body length. Aside from the pressure, which is exerted on the convex side of the axial flexions, a suction effect on the concave sides should not be neglected, as Blickhan (2007) has pointed out. The *body wave* travels faster rearward from the head towards the tail than swimming speed of the animal. Already the first excursion of the anterior trunk accelerates water caudally towards the tail (right side of drawing Fig. 3.1b). As reaction, the body is pushed forward. Streaming velocity along the body is increased by the flexions which take place in the trunk and proximal tail segments; the most caudal segments give them an additional kick. While the frequency of the flexions along the body cannot be increased, the caudally growing amplitudes comprise greater volumes of water and increase the rearward stream further. Behind the animal, vortices are left behind, which rotate about a vertical axis. In steady-state locomotion, the thrust forward must be equal to the resistance exerted (permanently) by the surrounding water against moving of the animal (Webb, 1975).

In most fishes, the dorsal and the ventral seam of fins is interrupted, so that several unpaired dorsal and anal fins remain. In trouts, as an example for the perciform swimming mode, detailed studies have been executed (Nachtigall 1977; Blickhan, 1990; Blickhan and Nachtigall 1991; Blickhan 2007), which yielded information about the streaming situation along the body, so helping to understand the arrangement of these fins.

The wavelengths can be measured in relation to body length. Typically one (or even a bit more of a) complete sinusoid wave is visible on the body. This method of swimming has been called anguilliform, referring to the movements of eels (Lighthill 1969, 1970; Lindsey 1978; Reif 1981b). The same style can be observed in lampreys, sturgeons, many small sharks, morays and swimming snakes. Several marine snakes possess the laterally compressed trunk and tail like that in fishes. Flounders and plaices show, if observed in side view, exactly the same pattern.

Aside from animals swimming with a relative wavelength of about 1, there exist forms, in which not the trunk is undulating but instead the ventral or dorsal fin seam, still in z-direction (e.g. *Amia*, *Gymnotus*). In these forms, it is no problem to perform more than one sinus swing in relation to body length. These undulations of the fins are produced by the relatively weak muscles which move the skeletal rods of the fins against the long dorsal and ventral processes of the vertebrae. A similar arrangement is also realised in skates and rays (see Chapter 3.4), where the margins of the huge pectoral fins, however, act in y-direction.

3.2.2 *Undulation Restricted to the Fluke: The Example of Rapid Epipelagic Swimmers*

Propulsion by undulation can be made more efficient by a *tail fin*, an element which enlarges the area exposed to water resistance at that part of the body, which is moved fastest. In a great number of fishes, the ratio wavelength/body length is smaller, less than 1, so that only part of one full swinging period fits to body length and so becomes visible. This swimming mode may be called perciform (after *Perca*, the perch Fig. 3.2b). Blickhan (2007) found for trouts very low energy requirements for the locomotion he calls "subundulatory". At the end of the variation range, the relative wavelengths are so long that not more than 1/2 fits on the body, of which alone the caudal third is flexed during oscillating movements ("carangiform" or "thunniform" mode, Fig. 3.2c). This mode is found in the fastest and most endurant swimmers.

The extreme morphological optimisation among the inhabitants of the open ocean was described by Fierstine and Walters (1968) in the case of tunnies: a stiffened tail peduncle is moved in one joint at the base of the peduncle and a second joint at its end, which allows the sickle-shaped fluke to assume the optimal angle for maximal propulsion. The peduncle has a slender shape, and so finds but low resistance when swinging from one side to the other. In thunniform swimming, the great mass of the unmoved trunk seems to provide the stabilising mass moment of inertia to balance the sideward-directed force components exerted by the fluke.

The more or less symmetrical and sickle-shaped tail fin (Fig. 3.3) of thunniform swimmers, (i.e. isurid sharks and mackerel-like fishes, tuna, marlin, swordfish, and also dolphins, rorquals, see Sub-Chapter 3.3.4, and Pacific sea cows), the "fluke", becomes the only organ of propulsion. If the streaming fluid has high velocities (in relation to the animal), the degree of efficiency of the propulsive organ is greatest, if the span of the fin (or its "height", i.e. extension in *y*-direction) is great, while its depth (length; extension in *x*-direction) is low. This spread ("*Spreizung*") is measured as one half of span width in relation to depth and assumes values of 3–10.[2] Because of its rather small area, such a fin is not exposed to outstanding water resistance and therefore permit very rapid oscillations. All these rapidly swimming forms possess the characteristic streamlined body shape: the head continues directly into the trunk; a visible neck is absent. Head and trunk are stout and contain by far the greatest part of body mass. Their cross section is often circular, because this shape combines the greatest mass with the smallest (friction-producing) surface. This body shape is generally known (since Prandtl, 1918 and 1919) and often recruited as an example for convergent evolution, because it occurs in widely different classes of vertebrates (Fig. 1.2 and Chap. 1).

[2]The same principle, by the way, is used by birds (Sub-Chapter 6.6.1) and discriminates the wind-using, extremely efficient "gliders" (gulls, bassets, albatrosses, swifts) from "soarers" (eagles, vultures, storks).

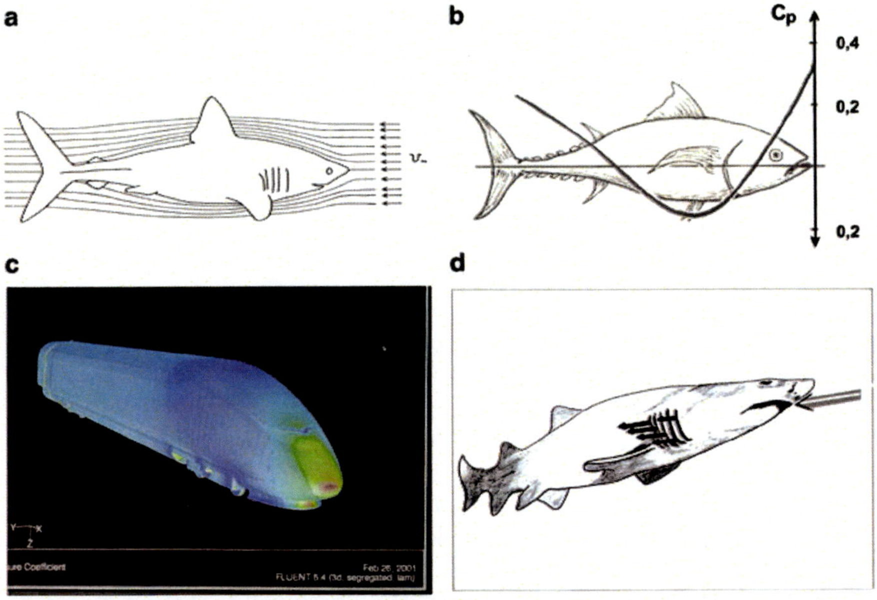

Fig. 3.3 (**a**) Streamlines around the model of a big white shark in a fluid. Based on experiments made by Ungerechts (1980). (**b**) Pressure acting against the surface of a moving tuna: high ram pressure acts against the head, decreasing rapidly to negative values at the greatest diameter and then again slowly growing friction resistance. Behind the head, where the gills are located in fishes, the surrounding fluid does not exert pressure, but suction. The drawing is an estimate based on data collected from the literature by Ungerechts (1980). (**c**) FE calculation of the surface pressure occurring in an ICE train. Negative pressure (=suction) occurs at that place, where the diameters of a lengthy body do not longer increase (DB-Mitteilungen, 2001). (**d**) Sand tiger shark in which stow pressure helps in ingesting O_2-rich water, while the O_2-depleted water is sucked out of the gills

3.3 Common Morphological Characters Depending upon Mechanical Necessities

3.3.1 Location of Gills

Figure 3.3a shows the experimentally determined streaming of a fluid around the model of a big white shark. This pattern is not changed if the model is given a flexed body axis (as in the typical oscillations of the fluke), and its overall streaming resistance does not become greater (Ungerechts 1980). The organ of gas exchange, the *gill apparatus*, has an optimal position: in Fig. 3.3b the compressive forces acting on the body surface of a tuna are shown. They are high at the most anterior part (stow resistance). Behind the greatest diameter, the pressure against the surface becomes rapidly smaller and changes its sign, switching to suction. This exactly is the place of the gill slits in sharks (Fig. 3.3d) or of the operculum in bony fishes. The correctness of this finding is confirmed by a FE calculation of the surface pressure acting on an ICE-train (Fig. 3.3c): where the diameters do not longer grow, the surface is exposed to suction. Obviously, the take-in of O_2-rich water is assisted by the stow pressure against the head of the moving animal, while at the outlet water is sucked through the gills.

3.3.2 *Why Persist Anterior, or Pectoral Fins, While the Posterior Pair Is Reduced?*

Water resistance can be kept low by reduction of all sorts of appendages. *Reduced are* the unpaired, as well as the paired, fins, and the surface of the body is smooth, without any structure offering resistance against streaming water (for more details about surface cover, see Hertel (1963, Chapters G – L, especially K; Reif 1978, 1981b; Nachtigall 1977, 2000; Ungerechts 1980). However, in all thunniform swimmers (sharks, mackerel, tuna, swordfish, marlin, cetaceans), the anterior pair of limbs persists in contrast to the posterior pair, and in ichthyosaurs the former is markedly larger than the latter. The whole extremity is equipped with a streamlined profile and can be moved actively – in contrast to the posterior limbs, which, if at all present, are rarely moved. The question should be raised *why* this is so.

The location of the *anterior limbs* immediately behind the head permits the water living animal a precise positioning of the head, to make optimal use of its sensory organs and of its jaw apparatus. This increases the ability to avoid danger, or to capture prey effectively. Movement of the head alone, without the heavier trunk, saves energy, as has been worked out by Preuschoft and Gudo (2005): If a small shark of 1 m length and a mass of 2 kg turns towards a potential prey appearing at its left side (Fig. 3.4a), the pivot of its full-body turn is the CoM at about ½ body length; its mass moment of inertia (J) is

$$J = m * l^2, \quad \text{or} \quad J = 2 * 25^2 * \frac{m}{2} = 1250 \text{ cm kg} \tag{3.3}$$

In case the shark's trunk axis is flexible, the head with sense organs and jaws may be re-directed without moving the entire body (Fig. 3.4b). Movement of the head is coupled tightly to the action of the anterior fins, if a direct bony contact exists between the shoulder girdle and the head – as is the case in many chondrostei and teleostei (Gudo and Homberger 2002). If the head alone is turned towards the prey, the mass moment of inertia, which resists being moved, is

$$J = 12.5^2 * \frac{m}{4} = 78.13 \text{ cm kg} \tag{3.4}$$

That is not more than 1/16 of the former value. In addition to inertia, water resistance must be overcome. The acting force F_w is

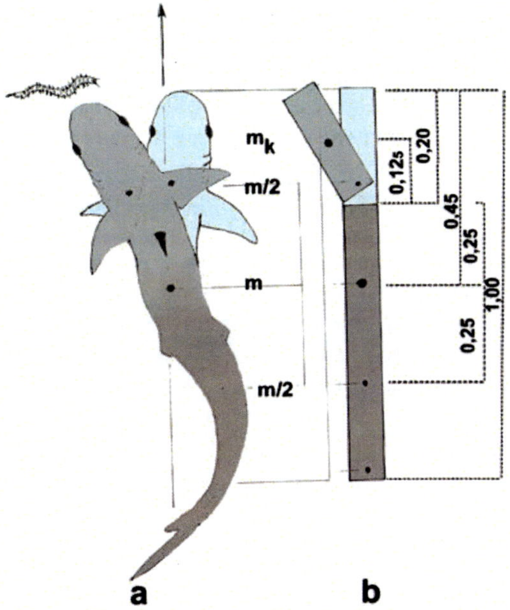

Fig. 3.4 (**a**) Small shark, like *Squalus*, or *Scyllium*. The orientation of the anterior body parts alone with head and gill apparatus toward a prey (**a**) is performed by the anterior, pectoral fins, which are coupled to the head by a skeletal element (Gudo and Homberger 2002). Preuschoft and Gudo (2005) have made a model calculation in which the shark's body was replaced by a beam of the same length (**b**) and about the same volume. The dimensions used for the model calculation are given on the right. If this beam is rotated about its CoM toward the prey, its mass moment of inertia must be overcome as well as water resistance against the movement. Movement in this case costs about 16 times as much (force) than head movement alone; water resistance is about 3 times greater than if the head is moved alone (see text, especially Eqs. 3.3 and 3.4)

$$F_w = c_w * A * q \tag{3.5}$$

where c_w is a form- and surface-specific factor ("Widerstandsbeiwert"), A is the exposed area, and q is the pressure of the water pushed aside. It consists of

$$q = \frac{1}{2} \frac{\text{mass}}{\text{volume of water} * \text{velocity of movement to the square}}. \tag{3.6}$$

Turning the stiff body about its CoM (radius of 45 cm behind tip of rostrum) in our example costs 146.2 units of force/unit area. Redirecting alone the head requires a higher angular velocity but a radius of not more than 20 cm, and alone the lateral aspect of the head offers resistance against water. The calculation in this case yields 46.2 units of force/unit area, that is, little more than one third of the first value. In addition, the most anterior part of the head, which is moved most rapidly, in many forms is flattened, so that it offers a smaller lateral plan. In most sharks, the rostrum is depressed; in garpikes (*Lepisosteus*) and luce (*Esox lucius*), as well as in crocodiles (see Sub-Chapter 2.3.6), the long snout (upper and lower jaws) is obviously flat in side view.

The above-described *rapid swimmers* among selachians and teleosts indeed make use of the option to place their head without trunk into a favourable position towards an aim. They seem to use their *anterior fins* mainly to adjust the height (on their *y*-axis) and the tail fin to determine the lateral excursion (on *z*-axis). In living sharks it can be seen nicely how the anterior fins change their position in relation to the trunk when initiating a change of direction. The secondary water-dwelling whales and sea cows also possess anterior, but no posterior, limbs. The forelimbs may control the lateral (*z*-direction) and the fluke the vertical (*y*-direction). This is the reverse of fishes. It may well be that the height of the peduncle in many whales allows it to replace the steering function of the tail fin in *z*-direction like in fishes. In several whales (beluga, narwhale, northern and southern right whales), the flukes deviate from the described outline, and their peduncle is less high. Aside from their lesser travelling speed, it is not known which special behaviours are connected with their deviating morphology.

3.3.3 Segmentation of Body Wall

The *segmentation* of the trunk wall, especially in fishes (Fig. 3.5), is a tricky problem of its own and has attracted much attention. This sort of musculature is commonly seen as an evolutionary old, "plesiomorphic" trait, maintained from more primitive, fishlike ancestors. It can be observed in the tail and trunk of fishes, newts (Omura et al., 2013, Carrier 1993) and crocodiles (see Sub-Chapter 3.3.3), the musculature of which is otherwise determined by the limbs and their girdles and hardly shows remnants of the primary segmentation of the trunk wall. The muscular segments are innervated by segmental nerves, which take their origins from the spinal cord, at the height of the respective muscles.[3] The spinal cord itself extends in reptiles caudally onto the tail (Anthoney, Guibé, both in Grassé 1970), while in mammals it ends in the "filum terminale" at the level of the lumbar or the first sacral vertebrae. The newly evolved (apomorphic) limb muscles, by contrast, receive their nerves from plexus in the lumbar or sacral region (Chap. 10.4).

The most "primitive" state can be seen in subdivisions more or less perpendicular to the long (or *x*-) axis, as in *Branchiostoma* and in young embryos of various forms. Such a segmentation has been observed in *Metaspriggina walcotti* from the Cambrian (see Chapters 1.4 and 2.1). All other fishes show a complex, three-dimensional folding of the borderlines between myomeres, the myosepta. They assume roughly

[3] In comparative anatomy, innervation of muscles or areas of the skin is commonly taken as very convincing information concerning the evolutionary or ontogenetic origin of the innervated structures and as a strong argument for their "homology". (See also Chaps. 1.3 and 10.4)

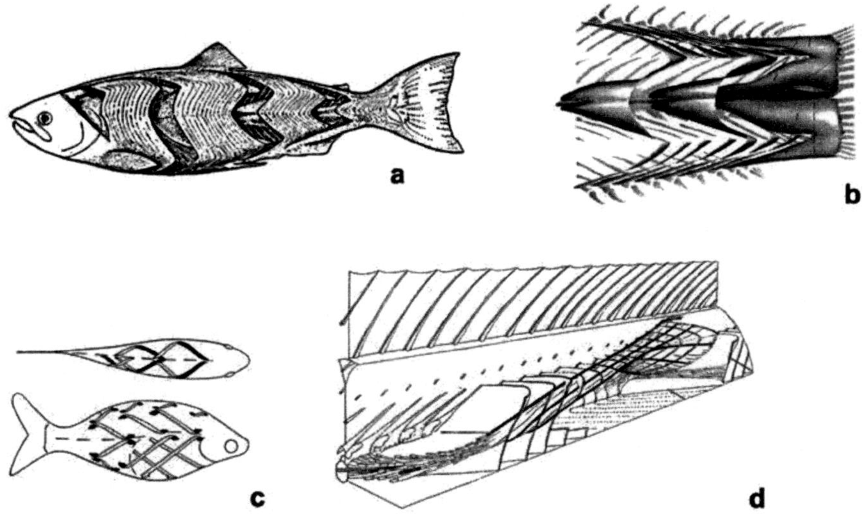

Fig. 3.5 (**a**) Arrangement of segmented trunk wall musculature in a perciform fish. (From Harder 1975). (**b**) Myosepta on the tail according to Gemballa (1995); (**c**) directions of muscle fibres across several segments, "muscle-fibre trajectories" as drawn by Alexander (1969); in top view (top) and lateral view (below); (**d**) three-dimensional arrangement of muscle fibres across segments ("Kontraktionsbogen") according to Gemballa (1995). Note that force-transmitting fibres are distributed over several vertebrae

the form of a W, the upper side of which is turned forward, the lower two angles rearward. The latter are located dorsal and ventral of the vertebral column. The greatest length in cranial direction have the myomeres close to the axial skeleton. In caudal direction, however, close to the surface, fibres of the myosepta intermingle with the connective tissue of the skin (Gemballa 1995), and back to the vertebrae (Bruna et al. 1989). In thunniform swimmers the myosepta stretch along the whole peduncle and so assume the role of tendons. As has been shown by Alexander (1969) and Gemballa (1995), the muscle fibres of subsequent segments can be traced over long parts of the body, and form a complex system of helices, by which the coherence of the muscle bundles is secured. This arrangement keeps the bundles together and counteracts any possible deviations of water pressure on both sides. Such pressure differences may give rise to deflections of the tail fin's greatest height and so reduce propulsion, while producing nicking of the trunk axis. The fibres of epaxonic and hypaxonic myosepta form a slidable lattice grate, which allow elongations on the convex side of trunk curvature. This arrangement leads to tensile forces which span several (up to 12) vertebrae.

The essence of muscle segmentation obviously is the possibility to activate only parts of the long strands of musculature consecutively in a coordinated sequence. A fine-grained segmentation of the lateral musculature can control and optimise the flow of water around the body (Blickhan 2007). The strongly curved myosepta, however, distribute the tensile forces exerted by short muscular segments over a number of skeletal segments. This gradual transmission of muscular force on the skeleton entails a smooth and gliding onset of movements. Nevertheless, segmentation and the pronounced curvatures of myosepta produce contradictory effects and remain an unsolved question.

3.3.4 Why Is the Tail Preferred as Propulsive Organ?

If propulsion concentrates on the rearmost part of the body, an undulating tail, a tail fin (or the posterior limbs; see Sub-Chapters 3.2.1, 3.2.2, and Chap. 6), the parts of the body cranial to the propulsor are *reliably compressed*; no tensile stresses occur. Compressive forces can be sustained by the skeleton without expenditure of energy. The compressive forces along the vertebral column up to now have not been estimated in anguilliform or perciform swimmers. In view of the waves travelling from head to tail, it must be expected that they are of roughly the same size along the whole length of the body axis (for thunniform swimmers see Sub-Chapter 3.2.2). In recent newts, tadpoles and crocodiles, as well as in more or less aquatic lizards, undulation is in fact more or less confined to the long and flexible tail (Fig. 3.6). Such a mode of locomotion seems to have been used as well by the primitive jawless or jaw-possessing vertebrates (ostracoderms) of the Silurian and Devonian. Only in few of these animals (e.g. *Cephalaspis*, *Hemicyclaspis*), the trunk seems to have been mobile enough to take part in the undulatory movement. This more extended mobility later became common in the placoderms of the Devonian.

The *low mass* of the tail keeps inertia against being moved within narrow limits.

Its *expanded lateral plan*, offering a large area against water (Fig. 3.6a), is often increased by a dorsal and/or ventral fin seam, which enlarges the area actively moved against water and at the same time impedes an exchange of compression and suction components which occur separately on both sides of the animal (see below). Unpaired fins (in sturgeons or sharks) or expansions of the horny skin (crocodiles) also seem to improve the efficiency of an undulating tail while keeping the resistance-causing area low. In the earliest land animals, a rather stout trunk is combined with a narrow (compressed) and high propulsive tail. A predominance of undulating tail movement during swimming is at least probable in these animals.

It has never been investigated whether the low-amplitude *motions of trunk*, neck and head are nothing than co-movements under the influence of the muscles which move the tail, or contribute to propulsion. In shallow water, or in swampy, or vegetation-filled "half-liquid" environments near the banks of slow-floating rivers or lakes, the trunk may perform undulating as well as limb-dominated, quasi-striding movements, while the tail undulates (own observations in crocodiles and newts).

In crocodiles, the long tail is exposed to considerable water resistance, which evokes cranially growing *bending moments* (Manter, 1940; Fig. 3.6b–d). These are countered by the segmented musculature and on the proximal section by the externally visible m. caudofemoralis, which is responsible for the breadth of the

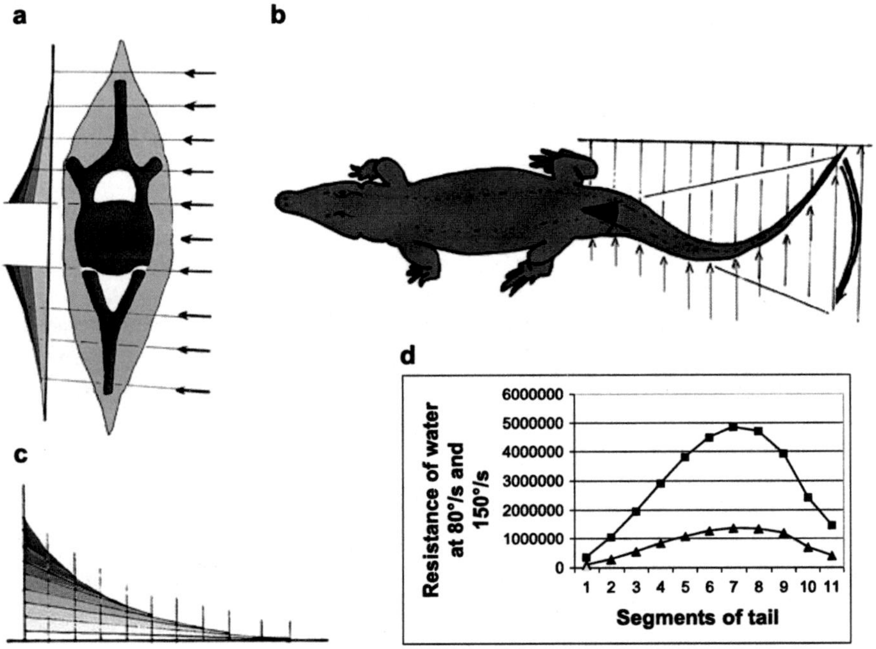

Fig. 3.6 (**a**) Schematic cross section through the tail of a reptile. Water resistance is equally distributed over the height of the tail, and the dorsal as well as the ventral sides are exposed to parabolically increasing bending moments in the x/z-plane. Skeletal elements are dark grey. Note in dorsal and ventral processes the increase of bending strength toward the centrum. (**b**) Tail of a crocodile during a strike to the left side (bowed, grey arrow). (**c**) If viewed from the top, the bending moments add up to a parabolic envelope of the graph (for calculation see Chapter 4.5). (**d**) Resistance of water against lateral moving of the tail grows with peripheral speed of the segments. It is calculated for angular velocities of 80 °/s (triangles) and 150 °/s (squares). Resistance reaches its maximum not at the fast moving tip of the tail, but a bit distal to the middle, where its lateral area is larger than at the tail's tip

proximal tail and inserts into the transverse processes of vertebrae. This limb muscle overlays and covers the segmental, seemingly primitive musculature of the tail.

Since the muscles of the body stem always exert their greatest *force component* along the x-axis of the body, they compress the axial skeleton (see above, Chapter 3.2). This compressive stress is sustained by the vertebral bodies, or "centra", as well as the intervertebral discs. A first estimate of these forces is attempted in Fig. 3.6. As an example, the tail of a crocodile is investigated during a simple strike to the left side, not during undulating movements. Water resistance is dissolved into a number of partial forces and shown in the form of several arrows. These partial forces evoke bending moments, which culminate at the base of the tail. If summed up, all these bending moments form an envelope with a parabolic outline (Fig. 3.6c). In proximal direction, the counteracting moments of the muscles grow because of their increasing lever arms (i.e. distance between the centre of cross-sectional area of muscles and centre of vertebral body) and because of the proximally growing cross sections of the muscles. This becomes particularly obvious in the m. caudofemoralis. The compressed and high cross section through the tail (Fig. 3.6a) is also exposed to water resistance, symbolised by several partial forces (=arrows). The resulting bending moments increase towards the axial skeleton and are balanced, by passive structures, connective tissue below the skin and the long spinal (dorsal) as well as hemal (ventral) processes of the vertebrae. Both show clearly their strength against bending by their laterally displaced contacts with the centra. While the dorsal processes are firmly fused to the vertebral bodies by their arches, the weaker ventral processes can swing in longitudinal direction because of the pair of joints by which they are attached to the centrum. Occasional imbalances between water pressures of either side (which may well be evoked by environmental conditions) may well twist the tail along its x-axis. Such twisting is countered by the above-described helical arrangement of muscular and collagenous fibres.

The calculation in Fig. 3.6d is based on a tail length of 90 cm; the lateral surfaces have been cut into 8 cm broad rectangles, the last one as a triangle. As speed of the movement, values of $\omega = 80°$/sec and 150°/sec have been assumed arbitrarily. Like in Eq. 2.4, peripheral velocity (v_p) is proportional to length, and therefore the speed of the distal tail segments is greater than that of the more proximal ones. Water resistance (W) is

$$W = A * \omega^2, \tag{3.7}$$

and therefore the forces in Fig. 3.6d come into action against the tail segments and must be overcome. Obviously these resisting forces do not reach their maximal values near the rapidly moved tip of the tail but are greater near the middle of the tail, where its lateral plan is larger. This, by the way, explains why a loss of the tail tip does not impede the movements of a crocodile too much.

3.3.5 *Strength Variation of the Axial Skeleton in Aquatic Mammals*

A divergent specialisation of *thunniform swimmers* (Fig. 3.7) is illustrated by the example of cetaceans. For propulsion, the fluke is moved up and down (see, e.g. in Buchholtz 2001a), not from one side to the other as in fishes. The propulsive movements in aquatic mammals, especially cetaceans and sirenians are described and explained convincingly in relation to the lateral undulations in fishes and reptiles by Gutmann (1994). Water resistance (F_{e1} and F_{e2}, respectively) is concentrated on the fluke, while the tail stock is narrow and high in the dorsoventral (y-) direction of propulsive movements. The propulsive muscles are not segmented, but instead the dorsal muscle system of land-living ancestors is strongly developed as m. extensor caudae, a slightly modified and strongly developed derivative of the autochthonous muscles of the back. Its ventral antagonist is the newly formed m. flexor caudae, which seems to be derived from the m. quadratus lumborum and the m. psoas. The muscles of the ventral abdominal wall (Slijper 1937), especially the m. rectus abdominis, change their direction and turn dorsally towards the vertebral column. This change of direction requires the transmission of a tensile force. The latter seems to be provided a fascia stretched out between the abdominal wall muscles and the flexor caudae. Figure 3.7a, e show that and why the proximal tail of a cetacean is exposed to particularly high compressive forces, which require more strength than is offered by the trunk and the more distal tail segments. Since the material (spongy bone, intervertebral discs) is the same along the entire axial skeleton, the cross-sectional area should be proportional to strength. Therefore we have measured the horizontal and vertical diameters of the vertebral bodies (Fig. 3.7b; Table 3.1) and calculated from these values the areas of ellipses as approximations to the cross-sectional areas (Fig. 3.8). Similar measurements have been published by Tinker (1988) and Buchholtz (1998, 2001a), but not interpreted in the same way. Easy to understand is the distal decreasing of the cross-sectional areas: because muscle or tendon fibres insert gradually into the vertebrate bodies, less and less force is transmitted along the axis of the animal in distal direction. Only a part of the fibres reaches the fluke where they fuse with the connective tissue constituting this soft-part element. In proximal direction, the cross-sectional areas also decrease, though not to the same extent. This part of the axial skeleton mainly transmits the longitudinal force produced by the tail fluke to the head, where the stow pressure acts, while the friction resistance of the streaming water grows gradually along the length of the body. In addition, the muscles have longer lever arms along the trunk than in the tail and therefore need to exert less force in longitudinal x-direction.

The only herbivores among permanently aquatic mammals are the *sea cows* (sirenians, Fig. 3.9). Among them, the Pacific dugong resembles externally the porpoises by its sickle-shaped fluke and its up and down oscillating movements. This is not the case in the Atlantic manatees (Hartmann, 1979), which possess a sort of fin seam around the posterior quarter of their bodies, rounded at its tip, and their movements are closer to undulation with waves travelling along the x-axis. Sirenians do not show the great height of the peduncle, which is so typical of many whales and dolphins. The muscles which take part in moving the fluke, however, are similar to those in cetaceans (Steiner and Rozak 2001) A rather inobvious divergence between the vertebral columns of sirenians on one and cetaceans on the other side, lies in the largest diameters being located in the lumbar, rather than the tail region (Fig. 3.10). While the conditions in the dugong are quite similar to those in the porpoise, shown in Fig. 3.7 (except the slightly more anterior maximum of diameters and of axial flexion), the use made of the more mobile forelimbs, especially in the manatee, requires more activity in the shoulder girdle muscles than in the forms mentioned before. All these muscles possess components in longitudinal direction, which exert compressive forces on the axial skeleton and require additional strength. This is visible in the relatively greater diameters of the thoracal vertebrae of sirenians if compared to cetaceans.

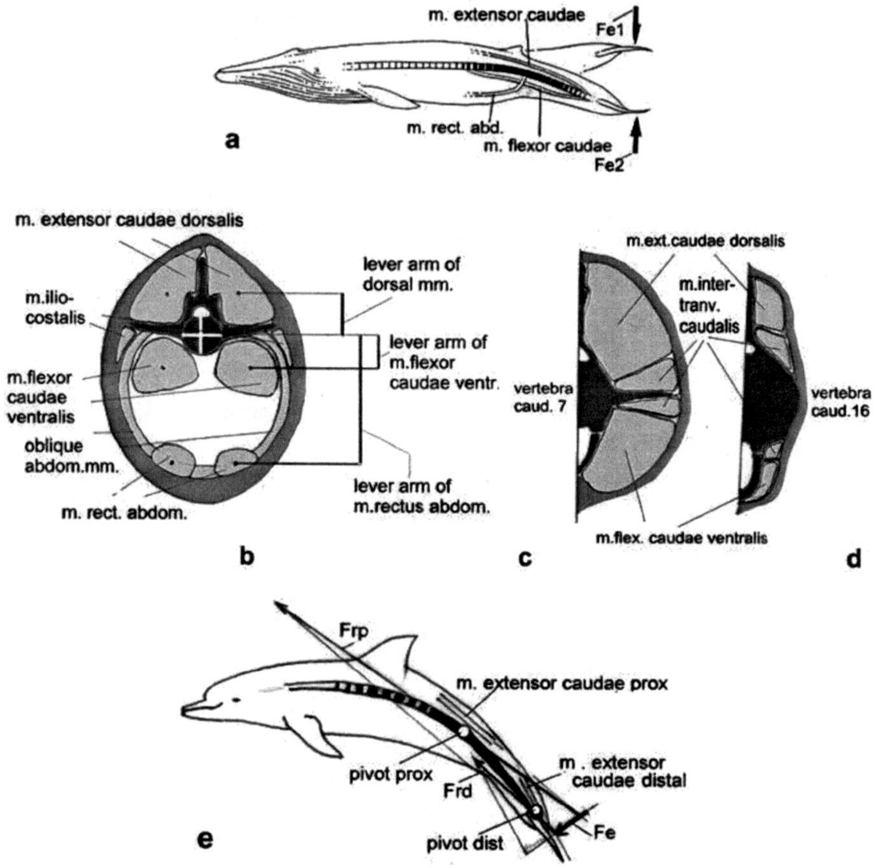

Fig. 3.7 Schematic lateral view of the muscles moving the tail in cetaceans. (**a**) In the absence of a pelvis, the ventral muscles of the abdominal wall curve dorsally, toward the vertebral column. The change of direction requires a force in the direction of the m. flexor caudae, probably through connective tissue fibres of a tough fascia. The ventral tail flexor and its dorsal antagonist have numerous connections to individual vertebrae, so that their force is distributed over several segments. (**b**) Schematic cross section through the posterior trunk of a cetacean. The strong ventral m. flexor caudae has no unambiguous homologue in other mammals, while its antagonist, the dorsal m. extensor caudae, is also present in land-living mammals. The white lines on the vertebral body indicate the diameters measured for estimating the cross-sectional area. (**c**) Right half of the cross section through the proximal tail, (**d**) right half through the distal tail. Drawn on the basis of data presented in Slijper (1937). (**e**) Activity of muscles and resulting compressive forces in the vertebral column during the upstroke in a dolphin. In the distal joint between the axial skeleton and the fluke (*pivot dist.*), the tendons of the m. extensor caudae balance the external water resistance F_e. The resultant of F_e and the force of the tendons, F_{rd}, is greater than the individual acting forces and passes ventral to the vertebral column (black). In the proximal joint (*pivot prox.*), at the basis of the tail peduncle, the load arm of F_e is longer and therefore counterbalanced by stronger muscles (which have not yet lost a part of their fibres by insertions). Consequently, the resultant F_{rp} is longer than F_{rd}. Resisting this greater force requires greater strength of the vertebral column. Along the trunk, little movement takes place, and the lever arms of the muscles are still longer than in the tail. This means lower compressive stress (indicated by white banding)

Table 3.1 Vertebral bodies of aquatic mammals in which diameters have been measured

Species	Number of measured trunk and tail vertebrae	Main propulsive organ(s)	Neck length, mobility
Beaver			
2 *Castor fiber (3)*	36, 39	Limbs, tail	Short, mobile
Otters			
Enhydra lutris (1)	27	Hindlimbs	Medium, mobile
Pteronura brasiliensis (1)	28 (40)	Limbs, trunk, tail	Medium, mobile
Seals			
Zalophus californicus (2)	28	Forelimbs	Long, mobile
Callorhinus ursinus (1)	27	Forelimbs	Long, mobile
Eumetopias jubatus (1)	27	Forelimbs	Long, mobile
Arctocephalus forsteri (4)	21	Forelimbs	Long, mobile
Otaria byronia (4)	20	Forelimbs	Long, mobile
Earless seals			
Hydrurga leptonyx (4)	21	Fore-/hindlimbs	Long, mobile
Phoca vitulina (2)	26	Hindlimbs	Long, mobile
Pagophilus groenlandica (2)	24 (28)	Hindlimbs	Medium, mobile
Erignathus barbatus (2)	28	Hindlimbs	Medium, mobile
Mirounga leonina (4)	21	Hindlimbs	Long, mobile
Odobenus rosmarus (1)	26	Fore-/ hindlimbs?	Medium, mobile
Sirenia			
Trichechus manatus (1)	43	Tail	Short
2 *Trichechus senegalensis* (4)	20(?), 36 (?)	Tail	Short
2 *Halichoerus dugong* (2) (4)	38	Tail (fluke)	Short
Hydrodamalis gigas (4)	55	Tail (fluke)	Short
Baleen whales			
2 *Balaenoptera acutorostrata* (1) (4)	46 (<50)	Tail (fluke)	Short
Balaenoptera borealis (1)	38	Tail (fluke)	Short
Megaptera noveangliae (4)	41	Tail (fluke)	Short
Toothed whales			
Physeter macrocephalus (4)	35 (+6)	Tail (fluke)	Short
Hyperoodon borealis (4)	32	Tail (fluke)	Short
Monodon monoceros (1)	45	Tail (fluke)	Short
Delphinapterus leucas (1)	49	Tail (fluke)	Short (mobile)
Lagenorhynchus obliquidens (1)	65–75	Tail (fluke)	Short
Phocoena phocoena 390 (4)	>40	Tail (fluke)	Short
Phocoena phocoena 232 (4)	54	Tail (fluke)	Short
Cephalorhynchus commersonii (4)	57	Tail (fluke)	Short

In several individuals, some vertebrae were preserved as fragments, or tiny in comparison to others, so that no measurements were taken. The total number of existing vertebrae is given in parentheses. Collections are indicated as follows: *(1)* Zoologisches Museum, Vienna; *(2)* Senckenberg-Museum, Frankfurt am Main; *(3)* Senckenberg-Museum Dresden; *(4)* Musée d'Histoire Naturelle, Paris

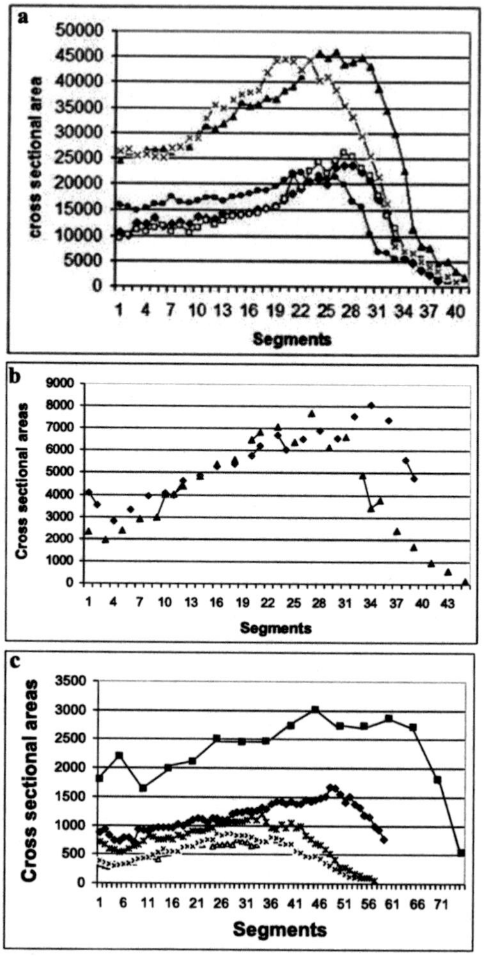

Fig. 3.8 Cross-sectional areas of vertebrae in cetaceans. (**a**) The "big" whales (crosses: *Physeter*; triangles: *Megaptera*; rosettes: *Hyperoodon*; diamonds: *Balaenoptera acutorostrata*; white squares: *B. borealis*) show similar variation of frontal areas along the body axis, calculated in cm^2. The maximum in toothed whales seems to be slightly more cranial, but always on the tail (=peduncle). (**b**) White whale (*Delphinapterus*, diamonds) and narwhale (*Monodon*, triangles) show rather large surfaces in the anterior part of the thorax. (**c**) Small, toothed cetaceans possess very large numbers of skeletal segments (which means that they have extremely flexible vertebral columns). In delphinids (*Lagenorhynchus* spec., black squares, specimen from Vienna, and black diamonds, specimen from Canada) the maximum is more caudal than in the smaller phocoenids (*Phocoena spec*., black triangles, 232; black rosettes, 390) and *Cephalorhynchus*, crosses)

Buchholtz (2001b) has observed in some Jurassic ichthyosaurs (*Stenopterygius*, *Ophthalmosaurus*) a "thunniform" morphology, especially a semilunar fluke, laterally mobile tail stock and a rigid trunk. The hindlimbs were small and "suggest" being "in a process of reduction and loss". The author measured diameters in *y*-

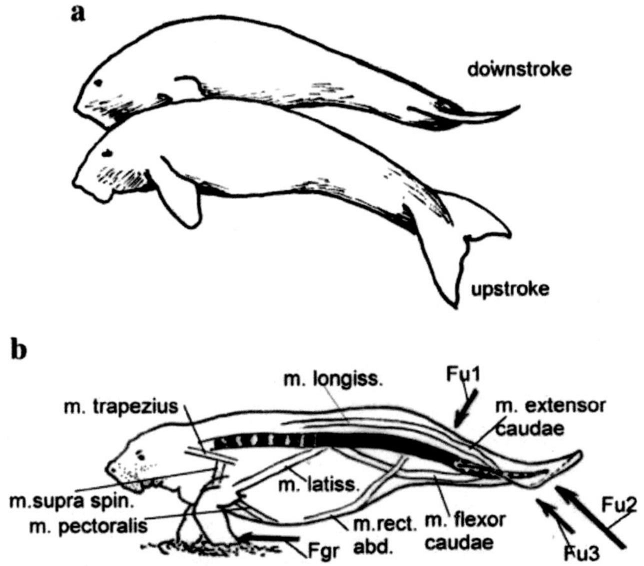

Fig. 3.9 (**a**) Pacific sea cows, genus *Dugong*, in two phases of tail movement. (**b**) Atlantic manatee, genus *Trichechus*, with schematic presentation of the major muscles and axial stresses on skeleton as in Fig. 3.7a, e. The anterior limbs are used more for bracing and propelling the body against the ground than in cetaceans and even in dugongs. Therefore, the muscles of their shoulder girdle have to exert more force, which contributes to compressing the thoracal vertebrae

Fig. 3.10 Cross sectional areas of vertebrae in Sirenia in cm^2. (**a**) Compressive forces act against maximal areas in the thoracal (*Trichechus spec.*, white squares, diamonds, triangles) or in the lumbar region (*Dugong*, black squares and diamonds), depending on the propelling organ (fin seam or fluke). (**b**) The extinct Steller's sea cow (*Hydrodamalis*) possessed a true, forked fluke, on which water resistance concentrates. In agreement with this trait the lumbar vertebrae are particularly strong, like in *Dugong*

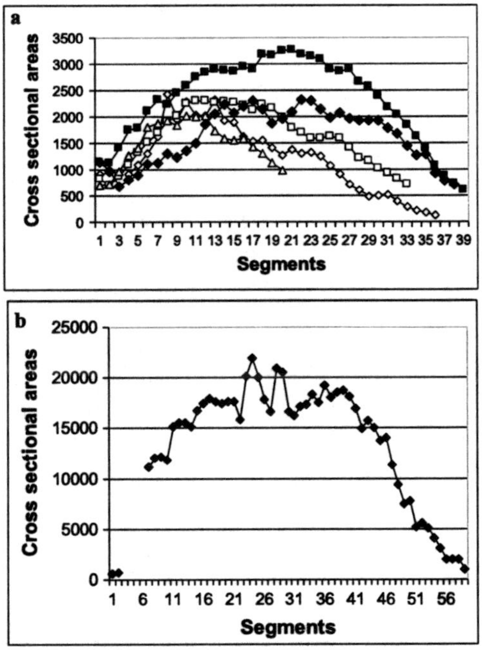

direction and lengths of centra in x-direction and found the highest strength and shortest centra in the tail stock. These traits are similar to those found in rapidly swimming delphinids, but not to the more perciform-like morphology in other ichthyosaurs. The author, like in her studies on cetaceans (2001a), connected morphology directly with "adaptations for extreme speed" on the basis of comparisons, without trying to unveil the (proximate) mechanical reasons.

3.3.6 Diverging Locomotion and Body Shapes in Pinnipeds

A very special locomotor mode can be seen in earless seals (*Phocidae*, including *Mirounga* and *Hydrurga*, Fig. 3.11; see also Gutmann 1994). These animals do not use the (reduced and short) tail, but instead alternatingly one of the hindlimbs for

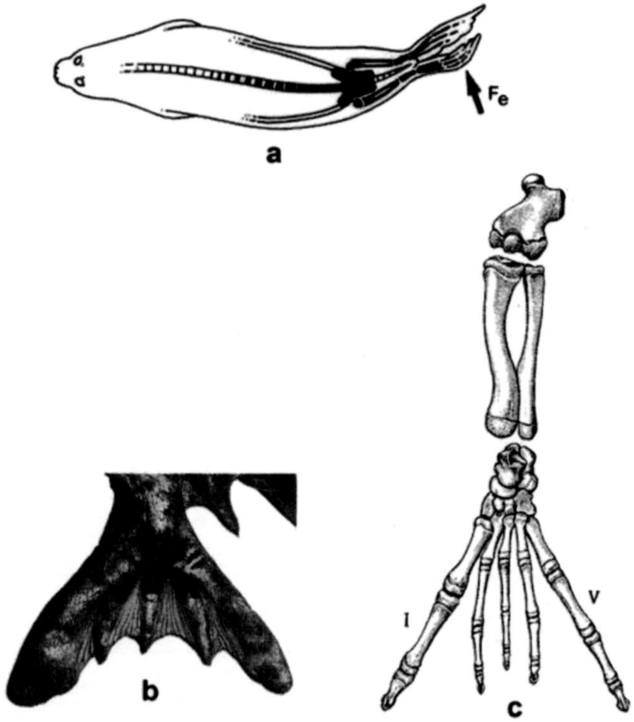

Fig. 3.11 (**a**) Schematic illustration of propulsive oscillation of the hindlimbs in an earless seal. Water resistance is evoked against the in x-direction (rearward) extended hindlimb on the (right) trailing side of movement. The hindfoot on the (left) leading side of the movement is folded. The force required to overcome water resistance is exerted by the strong adductor muscles of the hip. (**b**) Inflated left hindlimb of a newborn *Mirounga*, dorsal side. (**c**) Skeleton of the hindlimb of *Mirounga* (both after Starck 1979)

propulsion, so "imitating" the thunniform mode. It is the trailing pes, which is brought into action; the other one, on the leading side, is folded by adducting the slightly flexed toes in stereotypic form (not published observations of Kuschel). By this distribution of load, the (like in all mammals) strong adductors of the hip joint are recruited for propulsion, and the variable, less strong abductors are relieved. To perform their function, the hindlimbs are extended in caudal direction and have assumed symmetrical, fluke-like outlines (Fig. 3.11b, c). Their skeleton obviously shows the characteristics of a five-rayed terrestrial autopodium, but both rays at the margins (*I* as well as *V*) are reinforced, and much longer than those in the middle, so that the whole element exhibits similarities to the flukes of scombriform fishes, with its short depth and streamlined form of the cross sections (see Sub-Chapter 3.2.2). Cetaceans and sea cows arrive at essentially similar fluke shapes alone on the basis of connective tissue. It is not known whether the folded pes on the leading side has any effect on the streaming water (like the longitudinal keel in dolphins, whales, sharks) or two keels in marlins, swordfishes) on the distal peduncle. Anyway, the permanent posture of hindlimbs makes them useless for terrestrial locomotion.

For the oscillating movement, the *lateral body wall musculature* is recruited, which moves the pelvis plus hindlimbs in either direction. Like the bending moments which must be overcome, the force arms increase cranially, so that less force is required. The greatest forces are exerted at the lumbar vertebrae, where most movement takes place. Consequently, the lumbar vertebrae have the largest cross-sectional area (Fig. 3.12a) The massive bone structure of the pelvis has considerable bending strength by its own (especially if its breadth is greater than its dorsoventral depth!), and the hip joint is counterbalanced by the powerful adductors against water resistance, while the weaker abductors are not recruited (see previous paragraph). As will be shown below, the stressing of the vertebral column is quite similar when the earless seal is moving on land.

The anterior limbs of Phocidae with their strong claws resemble the paws of other carnivores and are used – on firm substrates, like ground or ice – in a similar fashion. When spreading the digits of their *forelimbs*, large webs are inflated, which are used effectively in changing the direction of swimming or to prevent a "rolling" rotation of the body about its *x*-axis. In travelling forward, the forelimbs are usually adducted tightly to the trunk (resembling the anterior fins of, e.g. tuna, which also spread only if needed).

In sharp contrast to Phocidae, their close relatives, the *Otariidae*, as well as the walrus, *Odobenus*), use for propulsion in water primarily their anterior limbs. Their up and down movements (Fig. 3.13a, b) have led to calling this non-axial propulsion "flying under water"[4] (Reif 1981b; Kuhn and Frey 2012).

---

[4]This attribution is popular and common, but does not really fit. The essence of flying is primarily maintaining height above ground – while aquatic animals must not spend much energy for maintaining their depth below the surface, because they have more or less the same specific weight as the surrounding medium.

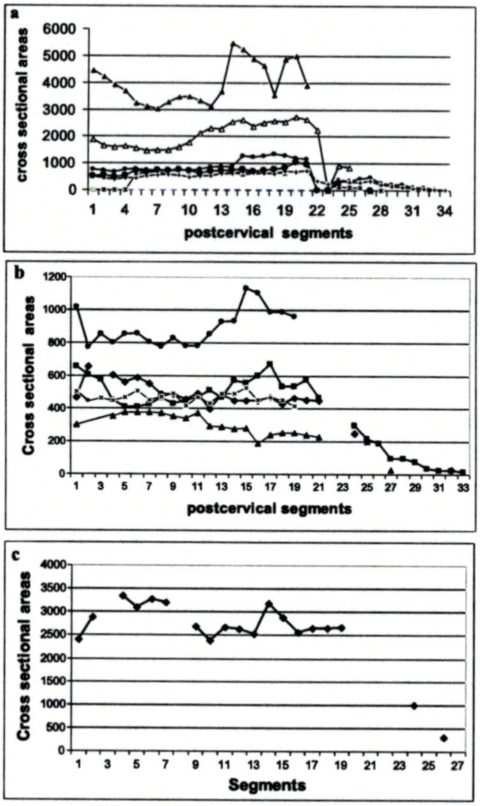

Fig. 3.12 (**a**) Frontal areas of vertebral bodies in mm^2 in earless seals, Phocidae, represented by sea elephant (*Mirounga,* black triangles, where several vertebrae are damaged, and measurements therefore underestimate the real areas); sea leopard *(Hydrurga,* white triangles)*;* bearded seal *(Erignathus barbatus, -•-•-);* and harbour seal *(Phoca spec., -◇-◇-, --)*. (**b**) Frontal areas of vertebral bodies in mm^2 in eared seals, Otariidae, represented by *Eumetopias, -◆-◆-*; *Zalophus*, -□-□-; *Callorhinus -▲-▲-*; *Arctocephalus* -×-×-; and *Otaria -◇-◇-).* In Phocidae (**a**), the cross sectional areas (and therefore the strength against compressive forces) increase caudally toward the pelvis (**a**), while in otariids (**b**), and in the walrus (*Odobenus rosmarus* (**c**)), the values are nearly on the same level along the whole vertebral column. This corresponds to their strongly developed musculature of the proximal neck, shoulder girdle and anterior trunk (Fig. 3.13e), needed for propulsion and the extreme flexibility of thier necks and trunks

The forelimbs, or anterior flippers, in seals have short proximal segments, while the long digits of the manus are elongated beyond the finger tips by fibrous "lepidotrichia", so that the rudimentary claws are shifted away from the outer margin to the dorsal surface. The outline of the flippers is long and slender, similar to a bird's wing or the fluke of a whale or fish with pronounced spread of >3/1 (Fig. 3.13c, d). They possess a streamlined profile without vaulting. Their wing-shaped outline implies a limited ability to abduct the rays from each other. The hindlimbs also

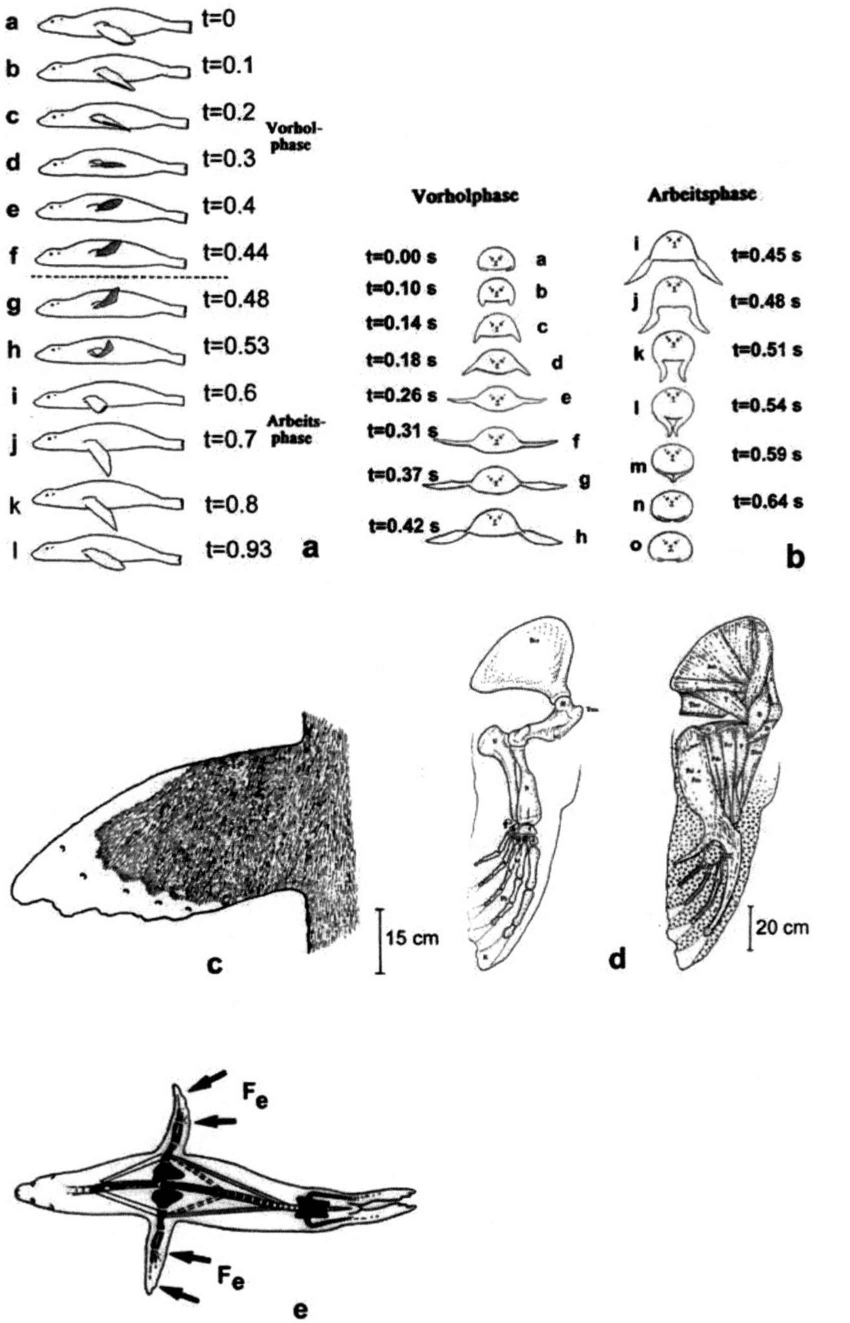

Fig. 3.13 Eared seal (*Zalophus*) swimming. Drawn after video recordings of Kuschel (1994). Left column (**a**) in side view (volar side of flipper grey), middle and right column (**b**) seen from in front. (**c**) Dorsal aspect of the forelimb of Steller's seal, ***Eumetopias jubatus.*** Note the hairless tip of the flipper, its cranial and caudal margins and the rudimentary nails. (**d**) Anatomy of the same flipper,

Fig. 3.13 (continued) on left the skeleton, on right the deep layer of muscles. Note the cartilaginous and fibrous elongations of the bony digital rays, the "lepidotrichia" (white). **(e)** Schematic illustration of propulsive movement of the forelimbs in an eared seal. Flippers (skeleton black) are moved against water resistance (black arrows) by the strong muscles of the shoulder girdle (double lines). These muscles exert high compressive forces on the axial skeleton, especially its thoracal and lower cervical sections (black)

have short proximal elements and a long pes; especially the webbed digital rays are long.

For propelling the body, muscles of the shoulder joint and especially those of the shoulder girdle are activated. The latter muscles span the cranial two thirds of the body stem, including the neck (Fig. 3.13e), where they compress the axial skeleton and evoke high compressive loads. In accordance with this requirement, the neck and thoracal segments of the vertebral column show large intervertebral contact areas (Fig. 3.12b, c). The statement of Ridgway (1972) and Ridgway and Harrison (1981) that the vertebral column reflects the pinniped mode of locomotion is confirmed by our results.

In addition, the body axes in all seals are remarkably flexible in lateral, even more so in dorsoventral direction. This flexibility requires a differentiated and strong lower neck and upper thoracal musculature, which also exerts considerable compressive forces along the vertebral column, so increasing the necessity of strong, that is broad and high vertebral bodies (Fig. 3.12). This flexibility can be – and in fact is used – only under the condition that a reliable bearing exists, against which flexion is performed. While this is a minor problem on land, where gravity presses parts of the trunk firmly on the ground, the stabilisation of the posterior against the anterior part of the body stem requires the existence of an anterior as well as a posterior pair of flippers (see also below, sea turtles). It should be noted that the frontlimbs in both families of pinnipeds in y-direction are attached to the middle of the trunk and stretch mostly sideward, rather than downward. This position excludes rotating moments in lateral view (see below, Chapter 5.4).

3.3.7 How Do Amphibious Mammals (Pinnipeds, Otters, Beavers) Move on Land?

Pinnipeds, otters and beavers spend parts of their lives on land, where their bodies are exposed to much greater loads than in water, because of the permanent action of gravity. Their body must sustain the occurring (greater) forces as well as those evoked during swimming. In order to understand their body shapes completely, the points to follow must be taken into consideration here and not in Chap. 4 or 7.

Among the *pinnipeds*, terrestrial *locomotion on land* differs not less than aquatic locomotion: while otariids use the same gaits as land mammals and keep their trunk like a beam supported on two pairs of supports (Fig. 3.14b), the earless seals rest

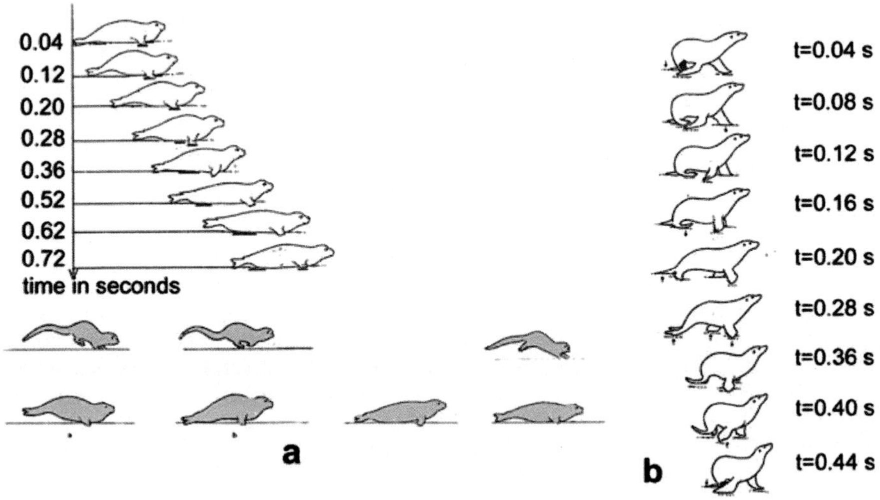

Fig. 3.14 Terrestrial locomotion in pinnipeds. (**a**) Harbour seal, from top downward. A body wave travels from the pelvic region toward the head – in contrast to the body wave in undulatory swimming, but in agreement with the "dorsomobile" locomotion of small mammals. This parallel is indicated by the two rows of animals in gray (see Chaps. 4 and 7). (**b**) Californian sea lion, walking with lateral footfall sequence. (From Kuschel (1994), unpublished

their whole body full length on the ground, without support on the hindlimb and little support on the frontlimbs. Just the cranial half (the long neck and the cranial part of the trunk) is sometimes lifted and placed in a more or less perpendicular direction (as is well known in sea elephants!). Their locomotion on land has attracted much less attention than that of otariids. Kuschel (1994) has confirmed that during phases of rest, the whole trunk of phocids rests on the substrate. To move forward, the earless seals contract alternatingly dorsal and ventral trunk muscles, which extend the trunk, throwing the anterior torso forward, or flex the trunk, so pulling the posterior part of the torso behind (Fig. 3.14a). This activity pattern in fact has very much in common with the bounding of "small mammals" (see Chap. 7), although it looks like a wave travelling from the tail forward to the head (that is opposite to the body wave in anguilliform swimming!). The first half is performed against the weight of the cranial part of the body stem; the second half needs less energy, because the weight of the caudal section assists flexion.

Really amphibious are two mammals belonging to different orders and having divergent lifestyles, namely, *otters* and *beavers.* Both spend a considerable time of their active life in (shallow) water, but as much on firm ground, exposed to gravity. While their external shapes are not too different (long, very flexible and rather slender body stem, covered by dense fur; short anterior and longer, powerful posterior extremities, the latter with expanded autopodia; long and a fairly massive tail), the modes of propulsion in water are in both forms quite variable (Fig. 3.15a, b). Forelimbs seem to be used primarily for changing directions. If possible, both, the

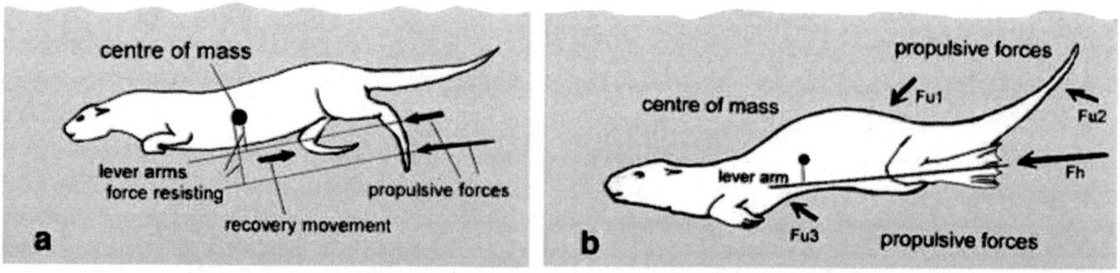

Fig. 3.15 European otter as example for semiaquatic mammals with an amphibious lifestyle. Propulsion in water can be performed in different ways: either by the limbs (in (**a**) shown in alternating sequence, water resistance acting preferably on hindlimbs) or by undulation in the x/z-plane of the body stem (**b**), which can culminate in either the hindlimbs (otter, sea otter) or the tail (beaver). This mode can hardly be discriminated from synchronous retroversion of both hindlimbs. Sea otters seem to spread out their hindlimbs in *z*-direction (so representing a sort of tail flipper) and propel themselves by undulation in the x/y-plane, in a way reminiscent of whales. Forelimbs are primarily used to change direction, hindlimbs primarily for propulsion. In swimming shrews (Wikipedia), the limbs are also in a sprawled posture, so that up and down movements of the head during propulsion are avoided

carnivore and the rodent, take advantage of submerged firm structures (soft ground, rock, subaquatic vegetation: roots, stalks, parts of fallen trees). In free water, the limbs can, as on firm ground, either be retracted alternatingly, or the hindlimbs are swung caudally in unison, synchronous with the dorsal extension of the trunk, like in the bound of terrestrial mammals. The limbs have short stylopodia and the joints can be strongly flexed. Shortness of stylo- and zeugopodia as well as flexed posture of the joints keeps load arms of water resistance short and the force required for movements limited (in spite of the disadvantage that short limbs do not allow very rapid peripheral speeds). Shortness of the limb segments also keeps down the rotating moments in the plane of movement. This is in most cases parasagittal (Fig. 3.15a; see also Chap. 7), but sea otters spread their characteristically large hind flippers horizontally in the x/y-plane of the body like cetaceans, so that the rotating moments exerted by each flipper are counterbalanced by the contralateral, if both sides act synchronously. Often the body stem performs complete undulations. In this mode, forelimbs are more or less adducted, the hindlimbs seem to act together with the tail – which in beavers has the well-known flat form, reminding of a manatee, and is hairless; in the giant otter of South America it is depressed. Both shapes enlarge the fastest-moving most posterior parts of the body, so increasing water resistance. Since ram resistance acts primarily against the head, and friction resistance acts along the whole length of the body, increasing with the caudally travelling muscle contractions, the compressive forces exerted by the muscles can be expected to assume the same height along the whole axial skeleton, with a tendency to become greater in the caudal segments. This obviously is the case, as can be derived from Fig. 3.16.

The greatest forces in these semiaquatic mammals act on the vertebral columns during locomotion on land, under the influence of gravity. They are very similar to those in "small mammals" (see Sub-Chapter 7.3.2), and their body shape must be able to sustain these internal forces.

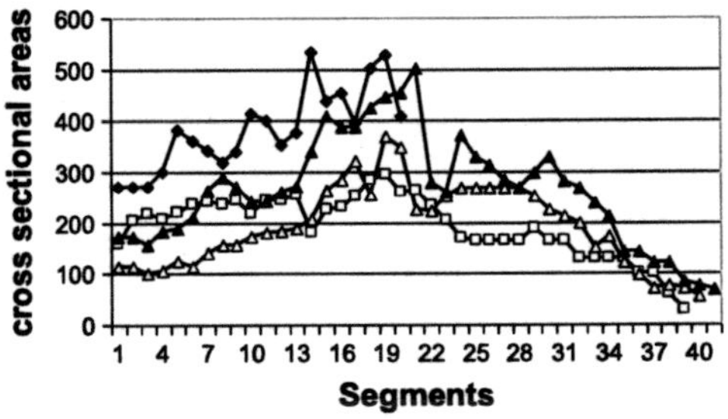

Fig. 3.16 In the semiaquatic otters (*Enhydra*, black diamonds, and *Pteronura*, black triangles) and beavers (white triangles and squares), the most caudal vertebrae of the lumbar section are reinforced, and the tail is also able to sustain high compressive forces. The remarkable strength of the lumbar vertebral column corresponds to its being loaded in terrestrial locomotion (see Sub-Chapters 7.3.2 and 7.3.3). In the proximal tail, behind the sacrum, the diameters of the vertebrae become smaller, especially in otters. In beavers, which possess a broad and flat, finlike tail, the reduction of diameters behind the sacrum is less pronounced

3.4 Do Alternative Technical Solutions Exist Among Aquatic Non-mammalian Vertebrates? Divergent Body Shapes and Locomotor Modes

Animals belonging to widely diverging classes, skates and rays (*batoidea*), a number of teleosts (most of them inhabitants of coral reefs and rocky slopes; the very large *Mola*, however, being epipelagic) and marine turtles present a completely different body shape. In sharp contrast to the above-discussed elongate bodies, they exhibit a *shortened x-axis,* while the greatest diameters of the more or less *disc-shaped bodies* may well be as great or even greater than the length of the body axis exclusive of the reduced tail (Fig. 3.17). These shapes are phylogenetically old and appear in several lineages: the batoidea are first documented in the Jurassic and have developed a variety of forms in the Cretaceous. They are clearly younger than the related selachians (which occur in the fossil record since the Carboniferous). Their depressed body shape corresponds to the arrangement of the eyes, the gill openings and the dentition, so that their morphology is commonly interpreted on the basis of its undoubted biological role as an "adaptation" (better: optimisation) to living on or near the ground. The essential characters have been – following the current opinion – just maintained even after transition to a pelagic (and no longer ground-dwelling) lifestyle in the eagle rays and mantas. Among bony actinopterygian fishes, disc-shaped forms occur very early, since the Carboniferous, and reappear independently in many evolutionary lines. The marine turtles are much younger; their oldest representatives appear in the Cretaceous.

In all these forms, propulsion is "paraxial", not necessarily accomplished by the body axis. Interestingly, angel fishes (squatina-like sharks) and guitar fishes (long-tailed skates), as well as the electric ray *Torpedo*, which all possess the morphological requirements for *both modes of propulsion* (axial as well as paraxial) prefer in

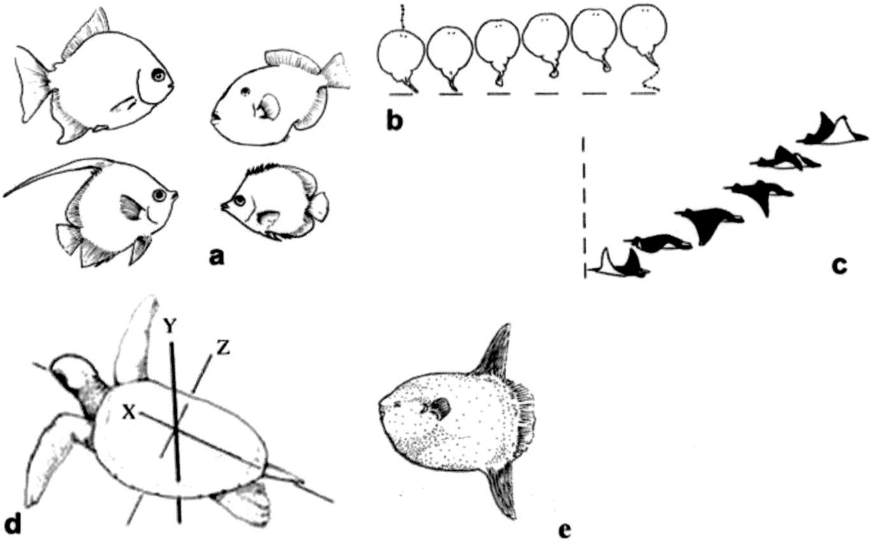

Fig. 3.17 Alternatives to elongate body shapes among aquatic vertebrates. All these forms have in common the concentration of body mass in two dimensions, one of them being the longitudinal *x*-direction and the stabilising expansion in *z*- (rays and turtles) or *y*-direction (bony fishes). The modes of propulsion vary considerably: (**a**) Highly manoeuvrable fishes dwelling in coral reefs. Redrawn from Grzimek (1970), top left, *Metynnis*; top right, *Ballistis*; bottom left, *Pomacanthus*; bottom right, *Chaetodon*. Locomotion depends upon independent undulatory movements of the paired anterior limbs and the unpaired fins of the caudal body part. (**b**) Electric ray, *Torpedo,* in six phases of locomotion seen from above. Propulsion is accomplished by undulation of the tail in *z*-direction and/or undulation in *y*-direction of the huge pectoral fin, which extends from the head along the trunk. (**c**) Six phases of locomotion in lateral view of a rapid swimmer of the open sea (eagle ray, *Myliobatis*). Dorsal side, black; ventral, white. Locomotion is due alone to up- and downstrokes of the pectoral extremity in *y*-direction. Note the triangular outline of the pectoral fin, which resembles the anterior flippers in seals and in toothed whales. (**b**) after Reif (1995). (**d**) Green sea turtle (*Chelonia mydas*) as example for all marine turtles with body axes. Turning is easiest about the y-axis. Propulsion is made alone by the forelimbs; hindlimbs are mostly kept without movements like steering rudders, but can be recruited to produce water resistance in any direction to assist rapid changes of direction. (**e**) Sunfish, *Mola.* (Modified from Norman and Fraser 1963)

travelling the undulating movements of trunk and tail (Norman and Fraser 1963, own observations). A tempting and simple assumption is that this preference is due to its energetic efficiency. Anyway, the method of undulating propulsion can be realised alone in water, that is a fluid environment which has about the same specific weight as the animal's body and a considerable viscosity. Energy requirement is not more than a small fraction of locomotion on land, where counteracting gravity requires great additional efforts. In most rays, the slender and whiplike tail is not designed for propulsion. For propelling the animal, the lateral fin seams undulate (Fig. 3.17) or are swung up and down. Minor modulations of the external margins allow precise control of the orientation in space. In those batoids, which have acquired a pelagic lifestyle (eagle rays and mantas), the lateral parts of the huge pectoral fins are laterally extended, resembling the wings of birds.

In many *teleosts*, flexion of the body axis in the *x*/*z*-plane is limited or confined to the tail fin. In these forms the anterior limbs as well as the unpaired fins take responsibility of locomotion, usually in cooperation with the tail fin, and control swimming direction and swimming speed precisely. In the sunfish, *Mola*, not alone the body axis is shortened, but the tail is replaced by fin seams along the posterior margin, and unpaired dorsal and ventral fins propel the body, mostly by alternating excursions.

Marine turtles also possess a short and rather broad body, which is totally stiff. They propel themselves primarily by the anterior limbs, while the posterior extremities mainly act in stabilising the position of the animal in space or assist in steering. Because mass moments of inertia are so low, an organ (flipper) to control the direction of the *x*-axis is particularly necessary.[5]

All these animals have in common a technical alternative to axial undulation. Our mechanical approach starts from the physical rule that *rotations of a body* suspended in a medium of about the same density normally take place about its centre of mass (CoM). If part of body mass is remote from this centre, the body resists being rotated more strongly, than if the whole mass is concentrated close to the CoM, because its mass moment of inertia M_i is given by

$$M_\mathrm{i} = \Delta_\mathrm{m} * y^2, \tag{3.8}$$

where y is the distance between the mass units (Mi) and the CoM.

For example, the long bodies of a shark or a crocodile possess very great mass moments of inertia against rotation about a vertical y-axis and a horizontal z-axis, but not around the longitudinal (x-)axis.[6] By contrast, a piranha, or a scalar, or a skate possesses but small mass moments of resistance against being rotated about the vertical y-axis or the horizontal z-axis, respectively. Both forms are quite stable against rotations about the horizontal x-axis. In addition, the resistance of water also impedes more than any other movement a "rolling" rotation about the longitudinal x-axis. This means that these fishes are able to manoeuver easily, while maintaining their orientation in space (greatest diameters vertical in the case of teleosts, horizontal in the cases of skates and turtles) does not require great efforts.

In turtles, where the trunk is completely rigid and the conical tail reduced, the short, disc-like body shape is also characteristic. The ventral surface is flat and the dorsal surface vaulted. Because water must stream faster along the vaulted side, and high speed of streaming leads to reduced pressure, swimming forward creates some lift (air stream and its consequences will be treated again in Chapter 6.6). The stabilising effect of the low and broad trunk is increased by sharp keels along the

[5]This makes the independent mobility of the head against the trunk useful, as shown in Chap. 2 "Head".

[6]This allows the death role of crocs, as described in Chapter 2.6.

lateral margins. All these traits are not confined to marine turtles; some semiaquatic and sweetwater forms (Trionychoidea and Emydidae) show the same characteristics. Just because of the low mass moments of inertia, the powerful frontlimbs alone cannot sufficiently control the orientation of the animal in three-dimensional space. Therefore, posterior flippers are required. In travelling, they are kept at an angle of 60°–40° against the ventral plastron and are actively moved alone in the (frequent) changes of direction. Aside from their versatility, sea turtles possess a rather long and mobile neck, which permits to control a fairly large volume without changing their body position.

In fossil, probably *marine crocodiles* and the *mesosauria* (some of which had true tail fins), the anterior limbs are reduced, while the posterior extremities are strongly developed and seem to have possessed large webs between the digits. This diverges from the role ascribed to anterior limbs as the organs controlling the head in fishes and marine mammals (Sub-Chapter 3.3.2) but falls in line with the preference of the tail or the hindlimbs as propulsive organs, which secure compression along the whole body as emphasised in Sub-Chapter 3.3.4.

A still unsolved riddle is presented by the *Plesiosaurs*, which are known only as fossils (see also Troelsen et al. 2019). The breadth of their trunks seems to exclude undulating locomotion, although the long neck and the tail have possessed the necessary mobility in bilateral (z-) direction. Flexion of the neck in dorsal direction was probably inhibited by the long neural spines. Their tails resemble in the height of neural and hemal processes (=y-direction) those of crocodiles. This suggests excursions in lateral (z-) direction, but not in dorsal or ventral (y-) direction. If so, it could be used during rapid lateral movements of the neck, for stabilising the trunk in space. This conclusion seems to be confirmed because extremely long necks are usually combined with long tails. Difficult to understand, however, is the existence of four obviously highly efficient paddles. The shape of the cranial limb is characterised by a spread of three (or more) in relation to one unit of depth and pointed tips as can be observed in rapid epipelagic swimmers and in birds. The ventral part of the shoulder girdle forms a massive plate, which offers origins for a powerful pectoral musculature. The dorsal part of the cranial girdle, however, seems weakly developed; the scapula is a short, upright plate, firmly connected to the joint and the ventral plate. All this fits to the imagination of the downstroke being the propulsive phase, like in marine turtles. Different, however, are the hindlimbs of plesiosaurs, which show the same slender and elongate form as their frontlimbs. They are not much shorter than the frontlimbs in most forms; in some species, however, they exceed the anterior extremities; the hindlimbs should also have been efficient propulsive organs. The connection between the dorsal pelvis and the vertebral column (ilium) is markedly weak – resembling the shoulder girdle. Several ideas have been developed concerning the cooperation of the limbs: Both sides in synchrony, anterior and posterior limbs at the same time, or alternating, front and hind alternatingly or in more complicated cycles as in marching on ground (Chaps. 5 and 7). Nevertheless, all these models leave open an unexplained remainder.

3.5 Essential Results

In an aquatic environment, the factor "weight" does not play an important role, but locomotion requires continuous input of propulsive forces. Resistance against propulsion is minimised by narrow diameters. Energetically inexpensive propulsion is possible by undulation of a lengthy body and can be improved by a compressed profile of the body. These physical conditions set very rigid limits to body shape, which must be elongate, and have rather small cross-sectional areas. Locomotor movements are produced by more or less longitudinal muscles, and these muscles shorten the body – unless shortening is excluded by a longitudinal, pliant, but not compressible notochord or vertebral column. Optimisation of aquatic propulsion leads to concentrating the propulsive forces at the posterior part of the body, so favouring the tail as the major propulsive organ, either undulating (crocodiles, hemichordates, tadpoles) or oscillating (epipelagic fishes, whales, dolphins). Details of general body shape depend upon the conditions of streaming fluids (position of gills), or on saving energy (maintenance of anterior extremities). Internal organisation of the body (skeleton, musculature), as well as its cover (dermal denticles in Chondrichthyes, scales and secretion of slime in most fishes, sponge-like skin in porpoises; see Reif 1981a), varies considerably. While in fishes (and on the tail also in crocodiles), the musculature is segmented, secondary aquatic mammals maintain large parts of their ground-adapted skeleton (structure of flippers and shoulder girdle, ribs, sometimes pelvis) and their musculature.

Alternatives Are Numerous. Many teleosts produce propulsion by means of undulatory movements of paired or unpaired fins, which are entirely independent from the body axis. A frequent alternative is propulsion not by undulatory movements of the body axis but instead by (symmetric) up and down movements of the pectoral limbs. This is realised in unrelated groups: skates and rays, marine turtles and eared seals. While the two former groups facilitate changes of direction at least in the horizontal x/y-plane by their short body axis, the latter change direction by their extremely flexible trunk and neck. Less frequent are imitations of the fluke in form of laterally (earless seals) or up- and downward moved and accordingly shaped hindlimbs (sea otters). Semiaquatic or amphibious mammals are exposed to greater external forces because of gravity and need body shapes suited to withstand stresses caused by moving on land. Analysis shows that their body shape suits the particular alternative of propulsion they have "chosen" and is also in agreement with the common stressing in aquatic conditions.

Chapter 4
Axial Skeleton and Muscle Arrangement in Terrestrial Tetrapods

Keywords Beam on two supports · Body stem · Bending moments · Shearing forces · Torsional moments · Arrangement of axial skeleton · Girdle skeletons · Trunk muscles · *Tiktaalik* · Posterior cantilever · Anterior cantilever

The external physical conditions for land-living animals diverge fundamentally from those with which water-living forms have to cope: gravity acts permanently and must be sustained in any time interval. On the other hand, there is no need to overcome the incessant resistance of water as soon as locomotion is performed. From the viewpoint of morphology, the first question to be answered is: What is the role of the axial skeleton under the permanent, everlasting influence of gravity?

4.1 Which Animals Are Called "Tetrapods" in the Present Context?

Most of the so-called "lower" tetrapods possess (in contrast to "fishlike" forms) a mobile neck, either short and bulky, or slender and elongate, and a slim trunk, the musculoskeletal parts of which surround a spacious cavity for the intestines, including the lungs, circulatory system and digestive tract, and a long tail, of which often the most distal part is dragged behind on the ground. Bony ribs are well developed and affixed at two bearings (capitulum costae and tuberositas costae) to the vertebral column. In ventrolateral direction, bony ribs transform at the rib's "knee" into the cartilaginous ribs which converge to the sternum. The "angulus costae" and the ventral aspect of the abdomen are usually left without bony cover. The body stem is supported by two pairs of limbs – which make the animals able to move on firm substrates. The rather uniform extremities will be treated in Chap. 5. This body shape occurs in various evolutionary lines: recently in caudate amphibians (urodeles), rhynchocephalians (tuatara), crocodilians and most lizards; fossil in Ichthyostegalia, Anthracosauria and Temnospondyli among so-called fossil "amphibians"

H. Preuschoft, *Understanding Body Shapes of Animals*,
https://doi.org/10.1007/978-3-030-27668-3_4

(Parrington 1967); and in subclasses of "Reptilia", namely, Pelycosauria, Therapsida, anapsids, in several Archosauria (and many extinct Lepidosauria).

The monitors (varanids) and other big lizards (*Iguana*, *tegu* and relatives) can be counted among the true *land dwellers*, though they are good swimmers (like the marine iguana *Amblyrhynchus* from Galapagos). They inhabit a variety of biotopes: hard stony ground, rocks, bushes, thickets as well as the above-described semi-liquid substrates. Some species live on trees (Sweet 2007). Their limbs, especially their autopodia, are mobile, and the existence of claws (see below, Chapter 5.4 and Chapter 7.2.1) permits hooking on even small rough substrates, like rocks or bark. All "lower" land-living animals have in common the low position of the body's CoM and a large area of support. Both characteristics entail a remarkable stability in standing and locomotion. In climbing, the heavy tail is sometimes curled around a stem and so approaches the CoM still closer to the substrate.

Many of the recent salamanders and newts, as well as most crocodiles, have an amphibious lifestyle in environments between water and land. These can vary considerably from free, sufficiently deep (more than three times body depth) water through shallow water (less than body depth) and *semi-liquid substrates*, characterised by algae, soft, submersed vegetation, or rather liquid mud forming a gradual transition to firm land or wet meadows.

The *oldest known forerunners* of land dwellers doubtlessly are water-living, fishlike animals, with lobate, sarcopterygian extremities. Beginning with the late Middle Devonian (385–380 million years ago), there appear fossils with a transitory morphology (of which we could analyse as an example *Tiktaalik* (Chapter 4.4)). From the *Middle Carboniferous*, however, we have as the oldest remains nothing than foot tracks (Voigt, 2012; Fig. 4.1) of *quadrupedally walking animals*. Fairly complete skeletons of true land-living forms are known from the *Late Carboniferous*, about 314 million years ago (Nyakatura et al. 2015, see also Chapter 5.1).

The *reasons for the transition* from living in water to living on land have been discussed at length. Under consideration have been overcrowding in drying ponds; the attraction by new, additional resources on land, where no competitors existed; and avoidance of predation pressure. All attempts are on the level of "biological roles" and therefore will not be detailed in this book. Our aim is finding out which among the structures documented by fossils are suited for locomotion in water as well as for maintenance of body shape and locomotion on land.

4.2 Why Is Trunk Morphology So "Conservative"?

The *arrangement of muscles and skeletal elements* belonging to the trunk is rather constant throughout all land-living tetrapods and differs widely from that in fishes. This constancy, or "conservatism" of axial skeleton and musculature of the body stem has often been observed, but not explicitly explained. As parts of the trunk can be seen the limb girdles, and because their bauplan varies among the classes, we will come back to the trunk in Chap. 5.

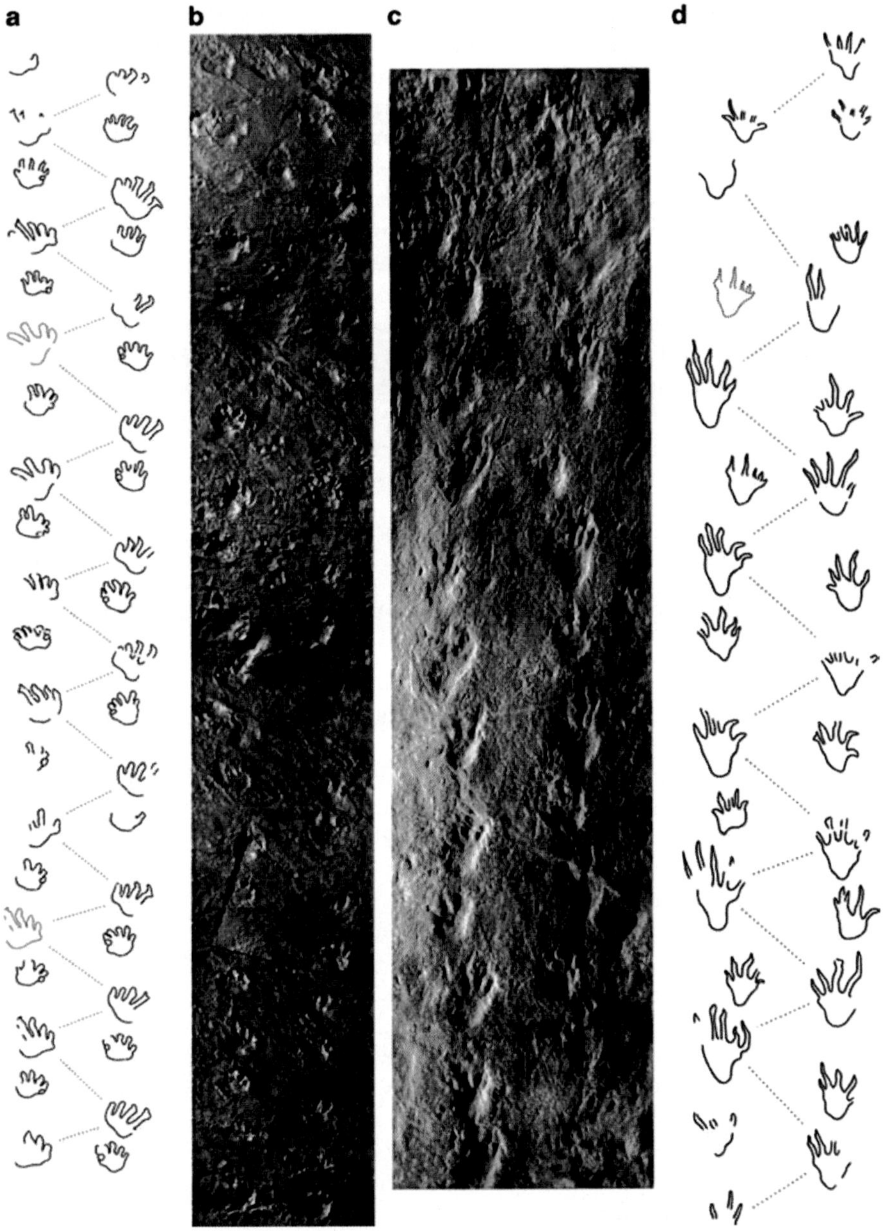

Fig. 4.1 Foot tracks of a four-footed animal from late Carboniferous named *Ichniotherium praesidentis*. (**a**) Drawing. The dotted zigzag line links the pes imprints, and dotted track outlines indicate the position of imprints, which must have existed, but are not preserved. (**b**) Photograph of a resin replica of several original fragments, (**c**) foot tracks of *Dimetropus* (late Carboniferous), plaster casts. (**d**) Drawing of the complete *Dimetropus* trackway as preserved on five blocks. Note the pointed digits, which seem to be equipped with claws. (From Voigt and Ganzelewski (2010). See also Chap. 5)

The close similarity across different orders and classes raises the question WHY this is so. An easy selling reason seems to be offered by the theory of evolution: descendants from ancestors which have possessed the same traits. Some of the morphological similarities look like early determination of particular variants, which have never been changed during evolution. Some authors are straightforwardly talking about early "setting of shifts" during evolution. Our major aim must be finding out whether such seeming "point settings" are simply due to mechanical necessities, which have enforced the evolution of the respective traits. Because these physical conditions and mechanic requirements hold true under all circumstances, we can expect principally similar morphological structures able to sustain the evoked stresses and behaviours which keep the stresses at a low level. Therefore, it seems worthwhile to search for compelling necessities or advantages for the "conservative" anatomical characters. In order to find such necessities, we have investigated in several studies monitor lizards and crocodiles as living and the only available representatives of this "primitive" body constructions among land-living animals. Large forms make more illustrative examples than small ones.

4.2.1 Physical Facts: Heavy Beam on Two Pairs of Supports

Basic for all further considerations are *physical facts*: while suspension in water requires not much energy for maintaining body shape, it needs, because of the medium's viscosity, energy for propulsion. By contrast, existence on land is under the permanent influence of gravity, which must be counterbalanced to maintain body shape, while energy requirements for locomotion can be kept on a rather low level, at least at slow speeds. The average absolute maximum speeds of aquatic animals are less than those in terrestrials (not to speak of flying birds!).

As soon as an animal leaves the water (the medium of about the same specific weight as the body itself), it has to withstand gravity. *Weight (= mass * gravitational force)* is distributed over the whole body; that means it attacks at all body segments. For our purposes, it is often useful to imagine weight as a force concentrated in the centre of mass (CoM) of the body or body part. The reaction force exerted by the ground (GRF), evoked by body weight, acts concentrated from below against the autopodia, where it is seizable for an analysis. Aside from the bending moments along the x-axis of the body stem, the joints of the extremities must be kept in equilibrium against gravity. All these necessities persist during locomotion – which itself poses its own demands, different from those in an aquatic environment (see below). Each action against gravity costs energy. Since gravity acts incessantly, those structures are most advantageous, which allow a resisting with minimal expenditure of energy, not alone in the trunk but also in the extremities.

4.2.2 *Body Stem in Primitive Tetrapods: Bending Moments, Shearing Forces*

In the "lower" *tetrapods* in the sense of Chapter 4.1, the elongate body can be compared to a *heavy beam, propped up on two pairs of extremities*. The anterior neck and head, as well as the posterior tail, can be understood as *cantilevers*, which are bent downwards by their own weight (Fig. 4.2a) as detailed in Preuschoft et al. 2012a and dealt with in the following Chapters 4.5 and 4.6. While the head + neck tend to bow downwards – as well as the tail – the part between the limbs, the trunk, tends to sag towards the ground between the supports, with its middle part deepest. If the trunk is heavier than the neck or tail, this dorsally concave bending will be more pronounced.

The tendency to deform under the load is quantitatively illustrated by *bending moments* (Fig. 4.2c), which are proportional to the shearing forces. The bending stresses increase towards the periphery of the body under bending and are rather small in its centre (Fig. 1.5c, f).

If a cut is made behind the head, or just in front of the anterior support, the head or head plus neck tend to fall downward, because internal forces not longer resist the *shearing forces* (Fig. 4.2b). If such a cut is made behind the anterior support, the head, neck and shoulder area will remain in their places, but the adjacent segments of the trunk will fall downward. The resisting internal forces change their sign. If we imagine sections farther behind the anterior support, about halfway to the posterior support, the shearing forces become smaller, pass a zero value and reverse their sign again in front of the posterior support (Fig. 4.2b). At the posterior bearing, the direction of shearing forces again is inverted; the tail segments tend to move downwards against the posterior trunk – opposite to the head and neck section.

The degree of bending the body stem depends largely upon the weights and lengths of its parts: the head and neck and trunk and tail. This influence is investigated by FESA in typical examples in Fig. 4.3. The first of these examples (a, b) shows the compressive (left) and tensile stresses (right) in a generalised low tetrapod.

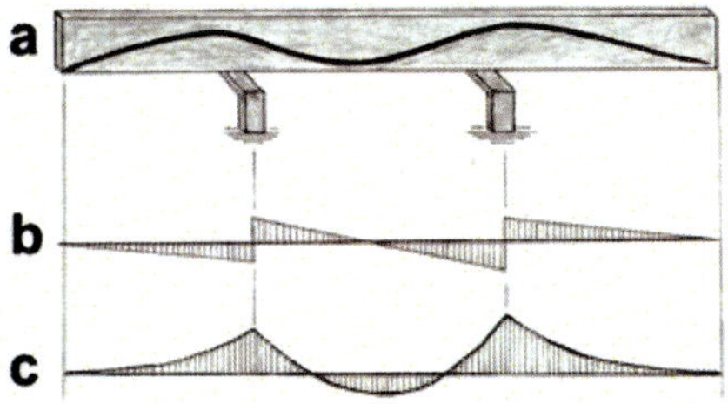

Fig. 4.2 A heavy beam, supported on two pairs of braces as a model of (**a**) "low tetrapod" with stress diagrams. The curved line on the beam is the expected deformation if the beam is not rigid: dorsally convex curvatures occur on top of the supports extending toward the "cantilevers" head + neck and tail, respectively, dorsally concave curvature between the supports, along the trunk. Diagram (**b**) shows shearing forces, (**c**) bending moments

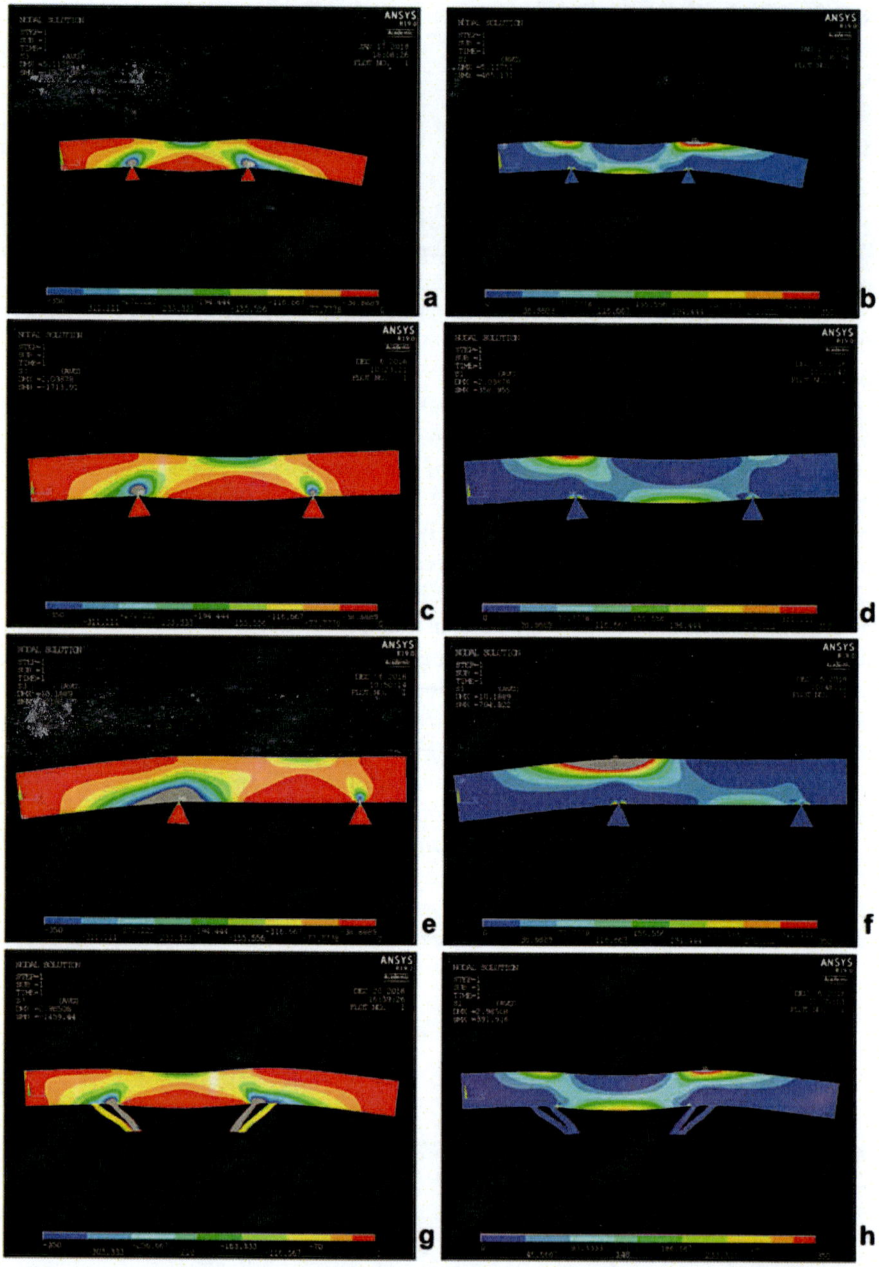

Fig. 4.3 FES Analyses of stresses in heavy beams resting on two supports as simplified models of tetrapods. The trunk is assumed to have two times the mass of the cantilevers: head + neck on the left and tail on the right. Although trunk length is kept constant, the stress patterns vary depending upon the length of the cantilevers. The left support (shoulder) is mobile in horizontal direction; the right (hip) is fixed. The deformation of the beam is emphasised. Compressive and bending stresses are indicated by colours. On the left: red means no compression: orange and yellow, moderate:

The second example illustrates a shortening or weight reduction of the tail (c, d), which is realised in several “mammal-like reptiles”. Increased head weight or elongation of the neck, combined with further reduction of the tail, has the consequences shown in e, f. The last example shows the influence of the position of limbs, as dealt with in Chapter 5.6.

4.2.3 *Structures to Resist Bending Moments and Shearing Forces: Axial Skeleton and Paraxial Muscles*

Of course the body must be able to sustain the permanent stresses. In fact, the typical arrangement of body structures is in agreement with the physical needs (Fig. 4.4a): the compression-resistant axial skeleton is on the dorsal side of the trunk – but in both “cantilevers”, tail and neck, shifted towards the middle of the cross section. At the attachments of the supporting extremities, there exist large bony plates as ventral parts of the girdle skeletons (see also Chapter 4.3). The bony shoulder girdle often extends more in cranial than in caudal direction, in agreement with the ventral compression of the neck basis visible in Fig. 4.3a. The tension-resistant muscles are strongly developed at the basis of the neck and the shoulder (aside from autochthonous dorsal muscles, the mm. trapezius and rhomboideus), as well as on the ventral side of the trunk (m. rectus abdominis). Likewise, there is compression at the ventral side of the proximal tail. The dorsal musculature in the same region (Fig. 4.3b) is also remarkably strong and becomes less so towards the tip.

Fig. 4.3 (continued) green and light blue, considerable: dark blue and grey, high compressive stresses. On the right: dark blue means no tension: light blue, moderate: green and yellow, considerable: red or grey, high stresses. The illustration is designed to show the influence exerted by neck or tail length on bending of the body stem under the influence of its weight. (**a**) Distribution of compressive forces in a “low tetrapod” with moderately long neck + head and long tail. (**b**) Pattern of tensile forces in the low tetrapod model. Note the extension of fairly high stresses toward the tips of the snout or tail and the low tension along the belly. (**c**) and (**d**) tail is shortened (or contains less weight), like in fossil therapsids. The bending of the trunk becomes more pronounced, and the stresses are higher. The arcuate arrangement of compressive stresses (yellow) corresponds to the rearward inclination of the shoulder blade and the forward inclination of the ilium. Tension along the ventral contour is nearly as high as at the base of the neck. (**e, f**) Compressive and tensile stresses in a model similar to a cursorial mammal with a long neck and a negligible tail. The highest stresses occur at the shoulder and indicate the major share of body weight, being supported by the frontlimbs. This corresponds to the long lever arms formed by the spinal processes, the sternum, and the low position of the vertebral column in the thoracal region. The lumbar segments of the axial skeleton and abdominal wall also receive high stresses. (**g**) and (**h**) “Low tetrapod” with frontlimbs inclined rearward and hindlimbs inclined forward. If these limbs are connected to the trunk by joints, the latter must be kept in balance by ties, and the ties reduce the bending moments along the trunk in comparison to (**a, b**). See also Chapters 4.2.1 and 5.6, as well as Fig. 5.13

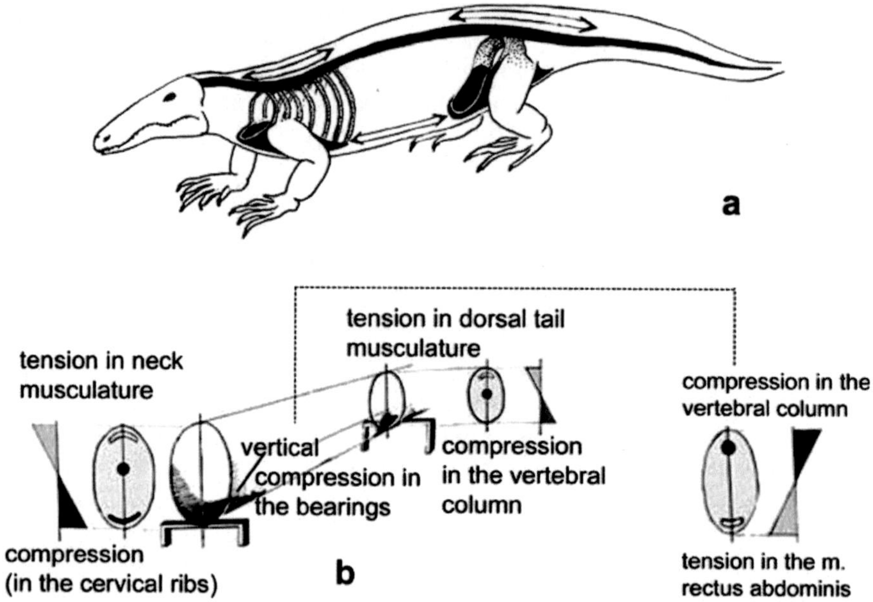

Fig. 4.4 Simplified model of a lower tetrapod while standing on all fours to illustrate the structures which resist bending of the body stem. (**a**) Distribution of stresses in the trunk in oblique view from the top left. Black, compression-resisting elements; white, tension-resisting parts. The former correspond to the axial skeleton and the ventral parts of the girdle skeletons, and the latter to the autochthonous dorsal (epaxial) muscular group and the m. rectus abdominis. Skeletal elements able to resist the greatest shearing stresses are stippled. (**b**) Distribution of stresses on selected cross sections through the trunk. (Modified from Preuschoft et al. 2007)

Contraction of the ventral m. rectus abdominis (required to resist bending in Fig. 4.3) exerts dorsally directed compressive force components. These are sustained by the ribs, which keep distance between ventral muscles and axial skeleton.

The structures resisting the shearing forces can either be considerable amounts of bone material (Fig. 4.4a; basal neck, pelvic region) or an array of +/− vertical compression-resistant rods, that is, spinal or haemal processes in the neck and tail, or ribs in the trunk, interconnected by oblique tension-resistant elements (i.e. interspinal muscles and ligaments, intercostal oblique muscles).

4.2.4 Torsional Stresses

A complete understanding of the trunk is only possible, if *torsional stresses* are taken into account. This was considered first by Carrier (1990) and Carrier (1993). This author argued on the basis of "propulsive forces", while in this book an approach through the rules of statics is preferred, as developed independently in Preuschoft

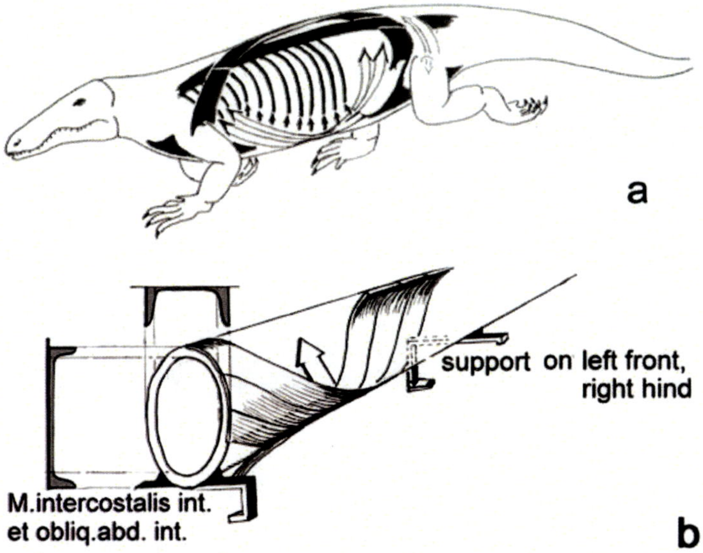

Fig. 4.5 (**a**) Simplified model of a lower tetrapod during locomotion to illustrate the structures which resist torsional stresses. Black is the flow of compressive force components, connecting the supporting limbs FL and HR, as well as the axial skeleton; white is the tensile component, the oblique muscles. The flow of compressive force follows the supporting forelimb, the structures of the shoulder girdle, axial skeleton and pelvis to the diagonal hindlimb. The tensile forces are in accordance with the oblique muscles of the ventrolateral trunk wall and m. iliocostalis. The activity of the oblique muscles of the trunk wall exerts force components which tend to approach the ventral to the dorsal side, but such a deformation is precluded by the ribs. (**b**) The trunk of the same animal, simplified as a tube. The diagonal flows of tensile forces and the concentration of torsional stresses near the periphery are emphasized, This concentration is particularly pronounced, if the resisting structures are rather stiff. (Modified from Preuschoft et al. 2007)

(1976) and Preuschoft et al. (2007). Torsional stresses concentrate along the periphery of a body while leaving its central part free of stress (Figs. 4.5b and 4.6a).

Precondition of locomotion on limbs is the ability to lift individual limbs from the ground (Fig. 4.5a). If one forelimb of the animal is without ground contact, and relieved from carrying weight, so that the body is supported asymmetrically, the stress pattern changes markedly. The trunk between the anterior and the posterior pair of limbs is now exposed to *torsion* (Fig. 4.5).

Torsional moments (M_t) are defined as

$$M_t = F * r, \tag{4.1}$$

where r is the distance of the force to the longitudinal (x-) axis.

Resistance against torsion (τ_t) depends upon

$$\tau_t = M_t * {y_{max}}^2 / I_t, \tag{4.2}$$

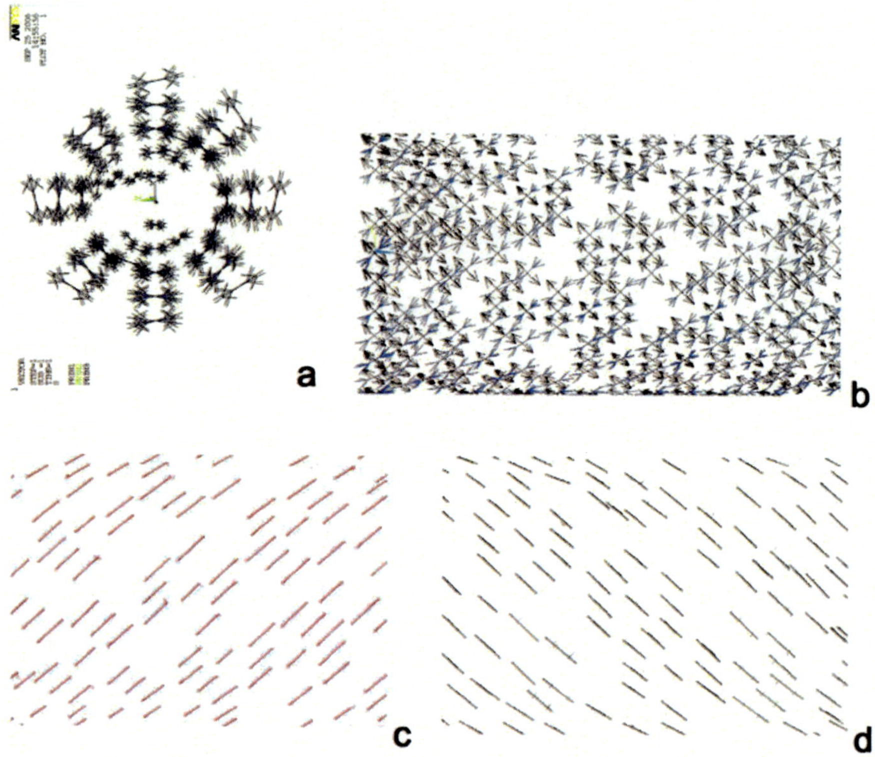

Fig. 4.6 Torsional stresses in the trunk of an FE model of a monitor lizard. (**a**) Trajectories on a cross section through a homogenous body, without particular rigidity close to the periphery. Note that they are greatest near the periphery and dwindle to zero values near the centre. (**b**) At the surface of a twisted body, compression (>--<) and tension (<-->) occur at the same places and are perpendicular to each other, as visible in side view. The magnitudes are indicated by the lengths of the arrows. (**c**) To show the calculated results clearer, the tensile components are redrawn in red (corresponding to diagonal muscles), and (**d**) the compressive components in black (corresponding to ribs). (From Preuschoft et al. 2007)

where M_t is the above-defined torsional moment, y is the distance from the x-axis, and I_t is the *area* moment of resistance as defined in Chapter 1.2.

Ventral between the forelimbs occurs compression in z-direction, opposed by tensile stresses on the dorsal side (Fig. 4.5). Viewed from above or below, the stress distribution in the trunk becomes asymmetric, shifted towards the supporting side. This is due to the superimposure of normal, longitudinal stresses with torsional stresses. As noted above, the latter concentrate near the periphery of the body (see also chapter 4.2.5).

If extremities are sprawled, *torsional moments* assume particularly high values, simply because their lever arms (r) are long. Torsional moments consequently are most prominent along the trunk between anterior and posterior limbs. They

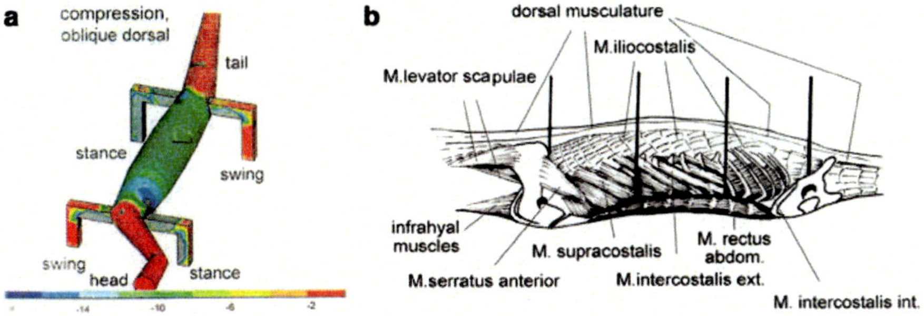

Fig. 4.7 (**a**) FESA model of a long-necked monitor lizard from above – left. The neck is turned to the left of the animal, and its head points to the right. Weight supported mainly on LF and RH. Torsional stresses superimpose bending stresses, leading to an asymmetric distribution. Remember that FESA results obtained from different models cannot easily be compared quantitatively. (From Preuschoft et al. 2007). (**b**) Trunk wall of a reptile (*Sphenodon*, taken from Guibé 1970). The dorsal (bony) parts of the ribs are orientated parallel to compressive components of torsion, and muscles parallel to tensile components. In reptiles, the m.iliocostalis is, like the oblique muscles of the body wall, more expanded on the flanks of the trunk than in mammals, where it acts as a part of the epaxial, dorsal muscle group. The ventral parts of the ribs and the m. rectus abdominis are dealt with in chapter 4.2.6

superimpose bending, so that the distribution of combined stresses is asymmetrical (Fig. 4.7a)

4.2.5 *Structures which Resist Torsional Stresses: Oblique Muscles and Ribs*

As noted above in Chapter 4.3 and shown in Figs. 4.5 and 4.6, torsional stresses become greater towards the periphery of the twisted element, the trunk. If structures of the periphery are strong enough (= possess a higher Young's modulus), torsional stresses leave the central parts of the cross section without any mechanical stress and so create the trunk cavity for the intestines. More or less circular cross sections through the trunk provide maximal torsional strength, in contrast to flattened or compressed cross sections (as, e.g. in the anterior thorax in mammals; see Chapter 7.7). A round cross section is indeed most frequent among lower tetrapods. The directions of torsional stress, obtained from the same model, are shown in Fig. 4.6. Tensile and compressive components cross each other at right angles.

The requirements of both are fulfilled by the typical arrangement of materials in the trunk of all tetrapods: the tensile components of torsion (Fig. 4.6) are sustained by the oblique muscles of the trunk wall (mm. obliqui abdominis externus et internus, mm. intercostales; Fig. 4.7b). In addition to these muscles, the m. iliocostalis in reptiles also acts as a tension-producing structure against torsion.

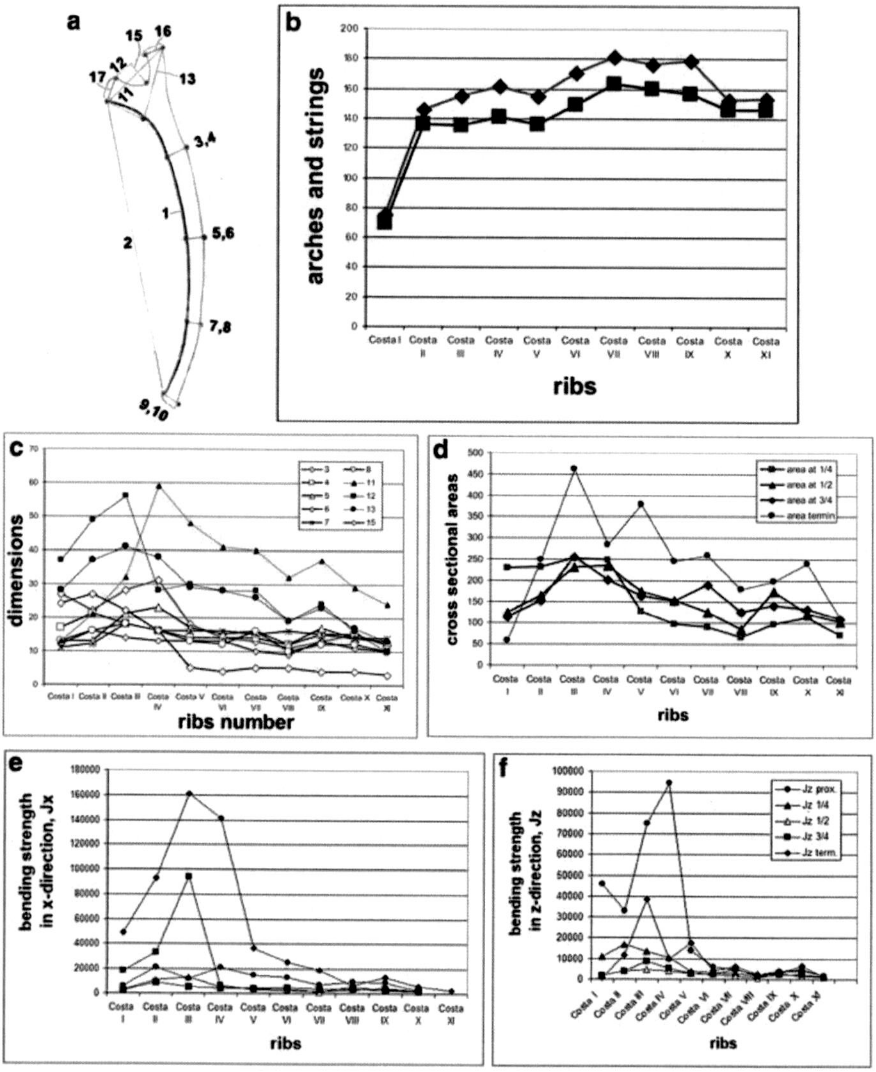

Fig. 4.8 An alligator is used as a representative of all archosaurians. (**a**) Rib of a crocodilian with the used measurements: 1: length measured on inner side of the bowed rib; 2: tendon of the arch; 3, 4 ; 5,6; 7,8: Diameters at 1/4, 1/2, 3/4 of rib length and 9, 10: terminal thickening of the rib, 11: external distance between vertebral joints, 12: internal distance between vertebral joints; 13: height of tuberculum; 14: length of collum; 15 depth of incision, 16, 17: breadth of joints 10: length of collum costae; 13: length of tuberculum costae; 16: width of joint surface on tuberculum, as a rough measure for bending strength. The values are plotted against rib numbers, from costa I through costa XI. (**b**) Bow-string relationship. Note the shortness of the first rib. (**c**) The values describing the spread between the tuberculum and collum costae are great (11–13; black symbols, thin lines), while the values describing the rib profile are close together (white symbols, heavy lines). In the huge fossil archosaurs can be observed a slight flattening of the ribs, like in cursorials (Chap. 7).

This muscle deserves special attention, because it is innervated by rami dorsales of the spinal nerves and therefore considered to be a part of the autochthonous trunk muscles in mammals. The muscle activities which are predicted here (chapters 4.2.3 and 4.2.5) as the results of theoretical calculations are realistic: they have been controlled empirically by EMG in highly specialised cursorial mammals. Carrier (1990) and Carrier (1993) have shown them in the dog and Kienapfel et al. (2018; see Sub chapter 7.7.3) in horses.

The compressive components of torsion also require resistant structures. A series of ribs, combined with intercostal (tension-resistant) muscles, are a perfect means to withstand the shearing forces which occur according to Fig. 4.4a. Likewise, the rodlike elements in oblique directions are perfectly suited to sustain the compressive component of torsion (as shown in Fig. 4.5). B*ony ribs* are indeed a characteristic of primary land-living animals and their secondarily aquatic descendants, while fishes possess commonly two series of (laterally and ventrally directed) thin, segmental elements of their skeleton. *Ribs* in general have rarely been the subject of studies. This may be due to several reasons: the confusingly great number of rather similar elements, their limited variability and the complete lack of ideas about their "function". If at all attempts are made in this direction, ribs are seen in connection with breathing. I do not intend to falsify these approaches (Pajor and Perry, 2007; Perry, 2010; Perry et al. 2011) but would like to add new information.

That ribs and the muscles of the trunk wall are involved in locomotion becomes clear by the EMG data of Carrier (1990, 1993) and Kienapfel et al. 2018, as noted above.

In large archosaurians, ribs are not very numerous: crocodiles have 11 and sauropod dinosaurs 8 or 9. Rib-bearing vertebrae come close to the pelvis; the ribless lumbar section of the vertebral column is short. The small number of ribs indicates shortness of the trunk between the supports, provided by front- and hindlimbs. A short trunk keeps the bending moments on a low level, but does not permit much mobility.

The individual ribs have characteristic and rather uniform shapes: the corpus costae splits towards the vertebral column into two branches, the medial of which can be compared to the collum and caput costae of mammals and the lateral branch to the tuberculum costae in mammals. Both are separated from each other by a deep incision (Fig. 4.8a). To the vertebral column, the bony ribs are attached through

Fig. 4.8 (continued) The varying symbols show the measurements, as indicated in the box on the right. (**d**) The cross-sectional areas, calculated as rectangles with the dimensions of both diameters on 1/4, 1/2, 3/4 of length and on the free terminal end, indicate compressive strength. Their values decrease from costa V caudally, i.e. from the caudal margin of the m. serratus lateralis. (**e, f**) Well marked is the variation of bending strength. Area moments of resistance (J) in longitudinal (x-) and lateral (z-) direction of the body, at proximal extension, 1/4 (diamonds) 1/2 (squares), 3/4 (squares), and terminal end of bony ribs (rosettes). The information shown here for the crocodilian is very similar to the information obtained from fossil sauropods, of which I have taken measurements from the mounted *Brachiosaurus* in Berlin and the *Diplodocus* in Frankfurt. Their ribs exhibit, like the construction of the shoulder, some similarity to cursorial mammals (see also Chapters 1.5, 4.3, and 5.8) large, well-developed joints at two points: the tuberculum costae[1] at the transverse process of the vertebrae and the caput costae at the vertebral bodies. This forked support gives them high stability in the plane of bifurcation (which is close to the *x*/*z*-plane) while permitting movements at right angles in *x*-direction. The more cranial position of the dorsal and more caudal position of the ventral joints favour moving of the ribs partly in caudal-dorsal direction. The variation of shapes along the trunk axis is shown by some of the measurements in Fig. 4.8. More important than the measurements themselves are for the present study the mechanically relevant strength properties. Compressive strength is proportional to cross-sectional area (Fig. 4.8d) and bending strength in longitudinal (*x*-) and transverse (*z*-) direction proportional to the area moments of resistance (Fig. 4.8e, f). Most obvious traits are the shortness of the first rib and the strength values (areas, *Jx* and *Jz*) against compression and bending.

The first rib appears to be not more than an insertion for the m. serratus, by which the trunk is hung up from the scapula. It does not provide a compression-resistant connection to the sternum, as is true in mammals. The other ribs, especially those in which the m. serratus inserts, are suited to sustain extraordinary bending moments (Fig. 4.8e, f, especially due to measurements 13 and 3). The compressive components of torsion seem to flow through the shoulder girdle, rather than the ribs. Shearing forces and bending and torsional moments in the trunk seem to be greater in reptiles than in mammals. This may be due to the long, weighty tails and requires the existence of compression-resistant elements along the trunk and the proximal tail. From the transverse processes of the proximal tail vertebrae originates the strong m. caudofemoralis, the contractions of which exert compressive forces in the vertebral column, the hip joint and the femur. Long diameters of the most proximal tail vertebrae, immediately behind the sacrum, are obvious in crocodiles and especially in the fossil dinosaurs.

4.2.6 *Sternum and Angulus Costae in Ventral View*

All tetrapods possess a (often neglected) skeletal complex at the ventral side of the trunk, namely, the deviation of left and right cartilaginous ribs (which form on either side the *costal arch*, arcus costae), both of which including, caudal to the short sternum, the rib angle, *angulus costae*. It is shown in the example of a crocodile in Fig. 4.9a. The common view is interpreting the ventral, normally cartilaginous parts of the ribs just as prolongations of the bony ribs, located dorsal to the "rib knee". Seen from a functional viewpoint, the costal angle and cartilaginous ribs seem to be something quite different from the bony ribs and deserve special attention. The persistence of cartilage instead of bone may well be due to the fact that considerable

[1]This term is derived from the ribs of mammals, in which indeed not more than a tuberculum is present, instead of the strong branch of a fork.

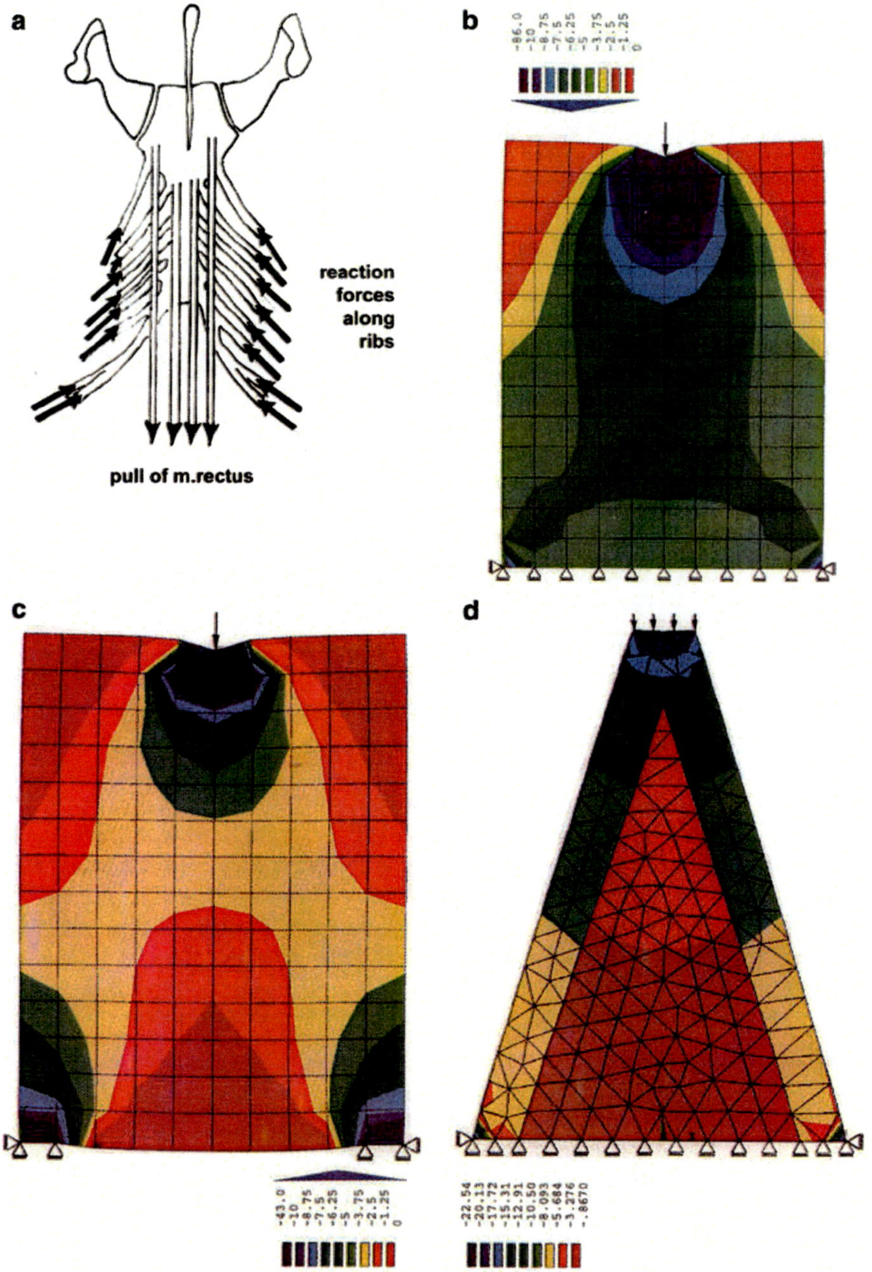

Fig. 4.9 (**a**) Sternum, cartilaginous ventral parts of the ribs, and origin of the m. rectus abdominis in a crocodile. The spoon-shaped, often cartilaginous cranial, "interclavicula" corresponds to the distribution of compressive stress caused by bending in Figs. 4.3 and 4.4. The m. rectus abdominis (white arrows) pulls the sternum caudally, and the force is transmitted to the ventral parts of the ribs,

movement of the ventral abdominal wall is unevitable (and so excludes ossification), because of breathing and the varying volume of the intestines. In addition, the forces acting in this region are not so great as in other parts of the skeleton. In Fig. 4.9a an attempt is made to understand the prevailing mechanical conditions: the m. rectus abdominis takes its origin from the sternum and the adjacent parts of the ribs. It produces a rearward-directed force, which is countered and supported by the ventral parts of the ribs. The flow of forces is illustrated by Fig. 4.9b–d with the aid of a two-dimensional FESA-model. The force flow in the plate already shows a clear preference to reach the caudo-lateral corners, while the middle receives lower stresses. Two ways of force transmission are shown as examples for further concentration of the stresses on both sides: the elimination of median support (c) and the reinforcement of the most highly stressed parts (d). The latter is in accordance with empirical observation. The result in both cases is the same: the cranial part of the ventral body wall is without compressive stress and consequently without skeletal support elements.

4.3 Extremity Girdles Seen as Parts of the Trunk

The *girdle skeletons* can be seen as parts of the trunk, and therefore they are mentioned in the present context, although they are also part of the extremities (Chap. 5). As can be seen from Fig. 4.3, the ventral parts of the shoulder as well as pelvic girdle are under compression in *x*-direction. In phases of locomotion, in which the body is supported by only one limb of a pair, the sprawled position of the stylopodia requires on the stance limb a contraction of the adducting muscles. These connect either the pubis/ischium with the femur or the rib cage with the humerus (Fig. 4.13b, c). These contractions compress the ventral side of the trunk in *z*-direction. Likewise, lifting of the swing limb of each pair also leads to compression in *z*-direction (Fig. 4.4). These varying, but always high compressive forces are sustained by ventral constituents of the girdle skeletons: coracoid and its derivatives, interclavicula in the shoulder girdle, pubic symphysis and ischium in the pelvic

Fig. 4.9 (continued) which are propped by small forces (black arrows) exerted on and by the ribs. The reason for this shape becomes visible in the two-dimensional FES model representing the ventral aspect of a terrestrial animal. **(b)** Is a plate resting on its lower (caudal) margin and loaded by a force concentrated on its upper (cranial) margin. The highest stresses (dark blue) occur near the force application and distribute toward the lateral sides of the plate (yellow, light green and dark green). The upper (cranial) corners are without stress (red). **(c)** The lower (caudal) margin of the same plate is supported only at its bottom corners, while loaded from above as before. This leads to stress concentrations at the lower corners, and to absence of stress in the middle, as well as the upper corners of the plate (red). In **(d)** the not stressed areas adjacent to the upper margin are cut off, and those areas found to carry most of the load are given more rigidity. In spite of the support offered along the entire lower margin, stresses flow along both sides of the triangular plate (blue, green, yellow) and leave the interior (red) unstressed, so forming the generally existing "rib angle"

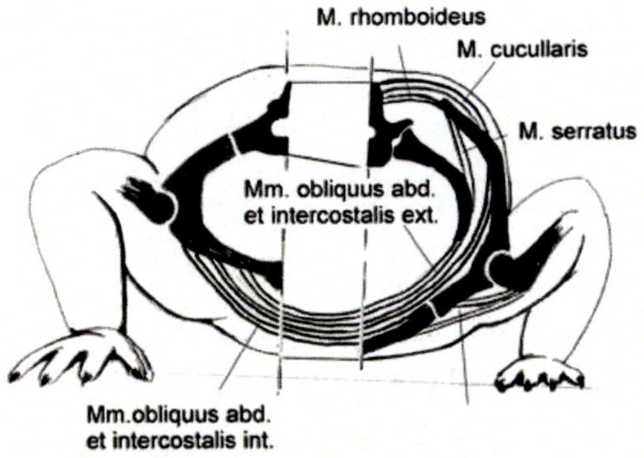

Fig. 4.10 Schematic arrangement of trunk wall musculature and the extremity girdles. Left half: cross section through the pelvic girdle with origins of the abdominal muscles from the pubic bone. Similar to the suspension on the frontlimbs by the m. pectoralis, the trunk is suspended on the hindlimbs by the adductors (running from the pubis or ischium to the femur). Right half: cross section through the shoulder girdle with insertions of trunk wall musculature and the shoulder girdle muscles which suspend the trunk on the forelimb. Modified from Preuschoft et al. 2007

girdle (Figs. 4.4, 4.10, 4.11, 4.12, and 4.13; see also chapter 7.7.4, especially footnote 22).

The dorsal parts of the girdles are responsible for supporting the (heavy!) trunk. The pelvic girdle carries trunk weight simply by a stiff, bony connection between the ilium and the vertebral column and by providing a rigid origin for the muscular trunk wall in form of the pubic bones (Fig. 4.10). The shoulder girdle, however, provides a (muscular) hanger for the thorax. This is best visible on transverse sections (Figs. 4.10 and 4.13). M. serratus lateralis connects the most dorsal part of the shoulder blade with lateral processes of the vertebrae (in lissamphibians, Fig. 4.13a) or with several, especially the short anterior ribs (in reptiles, Fig. 4.13b, c). The deviation of the muscle from the vertical permits transmission of compressive forces in z-direction from one side to the other, as occur in symmetrical support on both legs, and keeps the shoulder blades away from the trunk. In asymmetrical support on only one limb, however (Fig. 4.13c), mm. rhomboideus and trapezius (= cucullaris) must take over the tensile forces in z-direction.

We will come back to the limb girdles in Chapter 5.8, since the Bauplan of these elements varies fundamentally among the classes.

4.4 The Earliest Known Example: *Tiktaalik*

As an example of the transitory stage between fishlike, mostly aquatic and land-dwelling tetrapod animals, we (Hohn-Schulte et al. 2013) could analyse the newly described fossil *Tiktaalik*. This 382 million-year-old fossil (Daeschler et al. 2006;

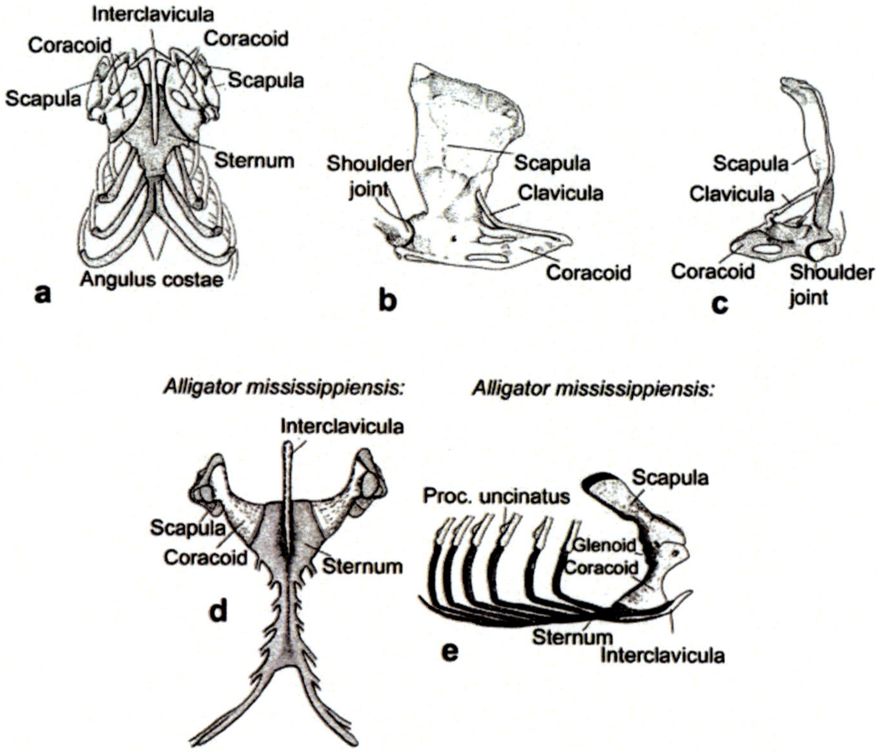

Fig. 4.11 Shoulder girdle of reptiles. (**a**) Ventral, (**b**) left side, (**c**) frontal views of a monitor lizard (*V. gouldii*). Movements of the forelimb rotate the coracoid along the fat line against the (grey) sternum (Jenkins and Goslow, 1983). Note the angulus costae, explained in Fig. 4.9. (**d**) Ventral, (**e**) right lateral view of the shoulder girdle of an alligator. (**d**) From Starck, 1978–1981, (**e**) from Bütschli (1921). The cartilaginous cranial process of the sternum is not shown, but the cartilaginous parts of the ribs (dark grey) and the "rib knee" are illustrated, as well as the coracoid and scapula. Note that the rib knees are located in the cartilaginous section of the ribs

Shubin et al. 2006 Ahlberg and Clack 2006, Fig. 4.14) is close to fishlike lobe-finned *Eusthenopteron*, *Gogonasus*, *Panderichthys* and *Elpistostega* from the Middle and the earliest land animals *Acanthostega* and *Ichthyostega* from the Late Devonian. It possesses well-preserved, rather strong-looking anterior extremities, which suggest the ability to lift the anterior part of the body above ground. Accepting this hypothesis of the above authors, we made visible the difficult to imagine stressing of the trunk including the girdle skeleton by means of the FES method (Fig. 4.15). The investigated model has the measurements of a well-preserved fossil individual and the mass distribution derived from these measurements (Fig. 4.15a). Head, neck and anterior part of the trunk are propped on the obviously strong forelimbs. The hindlimbs and pelvic girdle are not perfectly preserved and seem to be weakly developed. Therefore our model rests simply on the ground with its posterior part.

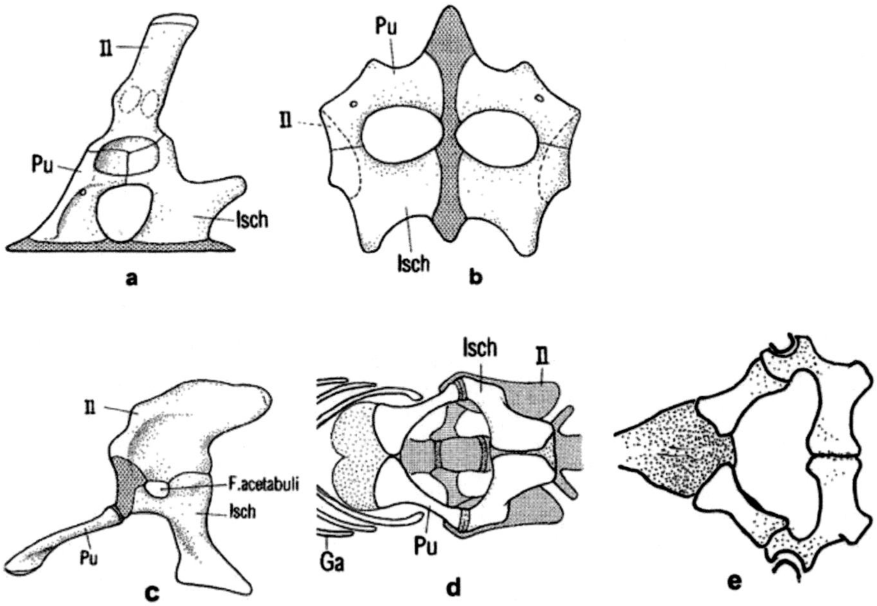

Fig. 4.12 Pelvis of reptiles. (**a**, **b**) Tuatara (*Sphenodon*) from left and ventral side, (**c**, **d**) Alligator (from Mickoleit 2005) from left and ventral side, (**e**) crocodile redrawn from Owen (1859) – bone is white, cartilage (and vertebral column on dorsal side in **d**) grey

Such a body posture can be observed in living mudskippers (teleosts), salamanders and lizards. If body weight is symmetrically supported by both forelimbs, the FESA-model shows high compressive stresses along the dorsal side of the trunk, especially in its caudal part and slightly cranial to the forelimbs. Along the ventral side of the trunk occur tensile stresses, as well as on the dorsal side above and slightly cranial to the forelimbs. Both stress qualities are acting longitudinally along the x-axis. If the body is asymmetrically supported be only one forelimb, there occur high compressive stresses in z-direction between both forelimbs, on the ventral side (Fig. 4.15b, c). Further aspects of *Tiktaalik*'s morphology are dealt with in Chapter 2.3.6 and Chapter 5.

In addition to bending occur *torsional stresses*. This combined stress pattern has been analysed with the aid of FESA (Fig. 4.15). As can be seen, the most strongly compressed area (at least in the model of homogenous material!) extends over a large part of the dorsum, covering not alone the vertebral column but also the dorsal parts of ribs and the lumbar transverse processes. Its asymmetry depends upon the asymmetric support during locomotion and changes its direction from step to step. On the ventral side, a corresponding asymmetry can be seen, which is related to the arrangement of the oblique abdominal muscles.

The distribution of compressive stresses corresponds to the arrangement of *compression-resistant skeletal elements*: ventral between and cranial to the forelimbs the massive ventral elements of the shoulder girdle, along the back structures

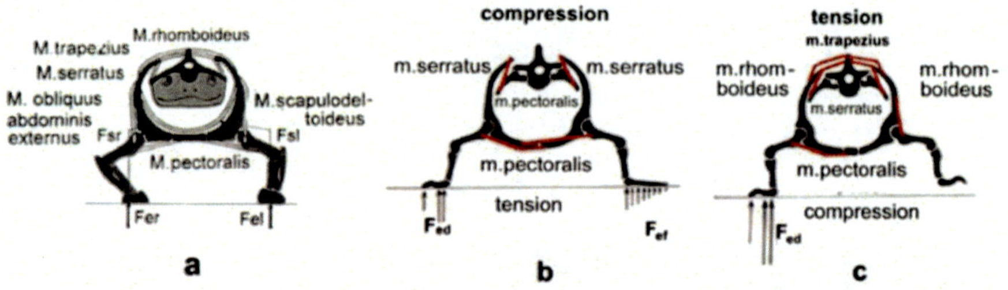

Fig. 4.13 Schematic cross sections through primitive tetrapods supported on two anterior limbs or only one of them. The arrangement of the muscles carrying the weight of the trunk is similar in lissamphibians (**a**, supported on two limbs) and reptiles (**b, c**). The vertebral column (and head, as shown in **a**) is hung up by m. serratus on both sides. Mm. rhomboideus and trapezius keep the scapula in its place. The intestines are wrapped by the muscles of the trunk wall, which originate in part from the medial side of the scapula. On the reader's right side (left of the animal), the forefoot is placed in a "wide gaged" posture far away from the midline, and the m. pectoralis is required for keeping the shoulder joint in balance. The forefoot on the left side of the reader is placed closer to the midline. This "narrow gaged" posture relieves the m. pectoralis but requires activity of the elbow extensors (and perhaps of the external muscles of the shoulder girdle). The same effect can be produced by inward rotation of the zeugopodium, so that the toes point medially (as can be observed in, e.g. frogs). In the cranial segments of reptiles (**b, c**), the same overall arrangement exists. (**b**) Loading in symmetrical stance on both limbs. The fixation of the trunk to the shoulder girdle is similar, tough not identical to (**a**). The shoulder joint is balanced by the m. pectoralis, on the left side as common among living tetrapods, on the right side by a muscle on the inner side of the coracoid plate (as inferred by Shubin et al. 2006). The autopodium on the right side of the reader has the shape of a flipper as in *Tiktaalik*, on the left side of a digited manus, as in modern reptiles. The diverging load application is emphasized on the left: parts of the GRF act against the carpus, metacarpophalangeal joint and tip of the digit; on the right side, GRF is distributed over the entire volar surface of the fin. (**c**) Asymmetric support on only the left limb. The fractions of GRF are two times those in (**b**); the manus on the right side is without ground contact, like the swing limb in walking. This not loaded limb must be kept in place by muscles, which are present in all living tetrapods: mm. trapezius (= m. cucullaris) and rhomboideus

representing an "axial skeleton". Fossil *Tiktaalik* lacks a complete vertebral column. This may be due to a notochord-like structure, at which just parts of the centra are ossified. Compression in longitudinal x-direction as well as the compressive components of torsional stresses can also and seem to be sustained by the broad, imbricate ribs. Strong ribs are a characteristic of *Tiktaalik* as well as all other terrestrial animals which so deviate from more fishlike forms. The flow of force in x-direction from one of the imbricate ribs to the other is possible if both are connected by tough fibres. The overlapping of the broad ribs allows distributing the compressive stresses over a large area of the dorsum – as is documented in Fig. 4.15a. This arrangement relieves the vertebral column or the notochord. In asymmetrical support, the ventral elements of the shoulder girdle receive high compression in z-direction, while the dorsal part is relieved. The model also shows in side view the compressive stress flow in the shoulder region to slope caudally. This corresponds to the rearward inclination of the scapula. The compressive components of torsional stresses seem to be counterbalanced by the ribs. Aside from being compressed at its dorsal side, the trunk is hung up from the dorsal margin

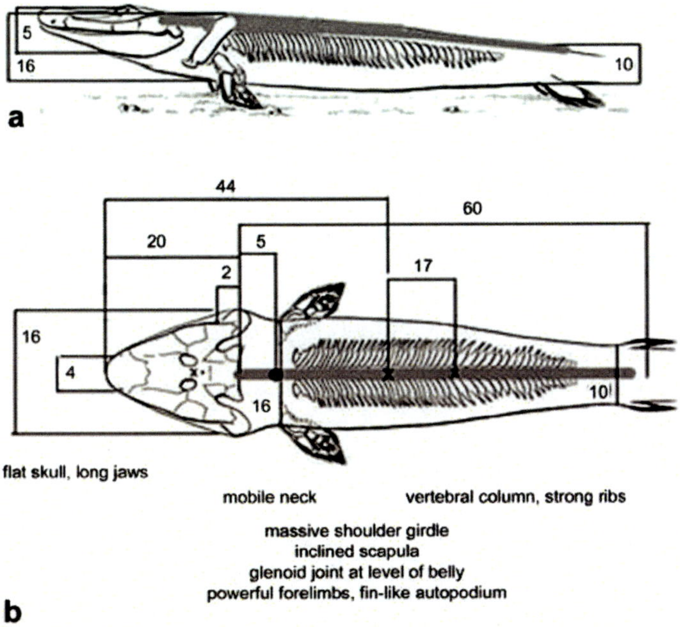

Fig. 4.14 *Tiktaalik*. (**a**) Reconstruction of the well-documented parts, with the measurements required for creation of the FE model. Note the flipper-like autopodia (see also Chap. 5), the sturdy ribs and the absence of a complete vertebral column. Maybe that this is due to a notochord-like axial skeleton, at which just parts of the centra are ossified. (Modified from Daeschler et al. (2006), and Shubin et al. (2006))

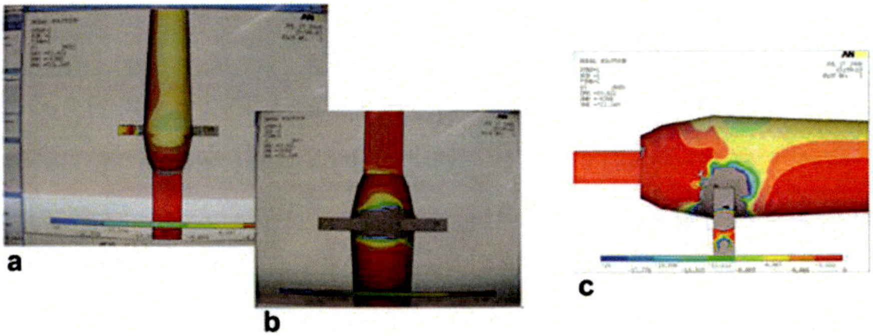

Fig. 4.15 (**a**) Dorsal, (**b**) ventral and (**c**) lateral aspect of the loaded FE model during asymmetrical support. The posterior part of the trunk is resting on the ground, alone the frontlimbs carry weight. The combination of longitudinal (*x*-direction) and transverse (*z*-direction) bending with increased load on the left frontlimb leads to asymmetrical compressive stress (yellow, green), on the dorsal side (**a**), especially between the shoulder blades. (**b**) High compressive stresses do also occur between the frontlimbs on the ventral side (blue, grey). (**c**) In lateral view, the stress flow in dorsal direction is inclined rearward, like the scapulae of tetrapods

of the scapula by the m. serratus, which has its distal insertion into the ribs (see Figs. 4.10 and 4.13). Most skeletal structures mentioned above (except the imbricate ribs) are present in all terrestrial animals. Their constant existence evidently is dictated by mechanical necessities. Interestingly, many, although not all, of these elements are preserved in their secondarily aquatic descendants.

The tensile stresses find their counterparts in *tension-resistant muscles*. These are, of course, not preserved in fossils. The musculature of the trunk, however, is very similar in all recent tetrapods. The constancy and conservativism of their arrangement give reason to conclude that the same muscles have been present in *Tiktaalik* as well. Along the *x*-axis of the trunk on the ventral side, the m. rectus abdominis can be expected, and dorsally the autochthonous musculature, reinforced in the shoulder and proximal neck region by girdle muscles derived from the extremities (see also Chapters 4.3 and 5.7). In transverse (*z*-) direction are powerful mm. pectorales spread between the limbs on the ventral side, while on the dorsal side the mm. rhomboideus and trapezius (= cucullaris in reptiles) take over tensile stresses. Torsion of the trunk occurs inevitably if support on the limbs is asymmetrical. Since torsional stresses concentrate near the periphery, they leave the abdominal cavity without mechanical stress. The tensile components of the torsional stresses (see below) can be sustained by the diagonal mm. obliqui abdominis and intercostales, as well as the m. iliocostalis. The compressive components of torsional stresses require the existence of strong ribs. The latter do not show up in the same form among fishes.

Tiktaalik seems to have possessed a rather flat body, although the possibility cannot be ruled out that this flat shape is just a deformation due to imbedding. Such a depressed body shape is by no means advantageous for a land-living animal, because the lever arms of ventral as well as neck muscles are shorter than in animals with a dorsoventrally deeper body (see also Chapter 7.7). A depressed body, however, keeps in shallow water a large part of the volume under water, so that it is nearly weightless and protected from drying. Whether the advantage of a deep or of a flat body is more important depends on the conditions under which the animal is living.

Aside from the bending moments along the *x*-axis, the joints of the extremities must be kept in equilibrium against gravity. The reaction force of the ground, evoked by body weight, acts concentrated from below against the autopodia. All these necessities persist during locomotion – which itself poses its own demands, different from those in an aquatic environment (see Chap. 5). Each action against gravity costs energy. Since gravity acts incessantly, those structures are most advantageous, which allow a resistance with minimal expenditure of energy, not alone in the trunk but also in the extremities.

4.5 Posterior/Caudal Cantilever: Tail

The body stem includes two *cantilever-like parts*, the tail and the neck plus head. Since the tail is easier to understand mechanically, it will be dealt with first. The long tails in low tetrapods are commonly seen simply as phylogenetic heritage. This view may be correct, but should not impede by a functional approach.

Bending moments in a heavy cantilever can be calculated by the following formula:

$$\Sigma M = F_{p1} * l_1 + F_{p2} * l_2 + F_{p3} * l_3 + \ldots, \tag{4.3}$$

where F_p stands for a partial mass, 1 for the most distal, 2 for the second segment from distal tail tip and 3 or the following for more proximal segments of the tail; l is the distance between the respective segment's CoM and the animal's hip joint. The same holds true for the neck, just in the opposite direction. The increase of bending moments is illustrated in similar form by Fig. 3.6b, c, (where a view on a tail from on top is shown, and instead of partial weights, the resistance of water acts against the tail).

In contrast to the situation in water-living animals (Chap. 3), the *tail* in land-living "low" tetrapods seems to mostly play a rather passive role – in spite of its well-developed musculature. As can be observed easily, at least the proximal tail is usually lifted from the ground, while the distal tail is dragged behind the animal. The faster the animal is moving, the longer the lifted proximal part of the tail. This variability makes a balancing of the tail by length-constant ligaments improbable. From the shallow scratches left behind by the distal tail, we may conclude that not much weight is transmitted on this way, while a larger part is balanced by muscle force. It is improbable that dragging the tail exerts more than minor resistance against being moved forward.

The use made of the sometimes impressively *long tails* or, in other words, the biological roles played by them in land-living forms are not well known. Definitely, the tails act as counterbalance of the long trunk plus neck and head in bipedal postures (Chapter 5.8), but it seems doubtful, if their length evolved just for this purpose. A body part of such a mass and weight of course has a considerable mass moment of inertia. This can be used for changing the body's position in space. According to Christian (1995), it is the means to perform rapid turns to both sides (Fig. 4.16): for a left turn, the heavy and long tail is moved into this direction. The lateral body wall muscles of the left side, which move the tail, also swing the head, neck and trunk to the left. In other words, the tail performs a countermovement against the anterior part of the body. Mass moments of inertia of both parts ($J_i = l_i^2 * m_i$) are equal; the tail's lack of mass in comparison to the trunk is compensated by its length (vide also Chapters 4.6 and 8.8).

Lateral tail strikes of crocodiles are really dangerous and in monitors at least surprising. Some dinosaurs of the past have possessed formidable weapons in the form of blunt and heavy clubs on their tails (*Ankylosaurus*) or spiked clubs (*Stegosaurus*). Like in water-living forms, the preferred tail movements seem to be lateral in *z*-direction, rather than up and down. An interesting idea has been proposed by Warsinger (2018): the author assumes (on the basis of fossil finds) the existence of a dermal sail, hanging down from the ventral surface of the neck, trunk and long tail. According to his calculations, such a sail would be suited to dissipate enough body heat to allow a huge sauropod to be active even in extreme environmental

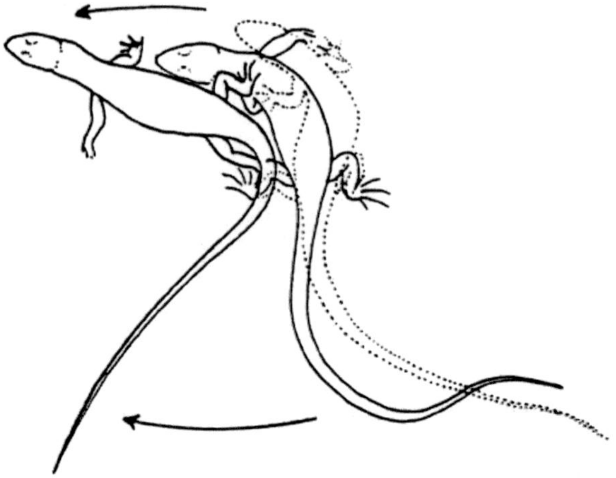

Fig. 4.16 Use made of the long tail for a rapid change of direction in a small lizard. If muscles of the left side (lateral body wall, tail) contract, they move at the same time the anterior as well as the caudal half of the body. The mass moment of inertia (J) of the short but heavy head, neck and trunk is counterbalanced by the less heavy, but longer, tail

temperatures. This role of a cooling device can be compared to the ears of African elephants.

Lifting of the tail, or just keeping it above the ground, is accomplished by muscles, which possess *long lever arms*, so that moderate forces are enough to balance tail weight. All muscles on the dorsal side act as dorsiflexors of the tail and also possess a strong component in x-direction. In view of the considerable weight and the energy requirements of muscles, passive structures (ligaments) may well have provided the necessary force – though this would depend on the lengths of the tension-resistant ligaments and therefore be relevant alone in particular, predefined postures. The length of lever arms can be derived from the long spinal processes (as illustrated in Fig. 3.6). The dorsal muscles on the most proximal segments also compensate the strong pull of the m. caudofemoralis, which has a ventrally flexing moment (see below). Complete control of the tail is only possible if muscles exist also on the ventral side – which indeed is the case. Their lever arm length is documented by the length of the *haemal processes*. Obviously these processes are less strong than those on the dorsal side. Allowance for changes of the length of the ventral contour in dorsoventral flexion is made by the joints between vertebral centra and the bases of the haemal processes. Both dorsal and ventral processes increase their length in proximal direction, following a linear progression. As discussed in Chap. 2, the diameters in the plane of bending must grow linearly, if a body of equal strength is constructed against parabolic increase of bending moments.

All structures producing tensile force possess longitudinal components and therefore exert compressive forces, especially on the vertebral centra – which therefore must be quite strong (that means provide large frontal surfaces; compare Chapter 3.3.4).

4.6 Anterior/Cranial Cantilever: Neck and Head

The other cantilever of the body stem consists of *neck plus head*. Forms with really long necks appear in the fossil record among the reptiles. All land-living forms possess a shorter or longer, well-developed neck. Among the living animals, the turtles, monitors[2] and birds are equipped with very long and mobile necks. In large mammals, the length of the neck is variable, in small mammals usually short.

In the past, long necks existed in aquatic plesiosaurs, some mesosaurs, pterosaurs and several dinosaurs, especially among sauropods and prosauropods (*Tanystropheus!*). All these necks consist of numerous (more than 20) segments. The large number of segments favours mobility. On this background, the divergence from mammals becomes obvious. Mammals sometimes possess quite long necks but, as is well known, only seven cervical vertebrae.

The shapes of *neck vertebrae* vary considerably. Aside from those with only short outgrows (e.g. in turtles, mammals and birds), vertebrae may have *long spinal processes*, which may even be split like a simple fork (some sauropods, like *Dicraeosaurus* or *Brachytrachelopan*). The distance between the two branches increases proximally towards the neck base. Above the shoulder, the distance becomes rapidly smaller until only one process remains in the anterior thoracic vertebrae. Long spinal processes provide, if connected by tension-resistant structures, long lever arms for balancing head and neck weight against the moments of gravity. This resembles the usual situation in the tails (see above). If long spinal processes are present, and balance is maintained by ligaments, however, movements in dorsal and ventral direction are restricted, though possible to both sides. In the necks of some large mammals, some bovids (bison, gaur) or horses, the spinal processes of the cranial thoracic vertebrae form an elevated withers from which the elastic nuchal ligament and muscles take their origin and span freely the distance to the head (see Chapter 7.7.1). This arrangement takes advantage of long lever arms but also leaves space for neck mobility. If only one "tie" along the midline is available, any deviation of gravitational or other external forces acting on the head or neck implies instability of the neck and must be compensated by additional muscle forces. Two ties, which deviate proximally, reduce the danger of such an instability – a fact well known from technical constructions. On the other side, however, passive ligamentous structures, while reducing energy requirements, impede or block active movement. Muscles, by contrast, can adapt their lengths. So the forked spinal processes are suited for reducing the high energy demands of horizontal neck postures but offer only limited possibilities for moving the neck. This particular shape suggests horizontal neck posture and movements preferably in lateral direction.

The active mobility of the neck is in contrast to the limited mobility of the tail. The biological role of long necks can easily be observed in many extant birds: neck

[2]The freely moved anterior part next to the head in snakes, by the way, is not the neck but rather the anterior part of the (rib-bearing) thorax! The neck of snakes is a quite short segment.

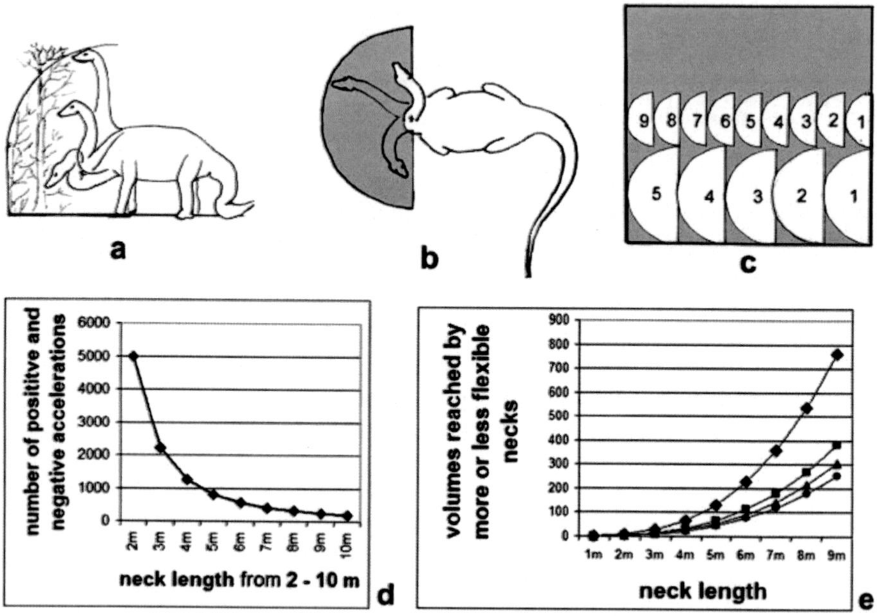

Fig. 4.17 The biological value of a long neck is the great reaching distance (**a**), in side view, and (**b**) in top view. This allows the animal to get hold of very distant resources (e.g. food, growing on high trees), without moving its total body mass. Aside from neck length, the length of the (front) limbs may contribute to reaching distance. Limitations of neck mobility (which indeed exist in the case of, e.g. giraffes, and are also discussed for sauropod necks), reduce the volume that can be controlled and harvested. The volume that can be accessed grows by the third power of neck length. (**c**) Top view on an arbitrarily assumed area which can be exploited by an animal. White semicircles show those parts which can be reached from one position, or "feeding station". In the bottom row, the neck is much longer than in the top row; the number of feeding stations is therefore smaller. A long neck allows harvesting the same area (grey square) with fewer changes of feeding stations. (**d**) The number of changes decreases with neck length following a hyperbolic curve, much faster than a linear curve. Obviously, each change of the feeding station costs energy for accelerating and decelerating total body mass; therefore, a long neck saves energy. If the neck has a certain length (e.g. 6 m), a further elongation does not pay. (**e**) Limitations of neck mobility entail a restriction of accessible volumes, as can be seen from the graphs below the diamonds, which shows a freely mobile neck; squares, triangles and rosettes show limited mobility. (Modified from Preuschoft et al. 2011)

length is a means to enlarge reaching distance of the head, for close inspection as well as acquisition of food, but without the necessity to accelerate and decelerate the whole body mass. Preuschoft et al. (2011) have quantified the selective value of a long, mobile neck in the admittedly extreme example of sauropod dinosaurs (Fig. 4.17a, b). The volume accessible from one point – or the "feeding envelope" (Shipley et al. 1996) – is defined by

$$V_a = 4 * \pi * r^3/12, \tag{4.4}$$

where r is the length of the neck. In an animal standing on limbs, the centre of the volume V_a, that is the basis of the neck, is lifted by the length of the forelimbs (h). Therefore, the volume of a half cylinder must be added to the above value:

$$V_c = \pi * r^2 * h/2 \quad (4.5)$$

Obviously, the lengths of the neck and the frontlimbs not alone increase the volume accessible from one feeding station (Fig. 4.17c); it also gives access to resources located high above the ground. In addition, an important effect of neck length can be made visible by projecting the situation into a horizontal plane (Fig. 4.17b). If we imagine a given area, say 1 ha (100 m $*$ 100 m), covered with ample vegetation. This area is harvested by large animals with necks of variable lengths of, e.g. 1–10 m. As soon as the accessible volume is used up, the animals must change their feeding stations. As illustrated in Fig. 4.17c, animals with short necks must make these changes more often (*top row*, 9 changes) than longer-necked animals (*bottom row*, 5 changes). Dividing the given area by the semicircles that can be controlled without change

$$N = 100 * 100/\pi * r^2 \quad (4.6)$$

yields the number of feeding stations N. In Fig. 4.17d, the number of changes is plotted against neck length. Each change to a new feeding station costs energy for accelerating and decelerating body mass. As is visible from Fig. 4.17d, very long necks are not as advantageous as those of less than, say 6 m. The extremely long necks do not reduce the number of feeding stations to the same extent as those of more moderate lengths do.

The *disadvantage of neck length* are the great bending moments (Fig. 4.18), especially if the neck is approached to the horizontal – as often illustrated and advocated by, e.g. Stevens and Perry, in several papers. None of the living, long-necked birds carries its neck in such a way, and the long-necked giraffes also keep their necks upright. The growth of bending moments from the tip of the snout to the base of the neck is more than linear (as detailed in Chapter 4.5); Preuschoft et al. 2011; Christian and Preuschoft 1996; Christian and Dzemski 2011). Head and neck weights reach so high values that the major share of body weight must be expected to be carried by the frontlimbs. However, the opposite is true: large sauropods leave behind deeper and larger foot than handprints (Salisbury et al. 2017).

The final step for a quantitative trade-off between advantages and disadvantages of neck length is only possible with detailed knowledge about the distribution in space of objects (mostly food) which are important for the animal.

Torsion of the neck becomes most obvious in the famous “death role” of crocodiles (Fig. 4.19a): after biting a victim, the crocodile rotates about its longitudinal axis, while the seized victim, with the jaws of the crocodile, remains in its position or at least offers resistance against being rotated. Strong oblique muscles connecting head and shoulder region keep the neck in balance. These muscles insert

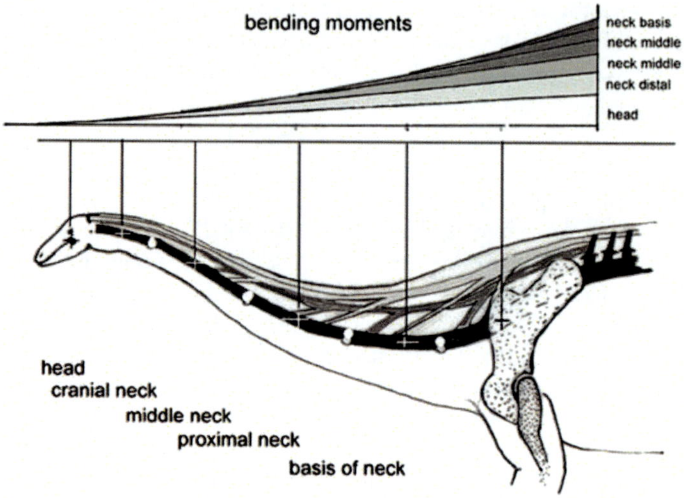

Fig. 4.18 (**a**) Anterior cantilever. (**b**) Bending moments in a long sauropod neck kept in horizontal posture. Bending moments caused by the head (white) grow linearly from the head toward the shoulder region. The arbitrarily assumed segments also possess their own weights, which exert bending moments (light, medium, dark grey). All of them are drawn one on top of those of the head. The addition of all bending moments leads to a parabolic increase of total bending moments toward the basis of the neck. To counterbalance the bending by weight, each segment needs its own tension-resistant ligament (light grey) and/or a tension-producing muscle (dark grey). The lever arms of the load are the horizontal projections of the distances between CoM of the head (or the simplified neck segments) and the base of the neck, hidden behind the scapula. (After Preuschoft and Klein 2013)

into outgrows of the vertebrae: spinal process (mm. trapezius, rhomboideus, multifidus cervicis) and at both sides the ventrolateral processes which are commonly interpreted as "cervical ribs" (with dorsal component, mm. levator scapulae, serratus profundus; with ventral component, mm. scalenus, costo-cervicalis). As can be seen from Fig. 4.19b, both processes possess enormous strength against being bent: the spinal process in the form of the two arches, surrounding the spinal cord, the so-called cervical ribs by their double contacts with the vertebrae (dorsal, diapophyses in contact with tuberculum costae; ventral, parapophyses in contact with capitulum costae).

Even without such a spectacular behaviour as the death role, the long necks of sauropods are exposed to torsion. Figure 4.20a–c shows three examples of a slender-necked sauropod like *Diplodocus* viewed from the top. The laterally flexed distal part of the neck exerts, due to its weight, a torsional moment on the proximal part as shown in Fig. 4.20d–f. The flexed part of the neck is longest – and heaviest – in Fig. 4.20a and shortest in Fig. 4.20c. The straight, proximal parts of the neck are subject to torsional moments of the constant size over the length of the part kept on the x-axis. Acquisition of food in the small-toothed sauropods probably was made (like in grazing and browsing mammals) by lateral movements of the head on the

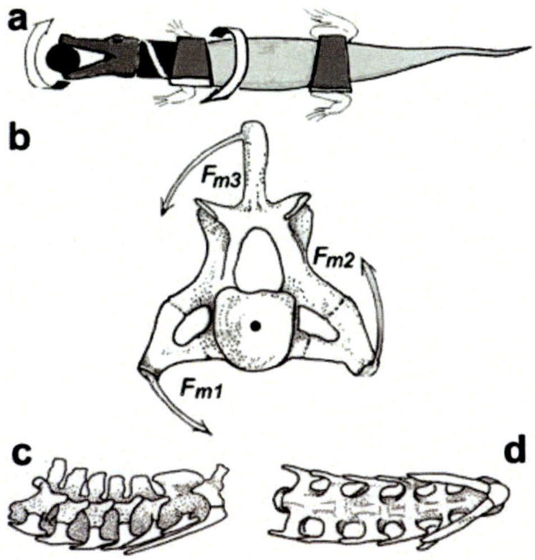

Fig. 4.19 (**a**) Torsion of a crocodile's neck (dark grey) during the famous "death role". The animal rotates actively about its longitudinal (*x*-) axis (white, curved arrow behind the shoulder). The orientation of the shoulder and pelvic girdles is emphasised by the boards (grey) attached to the dorsum. Rotation is opposed by mass moment of inertia of the prey and perhaps by its counteractions. While the teeth and jaws are anchored in the prey, the neck (dark grey) is exposed to torsion. The strong oblique muscles (white band, connecting the head with the sternum) insert into rather long lateral and dorsal outgrows of the vertebrae, as shown in (**b**). The seventh cervical vertebra of *Alligator mississippiensis* provides the long lever arms required for resisting the enormous torques exerted by the animal. In lateral (**c**) and ventral view (**d**), the ventral outgrows can be seen as "cervical ribs". (Modified from Preuschoft and Klein, 2013)

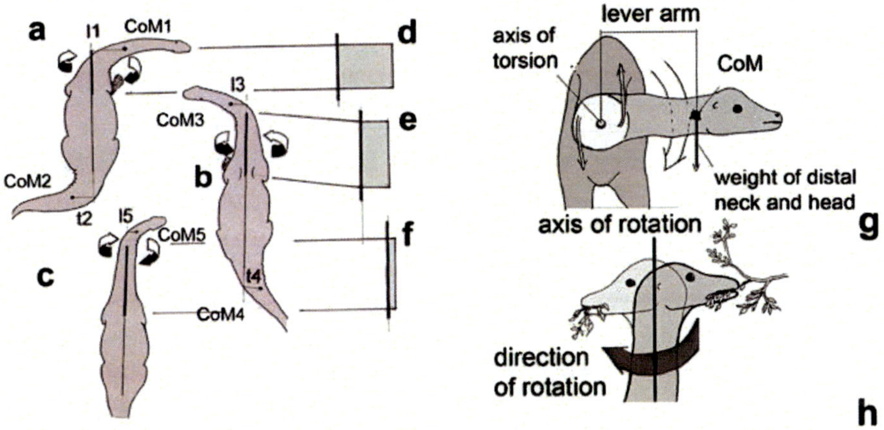

Fig. 4.20 (**a–c**) Torsion of a sauropod neck if flexed laterally. The longer the laterally flexed neck section, the greater the torsional moments along the forward extended, proximal section (hatched rectangles **d–f**). (**g**) A sauropod with horizontally kept neck from in front. It corresponds to (left) position (**b**) in *x*/*z*-plane, now in anterior view (= *y*/*z*-plane). The weight of the head and the distal section of the neck rotate the proximal section (white). (**h**) When ripping off twigs, leaves or fruit, the head may be rotated on the neck, here shown in vertical position. This also causes torsional moments in the neck. (From Preuschoft and Klein 2013)

neck (Fig. 4.20h). This again evokes (admittedly small) torsional moments along the neck.

Although the remarkable length of neck and tail gives the "lower tetrapods" great mass moments of inertia about a vertical (y-) and a transverse horizontal (z-) axis, and therefore high stability, they are able to *change their direction rapidly*, with the help of their long tails. How this is accomplished is shown in Fig. 4.16 for a small lizard as an example.

4.7 Morphological and Technical Alternatives: Turtles and Froglike Anurans ("Salientia")

There exist several forms, in which the technical problems are solved in a way different from what has been discussed:

Turtles possess, in the form of their carapax plus plastron, a sort of "external skeleton" like arthropods, which relieves the characteristic trunk structures, vertebral column and musculature of the body wall. The ribs are also not highly stressed by the mechanical conditions of quadrupeds but are involved in the construction of the armour.

Froglike anurans, the "salientia", recruit parts of the pelvic girdle, that is, the long, rod-shaped ilia, as longitudinal props instead of their extremely short vertebral column. These animals do not possess a tail (their name "anura" saying "tailless"!), a short neck and a rather large (and heavy) head. Their trunk is extremely shortened in x-direction and its diameters in y- and z-direction are long, so that torsion can be sustained without expenditure of great force. This may be the reason why ribs are reduced and the muscular layers of the abdominal wall poorly separated and thin. The advantages of this body shape will be discussed in Chapters 5.10.1 and 5.10.2.

4.8 Essential Results

Tetrapod locomotion on land is phylogenetically old. Because the body is elongate, and supported on two pairs of limbs, it is permanently subject to considerable bending and torsional stresses. Both stress qualities require an axial skeleton (like the phylogenetically older, aquatic forms), ribs and a particular set of muscles, which are maintained throughout all evolutionary changes (but not at all "anticipated" by their primary aquatic ancestors (the musculature of which is segmented)). This is the mechanical reason behind the repeatedly observed "conservative" trunk structure. The extremity girdles can be treated as parts of the trunk, and characteristic, widely spread traits are just adequate responses to the mechanical stress patterns. Many, though not all, of the traits typical for more advanced forms existed as early as the Middle Devonian in a fossil species (*Tiktaalik*) which is assigned to the sarcopterygian ancestors of land animals. Ancient and modern, "primitive" tetrapods preserve the long tail not alone in forms with an aquatic or amphibious lifestyle but also in true land dwellers. Nobody knows why this happens in reptiles, but not in the mammal-like reptiles. A new acquisition of diverse early land animals is the often remarkable length of the neck. Neck length has been kept in several forms that have returned secondarily to living in water. Regardless of the environment, it provides reaching distance for the jaws without the necessity to accelerate the mass of the trunk. Provided that knowledge is available about the used resources, the balance between energetic costs and expenses of a long neck can be calculated. Diverging from the above sketched body plan are turtles and the tailless lissamphibians.

Chapter 5
What Have the Extremities of "Lower Tetrapods" in Common? And Why?

Keywords Three-segmented limbs; Stylopodia; Zeugopodia; Pronation; Autopodia; Ground reaction forces; Flippers; Digits; Horizontal body waves; High walk

5.1 Which Animals Are We Talking About?

Most primitive tetrapods possess a rather slender trunk, neck of short or medium length and a long tail (see Chap. 4), all this in combination with sprawling limbs. Modern forms belong to lissamphibians (salamanders), archosaurs (crocodiles and alligators) and lepidosaurs (lizards, e.g. iguanids, monitors). The similarity and the old age of this body shape and the respective locomotor modes have also been emphasised by Reilly et al. (2006). Recently, Usherwood and Davies (2017) have collected empirical data on duty factors and footfall sequences in many species of amphibians, reptilians and mammals. Our own observations are concentrated on monitors and several other lizards (Christian 1995), as well as crocodiles (Preuschoft et al. 2007).

5.2 Gravity and Ground Reaction Forces, Rolling Off and Segmented Extremities

Land-living, tetrapod animals establish their first, permanent and force-transmitting contacts with their environments through the terminal segments of their limbs, the feet.[1] There are no arguments against, but many very good reasons in favour of,

[1]In view of the functional and morphological divergence between anterior and posterior extremities among, e.g. primates, and especially humans, the discrimination between "hands" and "feet" is common. This divergence does play an essential role neither in primitive four-footed tetrapods nor in more advanced quadrupeds. Nevertheless, it often is appropriate to discriminate between forefoot and hindfoot, without the allusion to the human condition, inherent in "hand" and "foot". Therefore the terms "manus" and "pes" are used. A general, though uncolloquial, term for both is "autopodia".

H. Preuschoft, *Understanding Body Shapes of Animals*,
https://doi.org/10.1007/978-3-030-27668-3_5

trusting that fossil early tetrapods had the same configuration (see, e.g. above, Chap. 4.2.).

It is the permanent *action of gravity*, which presses the animal downward, with a force proportional to body mass. This "weight force" evokes the ground reaction force, which must be resisted by the limbs.

During standing, the *ground reaction force* (GRF), opposite to the weight force, acts against the autopodia, exerting an "impulse" (force * time). If the supporting contact is interrupted (Fig. 5.2), the GRFs must become greater, because gravity acts permanently downward, without interruption, and its counter-impulses (force * time) acting at the autopodia must have the same value. The body is accelerated upward during contact intervals, and acceleration forces add to the weight forces. In more rapid locomotion, the body may be during time intervals completely without ground contact (see also Fig. 5.2 and Chapter 7). Between these phases of aerial suspension, GRF often grows to values of two times body weight – because the body is supported on a reduced number of limbs: often two, sometimes one or three. All this requires greater strength than in the flippers of aquatic animals, that is, reinforcement of extremity bones and their muscles as well as of limb girdles and their muscles. This enforcement of the limbs is combined with the development of digits (toes) instead of flippers (see below).

In many water-living animals, the limbs are fixed to the trunk near the middle of the vertical (*y*-) diameter of the body (Figs. 3.12, 3.13, and 5.2a, b), not shifted toward its ventral contour. Therefore, their movements evoke only limited rotating moments about the *x*-axis of the animal, which would lead to "rolling", and about the *z*-axis (which would lead to "nicking/pitching"). Land-living animals, however, can increase the height of their bellies above ground as well as functional limb length by inclusion of the shoulder blade, so that their shoulder joints are attached close to the lower margin of the trunk. In the hindlimbs, there exists a tendency to integrate the femur into the trunk wall, so that the free extremity also moves essentially about a pivot not in the hip, but in the knee joint, near the ventral margin of the trunk (see below Chapters 6 and 7).

Ground reaction forces in primitive forms like *Tiktaalik* obviously have long lever arms in *z*-direction, due to the breadth of the trunk and to the abducted stylopodia. This "wide-gaged", or "sprawling" (Figs 4.1, 4.7a, 4.10, 4.36, and 5.1) posture and locomotion is common among lower tetrapods. It is safe, because the area of support of the animal is wide, and therefore tilting laterally not at all probable. It requires, however, powerful torsion-resistant trunk muscles and contractions of the major adductors, the m. pectoralis and the mm. adductores femoris. Their activity leads to high compressive force along the ventral side in *z*-direction, to keep the shoulder and the hip joint in balance. A strong compression-resistant structure is needed and indeed present: the coracoid and its derivatives in the shoulder and the pubic plus ischium in the pelvic girdle (Figs. 4.1, 4.4, and 4.9). If GRF has a component in *z*-direction against the midsagittal plane its lever arms at the hip joint become shorter, and at the knee joint longer (Fig. 2.39c, showing, aside fro the lateral snap, also the change of lever arms).

Wide-gaged tracks are frequent, but by no means obligatory for animals with sprawling stylopodia (or a wide chest as in some mammals, like bears, or primates).

Fig. 5.1 Examples of living "lower tetrapods": Komodo dragon seen from in front (**a**) Redrawn from an illustration in FAZ 27.11.2014 RT and from behind on its wide-gaged track (**b**) thanks to the courtesy of R. Gössling). (**c**) Walking crocodile in side view. Note the diagonal support and the overall proportions which have probably been similar in the fossil tetrapods since the Carboniferous and Permian. From Preuschoft and Christian, 1999

This is perfectly documented in chamaeleons able to climb along thin branches, but can also be observed in amphibians and reptiles, or in bears which turn their forefeet often medially. In those cases (Fig. 4.13a), the torque in the shoulder joint is reduced, but instead a torque occurs in the elbow joint. This point should generally be kept in mind: any inclination of GRF has consequences for the torques exerted at the limb joints, by changing the lever arms of the intersegmental joints.

Since the whole-body CoM moves forward in locomotion, and each autopodium remains on the ground, the ground reaction forces also act first against the posterior part of each autopodium, then to its middle and last to its most anterior part. The term "roll-off" has become usual for this phenomenon.

5.3 Flippers Versus Segmented Limbs

According to the original descriptions (Daeschler et al. 2006; Shubin et al. 2006; Ahlberg and Clack 2006), *Tiktaalik* possessed lobated flippers and fins with numerous rays ("lepidotrichia"; Fig. 4.2). This is in line with all other fossils of the same time and evolutionary stage. The rather long fleshy pedicle gives the spread fin high peripheral speed (v_p – provided that the muscle moments are the same), because

$$v_p = \text{angular velocity} * \text{radius}. \quad (5.1)$$

Because (see Introduction, Chapter 1.2) water resistance grows with the square of (peripheral) speed, the area of the fin can be limited in size, without a decrease of efficiency. Pushed against water, the large areas provided by elastic fins evidently evoke much resistance and are therefore more efficient than other forms of autopodia. The relatively rigid fins of aquatic animals perform their most pronounced excursions in the joints closest to the trunk.

Such a flipper can also be placed flat on the ground and so carry a part of body weight (Fig. 4.13b and 5.2b). To fulfil this function, a *segmentation* is advantageous

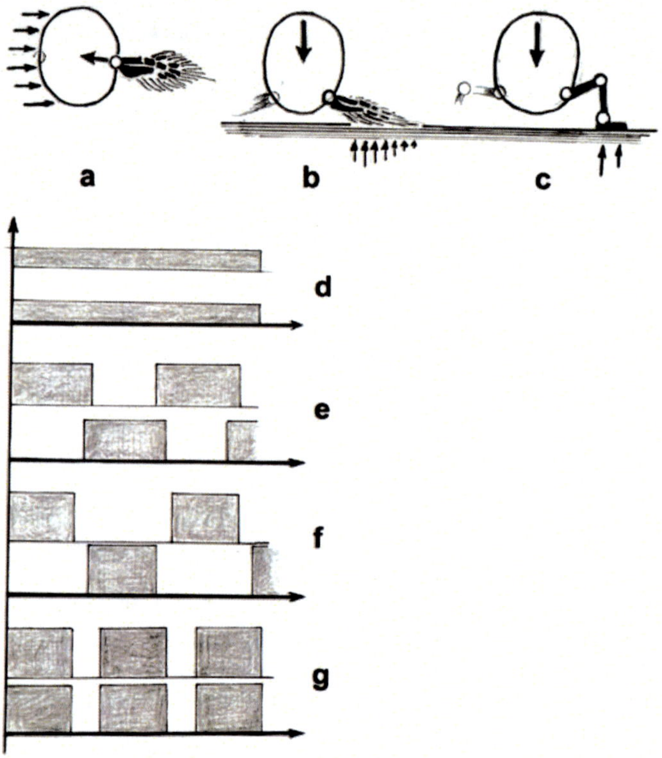

Fig. 5.2 Function of extremities. **(a)** The lobate flipper on left pushes the weightless body against water resistance to the left. Arrows indicate direction of forces: F_l, exerted by the limb; F_e, exerted by water resistance. Position of the extremity in the middle of the trunk distributes the resisting force evenly and avoids rotational moments about the *x*-axis. **(b)** Under the influence of gravity, the same body is on a firm substrate supported by the extremities, which are pressed against ground, each distributing a share of body weight (F_b) over the surface of the fin (F_e). **(c)** Mobility of the trunk against ground is further improved, if the extremity has three segments, the most distal of which, the autopodium, concentrates GRF (F_e) on a footprint. In **(b)** and **(c)**, lifting the body above ground is most efficient, if the extremity is near the ventral margin of the trunk. **(d)**–**(g)** Ground reaction forces (GRF, vertical) over time (horizontal), which means the impulses *P*, in standing **(d)**, walking **(e)**, trotting **(f)** and bounding **(g)**, drawn schematically for left and right limbs of a pair. The frequencies are assumed arbitrarily, so that the time axes are not identical. In standing, weight is distributed evenly on both limbs; striding is characterised by overlapping supports on left and right limbs, while each of them carries two times the weight as in standing. In faster gaits, ground contacts are interrupted, in trot and pace-trot shifting between the sides, in bounding synchronous for anterior or posterior limbs. The shorter the ground contacts, and the longer the intervals without, the higher become the GRFs

into stylo-, zeugo- and autopodium, separated by joints: shoulder, elbow and wrist or hip, knee and ankle. The ability to prop up its forequarters on the anterior limbs has been attributed to *Tiktaalik* by Shubin et al. (2006) because of the massive forelimb bones and the shape of the wrist joint, which indicates an ability to dorsiflex the joint. Transmission of GRF through a flipper-like autopodium is possible (as can be observed in seals; see above). It distributes the GRF evenly over a large area. On firm ground, however, the large contact area is superfluous and may even lead to lesions. Extremities used for support against the action of gravity, however, need a division into *segments*. Aside from the segment which articulates with the trunk (so-called *stylopodium*), a segment should be placed on the ground ("*autopodium*"). The intermediate segment ("*zeugopodium*") is needed to establish the connection between the others (Fig. 5.2c). A veritable disadvantage of flippers occurs, if the autopodium is *rolled off* on the substrate, that is, when the load on the respective foot is pushed forward (vide Figs. 5.3b and 5.6). For maintaining the shape of the long autopodium, or for maintaining equilibrium in its joints, muscle force is required, or the passive strength of the structure must be enforced. Both cost energy, and the former has the consequence of greater weight.[2] Under these conditions, a shortening of the digital rays is the appropriate means to keep energy expenditure on a moderate level, and reduction of the number of digital rays also may imply some reduction of weight without loss of strength.

5.4 Why Are Digits Directed Forward? Why at All Digits? Why Five Digits (or Less) and Not More? Why Hooves or Claws?

The possession of five digital rays (or less, but never more) is considered to be a characteristic of land-living tetrapods, in contrast to the multi-rayed limbs of primary water-living forms. It seems immediately plausible that fins are most efficient for swimming, if they are large, mobile within limits, streamlined and directed laterally. The position of the most primitive forelimbs with digital rays directed laterally in z-direction is abandoned in the majority of tetrapods, and this can also be observed very early during evolution. Already the wide-gaged foot tracks from the Carboniferous (about 320 MYr old) show the *digits pointing forward*, in the direction of

[2]In this context, it seems worthwhile to have a look at the eared seals, who also possess flippers. When on land, these flippers carry a large part of body weight. Contact to the ground is made with their entire volar surface, but its largest (distal) part does not contribute much to weight transmission and can even be lifted. Below the carpus, seals possess thick pads and horny skin (callosities) (Fig. 3.14c; from Kuschel 1994, unpublished). Especially if moving rapidly, the long digital rays are turned outward, so that the manus rolls off from its ulnar to its radial side, and the GRF does not get exceptionally long load arms. The disadvantage of such a shortening of functional limb length is outweighed by the advantage of saving energy when walking on ground and when swiming by the existence of long digital rays.

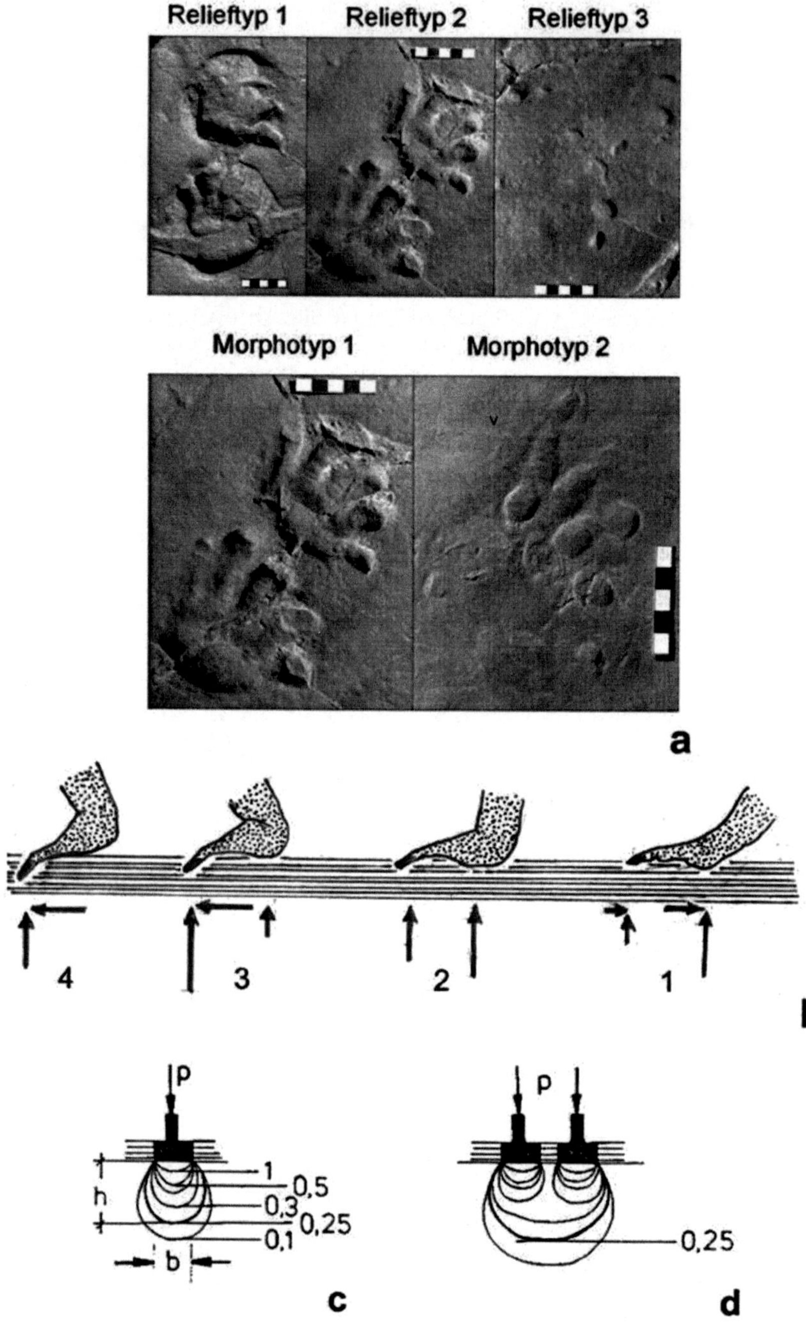

Fig. 5.3 (**a**) Selection of early footprints from the upper Carboniferous/Lower Permian in Thuringia, Germany. (From Voigt 2005). All (mostly five, sometimes four) digits are directed forward and imprinted deeper than the carpal or tarsal pad. Some of the early tetrapods have pointed digits, suggesting claws; others possess thick digital pads. (**b**) During walking, manus or pes are rolled off

locomotion and the body's x-axis (see above: Diadectomorpha; Voigt and Haubold 2000; Voigt et al. 2007; Voigt and Ganzelewski 2010). The foot tracks suggest the existence of three segments of the extremities (stylo-, zeugo- and autopodium). These assumptions are confirmed by the 30-MYr younger body fossils from Tambach/Thuringia. When confronted with this fact, most authors just contend themselves by referring to "evolutionary progress" toward living on land.

Ontogenetic development does not answer the questions raised in the headline. In embryos, the limbs appear as plates on a haft, where the latter at an early stage is subdivided into three segments. The most distal of them is first a uniform plate. The decay of cells in the interdigital spaces leads to a separation of the five digits (Fig. 5.4; Christ et al. 1993), which themselves show a proliferation of cells.

The answers to the above-raised questions (why all these traits are characteristic of terrestrial animals) seem to be contained in the fact that possession of toes rather than fins offers a number of advantages: the total length of a manus or a pes is determined by the length of carpus or tarsus, plus metapodium plus digits.[3] In standing, the position and structure of the autopodium are irrelevant. The sprawling of the stylopodia yields extremely large areas of support for the body, and these are still increased by directing the digits laterally. In walking, however, the autopodium

Fig. 5.3 (continued) from right to left. *First stage*: shortly after touchdown. Vertical force component (arrows) on rear part of palm or sole, moderate braking component against x-direction. *Second stage*: midstance, nothing but vertical force components, which now reach their maximum, distributed over vola manus or planta pedis (between the most posterior and anterior parts of autopodium, depending on its rigidity). *Third stage*: load is shifted toward the anterior part of autopodium. Aside from the vertical weight component, a forward-directed acceleration component is exerted to compensate for the braking at touchdown. Sudden flight, or rapid attack, depends on this component! It can be transmitted reliably, if the anterior autopodium is pressed deep into the ground. *Fourth stage*: immediately before liftoff, the vertical component becomes smaller, while the accelerating component remains as large or is even larger than in stage 3. To avoid slipping of the foot rearward, the most anterior margin should be inserted deep into the ground. This is greatly facilitated, if the GRFs are distributed over smaller areas by splitting the anterior sole into digits. At the tips of toes (black), internal stresses are concentrated and favour the development of hooves or claws. (**c**) Ground pressure under the influence of load p below the fundament decreases gradually; the areas of equal ground pressure are arranged like onion shells around the fundament. If more than only one fundament is close to each other (**d**), the "onion shells" of ground pressure flow together at greater depths. The individual toes are taken as "fundaments". The individual ground resistance of the single toes sums up to one bigger value. The number of five digits seems to be the result of a trade-off between the advantages of separation (small surface, high pressure) and combination of close-by fundaments (strength and high resistance against too deep insertion). The required strength of the toes does also set limits to a reduction of toe diameters

[3]The following hypothesis has the precondition that the most primitive tetrapods evolved in a biotope, which offered soft, compliant and smooth substrates for the toes. This indeed is in line with the fossil documents: the wet and swampy landscapes between water and dry land seem to have been inhabited.

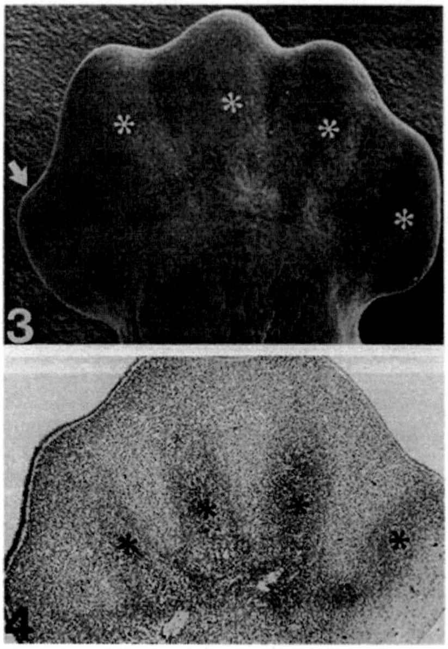

Fig. 5.4 Stages of embryonic development of toes. On top external aspect of a developing hand in a human embryo of about 18 mm, in which death of cells forms the interdigital spaces; on the bottom a section through the hand of a human embryo of about the same size shows the proliferation of cells forming the metacarpus. (From Christ et al. 1993)

is *rolled off* on the substrate (Fig. 5.3; Sub-Chapter 3.3.5) that means the resultant ground reaction force (GRF), which acts between the animal and the ground, travels from the point of the contact surface most caudal (in the x-direction of locomotion) to the point most cranial. This implies consequences: First, the longer the sole of the autopodium, the longer the time interval and the distance through which the foot can contribute to transmission of weight or control the body's movements and to equilibrate uneven structures of the substrate or to initiate changes of speed or direction. The orientation of the autopodia with toes forward elongates the potential "functional length" of the limbs, so that the maximal step length becomes greater. Second, those (caudal) parts of the autopodium which make the first contacts with the ground, as well as those (cranial) parts making the last contacts, are exposed to particularly high loads (Fig. 5.3a, b), depending upon stiffness of the segment. Length of the autopodia under any condition leads to high bending moments along the manus or pes. Balancing them requires high strength, and this is equivalent to increased weight of the limb under consideration.

Even if extreme loading of the anterior part of an autopodium is avoided, because body weight is taken over by, or is shifted to other supports, the ground pressure (=quotient force/area) assumes high values after touchdown and before liftoff. In striding or trotting locomotion, horizontal, braking or accelerating forces occur inevitably (as will be detailed in Chapters 6 and 7). Especially the accelerating force components, which allow rapid flight as well as sudden attack, can be transmitted to the ground reliably by inserting the anterior autopodium deeply into the ground. This is facilitated by splitting the anterior parts of the feet into *digits*, because their area is small and therefore the pressure (F/A) on the individual digit is high. In fact, numerous authors (e.g. Voigt et al. 2012 or Falkingham 2014) observed that toes in fossil, as well as recent tracks (own observations), are deeper impressed than other parts of the feet. In addition, the risk of slipping is distributed over several points. Even if some digits do not find secure contacts, others may provide reliable hold. Precondition, however, is a sufficiently great distance between the digits. If this distance (which depends on the properties of the substrate) is not reached, the effect of the anterior foot on the ground is the same as that of a continuous plate. The same argument holds true in cases of belly dragging (in the sense of Nyakatura et al. 2013 and Curth et al. 2014) where no vertical components are to be transmitted through the autopodia, but alone propulsive (=accelerating) forces must be exerted against the ground. From experiments in swimmers (Löhr and Ungerechts 1976), it is known that a slight spreading of the fingers increases water resistance of the hand. Similar information is provided by ground mechanics: the forces transmitted from a fundament on the ground do not only spread in vertical but also in horizontal direction to all sides (Fig. 5.3c, d). Areas of decreasing pressure can be compared with the shells of onions. In a rectangular fundament, for example, the stress σ is in a depth h of one and a half of its breadth b, not more than 25% of the maximal pressure below the fundament σ_{max}. If several fundaments are close to each other, the pressed zones ("onion shells") merge and reach greater depths than one coherent unit. Exact shape and diameters of these pressure zones depend on material properties of the ground, aside from the distance between the toes. The fact that tetrapods possess just *five digits* (in contrast to, e.g. *Acanthostega* with eight rays), at this time can be explained only in part by mechanics: evidently, the greater the number of digits, the smaller is the share of total GRF to be borne by the individual toe. This implies a less deep insertion into the ground (see above). The number of five digits seems to have fitted to the size of feet and the properties of the underground in early tetrapods. In more advanced forms, this number is often reduced to four (many rodents, carnivores), three (odd-toed ungulates, like tapirs or rhinos; most birds), two (even-toed Bovidae, Cervidae and Camelidae) or even one (horses). A reduced number of digits leads to an increase of bending moments along each toe and therefore require particular strength (which is equivalent to weight). A spreading of long digits, so that some of them deviate from the x-direction of locomotion, opens the option to transmit forces in different directions between animal and substrate without exposing them to extreme bending in mediolateral (z-) direction. If the external ground reaction acts along the toe, it is exposed to compressive force along its axis.

Anyway, the GRFs acting against the anterior part of the autopodium may assume high values. In the forelimb this is because of its position close to the whole body's CoM; in the hindlimb most marked during accelerations, when use is made of the above-noted "functional length". High GR forces lead at the (narrow) tips of the digits to particularly high stresses. These stresses meet at the *tips of the toes* in many animals with horny covers, that is, *hooves, claws* or *nails*, structures suited to resist extremely high loads. When moving on flat and rather soft ground, these terminal covers can be forcefully inserted into the ground (Fig. 5.3b) and so provide high resistance against being pushed rearward instead of propelling the body forward.

The roll-off of a manus or pes on the ground can easily exclude the toes and be confined to parts between the carpus or tarsus and the basal joints of the phalanges. A contraction of the toe flexors presses them against the ground and shifts the external GRF distally. Establishing an equilibrium at the basal toe joints requires muscle force and correspondingly strong muscles, located at the autopodium or zeugopodium. The longer the toes, the stronger – and heavier! – must be the flexor muscles (more details about toes and fingers will be given below under various headlines). Zeugopodia therefore become more massive and heavier. Tendon force and external load combine to a resultant, which must be sustained by the bony digit, if it shall not break under everyday loads. Shortening of the digits reduces the moments of the external force and leads to lower demands on muscles. This shortening can take place (like in the toes of humans in comparison to those of non-human primates) as a morphologically given trait or as a dorsiflexion of the basal and plantar flexion of the middle and distal toe joints. Such a "digital arch" can be seen in many tetrapods (e.g. rodents, carnivores; see also Chap. 8) and has been emphasised by Kümmell (2009) in the case of Permian synapsids.

Decisive for the stressing of digits are the *flexor tendons*, which keep the joints in balance against being dorsiflexed. These tendons in lizards and marsupials (*Varanus* and *Didelphis*, Landsmeer 1949, 1955, 1987, 1993) split from a common aponeurosis of the undivided flexor muscle. The (flexing) force exerted on each ray depends on the length and number of the collagenous fibres and is morphologically determined. Therefore, the deformation of the individual digits can be predicted under consideration of these data and the modulus of elasticity. This invariable determination is replaced in several higher evolved mammals (e.g. primates; see Chapter 8) by a longitudinal separation of the flexor muscles into a superficial and a deep layer as well as separate bundles to individual digits. Individual control of all joints is possible with this arrangement of muscles only, if not more than *three phalangeal segments* exist (see Chapter 8 on primates). A limitation of lateral toes to three, in the first ray only two segments can be observed in many synapsids (Dicynodontia, Therocephalia, pes of Cynodontia). Details of shape suggest the assumption that this arrangement has gained ground gradually against multi-phalangeal forms (Kümmell 2009).

After all, we may expect a *trade-off* between the number of digits, their length, distance between them (degree of spreading) and strength of the substrate on which the earliest tetrapods have been living. The existence of an optimum for the number of digits and their spreading is probable, but will depend on the properties of the ground. Nobody has yet searched for it, and so the number of five digits, each with only three phalanges, cannot be explained satisfactorily from a functional point of view.

5.5 Consequences of Sprawling Limb Positions?

5.5.1 *Zeugopodia and Abducted Stylopodia*

Among the most characteristic and most widely spread features of "lower tetrapods" is the sprawling posture of their limbs (Figs. 4.5, 4.10, 4.13, 4.16 and 5.1). It is in essential due to an abduction of the stylopodia in z-direction, while zeugopodia are, when loaded, kept in more or less vertical postures with elbows and knees in anterior view perpendicularly above the manus or pes. In the posterior limb, the orientation of the toes forward implies some (moderate) twisting of the lower leg. The toes easily maintain their forward direction, if the knee is slightly swung cranially, as in fact can be seen in most tetrapods. In the forelimb, however, such a rotation of the forefoot would result in a swing of the elbow joint forward toward the head. This is not realised and indeed would entail several mechanical disadvantages (see below). In fact, the elbow in most (except turtles) living forms has a tendency to point rearward. Combining a rearward-directed elbow with forward-directed toes is only possible if the stylopodium is twisted about its long axis, a movement called "pronation" (Fig. 5.6): the distal radius is rotated about the distal ulna, while the proximal parts of both bones maintain their position to each other.

Abduction of the stylopodia allows easy lowering of the body stem (for resting) as well as its lifting from the ground or keeping it at the same height in spite of uneven ground. Abduction also permits and facilitates rapid movements of the head plus neck, or the tail in lateral direction, without losing balance (compare Fig. 2.36). Elbows and knees are pointing primarily in lateral direction. Because the zeugopodia are but slightly inclined forward or rearward (Figs. 4.5 and 5.6), there are no great torques to be balanced by muscle contractions, neither in the elbow nor in the knee joint. Their preferred position keeps them more or less free from being bent and reduces stresses to compression alone. Empirical data for this posture in lizards and crocodiles have been provided by Christian (1995) and Preuschoft et al. (2007).

During locomotion, the regularly observed force component parallel to the ground in transverse (or z-) direction leads to shorter lever arm lengths of the GRF in the shoulder and hip joint while requiring activity of the extensor muscles of elbow and knee joints (Figs. 5.8 and 5.9). The sum of all torques at the big joints so becomes smaller than in a vertical orientation of the GRF (see Christian 1995).

At least at rest, the anterior autopodia are often *placed closer to the midline* than the elbows or knees (frogs, lizards). A rotation of the digits toward the midline (salamanders, frogs) also reduces the lever arm of the ground reaction force at the shoulder or hip joints. This advantage, however, is opposed by the disadvantage that torques in the elbow, knee and carpal/tarsal joint must be balanced by muscle force. The hindlimbs, by contrast, are often more abducted in resting postures than during locomotion, with toes pointing laterally. This posture widens the area of support for explosive takeoff, but may also well be a sort of comfort behaviour, namely, the consequence of stretching of muscles which are shortened and relaxing muscles extended in other postures. The distribution of torques over several joints, instead of a concentration at only one joint, requires a distribution of muscle mass on the free extremity. Neither possible advantages nor disadvantages of these alternatives have been investigated systematically.

Anyway, the more or less horizontally kept stylopodia need the action of very strong adducting muscles. In the frontlimb, this is primarily the m. pectoralis, which balances the shoulder joint. If the elbow points rearward, this muscle also gets a moment as protractor. In the hindlimb one of the strongest adductors is the m. caudofemoralis, which connects the transverse processes of the proximal tail vertebrae with the femur (Gatesy 1990). This muscle also has a considerable moment as a retractor of the hindlimb, especially if the knee points forward. Christian (2007) (see also Christian and Garland 1996) has estimated the muscle mass necessary to balance the sprawled extremities in monitors of increasing sizes. Because of the huge moments in the shoulder and elbow joints, the muscle force must increase – of course by increasing its physiological cross section. So the diameters of the upper arm (segment lo) eventually become longer than its length. Christian (2007) has drawn the conclusion that the fossil giant monitor *Megalania* cannot have exceeded essentially the size of a *V. komodoensis.* Preconditions of this calculation were the body posture and locomotion of living monitors. Anyway, the much bigger size of crocodiles is a strong argument against the premature conclusion that sprawling gaits are confined to small- or medium-sized animals.

5.5.2 *Locomotion on Sprawling Limbs: Why Are There Horizontal Body Waves in Lower Tetrapods?*

The oldest >300-MYr-old foot tracks (see Chap. 4, Fig. 4.1) are obviously wide-gaged and so suggest sprawled extremities – as well as the earliest body fossils (Diadectomorpha (see above) and Pelycosauria/Cotylosauria from the Latest Carboniferous and Permian). This fits well to common ideas about their derivation from semi-terrestrial forerunners (see Chap. 4.4).

Sprawling limbs provide a wide area of support, and the length of the trunk provides a long area of support (Fig. 5.5), so that these animals can move their CoM within wide limits without the danger of toppling over. In addition, the CoM is low above the ground, and therefore the animals have a considerable cross-country mobility. This is of outstanding importance in the case of, e.g. small lizards. Even minor elevations of the ground, stones, or exposed roots of low, ground-covering plants, which seem to be completely unimportant from our point of view, are, for a small animal, difficult-to-overcome obstacles. Larger representatives of this group often possess impressive abilities as climbers (Varanidae, Iguanidae, Agamidae), which can cope not only with thin branches but also with smooth, tall stems of high trees (Sweet 2007; see also Chap. 4). Morphological "adaptations" to climbing can be seen in sharp claws, suitable for hooking behind roughnesses of the bark, or in the mobility of their long toes. Geckos possess in addition special organs for adhesion

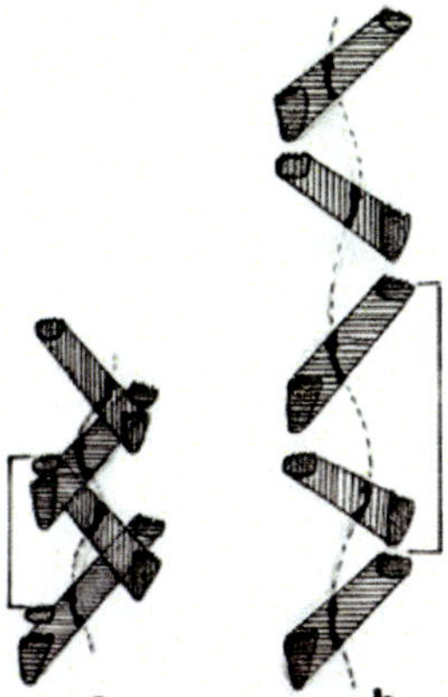

Fig. 5.5 Areas of support in a typical modern reptile (hatched horizontally or vertically). The broken lines show the pathways of the CoM, which are redirected during each stance phase (black arrows). The position of the CoM closer to the hind- than to the frontlimbs is chosen as in recent Lepidosauria (in mammals it is closer to the front- than to the hindlimbs, because of their heavy heads and reduced tails) (**a**) Slow, striding locomotion, in which hindfeet are placed behind the frontlimbs. This represents the same type of locomotion as those in Fig. 4.1. (**b**) Fast locomotion with "overstepping", which means placing the hindfeet in front of the anterior feet. Step lengths of left or right frontfoot are indicated by the vertical bar; in (**b**) step length is greater and includes a phase of aerial suspension. The width of the track is often narrower in faster locomotion

on their hands and feet. The overall body shape and position of limbs are maintained in these forms. Chameleons, however, usually sprawl their limbs in the same way as other reptiles, but place their autopodia very close to the median sagittal plane of the body when marching along slender twigs. As a means to counter imbalance in lateral (z-) direction, they are using the prehensility of their autopodia. This proves that very narrow-gaged tracks are also within the range of competence of reptiles and the width of tracks is obviously independent of the sprawling position of the limbs – a fact often overlooked.

The energetic costs for transportation of each kg body mass per unit distance are in reptiles not higher than in mammals of the same size (Schmidt-Nielssen 1984). The observed locomotor speeds are in small lizards (body mass < 1 kg) sometimes higher than in mammals of the same size (Christian 1995, 2007). The mass units of each limb follow during each foreswing an elliptic pathway (Figs. 5.1 and 5.8c). Because partial masses of limb segments must not be slowed down or accelerated like in the pendulous limbs of mammals, the limbs can be moved very rapidly without increased expenditure of energy (Christian 1995), so that high frequency of cycles is possible and in fact is used to increase speed.

Locomotion is performed by placing one or two "swing" limbs forward (Figs. 5.1, 5.8, and 5.10), while the others support the body in a "stance" phase (Fig. 5.8). In the majority of cases, the foreswing as well as the stance is performed by diagonal limbs at the same time. Limbs of the right side move in symmetry to those of the left. Fine-grained analysis shows an inobvious leading of the hindlimbs, which touch down slightly before the diagonal forelimbs. If support by one diagonal overlaps with that of the other, the so-called "striding" gait or "walk" is rather slow, but safe. The "diagonal sequence of footfalls" is maintained even if the ground contacts at higher speeds become shorter, so short (Figs. 4.1 and 5.2) that phases of free aerial floating occur. This characterises the "trot", which is per se less safe than the walk, though minor imbalances can be compensated by the rapidly following footfalls of the diagonal limbs. While high frequencies are used to make locomotion faster (see above), the stride length in living lizards is constant, and the excursions of the limbs are rather wide. The same of course holds true for the angles between the limb's functional axis and ground during the stance phases.

Springing gaits with asymmetric footfall sequences have sometimes been observed in smaller or younger crocodiles (Hutchinson, pers. comm.), never in other reptiles nor in amphibians.

As characteristic as sprawling limbs are the wave-like movements of the body axis in low tetrapods. They are commonly connected with the undulating movements of the body axis in primary aquatic animals. There exists, however, an important difference (Alexander, personal communication): while the undulatory waves in fishes and other water-living forms travel from the head on rearward (see above), the "knots" of body waves in tetrapods stay at the same places throughout the whole motor cycle, shortly behind the shoulder and behind the hip joint (Fig. 5.7b; Christian 1995; a deviating description of longitudinal waves has been presented by Ritter 1996). A consequence (but not the cause!) of these standing waves is the shift of the partial CoM toward the anterior or posterior stance limb (Fig. 5.7a). This reduces the lever arm of the load and so saves energy. Muscles which are frequently considered to be "propulsive", that is, the main retractors of the front- and hindlimbs (namely, mm. latissimus and pectoralis as well as the m. caudofemoralis, respectively), indeed exert forces at their origins along the body axis, which flex sections of the trunk axis laterally (Gatesy 1990). While in Fig. 5.6 the movements of the stylopodia in relation to the trunk are shown, Fig. 5.8 and Fig. 5.9 show schematically the left front and right hindfoot during ground contact. While these feet are fixed to the ground, the upper arm, if viewed from above, swings about the elbow and the thigh about the knee joint. Both segments (humerus and femur) of course maintain their lengths. The feet, if pinned down to the ground, force the body to move its shoulder or hip region away from the respective stance foot. The distance, over which the body is shifted, is the cosine of the excursion angle $*$ length of the stylopodium.

$$\text{Shift of the shoulder girdle} = \cos\alpha * \text{length}. \tag{5.2}$$

The force component in transverse (z-) direction is exerted against the body stem. The corresponding component of the counterforce against the ground reaction force (GRF) in Fig. 5.9d has indeed been observed by Christian (1995) and documented in many lizard species by McElroy et al. (2014) – especially in the hindlimbs. It becomes smaller, if the elbow or knee joints are less flexed – as in the "high walk" –

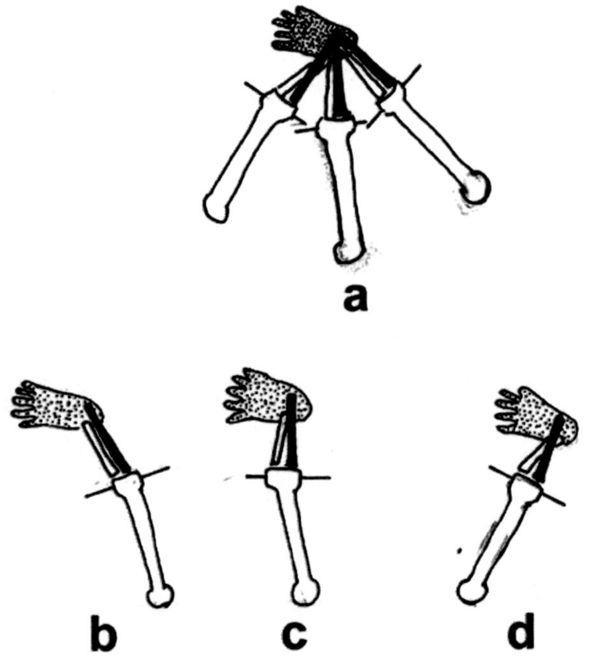

Fig. 5.6 Pronation of the anterior zeugopodium in walking to the left. (**a**) The movement of ulna (black) and radius (white), as well as humerus (white), around the standing manus (grey) in ground contact. The more the elbow joint is moving forward, and the more the forearm is inclined in the direction of locomotion, the more pronounced the twisting of the forearm bones. (**b-d**) zeugopodia, drawn separately in three phases of movement: (**b**) At touchdown, (**c**) midstance and (**d**) before liftoff. While swinging forward (from right to left), the twisting of the forearm bones is reduced

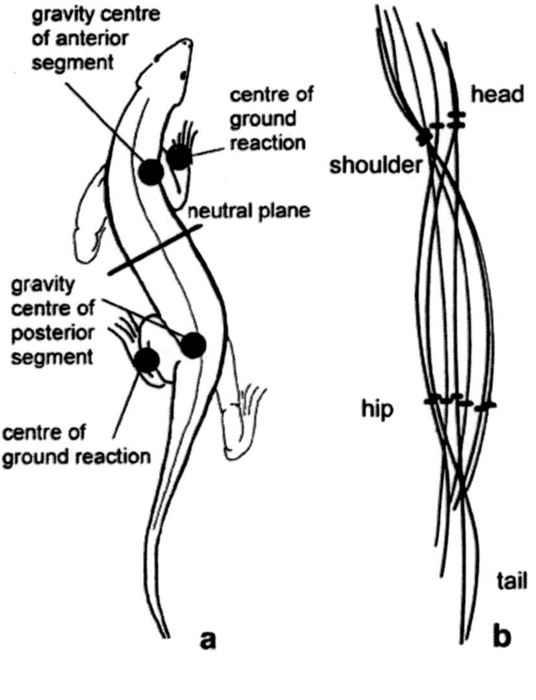

Fig. 5.7 (**a**) Walking monitor lizard, seen from on top. Cranial as well as caudal half of the body is curved around the supporting limb. The CoM of each half is approached to the supporting front- or hindfoot, so reducing the lever arm length of the load carried by the respective limb. The area of support in the illustrated phase extends between LH and RF (compare Fig. 5.5). (**b**) Superimposing the longitudinal axis of the body from several motor cycles shows the "standing wave" during locomotion. (From Christian, 1995)

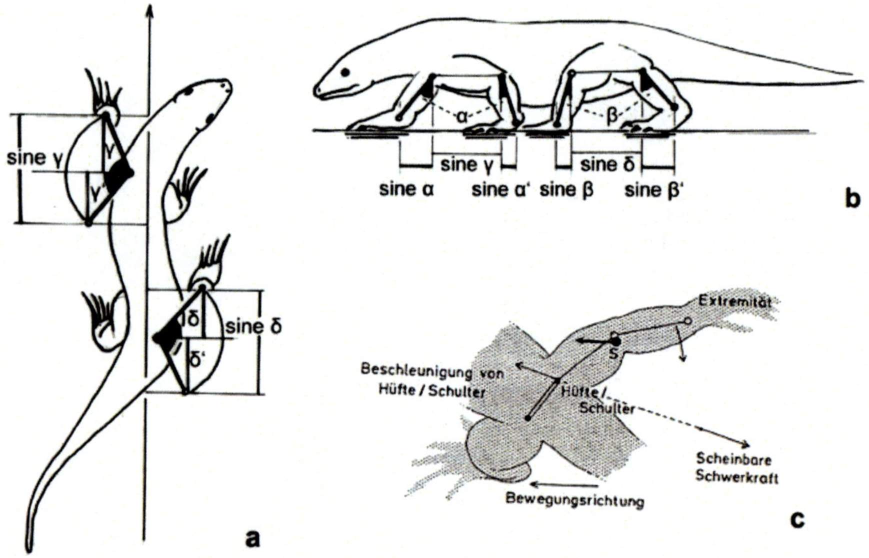

Fig. 5.8 Top view (**a**) and side view (**b**) of a schematised low tetrapod. Stylopodia are more or less moved through the angles γ and δ in a horizontal plane, and an inclination of zeugopodia is visible alone in side view as the angles α and β. Step length is the sum of excursions of the stylopodia in the glenoid and hip joints and of the zeugopodia in the elbow and knee joints ((**b**); see also Eq. 5.4). As visible in (**a**), the excursions of the joints follow the sine of the angles γ and δ between z-axis and stylopodium, the latter the sine of the angles α and β between the vertical and the zeugopodium. (**c**) Stage of the foreswing of right hindlimb to illustrate the elliptic pathway of the autopodium. The trajectories of the feet during foreswing is estimated in Fig. 5.1a (**c**) taken from Christian, 1995

and the force exerted pushes the body stem upward, rather than laterally (Fig. 5.9b, c). The upward movement can be used for improving the energy balance of locomotion (see below).

Locomotor speed (v) of a tetrapod is determined by.

$$v = \text{step length} * \text{step frequency}. \tag{5.3}$$

The same importance for economical locomotion (as step frequency) has step length. If step frequency is governed by the laws of the pendulum (see below in Chapters 6 and 7), it also depends on limb length. Raised frequency requires overcoming of the inertia of the swing limb. Frequency as well as length can be increased only by the input of much energy. As can be seen from Fig. 5.8a and b, the length of a step (s) is the projection of angular excursions of the proximal joint (α and β) in top view plus the excursions of angles γ and δ in side view:

$$\text{Step length } s = \sin\alpha * l_o + \sin\beta * l_o + l_u * \sin\gamma + l_u * \sin\gamma, \tag{5.4}$$

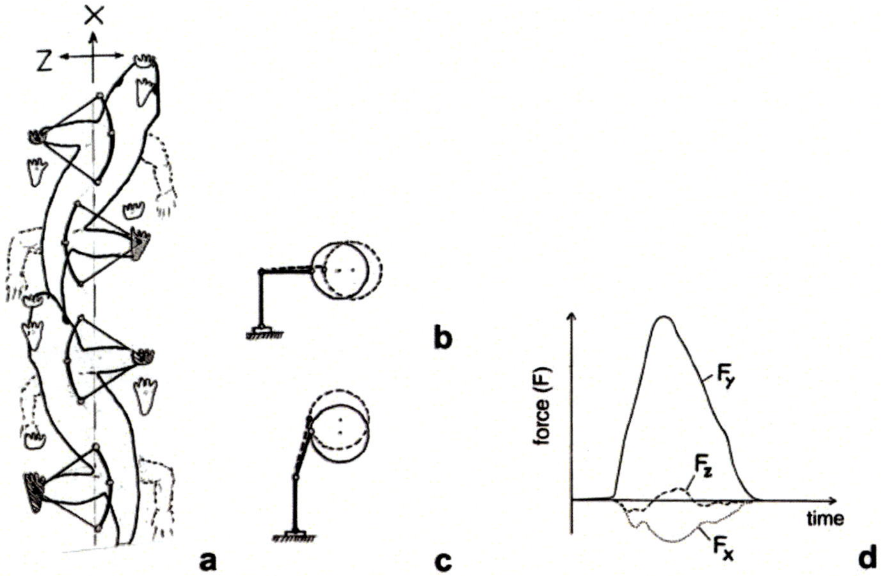

Fig. 5.9 (**a**) Two consecutive support phases of walking in a reptile. Because of the fixation of the autopodia to the ground, the +/− horizontal stylopodia swing the shoulder or hip joint away from the respective stance limbs, forcing the body stem to give way in the direction of the transverse force component in *z*-direction. The distance in *z*-direction between the elbow and knee joints and the median sagittal plane is greatest at midstance. This alternating giving way leads to the "standing wave" of the body axis. The effect is most pronounced if the excursion of the stylopodium is large. (**b**) Only if the stylopodia are kept close to the horizontal, the body stem is pushed to the other, swing side (broken outline). Because midstance phases of diagonal feet are reached at the same time, the body axis performs wave-like movements. (**c**) In the "high walk" with wider angles between stylo- and zeugopodium, the former pushes the body upward, not laterally. This allows making use of the stored potential energy by transforming it into kinetic energy as detailed in Chap. 7. Both movements can be neutralised by flexion or extension of the elbow or knee joint – what, however, costs muscle force! (**d**) GRF of a slowly walking monitor lizard measured by Christian (1995). Note the high transverse force component in z-direction, which is a consequence of shifting trunk mass to the other side

where the angles α and β are measured between x-axis of locomotion and stylopodium in the shoulder or hip joint and angles γ and δ between the vertical through the elbow or knee joint and the zeugopodium.

Aside from walking with sprawling limbs, the "high walk" (Figs. 5.8 and 5.9, see also Reilly and Elias, 1998) has been emphasised repeatedly in crocodiles which walk or run rapidly. Christian (1995), as well as own unpublished data, document that other reptiles (monitors and iguanids) also extend their elbow and knee joints when moving at high speeds. This extension lifts the head, neck, trunk and proximal tail higher above ground than is realised at slow walking speed. Their swing limbs are moved, as in slow walking, on an elliptic pathway (Fig. 5.8), in contrast to the pendulum-like fore- and back-swinging limbs of mammals. This swinging forward of the unloaded limbs is resisted by its mass moment of inertia and subject to the

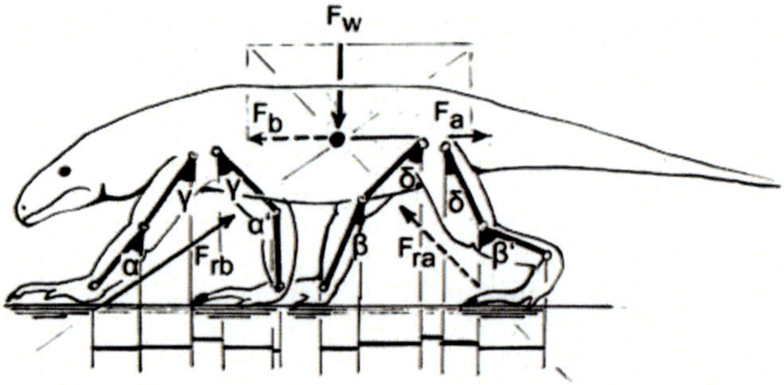

Fig. 5.10 Crocodiles as well as monitor lizards, when moving at higher speeds, extend their elbow and knee joints (angles α and β), so approaching their stylopodia to the parasagittal plane ("high walk"). This permits using flexion and extension of the elbow and knee joints (angles α and β) for increasing step length – or reaching the same step length with less excursion of shoulder and hip joints (angles γ and δ). Experience shows that shoulder and hip joints are moved somewhat forward and rearward. If this is neglected, and the same excursions in angles γ and δ are used as in Fig. 5.8, step lengths have the same size, and the lateral excursions of shoulder and hip are diminished. Energy can be saved because of the wider opened angles in the elbow and knee joints (angles α and β). Not only weight force (F_w) acts at the body's CoM in the middle of the trunk, but also inertia. In accelerating the body, e.g. from a resting position, inertia (F_b, solid arrow) tries to keep the body where it is, which means it is directed rearward. The resultant of weight and inertia runs toward the hindfeet, where its counterforce appears as GRF (F_{ra}, broken arrow). In braking, the inertia of mass (F_b, broken arrow) tries to maintain the speed of locomotion and therefore is directed forward; the resultant of both force components (weight and inertia) runs toward the frontlimbs, where it exerts the GRF (F_{rb}, solid arrow). Under the condition that the elbow points rearward, and the knee forward, both resulting GRFs do not pass too far from the joint axis of all joints (see also Chap. 7, Figs. 7.7b and 7.13)

laws of the pendulum. The velocity of the swing foot on an elliptic pathway, however, is not limited and can become very fast, so that step frequency can be very high.

The high walk attracts so much attention, since it resembles the locomotor mode of mammals. The reasons why the "high walk" is used remain enigmatic. At a first glance, the extension of elbow and knee joints seems to increase the "functional length" of the respective limb (shown in Fig. 5.10) and so seems to imply an increase of step length and higher speed without increasing of frequency. This seemingly plausible idea, however, is in contrast with the facts reported above: if the angles of excursion (α) are the same, the length of a step does not change, regardless of the plane, in which the proximal segments are moved, horizontal or vertical. In all cases, the elbow or knee joint moves over the same distance (s):

$$s = \text{length of stylopodium} * \text{sinus}\ \alpha. \tag{5.5}$$

To this value, the excursions of the zeugopodium must be added (see above).

As can be seen from Fig. 5.9b, c, the more extended limbs lead to lesser excursions in z-direction of shoulder and hip region, so that the lateral deviation of the body axis is reduced.

Another often-mentioned item is the saving of muscle force, or energy, because the limbs are less flexed. The saving of energy by extension of the limbs indeed can easily be observed in elephants (Ren et al. 2008) and is often taken as the decisive argument to reconstruct sauropod dinosaurs with extended extremities. The muscle forces necessary for maintaining the horizontal posture of the humerus and femur are really remarkable. The mechanical aspects are correct, but the argument leaves out of sight the fact that the earliest mammals were small and the data collected by Fischer and co-workers show convincingly that all small mammals are keeping their upper arm and their lower legs +/− horizontally during locomotion (Witte et al. 2002; Fischer and Blickahn 2006; see also Chapter 7). Although these elements are horizontal in a parasagittal (x-direction), not in a transverse plane (z-direction), the keeping of balance in the adjacent joints requires not less energy than the ancestral sprawling position.

5.6 Why Are Elbows Turned Rearward, But Knees Forward?

Own observations and video clips confirm the generally known fact that most squamata and crocodilians tend to direct their abducted elbows rearward and their (also abducted) knees forward (Figs. 5.8 and 5.10). In moderate expression, this can also be seen in amphibians (Urodela and Anura). This approaches the posture of limbs to that typical of mammals. This position is often reported in textbooks and sometimes discussed (see, e.g. Witte et al. 2002). The literature commonly states nothing but the assumption that locomotion is "improved" by this swinging of the limbs or that it is "energetically advantageous". A mechanically satisfying and complete explanation, however, is missing. One of its consequences is that flexing and extending the elbow and knee joints can be recruited for elongation of steps, whereas in the classical sprawled posture, only rotations about the stylopodia are used to swing the zeugopodia forward (compare Figs. 5.8b and 5.10).

At least a possible answer to the question, "why elbows and knees are pointing in different directions?", can be found in the deviating functions of front- and hindlimbs. Three situations are of importance (see also Christian 1995):

First during locomotion, the frontlimb is used for braking and the hindlimb for accelerating (Fig. 5.10). Doing so, the force flow in the trunk is mainly compressive, rather than tensile, and therefore easy to sustain by passive, not energy-consuming compression-resistant skeletal structures. If speed is to be increased, high horizontal forces in x-direction are produced, to overcome mass inertia. This inertial force is added to weight force, yielding a resultant directed from the CoM toward the hindfeet. If speed is to be reduced, a force in x-direction is produced contra mass inertia, which is superimposed by weight force, resulting in a GRF directed from the CoM toward the frontlimbs.

Second the same effect as braking and accelerating has any inclination of the support. Locomotion upward shifts a greater part of the load to the hindlimbs;

locomotion head-first downward shifts a greater part of the load to the frontlimbs (this will be detailed again in Chap. 8).

Third if a primitive tetrapod attempts to capture a prey, it may do this by lateral snapping (see above under *Tiktaalik*). Because of the inertia of head mass, sprawling frontlimbs are of advantage for lateral acceleration of the head (see Fig. 2.36b and Chapter 2). Likewise, the ripping off of plants also requires activity of the sprawled frontlimbs. But capturing a prey can also be accomplished by accelerating the whole body forward, which requires activity of the hindlimbs. After seizing the prey, the body must be stopped – again by the frontlimbs.

In all cases, the GRF has relatively short lever arms at the proximal, middle and distal joints of the respective limb (Fig. 5.10) – provided that the elbows are turned rearward and the knees forward. These postures clearly save energy.

A yet little observed effect of the tilting of elbows rearward and knees forward is that this posture allows them to remain below the trunk for longer time intervals and keep low the dorsally concave bending of the trunk: the inclined limbs must be kept in equilibrium at shoulder and hip joints, and the activated muscles contribute to dorsally convex bending (compare Sub-Chapter 4.2.1 and Fig. 4.3 as well as Chapter 7.2). Less bending of the trunk reduces compression along the axial skeleton and the muscle force in the m. rectus abdominis. The same effect on the hindlimbs is also brought forth by a reduction of the tail mass and tail length (see Fig. 4.3 as well as Chap. 7). If the tail is less long and less heavy, it does not longer provide upward bending moments in the posterior part of the trunk. The muscles necessary for keeping the hip joint in balance, however, replace this influence if the hindfoot is placed anterior to the hip joint. Both evolutionary trends (swinging the hindlimbs in a forwardly inclined position and reduction of the tail) are known and quite characteristic among the fossil forerunners of mammals (cynodonts). This point will be discussed again in Chaps. 6 and 7. The fossils, however, raise the hen-and-egg question: what was first and what is the consequence? The ultimate reason behind both developments may well be the increased agility of the mammalian ancestors in comparison to the stem reptiles. This perhaps has also included an improved ability to keep balance on very narrow tracks.

5.7 Why Is There No Bony Connection Between the Shoulder Girdle and Trunk in All Terrestrial Animals?

A basic problem of quadrupedal locomotion becomes visible in a simple experiment: imagine two persons carrying a ladder (representing the trunk) in the way shown in Fig. 5.11. Both press one rung tightly against the dorsal sides of their hip bones. The length of the ladder is constant, as is trunk length, because of the axial skeleton. The rigid connection between the two carriers forces them to walk at equal steps. Any deviation from synchrony of steps results in a strong thrust close to the CoM, which

Fig. 5.11 Experiment to demonstrate the consequence of twofold fixation of the trunk (represented by a ladder) to the supporting four extremities. Two persons are carrying a ladder, while each of them presses a rung tightly to his or her pelvis. The ladder's length is constant, like the length of a trunk. To avoid irritating pushes close to the body's centre of gravity, the two persons have to walk at equal steps. Alone if the two carriers walk at equal steps, they do not receive undue thrusts. A free choice of footfall sequence (or the coping with uneven substrates) is only possible if the tight contact between the ladder's rung and one of the carriers is abandoned, that is, uncoupling the body stem from one pair of limbs

disturbs the harmonic flow of movement. A free choice of the footfall sequence, however, is essential for coping with irregularities of the underground, for all methods of locomotion other than marching and for different sizes of the carriers (which is equivalent to diverging limb lengths). Only if the connection between the pelvis of one person (=pair of limbs) and his or her rung of the ladder is loosened, both carriers (=pairs of limbs) are free to choose their own rhythm. In a tetrapod, this uncoupling can be accomplished in different ways: lateral curvatures can change trunk length and so uncouple the movements of fore- and hindlimbs, provided that the length of body axis is sufficiently mobile. This seems to be realised in salamanders, in some lizards and in fossil diadectids. The most common solution among quadrupeds is hanging up the body stem at the shoulder blade by muscles (m. serratus lateralis, as shown in Fig. 4.4). In the swing limbs, the mm. trapezius and rhomboideus preclude a rolling downward of the trunk along the medial side of the scapula and a hanging downward of the swinging limb. These permit, like the supporting arms and hands of one carrier, some movement between the trunk and forelimb.[4] An element of the shoulder girdle, the scapula, extends from the glenoid joint in dorsal direction. The vertebral margin of this element is the insertion of those muscles which lift (in walking) an extremity from the ground and swing it forward, mm. rhomboideus and trapezius – or cucullaris. On the medial side of the scapula, the muscle takes its origin, which carries the vertebral column, the head and indirectly the whole trunk: the m. serratus. These muscles are very conservative among tetrapods. Giving them divergent, special names just camouflages their widely spread existence. The dorsal extension of the pelvic girdle, the ilium,

[4]Technical constructions (bridges) in which a beam is supported by two supports also provide some limited mobility in longitudinal direction (e.g. because of bending under load or temperature-dependent length changes).

establishes the rigid (and energy-saving!) contact between the limb and the axial skeleton.

That the anterior (shoulder) girdle is mobile, while the posterior (pelvic) girdle is firmly connected with the vertebral column, seems to be due to the necessity to initiate rapid flight by the hindlimbs (see Chap. 7.2). A rigid connection between the hindlimbs and the axial skeleton transmits the GRF without delay and without input of energy on the trunk. In braking, however, the GRF is transmitted by well-synchronised muscle activities, which permit some damping (see below Sub-Chapter 7.5.5).

The influence exerted by the position and muscles of the shoulder girdle has been investigated with the aid of the FE method by Hohn-Schulte 2010, 2013 (Fig. 5.18).

5.8 Bipedality in Lizards and in Dinosaurs

In many fast-moving lizards, the hindlimbs are elongated and therefore rather heavy. Hindlimb length is advantageous, because it allows great accelerations – what is essential for animals with the lifestyle of "escapers"/accelerators as detailed below in Chapter 5.9. These long hindlimbs – together with the long tails – shift the body's centre of gravity rearward, toward the hindlimbs (Christian et al. 1994a). At each step the long hindlimbs are placed close to or below the CoM (Fig. 5.12a-c). The longer these limbs, the greater the overall share of body weight they carry, since their lengths give them long stance phases. So the impulses (*P*) they contribute to locomotion become increasingly greater than those of the forelimbs, which at the same time are relieved, because the impulse of total GRF cannot exceed a value given by the permanently acting body weight (see above Chapter 5.2). Mainly because of their different own frequencies, the coordination between short forelimbs and long hindlimbs becomes increasingly difficult, the greater the deviation between both (Christian et al. 1994a, b). A solution for all these problems is lifting the frontlimbs entirely from the ground, especially during phases of increasing locomotor speed or accelerations. In fact, quite a large number of lizards run bipedally as soon they are in a hurry (about 1/3 of all available species, mainly Agamidae and Iguanidae; Christian et al. 1994b). In all these animals, the bipedal running does not seem to be connected with any particular known morphological trait (so-called adaptation) except hindlimb length.

For rapidly changing the direction of locomotion, these lizards usually recur to quadrupedalism and make use of the mass moment of their tail to turn their head and trunk into the new direction (Fig. 4.16).

The bipedal posture changes the pattern of bending moments along the body stem essentially (see Fig. 5.13, also Preuschoft and Christian 1999). This becomes particularly evident in the case of the bipedal dinosaurs, no matter whether they belong to the theropods or the ornithischians. The extreme forms seem to be reached in the often huge flesh-eating theropod carnosaurus, like *Tyrannosaurus* and its kin, which have reduced their frontlimbs to surprisingly small appendages (Fig. 5.13),

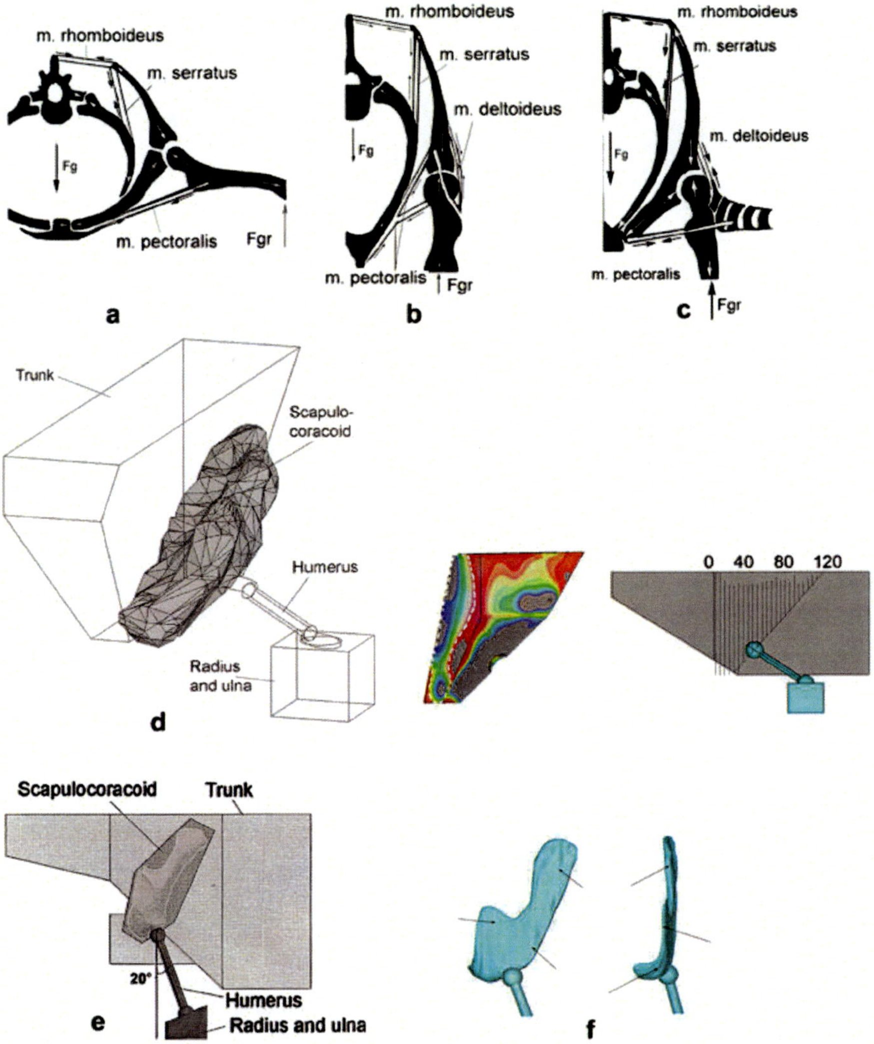

Fig. 5.12 Connection between the trunk and shoulder girdle in the thoraco-scapular junction. (**a**) Anterior view in a crocodile: m. serratus lateralis stretches between scapula and anterior ribs, shoulder joint balanced by m. pectoralis. (**b**) In cursorial mammals, by contrast, all ribs reach the sternum, from which the m. pectoralis takes its origin. It contributes to carrying body weight (F_g), but loses its adducting lever arm. F_g is identical to GRF. (**c**) Hypothetical situation in a sauropod. Scapula fused to the coracoid, extremely narrow shoulder region, ribs connected to interclavicular skeletal element, m. pectoralis acting as a strong adductor. (**d**) FESS model of a crocodilian scapula and coracoid after the second reiteration. White stands for the anterior trunk; grey is the skeleton of the shoulder girdle. (**e**) FESS model of the girdle skeleton of a sauropod. (**f**) Reduced model of scapulo-coracoid, seen from left lateral and from in front. Note the fork-like lateral aspect and the fusion of scapula and coracoid, both traits similar to the fossil (**f**). (From Hohn 2010, 2011).

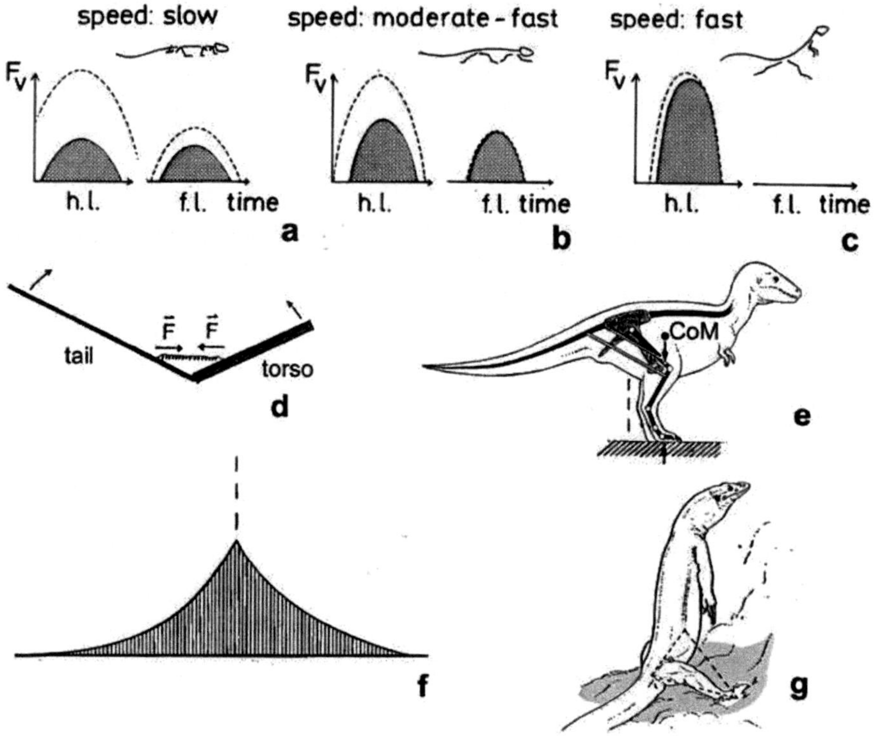

Fig. 5.13 The CoM of lizards (**a**) is close to the hip, because of their long tails; the hindlimbs, being often longer than frontlimbs (**b**), step easily below the CoM, so that body weight is carried to a large extent by the latter. In addition, the long hindlimbs can interfere with frontlimbs, especially because of their diverging pendulum periods. The way out is lifting the forelimbs entirely from the ground (**c**) by becoming bipedal. (**d**) Statics of a bipedally running lizard. The long tail counterbalances the trunk plus head; strong forces are required for dorsiflexion of the body stem. Both segments together provide high mass moments of inertia against rotating about *y*- and *z*-axis. The balance between the trunk and tail can be influenced by changing the inclination of both parts independently, so that their load arm lengths are inversely proportional to partial masses. (**e**) *Tyrannosaurus rex* as an example for obligatory bipedalism. Morphological adaptations to bipedality are obvious in fossil dinosaurs, regardless of their taxonomic position among saurischians or ornithischians. (**f**) Bending moments have their maximum at the hip joint, decreasing toward the head as well as toward the tail. (**g**) Upright standing in a monitor lizard is facilitated by enlarging the area of support by the tail. ((**a**)–(**d**) From Christian et al. (1994a, b, 1995), (**e**), (**f**) from Preuschoft and Christian (1999). (**g**) is a redrawing of a photo provided by H.G. Horn, owner of the animal

sometimes with large and sharp claws and sometimes with puzzling hooflike terminal phalanges, of which nobody can imagine their residual functions. In these animals, the trunk and neck are relatively short, and the vertebral column shows the greatest diameters of centra between the lower lumbar region, the long ilium strengthening the sacral region and the proximal tail – where presumably the strong m. caudofemoralis took its origin. The hindlimbs became very long and heavy, while the frontlimbs lose their importance for locomotion.

The mass moments of inertia of the fore-and-back-swinging, heavy hindlimbs exert a considerable rotating moment about a vertical (y-) axis. Of course, the length of the tail contributes much to facilitate bipedality, because it counterbalances the trunk plus neck plus head while shifting the CoM rearward. In addition, the long tail in unison with a nearly horizontally carried trunk provides the great mass moments of inertia necessary to keep the trunk in its place regarding the y- and the z- axis (Yamasaki 1990).

There exist many considerations and calculations concerning the speed reached by bipedal dinosaurs, which must not be repeated here. The importance of these potentials depends on a variety of poorly known factors: lifestyle, natural enemies, prey and environment. Less often considered are the questions concerning manoeurerability of these huge creatures. Nobody has yet investigated the conditions which must be fulfilled to turn or to brake such a giant. At a first glance, it seems that most of the necessary external forces must be derived from ground reaction forces (e.g. see Preuschoft and Gudo 2006), except those evoked by voluntary changes of body shape, connected with countermovements of other body segments (like in changing directions of lizards, as detailed above).

5.9 Alternative Body Shapes and Locomotor Modes Among "Lower Tetrapods"

5.9.1 Salientia

A technical alternative to the above-treated forms, which dominate our self-centred imagination (because we believe that we evolved from generalised tetrapods), is the tailless Anura or Salientia (Fig. 5.14). Fossils with traits of frogs seem to have existed in the Triassic (Poland, Madagascar) and are definitely known since the Jurassic and Cretaceous. In the Palaeocene and Eocene, all living families of anurans are well-documented (Rage 2012).

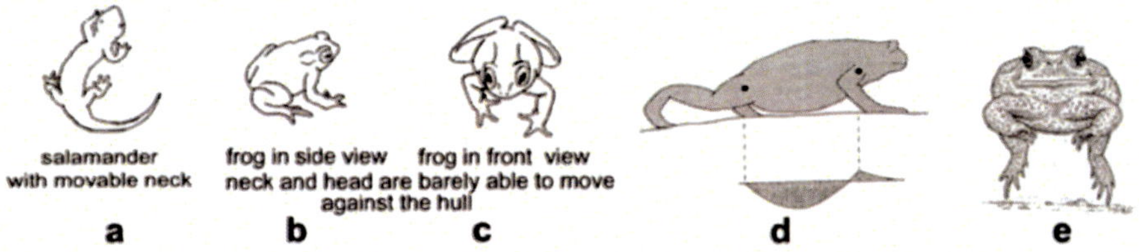

Fig. 5.14 Recent, preferably land-living lissamphibians. (**a**) As an example of a lower tetrapod, a newt or salamander, (**b**) a frog in lateral and (**c**) in anterior view, (**d**) a walking toad with bending moments along the diagonally supported trunk, (**e**) leaping toad, seen from in front. ((**e**) From v. Praun, unpublished, M.S.-thesis, 1997)

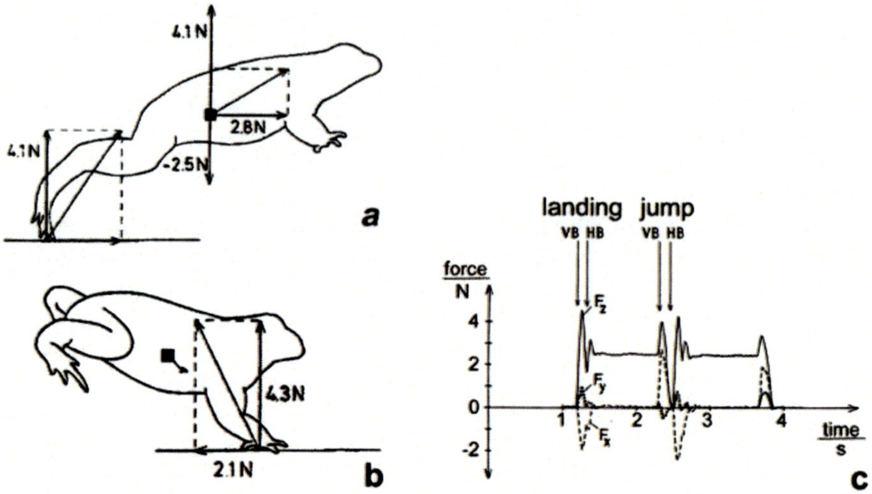

Fig. 5.15 (**a**) Side view of a toad during takeoff, (**b**) during landing. In both drawings, the resultants of the weight force and inertial force are shown. The measured GRF (**c**) does not coincide exactly with the calculated values, but also passes dorsal to the CoM. Consequently, in both approaches a marked rotating moment is exerted about the transverse *z*-axis through the CoM. This brings the lifted front- (**a**) or hindlimbs (**b**) reliably in contact with the ground. The longer the hindlimbs, the longer the GRF during takeoff, and the bigger the impulse (*P*). (**c**) the measured components of the GRF in a landing and immediately following takeoff. F_y is vertical (continuous line), and F_x horizontal (broken line). The horizontal force component in *z*-direction is minimal and not shown

After metamorphosis into the adult animal, they possess neither a tail, nor a neck, nor an elongated body stem.[5] The body stem is short; their extremities are principally similar to those of other tetrapods, as described above, whereby the hindlimbs are more or less elongated and accordingly strong. Their vertebral columns are extremely short (nine segments), and their ilia are long, resembling a pair of rods attached to both sides of just one vertebra, so elongating the vertebral column caudally to the hip joint and beyond. Dorsal to these elements exists a not paired "urostyle" as elongation and seemingly a derivative of the vertebral column. Several strong bundles of the autochthonous dorsal muscles take their origins from this element, running in dorsal-cranial direction, so transmitting stresses from the pelvis to the axial skeleton. This complex of traits is obviously related to their characteristic mode of locomotion, hopping or leaping, whereby some species, especially toads, prefer (like urodeles and reptiles) walking with a diagonal footfall sequence. During fairly long intervals, the animals remain supported on diagonal supports (Fig. 5.14d). In spite of this preference, they also can make use of hopping, if in a hurry (Figs. 5.14e and 5.15, see also Reilly and Jorgensen, 2011; Reynaga et al. 2018). A detailed comparison between both forms of

[5]In an earlier stage of ontogenetic development, as tadpoles, they function very much like a "normal", primitive chordate (see Sub-chapter 3.3.3).

locomotion is missing. Anyway, hopping is a sequence of consecutive leaps, and the individual leap is easier to analyse than hopping.[6] Taking off for a leap as well as landing after a leap means overcoming earth gravity and inertia of the resting (takeoff) or already moving (landing) mass. The forces necessary for both changes of speed combine with gravitational force to a resultant, which is inclined rearward during takeoff (Fig. 5.15a) and forward at landing. Consequently, it passes through the hind- or frontlimbs, respectively (Fig. 5.15b).

These limbs must exert the resultant force (F) against the supporting ground, and the longer the force acts, the bigger the impulse (P), which is transmitted to the animal's body. The bigger the impulse, the faster the movement and the higher (or wider) the leap. The force can act on the body not longer than contact with the substrate is maintained. This is the reason why long hindlimbs are of advantage for leaping and hopping. For making use of their full length, they should be folded to assume their shortest possible length. The length of the segments, however, leads to considerable rotating moments at the joints. The muscles necessary for overcoming these great moments lead to high compressive forces in the limb bones. Therefore, the limbs in leapers cannot be long without being heavy. Their mass is growing in proportion to the third power of length. A division into chains of many segments reduces the moments at the joints and distributes the necessary muscles on several segments (Witte unpublished Christian pers. comm). By developing the long elements tibiale and fibulare, the Salientia have acquired an additional segment of their limbs; the foot itself is also quite long.

The height, width and duration of a leap depend upon the factors takeoff angle (α) and speed (v) at takeoff. Since the animal is at rest before taking off, it must accumulate its speed by the acceleration (a). Acceleration is change of speed during time unit t, which means

$$a = 2 * \frac{\Delta s}{t^2} \tag{5.6}$$

The force (F), necessary for assuming the acceleration (a), depends upon the body mass (m):

$$F = a * m \tag{5.7}$$

This force gives the body an impulse (P), which is determined by

$$P = m * v = m * (a * t) = F * t. \tag{5.8}$$

The longer force is put into the body, the greater the impulse and the attained velocity (as mentioned above). The extended ground contact also allows correcting the direction and compensating irregularities of ground structure.

---

[6]The following has been developed first for primates; see Chap. 8.

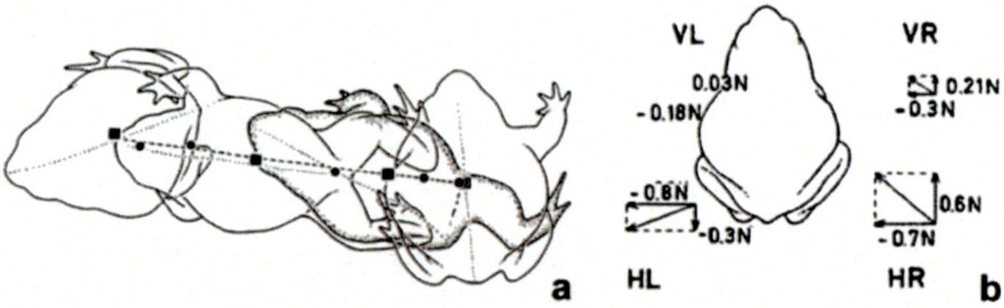

Fig. 5.16 *Bufo marinus* at takeoff for a leap to the left. (**a**) Four locomotor phases superimposed, with their respective CoM (black squares), and the *x*-axis of the body (stippled). The shortness of the body and the resulting concentration of its whole mass close to the CoM permit such a turn without being resisted by large mass inertia. (**b**) Ground reaction forces measured in the same animal. Regularly, the turn is initiated by the forelimb on that side, from which the animal gets away (in this case right *VR*). The major rotating GRF are exerted by the hindlimbs, which also serve to propel the leap (HR and HL). (From v. Praun, unpublished, MS-thesis, 1997)

The braking after short leaps is performed often with the aid of the frontlimbs (Fig. 5.15b). Their usual posture between moderate sprawling and rearward direction of the elbow leads to relatively short load arms of the GRF, but also to very short deceleration distances. The impulse (P) which must be overcome is

$$P = m * v, \tag{5.9}$$

whereby v depends on the (maximally reached) height of falling.

An interesting point is that Salientia possess the traits advantageous for taking off, but not for landing (see below on this page). The reason seems to be on the level of the biological role: acquisition of food, that is, capturing swiftly moving insects, requires urgently a rapid takeoff, as does the ability for very rapid flight to escape a predator. Landing, however, is not so decisive a manoeuvre. Frogs often land on their bellies, taking advantage of the elasticity of vegetation, or in water.

A concentration of the entire body mass near the CoM saves force for keeping the trunk in balance. This alone seems to be a good reason for abandoning the slender, elongated body shape of salamanders or newts and for assuming a short, compact body. Another point seems to play a still more important role: everybody who has tried to capture or to keep a frog or toad knows that these amphibians also possess the ability to direct their leaps laterally, even backward by about 180° of their body axis before taking off. This rotation about a vertical (y-) axis through the CoM is produced by a combination of GRFs, shown in Fig. 5.16. According to v. Praun (1997, unpublished), who studied *Bufo marinus*, this rather heavy toad (up to 1 kg) moves exclusively by short leaps, instead of walking. A rotation is always initiated by pushing the outer forelimb (on that side, from which the toad is moving away) laterally. For this purpose, the laterally directed elbow is perfectly suited, because one single twitch of the m. pectoralis and the elbow extensors is enough. The essential acceleration for the leap is provided, milliseconds later, by the hindlimbs. Here also single twitches of the extensors of the hip, the knee and the digits lead to a rapid and complete extension of the long hind extremities. Either hindlimb, the inner or the outer, may be recruited for transmitting the major impulse for accomplishing the turn, besides accelerating the body for the leap. Their choice seems to be influenced by the existing exact position of the hindfeet or by the chosen direction of the leap.

Anyway, the redirection of the leap becomes possibly quicker and less energy-consuming if the mass moment of inertia of the body has a low value. Exactly this is characteristic of the anuran body shape – at least at rest and before taking off. During its phase of free flight, the body assumes by its extended, long and heavy hindlimbs a considerable resistance against any deviation from the once chosen direction (see Fig. 5.15a), because its mass moment of inertia about the y- and z-axes is great and the loss of ground contact excludes any influence of an external force. In the case of toads, the problems of landing (see above on this page) are overcome by producing a considerable torque as long as ground contacts exist. As can be seen from Fig. 5.15a, b, the resultant GRF deviates clearly from the resultant of inertia and weight during takeoff and during landing. This creates a torque rotating the anterior or posterior legs, respectively, toward the ground. Another completely different way to escape the problems in landing is described below under "Primates" in Chapter 8.

In water, Salientia also propel themselves discontinuously but quite efficiently by stretching both hindlimbs, mostly at the same time. Instead of the GRF, the resistance of water acts against the wide and long webs expanded between the digits. Since the weight force is compensated by buoyancy of the water volume replaced by the animal, hindfeet must not be placed below the body but are abducted to both sides and moved in the x/z- plane by adduction in the hip joint. This keeps moments twisting about a z-axis (i.e. "nicking") at a low level or excludes it completely.

5.9.2 Turtles

An old and independent evolutionary line among the reptiles are the turtles (Chelonia, Fig. 5.17). Known since the Late Triassic, they are separated from other reptiles by several characteristics, among which the "anapsid" skull is given most emphasis by palaeontologists. More than other tetrapods (but similar to ankylosaurs and stegosaurs), they seem to have relied on passive means of protection, rather than the option to escape by locomotion. They have developed a solid armour (consisting of the dorsal carapax and the ventral plastron), covering the trunk. It consists of bony plates, its external surface being covered by derivatives of the skin, in most cases horny plates. The dermal armour is firmly connected to the vertebral column (ten trunk + two sacral vertebrae) and the eight ribs and to parts of the girdle skeletons. That means it possesses a stabile, self-sustaining (but immobile) frame for the trunk, against which alone the neck, tail and limbs can be moved.

The shell leaves uncovered ventrolateral incisions between the carapax and plastron which permit some mobility for the extremities. In spite of these incisions,

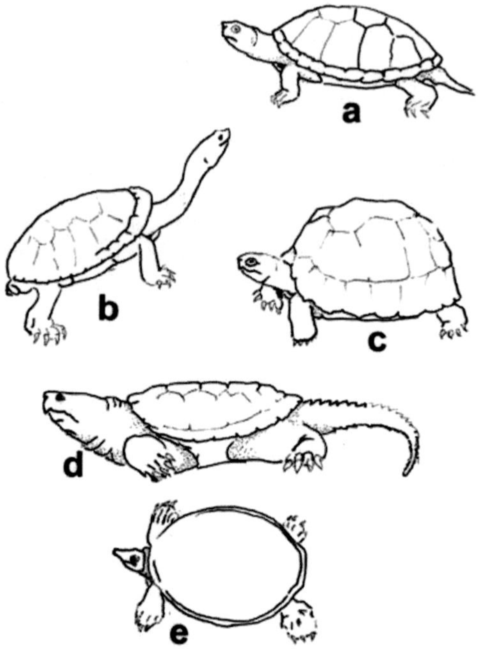

Fig. 5.17 Selection of turtles to show the variability in dependence of preferred biotope. (**a**) Amphibious European "swamp" turtle *Emys*, (**b**) semiaquatic Australian *Chelodina*, (**c**) terrestrial *Testudo* from Madagascar, (**d**) aquatic alligator turtle, *Chelydra*, (**e**) aquatic soft-shelled turtle *Trionyx*. (Redrawn and regrouped from Grzimek's Tierleben VI, 1972)

the animals cannot preserve the common angulation of the forelimbs (see Chapter 5.6), because the shoulder is located inside the armour and the upper arm therefore directed forward, not rearward as in other tetrapods. Just in order to reach the ground, the elbow joint had to develop an angle opening rearward (in ventrocaudal direction).[7] In the hindlimbs, the problem does not exist, because trunk axis and thigh include a forward (ventrocranial) open angle, which allows placing the hindfoot below the trunk.

The shapes of the autopodia vary considerably, depending on the mode of living and body weight. There are numerous transitions ranging from the slender and long propelling organs of the marine turtles through the broad, inflatable oar-like flippers of semiaquatic, sometimes soft-shelled forms (Trionychidae) to the poorly mobile, plump and claw-bearing feet of land-living turtles. As can be seen from Fig. 5.17d, the more aquatic animals keep their extremities preferably in lateral direction, while the more terrestrial forms extend their limbs more in ventral direction. The propulsive movements of mainly aquatic forms (marine turtles as well as soft-shelled trionychids) are

[7]Such an angulation is uncommon among quadrupedal animals, but uncommon angulations can also be found in a number of fossil theropods from the stock which gave raise to birds.

synchronous, while the less aquatic, swamp-dwelling forms usually prefer symmetric movements of front- and hindlimbs, comparable to the walk and trot.

The reason for these differences is sketched in Chaps. 3.3 and 4.4.

The *necks of turtles* are quite long (up to the length of the carapax; Fig. 5.17b) and mobile in all directions. Precondition for active mobility is the equipment with a differentiated musculature. The positions and insertions of the individual muscles resemble closely those in other reptiles, but many of these muscles take their origins from the inner surface of the shell (Guibé 1970). Therefore, their force arms can be rather long. A trait accompanying mobility of the neck, which can also be observed in birds and some mammals, is the absence of marked outgrows of the vertebrae.

The *tails* of mostly water-living alligator turtles (Chelydridae) may reach the length of the shell. In land-living forms, the tails are reduced and often hidden beneath the posterior carapax (Fig. 5.17a, c, f, e).

The biological *role of the shell* as a protection is evident, but it entails obvious restrictions. From a biomechanical viewpoint, turtles are characterised by the concentration of body mass near the CoM, and that means easy, energy-saving possibilities to rotate about the horizontal x- and vertical y-axis, while rotations about the z-axis are impeded. The rather flat, in dorsoventral direction depressed profile of the shell yields in a swimming animal a high degree of stability about the x- and z-axis, against rolling and nicking. Its advantages in shallow water have been explained above (see Chapter 4.4): it keeps more of the body immersed than a dorsoventral higher shape. All these morphological traits are most important for the varying forms of aquatic lifestyles, as discussed above (Chapter 3.4), but without any relevance for terrestrial forms. The highly vaulted shell of terrestrial turtles seems to contribute to strength of the armour.

The rigidity and breadth of the shell compulsory demand diagonal supports, when moving on land – like in other animals walking on wide-gaged tracks. This excludes everything except diagonal footfall of front- and hindlimbs – as is common among all reptiles (Christian 1995). Phases of free aerial floating (in trot-like locomotion) have never been reported. The long necks allow the clumsy animals to reach even distant resources. Moving the neck, however, evokes resistance because of mass inertia, and keeping it extended in any direction shifts the CoM cranially toward the head. Both forces must be counterbalanced by additional external forces – because the general shape of the animals implies low mass moments of inertia. A great reaching distance is of particular relevance for living on land. Additional components of external GRFs are easy to provide (Preuschoft and Gudo 2006). Although reaching distance seems to be less important, if the weight of the animal is carried by the surrounding water, and must not be moved against gravity, the tail can possess a mass moment of inertia, which compensates passive movements of the trunk, while the head is pushed rapidly toward a prey. (This interpretation is supported by the existence of long tails in those forms, which lie on the ground of ponds or slow-streaming rivers in wait for prey and try to capture it by a sudden head strike, like the alligator turtles.) In addition, movements of the tail may also evoke water resistance, which also may induce active while inhibiting passive rotation. The same function, however, can be and actually is performed by the hindlimbs.

5.10 Essential Results

As the name "tetrapods" suggests, the extremities play a dominant role in these animals. The palaeontologically documented transition from lobefins to segmented extremities is advantageous from a biomechanical viewpoint. The constant existence of three segments (stylo-, zeugo- and autopodium) permits sufficient freedom of the limbs from the trunk as well as lifting the body from the ground.

Autopodia on the ground are "rolled off" because the CoM of the animal and its most proximal joints are moved forward. This "rolling off' of the feet yields especially great step lengths if the autopodia are long and kept in x-direction. An attempt is made to identify the function of a splitting of the anterior border of the autopodia into digits: providing a possibility to exert rearward-directed force components against the ground. Their number (mostly five) seems to be the product of a trade-off between anatomical necessities (control of all digits individually), small areas of contact and ground mechanics (confluence of spreading compression zones of the underground at the individual "fundaments").

Locomotion on four limbs implies an interplay of braking and accelerating, regardless of the chosen gait. In braking, the forelimbs do most of the work, so that the trunk is under compression. Vice versa, accelerating by preferably the hindlimbs also produces compression of the trunk. The tendency of most, even the earliest, tetrapods to rotate their elbows rearward and knees forward brings the limb joints in an advantageous position to the GRFs in braking (forelimb) and in accelerating (hindlimb).

Anchoring the limbs to the trunks requires strong bony props or counterparts: the limb girdles (Fig. 5.12). They are part of the trunk and of the extremities as well. While their relation to the trunk was dealt with in Chap. 4, their relations to the free limbs are investigated here.

The first problem to solve is why in all tetrapods the connection between the anterior extremity and the trunk is mobile, while the posterior hindlimb girdle is firmly connected to the trunk. This characteristic permits to transmit accelerating forces immediately to the trunk mass and to move hind- and frontlimbs independently, not bound to equal steps.

In the shoulder girdle, two principally different variants exist: putting loads on the laterally spread forelimbs and so requiring a very strong pectoral muscle acting largely in z-direction. This is realised in reptiles and birds (Fig. 5.18a, b). Quite different is the situation in most mammals (Fig. 5.18c), in which external forces are acting against a manus below the shoulder joint. So they need a device for maintaining the angle between the shoulder blade and humerus, not an adductor. Alone arboreal mammals, like primates, have the same sort of problem and will be dealt with in Sub-Chap. 8.5.6.

The pelvic girdle of "low tetrapods" diverges principally from that in more advanced, tailless forms, because the most powerful retractor of the femur takes its origin in primitive forms from the proximal tail and just passes the pelvis without exchange of force. As soon as the tail is reduced, and therefore loses its enormous mass moment of inertia, new origins for the retractors of the femur (=extensors of the

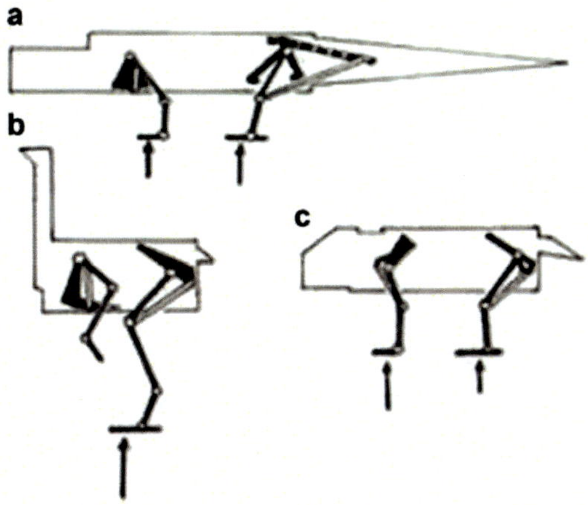

Fig. 5.18 Construction of body stem and limb girdles in ancestral and several recent reptiles (**a**), reprtile, (**b**) the bird, (**c**) the mammal. Total body masses are assumed to be similar in all cases; limb positions and muscles are arbitrary, though in line with empirical data. The major retractor of the hindlimb in reptiles (m. caudofemoralis) is replaced in the other classes by pelvifemoral muscles. The shoulder complex, in reptiles as well as in birds, is balanced by m. pectoralis, which acts in z-direction on the ventral part of the girdle skeleton. In mammals the shoulder is totally reconstructed: the activity of anteverting, rather than adducting, muscles are required to balance external forces. The skeletal counterpart of the anteverting muscles, the scapula, is used as an additional limb segment. Reptiles carry a larger part of their body weight on the hindlimbs than mammals (vertical arrows), in which the lack of a tail and the heavy head push the CoM forward. See also Fig. 4.6

hip joint) must be developed. Therefore, the origins of muscles are moved to the postcoxal part of the pelvis, and the latter needs additional strength (in birds the ischium, in mammals the pubis; Fig. 5.18c).

In the fossil history of reptiles appears often bipedality. The basic conditions for bipedal posture are keeping balance on only 1 foot, strength of hindlimbs and bending moments along the trunk with a maximum in the pelvis. The observation is emphasised that many of the living lacertilians use bipedal locomotion if in a hurry.

Technically diverging forms exist in turtles and anurans. In both groups, the body mass is arranged close to the CoM, so that rotation about a vertical x-axis and a horizontal z-axis is facilitated. This fact is used by anurans for catching prey and for rapid flight. In turtles, the shape of the carapax influences mobility: in marine and amphibious forms, it seems designed to limit nicking (about a z-axis) and rolling (about an x-axis). In terrestrial turtles, the highly vaulted carapax gives strength against external forces.

Chapter 6
Birds

Keywords Alternative use of forelimbs · Bipedality · Flightless birds · Gaits on ground · Locomotor speed · Mass distribution · Hopping · Running · Beak · Light-weight head · Length of neck · Hindfeet in swimming · Wings · Statics of body · Flying

6.1 General Characteristics

The vertebrate Bauplan has undergone a very particular evolution in the class of Aves. Irrespective of the enormous variation in colour, plumage, lifestyle, behaviour and size (living species from 0.010 kg to 150 kg; subfossil in Madagascar and New Zealand still larger – see Table 6.1), birds are rather similar in body structure (Romer-Frick 1966; Bock 1974; Oemichen 1950a, b, c): a compact trunk, long neck (often hidden in the plumage), small head with horny beak and folded front- and powerful hindlimbs. From a stock of small theropods evolved bipedal forms in the Jurassic (Chatterjee, 2017), some of them with similar postures of the forelimbs and some of them with dermal appendages which look like feathers. The long-tailed *Archaeopteryx* from the upper Cretaceous (Mayr 2013) was completely covered by feathers and possessed teeth.

Above all other specialisations, fossil and recent birds are *bipedal animals* (Fig. 5.18). Flying, as an additional mode of locomotion, is an ability which only some of them can dispose of. Several groups are excellent swimmers and divers (penguins, auks, divers/Gaviidae and grebes/Podicipedidae) and so have to cope with special conditions. These will be outlined under "alternatives". Avian bipedality is similar to that in fossil theropods, in which, like in the many species of ornithopod dinosaurs, the tendency towards carrying the major share of weight on the hindlimbs is driven to an extreme (e.g. reduction of forelimbs in tyrannosaurs). Though theropods are placed into the "saurischian", and the ornithopods into the "ornithischian" branch of dinosaurs, the former are considered to be the forerunners of birds. The living birds (subclass Neornithes) comprise two infraclasses, Palaeognathae and Neognathae, because of divergent composition of the upper jaws. These birds also show differences of their postcranial skeletons and their plumage. Most of the former are unable to fly, but the flightless penguins (as well

H. Preuschoft, *Understanding Body Shapes of Animals*,
https://doi.org/10.1007/978-3-030-27668-3_6

as several other recently exterminated forms) are nevertheless counted among the Neognathae. The ability or inability to fly is not a useful characteristic for taxonomy.

During the last few decades, a yet unknown wealth of information about the origin of birds from theropod dinosaurs has been uncovered. Essential "inventions" seem to have been the development of feathers and the loss of the reptilian tail, including the development of a "pygostylum". These points will be taken up again in Chapters 6.4 and 6.8.

Anatomical "specialisations" of Neornithes are the fusion of the clavicle from both sides to a largely independent "furcula" and the huge, often carinated sternum. In ratitae, the clavicles of both sides make contact independently with the flat, broad sternum. Instead of a tail, all living birds have a short, actively mobile pygostylum as the basis for the often long tail feathers. Aside from their characteristic pelvis (see below), birds are characterised by a rather short femur, the long tibiotarsus with the rudimentary fibula and an often long tarsometatarsus. The two most proximal tarsal bones of lower tetrapods in (in Chapter 5) and mammals (in Chapter 7) are fused with the tibia to a continuous, long, tube-like unit, while the fibula is more or less reduced. The distal elements of the tarsus are fused with the metatarsals to another tubular skeletal segment. They are moved against each other not in the ankle joint, like in mammals, but in a newly formed intertarsal joint (which, however, is often called "heel"). The tendons of the major muscles, however, have their usual positions on the caudal/posterior sides of the skeletal elements and function as flexors. The digits of birds make direct contacts with the environment. Their highly variable form can be explained only on the level of their biological roles, under consideration of their particular mode of living (see below, Chapter 6.6). As far as known to me, detailed attempts to identify the mechanical forces exchanged between the animal and its environment are rare; one will be mentioned below.

The uniformity of aves and the seeming simplicity of anatomical structure should not distract from the fact that some of the systems in birds are more efficient than those in mammals. Insulation of the body against temperature differences to the environment keeps average body temperature at about 40 °C, respiratory system allows reducing O_2-partial pressure from 100 mm Hg down to 40 or 24 mm Hg in the exhaled air (instead of 80 mm Hg in mammals). The brain is nearly without cortex, but in the relatives of crows (Corvidae), as well as in parrots, mental abilities have been observed which are comparable to those in monkeys and apes. The cavities of long bones are filled with air and therefore lighter than those of mammals which have never succeeded in replacing bone marrow (which has similar mechanical properties as water, but less weight) by air. Birds occupy nearly all environments existing in our world: although reproduction always takes place on "land", some forms spend nearly their whole life in or close to water (strictly marine are penguins, auks (= Alcidae) and the tube-nosed birds (= Tubinares, including mollymawks and shearwaters)). The gulls and their relatives may be included into this group, since they are spending much time above or on water. As "land-living" animals, birds inhabit the entire variety of biotopes from desert-like steppe to dense forest and from swamps to high mountains.

6.2 Head, Why So Small?

The most important ("prehensile") organ for grasping and making other close contacts with external objects in all birds is the *beak*; only few forms make also use of their feet (e.g. owls, parrots, birds of prey). Because the beak establishes direct contacts between the animal and its environment, beaks are among the most variable parts of the body. The variety of their shapes can be understood by the biological role, which it plays in the respective genus (see below). Anyway, the beak is part of a small head, which is mounted on a long, very mobile neck. The mechanical interplay between these traits should always be kept in mind.

The head is moved frequently and rapidly, and each movement is equivalent to an acceleration against the mass moments of inertia of the head:

$$J = \Sigma m_i * l_i^2 \tag{6.1}$$

Therefore J must be kept low, and all characteristics of a bird's head are indeed means to confine J to a low level. The brain of birds is "miniaturised". The eyes, however, as the most important sense organs, are rather large. Maybe the reduction of the olfactory organ also has taken place as a means to reduce weight. The most convincing restrictions, related to the head in its role are due to the role of the beak as the most important grasping organ. Birds cannot afford an apparatus for mastication, like in mammals. Neither teeth for mastication, nor their bony bearings, nor the muscles to perform movements for comminution of food objects can be housed in a mass-reduced head. The horny sheath of the beak is much lighter than these structures, but not suited for crushing great amounts of hard objects at the same time. Because of all these items, birds tend to take up preferably small objects by the forceps formed by their jaws or to comminute the objects to be ingested outside the mouth cavity (see also below). In order to produce bite forces sufficient to hold small, often slippery objects, the jaws need to be very narrow, so that the bite force is concentrated on small areas. In Fig. 6.2, the formal relations between beak length, bite force and contact areas between upper and lower branch (as measure for bite pressure) are detailed. The longer the beak, the smaller the bite force at its tip, but the decrease becomes less and less, above 7 or 9 cm, and it remains at nearly the same level.

Bite force is the moment of adducting muscles divided by beak length. The "sharpness" of the beak depends upon the ratio bite force/contact area. In most cases, the area of the lower branch of the forceps is smaller than the dorsal branch and therefore determines bite pressure, but in many birds, a movement of the head downward during "pecking" seems to increase the force on the upper beak. Anyway, the decrease of contact area acts opposite to the decrease of bite force with length (Fig. 6.2b). That means that high bite pressures may well occur even in long beaks, if their tips are pointed and their edges narrow. Thanks to its special form and sharpness, the horny beak is suited to reduce the sizes of swallowed portions outside the oral cavity (see above; e.g. in birds of prey, vultures, ravens, parrots). Another

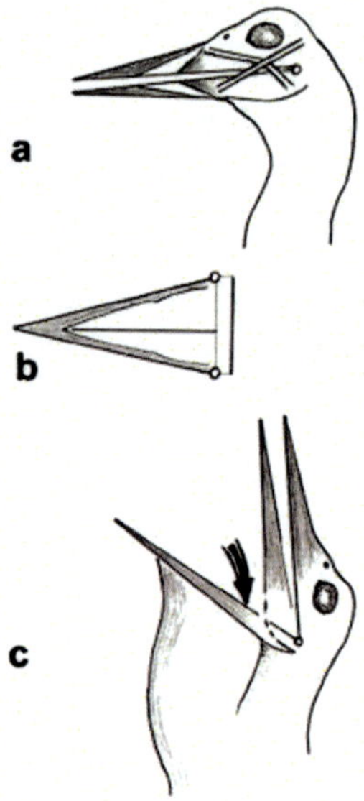

Fig. 6.1 Schematic illustration of a bird's head. (**a**) The most important adductors of the mandible are shown as double lines. Three beak lengths are illustrated: very short (grey), moderate (white) and long (grey). The longer the beak, the smaller the bite force at its peak, if muscle moment is constant. (**b**) Ventral aspect of the mandible with lower beak. The floor of the mouth has triangular shape, extending from the tip to a line connecting both mandibular joints. These distances set limits to the size of objects that can be swallowed. (**c**) Birds swallowing big objects (prey, curved arrow) often direct their head upward, in elongation of the cervical spine. The dilatable oral floor becomes visible, when the lower beak is at a wide angle to the upper. An extreme can be observed in pelicans. Bowing the branches of the mandible outward further dilates the passage and permits swallowing of prey which is larger than the distance between the mandibular joints

option is swallowing of surprisingly large, undivided lumps (e.g. owls, herons, storks, penguins). Whole prey animals, like fishes or crabs, or bones as well as fragments of them can be swallowed because of the extensible floor of the mouth and the pharynx (Fig. 6.1c). The maximal width between the mandibular joints of both sides (Fig. 6.1c) is not necessarily a limitation; it can be circumvented by bowing the branches of the mandible outward (as, e.g. in pelicans).

The arrangement of the most important muscles (m. adductor mandibulae from fossa temporalis to the middle part of the lower jaw and m. pterygoideus from the dorsal side of the palate to the posterior part of the mandible (Fig. 6.1a)) provides an active tie for those parts of both opposed parts which are exposed to the greatest bending moments. Whether the periosteum also provides a (passive) tie along the free jaws can be expected but has not been investigated. The (external) reaction force evoked by closing the upper and lower branches in most cases acts against the horny beak, while the muscle force acts on the skeleton. Both structures may well reinforce each other, but the transmission of stresses between them is not completely known –

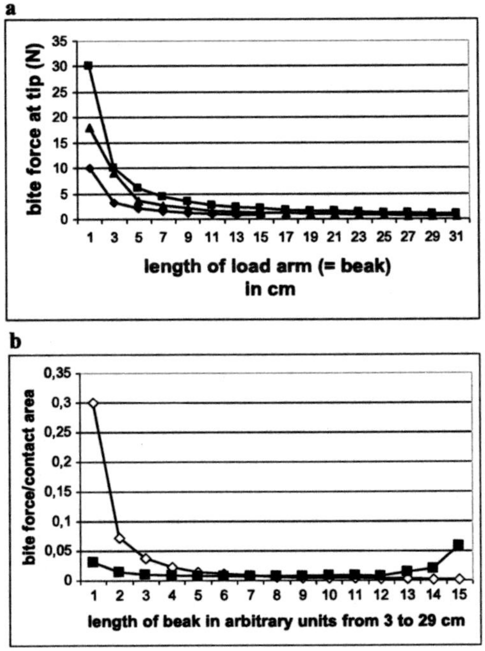

Fig. 6.2 (**a**) Bite force as a function of beak length, varying from 1 to 30 cm. Muscle moments are assumed to be 20 *N*, multiplied by lever arms (h_m) of 1.5 cm, 0.9 cm and 0.5 cm, respectively. Note the slow decrease of bite force above lengths of more than 7 cm. This somewhat surprising result is due to the changing ratio between constant force arm and growing load arm. (**b**) Bite pressure and "sharpness" of the beak are determined by the area of contact between the upper and lower beak. The contact areas are proportional to the width and length of the beak, and their influence is opposed to the influence of beak length. The ratio of bite force divided by contact area ("sharpness") is illustrated. Distally growing contact areas (= decreasing sharpness, white diamonds) leads to further reduction of bite pressure, while contact areas which become smaller distally lead to increasing sharpness (black squares), so compensating the loss of bite force

but see Homberger et al., 2001 on parrot beak, as well as W. Bock (1974). As shown in Chap. 2.1 and Figs. 2.5, 2.11, the strength of a homogenous body against bending depends upon the greatest tolerable stress σ_{max}. Its value is determined by material properties (horn or bone, respectively) and grows by the breadth (*z*-direction, perpendicular to bite force) and the square root of its height (*y*-direction, parallel to bite force). Therefore, it can be expected that the profile of a beak grows with the square root. Proximal to the hook (if present), the beak's contour comes close to the expected profile.

Most of the work on function of avian skulls dwells on the "kinetic" of the jaw apparatus (Bock, 1999) – which plays a subordinate role in the forms discussed above and no role at all in mammals. The value of movements of the lower beak is fairly well known in pelicans and nightjars (Bühler 1972, 1981: "butterfly/ring net"), and the value of the often-treated and systematically described kinetic of the upper beak has never been quantified, nor assigned to a special role. It just seems to be a means for increasing gape or for damping: the forces that occur in biting or pecking.

Better than by its mechanical function, the shape of the avian beak can be understood from its "biological role". This is diverse and often yields convincing and highly plausible reasons for characteristic shapes. In birds of prey, for example, the upper beak is formed as a hook that may be related to the possibility of making use of gravity when the bird "pecks". The ventral neck muscles, which flex the neck ventrally, act in synergy with gravity. In parrots, the hook formed by the upper beak is an additional prehensile organ, well suited for climbing, as long as the bird keeps the dorsal side of its head up. The contact pressure is increased by an adduction of the lower beak with the aid of the jaw muscles. This situation is similar to that in the prehensile hands or feet of primates (see Chap. 8): the upper beak corresponds to the lateral digits and their ability to transmit tension, the lower to the thumb or big toe which secure the grip by producing a counterforce and friction. In special cases, this allows even the transmission of rotating moments between animal and substrate (Homberger, 1986) as well as (Homberger and Brush, 1986) have investigated the complex movements by which parrots separate seeds or kernels from the fleshy mesocarp and have identified mechanical factors which are relevant for the shape of beaks.

6.3 Why Are Bird's Necks So Long?

The typically *long and actively flexible neck* of birds permits moving the prehensile organ (beak), together with sense organs (vision, taste, olfaction, touch), to any place within reaching distance of the neck. In other words, the long neck allows controlling a very large volume (Fig. 6.3a and see also Chapter 4.6) without moving the whole body mass. Neck mobility is impeded by its own weight and mass inertia. Because of the length of the cantilever "neck", weight causes considerable bending moments:

$$M = \Sigma m_i * l_i, \tag{6.2}$$

which must be balanced at the joints between neck vertebrae by muscle force. The biggest partial mass (m_i) is contributed by the head (see above). This also has the longest lever arm (l_i) and so exerts the strongest influence on static loading. In addition, external forces (like the weight of prey objects or other things lifted) increase bending moments. These and the muscle force necessary for balancing them are kept low by upright, vertical neck postures (Fig. 6.3b). Not alone the conditions of statics, but more so the mass moments of inertia (J) (see Eq. 6.1) become very great with neck length.

Fig. 6.3 Pheasant with various positions of the neck. The volume which can be accessed alone by moving the neck is surrounded by a heavy line. Lifting or lowering the trunk by movement about the knee or hip joints (white circles in the black skeleton) leads to a shifting of the neck base along the fat, curved line and therefore to an enlargement of the accessible volume (thin outlines). The bending moments along the neck are particularly great, if the neck is in a horizontal position, because the lever arms of segment weights are longest. In upright neck postures, the bending moments along the neck are small or zero (if no lever arm exists). The mass moments of inertia (J) against being moved, however, are the same in all positions of the neck! (**b**) A number of Canadian geese to show the most preferred neck postures

6.4 Trunk, Why So Short and Compact?

The common discrimination between "saurischian" and "ornithischian" dinosaurs refers to a characteristic of the *pelvis*, more exactly of the *os pubis*: the cranioventral inclination of the broad and large plate formed by the pubes of both sides in saurischians and the rather slender bony rod in ornithischians, which is inclined caudoventrally, parallel to or connected with the ischium. The latter is similar to the situation in birds. The knack is that the caudal insertions of the ventral muscles of the body wall are shifted rearward with this turn of the os pubis (Fig. 6.4b). While they assume a position in front of the femora in saurischians, they are placed in ornithischians as well as in birds clearly behind the femora. By this translocation, the abdominal cavity is extended in caudal direction, and the heavy intestines (digestive tract and urogential system) shift the CoM of the trunk rearward. This is of course more advantageous for bipedal posture than a CoM more in front, as in saurischians (Fig. 6.4a). Balance on top of the hindfeet was kept in the fossil theropods with the help of their long and massive tails. None of the living birds possesses a long, skeleton-supported tail, which, however, was present in their fossil forerunners. These possessed this long, unpaired extremity. In birds, it is replaced by the short, nearly weightless pygostylum (Fig. 6.4b).

The compact, ellipsoid *trunk shape* of birds gives rise to but low bending moments and mass moments of inertia (see also Hildebrand and Goslow (2004) on p.610 f.). In addition it is an advantage for warm-blooded, endotherm animals. It combines maximal (heat-producing) mass with minimal (heat-dissipating) surface area, aside from the extremely insulating plumage (see above). Heat losses take place alone at the not plumage-covered distal segments of hindlimbs and to a lesser extent

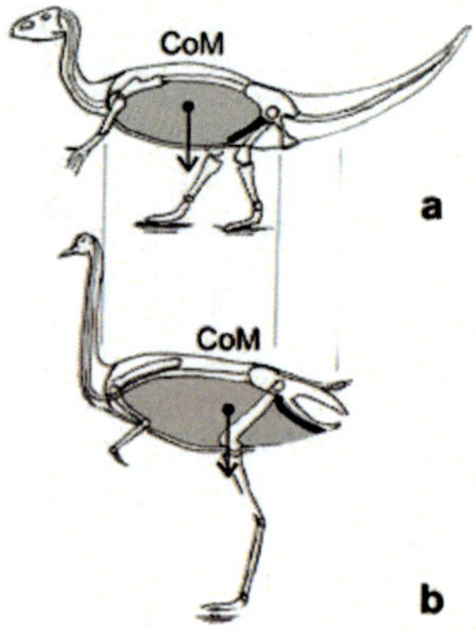

Fig. 6.4 The saurischian pelvis of, for example, *Plateosaurus* (**a**, drawn after the Tübingen specimen) confines the body cavity (grey) to the trunk in front of the hip joint (o). Bipedal posture is possible because of the counterbalancing, long and heavy tail. The pubes of an ostrich (**b**) extend rearward and widen the body cavity in caudal direction. So the CoM (black dot) is approached to the hip joint (o) without counterbalancing by the tail, largely because of the heavy intestines. The ostriches and their relatives possess, in contrast to Neornithes or modern birds, a bony connection between the pubic bones of both sides. Note the small, light-weight head on the long neck

at the widely extended, but feather-covered frontlimbs. The stiffness of the trunk in birds has been emphasised repeatedly in the literature.

In anterior view the tetrapod *pelvis* is interpreted as a ring (see Chaps. 4 and 7). In standing on two limbs, its ventral (pubic) part between both hip joints is exposed to tensile forces. In standing on one hindlimb alone (as occurs regularly in striding locomotion), however, the ventral part of the pelvis is exposed to compression (see also Chapters 4.2, 4.3 and 7.7.6). The pelvis of birds does not follow the same constructional principle. In flightless birds (ostriches and relatives), there exists at least a ventral connection between the pubic bones of both sides. It is, however, located far behind the hip joints, while the dorsal connection (through the synsacrum) is cranial to them (Fig. 6.4 and 6.7a). A comparison with a ring is not at all evident. Still bigger is the difference in the flying Neognathae, in which no bony connection between both sides exists at the ventral side of the pelvis. Instead, the transverse processes of numerous vertebrae are firmly fused to the ilium and expanded in dorsoventral direction, so being able to sustain very high bending moments in *z*-direction (Fig. 6.6), as are inevitable in asymmetrical support on only one hindlimb in walking and running (R. Fechner, unpublished). In symmetrical support (standing and hopping), no compressive stresses occur on the ventral side of the trunk but instead considerable tensile forces.

All this gives reason to imagine that the earliest ancestral birds were specialised to hopping and therefore could reduce the ventral connection between both sides. With increasing body size, and growing importance of striding and running (see below), evolution has not led to a reintroduction of the ventral connection, but to an

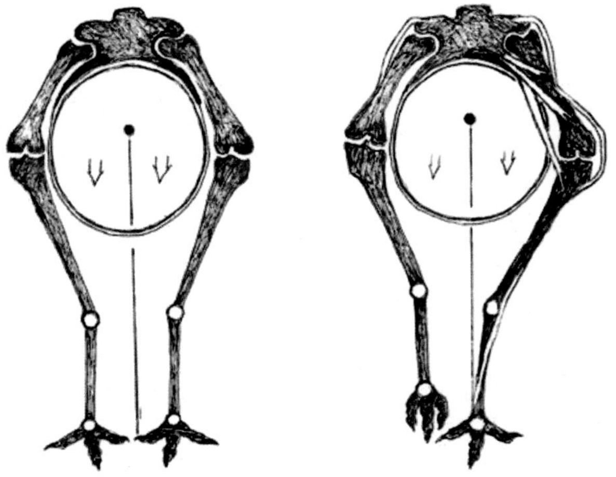

Fig. 6.5 Schematic projection of a bird's pelvis and hindlimbs into a transverse *y*/*z*-plane in anterior view. (**a**) In symmetrical support on two limbs, no matter if standing or hopping, tensile forces occur on the ventral side of the trunk, which are sustained by the muscles of the trunk wall. The weights of the intestines (white arrows) anyway require tensile forces in the abdominal muscles. The dorsal sides of ilium und synsacrum are exposed to compressive forces in transverse *z*-direction. (**b**) In asymmetrical support, during the stance phase of walking or running, compressive forces occur at the ventral side of the trunk, where a ventral connection between both sides of the pelvis (like in mammals at the symphysis pubis) is missing. The hip joints are kept in balance by the pelvifemoral muscles of the thigh (double lines), and the knee joints by the m. quadriceps, perhaps also by m. gastrocnemius. The muscles are shown alone on the stance side. On the swing side, the relatively weak flexors and abductors of the hip joint are recruited

alternative strengthening of the dorsal bridge, consisting of the ilia and the synsacrum. The length of the postcoxal part of the pelvis, which is so obvious in the heavy ostrich (Schaller 2008, p. 57), gives the extensors of the hip joint very long lever arms. This means that the required forces are low, whereas the induced movements are slow. The long muscle fibres of these muscles provide the considerable lifting height which is necessary for the wide excursions in hip and knee joints (Kostuj 2002).

In Fig. 6.7a, the body stem of a nonflying, ostrich-like bird, including the frontlimbs and the pygostylum, is replaced by a heavy, horizontal beam, which is completed by a long, slender, upright neck and a lightweight head, both kept in a "natural" position. Bending moments along the body stem increase rapidly along the trunk, reach their maximum at the hip joints and decrease steeply towards the most caudal part, the ischium-pubis complex. Caudal to the hip joints, bending moments are not so much determined by the weight of the segments. Greater influence is

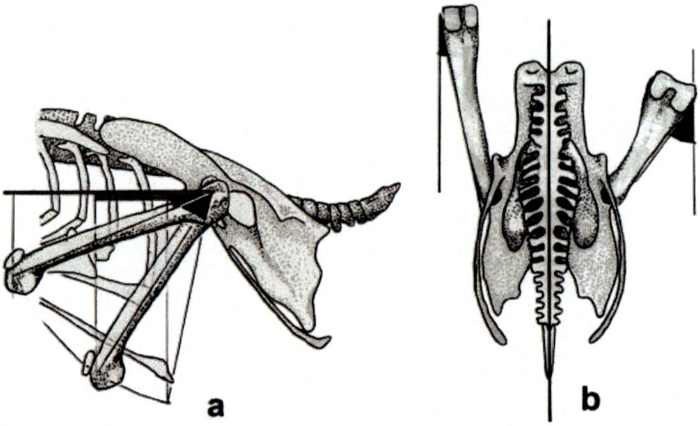

Fig. 6.6 Pelvis with femur of a bird (domestic fowl) seen from the left (**a**) and the ventral side (**b**). The foot in quails and tinamous is placed on the ground, while the femur is anteverted by 40° against the *x*-direction. At liftoff, femur and *x*-axis include an angle of less than 90°. The locomotor progress made is proportional to the cosine of the excursion angle * femur length. As can be seen from the horizontal bar through the hip joint, excursions of more than 80° do not yield much progress. (**b**) In dorsoventral projection, viewed from ventral side, the femur on the left side is shown at touchdown, on the right side near midstance. The angles of abduction are indicated by the filled angles between femurs and parallel to the *x*-axis. The femur of the ground-running birds moves in abduction angles of 10–20° (Stoessel and Fischer, 2012). The low bicoxal and high postcoxal breadths give the ischio-femoral muscles a nearly parasagittal direction slightly lateral to the hip joint, thus influencing the abducting/adducting moments. In all cases, the ischium is subject to considerable bending moments in top view as well as in side view and indeed possesses high bending strength in both planes, because of its platelike shape which is extended in an oblique direction between horizontal and vertical

exerted by the pull of the muscles which keep the hip joint in balance (see also above). Shortness of the trunk is per se a means to keep bending stresses low. A more upright trunk axis, as can be observed in many birds (e.g. penguins, woodpeckers, birds of prey, owls, parrots, Fig. 6.7b), also makes the lever arms of body segments shorter and so reduces the bending moments. These lever arms follow the cosine of trunk inclination against the horizontal. Therefore, a relevant energy-saving effect can be expected in marked erection of the trunk. The advantage of energy saving, however, is opposed by a disadvantage in running (see below). Examples for endurant standing during long periods are penguins. They can maintain energy-saving positions, since they do not run very often – in contrast to galliform birds or ostriches, which spent more efforts in balancing their horizontal body axis but have the means to stabilise their trunk in running (as detailed below).

In the simplified model, the dorsal margin of the trunk is under tensile, the ventral under compressive forces. The highest tensile forces are sustained by the high, boxlike profile of the synsacrum, the fibres of the muscle insertions and the tough periosteum. Cranial to the pelvis, the joints between the vertebrae must be kept in

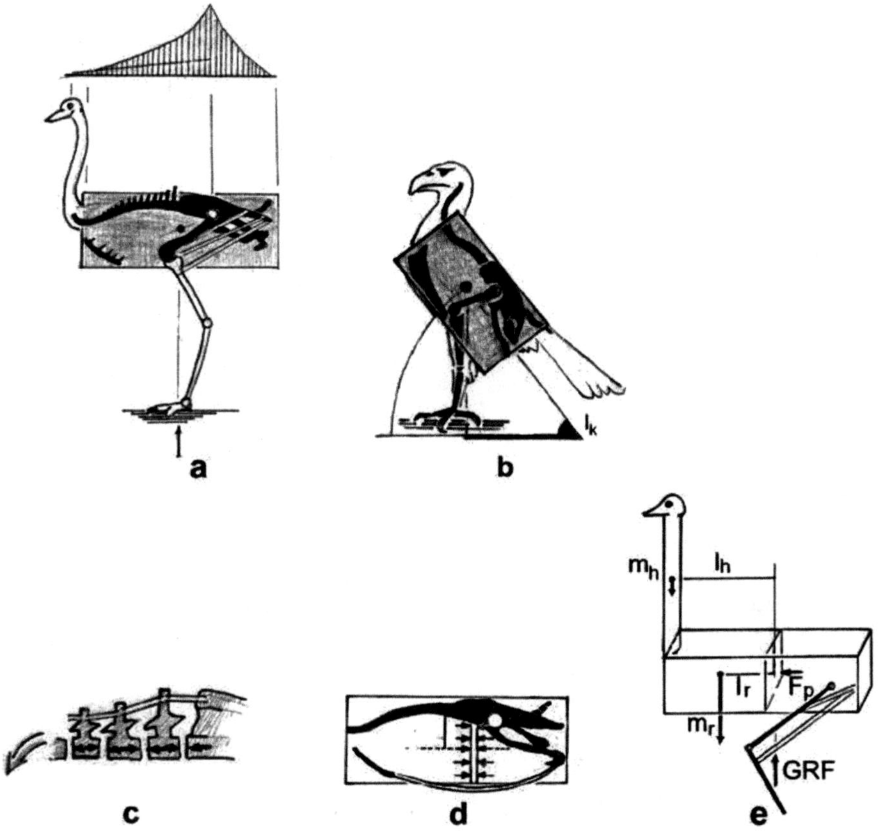

Fig. 6.7 (**a**) Statics of the trunk in an ostrich-like bird. The most important parts of the skeleton are shown in black, and the muscles balancing the hip joint are white double lines. The supporting hindlimb is drawn only for the left side. On top of the drawing, the bending moments are shown. The linearly increasing bending moments of the head are separated from the parabolically increasing moments of body weight by a straight line. (**b**) A steep inclination of the trunk axis implies shorter lever arms of segment weights, thus reducing the bending moments (like in owls, or birds of prey, or penguins). (**c**) Details required for establishing equilibrium in the vertebral column, exemplified in several presacral vertebrae of an ostrich. The weight of the anterior part of the body is pushing the trunk segments downward (curved arrow) and kept in balance by the dorsal muscles (double lines). Due to their caudally increasing force arms, these muscles exert caudally increasing moments in the intervertebral joints. Their pulling force presses together the vertebrae. (**d**) Trunk of the ostrich in (**a**), in which the intestines as a whole behave like a liquid filling of a "pneu". Their weight presses the trunk wall muscles downward, thus inducing forces in the abdominal wall and evoking hydrostatic pressure inside the trunk cavity. This pressure is distributed evenly over the entire cross-sectional area (arrows), but can be imagined to be replaced by the sum of partial forces (F_p in FAT) acting at 1/2 of the vertical trunk diameter. (**e**) Because this (simplified) force arm is rather long; already a small increase of intra-abdominal pressure (F_p in FAT) can contribute considerably to maintenance of equilibrium in the trunk. According to a rough estimate (see text), the mass of the neck (m_h) is 1/4 of the anterior trunk, and the total force necessary for stabilising the trunk, F_p is 1/2 of the weight force of the neck plus weight force of the anterior trunk, and is distributed over the entire cross section of the trunk, condensed to the force F_p

balance by the dorsal muscles (at least in the ratitae, which have a mobile vertebral column). The lengths of their lever arms can be approximated by the lengths of the spinal processes, which also grow in caudal direction, following a parabolic curve (Fig. 6.7c). The force (F_d) which the dorsal muscles have to exert in order to keep the trunk in its position can be calculated according to

$$F_d = M_B/h_F, \tag{6.3}$$

This force compresses the vertebral bodies and the intervertebral discs. Both elements have limited strength, and therefore the stresses (σ) within the vertebral column must be expected to be

$$\sigma = F/A = \text{const.} \tag{6.4}$$

The cross-sectional area must be proportional to the calculated force (see also Chap. 3), that is, it must grow faster than linearly in caudal direction. This in fact can be observed in ostriches and their relatives. In flying birds, the entire vertebral column is fused to a rigid rod, in which the individual vertebrae are often difficult to identify. In these forms, equilibrium between the vertebrae is not alone maintained by the force exerted by the dorsal muscles. In spite of this, the vertebral columns of large, mainly ground-living birds, like capercaillies (*Tetrao*) or bustards (*Otis*), show similar metric characteristics like those measured in ratitae.

The ventral margin of the trunk is not stretched by body weight, but compressed. These compressive forces are in flying birds sustained easily by the impressively voluminous sternum, which extends caudally to the height of the knee joints (Fig. 6.7b; see below Chapter 6.8). In spite of the length and strength of the sternum, there does not exist a force-bearing structure which connects the latter to the pelvis. Alone the intestines and its latero-ventral wall may act as a compression-resistant "pneu". This seems to be particularly important in the nonflying ostriches (Fig. 6.7d) because of their not so strongly developed sternum. The intestines have, because of their high water content, the mechanical properties of a fluid: not compressible and without form. Any contraction of the trunk wall muscles raises the hydrostatic pressure inside the abdomen. For the sake of simplicity, in Fig. 6.7e the trunk is replaced by a thick, heavy beam, cut in the middle. Its right ("posterior") half is supported by the hindlimbs as in Fig. 6.7a. Its left ("anterior") half rotates downward with a moment defined by 1/2 of its total masses m_h and m_r, their corresponding weight forces multiplied by their lever ams l_h and l_r. These moments are counterbalanced by the intra-abdominal pressure (F_p), acting on the lever arm h_t. This force acts distributed over the entire cross-sectional area and is imagined in Fig. 6.7e to be condensed to the force (F_p). The lever arm of this force F_p is 1/2 of the ventrodorsal diameter of the beam or the "pneu" in Fig. 6.7d. Because of its long lever arm, the force required is relatively small.

$$F_a * \frac{1}{4} \text{ trunk length} = F_p * h_t \quad (6.5)$$

$$\text{and } F_p = \text{pressure} * \text{Area of cross section} \quad (6.6)$$

Because the (small) force is distributed over the (large) cross-sectional area, a moderate increase of the intra-abdominal pressure exerts a moment large enough to counterbalance the bending moments. In more upright trunk postures, the bending moments along the trunk become smaller with the cosine of the angle between trunk axis and the horizontal (see above and Fig. 6.7b).

The above considerations suggest a ***bearing of the massive trunk plus thigh muscles*** at the knee joints. Indeed the femur in all birds is commonly kept in a pronounced anteversion (Figs. 6.7a, b). Kummer (1959a, b) has shown that bending moments in the femur are smaller in anteverted than in vertical postures. That means the necessary strength can be provided by less material. Another means to reduce bending moments is shortness of the femur. Bipedal stance and locomotion is only possible, if the feet are placed perpendicularly below the body's CoM. Although the above-described caudal expansion of the body cavity shifts the CoM rearward, the feet must be placed rather far forward. Without extreme (i.e. dangerous and energy-demanding) bending moments, this is possible only by flexion of the hip and thus placing the knee joint forward in cranial direction. The powerful muscles necessary for keeping the hip joint in balance against trunk weight are considered in the above model as part of the trunk mass, and the body is not balanced at the hip but instead at the knee joints.

In the anteverted position of the femur, any extension of the hip joint leads to a lifting of the body in y-direction. Lift follows the sine of the extension angle, multiplied by femur length. On the other hand, excursions in the hip joint imply only limited movements forward (in x-direction), because this movement follows the cosine of the extension angle. In domestic chicken, we found the angle between the horizontal and the femur to oscillate around 45°, which seems to be a compromise between lifting and propelling the body. In ostriches, however, we as well as Schaller (2008) found in most cases a more inclined position of the femur (or more flexion of the hip joint; see Fig. 6.7). Much greater than hip extension is the contribution of the knee and tarsal joints for increasing step length (see below).

In this context, a look on the phylogeny of birds promises additional evidence: Jurassic forerunners of birds are small theropods (dromaeosaurids) and forms like *Archaeopteryx* or *Jeholornis*, are characterised by a long tail and a three-radiated pelvis. The equilibrium in the hip joint was probably maintained, as in other reptiles, by strong caudofemoral muscles. This can be documented by the unproportionally great diameters of the proximal tail vertebrae. When the long tail was replaced by something like a pygostyle (e.g. *Confuciusornis*), these muscles found a new origin at the postcoxal part of the pelvis – which consequently lead to a birdlike shape of this element with enormous bending strength – as in Figs. 5.17, 6.6 and 6.7a – and to the fusion of ilium and posterior vertebrae to a synsacrum (as in *Confuciusornis*).

6.5 Bipedal Locomotion on Ground. Gaits. Speed. Why Hopping or Running? Why the Characteristic Mass Distribution on Hindlimbs?

Locomotion on two limbs can be performed in three gaits: *striding, running or hopping*[1]. The latter two gaits have been investigated by Christian (1999) on the example of the fossil mammal *Leptictidium*.

In the widespread *striding*, the swing limb makes contact with the ground before the stance limb is lifted, both support phases overlap (Fig. 5.2). In *running* (common in ratitae and Galliformes/Phasianidae, also in blackbirds or magpies) the swing limb is set down on the ground not before the stance limb has given up its ground contact. The mass of the body is supported only within an interval of the time and floats in the air without any external input between these stance intervals. This corresponds to the trot of quadrupeds. It should be noted, however, that birds (Andrada et al. 2015) and low tetrapods (Reilly et al. 2006) seem to make use of the characteristic CoM-mechanics (and consequently of elasticity) in walking as well as running, while mammals use spring mechanisms and the related CoM-mechanics preferredly in trotting (see below, Chaps. 7.3, 7.5). As Fig. 5.2 d-g shows, greater GRFs lead to longer phases of aerial suspension. *Running* of bipedalists can be seen as a sequence of leaps performed by only one limb. Both modes of locomotion, striding and running, have a rather continuous substrate as precondition. On discontinuous substrates, which offer support in different distances and at different heights, like the branches of trees, are individual leaps, or sequences of bipedal leaps the better option – the latter being equal to *hopping*. Both limbs are swung forward in unison and support the body against gravity. They also provide the impulse for the next following phase of aerial floating.

All modern birds use the *walk* as the *least energy-requiring gait*. When in a hurry on ground, small birds, up to the size of blackbirds or magpies, prefer hopping, often after an interval of running when their speed further increases. The heaviest bird I have seen hopping (in a TV presentation in December 2016; also Internet: oceanwide-expeditions.com) is the small rockhopper penguin *Eudyptes crestatus*, weighing 2.5 kg (although they seem to do it preferably on declining slopes; see also Sub-chapter 6.4.3). Larger birds decidedly prefer running over hopping. With growing speed, frequencies of cycles increase as well, but intervals of ground contact decrease. This means that GRFs become greater, and the swing limbs must be moved rapidly into their forward position. These conditions hold true for hopping as well. The swing forward takes place against the mass moment of inertia (J) of the moved limbs:

---

[1]The physical rules that govern striding and running have been identified first for bipedal humans, and therefore will be dealt with in Chapter 9.2).

$$J = m_i * {r_i}^2 \tag{6.7}$$

Why birds either *hop or run* is a fascinating question. Its solution depends on a series of basic conditions which must be treated first. An attempt to explain this divergence will be made below.

If a constant force (F) acts during a time interval (t) in the same direction on a body, an impulse (P, the product of force and time) is exerted, and the body is accelerated (a) to a velocity (v), which is limited by its mass (m). Consequently, the impulse (P) transferred to the body is

$$P = F * t/m, \tag{6.8}$$

and its velocity (v) depends upon

$$v = a * t \tag{6.9}$$

The impulse P can also be defined as

$$P = m * v = m * a * t \tag{6.10}$$

As long as mass is constant, the speed depends upon the product of force (F) and time (t) (Fig. 5.2). The available time interval is limited, since the limbs are losing their ground contact sooner, at higher speeds. Just in order to keep contact with ground for a rather long time, limbs must be long for efficient striding and running, not to speak of hopping (see also above, Chap. 4 and Chap. 5.9). Limb length is not alone a means to put energy into the accelerating body but also to brake slowly.

In all modes of locomotion on limbs, these may exert *components of GRF* in three directions): vertical (in y-direction) as the response to body weight, horizontal (following the x-axis) in the direction of locomotion (= accelerating) or against it (= braking) and horizontal at right angles to locomotion (in z-direction). The latter force component is largely suppressed in birds as well as in mammals. The more important of these components can be seen in the experiment illustrated in Fig. 6.8: a mass rotates about a bearing, which measures the reaction forces in two planes (x and y). The resultant of the vertical weight force (y) and the horizontal, braking or accelerating force component (x) is the ground reaction force (GRF). Since relaxed animals tend to be lazy, they always, in any gait, try to save energy, that is, they chose that excursion of their limbs, which implies the least muscle forces: the big joints of the limbs are kept close to the "functional limb axis" (Fig. 6.9a). In this position the lever arms of the load are short, and muscles can maintain equilibrium with a minimum of energy.

Since in hopping both hindlimbs act at the same time, the total force exchanged between animal and ground is twice as big as in running. The impulse therefore is twice as big and the width of the hop (= a cycle) about twice the length of a single step in running. However, the cycle of running consists of two steps, so that the sum is the same as before!

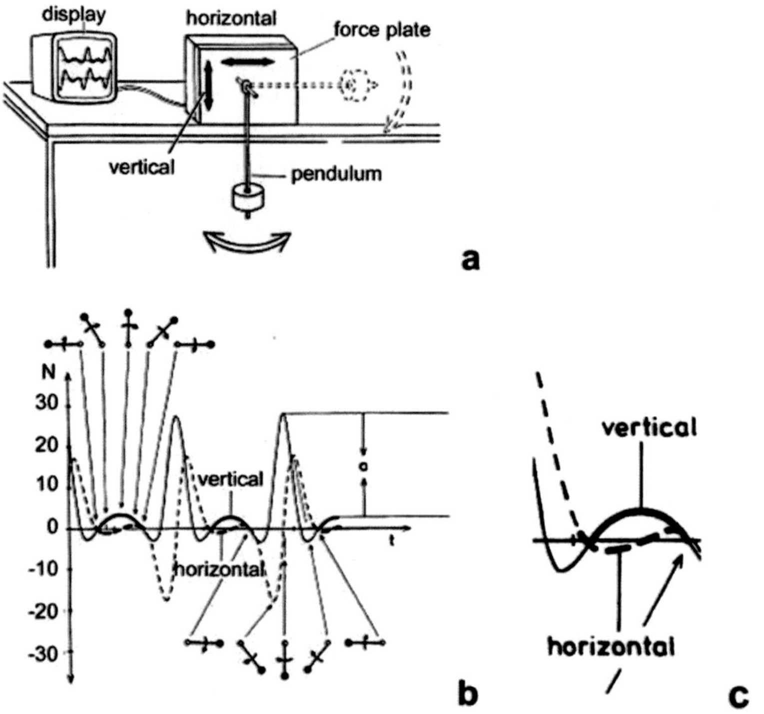

Fig. 6.8 (**a**) Pendulum mass, fixed by a stiff rod to its pivot on a vertically placed force plate. The mass is accelerated to such an angular velocity that it performs complete 360° rotations about its pivot without input of external force. (**b**) The forces at the pivot are measured in the plane of the force plate during two full rotations. Most relevant for our interest are the forces occurring while the mass is above the bearing (fat line on the left part of the graph). This part is enlarged in (**c**): Acting as an "inverted pendulum", the system exerts forces on the bearing which are equal to the GRFs below the foot of a walking or running animal: a vertical component, reaching its maximum while the load is perpendicularly above the pivot (continuous line), and a component in *x*-direction which has two maxima (broken line), changing their sign from minus (braking) to plus (re-accelerating) in the middle of the phase. If the mass is below the bearing, like a "suspended pendulum", the same horizontal force components can be seen as above. In the phase of suspension, forces at the bearing are greater, because weight and centrifugal force components sum up (while being opposite to and therefore subtracted from each other in the above-discussed "stance" phase). The big difference between suspended and inverted phases comes from the relatively high angular velocity which was necessary for getting a full rotation. (From Witte et al. 1991. Compare with Fig. 6.14

Locomotor speed (v) in any rhythmic gait is composed of

$$v = \text{cycle length} * \text{frequency}. \tag{6.11}$$

Both variables must be considered separately:

1st The longer the limbs, the greater is the distance covered and the length of the time interval, during which force can be transmitted between the animal and the substrate. Precondition for this direct relation is, of course, that excursion angles are kept constant. *Cycle length* is the distance wherein the body is moved over ground during one locomotor cycle and the excursion angles of the limbs are identical to the inclination of the limb against ground from touchdown to liftoff (Figs. 6.9a see also 7.4a,b).[2] In striding, the movement of the animal's CoM can be described roughly as movement on the periphery of a circle about the stance foot, the radius of which is "functional limb length", so that speed is proportional to the latter multiplied by the sine of the (above-defined) "functional limb axis". In running, however, phases of aerial floating are intercalated between the stance phases. Nevertheless, the relation between functional limb axis and GRF is the same (Fig. 6.9a).

Already the extension of the joints between the major segments leads to an elongation of functional limb length and so to an elongation of the steps. In all birds, the femur has a moderate length. Its most frequent position is in chicken about 45° against the horizontal, so providing a fair compromise between producing progress and lifting the body. In (running) ostriches, we have found angles of about 25–50° against the horizontal (Fig. 6.9a). Because lift follows the sine, but progression in x-direction the cosine of the excursion angle, the femur in these cursorial birds mainly acts to control the height of the body's CoM above ground. More efficient for progression in x-direction are the excursions of the tibiotarsus about the knee axis and of the tarsometatarsus in the intertarsal joint.

The interplay between braking and accelerating raises the question whether the impulses contained in the decelerated body can be stored and used for reacceleration. This depends upon the own frequencies of the moved body and of the elasticity of the limbs, more exactly of the tension-resistant muscle-tendon complexes. According to data collected by Alexander and his group (1988), the collagenous fibres of tendons alone are perfect means for *storing energy*. Rao 1998, Rao et al. 1995, 1996; Witte 1996, Witte et al. 1995b have focussed on the elastic elements within muscles and provided good arguments for the efficiency of titin. We have to come back to these items below (Sub-Chapter 7.5.6).

The advantageous length of the limbs, however, implies *disadvantages* as well. Aside from increased danger in accidents, where long limbs clearly are more exposed than short limbs, growing of length implies an increase of mass and a shift of the CoM in distal direction, therefore increasing its *mass moment of inertia* (J) and the time available for the foreswing (Fig. 6.10). In view of the phylogenetically prescribed and functionally given arrangement of the muscles, in birds primarily tarsometatarsus and tibiotarsus can be elongated. Fusion of independent elements to one unit allows reduction of those muscles which control the position and the movements of separate skeletal elements. This is of relevance for tarsal

[2]The excursion angle of the limb can be measured between the vertical and the "functional axis" of the free extremity and also between the horizontal ground and the above-defined "functional axis" (Figs. 6.8 and 6.9). It can be chosen arbitrarily (everybody can walk by taking shorter or longer steps), but in the vast majority of investigated examples, the excursion angle is chosen to minimise the sum of moments at the joints – which saves muscle force – and saving energy is a very good reason, why animals are lazy!

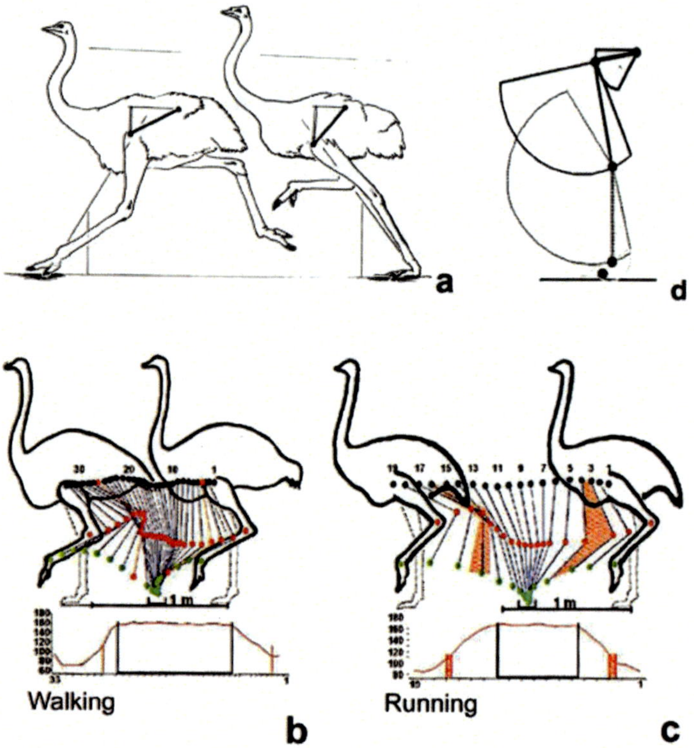

Fig. 6.9 (**a**) Ostrich in two phases of running. The limbs in both phases are approached to the GRF thin lines connecting the stance foot with full body´s CoM, which results from vertical and braking (left) or accelerating (right) horizontal force components. (**b**) Striding (with continuous ground contact), (**c**) running (with phases of aerial floating), both recorded with 25 frames/sec. The figures show the numbers of the frames. The pathways of the knee joint are marked by black dots, and the respective positions of the intertarsal joint by red dots. Lifting and progression of the knee joint (which depend upon the excursion of the femur) are indirectly visible: deep position means extension of hip joint; high position means flexion. The intertarsal joint (red) does not change its angle during carrying load in the stance phase, as emphasised in the graphs below each phase. The distance covered in x-direction is mainly determined by excursions in the knee joint; the excursion angles of the femur in the hip joint primarily lift the body, following the sine of the angle between x-axis and femur in the interval of maximal anteversion (about 25°) and maximal retroversion (about 50°). The progress in x-direction follows the cosine of the same angle. (**d**) Maximal excursions observed in a defleshed hindlimb of *Struthio*. (From Schaller 2008)

segments, but not for those muscles, which control the digits, since their position must be adapted to the respective substrates. The flexor tendons can be guided across the extensor sides of proximal joints, where they contribute to maintaining equilibrium against gravity. Because of the forces required from these muscles, their mass cannot become smaller. The only means left for reducing the mass moment of inertia (J) is the concentration of muscle mass on the proximal part of the respective segment. Exactly this is the reason for THE characteristic shape of the “lower leg”

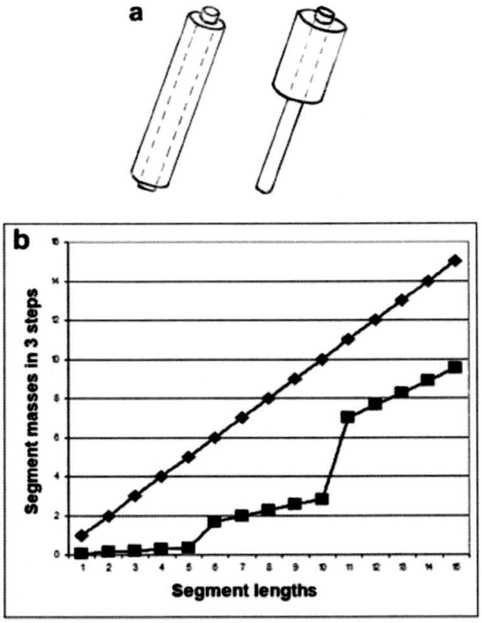

Fig. 6.10 Effects of proximal concentration of muscle mass on the lower leg. (**a**) On the left side, a rod, representing the skeleton (tibiotarsus) wrapped into a layer of muscle tissue with mass m_u. This is assumed to have about two times the mass of the stiff rod in its centre. On the right side the same muscle mass (m_u) is confined to the proximal half of the rod. (**b**) Volumes (as equivalents of mass) of the skeletal element plus musculature grow with increasing length, linearly in the case of cylindrical shape (diamonds), slower in the case of proximally concentrated musculature (squares). The diameters of the central rod are assumed to be 0.3 cm for segment lengths of 1–5 cm, 0.6 for lengths of 6–10 cm, and 0.9 cm for lengths of 11–15 cm. These estimates may be too low, but the general tendency becomes visible: the linear relationship between volume (that is mass) and length

in birds: while foot and tarsometatarsus consist of nothing than bones (for carrying weight), tendons (for controlling the joints) and skin (to wrap in the whole segment), the heavy muscles are confined to the proximal half of the lower leg, so forming the well-known *club-like shape*. The upper thigh, by contrast, which is not moved so much, is accompanied by muscles along its total length. The shifting of muscle mass proximally, towards the knee joint, can reduce the lower leg's mass moment of inertia (J) and so save energy. Figure 6.10 shows a simplified model of a lower leg. Its core is a rigid rod of given length (l) and mass (m_t), comparable to the skeletal element. Wrapped around this rod is the muscle mass (m_u). Mass moment of inertia of this (cylindrical) segment is

$$J_u = (l/2)^2 * m_u + (l/2)^2 * m_t. \tag{6.12}$$

If the muscle mass (m_i) is concentrated on the proximal half of the (club-shaped) segment, its mass moment of inertia (J_k) is

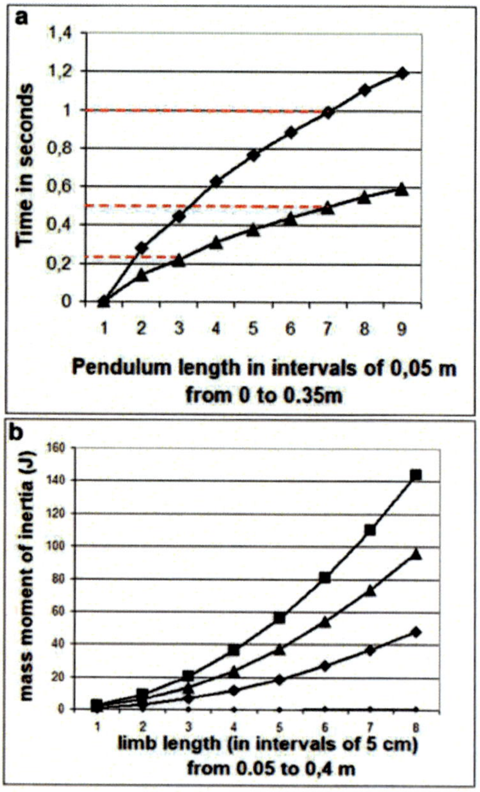

Fig. 6.11 (**a**) Numerical values of pendulum periods (*T*, diamonds) and time intervals available for swinging the limbs forward (*T*/2, triangles) for lengths of the limb pendulum increasing from 0 through 2 cm up to 35 cm, in steps of 5 cm. Horizontal bars indicate the estimated time intervals available for the foreswing of the legs: 0.25, 0.5 and 1 s. These values correspond to frequencies of 4 Hz, 2 Hz and 1 Hz. (**b**) Mass moments of inertia, which resists the active swinging of the hindlimbs forward. Drawn for varying limb masses: diamonds: 0.3 kg; triangles: 0.6 kg; squares: 0.9 kg

$$J_{\mathrm{k}} = (1/4)^2 * m_{\mathrm{u}} + (1/2)^2 * m_{\mathrm{t}}. \quad (6.13)$$

The ratio between both values is 4:1, independent from all other factors. The ratio of muscle mass/skeletal parts and segment length are without relevance; the energetic reward remains the same.

The active *swinging forward* of the unloaded limb is opposed by its mass moment of inertia (J), which must be overcome by muscle force (Fig. 6.11). The smaller the mass of the limb and the shorter the distance (l) from the pivot to its CoM, the smaller the muscle force necessary for swinging the limb forward and the shorter the time interval required for doing so. This influences the above-mentioned factor *frequency*. If the foreswing follows the *laws of the pendulum*, the energy requirement is at a minimum. Frequency of a pendulum is determined by the swinging period (T) and depends alone from pendulum length (l):

$$T = 2 * \pi * \sqrt{(l/g)} \quad [\text{in s}], \quad (6.14)$$

where g is earth gravity.

Pendulum length is the distance between its pivot (in many cases the hip joint, in ostriches according to the above illustrations the knee joint) and the CoM of the freely swinging extremity. If this distance (l) is short, the limb is moving forward faster than if (l) is long.

The shortening of pendulum length is the simple reason why all animals lift their limbs higher and angles between their segments become narrower if gaits become faster.

From the GRF in Fig. 5.2, it can be seen that the time span available for swinging both legs forward in hopping is alone the interval of the aerial suspension. In running, however, the time span for the foreswing of only one limb is much longer, ranging from liftoff through the stance phase of the opposite foot plus floating phase. This allows approaching the above-raised question, *why running is preferred over hopping* in larger birds.

The starting point for such an attempt can be a rather well-investigated extreme: running in the largest living birds, African ostriches. They have been observed running at a speed of 60 km/h or 16.7 m/s for 20 min, 80 km/h or 22.2 m/s for shorter distances (Schaller 2008).

Under consideration of the distribution of mass on the hindlimbs (see above), the length of the limb pendulum is proportional to limb length, so that the time needed for swinging the limb forward grows with limb length. According to Schaller 2008, the length of the hindlimb pendulum is in an ostrich roughly 35 cm (See Fig. 6.12a), and the swing interval lasts about 0.22 s (= 4–5 Hz). Stride frequency in walking is about 1 Hz (i.e. a cycle lasts about 1 s), in running with the moderate speed of 3–4 m/s about 2 Hz (= two cycles per second). In both gaits the available time interval is much greater than necessary for a passive and energy-saving foreswing.

In order to find a generally valid, mechanical rule, it seems appropriat to recur to the pendulum period (T, which determines the duration of the swing phase, see above). If plotted against limb length, T (and this also means duration of the swing phase), follows a root function (see above, and see Fig. 6.11) with gradually dwindling inclination. With increasing length, it comes to cross the value demarcating the time available for swinging the limb forward. Below this value, the time interval available for the foreswing is long enough; above this value, the available time interval is shorter than the foreswing itself. This time interval is shorter in hopping than in running (see above); therefore, at least two of these demarcations are to be considered, one for hopping and one for running. Both gaits can be performed with various frequencies. For a calculation, we need the (pendulum) lengths of bird legs (between say 2 cm and 35 cm) and the frequency of hops or, better, the time of aerial floating without ground contact (unknown. Estimated frequency decreasing from about 4/s to 1/s). The lengths of the pendulum also represent size.

This would be a very good reason why hopping is limited to smaller body sizes and why larger birds prefer running over hopping. In the case of ostriches, their long limbs per se allow them very fast locomotion, but their total body mass seems

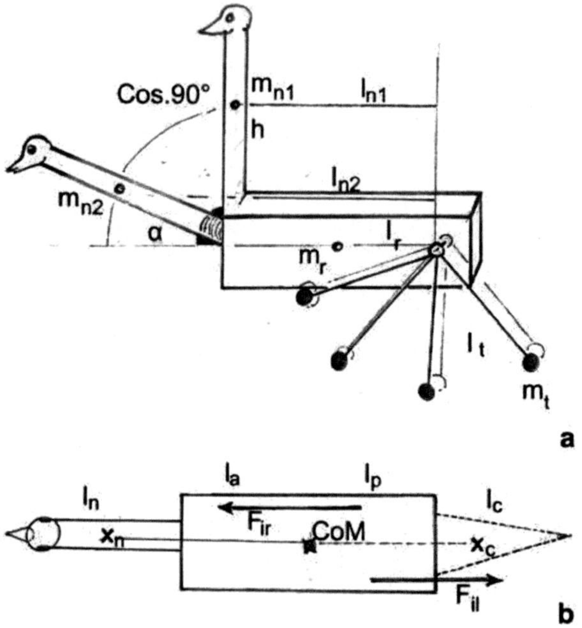

Fig. 6.12 (**a**) The three-dimensional situation is seen from its top left side. Both hindlimbs are swung rearward during stance and forward during swing phase and exert mass moments of inertia (for each limb $J_t = m_t * l_t^2$ where m_{n1} stands for the weight force of the upright, m_{n2} for the forward extended neck plus head). Rotations about a *horizontal axis* through both hip joints (= "nicking" of the body stem) is opposed by the mass moments of inertia of the trunk ($J_r = m_r * l_r^2$ where m_r stands for the weight force exerted by the trunk). Stretching the neck + head in x-direction increases this value by $J_n = m_{n2} * l_{n2}^2$, where $l_{n2} = l_{n1} + h * \cos \alpha$, and the angle α characterises the degree of extension. Since the mass moment of inertia follows the second power of its lever arm, even a light-weight neck may exert considerable influence if stretched forward. (**b**) Viewed from above, a running bird "yaws" under the influence of the rearward (left side of animal)- and forward (right side)-directed inertia of the swing limbs about a vertical y-axis through the CoM. The inertial forces of the limbs are applied to the hip joints (arrows). The moments exerted by these forces may diverge in size and must not act at the same time. Rotations about a vertical y-axis can be suppressed by the mass moment of inertia of the trunk and by stretching out the neck in horizontal direction. This is the reason, why running birds commonly cannot afford the energy-saving upright posture of the neck. In bipedal dinosaurs the tail (broken outline) yields considerable mass moments of inertia. The black dot • stands for CoM (= pivot of rotation); x_n demarcates the centre of head and neck mass; x_c stands for centre of tail mass (existing in dinosaurs, not in birds!); l_n is the lever arm of neck plus head mass, l_a the lever arm of anterior half of the trunk, l_p of the posterior half of the trunk and l_c of the tail (only in dinosaurs, not in birds!). Remember that the distances l must be raised to the second power to get the mass moments of inertia (after an idea proposed by Yamasaki)

too large to be accelerated so much as is necessary for longer floating phases in hopping.

Another argument, why hopping is limited to small birds, seems also quite plausible: because the acceleration for each individual hop depends alone from the *available force*,[3] if mass remains constant. The available force (F_a) is greater in relation to mass as long as an animal is small (see Fig. 1.8).

In fowl-like birds (mainly runners but able to fly), and the above-mentioned rockhopper penguins, the available force of the hindlimbs is probably restricted by the massive pectoral musculature which moves the wings. The nonflying ratitae do not have this problem, and therefore their hindlimbs may occupy a greater share of total body weight and be more powerful than in a chicken-like bird.

It may well be that sufficiently long phases of aerial floating can be reached only if both hindlimbs together accelerate the body. More acceleration means longer airborne phases and greater distances covered. It should be kept in mind, however, that an array of individual jumps (as hopping) is not a very fast mode of locomotion, because it hardly exceeds the speed during flight, which is in small prosimians (Günther, 1985), as well as in large horses (Preuschoft et al. 1988; see Chapter 7.5) not more than 5–6 m/s. Really fast animals therefore do not exploit possible widths, but put in additional acceleration (= energy) within very short time intervals to gain more speed.

Both attempts to explain on the basis of mechanics, why large birds run rather than hop, are in conflict with the fact that *kangaroos hop* and do not run, although their body size is about the same as in large running birds (Table 6.1). Probably the reason for the divergence between locomotor modes of kangaroos and ostriches must be sought in the possession of a tail (see below) or in evolution: like all mammals, kangaroos evolved from other ancestors than birds and therefore have divergent anatomical preconditions – and divergent programmes for locomotion – whereby both of course are subject strictly to the physical conditions. Both possibilities will be dealt with below, in Chap. 7.3.

In running, the alternating passive fore and back swaying of the hindlimbs and still more so their active acceleration and deceleration at the beginning and end of the foreswing lead to considerable *torques about a vertical (y-)axis* (Fig. 6.12b). These must be suppressed unless the body starts uncontrolled "yawing". An energetically inexpensive/cheap means to do this is using the mass moments of inertia inherent to the body stem. If this is great enough, no energy is required to exclude disturbing and probably self-enforcing rotations.

The body part with the greatest mass, the trunk, in birds is stout and therefore has limited mass moments of inertia, a tail is lacking, and neck and head comprise little mass. As mentioned repeatedly, mass moments of inertia grow with the square of lever arms, and therefore even a slender neck, if only long enough and extended in the (x-)direction of locomotion, may well provide the mass moment of inertia necessary to counterbalance the moment exerted by only one hindlimb (compare the data in Footnote 3). This can be observed in Galliformes: when running, their

[3]In analogy to our data on human high jumpers and jumping horses, the acceleration of an ostrich should be about 2–4 g where the exact value depends from width or height of the respective leap or stride. Günther et al. 1991 have found in *Lemur catta*, weighing 2.4 kg, accelerations of 3.8 m/s^2 and in *Galago moholi*, weighing 0.25 kg, 5.1 m/s^2, both leaping over 2–2.5 m (see below, Chapter 8.8). This may be within the range of what birds can perform.

Table 6.1 Body mass of several hoppers (mammals) and runners (birds)

Rodents	Body mass in kg
Kangaroo mice (*Microdipodops*)	0.017
Kangaroo rats (*Dipodomys*)	0.035–0.180
Jerboa (Dipodidae), including *Jaculus*	0.040–0.075
Mongolian jerboa *Allactaga sibirica*	0.080–0.140
Great jerboa *Allactaga major*	0.280–0.480
Springhare (*Pedetes*)	3–4
Kangaroos	
Rat kangaroos (*Hypsiprymnodon*)	0.36–0.68
Potoroos (*Potorous*)	0.66–2.2
Hare wallaby (*Lagostrophus*)	1.2–4.5
Swamp wallaby (*Wallabia*)	7–27
Bennett wallaby (*Macropus*)	15–20
Grey and red kangaroo (*Macropus*)	17–85
Galliform birds	
Chinese painted quail	0.045
Turkey	22
Nonflying, paleognathic birds/ratitea	
Kiwi (*Apteryx*)	1.5–4
Fossil Moas (*Dinornis, Pachyornis*)	180–250
Subfossil ostriches from Madagascar (*Aepyornis, Pachyornis*, etc.???)	350–400
African ostrich (*Struthio*)	100–150
Nandu (*Rhea*)	20–25
Emu (*Dromaeus*)	40–55
Kasuar (*Casuarius*)	up to 85

Collected from Grzimek 1988, Schilling et al. 1983 and Wikipedia

trunk axis becomes less inclined, and they stretch their neck full length horizontally. The lowering of the anterior part of the trunk and the stretching of the neck move the greatest length into the (horizontal) plane of yawing, so maximising the lever arms. In ratitae, the obvious length of the postcoxal pelvis may be an additional means to increase the trunk's mass moment of inertia.

In a tail-bearing bipedal dinosaur, the balance is easier to establish because of the long, heavy tail. In a first approximation, the values are

$$\begin{aligned} J_{\mathrm{d}} = {} & \text{trunk mass} * \tfrac{1}{2}\,\text{trunk length}^2 + \text{neck} + \text{head mass} * \tfrac{1}{2}\,\text{neck length}^2 \\ & + \text{mass of tail} * \tfrac{1}{3}\,\text{tail length}^2 \end{aligned} \tag{6.15}$$

6.6 Why Are Three or Four Toes So Common?

The toes (digits) in all birds are, aside from the beak, the body segments used for establishing contacts with the environment. Their highly variable shapes can be explained alone on the basis of their lifestyle, preferred substrates and the forces

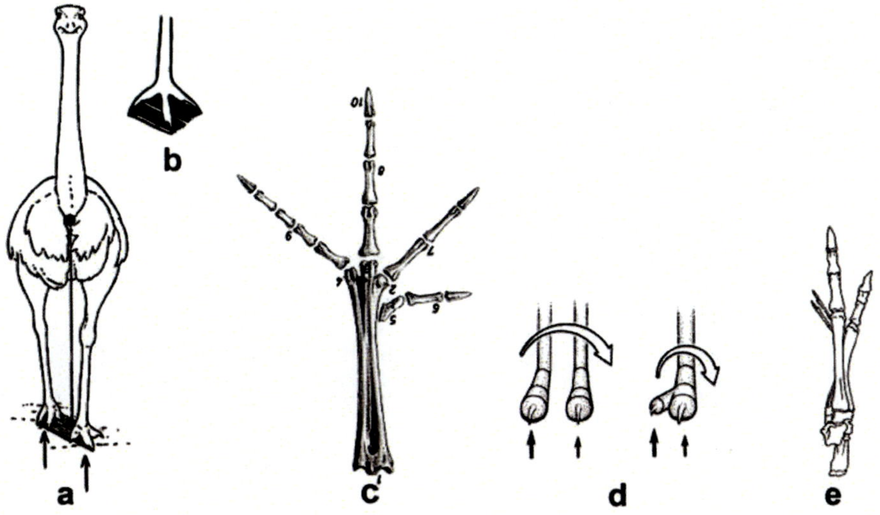

Fig. 6.13 (**a**) Walking nandu (*Rhea americana*), seen from in front. The CoM is perpendicularly above the area of support, which is delimited by the two feet on the ground. (**b**) Three toes provide a rather large area of support even if the animal is carrying its weight on only one foot. (From Preuschoft and Christian 1999). (**c**) Skeleton of a four-toed foot in chicken (*Gallus domesticus*, from Nickel et al. 1968). Note the melting of the parallel tube-like bones of the tibiotarsus to one element and the varying number of phalanges. (**d**) Balancing body weight above only one toe is difficult. As long as both one-toed feet are supporting the body: no problem. Minor imbalances, e.g. wind pushing the animal to its right side, can be countered by exerting higher GRF (left vertical arrow) on the right foot. This produces a moment (curved arrow), shifting the body back to the left. If only one foot is on the ground, as obligatory in walking or running, the restoration of balance is possible only with the aid of at least one lateral toe (Christian 1999). (**e**) The feet of typical kangaroos (*Macropus*) also show a rather strong digital ray lateral to the primarily weight-bearing third toe in spite of their bipedal hopping - as shown in **d**, left. (From Jouffroy and Lessertisseur 1979)

occurring during species-specific usage. Their biological roles (standing on ground, running, standing on twigs, climbing, propelling the swimming bird) determine the number and position of *digits*, as well as the shape of their claws at the terminal phalanges (which may be blunt and short, often nail-like, or sickle-shaped, long, often sharp). Most of the *ground-living forms* use *three digits* (according to comparative anatomy, the rays II, III, IV) for transmission of weight to the substrate. A fourth digit may exist, but not be used to transmit body weight (Phasianidae). Three widely separated toes indeed provide a considerable width of the area of support to balance the CoM above it (Fig. 6.13a, b). A tipping to the medial or lateral side can be impeded by pressing the respective toe against the ground. During roll-off, not alone the median toe does transmit GRF, but also the lateral and medial toes can take over considerable forces. The same arrangement of toes existed in bipedal dinosaurs, as can be seen from numerous fossil foot tracks. These animals were often heavy, weighing several tons, and document that this sort of support works independent

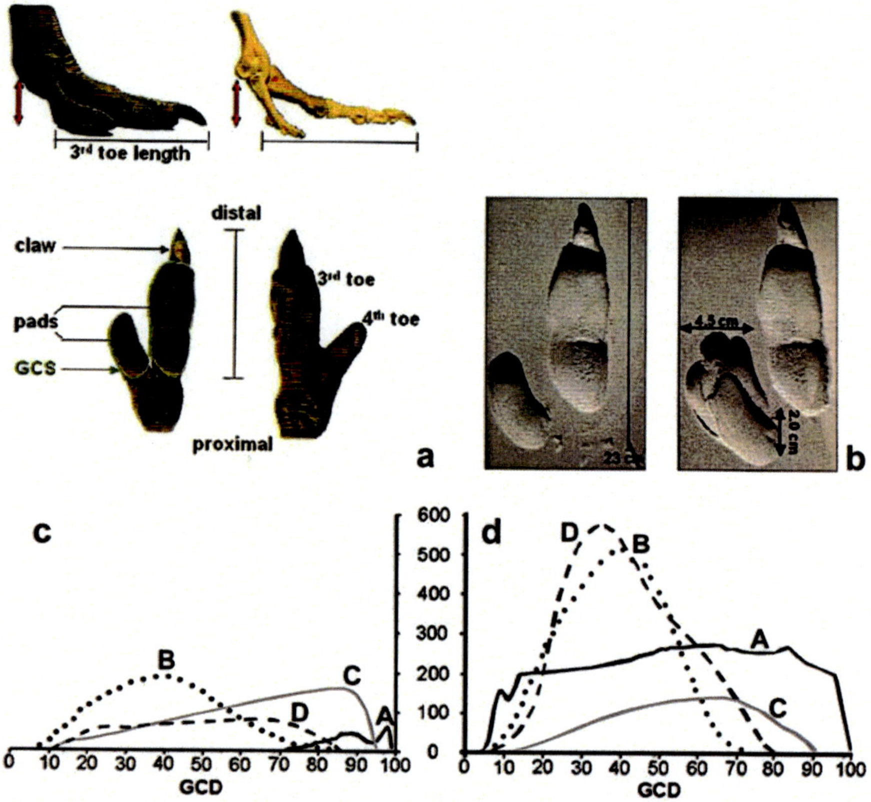

Fig. 6.14 Two-toed foot in the African ostrich (*Struthio camelus*). (**a**) Seen from lateral side on *top left*, the skeleton of the same foot on *top right*. *Below, on left side* plantar, *on right side* dorsal aspect of right foot in action. (**b**) Two footprints: *left* walking, *right* running. Note the repeated contact made by the lateral toe. (**c, d**) Changes over time of those forces, which act on selected segments: A below the third claw, B below fourth toe, C below the third terminal phalanx, D below the third joint between the middle and terminal phalanges. Compare with Fig. 6.9. (From Schaller 2008)

from body size. In swimming birds (anatids, grebes, divers), three toes provide by spreading an optimal area to produce high *water resistance*.

Like in mammals, the number of toes can be reduced beyond three (Fig. 6.14a). This may yield some (limited; see also in Sub-Chapter 7.5.3) saving of weight and so contribute to minimising the mass moments of inertia. More important seems to be the loss of ground contacts of shorter toes, once the longer has become sufficiently strong. Reduced ground contacts entail a reduction of external forces. The extreme among birds is the African ostrich, in which the third toe is very strong and carries most of the GRF (Fig. 6.14b). The lateral fourth toe is clearly weaker. (A similar configuration exists in kangaroos (Fig. 6.13e).)

The question should be raised, why less than two toes have never been observed. Christian (1999) has noted that ostriches (like all birds) have but weak abductors of the hip joint, so they must make sure that in walking and running, the stance foot is not placed at the "wrong" (i.e. opposite) side of the medio-sagittal plane in which the animal moves. The passively adducting moments in the hip joint could become greater than can be balanced by muscle force. Although such a mistake can be corrected during the stance phase of the following stride (especially if stride frequency is high), an additional component of the GRF applied on the lateral side through the fourth toe is a considerable help (Fig. 6.13d). Schaller (2008) has investigated the use made of the toes in ostriches experimentally (Fig. 6.14b–d). She found the lateral toe abducted actively before touchdown, making first contact with level ground and carrying a surprisingly large part of the GRF. At higher speeds, the GRF concentrates more and more on the stronger median toe. All her observations fit perfectly to Christian's theoretical considerations. The above-mentioned kangaroos do not have, because of their bipedal hopping, the problem of keeping balance on only one foot.

An additional, *rearward directed toe* (as in Fig. 6.13c) in ground-living forms can transmit braking force components on flat ground or exclude a tilting of the animal rearward. In climbing on branches, the fourth toe provides a counterforce against the grip of the anterior toes – comparable to the thumb in mammals. The "gripping forceps" is morphologically and functionally separated from the free limb, which includes the tarsometatarsus. A similar arrangement can be found in arboreal primates (see Chap. 8.3). The prehensile function is emphasized in the often climbing parrots by an additional toe being turned rearward, and so securing an equilibrium of external forces when grasping (comparable to chameleons). Birds of prey and owls need for capturing prey the possibility of exerting force concentrical from all sides and therefore need a strong fourth digit.

6.7 Alternative Use Made of Hindlimbs. How and Why Do Hindfeet Differ Among Swimming and Diving Birds?

Generally known specialisations of many birds are the large webs connecting the three anterior digits (e.g. ducks, geese and swans, as well as divers). In coots (Rallidae) and grebes, there exist instead of the webs large flaps on both sides of the digits (Fig. 6.15a, b). The existence of these feet coincides with the mode to swim of the respective forms, but their particular propensities have not yet been explained in detail.

Most commonly known waterfowl propel themselves continuously by alternating movements of the hindlimbs, similar to those in running. Continuous propulsion, regardless by which sort of feet, is an advantage in a viscous medium, because it avoids losing speed during the intervals between propulsive strokes, which must be regained. The birds mostly remain at the surface, with trunk axis more or less parallel to the surface, while only a part of the body is immersed. The neck is kept upright or

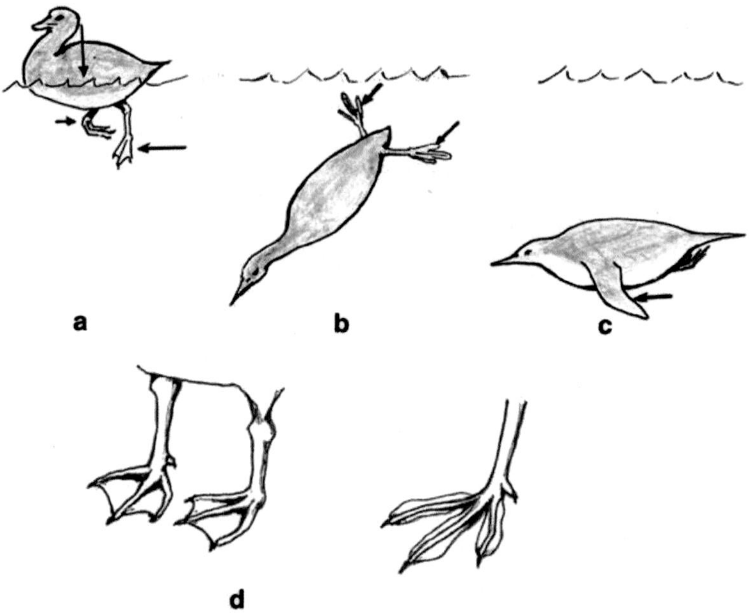

Fig. 6.15 (**a**) Ducks, geese, swans and other birds which swim on the surface of water propel themselves by alternating or synchronous retroversion of their feet. While the expanded webs cause considerable water resistance, the forward movement of the folded foot finds less resistance. The resultant propulsive force rotates the head and neck upward. This rotation is countered by body weight, which rotates the head and neck downward. (**b**) Diving birds (divers, grebes, cormorants, anhinga, coots) propel themselves by laterally abducted hindlimbs. The feet are often equipped with *flaps* on both sides of digits instead of webs. Both hindlimbs can be retracted in rapid locomotion forward more or less in unison. For changing directions, the feathered tail (if present) may be used aside from varying propulsion by the hindlimbs. Note the posterior position of the limbs near the caudal end of the trunk. (**c**) Swimming penguin, using the forelimbs for propulsion and the rearward positioned, webbed hindfeet mainly for steering. Major external forces, without water resistance, are symbolised by arrows. (**d**) Typical webs and vestigeous ray I in an anatid foot, flaps in coots (*Fulica atra*); similar, but without claws and first toe, in grebes (*Podiceps*)

recurved. In acceleration, for example, in taking off for a flight, the neck is stretched forward and so increases the body's mass moment of inertia – as detailed above. If very rapid acceleration is required, both feet may by retracted at the same time. Water resistance (a force produced by the feet) is directed horizontally, somewhat deeper than the ventral contour of the swimmer. The so produced moment tends to rotate the bird's body head upward, tail downward. It is countered by gravity, which can be imagined as concentrated in the body's CoM, acting at right angles to the surface (Fig. 6.15a).

The more or less two-dimensional biotope at the surface of water can be enlarged to a three-dimensional space, if the birds are able to dive. For diving, that is, swimming in a three-dimensional space, the fore-and-aft movements of the propulsive feet below the belly are not optimal, because of the head upward rotating moment – against which the counterforce is missing in the absence of gravity.

This unfavourable moment does not occur, if the propelling feet are producing thrust at the height of the CoM (same thing as above, Sub-Chapter 3.3.7, under "Otters and Beavers"). Precondition for doing so is a position of tibiotarsus and tarsometatarsus in a parasagittal *z*-plane (Fig. 6.15b). Their movement must not be restricted by the space-demanding trunk and therefore the hindlimbs are attached, for example, in divers very close to the rearmost part of their trunk. The postulated movements (and the required morphology) can be observed in penguins, divers, grebes, cormorants and gannets that are birds which spend much time below the water surface while diving. In all of them, the tibiotarsus is rather long[4] and the tarsometatarsus rather short, but long enough to keep the foot clear from the trunk and the short tail. Both segments are kept (like in flying) nearly parallel to the longitudinal axis of the trunk, in penguins not alone when swimming but also when standing or walking on ground.

6.8 Frontlimbs and Flying

The most observed and frequently discussed characteristics of birds are the transformed *anterior extremities*, forming *the wings* which permit flying. The *flying* of birds has been and still is the subject of numerous studies, most of them focussed on one of two problems: stream mechanics and evolution of the ability to fly. The first aspect is summarised in Hildebrand and Goslow (2004, pp. 588 ff.). I have nothing to contribute to these items. Instead, possibilities to understanding body shape are opened via the "internal mechanics" of the bird's body. This is presented here as a sketch; a thorough study based on empirical data is not known to me. Baier et al., 2007 and orally in Paris, 2008, has presented wonderful animations of movements and the respective changes in position of the skeletal elements. The following rules therefore are probably known, but not described in detail.

Like fossil pterosaurs as well as living bats, birds make use primarily of their forelimbs for *flying*. The easiest way to see and understand the functional conditions of flying is gliding or soaring: two variants are realised, as illustrated in Fig. 6.16a. For both purposes, the forelimbs, equipped with strong, mechanically resistant and effective feathers, and the "wings" are abducted and unfolded.

If the velocity of the streaming air is predictably and reliably high (as on the open sea and near the seashores), a wide "spread" and short "depth" of the wing is of advantage. This is the same situation as mentioned above for the flukes of fast swimmers among fishes and whales. It explains why, e.g. albatrosses, gulls and gannets possess outstanding spread (= length) but very limited depth (= breadth) of their wings. Different are the external conditions for soarers (vultures, eagles and

---

[4]Visible in the living animal is only the tarsometatarsus and the foot, while the tibiotarsus is completely wrapped in the skin covering the trunk. The distal limb segments poke out of the compact, streamlined body close to or behind the short tail. Therefore their name "Podicipedidae" for divers.

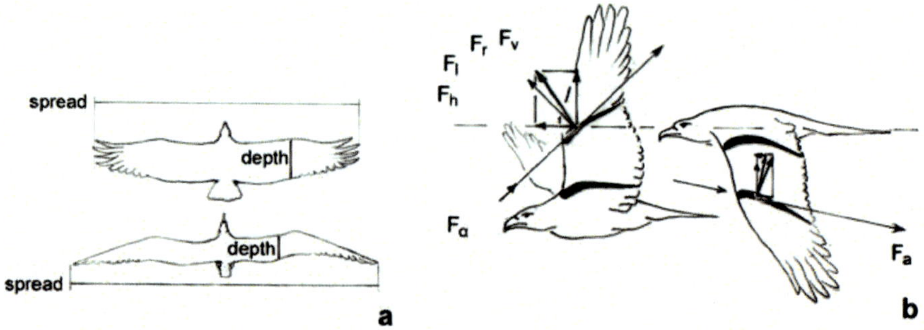

Fig. 6.16 (**a**) Soaring (top) opposed to gliding bird (bottom). "Spread" and "depth" of the wings are indicated. The bigger the ratio between 1/2 spread and depth, the better are the wings suited for gliding, and the smaller the ratio, the better the wings are suited for soaring. See also Sub-Chapter 3. 2.2. (**b**) Wing profiles in active "rowing" flight. *Left*: downstroke, *right*: upstroke. Two approximate profiles of the wings are drawn in black. The forces acting on these profiles (F_a) are broken down into vertical (carrying, F_v) and horizontal (propelling, F_h) components. F_l is the lift produced by the air stream (F_a) along the wing's profile, which, added to the air resistance, yields the resultant lift, F_r. (Redrawn after Herzog 1968)

relatives, storks): the speed difference between the surrounding air and the bird is low; the streaming velocity is moderate but may well possess an upward component. To make optimal use of such conditions, a limited spread of the wing but great depth is of advantage (Fig. 6.16a; Herzog 1968).

In both variants, the plumage and feathers give the wings a profile (Fig. 6.16b), which produces a lifting force component. It compels the streaming air to flow faster on the vaulted upper than on the hollow or flat lower side. The faster the stream, the lower the pressure exerted against the wing, both being inversely proportional to each other. The low pressure against the upper side in combination with the opposed high pressure against the lower surface (Fig. 6.16b) lifts the wings and – if these are firmly connected to the trunk – the whole body.

When the velocity of the air stream is too slow to carry the bird, it produces the required velocity of air stream itself, by beating the wings up and down in *active*, "*rowing*" or "*flapping*" *flight*. The (rapidly moved) hand wings make the greatest propulsive effect, because their speed (V_{peri}) grows with distance from the pivot ("radius"; see above, Chap. 5.5)

$$V_{peri} = \text{angular velocity} * \text{radius} \qquad (6.16)$$

(Nachtigall 1977, Bilo and Nachtigall 1977). Because of their long lever arms (= "radius"), hand wings are moved rapidly, but cannot carry much weight force. As shown in Fig. 6.16b, a bird rotates the profile of its hand wings to produce a resultant lift not alone in vertical but also in forward direction, preferably during downstroke. Arm swings produce the major part of the lifting force which carries the body. The great mass of the trunk is suspended by the m. pectoralis from the upper arms (Fig. 6.17). It takes its origin from the carina of the sternum. If muscle force is greater

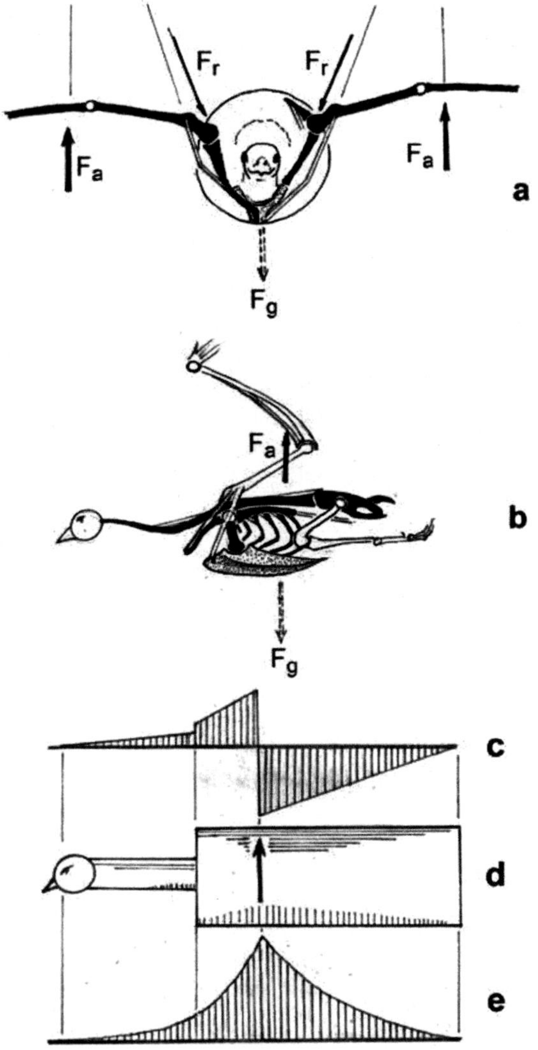

Fig. 6.17 (**a**) Anterior view of a flying bird. Humerus, proximal part of the forearms and ventral elements of the shoulder girdle are shown in black. On the left side of the reader (right side of the bird), the furcula (black) and sternum (stippled) are also shown. On the right (left side of bird), the scapula and coracoid are drawn in black, and the sternum is stippled. The projections of most important muscles are indicated as double lines. The scapula is fixed to the rib cage by the m. serratus. The lifting forces are distributed on the wings and can be represented by their resultant (upright arrows F_a), which is concentrated on the centre of the wing area. It is comparable to GRF in animals on the ground and opposed by body weight (F_g downard directed, white broken arrow). Equilibrium in the shoulder joint exists, if the resultant (F_r oblique, downward directed arrows) of both forces, F_a and F_m (the latter exerted by the m. pectoralis), passes from their intersection (not shown) to the shoulder joint. These resultants (F_r) are close to the most proximal parts of the humerus and fall nearly in line with furcula and/or coracoid. (**b**) Side view, conventions as in (**a**). From the body's CoM, body weight F_g (indicated by broken double line) acts downward. It falls in line with the lifting forces F_a of both sides. The bracing of the glenoid joint in ventro-cranial

than necessary for equilibrium in the shoulder joint, the wing is swung downward. Deeper than the pectoral originates the m. supracoracoideus (= m. pectoralis profundus of mammals) from the ventral surface of the sternum. Its tendon passes the glenoid joint on its dorsal side. Therefore, this muscle is responsible for the upstroke (Hildebrandt and Goslow 2004; Wikipedia). Both muscles have highly elastic tendons[5]. For a balanced body position during flight, the resultant of all lifting forces must be located at the height of the body's CoM: the resultant of the lifting forces should ideally pass through the body's CoM. If the resultants of lifting forces and weight forces do not coincide, the bird has two options: changing the form of its wings (most often by flexing the carpal joint) or correcting its balance by the tail feathers. They may be flexed downward and so create additional lift at the caudal part of the body, or tail feathers may be flexed in dorsal direction and so press the tail downward.[6] Both manoeuvers, however, imply a disadvantageous braking component, directed against progression and influence the bending moments along the trunk (see below). This disadvantage is probably avoided by the flexion of the free wings, which can be observed frequently.

If the body stem of a bird during flight is modelled as a *homogenous beam*, it is exposed to compression along its ventral and to tension along its dorsal margin (Fig. 6.17d,e). This is the same situation like in bipedal posture, though its maximum is different: shoulder joint rather than hip joint. In both, bipedal posture as well as flying, and shortness of the trunk in combination with great diameters keep the bending stresses low. In many flying birds, the neck is carried in an S-shape, so that the lever arm of head weight is shortened and not much force is needed to keep it in

---

Fig. 6.17 (continued) direction by the furcula is evident in both views, lateral and frontal, as well as in ventro-caudal direction by the coracoid and in dorso-caudal direction by the scapula. The latter is kept in its position relative to the trunk by the mm. serratus (ventral) and rhomboideus (dorsal). **(d)** Body stem (head, neck and trunk) of a flying bird in side view, replaced by cylinders of varying diameters. The rearward extended hindlimbs are neglected. Tensile stresses (dorsal, horizontal hatching and compressive stresses (ventral, vertical hatching) reach their maxima at the shoulder region, where the body is suspended from the forelimbs. **(c)** Shearing forces. **(e)** Bending moments. An imbalance between weight force and lifting forces can be compensated by downward inclination of the (nearly weightless) tail feathers – which, however, has also a braking effect

---

[5]This is unusual (most tendons are rather stiff, for precise transmission of precisely controlled muscle contractions against a solid substrate). The flight muscles, however, work against the "soft" and widely distributed resistance of air, by means of the also soft and compliant feathers. Probably this "damping" device is an additional means to exclude abrupt movements of the wings, which might disturb the air flow, and to adapt the movements of the wing to inhomogeneities of the airflow. Force transmission by an elastic tendon also allows the muscles to work permanently with minimal energy expenditure and to avoid energy-requiring corrections of muscle pull.

[6]This sort of "steering" can also be performed by a feathered long tail as was present in reptilian forerunners of birds. Nevertheless, it implies the disadvantages of increased air resistance and additional weight!

balance. In other forms, however, *neck* and head are stretched out in x-direction, so producing bending moments. Both segments contain little mass, and therefore the bending moments along the neck are not extremely high, even if the neck is extended. This is in contrast with the tendency among all birds, to keep their necks close to the vertical as long as no necessity compels them to deviate from this energy-saving neck posture. It seems worthwhile to investigate whether the head in storks, flamingos, geese, swans, etc. possesses a profile that can produce lift, so that the head rests comfortably on a cushion of air.

Compressive stresses along the ventral side of the *trunk* can easily be sustained by the massive sternum, in addition to the (dominating) compressive forces derived from the pull of the mm. pectoralis and supracoracoideus, the fibres of which are more or less parallel to the x-axis. The tensile stresses along the trunk's dorsal margin are, like in bipedal standing, sustained by the ilium plus synsacrum and their periosteum. Where bending during flight reaches its maximum, the slender and long scapula is located on both sides parallel to the trunk's longitudinal axis. Parallel to this bony element, there exist tension-resistant structures along the dorsal margin of the trunk in x-direction: the scapula is covered on both sides by muscles, the most obvious function of which seems to control the glenoid joint (this will be detailed below, Chaps. 8 and 9), but all of them exert longitudinal force components. The number of their fibres and subsequently the longitudinal force components decrease from cranial to caudal. The most caudal part of the scapula is tied rearward by a large fan of fibres of connective tissue spreading in caudal direction and so distributing the transmitted force.

Where bending moments assume their greatest values, the *shearing forces* also have their maximum. They are opposed by the ribs, which keep the distance between the sternum and the vertebral column. Therefore, compressive stresses along the ribs can be expected. Since birds possess a clear-cut joint at the "knees" of the ribs, the angle between sternal and vertebral parts may be narrowed. Such a movement compresses the thorax and entails an expiration. The rib knee is controlled and opened against the forces tending to compress the thorax in dorsoventral direction by the mm. serratus and obliquus abdominis externus. Both muscles have very long lever arms in form of the obvious and often discussed processus uncinatus.

As can be seen from the anterior view (Fig. 6.17a), the mass of the body stem is suspended by the m. pectoralis from the humeri. Muscle force and the lift derived from air resistance combine to a resultant which passes through the shoulder joint. Depending from the position of the wings, and from the exact size of the lifting force, the direction of this resultant may vary. In most cases it probably shows a ventro-medial direction, fitting to the inclined position of the humeral head and to the strong development of the coracoid, which transmits the force of the pectoral muscles to the sternum. If wings are in a deep position, the joint force may well have a dorsal component, which acts against the scapula. The same holds true for components of the joint resultant in caudal direction, while joint forces directed forward, are passed on to the furcula. Because fibres of the m. pectoralis completely surround the humeral head, it seems possible that the joint force can be redirected independently from the external conditions to fit to the existing skeletal elements.

The most cranial part of the shoulder blade is curved ventrally to form the (above-mentioned) dorso-caudal brace of the glenoid joint. At the concave (latero-ventral) side of this curvature, the ventrally directed fibres of the m. serratus change their insertions. The *circles of forces* shown in Fig. 6.17a, formed by lift against wings, body weight, pectoral and serratus muscles are closed through coracoid, proximal humerus and scapula or furcula, respectively. Aside from the glenoid joint, the latter two skeletal elements do not have other than muscular contacts with other parts of the trunk.

6.9 Alternative Use Made of Forelimbs Among Birds. What Do Flightless Birds with Their Wings?

Forelimbs can also be used for *propulsion in swimming* and diving (penguins, auks, gannets, shearwaters). Those birds able to fly (all except penguins), use their forelimbs in water like in air, but shorten the lever arm of the wing by flexing their carpal joints. This confines water resistance to the arm wings; hand wings seem to be trailed behind the up- and downstrokes. Narrow, long wings are well suited for diving, because they have rather small surfaces – enough to propel the body in the tough medium (factor ρ in Reynold's number) – but evoking not too much resistance against being moved.

Penguins have chosen a different way: their forelimbs do not consist of feathers forming wings but of skeleton and connective tissue, covered by skin. The shortness and moderate depth of these extremities is in line with the toughness and specific density of water, which makes movements much harder than movements in air. They also do not need the lift necessary for keeping the animal airborne. Therefore they do not need a profile like flying birds; the effect of penguin "wings" is best compared to the forelimbs of eared seals or to the flukes of fishes and whales.

In the recent land-living flightless birds (ratitae), the *forelimbs are poorly developed* in comparison to the powerful hindlimbs, but still present, and in ostriches and nandus rather strong, equipped with a considerable musculature and, because feathered, quite obvious. A comprehensive study on the use made of the forelimbs in cursorial birds is still missing, though planned (Schaller). Usage of "wings" in courtship is known, as well as in demonstrative behaviour, and as "shield" in intraspecific conflicts. All this, however, does not evoke stresses of considerable height, which are repeated regularly, at least often enough, to justify a morphological specialisation (for more details, see Chap. 10).

Some observations of ostriches and nandus yield starting points for predicting stresses acting regularly on the "wings": in fast running, they are able to *change direction* rapidly. The ability to stop suddenly high-speed running is an obvious necessity. Both is problematical, in view of the considerable mass of the animals and their speed of about 17 m/s (= more than 60 km/h), not to speak of the height of the CoM above ground, which gives tilting forces an enormous moment. When wheeling around, the inner wing is regularly spread, and in braking manoeuvers, both wings are inflated (Fig. 6.18), even moved against locomotion. At the same time,

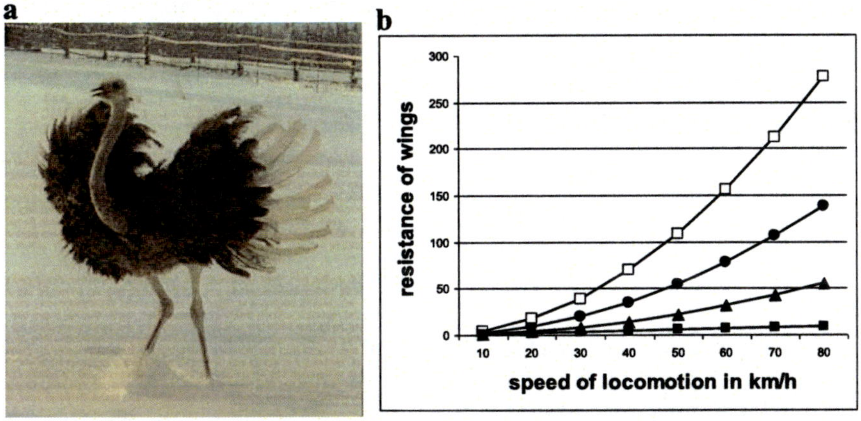

Fig. 6.18 (**a**) Ostrich, braking from running at medium speed. Note the spreading of the wings taken from (Schaller, 2008). (**b**) Air resistance *W* (rosettes) against both wings depends on the square of running speed (between 10 and 80 km/h). Therefore, the braking impulse of both wings (white squares) and the moment which changes the direction (triangles) grow faster than linearly with increasing speed. The deceleration of running velocity within the first seconds is between 1% and 9% (black squares), depending on the original speed. This model calculation is based on estimates of the smallest exposed area, corresponding to the black feathers. If the exposed area is larger (corresponding to the white feathers), the deceleration grows linearly with increasing area. Movements of the wings against running direction entail an increase of the factor speed. They have been observed, but are not included in this calculation

the trunk is lowered by flexion of the limb joints, and so the CoM is approached to the ground (observations of Schaller, not published). While high braking forces are applied to the feet, the moment of the inertia, which tilts the body forward, is reduced.

Each wing is exposed to air resistance (*W*), a force which depends upon the square of running velocity (v^2), area of the wing (*A*),[7] multiplied by density of air $\rho/2$ (= 0.6 kg/m^3) and the cw-value of a plate (= 1.17, values taken from Papenfuß 2004). Written as formula, this is

$$W = v^2 * A * \rho/2 * \mathrm{cw} \tag{6.17}$$

The dimension of the result is kg m/t^2, that means a force, measured in Newton (N). Values are plotted in Fig. 6.18b against running speed. The braking force on both wings decelerates running.

[7]This area is estimated on the basis of N. Schaller's data to be 0.40 m^2. It comprises nothing than the black feathered part of the wing. The white feathers at the margins do not form a continuous area but fan out. Without experimental results obtained in a streaming channel, we do not know their influence on streaming resistance. If they contribute, the resisting area and the resisting force may easily grow by factor two or more.

The *braking* force (*W*) acts against mass inertia of the moving body. The effect of wings on braking can best be determined by the impulse (*P*). The running ostrich may have an impulse (*P*):

$$P = (F * t) \text{or} \, m * v(100\,\text{kg} * 13.7\,\text{m/s} = 1370\,\text{kgm/s}).$$

Each of the two wings exerts a braking force (see above) of at least 55 N, together 110 N. The ratio resistance/impulse (*W*/*P*) is 0.08. The running speed of, say, 50 km/h or 13.7 m/s is reduced during the first second after spreading of the wings by 8%. This does not seem an impressive effect. However, if the wings are moved forward, opposite to the direction of running, their relative velocity must be added to locomotor speed, and the impulse *P* inherent to the moving mass remains the same. In contrast, if the ostrich runs faster, the braking effect by air resistance becomes more pronounced, but at the same time, the impulse of the moving mass becomes greater. It may well be that the black feathered part of the wing underestimates the resisting area and that the white feathers are a better guess. If so, deceleration becomes greater by factor 2. Under these more favourable conditions, the ratio between *W* and *P* becomes greater, and the deceleration during the first second after wing-spreading may reach nearly three times as much, namely, 24%.

During braking, air resistance (*W*) is applied to the body at about the height of the CoM, so that the ostrich must not take care of balance-disturbing moments. If only one wing is spread, it exerts a turning moment of $W *$ lever length, which is the distance between centre of area of the wing and CoM. This distance is in ostriches about 0.40 m.

6.10 Essential Results

All living birds are equipped with a *beak*, rather than teeth-bearing jaws – despite of their ancestry. Although the length of the beak entails low bite forces at the tip, the small contacts between upper and lower beak lead to considerable bite pressures. The beak is part of a surprisingly *small head,* in which especially the brain is miniaturised. The low weight of the head permits being moved rapidly on a long and extremely *mobile neck.* The three characteristics in combination permit free mobility of the primary prehensile organ (the beak), by which birds make first, exploratory contacts with the environment, or as the tool or for taking in food.

All birds are bipedal animals, and most, but not all species are able to fly. In bipedal postures, the *trunk* of birds is supported at the hip (or knee) joints, in flying, however, at the shoulder. The distribution of compressive and tensile stresses is similar in both cases, but not identical, and the major resisting structures are discussed. A very short, compact trunk approaches shoulder and hip and so allows balancing the trunk at the hip as well as at the shoulder joint. In contrast to their presumptive forerunners (theropod and that means saurischian dinosaurs), birds have expanded their trunk cavity behind the hip joint by re-orientating their pubis. This leads to shifting the CoM rearward. The avian ischium is widely divergent from that in tail-possessing theropod saurischian dinosaurs, simply because no long tail exists, and therefore the muscles that retract the femur need new origins behind the hip joint. Both sides of the avian pelvis are not in bony contact with their pubes (in ostriches there exists a contact far behind the hip joint). Instead, the bony structures of ilium and sacrum possess extreme bending strength in the y/z-plane (see below).

On firm ground, the bipedal locomotion takes place in one of *three gaits:* walking (= striding), trotting (= running) and hopping. The mechanical characteristics of all three gaits have been investigated first in the case of primates (Chapter 8.7), especially humans (Chapter 9.2 and 9.3), but the requirements and conditions determine body shape in bipedal birds as well. Locomotor speed depends upon cycle length and frequency. Cycle length is influenced by limb length. Frequency is determined by "pendulum length", and this can be influenced by an appropriate concentration of segment mass on the proximal parts of the segments. Especially the heavy musculature can be located proximally, while the "passive" materials, bones and tendons must necessarily reach down to the toes. The feet of birds are exceptional light-weight structures.

While large birds run, *hopping* is preferred by small forms. This seems to be a consequence of the notorious advantage of small animals concerning muscle force (Chapter 1.4). The time needed for swinging a limb forward is longer in running than in hopping and so favors running over hopping in larger animals. The lack of a ventral closure of the pelvis in birds (see above) is compensated by extremely strong and bending-resistant dorsal elements: sacrum and ilium. The morphological evidence, together with simple statics, suggest the hypothesis that recent birds are descendents of hopping forerunners.

For moving on flat ground, three *toes* are an advantageous and safe arrangement, which provides ample area for balancing body weight. A fourth toe can be a means for braking, or for compensating external forces acting in caudal direction. For reliable gripping hold of objects, or in climbing, as well as in seizing prey, the development of four toes is necessary.

We cannot contribute new information to the mechanics of *flight*. The construction of the shoulder girdle is in essential more similar to that in reptiles, than to mammals (see Fig. 5.17).

The ability to fly is not present in all groups. Not-flying birds may use their *wings as flippers* (penguins) – as do flying forms (auks, gannets). Ground-living ostriches and their relatives seem to use their frontlimbs as *air-brakes* and for facilitating changes of direction. The basic, but admittedly preliminary, information for this assumption is presented.

Chapter 7
Land-Living Mammals

Keywords Autopodia · Claws · Parasagittal limb movements · Small mammal locomotion · Body stem · Symmetric gaits · Footfall sequence · Asymmetric gaits · Dorsomobility · Dorsostability · Tails · Cursorial mammals · Spring mechanisms · Tracks · Body construction · Bending moments · Shearing forces · Torsion of trunk · Rib cage · Weapons in combat

7.1 General Characteristics

The *living mammals* comprise a wide variety of shapes. This corresponds to the diversity of locomotor propensities: walking, running on four, leaping, hopping on two limbs, fossorial digging, swimming, diving, flying, etc. The size of land-living mammals varies between 0.01 kg or 0.007 kg (shrews of genera *Sorex* and *Crocidura*) and 5000 kg (= 5 tons; African elephant, *Loxodonta africana*). Water-living mammals (cetaceans, *Balaenoptera musculus*, blue whale) are with nearly or above 100 tons the largest animals that ever existed. The external surface of most mammals is covered by a pelt, but this pelage can also be missing or reduced to randomly distributed isolated hairs. The tail may be missing completely or be a passively dangling appendage, or a long and actively mobile extremity; it can even be the major propulsive organ, like in aquatic cetaceans and sirenians. The short or long limbs are always divided into the same segments, although they can be specialised as organs for propulsion in water, prehension on trees, burrowing underground or wings for flying. The greatest variation is found in the autopodia. Head shape also varies enormously, while a trait common to nearly all mammals is its considerable mass. The differentiation of the teeth and the dentitions is an outstanding characteristic among all animals. It is closely related to the general tendency widely spread among mammals: to comminute food in the mouth cavity between the teeth, rather than outside the jaws.

Egg-laying *monotremes* (order) or *prototheria* (subclass) obviously represent an isolated evolutionary line since the Early Jurassic. All presently existing forms (duckbills and spiny ant-eaters) possess sprawling anterior (though not the posterior) extremities, and their shoulder girdle follows the scheme described above for

H. Preuschoft, *Understanding Body Shapes of Animals*,
https://doi.org/10.1007/978-3-030-27668-3_7

reptiles. The phylogenetically oldest mammals from the Jurassic were small, although their therapsid forerunners from the Permian and Triassic reached medium sizes. The characters which discriminate mammals from the (older) reptiles and the relatives of birds are the "secondary" temporomandibular joint (and probably a divergent arrangement of mandibular adductors); brains with an expanded, stratified cortex; an isolating pelage (rather than plumage; see Chap. 6.1); dominance of the olfactory system; and the related parts of the brain, as well as a locomotor apparatus, which diverges from that of reptiles. The today dominating groups, *metatheria* or *marsupials* and *eutheria* or *placental mammals*, seem to have evolved in parallel since Mid-Jurassic and are continuously documented since the Late Cretaceous (Maier 2008). Older than or at least as old as these subclasses are several, completely vanished mammals of taxonomically different, uncertain and much debated groups as early as the lower Jurassic (subclass *Allotheria*, e.g. Trituberculata, Multituberculata, Docodonta).

The attribution of fossils to these taxa rests primarily on *dental and jaw traits* and is difficult to memorize and to evaluate for non-specialists. All remains of postcranial elements known up to now from Jurassic and Cretaceous times (e.g. *Juramaia, Henkelotherium, Morganucodon, Ukhaatherium, Haldanodon*) and from the earliest Tertiary indicate the same general *body shape* as that which is widely spread among recent insectivores, rodents, lagomorphs and marsupials (among the latter all inhabitants of the New World): *small size*; cylindrical, rather long trunk with short neck; head often with protruding, pointed nose; short limbs; the proximal segments of which seem to be closely attached to the "trunk"; and a sometimes long, sometimes short, in comparison to reptiles slender tail (Fig. 7.1c). Their dense and long-haired pelage covers a loose skin, which can easily move on the underlying structures. This makes observations of the essential movements of proximal limb segments difficult (but see Chapter 7.2). The sketched combination of morphological traits appears in various subclasses and orders. Therefore, this morphology is generally considered as "the primitive condition", conserved in several evolutionary lines. In this book, however, we should start off from the expectation that these body shapes are highly "adaptive". To test this hypothesis, we have to search for selective advantages connected with the existence of particular traits.[1]

Some of the locomotor types, into which mammals have evolved, have been investigated in detail. These are in the focus of the following text, which deviates in some respects from the scheme maintained in the previous chapters.

7.2 What Do "Small Mammals" Have in Common?

Because of the convergence of overall body shape in *small-sized mammals* (Fig. 7.1), we have to consider first the mechanical conditions in these animals. The first author to investigate the locomotion of this group systematically was M.

[1]The present variability of divergent, highly specialised and often very large forms (see Sub-Chapter 3.2.7 and 7.5) has not evolved earlier than the Palaeogene or early Tertiary.

Fig. 7.1 (**a**) Prototheria. Spiny anteater (courtesy of K. Kienapfel). (**b**) Examples of “small” mammals from various orders, drawn to about the same size. *Left row* from top downward: carnivore, marten; insectivore, *Elephantulus*, hedgehog; rodent, mouse. *Right row*: rodent, ground squirrel; carnivore, mongoose; hyracoid, rock hyrax. Note the position of the limbs and the curvature of the back, both characters diverging from reptiles. (Redrawn and adjusted from various sources)

Fischer (1994a, b). He used the appropriate cineradiographic method, which allows observing the movements of the skeletal elements that are hidden in the skin and pelt. His innovative first study was made on the rock hyrax (*Procavia capensis*); subsequent investigations, made by his co-workers at the Special Zoology and Phyletic Museum in Jena comprise pika, *Ochotona rufescens*; common rat, *Rattus rattus*; kowari, *Dasyuroides byrnei*; shrew-like opossum; *Monodelphis domestica*; tree shrew, *Tupaia glis*; wild cavy, *Cavia aperea*; cui, *Galea musteloides*; domestic cat (*Felis sylvestris* f. *domestica*); tamarin (*Saguinus oedipus*); squirrel monkey (*Saimiri sciureus*); brown lemur (*Eulemur fulvus*), mouse deer (*Tragulus javanicus*); and, for comparison with larger hooved animals, also goat (*Capra hircus* f. *domestica*). The above listed small experimental animals had body weights of 0.15 kg and 3 kg and showed the same locomotor pattern (Witte et al. 2002). The kinematics of the free limbs, as well as flexion and extension of the trunk, has been documented quantitatively. Special attention was paid to the movements of the

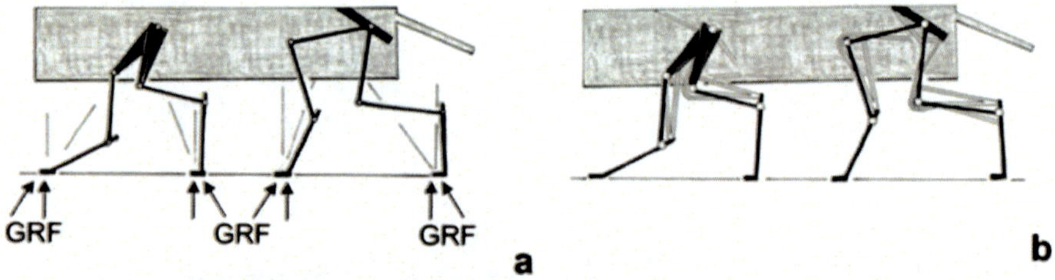

Fig. 7.2 In all "small" mammals, the movements of the limbs during locomotion are similar, and the joints are strongly flexed. The scapula is seen as the most proximal limb segment; the pivot of the forelimb is assumed (not entirely in accordance with observations!) to be at the vertebral margin of the scapula. In (**a**), the segments of the front- and hindlimbs are shown together with approximate GRFs after touchdown and before liftoff (arrows). Between these stages, the second segments from proximal (upper arm or lower leg, respectively) are swung from a nearly vertical into a more or less horizontal position. The upper thigh, by contrast, takes over loads in a nearly horizontal and is swung into a nearly vertical position. The GRFs (**a**) can be vertical or inclined, depending upon the speed of movement. (**b**) In all load cases, the m. triceps brachii and triceps surae must be activated (double lines), the biarticular parts of these muscles forming the repeatedly discussed "parallelograms", together with the respective skeletal elements. The muscles extending the shoulder and knee joints are activated in most – though not in all – positions and their activity varies, depending upon inclination of the GRF and posture of the limb. The horizontal positions of segments during much of the stance phases lead to high rotational moments and therefore require considerable muscle forces for balancing the joints, probably not less than the sprawled limb position in reptiles

shoulder blade on the thorax. The following considerations rest largely on the facts found in these studies.

Thanks to the extensive studies made in Jena (previous paragraph), general statements can be made about their characteristics (Witte et al. 2002): In side view (Figs. 7.1, 7.2), the limb joints are kept in strong flexion throughout most of the cycle. The upper arm as well as the lower leg is kept nearly vertical at touchdown and nearly horizontal before liftoff (Fig. 7.2). The thigh, however, varies between horizontal and vertical during the stance phase. The GRFs vary, as shown in Figs. 7.2 and 7.5, depending on speed, between inclined (during fast locomotion) and vertical (in standing). Measurements of GRFs are described in Chapter 6.5 and illustrated in Figs. 6.8 and 6.14. The major difference from the sprawled limb posture in reptiles is not the length of lever arms of GRF, but their shifting between segments.

7.2.1 *Autopodia, Claws*

The permanent or at least long-lasting and often repeated contacts with the substrate are established by the *autopodia*.[2] Among small mammals they are quite variable in

[2]In the sense of their function, but not in the sense of comparative anatomy, autopodia are the "paws". If equipped with soles, the best fitting term in the case of small mammals may be "paws".

accordance of their respective environments. Examples exist for wide variation: Morphologically complete *paws* placed with their soles on the substrate are called *plantigrade*. If alone the digits plus distal heads of the metatarsals come in contact with the underground, this is called a *digitigrade* position. Witte et al. (2002) and Fischer and Blickahn (2006) emphasize that digitigrade hind autopodia add an additional segment distal to the ankle joint, to form the three-segmented extremity, which is most typical among mammals. Digitigrade position of the front autopodia just leads to an elongation of the distal segment, while an additional segment is formed by the mobile shoulder blade (see Sub-Chapters 7.2.2 and 7.3.2). Very complex are the *prehensile organs* in arboreal marsupials (e.g. *Didelphis*, *Phalanger*, *Dendrolagus*), mice and squirrels and especially in primates (see Chap. 8.3). Prehensile paws are also present in some other classes (chameleons, birds, especially parrots and birds of prey). The also highly specialised devices for digging, for example, in moles, deviate widely from the conditions sketched here, and the wings of bats are completely different constructions. Both are not investigated in this book.

The terminal phalanges in most small mammals are armoured by *claws*. The highest external forces either act against the tips of the claws or are distributed along their length. Unambiguous and therefore well suited for an analysis are the specialised claws of cats (*Felis*, *Panthera*, *Neofelis* and relatives; Fig. 7.3; see also Homberger et al. 2007, 2009, Bragulla and Homberger 2009), which are similar to the claws of some martens (Mustelidae), civets (linsangs, Paradoxurinae) and the malgache fossa (*Cryptoprocta*). These pointed claws are used for (climbing) locomotion, as well as for intra- or interspecific fights and for capturing prey. They are brought into play by active flexion of the distal interdigital joints. If the deep digital flexor is relaxed, the claws are "retracted" (better, dorsiflexed) into their "pockets", by means of elastic ligaments. The narrow and high cross sections of the terminal phalanges and the horny claw sheaths are well suited to withstand the bending stresses which occur, as soon as the claws get hold of any substrate, enemy or prey. The fixation of the keratinous sheath requires the development of an outgrow of hood-like shape on the dorsal side, the well-known claw hood. The material properties of the horny claw are also in accordance with the mechanical needs (Anonymous 1995).

7.2.2 *Parasagittal Limb Movements*

Except monotremes (see above), and the highly specialised moles and aquatic pinnipeds, all (adult) terrestrial mammals diverge from reptilians[3] if performing locomotion on ground: Limbs are moved *parallel to the median sagittal plane*

[3]So-called reptiles usually are considered to be more or less ancestral, simply because they appear earlier in the fossil record than the fully developed mammalian type.

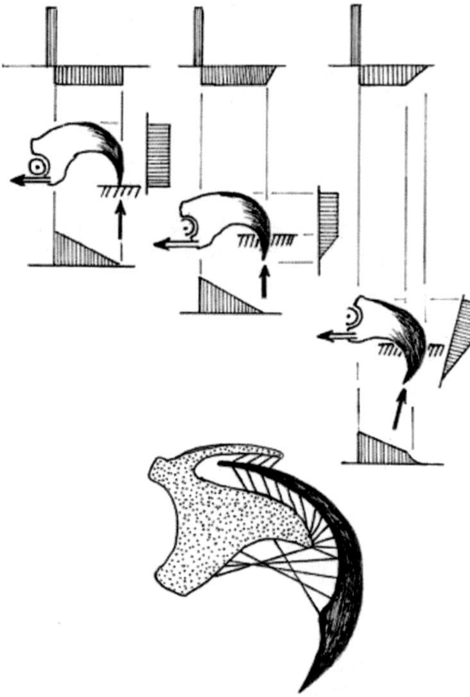

Fig. 7.3 Claw of a cat as a highly specialised example of the claws not alone in small mammals. (**a**) The claw is pressed actively against the substrate. Depending upon the resistance properties of the substrate, the tips are inserted slightly (left), moderately (middle) or deeply (right) into the substrate. The sections of the claw then are exposed to shearing stresses (top row of diagrams), compressive stresses (middle, on the right side of each claw) and bending moments (below the claws). The diagrams are not drawn to the same scale. The highest stresses occur if the tips are not inserted deeply, since the length of the claw defines the lever arm. (**b**) The forces act primarily on the keratinised, horny claw and afterward are passed on to the terminal phalanx. The shape of the latter allows, as well as that of the claw itself, resisting high bending moments alone in the dorsoplantar plane. Aside from shape, the material properties of the horny claw sheath are suited to resist the stresses (compression-resisting keratin indicated by dark grey). It is anchored to the skeleton by means of collagenous fibres. The hood at the dorsal basis of the terminal phalanx is a means to connect the claw with the skeleton. The fibres at the volar side continue the fibre direction of the flexor tendon and are partly produced by the deep layers of the dermis, partly of tension-resistant keratin. The quality of horn substance was determined by Homberger et al. (2007). Existence of the tissue fibres which keep the claw in place has been confirmed by Homberger (oral comm.). The role of cutting edges at the concavity of the claws is described in Chap. 2 under "Fang Teeth"

(*x*/*y*-plane; Figs. 7.1, 7.2 and 7.4) below the trunk, not at its sides (as detailed in Chapter 5.5). The parasagittal movements of the extremities include the movements of the shoulder blade and lead to narrow-gaged trackways. The changed position of the limbs and the corresponding deep-reaching reconstruction of the shoulder and hip region occur in combination during ontogeny (Strotdrees 1998, Klima 1987; Preuschoft and Gudo 2006) and in phylogeny probably since the late Permian to

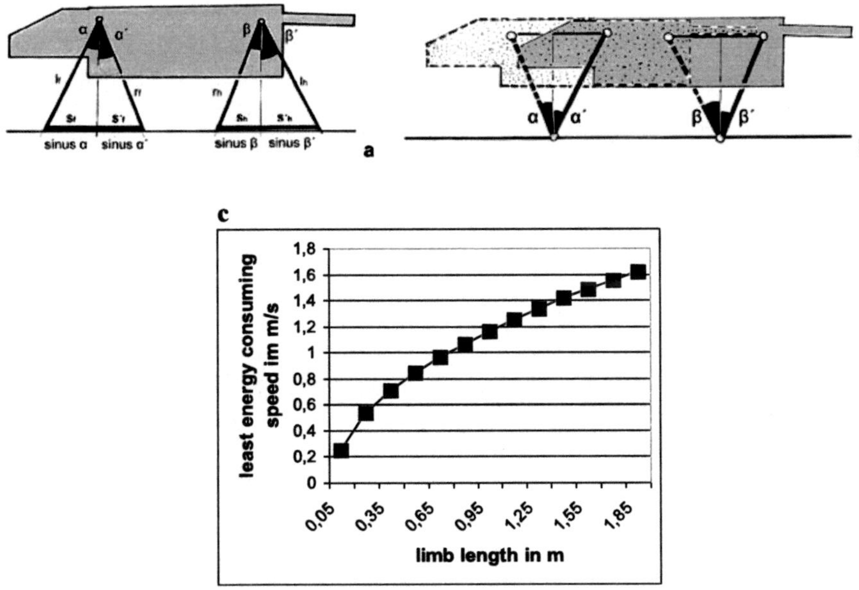

Fig. 7.4 Factors which influence walking speed. (**a**) Movements of "functional limb axes" in relation to trunk. (**b**) The same movements of the left forelimbs and right hindlimbs in relation to ground. The inclination of limbs shifts the body stem from the initial position (continuous outline) to the new position (broken outline). Step length (*s*) in (**a**) as well as in (**b**) follows "functional limb length" multiplied by the sine of the excursion angles α, α' and β, β'. (**c**) Velocity of striding with minimal expenditure of energy is determined by own frequency of the pendulous foreswing. Excursions ($\alpha + \alpha'$, and $\beta + \beta'$) are assumed to be 25°, functional limb lengths grow from left to right

Middle Triassic, so suggesting a causal relationship. Data presented recently by Gaetano et al. (2018) make a gradual change probable: The authors describe postcranial remains of cynognathian cynodonts, with a shoulder girdle comparable to that in mammals. The proximal parts of the humerus and the femur, however, allow pronounced abduction, besides ante- and retroversion. The parasagittal limb movements imply a change from moving the trunk in a horizontal (*x*/*z*-) plane into a vertical (*x*/*y*-) plane (Fig. 5.9b, c; Witte et al. 2017). This behavioural trait can hardly be derived from the skeleton. While the shoulder muscles of reptiles can more or less be identified in mammals, the muscles of the caudal limb girdle do not *longer* take their origins from the axial skeleton but alone from the (remodelled; see Chapter 5.8) pelvis. This change facilitates resorption of the tail – and that means a saving of dead weight. As in other cases, however, the hen-and-egg question remains unsolved: what was first, changes in mode of living or morphology?

7.2.3 Why Are the Limbs Moved in Parasagittal Planes Below the Trunk?

Up to now, no really convincing, straightforwardly mechanical reasons have been identified to be responsible for these characteristic *locomotor movements* and the corresponding morphological traits. Christian (1995) denies completely that reptilian locomotion can be explained on the basis of the "inverted pendulum" (see above, Chapter 6.5) and sees its use as a peculiarity of mammals (and birds). Christian states that the sprawling posture of reptilian limbs is only advantageous for limited body sizes, whereas the parasagittal position of the extremities and the use of the inverted pendulum are the essential condition for really endurant locomotion of large-sized animals over long distances. The eyeball approach makes probable the increase of "functional limb length" by changing sprawled position into parasagittal movement, but Chapters 5.5.2 and 5.6, Figs. 5.8 and 5.10 raise doubts against this explanation. Parasagittal movements instead of sprawling are definitely not simply a means to save muscle energy: this has been postulated but holds true only for limbs with extended joints (elephants, as well as large dinosaurs), not for the more frequent and phylogenetically older "small mammals" (sensu Fischer 1994a, b (Chapters 5.5 and 7.2)). These animals keep the upper arm as well as the lower leg in +/– horizontal positions, incurring the need of exerting great muscle forces (Fig. 7.2). The shoulder and hip joint, as well as the elbow and knee joint, are exposed to considerable rotational moments, which require muscle force to be kept in balance. The commonly overlooked point is that the rotating moments of GRF are not per se smaller in "primitive" (= small) mammals (Fig. 7.2) than in the sprawled limb postures typical of reptiles. From a strictly mechanical viewpoint, there is no difference between upper arms sprawling in the x/z-plane or held parallel to the x/y-plane of the body (compare Figs. 5.8 and 5.10 with Figs. 7.1 and 7.2). As long as the segment is kept at right angles to the dominating weight force (derived from gravity), its joints are exposed to considerable rotational moments. No essential difference exists between the moments acting on a reptilian thigh or a mammalian lower leg if both are in more or less horizontal position – regardless of their orientation to the longitudinal (x-) axis of the animal. In relatively "small" animals, the necessary forces are available (see Chap. 1, and repeatedly elsewhere). The anatomical structures which keep the joints in balance, however, differ in reptiles (see Chaps. 4.3 and 5.8) and mammals. One fact becomes obvious in parasagittaly moved limbs: The influence of their functional length on speed (Fig. 7.4).

The loss of rhythmic horizontal translations of the CoM in z-direction (in reptiles) during striding and running seems to save energy, although measurements of energy expenditure, which have been made by Schmidt-Nielson and others and summarised by Christian (1995), show that reptiles and mammals need the same energy for transporting a constant unit of body mass at a given speed.

One point can be taken for sure: the sprawling posture of abducted limbs entails marked torsion of the trunk, which requires strong, resistant structures, and that means weight. This disadvantage can be reduced by parasagittal supports, which evoke smaller torsional moments and therefore allow the saving of weight of the trunk wall. However, the influence exerted by the mammal-specific postures of the limbs on trunk statics (see below, Fig. 7.3) has never been analysed in detail and remains an open question to be answered in the future. It may be that walking on a narrow-gaged track also implies a *biological advantage*: less disturbance of surrounding vegetation (the energy transferred to the surrounding must be provided by the animal). Definitely, parasagittal locomotor movements facilitate keeping the (heat-producing) trunk clear from the (colder) ground (Fischer, pers. comm.). At present, the really important question *why* must be left open (see Chap. 10).

7.2.4 Are There Reasons for "Small Mammal Locomotion" on the Level of "Biological Roles"? Strategy for Survival

To the functionally defined group of "small mammals" belong, aside from the above-mentioned fossil and recent forms, most insectivores, rodents and several carnivores, like martens, viverrids, as well as hares and relatives (Lagomorpha). In all these forms, their strategy of survival rests on explosive *acceleration*, unforeseeable changes of direction or rapid attack over short distances. Two essential properties (which depend upon their small size, as has been explained above in Chap. 1.5) contribute to this ability: first, moderate mass moment of inertia and, second, a favourable ratio between mass and muscle force see also Sub-Chapter 8.4.5. A question of survival or death is not always and not alone the ability to accelerate rapidly but also the absolute velocity over long distances, for instance, in order to search new resources; attack on alert, fast running prey; or flying to escape endurant predators. Often the simple fact is overlooked that the substrates on and the environments in which small mammals move are always uneven in relation to their body size. Roots, stalks and rhizomes, not plane ground are important *obstacles* for, e.g. mice or rats, while we, as large animals, do not even perceive them, simply because of our size. The horizontal position of one limb segment on one side indeed entails long lever arms of the GRF and high moments of the GRF. On the other hand, it allows adjusting the functional limb length (see Fig. 7.4) to maintain a constant height of the CoM above ground, when overcoming the obstacles without the need of lifting and controlled lowering of the body mass (Fischer 1994a, b). The same method of coping with height differences between the heavy trunk and substrate, by the way, exists in the sprawling limbs of reptiles. Aside from this advantage, the generally existing surplus of muscle force in small animals (Chaps. 1.5; see also Figs. 1.6 and 1.7) anyway allows to assume positions and to perform locomotor modes which are not primarily energy-saving.

7.2.5 Body Stem: Head, Neck, Trunk and Tail

In the *body stem* of all, especially the "small" mammals, occur shearing forces and bending moments which have their maximum at the shoulder, not at the hip region as is common among reptiles (Figs. 4.3 and 7.5; Chaps. 7.5 and 7.7), mainly because of their heavy head and their slender, light-weight tail. The anterior as well as the posterior limbs are subject to requirements different from those in reptiles, and the girdle skeletons plus their muscles therefore follow a design that clearly differs from that in reptiles (Chap. 5).

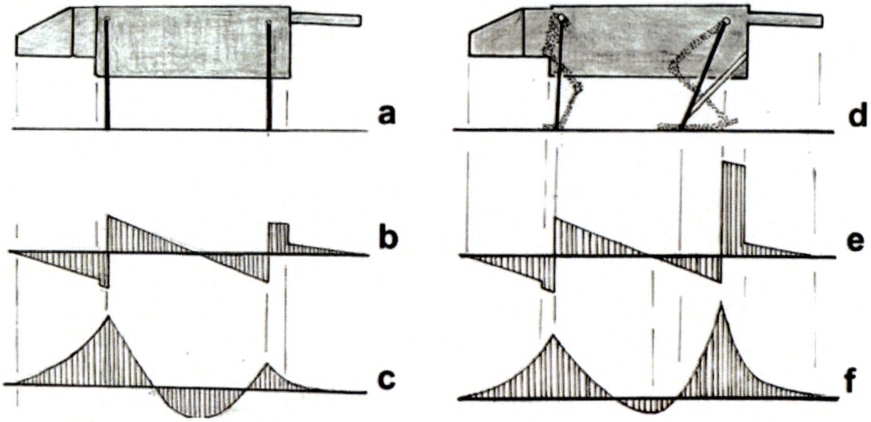

Fig. 7.5 Statics of the body stem in a small mammal. (**a**) Schematic outline to show mass distribution, (**b**) shearing forces, (**c**) bending moments. Bending moments which stretch the dorsal side are drawn above the baseline, although conventionally called "negative". Bending moments stretching the ventral margin, by contrast, are drawn below the baseline although conventionally called "positive". (**d**) Placing the hindfeet in cranial direction changes the shearing forces (**e**) and the bending moments (**f**) essentially. In a crouched posture, the load arms of the trunk are shorter. The influence of limb position on trunk statics is also illustrated in Fig. 4.3. The frontlimbs and especially the hindlimbs are illustrated in the characteristic angulation (stippled), and their "functional limb axes" are indicated as heavy bars

A characteristic of mammals is the heavy *head*, which houses not alone a rather big brain but also the highly developed and therefore heavy masticatory apparatus, which permits breaking down food into small particles inside the oral cavity. In the majority of "small mammals", the *neck* is mobile but short, so that it does not have much influence on the bending moments along the body stem.

Not much attention has yet been put on the *influence of limb position* and limb length on the *statics of the trunk* (Fig. 7.5). If the hindfeet are placed cranial to the hip joint (Fig. 7.5d–f), muscle forces are required for balancing the latter. These downward-directed forces must become greater, if the limb is longer. As the muscles induce a dorsal vaulting of the posterior trunk axis, the positive bending moments in the sacrum and lumbar region become greater.[4] The inversion of the bending pattern can be kept in limits, if the characteristic crouched posture (Figs. 7.1 and 7.5g) is assumed. The bending moments along the trunk are smaller, because the load arms of body segments become shorter. In addition, the forward-displaced hindlimbs also take over a larger part of body weight than those placed caudally and so facilitate rapid acceleration.

Because of the purely muscular connection between the *frontlimbs* and the body stem, the static situation is less clear-cut, though in essential the same. If the forefoot

[4]The same effect on bending moments is exerted by a long, heavy tail (as in reptiles). If, by contrast, a formerly long tail is reduced in length or mass, the negative (dorsally concave) bending in the lumbar region is increased (see also Fig. 4.3).

is placed rearward, behind the shoulder girdle, the positive bending moments on top of the shoulder blade become greater, because of the downward-directed force components of muscles which keep balance (mm. brachiocephalicus, supraspinatus, anterior part of pectoralis). The negative bending moments along the cranial trunk section become smaller. Exactly the same influence on the body stem comes from an elongation of the neck or an increase of head weight. By contrast, a placement of the forefeet cranial to the shoulder requires an activity of retractor muscles like m. latissimus. This increases the positive bending moments along the trunk behind the maximum at the shoulder blade.

One of the major outcomes of the mammalian shoulder are the free translatory and rotational movements of the *shoulder blade* on the rib cage, in the scapulo-thoracal "joint". The scapula so obtains the characteristics of an additional limb segment (Fig. 7.3; Fischer 1994a, b; Witte et al. 2002).

In anterior view, the shoulder blades of both sides approach each other in z-direction; the *anterior thorax* is obviously narrow. So the CoM is always close to the glenoid joints, and lever arms of body weight in asymmetrical support are short. Consequently, the tilting moments during asymmetrical support on one forelimb are low (more details below in Chap. 7.5).[5] In side view, however, there exist impressive rotational moments in the limb joints.

7.3 Mechanics of Locomotion on Four Limbs. Gaits, Footfall Sequences

The regularly repeated sequences of movement cycles during locomotion are called *gaits*. If both limbs of the anterior and posterior pair perform the same movements at different times, the gait is called "*symmetrical*" (walk, amble, pace-like walk, trot, pace-trot). If both limbs of a pair perform different movements at the same time, the gait is called "*asymmetrical*". These gaits are similar to sequences of short jumps and therefore also are called "springing gaits". A major role in defining gaits plays a time variable, namely, the footfall sequence (see below).[6] With the loss of sprawled limb posture, there is no longer an urgent need for a *diagonal footfall sequence* on more or

[5]A change in shoulder morphology (but not so much in shoulder positions) seems to have taken place repeatedly during evolution of several classes: sauropod dinosaurs (Hohn 2011), mammal-like reptiles (*Scymnognathus parringtoni* in Tübingen); and mammals, especially cursorials (Chap. 7.5).

[6]Footfall sequence can be identified best, if the ground contact of right hind (RH) is taken as a starting point, and the next footfall is observed, which may be either LF or RF. The first is the diagonal footfall pattern, common in reptiles, and the latter the lateral footfall sequence, as preferred by many mammals.

Aside from symmetrical/asymmetrical, the gaits are grouped by many scientists as "out-of-phase" or "in-phase". The former term relates to both hindlimbs being moved in counterphase to both forelimbs; the latter term emphasises that both hindlimbs are swung forward nearly in phase or synchronous to both frontlimbs.

less level ground. Because the tilting moments are small during narrow-gaged striding of mammals, the next ground contact after that of the hindlimb can (but must not) be made by the frontlimb on the same side. The so established "*lateral footfall sequence*" is in fact used by the majority of mammals. A characteristic of lateral sequence is the quick change between supports on three, two and even one limb (as illustrated in Gray 1968). This rapid change yields safety, because the next following support pattern can be used to compensate undue deviations of the CoM from the optimal, stabile pathway above the narrow trackway (see also Sub-Chaps. 7.5.5, 7.5.6 and 7.5.7). Hildebrand (1965) has emphasised the observation that horses at very slow speeds change to a diagonal sequence and explained this by avoiding the support stages on only two limbs of the same side, which is connected with lateral footfall sequence. While in sprawling locomotion an impulse in *z*-direction is difficult to avoid, it must not occur in each step of narrow-gaged mammalian locomotion. In addition, the lateral sequence opens access to fast modifications of the walk, like the pace-like walking and tölt (= running walk; see Sub-chap. 7.5.5). Both footfall sequences have been observed in vervet monkeys on ground (Vilenski and Moore 1992). When walking at slower speeds, they tended to the primate-specific diagonal sequence; at higher speeds they preferred the lateral sequence, in which the stance phases are shortened for the sake of swing phases. This point will be taken up again in Chaps. 7.5 and 8.4. In a recently published study, Usherwood and Davies 2017 have compiled empirical data from a great number of amphibians, reptiles and mammals on duty factor and time shift between hind- and forefeet. Their data are in essential agreement with the ideas proposed here. The authors found that most animals at slow speeds make use of the energetically most efficient combination of these factors.

7.3.1 Striding: Speed, Ground Reaction Forces, Excursion Angles, Lengths of Limbs

Regardless of the chosen gait, or footfall sequence, the speed of locomotion is determined by the distance covered during each locomotor cycle (s) multiplied by cycle frequency (inversed value of T). This has been explained in detail in Chap. 6.4. For step lengths see Eqs. 5.3, 5.4, and 6.11; for the duration of a pendulous swing (T), see Eq. 6.14.

Locomotor speed is determined by the frequency, and that means duration, of the foreswings. Effective means to shorten the foreswings are minimising pendulum length, high *lifting the limbs from the ground by flexion* of the joints (Sub-Chapter 6.4.1), as well as reducing the mass of the free joint. This indeed can be observed often in mammals.

Figure 7.4a shows the approximate positions of the "functional limb axes" in a walking "small" mammal. As indicated by the heavy lines, the limb axes deviate from the vertical through the most proximal joint, including excursion angles α and α', as well as β and β'. The sine of these angles, multiplied by "functional limb length" (*r*, *l*), is about one half of step length (*s*). Progress is made, if the feet are anchored firmly to the ground, while the most proximal joints are moving (b). In Fig. 7.4c, speed on the lowest possible energetic level is plotted against functional limb length. Obviously, short limbs yield less speed than longer ones, although a limb elongation from 0.05 m to about 0.5 m is more effective than elongations to more than 0.5 m.

The above-described force components for acceleration and deceleration act during each locomotor cycle. Because these cycles are performed so frequently, even minimal savings of energy sum up to considerable effects because of the frequent repetitions of cycles, in view of the active lifestyle of mammals.

Measurements as well as calculations of GRFs (Kimura et al. 1979, Demes et al. 1994; Dan Schmitt 1998, 2003a; Lee et al. 1999; Fischer and Lilje 2011; Preuschoft and Fritz 1977; unpublished own measurements) have demonstrated a constant pattern of the *components of GRF* (Figs. 6.8 and 7.5). As has been shown in Sub-chap 6.4.1 for bipedally striding birds, this pattern of GRFs is a physical phenomenon, which occurs in any "inverted" pendulum with a rigid chord. If the components in *y*- and in *x*-direction are combined, their resultant GRF falls more or less in line with the generalised "limb axis" (Fig. 7.4a, b), which indicates the inclination of the *free extremities* (see also Fig. 6.8 and Sub-Chapter 7.5.4). In other words, the inclination of the limb axis is chosen to conform with the respective GRF, and this depends upon locomotor speed and body weight. What has been shown for bipeds holds also true in quadrupeds, where front- and hindlimbs usually carry different shares of body weight. As a consequence, the joints are kept close to the functional limb axes, so that the rotating moments at the carpal, ankle, elbow and knee joints are comparatively small. In very "large" animals, like the huge sauropods, as well as in elephants, the rotating moments in elbow/knee joints and in the shoulder/hip joints are additionally kept on a low level by permanently extended joint postures.

Because the inclination of the limb axis, or the GRF, is chosen to keep moments at joints low (see previous paragraph), the shear *length of limbs* obviously entails greater step lengths and therefore yields greater speed. Limb length, however, also includes the danger of lesions and requires extra muscle forces for controlling them. Since long limbs are advantageous for reaching high locomotor speed, a trade-off is to be expected between speed and risk of injuries. Its result depends from many nonmechanical factors which are connected with the lifestyle and environment of the respective animal. Because of the dependence of speed from frequency (v = frequency $*$ stride length), as explained above (Chap. 6) and mentioned again in Chap. 7.5, flexed postures of the limb joints are equal to a vasting of limb length – with the consequence of shorter step lengths and therefore slower locomotor speed in symmetrical gaits. In view of their influence on speed, the limbs of small mammals must be expected to be gracile and slender, with the strong muscles concentrated on the most proximal segments. This is indeed the case in nearly all forms, which habitually move on ground (Sub-chap. 6.5). Total limb *length*, however, remains limited, because of several reasons:

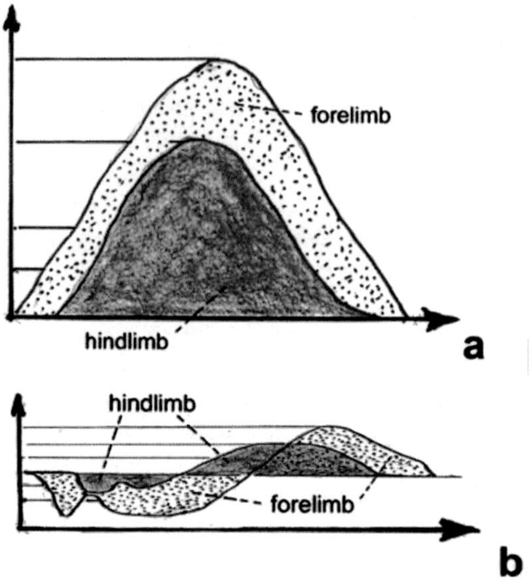

Fig. 7.6 Schematic presentation of the components of the GRF in a quadruped (dog as an example, after Lee et al. 1999, redrawn and modified from Fischer and Lilje 2011). Stippled, frontlimb; grey, hindlimb. (**a**) The vertical (y) and (**b**) the horizontal components in x-direction. In steady-state locomotion at constant speed, the forelimbs sustain a greater share of the load (stippled) than the hindlimbs (grey). The areas below the graphs outlining the braking and the reaccelerating impulses (**b**) should be identical in steady-state locomotion. The frontlimbs contribute more to braking; the hindlimbs begin earlier to push the body forward. This is typical for most (not all!) mammals (see below, under primates). According to our data on horses, the peak GRFs exerted by individual limbs can vary up to 40% in both directions, from one step to the next, depending on the ground, without influencing the rhythm in time. Ground reaction forces are also shown in Figs. 6.8, 6.9, 6.14 and 7.3

Short limbs require less force for controlling their positions and movements and are safer in cases of accidents, because they do not offer long lever arms for unphysiological and potentially dangerous external forces. That injuries play a considerable role is beyond doubt but extremely difficult to determine quantitatively. An example is given in Chap. 8.2. Short extremities also seem useful for crawling through narrow passages and burrows – but the long-legged spring hares and dipodids dig vast subterraneous burrow systems, in which they spend much time! The fundamental advantage of short limbs requires providing solid arguments for the existence of long extremities, which must be weighed in any special case against their disadvantages.

One outcome of the flexed limb joints is the ability to *accelerate rapidly*, without any delay, just by extending the joints. This is of outstanding importance for instantaneous flight or sudden attack. Another consequence of the zig–zag position of the limb segments is that stance phases and step lengths are relatively longer than in large forms with extended joints, which, in addition, prefer narrower excursion angles (see also in the following paragraph and p. 224 ff. in Witte et al. 2002). By flexion of the big joints alone at midstance, the up-and-down movements of the CoM are suppressed, and the virtual length of the "inverted pendulum" becomes greater, since it rotates about a virtual pivot below the underground. Limbs with joints, in combination with elastic tension-producing structures, act like springs and can be used for recovering energy by elastic recoil. The resonance properties of the limb springs can be modulated within wide limits by recruiting few or many compartments of the muscles and so adjust themselves to environmental conditions.

The *elastic properties of really existing limbs* have been determined and compared by Witte et al. (2002) in small and large mammals: Small mammals have comparatively long muscle fibres and short tendons; therefore they are able to modify the elastic properties to a large degree. This is in agreement with the well-known surplus of force in small animals as mentioned in Chap. 1.5. Because of the often flexed position of the limb joints, considerable changes of muscle length are necessary in small mammals. The existence of ample muscle force per unit body weight allows small mammals to modulate the elastic properties of their limbs by means of muscle tension (Witte et al. 2002). Large mammals, by contrast, possess short muscle fibres in combination with long tendons. Marked pinnation allows numerous fibres and therefore high forces but only minimal shortening. The elastic properties of collagenous tendon fibres are characterised by stiffness and high own frequencies. The latter can be modulated by muscle-induced pre-stretching only within narrow limits. As a consequence, "large" mammals can hardly change the frequency of gait cycles and therefore prefer changing cycle lengths. This increase of stride length in fact becomes visible in movie cines (Preuschoft 1985, Streitlein and Preuschoft 1987) and trackways (Kienapfel et al. 2014). "Small" mammals, however, can increase their locomotor speed better by changing the frequency of cycles. This indeed is what they do.

Because of the short absolute lengths of their extremities, most "small" mammals do not reach speeds in the symmetrical gaits, walk and trot, which allow them to compete with "larger" forms, no matter whether these are reptiles, birds or mammals. While a mouse covers in the walk with minimal expenditure of energy and excursion angles of 25° about 0.2 m/s, a human or a horse under the same conditions makes 1 or 1.3 m/s, that is five to six times the distance within the same time.

7.3.2 *Newly Invented Gait: Limb Pairs Moving in the Same Phase, Flexing and Extending the Trunk*

The way to escape the disadvantages of inherent absolute slowness in small mammals was the "invention" of a faster gait, in which both limbs of a pair do not perform the same, just time-shifted movements. Instead, short jumps are made in sequence: either both forelimbs or thereafter both hindlimbs are swung forward and placed side by side (symmetrical "*bounding*", Figs. 7.7 and 7.8), or the frontlimbs are placed one behind the other like in trotting, and the hindlimbs swung forward simultaneously

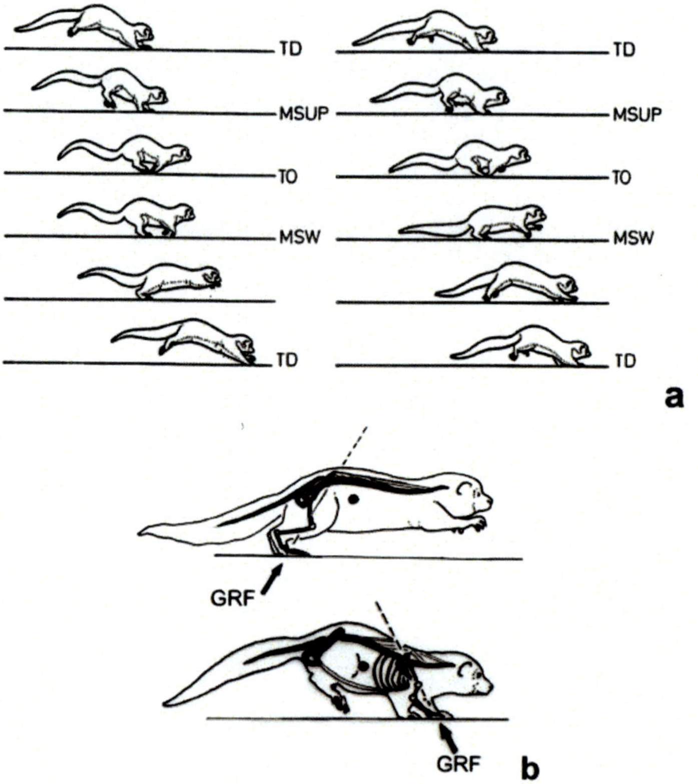

Fig. 7.7 Newly "invented" gait in "small mammals", *Cheirogaleus medius* as an example. (**a**) *Left column*: symmetrical bound. *Right column*: asymmetrical half-bound. Two phases of aerial floating may occur, in collection (third from top) and in extension (last but one). (**b**) Flexion and extension of the trunk with vertebral column and pelvis (black) and active muscles (double lines). On top the maximal stressing during acceleration of the body stem forward-upward, on bottom highest loading of forelimbs after touchdown. May be that the m. psoas is underestimated in these drawings. The functional limb axes intersect dorsal to the back, so rotating about the same pivot ("high fulcrum", according to Fischer 1994a, b). Distance between joints and the prolongation of GRF is responsible for the lever arms of loads and therefore for the size of rotating moments (redrawn from Nieschalk, 1991)

(asymmetrical "*half-bound*", Figs. 7.8 and 7.14). Ground contact may (but must not) be interrupted two times in each cycle, at the end of the foreswing of the hindlimbs by lifting the former stance (front-) limbs before touchdown of the hindlimbs ("collected" floating phase) or at the end of the foreswing of the frontlimbs, while the hindlimbs are trailed behind the body ("extended" floating phase). A third variant of these "*asymmetrical*" or "*springing" gaits* is that not alone the forelimbs but also the hindlimbs are placed on the ground one after the other (*gallop* – in horses usually called "canter" as in Fig. 7.16). This variant permits extended ground contacts and is

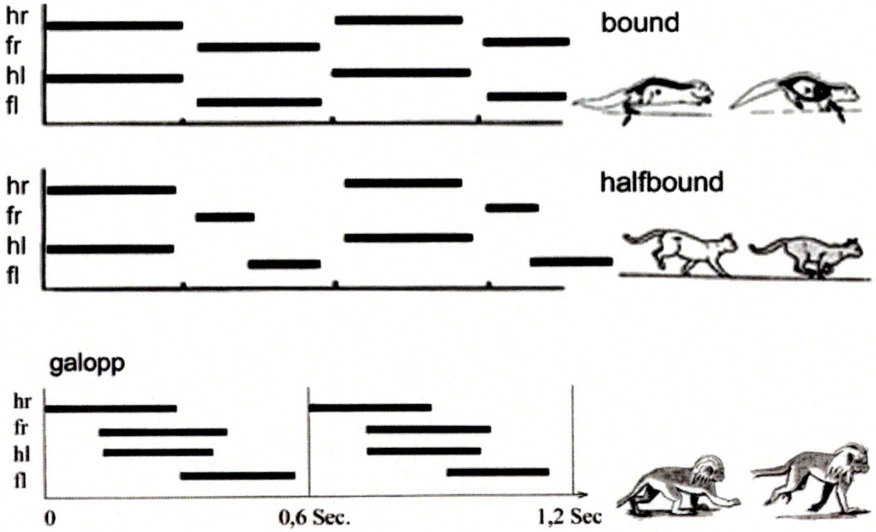

Fig. 7.8 Footfall sequences in springing ("asymmetrical") gaits of small mammals. From top downward: symmetrical bound (*Cheirogaleus*) and asymmetrical half-bound (domestic cat), right-lead gallop (*Saguinus oedipus*)

preferred by large mammals and primates.[7] The special advantages of galloping over bounding will be discussed in Sub-Chapters 7.5.5 and 8.6.

The *frontlimbs* do most of the braking (Figs. 7.7b and 7.10) and so contribute much to pre-stretching the "spring" formed by the trunk during the foreswing of the hindlimbs. In re-accelerating the body, the elastic recoil of the dorsal muscles can be used, while alone the *hindlimbs* are on the ground (also Sub-Chapters 7.5.6). In this phase occur the greatest stresses (Figs. 7.7, 7.26 and 7.27).

The phylogenetically innovative gait in "small" mammals is also useful for overcoming the numerous, from our human viewpoint, minor obstacles on the paths of small animals (mentioned in Chap. 7.2.4). Such obstacles can be crossed not alone by climbing but also by jumping, which is easy to perform during bounding or half-bounding.

There are, however, several "small" mammals which *prefer trotting* (e.g. Kowari, *Dasyuroides*) *or walking* (e.g. hedgehog, *Erinaceus*) rather than bounding.

The distance covered during each cycle of a "springing" or "asymmetrical" gait is particularly long, if not alone the limbs are moved, but also *the trunk performs pronounced flexions* and *extensions* in the sagittal *x*/*y*-plane (Fig. 7.7). Fischer

[7]In Fig. 7.16 horses are shown as examples. Arms (1998) has observed and documented the "canter", or gallop in the small New World monkeys *Saimiri* und *Saguinus* in the form of footfall sequences, but not illustrated. By contrast, bound and half-bound occurred very rarely in the observed animals and only in some individuals. The same has been observed in other, larger primates (e.g. Morbeck, M. Rose 1979; see also Chap. 8).

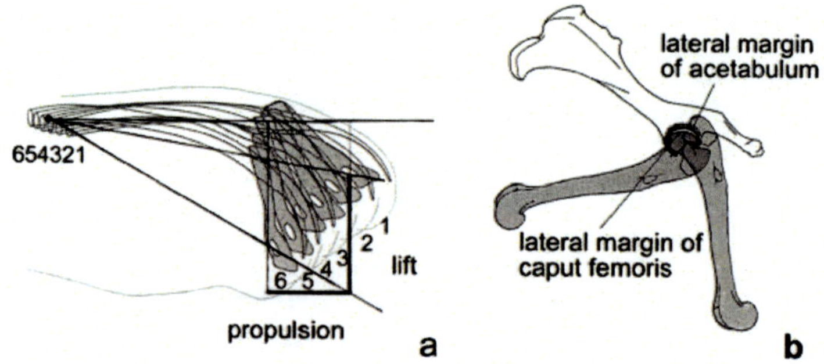

Fig. 7.9 (**a**) Six stages of flexion of the trunk in bounding *Procavia*. The thin line shows the trunk contour in extended posture; the superimposed drawings 1–6 indicate the dorsal and ventral outlines of the vertebral column plus pelvis in six phases of excursion. The tuber ischiadicum and the hip joint cover a considerable distance. It is easy to see that the propulsion in horizontal direction of the hip joint is proportional to the cosine of the angle between the horizontal and a line connecting the pivot with the hip joint, while the height difference in the vertical follows the sine of that angle. (**b**) Pelvis of the same hyrax with femur in the phase (6) of maximal flexion, with superimposed extension of the femur during phase 1 in lateral view. (From Fischer 1994a)

(1994a, b) has measured in the hyrax excursions of 30°–80° (Fig. 7.9), and my admittedly anecdotical measurements on drawings of hares in Grzimek 1988, yielded angular excursions of about 80° during a locomotor cycle. The angle between pelvic region and anterior part of thorax in a hopping kangaroo varies in Fig. 7.13b from 93° to 150°, that is, nearly 60°. Fischer, although he took movements of the posterior thoracal segments into consideration, has put most emphasis on the movements between the lumbar segments of the vertebral column. Simple eyeball observations of several "small" mammals, as well as the anatomical facts visible in skeletons, suggest that the most pronounced excursions take place in many forms between the caudal thoracal and the cranial lumbar vertebrae. These segments are usually the shortest and therefore allow the most marked flexions. Even moderate excursions of neighbouring vertebrae add up to a considerable angulation of the pelvic segment against the thoracal spine. The flexion of the vertebral column swings the hip joints forward, so that the total excursion of the hindlimbs becomes much greater (according to Fischer 1994a, b, up to 40%) than flexion of the hip joint alone. The pivot of the excursions of the hindlimbs is dorsal to the animal's contour (Fig. 7.7b). Likewise, the "frontlimb pendulum" of small mammals has its pivot dorsal to the back. Nieschalk (1991) has observed in *Cheirogaleus* the same pivot for the "pendulum chords" of front- and hindlimbs (Fig. 7.7b). Fischer (1994a, b) characterised this as the principle of "high fulcrum", typical of all small mammals. The horizontal translation, combined with a rotation of the scapula on the thorax, swings the glenoid joint and with it the humerus forward. (The latter effect also holds true for symmetrical gaits (Fig. 7.2).) Together with the wide excursions in the shoulder and hip joints, the proximal segments so are responsible for the distance covered during each cycle, while the distal joints elbow, carpal, knee and ankle do not contribute as much. The permanently flexed positions of shoulder, hip, elbow and knee joints are not in conflict with this general conclusion.

A springing locomotion, similar to the bound, has also been observed in small crocodiles, back in the 1960s (*Crocodylus johnstoni*; Gans, pers. comm.). According to more recent observations of Hutchinson (pers. comm.), bounding is fairly widespread among younger, smaller individuals of various crocodilian species, but not used by older, bigger individuals.

The flexing and extending of the trunk in springing gaits, permits recruiting the *musculature of the body stem* for propelling the body during hindlimb stance and for swinging the hindlimbs forward. This is illustrated *in* Fig. 7.7b. Locomotor speed (v) is proportional to the peripheral velocity (v_p) of the most distal limb segment in relation to the trunk. It depends – as in Eq. 2.4 – on angular velocities (ω_i) times lengths of the segments (l_i), and v is defined as

$$v = \Sigma\omega_i * l_i * \text{frequency} \tag{7.1}$$

The greatest stresses occur in the axial skeleton when the trunk is accelerated by the dorsal muscles against gravity (of body weight) and mass inertia in propulsion. Alone the length of the trunk leads to considerable mass moments of inertia, which resist the acceleration of the cranial trunk section. Muscle force required for balancing and overcoming inertia must increase caudally toward the pelvis, and this evokes high compressive stresses within the more caudal vertebrae.[8] The structures must sustain the stresses evoked in daily life. In fact, the lumbar segments of the vertebral column show the greatest diameters. In the stance of frontlimbs, during the foreswing of the caudal part of the trunk and the hindlimbs, however, movement is facilitated by synergetic action of gravity and inertia. Therefore, the muscle forces exerted by the ventral trunk wall and the m. psoas must not be as high as during the before-mentioned accelerating phase.

The *pelvis* in mammals has a rather uniform shape, combining long, rod-shaped ilia with shorter ischia. Because the forerunners of mammals lost their long reptilian tails (Chap. 5.8 and Fig. 5.17), the hip-balancing caudofemoral muscles were forced to find new proximal insertions on the pelvis, more exactly on the ischia rather than the proximal caudal vertebrae. The pubic bones look like a console, which the ischia of both sides have in common for support against the pull of the very strong ischiofemoral and ischiocrural muscles. These bony elements, however, get their stability by propping the ischia of both sides on the fused pubic elements of both sides right in the midline of the body. A part of the bending stresses may be relieved

[8]This has been mentioned in Sub-chap. 3.3.2: The *"amphibious" or semiaquatic otters and beavers* are exposed, when on land, to gravity, like other terrestrial mammals. Their locomotion on land also is basically similar to that in the terrestrial forms. Because of the permanent action of gravity, the stresses inside their bodies will be greater when moving on land than in water. Because in amphibious otters and beavers the caudal segments of the vertebral column must be counterbalanced by the dorsal and the shoulder girdle muscles against gravity, they need greater strength (=diameters) of the most caudal vertebrae.

by a tension-resistant tie between the pubic bones and the sacrum. The dorso-caudal inclination of the ischia, which is frequent in large mammals (see below in Chap. 7.5), is not obvious among small forms. Surprising is the shortness of the iliosacral connection, which includes even in large bovids or horses not more than a single sacral vertebra (in carnivores often two), although a much greater number of vertebrae is fused to a long, rigid os sacrum. The reduction of the strong and massive tail to a less heavy appendage doubtlessly facilitated the acquisition and development of the springing gait. The ventral closure of the pelvic ring in mammals deviates clearly from the situation in birds (Chap. 6.3).

7.3.3 *Why Do Small Mammals Flex and Extend Their Body Axis in Bounding While Large Mammals Keep Their Trunks Unmoved? Does a Borderline Exist?*

As noted above, locomotion powered by flexing and extending the trunk in a vertical plane is largely confined to "small" mammals. The first author who emphasised that "small" mammals flex their body axis during bounding and half-bounding more than larger ones, while the biggest perform the canter with a nearly stiff trunk, was Slijper (1946).

A first attempt to find out *why dorsomobility is restricted* to rather small mammals was made by Preuschoft and Franzen et al. (2012): head, neck and ribcage have about the same length as the lumbar section and pelvis. As simplified in Fig. 7.10, the trunk is extended and flexed about a pivot at mid-length. Seen from a technical viewpoint, both excursions are angular accelerations, either of the anterior half against the posterior or of the posterior half against the anterior. As mentioned in Sub-chap. 7.3.2, extension is caused by muscle force and opposed by gravity and inertia, while flexion is induced by muscle force and facilitated by mass inertia and gravity (Fig. 7.7). The extent of movement is determined by the duration of the locomotor cycles. (The forces required have been calculated and illustrated in Preuschoft and Franzen (2012).)

Flexion and extension take place between the vertebral bodies (centra) in the intervertebral discs (Fischer 1994a, b; Christian 1999; Preuschoft and Franzen 2012; Jones and Holbrook 2016). Active are the dorsal musculature (extension, short lever arms) and the longitudinal muscles of the ventral wall (flexion, long lever arms) and m. psoas. The activity of the ventral muscles has been documented by EMG repeatedly for various mammals (Fifer 1987, Schilling & Carrier, Kienapfel 2014 and Kienapfel et al. 2018); the m. psoas is very difficult to reach by EMG electrodes. The ventral muscles have longer, and the dorsal muscles shorter lever arms. The speeds of trunk extension and flexion vary inversely and linearly with length of the lever arms, while accelerations become smaller with length. Because of the given (and limited) velocity of muscle contractions, the extension of the trunk is performed faster than flexion, although the former requires more force. (The lever arms of the

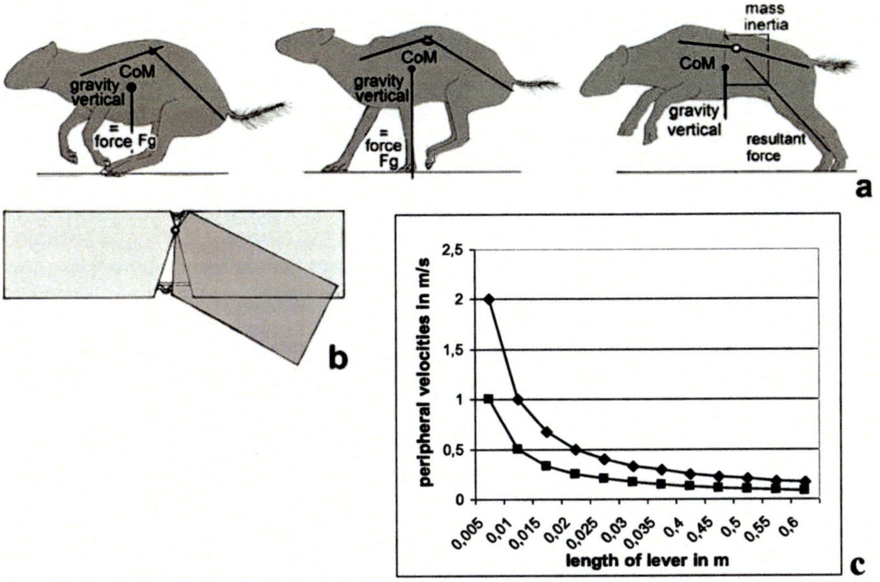

Fig. 7.10 Flexion and extension of the trunk in the half-bound. (**a**) Three phases in a living *blue duiker* (from Preuschoft and Franzen, 2012). (**b**) Schematic presentation of two extreme trunk positions. The vertebral column is not in the middle, but displaced dorsally. The lever arms of dorsal muscles are shorter than those of their ventral opponents. Therefore the dorsal extension needs more force but also can be performed more quickly than ventral flexion. In addition, earth gravity must be overcome for extending the trunk, which also costs additional force. The relation between mass and available force is shown in Fig. 1.9b. The deficiency of force is an argument against trunk flexion/extension in larger animals. (**c**) To accelerate one half of the body stem against the other, the force F_g is necessary. This force is $F_g = \text{mass} * a$, where acceleration (a) is assumed to take place on a straight pathway of the CoM tangential to the periphery. This clearly is determined by the angular velocity of the movement multiplied by its radius. The velocity of movement is greater if lever arms are short, and it must be in accordance with the time interval available (= inverse frequency; that is something between 0.05 s (in mice?) and 0.25 s in horses). Long lever arms can compensate missing forces, but with longer lever arms the speed of the resulting movement also becomes smaller!

m. psoas are of similar length as those of the dorsal muscle group, and so this muscle flexes the trunk more quickly.) To be of use for locomotion, flexion *and* extension must be performed during each cycle. Cycle frequency is, in the size order of duikers, about 8/s or 6/s. In (large) horses, cycles are repeated two or three times each second – if going at high speed (that is between 2 and 3 Hz; see 7.5.5), the time available for extension therefore is roughly one half of cycle duration (0.125 or 0.166 in small and below 0.33 s in large animals). Velocity of muscle contraction depends from the load to be moved ("Hill-graph" according to Astrand et al. 2003, p. 67; and Fig. 1.4e) and increases from 0.1 to 4 m/s, if the load decreases from 65 to 5 units. If a muscle between two neighbouring segments shortens with a velocity of 2.5 m/s (where its impulse P has its maximum), the excursion of the mobile segment depends from lever length. Under assumption of this value, Fig. 7.10c shows the angular velocities for force arm lengths of 0.05 m, 0.1 m, 0.15 m, etc. to 0.60 m. The long lever arms slow down movements in large cursorial mammals, where the lever arms of the dorsal muscle group seem to be too long to extend the trunk within the time interval available during rhythmic flexion/extension (dorsomobility). In spite of all this, however, horses perform (minor) flexions/extensions of the trunk with the frequency of locomotor cycles (see below).

It can be concluded that first the advantage of energy-saving by long lever arms has stronger influence on shape and behaviour than the advantage of rapid locomotion or second increasing body volume in large mammals makes the lever arms too long to permit in dorsomobile gaits the necessary excursions. The above-noted observation (Sub-chap. 7.3.2), according to which small crocodiles are more prone to bounding than larger and heavier individuals, also underlines that small size provides the animal with a surplus of muscular force, which is lacking in larger animals (Chap. 1.4).

The recruitment of trunk muscles is maintained in fairly large animals and so seems to be highly efficient. While the limits of "small size" are somewhat around 3 kg, marked flexion and extension of the trunk can be observed in animals of >100 kg. A *sharp borderline* between dorsomobility and dorsostability has never been found; the transition seems to be roughly around or above 200 kg. Pronounced *trunk mobility* is not combined with the flexed limb joints typical of "small mammals", as discussed above. Instead, quadrupedal walking on long, extended limbs, as typical of medium-sized and large mammals (see below), can often be seen in combination with dorsomobility (see also Sub-chap. 7.5.5).

If the lumbar vertebral column is used as an additional segment for "dorsomobile" locomotion (Figs. 7.9 and 7.10), its movements require obligatory morphological specialisations. The *first condition* is a powerful, differentiated and pinnated dorsal, epaxial musculature. Indeed the *m. iliocostalis* in mammals is positioned (in contrast to its oblique location on the sides of the trunk in reptiles, see Sub-chap. 4.2.5) dorsal to the vertebral column and therefore acts in combination with the other epaxial muscles as a synergist for extension. The dorsal muscles are located between the *spinal and transverse processes of the vertebrae*. If the latter are inclined ventrally, the dorsal musculature passes on both sides of the vertebra and so gets an additional moment in the horizontal z-direction. This allows exact control of lateral excursions of the long lumbar section. Such a lateral excursion is provoked, for example, when only one hindlimb pushes the body forward. The relevant component of the GRF is transmitted at the hip joint of the same side, slightly lateral to the midline. Mass inertia of all body segments cranial to the pelvis then leads to lengthening the contralateral side. Such a flexion toward the active hindlimb is excluded by the action of the dorsal muscles, if they are arranged at the sides of the centra. The characteristic *cranial inclination* of transverse as well as spinal vertebral processes (Slijper 1946; Kummer 1959a, b; Christian 1999; Jones 2015) approaches them to the direction of the muscular pull and leads to a reduction of bending and increase of compressive stresses.

The *second condition* is that *mobility* of the long lumbar section in the sagittal plane is not restricted by ribs or very long spinal processes. It is especially the great number of the lumbar segments and the low dorsoventral diameters of vertebral bodies that sum up to considerable excursions of the pelvis (Fischer 1994a, b; Christian 1999). The latter author also connects the short connection between ilium and sacrum with dorsoventral mobility (see also below). The more caudal lumbar segments are confined to dorsoventral movements. The "antiklin" position of the spinal and transverse processes exposes them more to compressive and less to bending stresses, as noted above.

The *morphology of the lumbar vertebral column* in two groups of mammals was compared morphometrically and with statistical refinement by Jones (2014 and 2015 in press; see also Jones and Holbrook 2016). The authors took Slijper's (1946; see above) observation as a starting point. Among bovids, as well as felids, there exist large forms (ca. 650 kg or 200 kg, respectively) and small ones (about 2 kg). While all felids are more or less "dorsomobile", the large bovids are "dorsostabile", in contrast to the smaller species. Jones (2014, 2016) found in the felids, especially the smaller ones, the slender transverse processes inclined cranially and ventrally, the likewise slender dorsal, or spinal processes inclined cranially (in contrast or "antiklin" to those of the more cranial vertebrae) and the small intervertebral joints orientated parallel to the sagittal plane, so that primarily flexion and extension are possible while lateral excursions are blocked. Among the smaller bovids (below 200 kg), the inclination of the transverse processes was similar to that in felids, but all vertebral processes are less slender and more expanded in longitudinal direction in bovids than in felids. The most obvious difference between bovids and felids was the longer and more numerous centra in the latter group, with the dorsoventral diameters shorter than those in the medio-lateral *y*/*z*-plane. In all bovids, the lumbar vertebrae are shorter and less numerous than in felids. While the increase of size in both groups implies similar morphological changes, other characteristics of bovids and felids remain constant throughout the full range of sizes. In perissodactyles, Jones and Holbrook (2016) found contacts between the transverse processes of the most caudal vertebrae. These obviously provide a stiffening of the lumbar segment against lateral bending (see also Sub-chaps. 7.3.3 and 7.7.5).

7.3.4 *Why Are Hindlimbs Often Elongated? Jumping, Hopping and Bipedalism*

A characteristic of many small mammals is the *great length of hind* – in relation to their shorter frontlimbs. This has been observed in animals which excel by their abilities in jumping or hopping, the latter often as bipedalists on the hindlimbs alone. Long hindlimbs, however, are also present in many "small" mammals, which are not renowned as jumpers or hoppers. Elongation may take place *in all segments* (Fig. 7.11), but in most cases outstanding length is confined to distal segments, usually the metatarsals. In "prosimian" primates, however, the elongation takes place in the tarsal elements (See Chapter 8). In jerboas (Dipodidae), the metatarsal bones

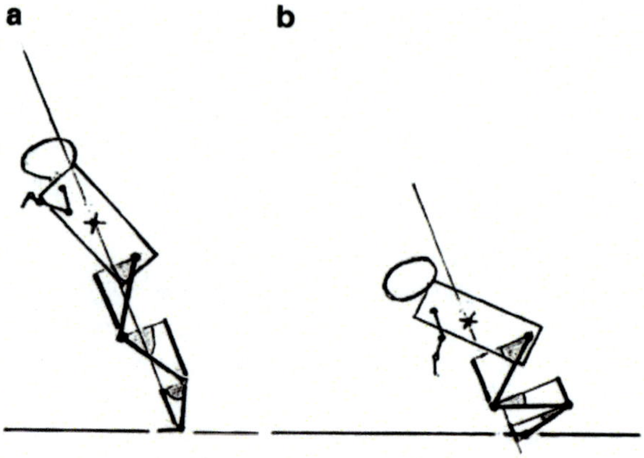

Fig. 7.11 An animal's body is accelerated alone by stretching the hindlimbs. (**b**) is an early phase of the takeoff; (**a**) is a late phase. (The line connecting the feet with the CoM gives the direction of movement, not the GRF! Instead, the latter usually passes dorsal to the CoM, producing a nose-downward rotation.) The acceleration distance grows with length of segments (heavy lines between the joints) and the sine of the excursion angles between the direction of movement and the femur, tibia and foot (heavy lines parallel to the direction of movement). Obviously, long extremities maintain ground contacts longer than short ones and so produce the greater impulse (*P*)

II, III and IV are completely fused except a short section near their distal (digital) joints – a configuration resembling that in birds (Chap. 6.1). Rays I and V are reduced in various degrees. In kangaroos, the long metatarsals IV and V are strongly developed, while the rays I, II and III are rudimentary or absent (Jouffroy and Lessertisseur 1979). The mechanical reasons for elongation of particularly the distal segments are explained in Chap. 6.5.

The ultimate reason for the development of *hindlimb length* is mainly a matter of *biological role*: small animals are permanently endangered by attacks of larger ones. To escape, they need the ability of rapid, explosive flight or of suddenly changing their direction. In terms of mechanics, they need the ability of extreme *acceleration, either forward* or into a newly chosen direction, when already moving. The same ability can also be used for capturing prey (for instance, in *Tarsius*, Niemitz 1984). Three traits are decisive, because of mechanical reasons: first small mass, which can be accelerated by moderate forces; second their surplus of muscle force (as explained in Chap. 1.5 and mentioned repeatedly); and third the long hindlimbs permit exerting force against the respective support during long time intervals, because they do not lose ground contact so soon as short limbs do (Fig. 7.12a). In specialized leapers, the ratio force arm/load arm is shifted toward the latter (as in *Salientia*, Sub-chap. 5.9.1), so that the distal parts of the elements reach very high velocities. This, however, requires remarkable muscle forces. These can be provided easily by small but hardly by large mammals.

Explosive acceleration with the aid of hindlimbs can – depending upon environment and lifestyle – be used for isolated leaps (like in several primates; see

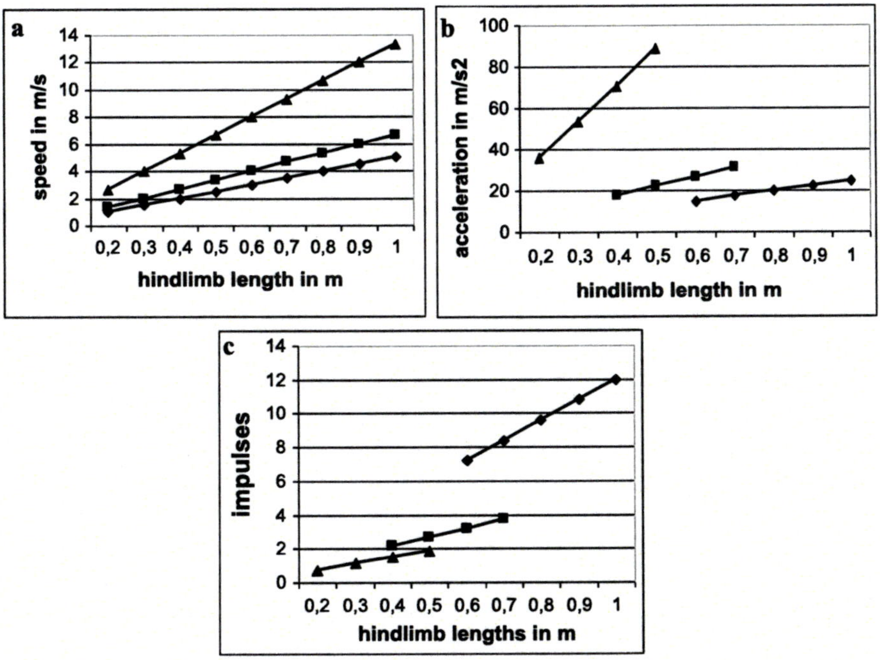

Fig. 7.12 Great hindlimb length and low mass are advantageous for leaping. (**a**) Velocities (*v*), (**b**) accelerations and (**c**) impulses (*P*) during takeoff plotted against limb lengths. Values are calculated for body masses of 0.28 (triangles), 0.8 (squares) and 2.4 kg (diamonds). In (**b**) and (**c**), only those values are shown, which are realistic for the respective hindlimb lengths. If angular velocity is given, and if proportions between force arm of muscles and load arms at the distal parts of the segments remain the same, the length of limbs increases the distance covered during each time interval, so that the velocity (*v*) of the entire animal becomes greater

Chap. 8.4.5), cyclical hopping (like in kangaroos, kangaroo rats, jerboas, springhares) or sudden changes of direction (all small mammals).

Alone the length of the hindlimbs shifts the animal's *CoM caudally*, so that the hindfeet automatically are placed close to or below the CoM. This relieves the forefeet (Fig. 7.3). The faster the locomotion, the more pronounced is this effect, because hindlimbs are placed farther forward. The stance phases of the hindlimbs become longer with their length (see above, Chap. 5.9). Since the sum of all GRFs multiplied by their contact times cannot be greater than the permanently acting body weight (detailed in Chap. 5.2), the increase of load on the hindlimbs is equal to a decrease of load on the frontlimbs. In addition, the coordination of long hindlimbs and short forelimbs is difficult, because of the diverging pendulous own frequencies during the foreswing. The solution of all problems is carrying body weight and propelling the body alone by the hindlimbs and relieving the forelimbs entirely from locomotion – as in reptiles (Chap. 5.8, Fig. 5.10).

Transition from quadrupedalism to bipedalism has taken place repeatedly in various orders (kangaroos, springhares, jerboas). All mammals have maintained the *mobility of their trunks*, and all living forms have chosen to hop rather than run,[9] the latter in contrast to large birds. Several bipedalists, for example, *Dipus* and *Allactaga*, when going at high speed, put their hindlimbs separately on the ground (Gambaryan 1974), not simultaneously as kangaroos do. The time shift between the leading and the trailing hindpaws can be so large that an additional phase of aerial floating occurs. The remarkable speed is reached because the impulse (P) is great, thanks to the length of the hindlimbs. The longer the time span (t) during which a force (F) is applied to a body, the greater the impulse P and the speed of locomotion. In addition, the prolongation of ground contact allows better control of direction and adaptation to inconsistency of ground.

$$P = F * t \tag{7.2}$$

The same principle is applied by large mammals when preferring the gallop instead of bound (see below). The time shift between hindlimb supports (as in *Allactaga*), however, is only possible, if sufficient force is available in each hindlimb to sustain the GRFs when coming down after the aerial floating and to re-accelerate the body. In small animals this is not a problem, due to the repeatedly mentioned surplus of force. Kangaroos do not seem to make use of the advantages of time shift. This may perhaps be due to their mass, which is too great in relation to muscle force to control the GRFs by only one limb; and it may also depend upon the use made of the tail for counterbalancing the hindleg's mass moment of inertia (see Chap. 7.4).

As a locomotor mode for longer distances, *hopping* is not by itself a very fast locomotion, because mammals become airborne for covering considerable distances of several meters at speeds of 5–6 m/s. This holds true for small (*Tarsius*, *Galago*) as well as large (horses) animals. The takeoff speed of course becomes less during the flight phase. Sequences of such jumps then lead to a locomotion with only moderate speed (see Günther et al. 1991). In order to reach really high velocities, the potential distances of free flight should not be exploited fully; instead an additional impulse must be put in repeatedly during each cycle. If this is realised, a small mammal like *Allactaga* can reach the speed of a galloping horse, 48 km/h (= 13,3 m/s; Wikipedia, Gambaryan 1974), or the still smaller *Dipus* can wander a distance of 14 km in one night (Wikipedia).

The high GRF must be counterbalanced at the limb joints. In contrast to static loading, it does, during leaping, not act vertically upward but inclined forward (at takeoff) or rearward (at landing). The positions of the limb segments are chosen to distribute rotating moments on all joints (Chap. 5.6; Figs. 5.12, 7.7, 7.12, and 7.15) so yielding a minimal sum. According to a model calculation of Witte (unpublished, see also Christian 1995), this is favoured by a greater number of joints. Their number, however, is limited by the subdivision of the limbs into always the same segments, which is dictated by evolution. Mammals recruit the shoulder blade as a mobile segment of the frontlimb (Fig. 7.2) and often the metatarsus of the

[9]The only exception are humans (see Chap. 9.2).

hindlimb (see also Chap. 7.2). Only few mammalian genera develop – in contrast to birds – an additional intertarsal joint (*Galago*, *Tarsius*).

The *pelvis* of leapers and hoppers is not significantly elongated. Kaup (1973) found in bipedal rodents only a minor elongation of the postacetabular section, ischium and pubis, if compared to non-bipedal forms. Like other small mammals, the leaping and hopping forms seem to make use of their lumbar vertebral column, rather than the ilium for propelling the body (see above, Sub-chap. 7.3.3).

7.4 What Are the Advantages or Disadvantages of Having a Long Tail?

All bipedal hoppers and the small forms among the leapers possess *long tails*. During resting, a strong tail can provide a large area of support (= high security) by being propped on the ground, as in kangaroos (Fig. 7.13a), dipodids, tarsiers and some birds (e.g. woodpeckers). More important for tail length are mechanical reasons which play a role during locomotion: for taking off, the long and heavy hindlimbs are stretched out rearward, trailing behind the trunk. For landing, the strong and heavy hindlimbs are needed in front of the trunk, to brake the impulse after long jumps or to prepare the next hop. The rotation is difficult to perform, because during aerial floating no external forces can act from outside on the body in order to initiate the rotation. Alone the opposite movement of another part of the body can do this, either the frontlimbs or the long tail (Fig. 7.13b). The mass moments of inertia of the counteracting tail (J_t) must be great enough to overcome those of the hindlimbs (J_h). According to Eq. 3.3, it is because

$$J_t = m_t * l_t^2, \text{and } J_h = m_h * l_h^2 \tag{7.3}$$

that even a slender tail (as in bipedal rodents) can have a large mass moment of inertia, if it is long enough (compare Sub-chap. 8.4.5). The heavy tail of, e.g. kangaroos must not be as long, because it contains more mass. On the other hand, the inertial resistance of the hindlimbs can be minimized by their low mass and by approaching their mass to the body's CoM by flexion of the joints. The *low-weight construction* of the hindlimbs has been mentioned repeatedly above, most detailedly in Chap. 6.5, and the crouching of the whole body to facilitate being rotated has been analysed first by Peters and Preuschoft et al. (1979) for tarsiers (see also Dunbar (1994) for leaping monkeys) and will be taken up again in Chap. 8.8. The mechanical conditions which must be fulfilled by the free extremities are the same in walking, running (trotting), hopping and leaping (see also Günther et al. 1992). Therefore, the existence of light-weight and long limbs cannot be taken as a proof for the preference of one or the other locomotor mode (see also Christian 1999).

Fig. 7.13 (**a**) Kangaroo standing (not sitting in a strict sense!) with feet in full length on the ground plus additional support by the tail. (**b**) Kangaroo hopping in four phases, with changing curvature of the back and tail (broken lines) as indicator of flexing the vertebral column. Only the balls and the digits make contact with ground; the heel is lifted. (Modified from Preuschoft and Christian 1999; also Gambaryan 1974)

Of course a *long tail* is a *vulnerable part* of the body: trailing behind movements, it may well be captured and hurt by an enemy. This point is illustrated in the very frequent events, where female monkeys seize the tails of their explorative offspring. A long and slender tail may also freeze in cold nights and get lost (Sauer 1960s, bushbabies). This problem is less of a danger if the tail is massive, like in kangaroos – but in these animals, it increases total body weight, being something like otherwise useless ballast.

7.4.1 *Why do Kangaroos Hop and Not Run?*

The above (Chap. 6.5) made attempts to explain on the basis of mechanics, why large birds run rather than hop, are in conflict with the fact that *kangaroos hop* and do not run, although their body size is about the same as that of large running birds

(Table 6.1). This contradiction deserves attention. A simple answer would be that evolution in the class mammals has started off from ancestors different from the class Aves. Maybe that the reason for the divergence between locomotor modes of kangaroos and ostriches must be sought in their respective evolutionary history: like all mammals, kangaroos indeed evolved from other ancestors than birds and therefore have divergent anatomical preconditions – and divergent programmes for locomotion – whereby both of course are subject strictly to the physical conditions. An answer on the field of mechanics would be more satisfying. Some aspects are proposed below:

A characteristic trait of kangaroos is their long and heavy tail. It is moved up and down during each cycle and counteracts the mass moments of inertia of the synchronously forward swinging hindlimbs (see Fig. 7.13b). In a running bipedalist, the heavy tail would as well stabilise the trunk, as it does in bipedal dinosaurs. Birds do not possess such a long tail; the only segment they have for counterbalancing a forward-downward movement of the trunk during foreswing of hindlimbs is their long neck with the attached head (Fig. 6.12b).[10] In addition, the mass moment of inertia of the head plus neck in an ostrich is not sufficient for counteracting the mass moment of inertia of both hindlimbs.

Swinging the hindlimbs forward requires a time span. If the available time interval is too short, kangaroos have two options: they can either extend the phases of free flight (what makes cycle lengths greater) or accelerate the foreswing by muscle activity. Both options cost energy but have never been investigated systematically. Only one attempt has been made by Janis et al. 2014, to identify – on the basis of morphological comparisons – a gigantic fossil kangaroo as a not hopping runner.

Being *mammals*, *kangaroos* can produce additional impulses for propulsion by extending their vertebral column. This is impossible in birds, and for bipedal runners, movements of the vertebral column in the sagittal plane are worthless. Therefore, the body axis of birds, especially the lumbar section between the rib-bearing thorax and the pelvis, is shortened and rigid and the synsacrum extremely long – all these items being means to sustain the static bending moments along the trunk with the help of passive, no-energy-requiring structures.

7.4.2 Alternative Shape and Locomotion: Hares and Relatives

A notable exception from the rule that long hindlimbs are combined with hopping in smaller mammals is presented through the hares and their relatives, the *Leporinae*

[10]Rhythmic up-and-down movements of the head instead of the tail would disturb reception of optical and acoustical stimuli (a fact leading, e.g. to the fore-and-aft movements of the head in walking pigeons (Frost 1978, Troje and Frost 2000).

Fig. 7.14 (**a**) Hare sitting, supported by its feet and the bottom of trunk (probably ischium). At least from European leporinae, it is well-known that they often lift their forelimbs from the ground to stand on their hindlimbs, whereby the feet may be in contact with the ground along their full length (plantigrade), so providing a large area of support, above which the *external equilibrium* can be maintained easily. True "standing", with only the balls and phalanges of the feet in contact with ground, can also be observed. (**b**) Foot tracks of a hare moving from right to left in slow half-bounding, for example, during feeding. Hindlimbs are placed on the ground behind the forelimbs, and the full lengths of the paws are in contact with the ground. (**c**) Foot tracks at high-speed half-bounding. The hindlimbs make contact with the substrate alone by the balls and the digits; heel is lifted. Hindlimb tracks are clearly in front of the frontlimb tracks. This "overstriding" is only possible after the frontlimbs have exerted GRFs and are lifted again, while the hindlimbs are swung forward. This is equivalent to a "collected" phase of aerial floating (**d**, **f**). The "extended" floating phase is shown in (**e**). An impulse rotating the nose downward is produced by the hindlimbs in all forms so far investigated. (Redrawn from various sources)

(subgroup of Lagomorpha).[11] The hindlimbs of all representatives are rather long; their frontlimbs are, especially in the genus *Lepus*, also long, though not to the degree of the hindlimbs. Their tails are mechanically negligible. At rest and during slow locomotion, the hindlimbs as well as the vertebral column are strongly flexed. Essential for survival is that flexion of the hindlimbs and the trunk keeps the animals permanently prepared to start off like a lightning, as well as the famous changes of direction. During very slow locomotion, rabbits sometimes use quadrupedal walking; the trot seems to be unknown to all Leporinae. At slightly higher speed, for example, when moving forward during feeding, they change to the bound or half-bound (Fig. 7.14b). The half-bound is typically used when the animals relaxedly

[11]The so-called springhares (*Pedetes*) of about the same size are indeed bipedal hoppers. They, however, do not belong to the lagomorphs but to the rodent group Anomaluromorpha and possess a long tail!

change their place, and in rapid flight. The difference is mainly that at slow speeds, the hindlimbs touch the ground behind the frontlimbs, while at high speed the hindlimbs are placed far in front of the footprints of the frontlimbs ("overstriding", Fig. 7.14c). Of course, the longer the distances covered in a cycle, the higher the speed is provided that frequency is constant. Hares are said to reach speeds of 80 km/h (= 22.2 m/s; Schneider 1988) and definitely can maintain speeds higher than those of (much larger) dogs for considerably long distances. During rapid movement, the contact areas become much smaller, because alone the distal parts of the hindpaws (heads of metatarsals, digits) contact the ground in a digitigrade fashion. Imbalances must be compensated during the next ground contact by placing the feet slightly into the direction of the "wrong", aberrant trajectory (see also Sub-Chapter 7.5.7). The very pronounced "overstriding" of hindlimbs over the track of frontlimbs at high speeds is only possible, when the hindpaws are touching down after the liftoff of front paws (Fig. 7.14e, f) and the CoM is shifted forward, so producing a phase of "collected" aerial suspension extended.

7.4.3 *Why Don't Hares Hop Bipedally, Like Kangaroos or Jerboas?*

Like in the above-treated kangaroos, the foreswing of the heavy and long hindlimbs requires the input of a moment which rotates the animal in the opposite direction. Diverging from kangaroos, however, hares do not possess a long tail. Small enough (1–6 kg) to be hoppers, hares and their relatives are lacking the long tail and are therefore not able to rotate their trunks forward-upward while airborne. Instead of a tail movement, they put their frontlimbs on the ground to stop the forward-downward rotation of their anterior body half and to give it an upward-directed impulse. To brake the nose down rotating moment produced by the strong hindlimbs, they make use of the impulse exerted by the frontlimbs. This implies that the animals perform two phases of aerial floating: "extended" before the forefeet touch the ground (Fig. 7.14) and "collected" after this ground contact. Their frontlimbs must make contact before the hindlimbs are swung forward. This indeed can be seen in Fig. 7.14e, f and is confirmed indirectly or implicitly by Kuznetsov et al. 2017. The latter authors focussed mainly on the rapid change of direction during full-speed running, which is a well-known and vital ability for hares. Their method was, like in our studies on horses, high-speed cinematography. For sudden cornering, hares reduce the time intervals of aerial floating and increase the intervals of ground contact, so that greater impulses can be transmitted from the ground to the animal. For a turn of about right angles, the hare needs not more than two motor cycles. The turn is initiated (like in our toads, Sub-Chapter 5.10.1) by the external forelimb, followed by an outward swing of the posterior half of the body. This means that only one half of the mass is accelerated by the frontlimb into the new direction. Next phase is ground contact of both hindlimbs; at the same time the anterior half of the body is redirected to fall in line with the posterior half (see also end of Sub-Chapter 7.5.5).

7.5 Large Cursorial Mammals

7.5.1 How Do "Cursorial Mammals" Look Like? Strategy of Survival

Many among the large mammals show a marked similarity of their morphology, regardless of their taxonomic position. The animals are characterised by:

- A short, massive trunk, which may be stiff or dorsomobile
- Often (though not always) a long neck
- Slim, often short, sometimes long tail, often decorated with colour or long hair
- Long and slender free limbs, digitigrade or unguligrade, often with reduced number of digits
- Very muscular proximal segments of the extremities, which more or less seem to be integrated into the trunk wall

The characteristic morphology in combination with the ability to endurant wandering and to fast running has lead since long time to classifying mammals belonging to different evolutionary lines as "cursorials". The above-listed characters make them endurant travellers over long distances without becoming too slow, when using the low-energy mode of striding. Their size allows them to neglect the uneven structure of grass- or herb-covered ground. As is well-known, these traits also allow the cursorial mammals to run very fast. Most *cursorials* are *herbivorous*, like the even-toed deers (40 to >250 kg, moose up to 800 kg), antelopes including goat and sheep (30–500 kg), giraffes (up to 1 ton), camels (<1 t) and odd-toed horses (300–800 kg and more). The essential biological role of cursoriality, which plays a considerable role in everyday life, may be more the ability to cover long distances without spending much energy for finding rich feeding grounds or water, rather than occasional rapid flight to escape a predator. The cursorial *carnivores* (dogs of 30–60 kg more than cats of 2–200 kg) use the walk or the trot when searching for prey and, when attacking, however, their swiftest asymmetric gait. Carnivores do not need the voluminous digestive tract of herbivores, so that their trunks can be, and in fact are, more slender than the bulky and compact trunks of herbivores are. Cursorials occupy terrestrial biotopes, which extend mainly in the two dimensions of ground surface. Herbivorous cursorial mammals do rarely use the third dimension, but do not refuse from living in mountainous regions. This is in contrast to small mammals which often make use of three dimensions.

The development of cursorial habits and morphology seems to be tied to the spreading of the vast, rather dry *grassy plains* in the Neogene (Miocene, Pliocene; see Simpson 1951/1961, MacFadden 1992). This connection is confirmed by the fact that acquisition of the cursorial traits among herbivores are often coupled to the acquisition of specialised grinding dentitions, well suited to comminuting fibre-rich grasses. *Predators* from two families (Canidae and Felidae) followed the herbivores on their evolutionary pathway. In spite of their common cursorial characteristics, story-telling differences remain between carnivores and herbivores.

7.5.2 Are All Large Mammals "Cursorial"? Does a Borderline Exist Between "Small" Mammals and "Large" Cursorials?

Large, ponderous and long-legged animals existed since the Paleocene in the New World: Paenungulata (after Simpson 1945) or Amblypoda (after Thenius 1969, pp. 574 ff.), like *Coryphodon* – but they did not possess all characteristics of cursorial mammals. Likewise, rhinos and their extinct relatives (*Baluchitherium*, *Indricotherium*), which lived in the Oligocene and Miocene, have been huge (estimated 15 tons), but probably not cursorial. Not earlier than in the course of the Miocene, the phylogenetically older odd-toed hooved mammals from the superfamily Equoidea developed their cursorial qualities, and the even-toed hooved mammals radiated as a new, in itself diverse group. Both had less specialised forerunners in the Oligocene.

Suiformes that means pigs and their kin (weight 14–40 kg in pecaris to nearly 200 kg in wild boar) are normally not counted among the cursorial mammals, although they have many traits (compact trunk, slender and rather long limbs) in common with cursorials.[12]

Pigs do not possess "weapons" on top of their heads; instead they use self-sharpening canine teeth. When feeding, they employ their long head as a sort of plough for opening and then investigatig the upper strata of the ground. This imposes considerable external forces on their heads. The bending moments, to which the necks are exposed as in *pecora*, act along a short vertebral column of the neck, controlled by very strong musculature. Suid autopodia consist of four hooved digital rays, of which the second and fifth are shorter than the third and fourth and not necessarily in contact with hard ground. Pigs are using the same gaits which are known from cursorial mammals.

Taxonomically close to suids, the *hippos* are more ponderous than pigs (about 2 tons) but use the same gaits as cursorials. In several subgroups of the ruminants are also genera, which are not typical cursorials (e.g. *Bos*, *Bubalus*, *Bison* as well as musk ox (*Ovibos*) and takin (*Budorcas*)). These animals are often rather large (near 1000 kg), but most of them lag behind typical cursors if mechanical criteria are used. The small size of some subspecies (anoa, 150–300 kg) or the red African buffalo

[12]In contrast to most pigs, the warthog lives in the open savannah and in fact shows more pronounced cursorial traits than other members of the group.

subspecies does not make them obviously more cursorial. Large forms, which are not typical cursors but rather "*graviportal*", are also present among the Perissodactyla, *tapirs* (<300 kg) and *rhinos*, which belong, with up to 2 tons, to the heaviest land-living mammals. *Elephants* have many traits in common with cursorial mammals (Chap. 7.9), and the symmetrical modes of locomotion are quite similar in both groups. Nevertheless it sounds quite uncommon to count them under "cursorial mammals".

There is *neither a sharp borderline* between "small" and "large" nor between "cursorial" and "non-cursorial mammals". Usage of the symmetrical striding locomotion and dorsomobility in bounding, half-bounding or galloping (Fig. 7.16) does not exclude each other. Obvious and widely known examples are relatives of the small guinea pigs (Caviomorpha) from the New World: mara (*Dolichotis*, weighing 8–9 kg and having a length from head to ischium of 0,7–0,75 m) and capybara (*Hydrochaeris*, weighing 50–70 kg and exhibiting a body length of 1,0–1,30 m) stick out of the usual small mammal image by their long, extended and little flexed limbs and their marked preference of striding like hooved mammals. In rapid flight, however, they *flex their trunks* in the same "*dorsomobile*" way as "small mammals". The same can be observed in many small even-toed hooved animals, such as mouse dear (*Tragulus*, 2–8 kg); didik (*Neotragus*, weighing 2–6 kg); small cervids (like *Capreolus*, 15–50 kg); small bovids like duikers (*Cephalophus*, 4–20 kg), gazelles (*Gazella, Litocranius, Antidorcas, Antelope*, 15–60, rarely 80 kg) and the fairly large yellow-backed duiker (*Cephalophus silvicultor*) with a weight of 80 kg; and the tapirs with 200–350 kg as well as sheep (*Ovis*) and goat (*Capra*, ca. 50 kg). Since many of these forms stand at their hips higher than at their shoulders and often carry their heads in a deep position, their body shape has been interpreted from the viewpoint of an assumed biological role as "adapted" to slipping through dense undergrowth (Boeker, 1935). This interpretation seems to be a bit short-sighted, in view of the pigs, which are not "adapted" this way, but in fact spend much more time in dense undergrowth, than gazelles, sheep, goats or roe deer.

7.5.3 Free Limbs, Autopodia, Limb Proportions

All cursorial mammals are characterised by their *slender and long free extremities* (Sub-chaps. 7.5.1 and 7.5.2 and Figs. 7.15 and 7.16). Especially the distal segments of the limbs undergo elongation. Zeugopodia, carpus/tarsus, metapodia and digits are involved in varying degrees. Extension of the joints leads to low rotating moments in the joints and at the same time contributes to functional limb length. The essence of all these proportional changes is giving the free limbs their characteristic great length. The advantage of limb length is evidently the elongation of step length, the latter being a sine function of the excursion angle (α) times segment length (Chap. 7.4; Figs. 7.5 and 7.15c–e). In Sub-Chapter 7.3.1, it has been shown that large mammals increase locomotor speed mainly by increasing step or cycle length.

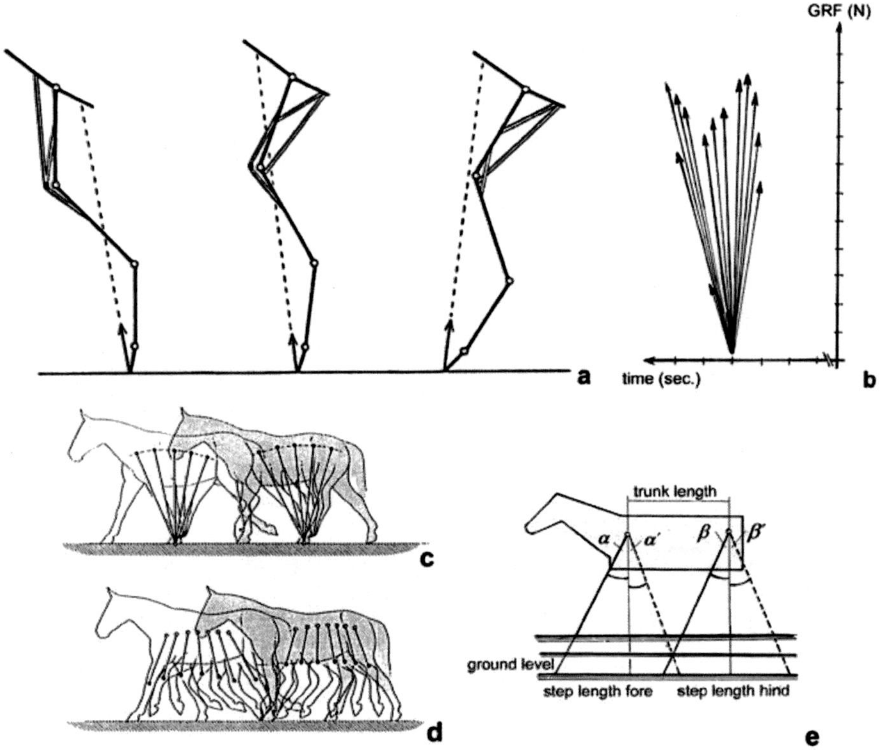

Fig. 7.15 Conditions of quadrupedal walking. (**a**) Hindlimb (sheep) from right to left in three subsequent stages of the stance phase of walking. GRFs (arrows) and their prolongations (broken lines) are measured values, and the muscles recruited for balancing the joints are shown as double lines. (**b**) Rotation of the recorded resultant GRF of the same hindlimb about the supporting hoof from right to left. (**c, d**) Walking horse as drawn in Preuschoft et al. 1994 a, b. During stance (**c**), a part of trunk mass rotates like the mass of an inverted pendulum about the joint closest to the ground (white outline, frontlimb; grey contour, hindlimb). During the swing phase (**d**), the frontlimb (white) as well as the hindlimb (grey) swings forward like a suspended pendulum. Length of the limb pendulum is from its pivot close to the trunk to the free limb's CoM. (**e**) Influence of limb lengths on step length and overstriding. Variation of limb length is indicated by different ground levels (hatched). If excursion angles are constant, the length of the limbs dictates length of steps (s) and therefore walking speed (see Eqs. 7.8 and 7.9). If trunk length is not changed, and limb length increased (lowest ground level), the hindlimbs step over those of the frontlimbs. This is not the case on the medium ground level: The decisive factor for overstepping is the ratio limb length/trunk length. If limbs are short in relation to trunk length (uppermost ground level), the hind footprints do not reach those of the forelimbs. Remember that speed is step length ∗ frequency. See also Fig. 5.8 and Fig. 7.5

The commonly known differences of autopodia have led to classifications of semiplantigrade, digitigrade or unguligrade extremities in contrast to the usually shorter plantigrade limbs (Fig. 7.17; also Chap. 7.3). Not so commonly realised is the fact that all variants must provide sufficient strength for sustaining the ground reaction forces. The autopodia have to cope with the conditions given by lifestyle,

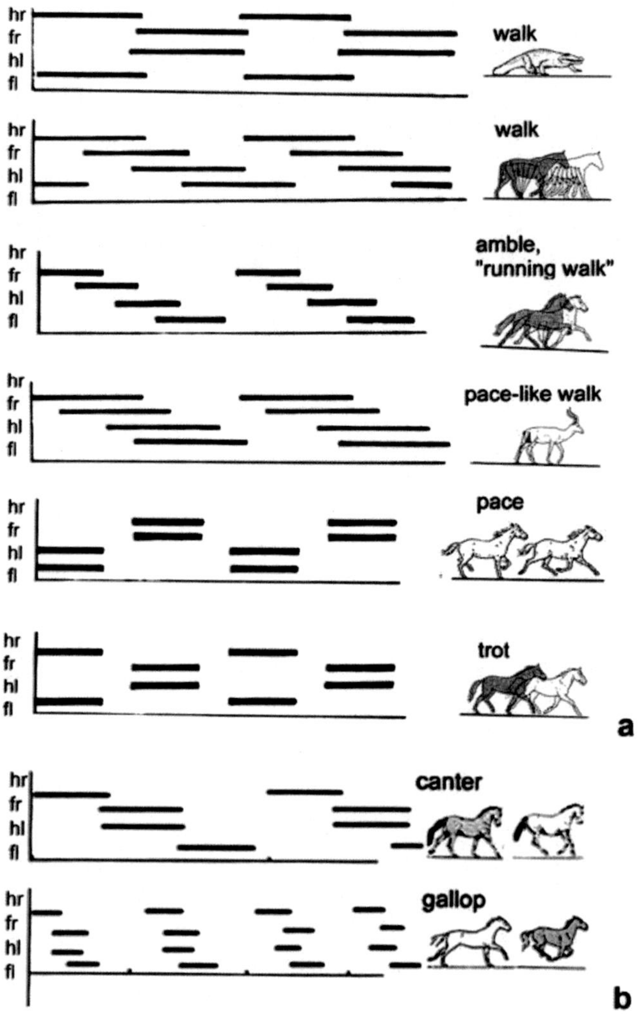

Fig. 7.16 Symmetrical (**a**) and asymmetrical gaits (**b**) of cursorial mammals in comparison to diagonal support of a crocodile. Note the shortness of the cycle length in pace-like walk and amble compared to four-beat walking

especially the substrate on which the animals live (Fig. 7.18). Therefore, the number, position and mobility of digits depend from their respective biological roles, which vary considerably. Among cursorial animals, *two types prevail*: the paws of carnivores are digitigrade, and hooved animals in a wide sense unguligrade. In the former (*digitigrade*) group, the paws are not alone used for transmitting GRFs but also for gripping hold of prey or for fighting (for more details on claws, see Sub-Chapter 7.2.1 and Fig. 7.3). For these purposes, they need to have freedom of mobility and a more complicated foot construction, with a differentiated musculature. Weight is

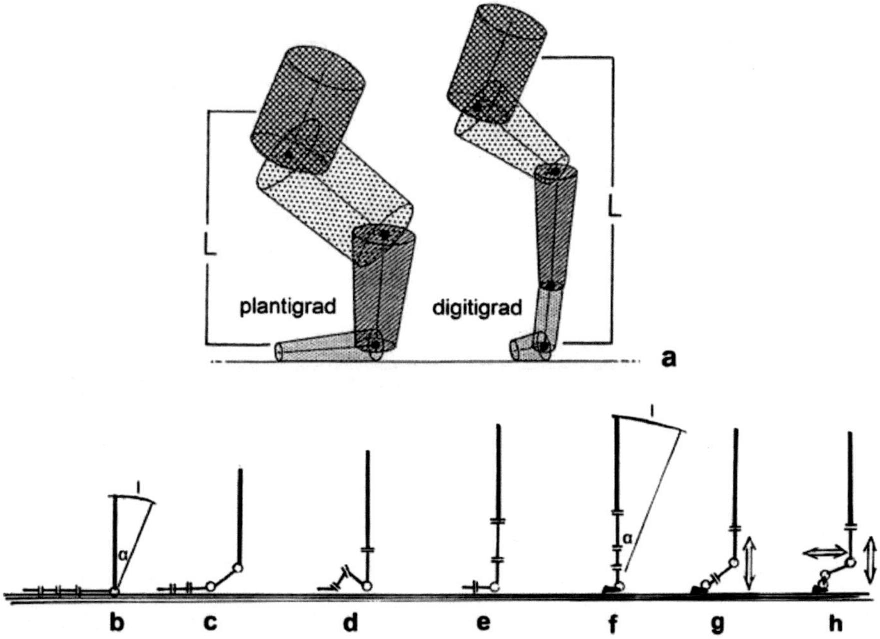

Fig. 7.17 Elongation of functional limb length in cursorial mammals. (**a**) Obvious is the divergent mass distribution shown in two related carnivores, bear and dog. The grey cones and cylinders represent the volumes (=masses) of the respective segments. (**b–h**) Examples of the basic structure of metapodia and phalanges. From left to right: Plantigrade (otter, **b**); semiplantigrade hands in bears and the majority of mustelids (**c**); digitigrade in two variants: dogs and cats (**d**), compared to camels (guanaco, **e**). Unguligrade autopodia transfer loads by only the modified terminal phalanx (**f**). Instead of maximising limb length, they develop a spring mechanism which permits damping either up and down (**g**, horse, giraffe?, cattle?), or up and down as well as fore and aft (**h**, even-toed hooved mammals, most pronounced in rock-climbing goats, chamois or klipspringers). The distal elements of even-toed hooved animals may also assume a spread-out shape (see Fig. 7.18g). All metapodia and respective autopodia are drawn to the same scale, to emphasize the effect of posture on limb elongation (l). When carrying weight during stance phases, most joints are fixed (=) by tendons. A damping spring mechanism requires movements in more than one joint. The extremity swings about the joint(s) closest to the ground (°) as an "inverted pendulum". The distance covered during each step is proportional to the sine of the excursion angle (α) times length above ground (l), as shown in (**b**) and (**f**). Obviously, length (l) either makes the steps longer or allows narrower angles (α) to reach the same step length

transmitted to the ground largely through a ball protecting the four capitulae of the metatarsals. In order to keep the roll-off distance short, the digits are hyperextended in the proximal and flexed in the middle joint; the distal phalanges with their claws are in some carnivores (felids, as well as several non-cursorial martens and genets) not in contact with the substrate, unless the animal uses them deliberately. In the second (*unguligrade*) group, the number of digital rays is reduced and the ground contact confined to the plantar surfaces or tips of the phalanges. Again, the lever arms of GRF are kept short by reduced length of the phalanges, of which a varying

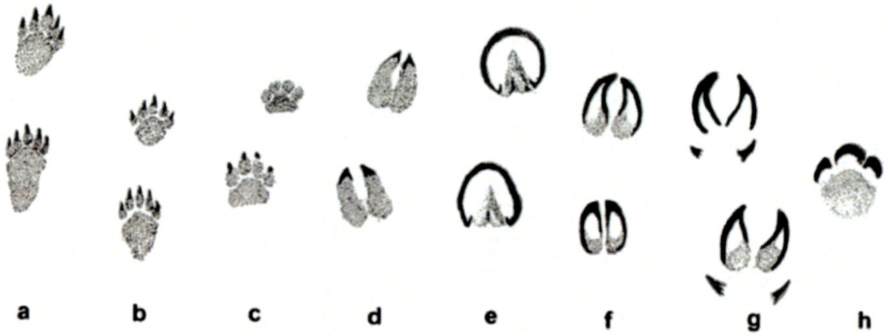

Fig. 7.18 Plantar aspects of autopodia, similar to Fig. 7.17. Claws, hooves and other cornified structures are black (indicating greatest stability) or white (plates parallel to, but not in contact with hard, level ground). Elastic cushions are stippled – even if covered by cornified skin, as in (**e**–**g**). Note the divergence between camels (**d**) and even-toed ruminants (**f**) and the similarity between odd-toed (**e**) and even-toed (**f**) hooves. (**g**) shows examples of even-toed hooved mammals spending most of their time on soft, deformable snow or swampy ground (reindeer, elk, the swamp-living antelope *Tragelaphus spekii*). Contact areas are enlarged by spreading the third and fourth digits and by putting load on the dewclaws. A similar arrangement is realised in three-toed perissodactyles (tapirs, rhinos, **h**). Transitions between (**a**–**c**) as well as between (**f**) and (**g**), are fluent. ***a*** Otter: on top forepaw, on bottom hindpaw. (**b**) Most martens, bears. (**c**) Top, cat; bottom, dog. (**d**) Guanaco. (**e**) Horse: top, front hoof; bottom, hind hoof. (**f**) Top, cattle; bottom, red deer, both on hard ground. Very similar are the footprints of, e.g. giraffe, wildebeest and oryx. (**g**) Top: reindeer; similar are footprints of elk (=moose) or the swamp-living antelope *Tragelaphus spekii*. Bottom: wild boar. (**h**) Tapir; very similar is rhino

number can be placed on the ground: two digital segments of two rays (camelids, Figs. 7.17 and 7.18; König et al. 2014) or one terminal digit (odd-toed horses) and terminal digits of two rays (even-toed giraffes and several large antelopes). Some bovids (e.g. *Tragelaphus spekei*; see Köhler 1993) and several cervids (e.g. elk, *Alces*, and reindeer, *Rangifer*), as well as suids, place their rather long terminal phalanges on the ground and the also widely splayed terminal phalanges of the second and fifth ray, the dewclaws, as well. They so provide the rather large contact area suited for moving on soft, swampy ground or snow (Figs. 7.17 and 7.18). All these variants are equipped with a very similar apparatus of tendons and ligaments (Köhler 1993; Leinders 1979).

It may be noticed here that weighty, graviportal *non-cursorials*, like tapirs, rhinos and hippos, modify the above-described unguligrade morphology. Like the digitigrade carnivores, tapirs, rhinos and hippos have developed a cushion underneath the capitula of their three or four metatarsals, which takes over a large part of total GRF. The phalanges seem to rest either on this cushion or directly on the ground, the most distal segments being covered by a hoof (Figs. 7.18h and 7.19a). This construction is reminiscent of the feet in elephants (Weissengruber et al. 2006; Panagiotopoulou et al. 2016), where the cushion has elastic properties, due to the resilient envelop of the not compressible filling. It also offers a large contact area on soft and moist ground. The tapir gave us an opportunity to estimate the use made of less developed digits. It probably also holds true for the dewclaws of even-toed mammals (Fig. 7.19b, d).

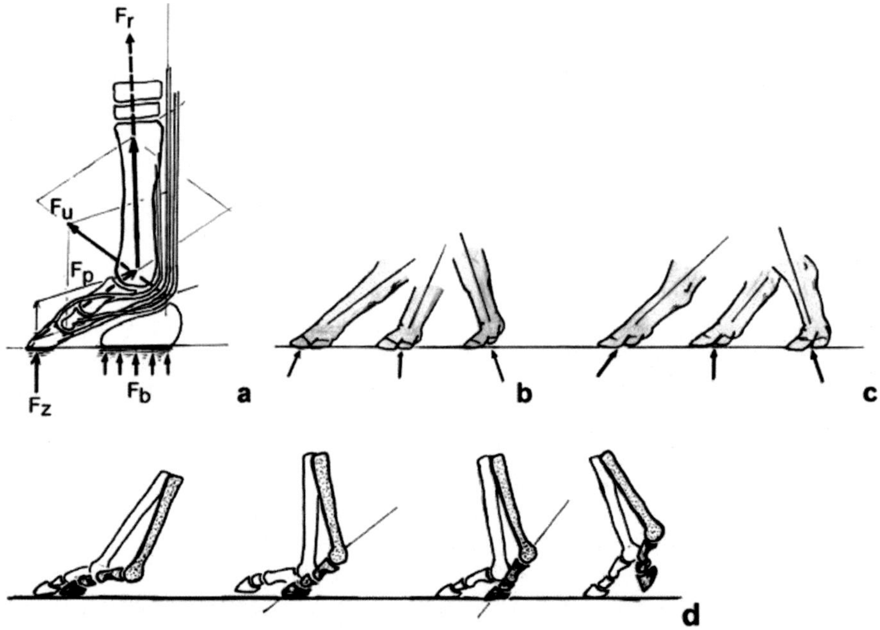

Fig. 7.19 (**a**) Construction of the foot in tapir or rhino. A considerable part of GRF is transmitted through the elastic cushion directly into the metapodial; only a part of total GRF dorsiflexes the third toe and must be counterbalanced in the basal toe joint by the flexor tendons. The resultant (F_r) of GRFs (F_z and F_b) and the redirectional force of the tendons F_u often deviate proximally from the metapodial. This requires proximally increasing bending strength of the latter (Mtp III not illustrated). (**b**) Three stages of the frontlimb stance phase in a tapir, (**c**) of the hindlimb stance in the same tapir as an example, with estimated direction of the GRF (arrows). In both limbs from left to right: shortly after touchdown, midstance and before liftoff. (**d**) Skeleton of metatarsus and phalanges in four stages of roll-off, from left to right. Toes II and IV (stippled) establish the first ground contact and are lifted before the strong median toe loses ground contact. Similar is the situation in even-toed hippos, reindeer and elk and also in carnivores, where just the median, third toe is doubled by the two rays IV and III

Aside from modifications of the ground contact, there exists a *spring mechanism* in the autopodia of hooved mammals (Fig. 7.17). This can be used for damping extreme or undue external forces, as has been analysed by Preuschoft and Fritz (1977), as well as for storing energy, as has been worked out by Witte et al. (1995a, b; see also Figs. 7.17g and 7.27) in horses or described by Köhler (1993) for even-toed ruminants (Fig. 7.17h). The movements of the fetlock joints in horses are restricted to up and down (y-direction), and Köhler's "type A" functions in a similar way, allowing a vertical up-and-down movement of the metatarsal against the ground. Köhler's "type B", however, keeps the interphalangeal joint in a flexed position and can move its hooves as well in fore and aft, that is x-direction. The latter arrangement seems to be most pronounced in rock-living goats. Both variants of springs are controlled by passive, ligamentous structures. The spring mechanism in horses is a means to cut off the peak impact of GRFs at touchdown on flat ground, where some sliding in x-direction commonly occurs (Falaturi 2000). The two-dimensional spring in mountain-living even-toed mammals allows in addition a damping of the horizontal force components in x-direction, so that the hoof, when loaded, does not move out of its once reached, secure place.

The *proportions of limb segments* vary among species (or perhaps families) but seem to be very constant within species. According to Fischer and Lilje 2011, they are very constant in domestic dogs in spite of their divergent sizes. This constancy, and its similarity to wild living species of canids (Wayne 1986a, b), is more interesting, since domestic dogs are bred in widely different body shapes, which all maintain the same relationships between lengths of segments. This lack of variation may be interpreted as indicator of functional relevance of the length proportions or as a remaining document of phylogeny.

Among horse people the experience rule circulates that "short" forearms and lower legs combined with "long" metapodia favour "high" lifting of the hooves and relatively short steps, while "long" forearms and lower legs in combination with "short" metapodia favour low lifting of hooves and long steps. Lilje (2000) in her unpublished master's thesis analysed the proportions (including the scapula) statistically and checked the relevance of relative length measurements as predictors of phylogenetic position, size, habitat and kinematics. She observed a more marked dependence upon phylogeny (on the level of families) than upon habitat. The dependence upon size was unclear, and a relationship to kinematics could not be identified, primarily because of the lack of kinematic data. The latter point has also been investigated by Heckner-Bisping with a not really satisfying result. It seems that animals living in areas with many natural obstacles (dense undergrowth, fallen branches, uneven ground) tend to lift their limbs higher than those living on open plains. All these attempts to understand limb proportions are on the level of biological roles.

Some *biomechanical facts* may theoretically be responsible for a shortening of the zeugopodium and lengthening of the autopodium and may justify the opinion widely spread among horse breeders and horse riders: if the rotational impulse (L) for anteversion of the limb exerted by the flexors of the elbow and knee joints is assumed to be constant, the excursions of the mentioned joints are the more marked, the shorter the respective zeugopodia are. This leads to a higher lifting of the carpal or ankle joints than in animals with the opposite morphological specialisations. At the same time, the inertial resistance offered by longer metapodia is greater than is the case in shorter segments. The marked lifting of carpal and ankle joints can indeed be observed in a number of cursorial mammals.

Less known than the advantages of total limb length is the selective advantage contained in *reducing the number* of digital rays.[13]

The true advantage of reducing the number of digital rays can only be found by looking at the mechanics of locomotion. As detailed in Chap. 6.5, Fig. 6.11 and as

[13]Too many authors seem to seek the causes for reduction of the elements making ground contact in factors like "friction" or "contact area". These factors, however, play a very subordinate role in terrestrial locomotion. Immortal seems to be the idea that reduction of the number of digits per se leads to a saving of limb mass – which is illogical, as will be detailed below.

will be reconsidered in Chapter 9.2, the only way of reducing the resistance of inertia of the long limbs is reducing and concentrating their masses. The minimum limb mass is determined by the summarised cross sections of bones, which have to sustain the occurring compressive forces, the tendons which control the joints, and the skin in which these parts are wrapped in. If the number of rays is reduced, the remaining parts must increase their strength, or their diameters, which is equivalent to mass (as long as length is a given factor). Reducing the number of rays allows the elimination of all "intrinsic" muscles of the autopodium and of all other devices for controlling the numerous elements. Muscles combine rather low strength (Sub-Chapter 1.1.1) with great weight, which is proportional to fibre length (providing "lift" of the muscle) and fibre number (providing "force" of the muscle). It is the reduction of muscles which leads to the effect of less weight on the distal segments (compare Chap. 6.5, Figs. 6.10, 6.11). The remaining structures are confined to few movements and can be sufficiently controlled by ligaments and the remaining digital flexor and extensor muscles. These muscles are located on the forearms and lower legs, and their muscles do not possess long fibres (see Sub-Chapter 7.3.1 and Witte et al. 2002), so that they do not occupy great volumes (= weight) in spite of their great physiological cross sections (= force). In horses, for example, the tendons pass through the entire length of the muscles and insert beside these collagenous structures into the bones of the skeleton. The strong muscles provide minimal lift, but great force. This indicates their essential function: to modulate the tension on the tendons, for example, in order to influence its own frequency in longitudinal oscillations. The powerful, heavy musculature of the upper segments (thigh, hip, upper arm and shoulder) is shifted proximally to both sides of the trunk, where it contributes only a small part to the mass moment of inertia during the foreswing. For many purposes, shoulder and thigh muscles can be considered as a part of the trunk.

7.5.4 How Do Cursorial Animals Move?

A complete understanding of the morphology in cursorial mammals is only possible under consideration of their locomotion. Large and long-legged mammals can frequently make use of the low-energy *walk* as the most preferred gait (see below).[14] No matter which gait they use, their outstanding endurance allows them to travel over very long distances within rather short times.

[14]All of our thinking about quadruped (cursorial) movements is more or less "*hippomorph*", because one species within the non-taxonomic group of cursorial mammals, the *horse*, is used since thousands of years by humans because of its locomotor abilities. In fact, horses have influenced the course of world history (Clutton-Brock 1992). They are an outstanding and thoroughly investigated example of cursorial mammals. In view of the widespread knowledge about and familiarity with our most preferred means of transportation, either tied by a harness to a chariot or under the saddle, it is not surprising that all our ideas about locomotion on four legs are biased toward horses. This begins with the names of gaits: walk, toelt, amble, pace, trot, canter and gallop. Because of their large size, horses must exploit all opportunities of optimisation (Simpson 1951/1961) and so have evolved to extremely

Of course the long tubular bones of cursorial extremities are highly endangered to fracture, because the not foreseeable, often high forces, which are applied in an accident to the body, may have long lever arms (Chap. 8.2). Shortness of the limbs increases safety, and therefore there must be good reasons for the evolution of long limbs. In the absence of such reasons, short length is to be expected.

In each locomotion on extremities occur forces between ground and animal (GRF) in three directions: x, y and z. Among these components of total GRF, that in z-direction is less marked in mammals and birds than in reptiles, in accordance with the limb movements in parasagittal planes (Figs. 5.1 and 5.7, 6.8, and 7.6). The parasagittal pattern of GRF components can be observed in bipedal birds and in both pairs of limbs of quadrupeds, due to the decoupling of anterior limb pair from the posterior pair in the shoulder (Chaps. 5.7 and 7.2.2; Fig. 5.9). The impulses transmitted contra x-direction (= braking) should in steady-state locomotion be equal to those in x-direction (re-accelerating). This indeed is the case in general, although these components of GRF are rarely noticed and difficult to verify quantitatively. An exception is Kawano and Blob (2013, 2021), who have presented the exact data.

7.5.5 *Modes of Locomotion, or Gaits, in Cursorial Mammals*

The common definitions of gaits rest largely on time variables; the spatial variables are not much observed. This is in contrast to the fact that the latter are documented in the form of tracks (Sub-Chapter 7.5.6).

The most frequently used gait of cursorial mammals is the striding *walk*, in which large mammals proceed 1–2 m/s. This locomotion is safe against disturbance of external equilibrium and does not consume much energy, so that it allows covering very long distances.[15] Our attempts to understand shapes are focussed primarily on the most often used gait (as in Pennycuik 1975). The travelling over very long distances in fact requires largely the same morphological characters as the ability for

efficient "master cursors" (following Hildebrand 1965). Horses are readily available experimental animals, large and strong enough to carry sensors and other equipment without disturbance of their movements. They are extremely cooperative, and their gaits can be chosen and modulated (within limits) by a rider, so that permanent control is possible, much more than in most other animals. Many of our own experiments were made on horses, and we are very sure that their results are valid for other cursorials as well and can help to understand the typical body shape.

[15] A look at the daily life in the African grasslands or on the prairies and steppes of the northern hemisphere shows why the energy-saving, not absolutely fast gait is so important: The large cursorial mammals spend many hours marching over remarkable distances in order to find enough food. They also have to drink, and water supply is often restricted to rare water holes. To reach them also requires covering long distances with low-energy expenditure. Both are common and unobvious necessities, but they decide on survival or death more often than the rapid flight to escape a predator. It is misleading to ask alone which speed can be attained, the same importance has the biological role played by endurance.

rapid flight. The following biomechanical analyses bring to light why the most rapid locomotion requires essentially the same shapes as does the slower but endurant walking.

Characteristic of the striding walk are *stance phases* of more than 50% of cycle duration (Fig. 7.16) and a passive foreswing of the unloaded limb which makes use of gravity (Fig. 7.15). Least energy is consumed, if the forward swinging front- or hindlimbs follow the laws of the pendulum. The pendulum period T determines step frequency (Eq. 6.5; Preuschoft and Demes 1984, Witte et al. 1995a, b). It can be influenced by shortening the "limb pendulum" (= distance between pivot and CoM of the limb). This is the reason why all cursorial mammals lift their swinging limbs higher, the faster they go. Flexion of the swing limb raises the limb's CoM and shortens the distance (l). Reduction of pendulum length (l) is also the reason for the slender build of the long, free extremities, in which the (heavy) muscles are concentrated close to the trunk (see also Sub-Chapter 6.5).

Walking speed (v), like in any gait, is, according to Eq. 5.2, the product of step length (s) and frequency, and also

$$v = s/t_s, \qquad (7.4)$$

where t_s is the time required for taking one step or stride (= the inverse of frequency). From Eq. 7.4 follows that step length (s) can be defined as the product of speed and t_s. Step length (s) is also determined by the sine of the excursion angle times length (Eq. 5.4). The longer the limb, the greater the step length, as long as the excursion angle against the trunk is kept constant (Figs. 5.6, 5.8, and 7.15).[16] The other way round, if the limb is long (see also in Chap. 9), hip and shoulder joints, and with these most proximal joints also the trunk, move farther forward when swinging as an inverted pendulum about the joint next to the ground.

The linear relationship between step length and speed has been documented in several studies (Van der Sluijs et al. 2010, llama; Dusek et al. 1970, horse; Pennycuik 1975, wildebeest, Thompson's gazelle; Dagg 1974, Gauthier-Pilters and Dagg 1981, dromedary; Preuschoft et al. 1985, horse).

The high walking speed, which is produced by *great limb length*, is a convincing reason for newly borne foals or calves to possess their generally known disproportionally long extremities. These allow great step lengths, so that even very young individuals can keep pace with the adults.

Because of the lengths of the extremities (and consequently the steps), at higher speeds the hindfeet are placed to the ground in front of the forefeet prints. This *overstepping* (Fig. 7.15e) has often been misinterpreted as indicator of a special gait. In fact, it is nothing than the product of a combination of great limb with short trunk

[16]The choice of the excursion angle is explained below. It can be measured between the functional limb axis and the vertical, as well as the horizontal (= ground), or between the vertical through the most proximal joint and the functional axis of the limb at touchdown (α) and before liftoff (β). See also next page.

length, more exactly of the distance between shoulder and hip joint. In the best known "pacers", camels and giraffes, the limbs are in fact extremely long and the trunk short. Advantageous and often visible in overstepping is lifting the forefeet from the ground shortly before the hindfeet touch down.[17] The shorter the time lag between ground contact of hindfoot and the ipsilateral forefoot (that means the closer the walk is to pacing), the less danger exists that both limbs of one side interfere. Many authors consider avoiding of interference as the only reason for pace-like walk and ignore the advantage of higher walking speed (as shown in Fig. 7.16 and explained below).

The *inclination of a limb* at touchdown or liftoff can be chosen arbitrarily (of course you can walk by taking longer or shorter steps), but only if the inclination of the limb falls in line with the direction of the resultant GRF, it is really energy-saving (Fig. 7.15a, b). Any deviation of GRF from the limb axis leads to bending moments along the limb and between the limb and the trunk, which must be balanced by muscle activity, and therefore is less comfortable for the animal. The extension is particularly evident in the distal joints, carpal and tarsal, and elbow and knee joints, while shoulder and hip joints preserve some flexion. The simple model of a striding animal (Fig. 7.15), which has been used repeatedly above, finds its most pronounced expression in cursorial mammals.

Walking speed in quadrupeds does not alone depend upon the duration of the foreswings but also on the arrangement of foreswings over time. Equids, for example, walk by aligning the swing phases of all four limbs. The time shift between touchdown of the hindfoot and the next (ipsilateral) forefoot is 25% of cycle duration (general knowledge among horse breeders and riders; for zebra see Usherwood and Davies 2017). Putting down the hooves at equal intervals yields the well-known regular four-beat pattern. If higher speeds are chosen, the time shift between hind and fore may decrease to about 20% of cycle duration. If the time shift diminishes to 19–11% (Old World camels, depending on speed) or 15–5% (New World camels, also depending upon speed), the duration of the whole cycle becomes shorter, and therefore marching speed higher, although the duty factor remains at >50% of cycle duration. Because of the reduced time lag between hind- and forelimb, the weighty body stem is supported during longer intervals on the limbs of one side, so producing the image of *pace-like walk*, which is well-known for camels, but as well used by giraffes or other cursorials. The footfall sequence changes from regular four beat to two double beats, similar to the pace trot (Fig. 7.16). An easy to observe example is large dogs, when having a walk with their human keepers. They cannot follow a human when walking but are too fast when using the dog's trot (see below). Although at a first glance balancing on two limbs of the same side seems a challenging task, it is easier than suspected (see below).

Lateral footfall sequence, in which the ipsilateral forelimb is put down after the first hindlimb of each cycle, is often preferred by cursorial mammals on the expense of the diagonal footfall sequence as preferred by reptiles (Chap. 5.5). A tricky

[17]Rapidly trotting horses in harness races put down their hindlimbs lateral to the forelimbs of the same side.

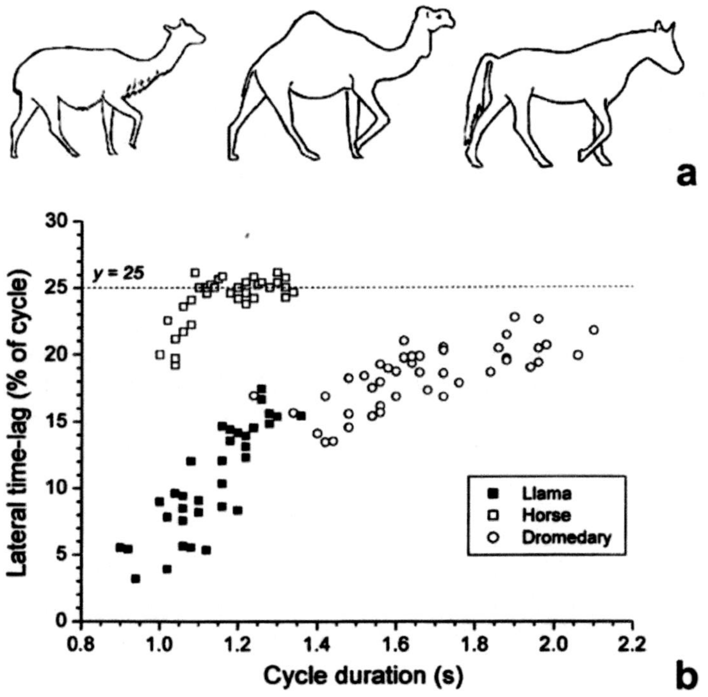

Fig. 7.20 (**a**) Pace-walking in llama and dromedary, compared to the four-beat pattern in horses. (**b**) While horses go at similar cycle durations with 25% time lag, New World camels use time lags of about 10–12%. Old World camels show rather long cycles, probably because of their very long "limb pendulum". The illustration does not show speeds, but among animals of the same size, the shorter cycles are usually combined with higher speeds (see Fig. 7.23). The shorter time lag in llamas yields shorter cycle duration and, in spite of their smaller size, considerable walking speed. (From Van der Sluijs et al. 2010)

question is *why* this preference is so widely spread. First the diagonal support is not required, because the supporting limbs are placed on the ground very close to the median sagittal plane, so that only small laterally tilting moments occur in z-direction (leaving behind narrow-gaged tracks). Second the area of support is larger during long periods, and therefore safety is greater in the lateral, than in the diagonal sequence walk (Fig. 7.21b; Hildebrand 1965). Third the lateral sequence opens access to variants of the walk, like amble (= toelt) and pace-like walking, which all include an increase of speed (see above). Usherwood and Davies (2017) have worked out that the preferred footfall sequences are energetically most efficient.

Stride frequency and therefore marching speed can also be increased by shortening the stance duration, without altering the four-beat pattern: in the *amble* or *toelt* (Fig. 7.16), the foreswing occupies more of the cycle duration than the stance phase, so that the duty factor remains less than 50%. Footfall sequence is the same as in the walk; the four-beat pattern is maintained, but the step frequency in the amble is higher than in walking. Because of its frequency, the amble is faster than the walk. It permits speeds which are clearly faster than the normal walk and can be raised to the speed of the extended trot or slow canter. Its energy consumption seems to be greater than in walking, because the swing limbs move quicker than a pendulum would do: the high frequency requires shortened swing phases of the limbs. This is only possible if additional energy is put in to accelerate them against mass inertia, and the pronounced lifting of hooves also costs more energy than walking or trotting.[18] When getting tired, many ambling horses tend to fall into the trot.

Shortening of the duty factor leads to interrupting ground contact for a short interval, and so the lengths of steps can be increased further, without changing the coincidence between excursion angle and the components of GRF. In fact rapid amble contains intervals, in which none of the forelimbs or none of the hindlimbs has contact with the ground. This is the case in rapid "toelt" of Icelandic horses, in the "running walk" of Tennessee Walkers, in the "rack" of American Saddle horses, or in elephants (Hutchinson et al. 2006; Christian et al. 1999). In view of the time shift between hind and fore, these airborne phases are restricted to either the fore- or the hindlimbs, not simultaneous as in the trot or pace trot. This raises a biomechanical problem: that pair of limbs which is not in contact with the ground moves farther forward than the other pair. The distance between shoulder and pelvic girdle becomes longer or shorter during one cycle. A long-legged cursorial may compensate this problem by modulation of the excursion angles or by more pronounced translation of the scapula on the thorax. Distribution of floating phases on front- or hindlimbs, respectively, leads to less acceleration of trunk mass upward than in the trot.[19] The gait called amble, running walk, walk and toelt is by no means restricted to horses but can be seen as well in other cursorial mammals as well as in elephants (see above).

If the time lag between footfall HR and FR becomes clearly shorter than 25% of cycle duration (according to Hildebrand 10%), the *lateral sequence walk* becomes the *pace walk*, similar to the pace trot, because phases of support on ipsilateral limbs become more obvious and longer than in the walk (Kienapfel et al. 2014). This "pace-like walk" leads to shortening of cycle durations and therefore to higher speed. It can be seen frequently in New World camels or, as noted above, in large dogs. (When leached, they are too slow in the walk and too fast in the trot.)

Synchronous with the foreswing of the frontlimbs, a "*nicking*" of the head downward and slightly to the side of the respective swing limb takes place. This is because the foreswing is initiated by a contraction of the m. brachiocephalicus (e.g. Kienapfel et al. 2014). The muscle spans the whole length of the neck, and

---

[18]Experienced long-distance riders confirm that ambling horses are not optimal for these exhausting events, because they are getting exhausted early.

[19]This is delightful for the riders, who are less shaken and therefore sit more comfortably on horseback although their speed is the same. Ambling is known on all continents and often an outstanding characteristic of special breeds. Only in West and Central Europe it is seen as an absolute no go since Baroque times – although it was highly estimated in Late Medieval and Renaissance times (see, e.g. Shakespeare: "As you like it").

its longest fibres insert into the lower jaw while taking their origins (together with the bulk of fibres) from the humerus, close to the glenoid joint. Muscle force is of course applied to both insertions, so that an impulse *P* is exerted on the heavy head and long neck as well as on the forelimb. In the trot, this nicking cannot be observed. This may be due to the short time available – a characteristic increase of the force on the reins can at least be taken as an indication that some impulse is exerted in the same direction. In the canter this nicking becomes part of the up-and-down movement of the anterior part of the body (see below, under gallop).

A marked increase of locomotor speed is usually accomplished by changing into the *trot* (Figs. 7.16 and 7.19).[20] This gait is characterised by first a divergent, two-beat footfall sequence, either diagonal or lateral – then called "*pace*" in a strict sense (see following page) – second the phase of free floating between the half cycles (= steps) and third the use made of *elasticity* of anatomical elements (see below, end of this Chapter). The duration of stance (= duty factor) of each limb is less than 50% of cycle duration. Step lengths in the trot (or pace) are usually longer than in the walk and make the trot faster than the walk. Both components (in *x*- and *y*-directions) of the GRFs become greater with increasing speed. As long as excursion angles are kept in line with the direction of the resultant GRF, the step lengths can be increased alone by intercalation of *floating* phases, in which no ground contact exists. Diagonal front- and hindlimbs (in trotting) – or both limbs of the same side (in pacing) – are swung forward at the same time, so that the body is in a phase of free floating. The swinging limbs so have time for being swung forward. In addition, for accomplishing the foreswing under the laws of the pendulum, the "limb pendula" are shortened by flexion of the joints (so that their pendulum period becomes shorter). The stance limbs, by contrast, are exposed to greater forces, as shown in Figs. 5.2b and 7.22, because the interval of support against gravity is shortened.

In order to classify the confusing *multitude of symmetrical gaits* in horses, Hildebrand (1965, 1966) has plotted the "stride interval that frontfoot follows hind on same side", that is, the time lag of front- to hindfoot, against "the stride interval that each hindfoot is placed on the ground" or the duty factor (Fig. 7.21). We think that our classification of gaits puts more emphasis on mechanical reasons (and possible pathways of evolution), although it is not at all based on such a comprehensive amount of data as Hildebrand's scheme.

Accelerations comprise the whole body of the horse (and also that of the rider!). Neither head plus neck nor the trunk of a trotting horse performs any visible internal movement. Measurements of the forces acting along the reins not alone of riding but also of carriage horses (Preuschoft 1990a,b, Preuschoft and Gottstein 1999) showed clearly marked peaks for each half cycle. As mentioned above, it seems that the activity of the m. brachioradialis exerts impulses for nicking of the head, which are too short to evoke a visible movement.

[20]An exception are New World camelids, which seem not at all to use the trot or pace trot (Van der Sluijs et al. 2010, while Old World camels prefer the symmetrical pace trotting much more than the asymmetrical gait "gallop".

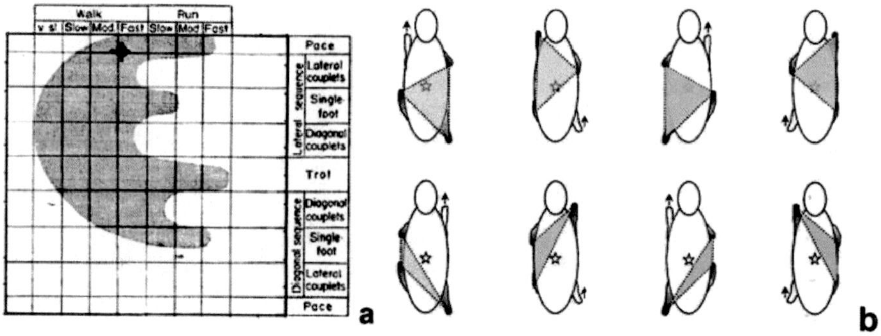

Fig. 7.21 (**a**) Hildebrand's often cited illustration of symmetrical gaits, supplemented by the Icelandic horses we have observed. Horizontal axis: relative stance duration (= duty factor) in % of cycle duration. 50% indicates the transition of walk to trot or pace trot ("run"), respectively. Vertical axis: time lag between hind right and the next extremity touching the ground, also in % of cycle duration. (**b**) The triangular areas of support change during each cycle from left to right. Top row, lateral sequence; bottom row, diagonal sequence. In the former, the CoM falls constantly within the larger area of support, while in the latter, balance is less safe. (Modified from Hildebrand 1965)

The short ground contacts of diagonal limbs in the *trot offer* only *limited opportunities* for correcting direction, uneven ground and inclination of ground. The frequency of repetitions, however, seems to compensate this shortcoming. Precondition for making use of the trot as an energy-saving form of locomotion is relatively even, smooth and continuous ground with few and low obstacles. For the *pace trot*, the same holds true as for the trot. Pace trot can be observed clearly in Old World camels and in European horses rarely, while in the United States, harness racing is also done by pacers. The anatomy of cursorial mammals is difficult to explain on the basis of pacing; it rather seems to have been developed on the basis of walking or trotting. Obviously, the existing body structures are strong enough to sustain the divergent loading in the pace, where torsion is minimised. The load arms of the weight force in z-direction are short in relation to the supporting front- and hindlimb and therefore do not evoke more than minimal torsional moments. A marked lateral swaying of the body's CoM, which is often postulated for the gaits of camels, has never been recorded. In fact the large cursorial mammals always move on very narrow tracks of about 1,5 times the breadth of a single foot- or hoofprint. Because the animal's CoM is 1 m or 1.5 m above the ground surface, their mass moment of *inertia against lateral tilting* is very great. In addition, the rapid changes of the area of support (Fig. 7.21b) provide an effective means against lateral excursions (Fig. 7.22).

At high speeds, the dominating *asymmetrical* or in-phase *gait* of cursorial mammals is the "*gallop*" (called in the case of horses the "*canter*"; Figs. 7.16, 7.23, and 7.24b). The gallop also permits frequent re-accelerations, so that speeds of more than 18 m/s are reached. While in small mammals a time shift between the first and the second ground contact of the frontlimbs leads to bound or half-bound, the time lag

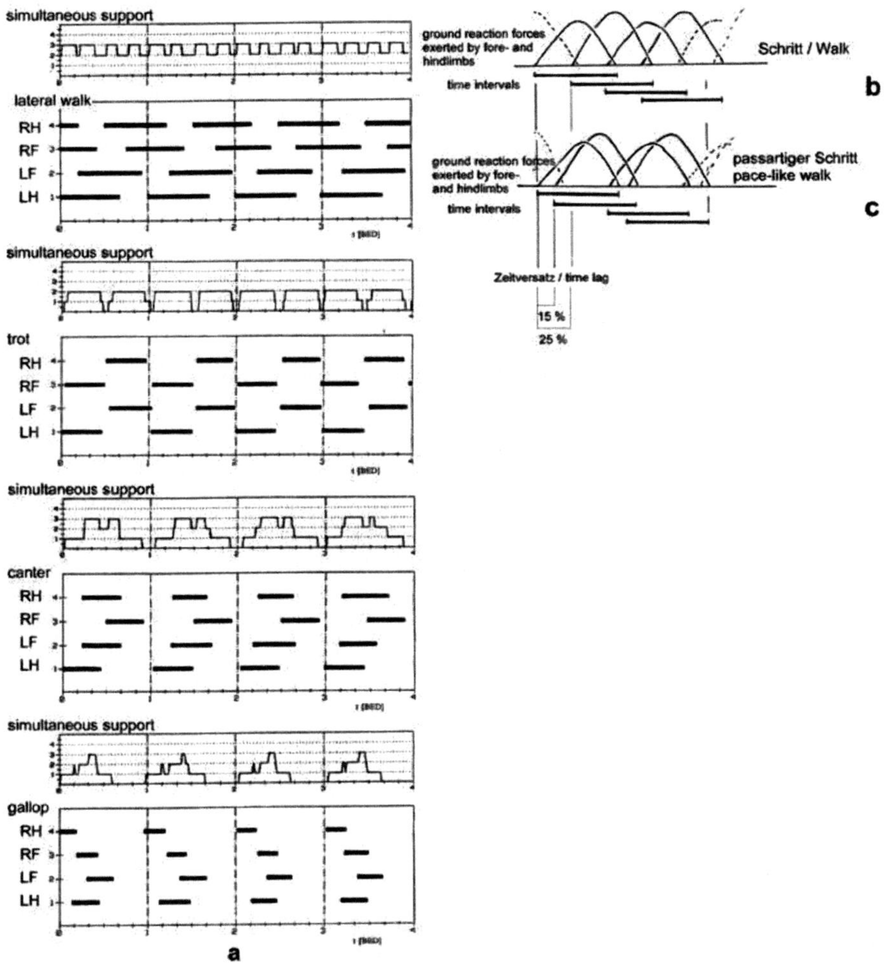

Fig. 7.22 (**a**) Support of body weight by one, two or three limbs in various gaits of horses as an example for cursorial mammals. From top downward: lateral walk, trot, canter, gallop. In the canter, frontlimbs and hindlimbs are placed on the ground one after the other, so that support is given at least by one limb to the trunk mass over a relatively long time interval. (From Witte et al. 1995b). (**b**) and (**c**) Vertical GRFs of all four limbs over time (schematised). (**b**) The four beat walk of horses with a time lag of 25% between hind and ipsilateral front; (**c**) the pace-like walk of New World camels with a time lag of 15%, which leads to a shorter cycle duration and so to higher speed

between "leading" and "trailing" foot of a pair is greater in the canter or gallop and occurs not alone in the frontlimbs but also in the hindlimbs. The position of the limb pairs reminds of that in striding, where in the anterior and the posterior pair the same (left of right) extremities may lead (lateral gallop) or the lead may change with each cycle (rotary gallop). The second variant seems to be used especially on uneven ground (Heckner-Bisping; a detailed investigation of "transverse" versus "rotary" gallop can be found in Biancardi and Minetti 2012). The placing down of the feet

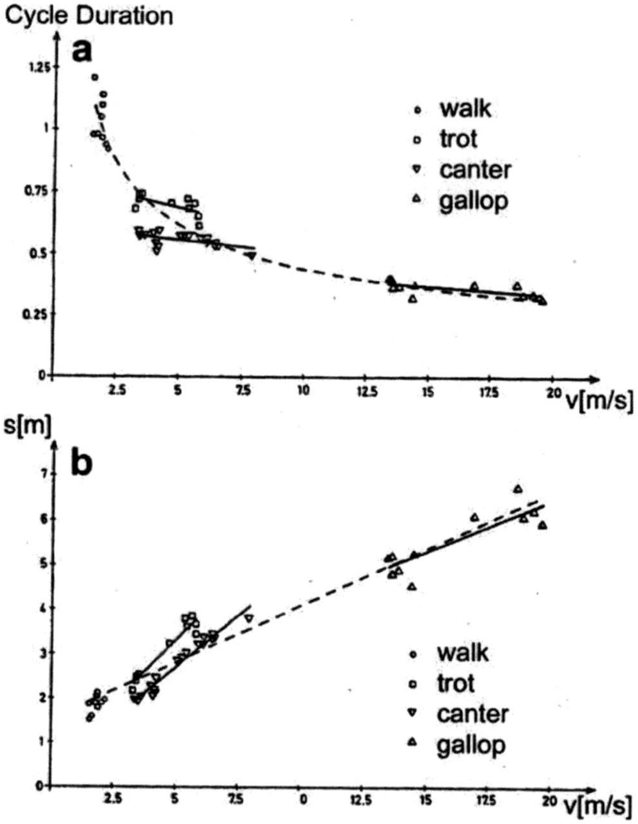

Fig. 7.23 With increasing speed, the duration of the cycles becomes shorter in all gaits (**a**), and at the same time the distance covered in each cycle becomes longer (**b**). The trot and the canter overlap in horses, but the regression lines of both gaits differ. The "extended gallop" of dressage horses is much slower than the "canter" of race horses (a three-beat gait of 12.5–15 m/s = 48–54 km/h), while the "gallop" of racehorses is a two-beat gait, in which the horses reach 18 m/s = 65 km/h (from Witte et al. 1995a)

follows in both cases a three-beat rhythm, unless the gallop is extremely fast. Disrupting the ground contacts of a limb pair leads to elongation of overall contact time. The greatest forces are transmitted between ground and the animal's body while two feet are on the ground, giving either diagonal support on contralateral front- and hindlimbs or lateral support on the limbs of the same side. The extended ground contacts allow adapting the transmitted forces on uneven ground or on grounds with variable rigidity. In horses we have observed variations of isolated GRFs up to 40% in both directions of the average values, while the accelerations of the trunk and the time variables did hardly change. The two- or even three-limb supports converge to a rather safe mode of locomotion, which also allows to jump over obstacles without changing the rhythm. Depending on the mass of the animal,

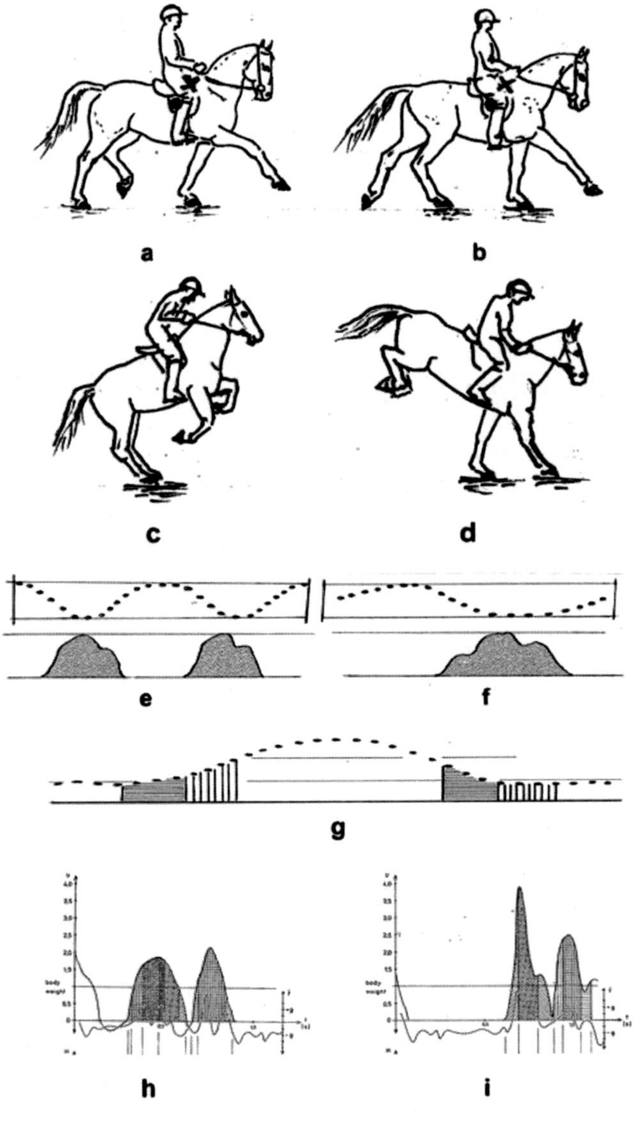

Fig. 7.24 Locomotion of horses. The image above is labeled to demonstrate: (**a**) Horse with rider in the trot, (**b**) in the canter, (**c**) in takeoff and, (**d**) landing after a jump (**d**). Letter (**e**) represents a schematic presentation of the CoM's pathway during a cycle of the trot (dotted line). In contrast to the walk (Fig. 7.15c and 7.15d), the highest point is reached during the floating phase and the (measured) vertical accelerations of a point (X) near the CoM (dot) in trot at "medium" speed (continuous lines surrounding grey areas). Letter (**f**) represents The same for the canter. The full cycle comprises up-and-down movement of the trot and in the canter, and the corresponding acceleration. Note the time lag between acceleration and movement. The accelerations reach 2x gravity and the lift is nearly the same in both gaits. The "saddle point", at which we had attached our accelerometer, is indicated by a cross (X), the estimated CoM of horse plus rider by a dot. Letters (**g, h**) demonstrate the schematic presentation of the CoM's pathway during the jump,

each cycle contains one or two phases of suspension. In smaller forms the extended phase of suspension is quite obvious and longer than the "collected" suspension with anteverted hindlimbs that "cross" the forelimbs. These extended phases of suspension make the gallop in the famous cheetah (*Acinonyx jubatus*) and in less heavy hooved animals (*Gazella subgutturosa, Capreolus capr., Rupicapra rupic., Capra caucasica, Saiga tatarica, Ovis orientalis*; Gambaryan 1974) similar to the bound of "small" mammals, diverging alone by the different footfall sequence. Very large mammals, not alone horses but also the giraffe, *Giraffa*; American buffalo, *Bison americanus*; eland, *Tragelaphus*; and reindeer, *Rangifer tarandus*, normally do not show the extended suspension (Gambaryan 1974). Galloping race horses (Figs. 7.16 and 7.22) reach their extreme of speed by disrupting the diagonal support in the gallop, so that not three beats but rather two double beats can be heard, because front- and hindhooves are put on the ground with a short interval. In this gait horses race over 1000 m, 2400 m or 3000 m. In our empirical work, we have never observed an extended phase of aerial floating, but Gambaryan (1974) states (with an illustration) that it indeed occurs in extremely fast sprints over very short distances of less than 1000 m.[21] Due to the rhythmic and rather slow accelerations visible in Fig. 7.24d, the intestines are also accelerated and so compress or dilateate the lungs (Young et al. 2013, Bramble and Carrier, 1983).

The *high frequency* of footfalls entails very short and therefore extremely high GRFs. The limbs of all cursorial mammals rotate in the gallop during the stance phases like an inverted pendulum about a pivot in the joint closest to the ground. The "retraction" of the limb during stance in essential takes place passively. Only moderate forces in x-direction are required for re-acceleration to compensate braking. During the recovery phase, the limbs swing forward as a suspended pendulum like in symmetrical gaits. The swing in rapid locomotion, however, requires additional input of energy, first because of the flexion of the joints to shorten the pendulum chord and second to accelerate the pendulous limb against its mass moment of inertia.

The frontlimbs are swung more forward and lifted from the ground earlier in the stance phase than the hindlimbs, in which the excursion rearward to the vertical is

---

Fig. 7.24 (continued) which includes both phases (broken line) and measured vertical acceleration (continuous lines). The forces evoked are calculated; hatching indicates the front (horizontal) or hind (vertical) hooves on the ground. Note that the forces in takeoff as well as landing are distributed on frontlimbs as well as hindlimbs. (Redrawn from Streitlein and Preuschoft 1987; Preuschoft et al. 1987). During everyday locomotion (trot, canter) about two times earth acceleration are acting, in extreme situations (**g, h**, jump over an obstacle of 1.6 m height) 2.5 and 3.7 times earth gravity. The acting force is mass times measured acceleration (Fig. 1.9). (Redrawn from Preuschoft 1989a; Preuschoft and Hüllen-Kluge, 1987)

---

[21]Own observations, as well as comparisons of outstanding performance in various disciplines of equine sports, suggest that the riders do not essentially influence the pattern of movement in horses but control direction and speed strictly.

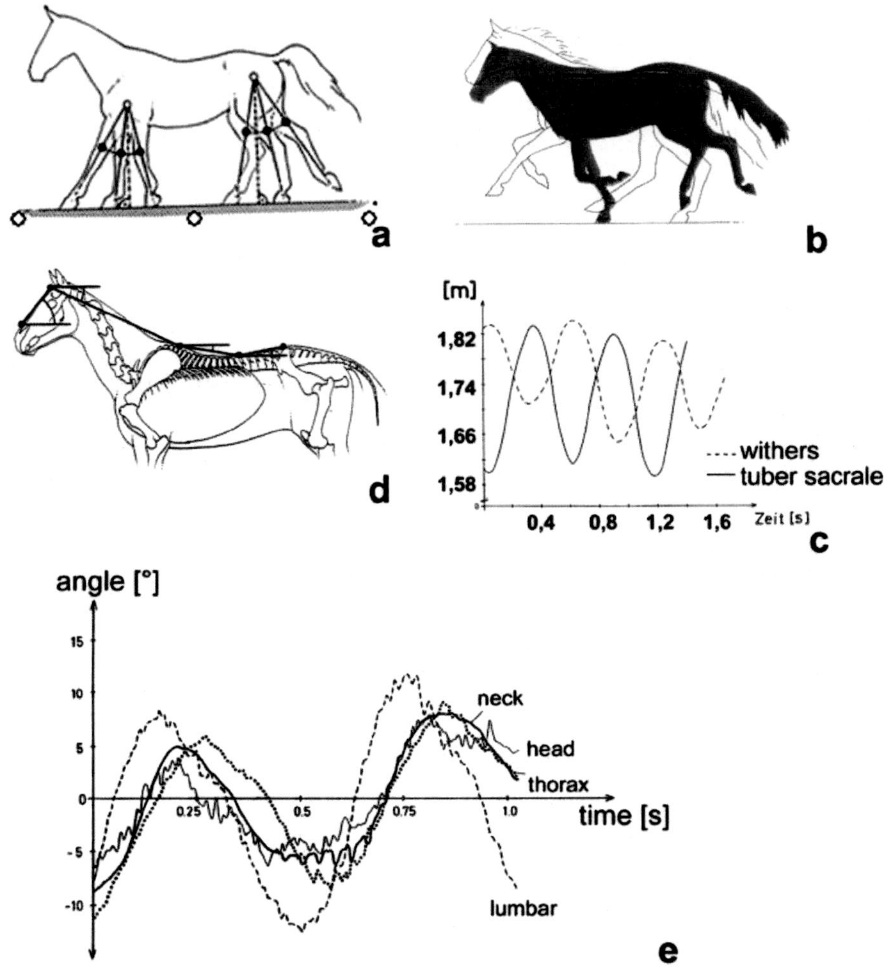

Fig. 7.25 Kinematics of the canter. (**a**) More obvious than in other gaits, the excursions of the forelimbs in front of the vertical are wider than those behind. The inverse is true for the hindlimbs, which are moved more rearward than forward. Several swinging movements are superimposed in the canter: (**b**) and (**c**) rotation about the CoM lifts and lowers withers and croupe alternatively. (**d**) A limited flexion and extension of the trunk takes place in horses, although they are "large" mammals with "dorsostabile" movements. We documented these movements by angles between the horizontal and segments of the animal's body stem. (**e**) These angles change rhythmically and are limited to about 20°. In *x*-direction, the hip joint moves only 10% of the length of lumbar plus pelvic segments. In vertical *y*-direction, however, the displacement follows the sine of the excursion angle; that means 30–40% of the lengths of lumbar plus pelvic segments. (From Loitsch 1993)

more pronounced than the excursion forward (Loitsch 1993).This more forward excursion of the frontlimbs and the more rearward excursion of the hindlimbs is obvious in galloping (Fig. 7.25a) but also exists in other gaits (trot; see, e.g. Streitlein and Preuschoft 1987) and has been observed in reptiles (see Chap. 5.6 and Kawano and Blob 2013). In addition, the total excursion angles of the forelimbs are more acute than those of the hindlimbs. This is correlated with the greater duty factor of the hindlimbs in comparison to the frontlimbs and is due to the greater share of body weight carried by the frontlimbs (Biancardi and Minetti 2012), which keeps the GRF in a more upright position. During steady-state locomotion at level ground and with constant speed, *the braking and re-accelerating force components* should exert the same impulses (*P*). A detailed comparison between GRFs of front- and hindlimbs in cursorial mammals unveils the braking force components to be larger in the forelimb and the accelerating force components larger in the hindlimb (Fig. 7.6). Bramble and Carrier (1983, also Carrier, 1996) connected the interplay of breaking and re-acceleration with breathing, since the foreward-moved intestines compress the lungs during the breaking and enlarge their volume during re-acceleration. This is perfectly in line with the experience of riders, who know that frequence of breathing increases remarkably with the change from trot or walk to canter.

Changing direction is difficult in large mammals (compare Sub-Chapter 7.4.3, hares), because of the inertia of their big mass and the height of their CoM above ground. Any turn leads to a centrifugal force, which is the greater, the faster the animal moves. The point, where horizontal force components in z-direction can be transmitted, are the hooves, and this requires frictional resistance. Turning in polo horses and race horses has been investigated by Tan and Wilson (2011). The former take sharper curves, while the latter go at higher speeds, since the racecourses are but lightly bent. The authors found that both polo and race horses kept the centrifugal force components within limits by considerably slowing down. Polo ponies go with horizontal speeds of up to 12 m/s and reduce it in exact accordance to the limits set by friction. At lower speeds, angular velocity becomes smaller and vice versa. Race horses, going at speeds of about 18 m/s on straight lines, reduce their speeds to 16 m/s or even less than 14 m/s around the bend.

Overcoming obstacles by *jumping* in cursorials and especially in horses is developed from the gallop. Even forms which otherwise do not show an "extended" suspension (horses!), the phase equivalent to extension is prolonged and other phases of movement are just slightly modified (Fig. 7.24). The accelerations and decelerations are divided over hind- and frontlimbs, while lowering of the head and neck in combination with lifting the extremities allows crossing the obstacle without lifting the CoM to the true height of the latter. The phase of free flight is relatively short, because the hindlimbs are in contact with the ground nearly to the highest point of CoM's trajectory. The velocity at takeoff in jumping is about 5 m/s, that is, less than 1/3 of maximal speed. For more details, see Preuschoft et al. 1987.

The swings of the limbs in the *canter* are superimposed and facilitated (Loitsch 1993) by several swinging movements of the trunk (Fig. 7.25, see also Townsend et al. 1983): first a rotation about a horizontal z-axis through the CoM lifts the cranial segments upward (during foreswing of the frontlimbs) and then the caudal segments upward (during the foreswing of the hindlimbs). Second the whole body performs pendulous swings of a few degrees fore and aft about an axis dorsal of the back. These swings can be shown only qualitatively, and we have been unable to determine its pivot. Third the trunk is flexed in the sense of a ventral flexion and back to dorsal extension. This is similar to the trunk flexion and extension in small mammals

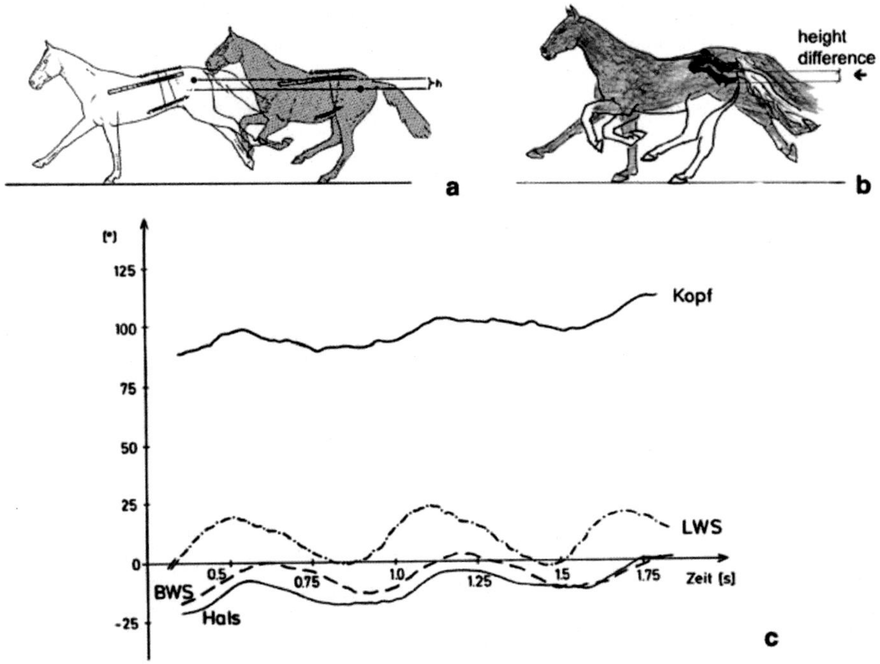

Fig. 7.26 Spring in the trunk of a horse. (**a**) In galloping, or cantering, the trunk is flexed and extended like a bending spring. The activity of the dorsal and ventral muscles has been documented by EMG. The height of the hip joint above ground changes rhythmically during each cycle by more than 20 cm. (**b**) Changes of angles against the horizontal, like in Fig. 7.25d, e. In the canter, excursions of lumbar plus pelvic against the hardly moved thoracal segments are about 20°, in the gallop some degrees more. The neck's participation in the swinging of the thoracal segment is restricted; the head ("Kopf") is kept at the same height, independent from the movements of the trunk. (**c**) Horizontal excursions of the hip joint are minimal, while vertical excursions of the hip joint in the canter are about 2∗20 cm/sec^2, which is more than 4 G. Consequence is an acceleration of the pelvis upward by about two times earth acceleration. Due to the lifting of the hip joint, the hindlimb pendulum therefore swings much faster forward than under the influence of G alone, the increased gravitation makes the foreswing of the hindlimbs faster. (Combined from Loitsch 1993; Witte et al. 1995a, b)

(see Sub-Chapter 7.3.3). It takes place in the lumbar segments and is most pronounced in front of the sacrum and pelvis. There are good arguments for the assumption that these movements are initiated by elastic stretching and recoil of not active muscle-tendon systems ("trunk spring", Fig. 7.26). According to Fig. 7.25b–d, the lifting and lowering of the hip region in y-direction follows the sine of the excursion angle; the distance in x-direction, however, follows the cosine. Because the flexion and extension are in large mammals not more than 20° or 25°, movement in the vertical y-direction is much more pronounced than in x-direction.

The gain of distance is not more than 4–13% of the distance covered in each cycle. The upward movement of the hip is very fast and takes place with an acceleration of about 2 G. This means that the hindlimbs are swung forward in a field of increased gravity, and their passive swing takes place in 0.75 s, instead of 1.06 s i.e. 1.4 times faster without the influence of being accelerated upward. Since it is the foreswing, which determines the frequency of cycles (in cantering horses 2 Hz, in galloping up to 3 Hz), the flexion and extension of the trunk permit higher speeds.

7.5.6 *Elastic Resilience of Structures and Spring Mechanisms Within the Limbs*

No matter in which gait, for any phase of aerial floating, the mass of the body must be accelerated upward; Streitlein and Preuschoft (1987) have measured in trotting German Warmblood horses about two times earth acceleration and height differences of the body's CoM of 6–12 cm (Fig. 7.27). This is the value by which the

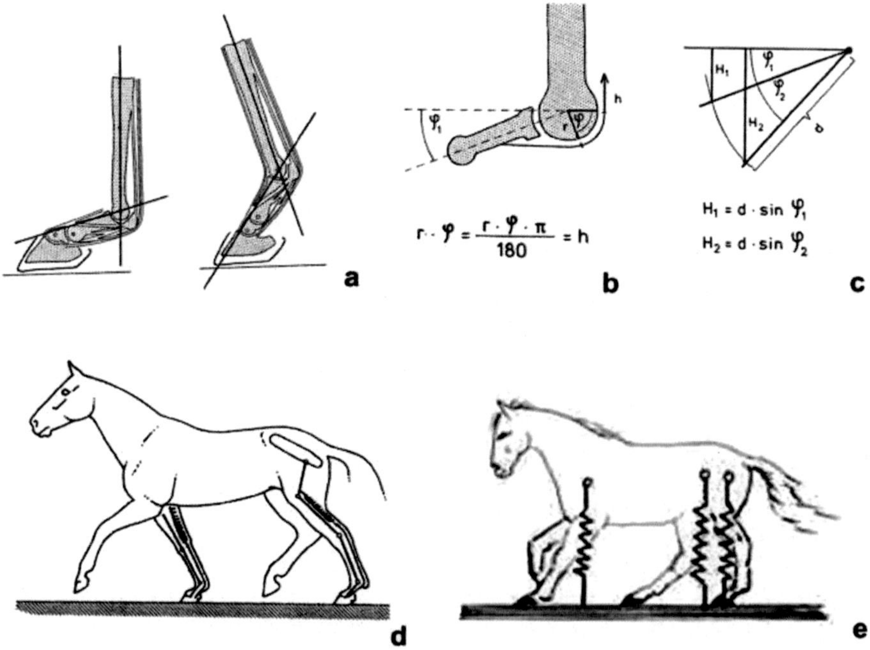

Fig. 7.27 Spring mechanisms in the limbs of horses. (**a**) Only the elastic stretching of the collagenous tendons by about 4 cm entails a swing of the fetlock bones (= digits) of 70° or 75° (**b**). If the lengths of the two proximal phalanges are 10 or 15 cm, this excursion is equivalent to a vertical lifting of the fetlock joint by 6–12 cm (following the sine of the angle φ, **c**). This value is close to the lifting of the horse's CoM in the trot, as observed by Preuschoft and Günther (1994). (**d**) During the stance phases of the trot, the springs of two limbs brake and reaccelerate body mass upward in *y*-direction (From Witte et al. 1995a, b). In the canter (**e**), at least two, or during most of the time three limb springs come into action (before and after diagonal support), in addition to the bending spring in the trunk (as shown in Fig. 7.26)

fetlock in horses lifts the animal; if observed and measured excursions are considered. The driving force seems to be provided by a *spring mechanism*. The flexor tendons of horses have a length of about 1 m. If stretched by the weight share carried by the respective limb (Fig. 7.27a), they get a length of about 0.04 m more (Dimery et al. 1986; Lochner et al. 1980, however, found strains of clearly more than 10% of length at rest under very high loads!). The first mentioned length increase leads to the dorsiflexion of the fetlock joint which allows the CoM to fall 6–12 cm deeper than during the swing phase of aerial floating (Fig. 7.27a–c). The elastic extension and recoil of a tendon or muscle-tendon complex is an efficient means to save energy (Biewener, 1998) and permits trotting over very long distances. Regarding the question which morphological elements provide the elastic spring, the opinions diverge somewhat. Alexander 1975 on kangaroos, and Dimery et al. (1986), Alexander (1988), Alexander et al. (1985), Goodship (1993) on horses and Dimery et al. (1986) on camelids hold *collagenous fibres responsible* for elastic recovery of energy. The elastic properties of collagen indeed fit to the rhythm of locomotion in these animals (see above). In our opinion the own frequency of collagenous fibres is too high to explain alone the movements of large mammals. As detailed by Rao et al. (1995), the toe flexors of cursorial mammals form muscle-tendon complexes, in which the *titin components* are less rigid (= have a lower Young's modulus) than the collagenous fibres of tendons. Therefore, their own frequencies are lower than those of tendons alone. Witte et al. (1995a, b), have measured the load exerted on a single limb of a standing horse by means of a force plate and at the same time the angular changes of the fetlock joint. They found an axial spring stiffness of 50 N/mm elongation of the flexor tendon. This would be too compliant for locomotion. They concluded that the stiffness (= Young's modulus) of the entire muscle-tendon complex during locomotion is modulated by contraction of the flexor muscles to 32 N/mm. With this value, and if the horse's body is supported on two springs (one front and one hind, like in Fig. 7.27a, b), the cycle duration can be calculated (Witte et al. 1995b, S. 268):

$$T_{2\ \text{springs}} = 2\pi\sqrt{}\text{m}/\text{c} = 2\pi\sqrt{}585\ \text{kg}/2 * 32\ \text{N/mm} = 0.60\ \text{s}, \qquad (7.5)$$

where T is the duration of one swing (or step), m the mass of the horse under investigation and c the stiffness of the spring (= forelimb, measured in N/mm). The time interval of ground contact and transmission of force (D) is

$$D = T_{\text{contact}}/\text{cycle duration}. \qquad (7.6)$$

In our gait analyses, cycle duration in the medium trot was 0.6 s, and D was 0.84. If the quotient is included into the two-spring model of the trot, the time interval required by one cycle can be calculated as

$$\text{Cycle duration} = T_{2\ \text{springs}}/D = 0.60\ \text{s}/0.84 = 0.71\ \text{s} \qquad (7.7)$$

This value is fairly close to the really observed average cycle duration of 0.69 s in the investigated gait. The calculation confirms that in fact alone the two "leg springs" are used for recovering energy.

Witte et al. (1995a, b) have also calculated the cycle duration in a model supported on three springs, having the same spring rigidity as assumed above. As can be seen from Fig. 7.22, the body is supported in the canter cycle repeatedly on three limbs. These phases are responsible for the safety and absence of sensitivity against uneven ground of the slow canter. The calculated cycle duration was 0.56 s, while we have measured a value of 0.55 s. The model does not take into consideration the "spring" provided by trunk flexion (Fig. 7.26a), as was first proposed by Alexander (1988), who held the strong aponeuroses of the dorsal (longissimus) muscle group responsible for elastic storing of energy (in accordance with his results on limb tendons). The relation between compliance of tendon and muscle fibres was also dealt with by Griffiths, 1991.

7.5.7 *Tracks: Support Patterns in Space*

While the common definitions of gaits rest on time variables, the *spatial variables* are not much observed, although they are documented in the *tracks* and can be preserved as fossils, so that an analysis may yield primary information about the behaviour of an animal which does not longer exist. To get a reliable basis for the relation between gaits and tracks, Kienapfel et al. (2014) have taken measurements on tracks of horses in defined gaits. Because of the general similarity of their gaits, the results are probably relevant for many, if not all, cursorial mammals, at least for hooved mammals. When looking at the trackways, the most obvious fact is the even (and surprisingly dense) distribution of footprints over the runway in all gaits. This gives the body weight continuous support and allows input of energy (by increasing the force components in x-direction) to maintain or change speed. Regardless of the footfall sequence and the "gaits" derived from them, the covered distances at higher speeds become longer, while the duration of the cycles becomes shorter (because of the higher stride frequency (Fig. 7.23): the number of cycles on a runway is smaller in faster locomotion. The width of the tracks in z-direction is narrow, in most cases less than two footprints side by side (Fig. 7.28), and minor deviations from the straight direction of locomotion occur regularly in all gaits. These deviations seem to be the means by which imbalances of the body over the supports are compensated.

Easy to discriminate are the *tracks of symmetrical* (Fig. 7.28a–f) from those of asymmetrical gaits (Fig. 7.28h, i, k). In walk and trot, the hindhooves are put behind, on top of or in front of those of the forelimbs. This "overstepping" clearly depends on the lengths of steps, which covaries with increasing speed in the respective gait, but its occurrence varies individually. The ipsilateral hoofprints often are in pairs, that is, distances between front- and hindhoof are shorter than the prints of the next following (mostly the opposed) sides. At the transition from walk to trot, the overstepping disappears and comes back at greater trotting speeds. The number of

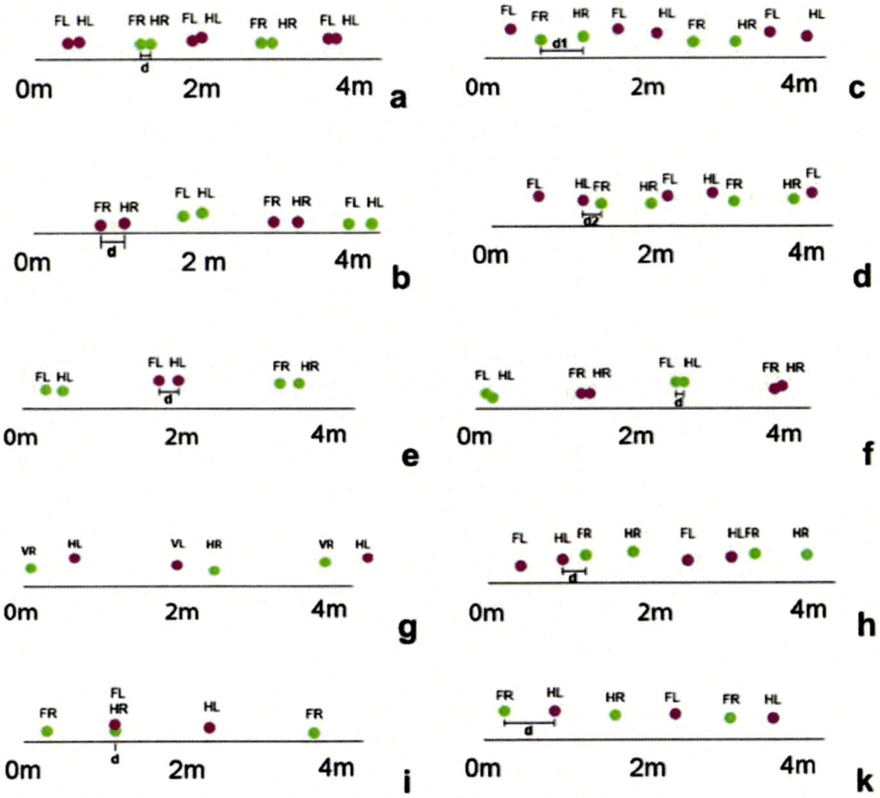

Fig. 7.28 Foot tracks of horses in selected gaits. Locomotion is from left to right. The attribution to the limbs was made on the basis of synchronous video-recordings. Right hoofprints green, left hoofprints pink. **(a)** Slow walk, **(b)** fast walk of a German warmblood; **(c)** fast and **(d)** slow tölt (amble) of an Icelandic horse. Overstepping (d1, d2) is more marked than in the walk. **(e)** Fast (extended) trot in a German warmblood, **(f)** slow (collected) trot of the same individual. **(g)** Running pace of an Icelandic horse. It looks similar to the fast trot, but the paired prints are from contralateral sides, so that the seeming "overstepping" is performed by contralateral rather than ipsilateral hooves; **(h)** very slow right lead canter (German: Galopp) of a Paso Fino, **(i)** left lead canter of a German warmblood; **(k)** right lead canter at relatively fast speed of a medium-sized German saddle horse. Overstepping of contralateral hooves is indicated by d. (Rearranged from Kienapfel et al. 2014)

prints in the trot is clearly smaller than in the walk, simply because step lengths are greater.

In the *toelt*, the hindhooves are usually placed in front of the fronthooves. The faster the amble, the more marked the overstepping. Toelt at very high speed leads to pair formation of contralateral prints, which makes the tracks similar to those in the pace. Characteristic of the *pace trot* are the long distances between the pairs of prints (= long step lengths) as in the trot, although the pairs are not made by ipsilateral but rather by contralateral hooves.

Completely different are the *tracks* left behind by horses *in the canter* (Fig. 7.28h, i, k). Often the hindhooves are slightly displaced to one side of the anterior hooves. In spite of the higher velocity than in the trot, the single prints are evenly distributed and at surprisingly short distances from each other. The longest distances between footprints come from the diagonal support, on the trailing hind and leading forelimb. As noted above, Gambaryan (1974) observed that small hooved mammals (*Gazella subgutturosa*, *Capreolus capr.*, *Rupicapra* rupic., *Capra caucasica, Saiga tartarica, Ovis orientalis*) interrupt ground contacts during their extended suspension phase. Such an interruption usually cannot be afforded by heavier mammals, like horses. The frontlimbs as well as the hindlimbs are also put down on the ground in a striding position.

Whether gaits can be identified on the basis of tracks depends from the knowledge of size of the trackmaker and its limb length. The first unknown variable (size) is roughly proportional to the size of the footprints – whereby the structure of the foot must be taken into account. In order to make estimates of limb length and trunk length, at least a vague idea of the trackmaker's body shape is necessary. If left and right footprints can be discriminated, and at least five prints in a row are available, the estimation of the gait should be possible.

7.6 High Stressing Aside from Locomotion

In contrast to the similarity of essential characteristics of locomotion among cursorial mammals, there are clear differences in *lying down* and *getting up* from resting positions (Fig. 7.29). Because of the length of limbs in cursorials, it is by no means an easy task to lift the mass of the body stem on about 1 m or more or to lower the mass in a controlled way. Alone getting up implies physical "work"; in lying down, however, the "fall" of body mass from a considerable height must only be slowed down under control of the limb position. Both movements imply the exertion of considerable forces, especially in those phases, in which the limb joints are strongly flexed and the muscles have to act against gravity.

Lying down seems not to be really difficult for the *cursorial carnivores*, because of their overall flexibility. It may begin with hindlimbs as well as with frontlimbs. Often they pass from a sitting posture on their flexed hindlimbs to lying by just sliding forward or by flexing their forelimbs. In other cases, canids as well as felids lower their CoM slowly while performing several rotations about a vertical axis, until they let themselves fall from low height. Getting up is frequently initiated by an extension of the frontlimbs forward, which already take over a great part of the whole body weight. Often, though not always, the hindlimbs are placed underneath the trunk by simply rolling the posterior trunk toward the strongly flexed hindlimbs. Forelimbs then are pulled backward, and all extremity joints are extended to lift the body stem. This action is against gravity, but the joints of the carnivores anyway are partly flexed during locomotion, so that the required muscle forces are not obviously higher than in locomotion.

Fig. 7.29 Lowering down, resting and getting up (from left to right) in odd-toed horses (**a**) and even-toed ruminants (**b**). The transitions from relaxed resting postures to standing and active locomotion, as well as the reverse, are a challenge for cursorial mammals, because the joints of their long limbs are exposed to rotating moments much higher than in locomotion, admittedly for only short intervals. In the case of camels, this has been described by Gauthier-Pilters and Dagg (1981)

The large herbivore, hooved mammals, however, are much more restricted in their mobility. The families have developed different techniques to overcome the mechanical problems (Fig. 7.29): *camels* first of all flex the carpal joints and thereafter the ankle and knee joint of the hindlimbs and finally the elbow and shoulder joints. They use to rest in symmetrical posture on the flexed limbs and the ventral side of the trunk. Old World camels possess conspicuous calluses at their preferred ground contacts. Getting up in camels also follows a fixed sequence: first the forearm is placed into the vertical, then the hindlimbs are extended, and the last movement is lifting from the carpal joints on the forefeet. *Horses* lower themselves in a way similar to that described above for carnivores: they become, with extended and laterally bowed neck, lower and lower by flexion of the carpal, elbow and perhaps shoulder, hip, knee and ankle joints. The carpal joints make first contact with the ground. The last phase is falling down from the lowered position. Their most frequent resting posture is with flexed forelimbs, symmetrically below the thorax, while the pelvis with (flexed) hindlimbs is tilted laterally, so that both hindlimbs are lateral to the belly. For getting up, horses first lift the anterior part of the trunk on one and then both extended forelimbs to a posture comparable to the sitting of a dog. The shifting of the body's CoM over the laterally placed hindlimbs is only possible by a powerful swing of the neck and the fixation of the hindhooves against sliding on the ground. *Even-toed cursorial* mammals normally rest in a similar posture as that assumed by horses. They usually first flex their carpal joints to get deeper and then place their hindlimbs forward while flexing slowly its joints. For getting up, Bovidae and Cervidae begin with the hindlimbs, which are placed far forward underneath the belly. This opens the flexion angles of elbow and carpal joints so much that the weight of the head, neck and anterior trunk rests on the carpal joints and the anterior side of the metacarpus. One frontlimb after the other is extended by placing the hoof forward, before the trunk is lifted by retroversion of the (extended) forelimbs (own observations, Nickel et al. 1968).

7.7 Stress Patterns and Their Relation to Shape in Cursorial Mammals

Stress patterns are identified in this book as the decisive factors for morphological structures. Therefore, it makes sense to investigate detailedly the distribution of stresses which act repeatedly, or during long periods, and compare them with the resisting body structures.

7.7.1 *Bending of the Body Stem*

The *body stem* of a typical cursorial mammal, for example, a horse, is exposed to gravity, which obviously tends to bend trunk and neck (Figs. 4.3f and 7.30a–c), preferredly in lateral view. The body stem can be replaced by a beam of varying diameter, the sections of which contain different weights (Fig. 7.30b). Shortness of the trunk keeps bending moments on a low level. In the body of a horse, the bending stresses (Fig. 7.30c–e) correspond to the arrangement of resistant tissues: the vertebral column sustains compression, and the tensile forces on the opposite side are provided by the dorsal or ventral musculature or no-energy-requiring ligaments or fascia. The dorsoventral diameters provide long lever arms, so that low forces are sufficient to maintain equilibrium. The location of the vertebral column on the longitudinal section corresponds to the regions where compressive forces are highest (Fig. 7.30c): in the lumbar and posterior thoracal region near the dorsal contour and in the anterior rib cage and proximal neck region approached to the ventral margin. Along the ventral margin of the thorax, a thick, strong rod (the sternum) is present in addition. While the vertebral column plus intervertebral discs are shown as one (black) line, in reality the intervertebral joints between the neighbouring centra must be kept in equilibrium, and this permits calculation of the required forces. The intestines as a whole nestle into the not highly stressed trunk cavity. In the space limited by the muscular diaphragm, pelvic floor and ventrolateral abdominal wall, they form the not compressible filling of a “pneu” or hydraulic cushion. This has relevance in side view (see below) and also keeps the muscles of the belly at distance from the vertebral column.

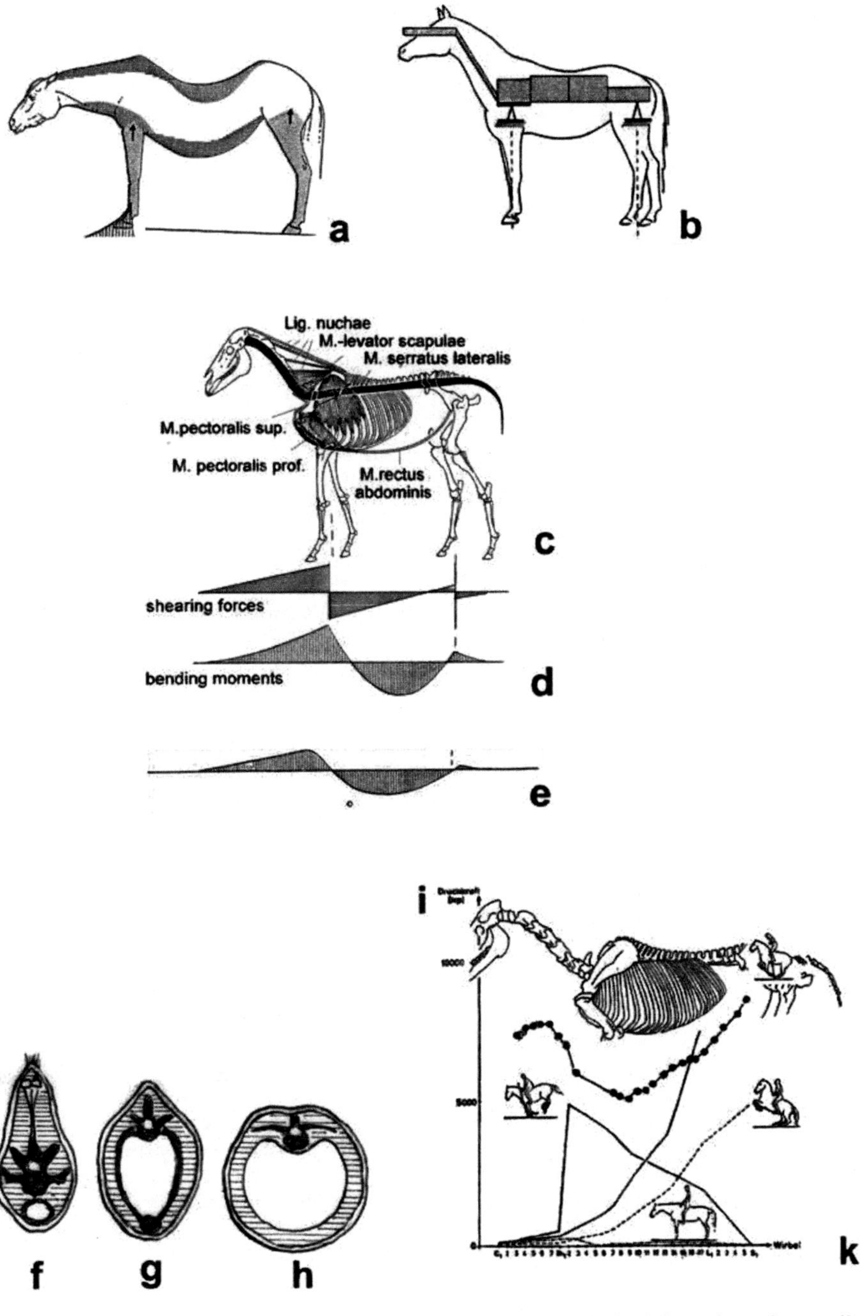

Fig. 7.30 (**a**) Under the influence of weight, the quadruped's body is deformed, so that tensile forces occur along the dorsal side of the neck, cranial thorax and the ventral margin of the belly, while compressive forces are evoked between the frontlimbs and along the dorsal side of the trunk. (**b**) Outline of a horse seen from its left side. The weights of body segments are indicated as grey rectangles. (**c**) Horse skeleton with tension-resistant structures that keep the trunk in its position.

The rather long neck is bent by its weight downward, and the bending moments are increased by the weight of the heavy head. Consequently, the compression-resistant vertebral column is located near the neck's ventral margin; and the tension-resistant ligamentum nuchae and dorsal neck muscles near the dorsal contour. Like birds (see Chap. 6.3), most hooved mammals (but not horses nor the heavy and short-necked members of genera *Bos*, *Bubalus* and *Bison*) keep at least the cranial half of their long necks vertically, so reducing the lever arms of weight and the muscle force which is necessary for balancing bending moments.

The tensile forces exerted by the muscles (including the above-mentioned m. iliocostalis) which keep the vertebral column in balance determine the compressive forces on the vertebrae. This is calculated for four critical postures or phases of locomotion, which represent typical load cases in Fig. 7.30i, k. Remarkable are the low stresses during standing, which the body must sustain during the whole lifetime. The greatest loads occur during takeoff for and during landing after a jump, because the (measured!) accelerations are between 2 and 4 G. Lifting of both frontlimbs from the ground for several seconds leads to static conditions, exerted without accelerations alone by earth gravity. This can be observed occasionally in free-moving horses and is induced in the "levade" of dressage horses in a particularly deep trunk position. All these loads must be sustained by the body, in the interest of survival. The strength of the vertebral column is proportional to the cross-sectional area of bony vertebrae and intervertebral discs (as mentioned above repeatedly). In fact, the greatest, normally occurring stresses and the anterior areas of the vertebral bodies change in proportion: from segment Th1 to Th7 or Th8, they decrease and increase again steeply toward L6 and the sacrum. Minor deviations may well depend from the fact that we have used locomotor data from trained athletes, but measurements on the vertebral column of a Przewalski's horse which has spent its whole life in a zoo. Jones and Holbrook (2016) put emphasis on two points: the extension of the cross-sectional areas of lumber vertebrae in lateral *z*-direction, while the diameters in dorsoventral *y*-direction are moderate and especially on the fact that the transverse processes of the most caudal lumbar vertebrae possess a synovial joint, so that the compression-resisting surface is

---

Fig. 7.30 (continued) The tensile stresses evoked by bending are sustained by muscles or ligaments (double lines); the reactive compressive stresses are sustained by the vertebral column plus sternum (black) below c Shearing forces, (**d**) bending moments along the body stem. Distribution of the supporting forces over a considerable distance along the trunk axis cuts off the high peaks of the bending diagrams (**e**). (**f–h**) Schematic cross sections through the horse's body stem (**f** at base of the neck, **g** behind the shoulder, **h** middle of belly) suggest the extent to which the general body shape is suited to withstand the occurring stresses with a minimum of material. The tension-resistant structures have rather long lever arms, so that low forces are sufficient to keep the body stem in equilibrium. (**i**) Skeleton of the body stem. (**k**) The compressive load on the vertebral column is calculated for four selected phases of movement. Counterclockwise beginning on the left side: landing after a jump, relaxed standing, lifting the forelimbs with full weight on hindlimbs ("levade"), takeoff for a jump to overcome an obstacle. The cross-sectional areas of the vertebrae (dots) are nearly parallel to the greatest of the occurring stresses. (Redrawn and rearranged from Preuschoft 1976, 1993; Preuschoft et al. 1994c) enlarged beyond the surface of the centra. This osteological fact fits nicely to the myological observation that horses have a massive extension of the m. glutaeus medius, which crosses the crista iliaca to fuse with the dorsal (longissimus) muscle group. If the hindlimbs are placed on the ground far forward, approaching the vertical through the CoM, very strong muscle contractions are required to keep balance in the hip joint, as well as in the iliosacral joint. Contractions of these muscles of course lead to high compressive forces in the most caudal part of the axial skeleton.[22]

7.7.2 Shearing

If the animal's body in Fig. 7.30d is thought to be cut in front of the shoulder, the head and neck will tend to fall downward, unless internal forces counteract. If such a cut is made behind the shoulder, the head, neck and shoulder complex will remain at their place, and the thoracal plus lumbar section will fall downward, unless internal forces keep balance. Responsible for these effects are the so-called shearing forces. They depend on the relative weights of the body segments and are shown along the trunk in Fig. 7.30c. They reach a maximum at the shoulder, where they change their sign. If the cut is made in front of the pelvis, the anterior part will maintain balance on the frontlimbs and the posterior part on the hindlimbs: the shearing moments are zero. Behind this "neutral plane", the anterior part of the body will again fall downward, while the posterior goes up – like in the neck. The shearing forces determine bending moments, because local bending moments (M) are proportional to products of distances to bearings and local shearing forces (F_q).

$$F_q = q * (l/2 - x) \text{ and} \quad (7.8)$$

$$M_x = q * x(l - x)/2 \quad (7.9)$$

And in a cantilever:

$$F_q = q * x \quad (7.10)$$

$$M_x = q * x^2/2 \quad (7.11)$$

where q is load per length, l is distance between the bearings (i.e. equivalent to trunk length) and x is the distance between the neutral plane and the nearest bearing.

A perfect construction, in engineering known to withstand shearing, and to provide the internal forces necessary for sustaining them, consists of a dense sequence of compression-resisting rods, connected by tension-resistant fibres in

[22]The same osteological traits seem to be present in small-bodied fossil Perissodactyla, but not in large-bodied forms. This distribution is in line with the assumption that small – and perhaps swiftly moving – forms like *Mesohippus* had the same mechanical problems as modern horses.

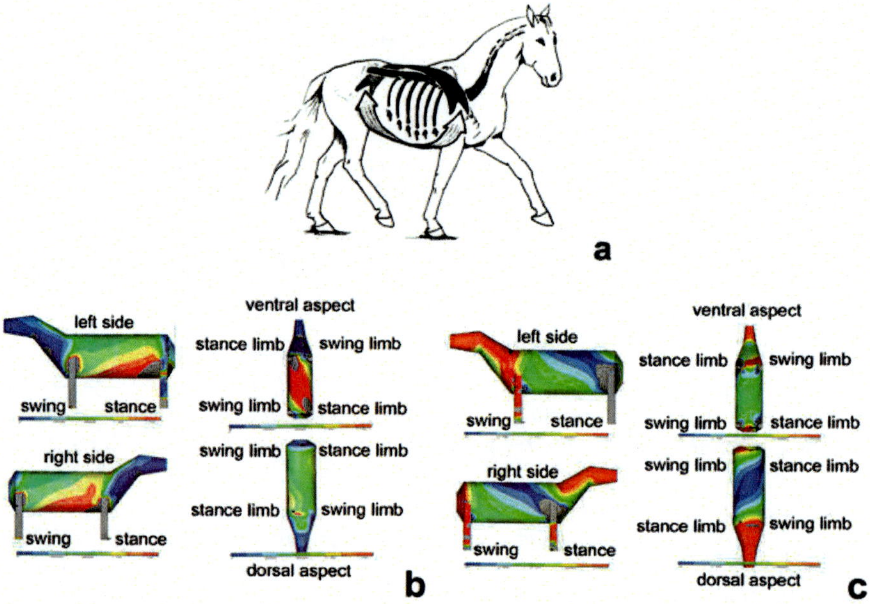

Fig. 7.31 (**a**) If supported by diagonal limbs (like in trotting), the trunk between the front- and hindlimbs is exposed to torsion. Black arrows show compression and white or grey arrows tension. (**b**) FES analysis of the tensile stresses in diagonal support, (**c**) FES analysis of the compressive stresses in diagonal support. The models show von Mises stresses: Torsion is superimposed on bending, caused by the weight of the horizontal body stem. In side view, the highest compressive stresses correspond to the inclined shoulder blade and the inclined ilium. See also Figs. 4.6, 4.7, 4.8, and 4.9 in Chap. 4

both oblique directions, anatomically the ribs plus their intercostal musculature, which together form the thoracal wall. Where the shearing forces are low or absent (lumbar region), such a structure is not required and indeed does not exist.

7.7.3 *Torsion*

Regularly repeated phases of asymmetrical support are inevitable during locomotion. If only one of four limbs is lifted from the ground, there occur *torsional moments* (as in reptiles, Sub-Chapter 4.2.4), especially in diagonal stance. In cursorial mammals, these torsional moments are low, because of the narrow distance between weight force and the supporting limb (Fig. 7.32a, b). Nevertheless, great body weight and the high accelerations acting at higher speeds make torsion a relevant factor (Fig. 7.31a). In the case of horses, in the trot we have measured regularly two times earth acceleration – that means the horse does no longer weigh 500 kg but during each support phase 1 ton. For sustaining torsional moments, a

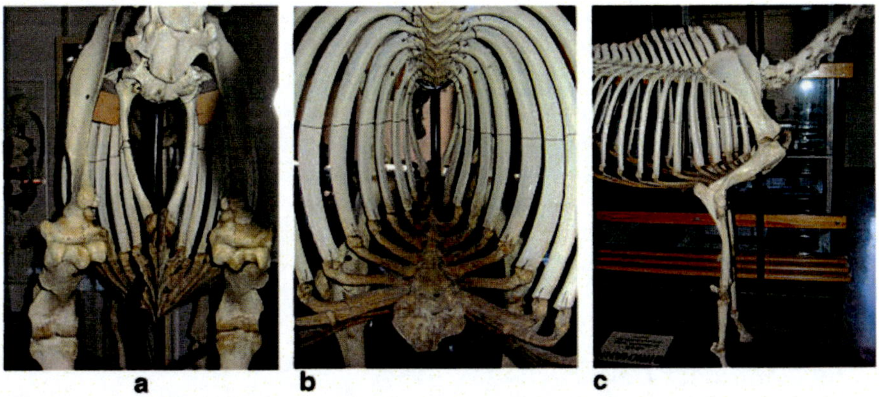

Fig. 7.32 Rib cage of a horse, seen in cranial (**a**), caudal (**b**) and lateral view (**c**). Note the massive, partially cartilaginous sternum and the caudally increased inclination of the bony ribs. Skeletons mounted in the Anatomical Institute, Tierärztliche Hochschule, Hannover

circular cross section gives the oblique muscles of the abdominal wall as well as curved ribs long lever arms (Fig. 7.32b). Torsional and bending stresses in side view sum up to the stresses shown in the FES -Analysis (Fig. 7.31b, c), which have their greatest values at the surface and so are visible in the models.

The directions of *stress trajectories* (e.g. Figs. 4.7 and 4.8 in Chap. 4.2) explain three typical and widely spread morphological traits: first the caudal inclination of the long, narrow shoulder blade and second the cranial inclination of the ilium (see also Figs. 7.31 and 4.3). Both elements are roughly in line with the compressive stress components. Third torsion compresses one side of the trunk where it is sustained by the ribs and stretches the other side, where the oblique muscles of the trunk wall are suited to counteract tension (Fig. 7.31a). The existence of muscle activity to counter torsion, especially in the diagonal support phase, has been demonstrated by EMG in trotting dogs by (Ritter et al. 2001; as well as Fife et al. 2001; Schilling and Carrier, 2009). The activity of the superficially accessible mm. rectus abdominis and obliquus abdominis externus in horses has been documented by surface EMG (Kienapfel et al. 2014, 2017/2018). *Torsional stresses* are on a low level, if the torsional moments are sustained by a cylinder with long diameter (Sub-Chapter 4.2.4). From a mechanical viewpoint, this is a good reason, why the posterior part of the thorax as well as abdomen are wide, round and therefore voluminous (Fig. 7.32b). The narrow anterior thorax, however, keeps the lever arms of the supporting forelimbs short and the torsional moments low. The latter trait is in full contrast to reptiles (see Sect. 4.2.4 and 4.2.5, Figs. 4.5, 4.6 and 4.7).[23]

[23]The fact that the varying cross sections through the trunk are visible in the skeleton, especially from the caudally increasing curvature of the ribs, should not lead to the conclusion that one morphological trait (curvature) is the reason for the other (rounded trunk)!

During the late stance phase of the hindlimbs, the cranially directed component of the GRF is applied to the hip joint, somewhat lateral to the midline. As described above for small mammals, the inertia of all segments cranial to the pelvis leads to bending of the vertebral column concave to the active hindlimb in top view. A veritable lateral flexion is excluded in the cursorial mammals by passive structures, not by musculature as in small mammals (see above). The expansion of accessory joints between the transverse processes of the caudal lumbar vertebrae is a means to stiffen the lumbar part against this sort of horizontal bending (Jones and Holbrook 2016). It also allows sustaining of torsional moments.

7.7.4 Shoulder Construction

In the skeleton of a cursor, if *viewed from in front* (Fig. 7.32a), the most obvious item is the above-mentioned narrowness of the rib cage (= moderate breadth between the shoulder joints). The short horizontal distance between the supporting limb and the CoM or the short load arms does not evoke exceptional torsional moments, in spite of the high forces. The narrow anterior part of the thorax is particularly pronounced in large animals (horses, Bovinae, elephants)[24] and in contrast to the more circular outline of the trunk's cross section in the segments behind the shoulder region (Fig. 7.32a, b).

The suspension of the weighty of ponderous trunk from the frontlimbs is illustrated in Fig. 7.33. Two load cases occur regularly under all-day conditions: symmetrical support on both frontlimbs, as in standing, and asymmetrical support on only one limb, as during locomotion. These must be analysed in detail to understand the details of shape of thorax and anterior ribs.

In *symmetrical support during standing* (Fig. 7.33a), the trunk is hung up by the mm. serratus and the m. pectoralis profundus on both sides. The first muscle inserts into the dorsal margin of the scapula on one side and the ribs on the other. The second muscle connects the sternum with the shoulder blade and the humeral head. For the sake of simplicity, body weight is assumed to be distributed evenly on both muscles. The ventral margin of the thorax is exposed to a horizontal tensile force, while the dorsal side is compressed in *z*-direction. Although the pectoral muscles have a horizontal component, they do not cross the midline. As a bridge between

[24]Hohn (2011) has demonstrated on the basis of shoulder girdle mechanics that the huge sauropods of the Mesozoic also had a strongly compressed thorax and the shoulder joints were surprisingly close to each other. The scapulocoracoids of both sides seem to have been in touch with a probably cartilaginous median element and the anterior, slightly concave margins of the often preserved "sternal plates". These not entirely ossified (and therefore not preserved) plates extended caudally and established a connection to the ventral tips of the ribs, like in other reptiles. Hohn's results coincide with my own measurements on sauropod ribs and my (unpublished) mechanical analyses of the reptilian shoulder girdle. The extreme narrowness of the sauropod shoulder is biomechanically convincing but rarely perceived and yet rarely presented in mounted skeletons or reconstructions. Shoulder construction in sauropods does not include, however, the free mobility in *x*-direction of the shoulder blade on the thorax, which is so characteristic for mammals.

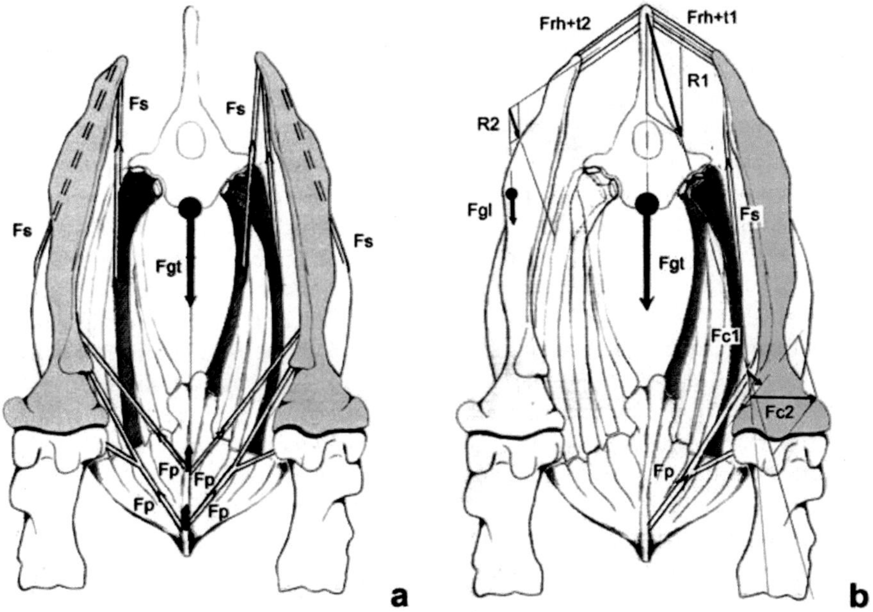

Fig. 7.33 Cranial views of the shoulder region of a horse. (**a**) Symmetrically supported by both frontlimbs, (**b**) asymmetrically supported by only the left frontlimb. The body stem is hung up on the most proximal segment of the upright standing frontlimb (scapula) like on a pillar. Weight force (F_{gt}) is transmitted by the strands of the m. serratus (F_s) and by the m. pectoralis profundus (F_p). We assumed the same share of body weight on both muscles. In two-limb support (**a**), the dorsal side of the trunk, especially both scapulae, are compressed in *z*-direction, the ventral side, however, is stretched. Internal forces are sustained mainly by masses of connective fibres and the m. transversus thoracis. By contrast, asymmetrical support (**b**) requires on the dorsal side tensile elements (mm. rhomboideus and trapezius, F_{rh1} and F_{t1}) and on the ventral side compressive structures: the cartilaginous parts of the anterior ribs and the sternum. In (**a**) the 1st, 3rd and 4th ribs are emphasised in grey, in (**b**) the 1st and the 4th rib. These ribs are exposed to compression, and their straight form is suited to sustain it. The weight of the not supporting/weight-bearing swing limb (F_{gl}) is also carried by the dorsal shoulder muscles (rhomboideus and trapezius, named F_{rh2+t2} and F_{t2})

both sides, there exists a strong m. transversus thoracis and a mass of horizontal collagenous fibres, acting in the same direction (not illustrated). The anterior fibres of the m. serratus have a tendency to swing the dorsal margins of the scapula medially. Although the caudal strands of the m. serratus (costa V, VI) may counteract this tendency (because of the curvature of caudal ribs, the insertions are lateral to the shoulder blade in Fig. 7.33a), there is the danger that the dorsal shoulder blade swings medially. This possibility is countered by the points of attack of the m. pectoralis profundus, which are distributed on the distal part of the scapula, the joint capsule and the humeral head, so blocking a lateral deviation of the shoulder joint. The most cranial ribs (I II, II, possibly IV in a horse) are exposed to considerable compressive force between their capitulae in contact with the vertebral column and the insertions of the m. serratus and to a lesser degree also between their dorsal (tubercula costarum) and their ventral connection to the cartilaginous sections. Their straight shape fits to loading in vertical y-direction and confines bending to their most dorsal sections (see below).

In *asymmetrical support* on only one frontlimb (Fig. 7.33b), the entire body weight (during locomotion multiplied by accelerations of 2 G, perhaps much more) is carried for a short time interval on one side, so exposing the described structures on the support side to very high loads. Different from the symmetrical stance, the trunk now tends to roll downward along the stance limb. This implies tensile forces on the dorsal and compressive forces on the ventral margins of the trunk. The former are countered by the mm. rhomboideus and trapezius, which not alone prevent the rolling downward of the trunk but also carry the weight of the entire frontlimb during foreswing. The activity of the muscles has been shown in the dog (Carrier et al. 2008). The compressive forces in z-direction are sustained by the sternum. The suspending muscles also press the trunk against the shoulder blade, especially the joint region. A giving way of the shoulder joint is countered not alone by the insertions of the m. pectoralis profundus (as above) but also by the mm. infraspinatus and supraspinatus, which must be active anyway to maintain joint equilibrium in side view.

For resisting the combined *tensile stress components* (derived from bending and torsion), the oblique abdominal and intercostal muscles are perfectly suited. The *compressive components* of torsion, as well as the *shearing forces,* are sustained by the bony ribs (see Sub-chaps. 4.2.5 and 7.7.5).

7.7.5 Rib Cage and Individual Ribs

Ribs, as well as their combination to a *rib cage*, have not attracted much attention by researchers. Since P. Schmid (Sub-Chapter 9.2.5), only Kagaya (2008) and Fujiwara et al. (2009) have approached the problem. While Kagaya focussed on primates (Sub-Chapter 8.5.6), Fujiwara pursued similar aims as those of this book and tried to understand the mechanic loading of ribs. Our own starting point (Preuschoft et al. 2005) has been the observation that the ribs of large hooved mammals look surprisingly similar. This similarity was documented for a small sample of cursorials with the aid of measurements (Fig. 7.34 and Table 7.1). In spite of the fact that horses possess 18 pairs of ribs, the even-toed ruminants only 13, camelids 12 and okapi 14,[25] the shapes of the ribs, as well as their particular strength properties, are quite similar in all included species.

[25]The similarity of rib shape in mammals of different sizes seemed to justify calculating "average rib dimensions" for cursorials. To facilitate comparisons, we have reduced the number of ribs in the case of horses by considering only ribs number 1–8, 10, 12, 14, 16 and 18. In camelids, we have inserted costa 10 two times. In addition, we have multiplied the values of smaller animals with a proportionality factor: to make the values comparable, we took the presacral length of the vertebral column of the trunk (caudal surface of last lumbar-cranial surface of first dorsal rib-bearing vertebra as a convenient and easily available reference).

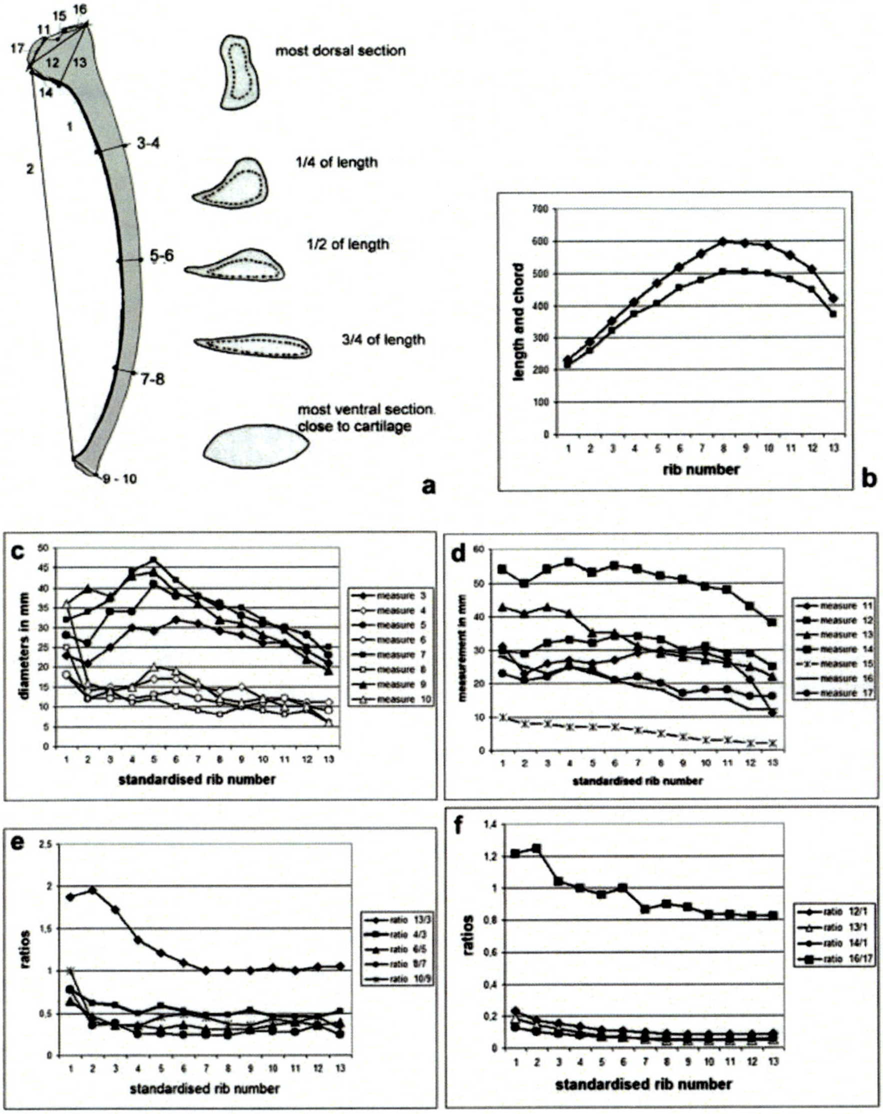

Fig. 7.34 Morphology of ribs in "averaged" cursorial mammals. The diverging rib numbers were "standardised" as in Footnote 25. **(a)** Cranial view of and cross sections through a rib of a giraffe as example for all cursorial mammals. The numbers demarcate the measurements. **(b)** Arch and chord characterise the length and curvatures of the ribs. The ratios become greater in caudal direction, but this is camouflaged by the shortness of the most posterior ribs. **(c)** Diameters in x- and z-direction of the cross sections as indicated in **(a)**. Obvious is the flatness of the ribs, documented by the marked differences between the x- and z-measurements. **(d)** Diameters of caput and costovertebral joints; **(e)** proportions of diameters in x/z-directions. Note especially 13/3, which indicates the high strength in z-direction. Except the first one, all ribs of cursorial mammals are flattened, as visible in **(c)**. **(f)** Proportions of joint surfaces indicate the strength of the costovertebral junction. The values in relation to rib length decrease rearward. The smaller values on the vertical axis do not indicate sustaining of high bending moments. The ratio 16/17 indicates more dorsally flexing moments in the cranial and relatively more ventrally flexing moments in the caudal ribs, as shown in Fig. 7.35b, d

Table 7.1 Mammals of which ribs have been measured

Common, *Scientific* name	Location	Presacral length of vertebral column in mm	Factor	Number of ribs
Camel, *Camelus dromedarius*	Institute d'Anatomie ULP Strasbourg	1310	1	12
Vicunya, *Lama vicugna*	Institute d'Anatomie ULP Strasbourg	555	2.324	12
Chamois, *Rupicapra rupicapra*	Institute d'Anatomie ULP Strasbourg	493	2.616	13
Cattle, *Bos taurus*	Anatomical Inst. ULP Strasbourg	1123	1.149	13
Horse, *Equus caballus*	Institute d'Anatomie ULP Strasbourg	1146	1.125	18
Okapi, *Okapia johnstoni*	Senckenberg-Museum, Frankfurt	Unknown	Unknown	14

From the most *cranial five or six ribs*, strands of the m. serratus take their origin, on which the trunk is suspended from the supporting limb (Fig. 7.35b). The high and permanently acting forces of the m. serratus bend these ribs in the *y*/*z*-plane laterally upward, abducting them from the median *x*/*y*-plane. A movement, however, is excluded by the tough costovertebral ligaments and perhaps to some degree by the "cartilaginous" ribs on the ventral side. The bending moments acting on the ribs increase from the muscle's insertions toward the vertebral column, where the ribs are – as mentioned above – tightly fixed by the tough ligaments of the joint between vertebra and rib. The necessary strength is provided in the dorsal sections of the ribs by their great diameters in transversal (*z*-)direction (as shown in Figs. 7.34e, f, 7.35b, and 7.36a). The ventrally decreasing strength of these ribs is reflected by their short diameters in *z*-direction (Fig. 7.36c). In addition to bending moments, the anterior ribs are exposed to compressive stresses, in their dorsal parts because of the m. serratus, in their ventral part because of the pectoral muscles as well as the m. rectus and the ventral neck muscles like m. brachiocephalicus (Fig. 7.35a, b). All these muscles exert considerable components in vertical (*y*-) direction. The orientation of the *ribs in side view* (Fig. 7.35a), as well as their cross-sectional areas (Fig. 7.36a), is in accordance with the compressive forces. The corpus costae is remarkably flat and extended in *x*-direction (Figs. 7.35 and 7.36). While their cross-sectional areas remain constant, their bending strength in *x*-direction assumes high values along most of their lengths (Fig. 7.36b). Being fixed along most of their length, the most dorsal sections of the ribs are exposed to longitudinal forces in *x*-direction applied to the capitula and tubercula costae. As shown in Fig. 7.35a, these forces evoke bending moments which increase in ventral direction.

The *posterior, more caudal costae VII–XII* (in even-toed ungulates, or *VII* to *XVIII* in horses) have to sustain lesser shearing forces (as shown in Fig. 7.35c) and are not exposed to compression between the insertions of the weight-bearing

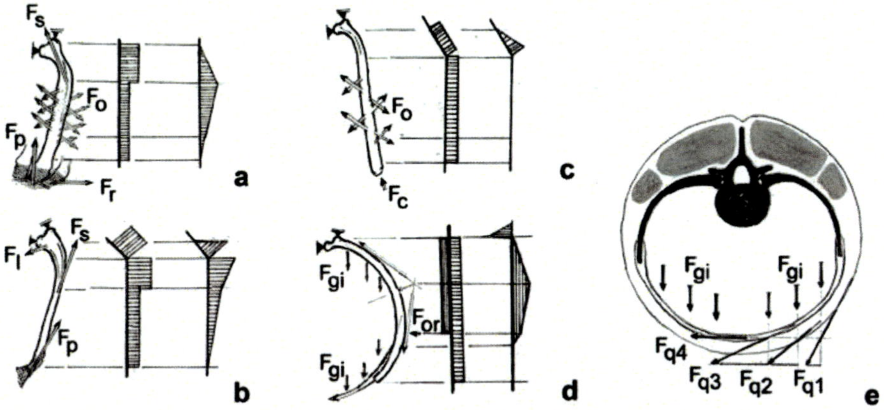

Fig. 7.35 Mechanical forces acting on ribs in lateral (**a**, **c**) and anterior view (**b**, **d**), and stresses evoked by these forces. The diagrams next to the rib show shearing forces; the diagrams on right show bending moments. (**a**) and (**b**) are cranial ("true") ribs, directly fixed to the sternum, (**c**, **d**) is one of the caudal ("false") ribs, with indirect and mobile contact to the sternum. In side view (*x/y*-plane, **a**, **c**), the individual rib is imbedded in the intercostal muscles; the forces exerted by them are distributed along most of the dorsoventral lengths. Muscles act alternatively in one of both directions. Only the most dorsal rib sections are free from insertions of intercostal muscles. The cranial ribs are subject to bending moments, caused by the joint forces applied to the capitula. In frontal view (*y/z*-plane, **b**), the mm. serratus lateralis and pectoralis profundus exert high forces, while the upward directed resultants of m. rectus and the ventral neck muscles are smaller. Both groups of muscles tend to swing the rib in latero-dorsal direction, so that bending moments increase toward the vertebral column. Consequently, the ribs are bent in both planes, but more so in the frontal y/z-plane. (**c**) Only parts of the oblique abdominal muscles insert directly into the corpus of a caudal rib, and the force transmitted through the cartilaginous rib sections (upright arrow F_c) is small. Like in (**b**), the section close to the vertebral column is exposed to bending moments in side view. (**d**) In anterior view, the weight of the intestines can be imagined as distributed over several points (like in **e**). It stretches the muscular abdominal wall, and these muscles exert tensile, downward directed forces on the ribs, which reduce the compressive force components shown in (**a**, **b**, **c**). The m. obliquus abdominis externus also exerts a redirectional force from lateral to medial. It is shown as only one redirectional force (F_{or}), **d**. (**e**) Schematic cross section through the trunk of a cursorial mammal at one of the last rib-bearing segments. The weight of the intestines (represented by short arrows) is distributed over the entire cross section and symbolised by six downward directed arrows (F_{gi}). It leads to tension in the abdominal body wall, which changes its direction gradually from vertical near the distal tips to horizontal near the linea alba (F_{q1}–F_{q4}), where it meets the same (reaction) force from the other side. The resulting tensile stresses are vertically hatched in (**d**). Tension in the body wall is redirected by the weight of the intestines, so that its horizontal components grow with the cosine of the angle between the horizontal and the respective part of the abdominal muscles. The local resultants F_{q1}–F_{q4} fall in line with the respective parts of the muscles

mm. serratus and pectoralis on one side and their connection to the vertebrae on the other. By contrast, these ribs have to sustain tensile forces, exerted by the m. transversus abdominis and the other ventral body wall muscles (Fig. 7.35d, e). These muscles encircle the mass of the intestines, which can be seen from a mechanical viewpoint as the liquid filling of a pneu. The weight of the intestines

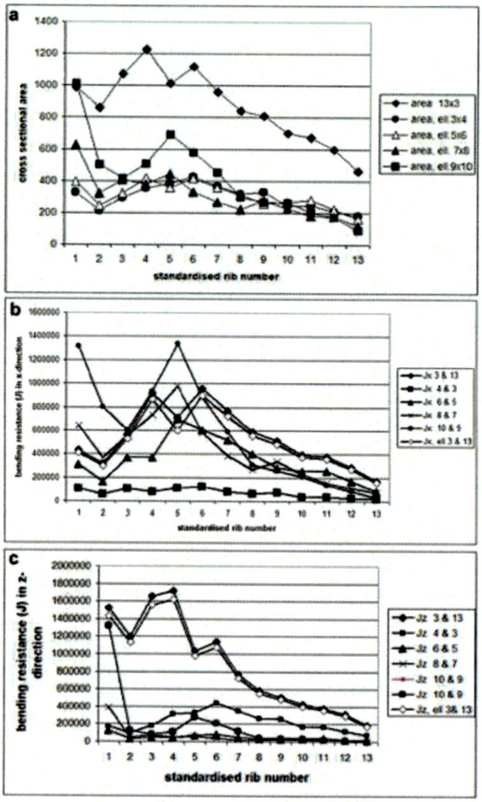

Fig. 7.36 Strength properties of cursorial ribs. (**a**) The cross-sectional areas indicating compressive strength are calculated as ellipses with the diameters shown in Fig. 7.34. Except the most proximal sections, the compressive strength decreases slowly along their entire length, in agreement with the stresses. (**b**) Bending strength of the ribs approximated as rectangles (black symbols) and in one example as ellipse (white diamonds). In *x*-direction exist maxima in costa I and ribs number II–VI; caudal to costa VII, the strength decreases gradually rearward, but still the values are high. Surprising and unexplained are the low values in costae II and III. (**c**) In the transverse *z*-direction (=frontal view), bending strength is very high in the most dorsal sections and then descends to the low values which can be expected from their dimensions (Fig. 7.34). The patterns in costae II–VI or VII, which show high bending strength in *z*-direction, are in accordance with Fig. 7.35c. The outstanding bending strength of the most dorsal parts of the ribs also implies high compressive forces in the vertebral joints (measurements 16, 17). Consequently, the diameters of these joints in Fig. 7.35d are greatest in the most cranial ribs and decrease rearward. Note the similarity between area moments of inertia calculated as ellipses (white diamonds) in comparison to those calculated as rectangles (black diamonds)

(F_e) is in the size order of 25% of total body weight (Grand 1977) and distributed over all abdominal segments, so that in quadrupeds, each rib carries a share of the intestine's weight. Consequently, the posterior ribs are bent in the transverse y/z-plane not toward the lateral side but instead toward the median plane (Fig. 7.35d, e) by the downward-directed pull of the ventrolateral muscles.[26] The share of weight force F_{gi} is suspended on both sides, where $F_{gi}/2$ pulls downward.

$$F_{gi} = F_g/n \tag{7.12}$$

where n is the number of posterior, non-sternal thoracal and lumbar segments: 12 in horses; 7 in bovids, deer and giraffe; and 6 in camelids. The tensile forces on the cross section change their directions (Fig. 7.35e): first, they are directed ventrally in y-direction. Then, the muscle pull assumes gradually increasing horizontal components in z-direction, following the cosine of the angle between horizontal and a tangent at the curved abdominal wall, as shown in Fig. 7.35e. Near the median plane, muscle pull is horizontal in z-direction (see also Chap. 4.2 and Fig. 4.6). The varying support during locomotion leads to varying contractions of the oblique and straight trunk muscles (Figs. 7.30 and 7.31). In all cases, the tensile stresses created in the ribs by the abdominal muscles are opposed to the compressive stresses evoked by torsion. Although their sum is yet unknown, we can be sure that the net compressive load on the "false" ribs is much smaller than that on the more cranial "true" ribs. Aside from being stretched by the mass of intestines multiplied by acceleration, the caudal ribs are compressed by several muscles: intercostal, diaphragm and serratus dorsalis. None of these muscles has the strength of the m. serratus lateralis and the m. pectoralis profundus. As a consequence of the reduced loads acting on them, the posterior ribs possess markedly less compressive strength than the cranial ones, as can be seen from their smaller cross-sectional areas (Figs. 7.34c, 7.35a, c, and 7.36a).

Following the conventional, descriptive anatomy and embryology, the bony ribs are continued from their costochondral junctions or "knees" by cartilaginous tissue (Figs. 7.32 and 7.35a). To match the lower strength of cartilage in comparison to bone, the most distal cross sections of (bony) ribs are a bit larger than those on more dorsal levels. The real diameters of the cartilaginous ribs are not visible in the common dry skeletons, since cartilage shrinks more by losing more water (and that means volume) than bone. From a mechanical viewpoint, however, the costal cartilages plus sternum are a structure of its own, loaded by the m. rectus abdominis and kept in its place by the mm. scaleni, as detailed in Fig. 4.12, in Sub-Chapter 4.2.6.

The mechanical conditions discussed above are primarily derived from theoretical considerations, being deductive arguments. To *test the postulates experimentally* was a difficult task, in particular since we felt responsible not to inflict damage to our experimental animals. Therefore we have taken the predicted activity of neck and trunk muscles in special phases of locomotion as a starting point for testing our hypotheses. The trunk muscles should be active at least when the tensile stresses reach high values. Aside from muscles there exist tough fascia or aponeuroses (abdominal wall, caudal section of dorsal musculature) or ligaments (ligamentum nuchae). So it may well be that energy-requiring muscles are not activated during

[26]This is contra Fujiwara's (2016) most recent interpretation!

long-lasting low-stress situations. The activity of muscles can be controlled by *electromyography* (*EMG*). Unfortunately, the electrical activity is not very tightly correlated with the exerted muscle force. This is particularly true when the non-invasive (and therefore preferred) surface electrodes are used. Muscles which are responsible for maintenance of body shape are often located near the surface, immediately underneath the skin, and therefore can be controlled. According to our results (Kienapfel 2014, Kienapfel et al. 2017/2018), horses show in standing and in the walk only minor activity of the trunk wall muscles. In the stance phases of trot and in the canter, when the body regularly is accelerated by two times earth gravity (Preuschoft and Fritz 1977), the mm. rectus abdominis, obliquus abdominis externus and internus are reliably active. In the neck and shoulder region, where the head-neck position is induced or strictly prescribed by the rider, the predicted influence on the activity of the m. semispinalis cervicis has been recorded. The m. brachiocephalicus serves clearly for protraction of the frontlimb, but its activity also is correlated with the expected mechanical requirements in the body stem. Our results are in agreement with the EMG literature (Ritter et al. 2001), which, however, was rarely focussed on the muscles of the body stem but instead on limbs and gaits.

All groups of cursorials have in common the narrow *shoulder girdle* (see above) and the mobility of the shoulder blade on the trunk (Fig. 7.2). A notable difference between carnivores and herbivores is that the former, especially the cats, allow the trunk to hang down from the vertebral margin of the shoulder blade on the stance side. This leads to the effect that the mm. rhomboideus and trapezius get strong vertical components opposite to weight force in (y-) direction and so take part in carrying the trunk, while this task is fulfilled in hooved mammals alone by the mm. serratus and pectoralis profundus (vide Fig. 7.33).

7.7.6 *Pelvis*

The *pelvis* of land-living mammals is rarely investigated and poorly understood from a mechanical point of view. For general aspects see Sub-Chapter 7.4.2. Since our knowledge is so scanty, I can offer not more than some fragmentary considerations regarding several traits. In all groups of cursorials (even-toed and odd-toed hooved animals, carnivores), the pelvis has a similar shape in side view. Especially in carnivores, the pelvis is narrow and so keeps the torsional moments in asymmetrical (diagonal) support low – comparable to the shoulder. The *ilium* has a more or less rod-like shape; alone its most cranial border (the crista iliaca of human anatomy) is broader in herbivores than in carnivores. The ilium is connected to only one or two vertebrae. This shape does not provide impressive strength, but the deficiency of strength can be understood after a look at the major forces acting on it. Figure 7.37 shows a lateral view of the pelvic region in a horse as an example for all cursorial mammals. Two phases of a step are investigated: the hindhoof placed forward like in

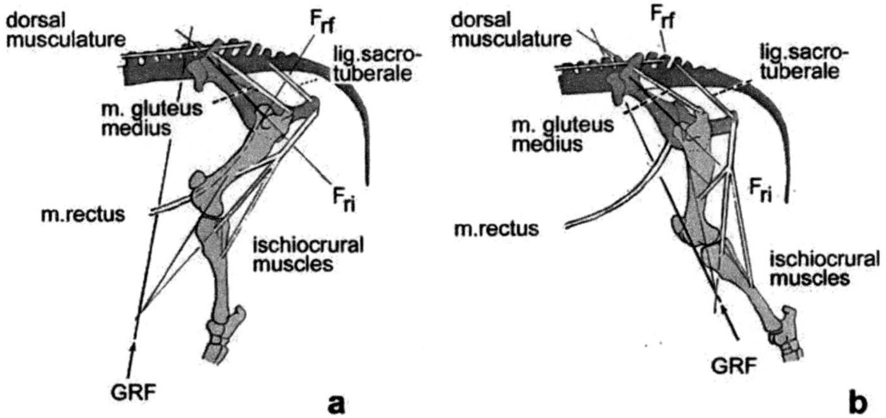

Fig. 7.37 Pelvis of a horse viewed from left side, with skeleton and most important muscles. Two load cases are shown: (**a**) Early stance phase, hindhoof placed moderately forward. At midstance, if GRF assumes a vertical direction, it passes between knee and hip joint. The resultant F_{rf} coincides with the femur. The ischium, from which the ischiofemoral and the ischiocrural muscles take their origin, is inclined dorsally and tied to the sacrum by the ligamentum sacrotuberale latum. So the ischium has roughly the same direction as the resultant of all forces acting on it. To keep the hip joint and the lumbar part of the vertebral column in balance, the dorsal hip extensors and the dorsal epaxial muscles are required. The resultant (F_{ri}) of all hip extensors and the GRF passes through the hip and the iliosacral joint, so coinciding with the long, rod-shaped ilium. (**b**) End of stance phase, hindhoof directed rearward, GRF with propulsive component. The hip joint is kept in balance partly by the hamstrings and partly by the dorsal hip extensors. Again the ilium is stressed by the resultant (F_{ri}) of the GRF and the dorsal hip muscle (= m. gluteus). This resultant (F_{ri}) passes through hip and iliosacral joint. Therefore, no or only small bending moments occur along the ilium and nearly no rotating moments in the iliosacral joint. The resultant (F_{rf}) of the hip extensors and the GRF is in line with the femur. See also Chap. 8.12, Fig. 8.51

the trot or in the canter and the hindhoof pressed rearward against the ground like in the final stage of a support phase or in the takeoff for a jump. In all stages, powerful contractions of the hamstring muscles and of the m. gluteus medius are required to balance the hip joint. The resultants of all acting forces fall more or less in line with the femur and with the ilium. Both elements are therefore exposed primarily to compressive forces, not much to bending. A rotation of the pelvis in ventro-cranial direction about the iliosacral junction under the influence of the thigh muscles or the abdominal trunk wall is excluded by the widely spread ligamentum sacrotuberale latum, as well as by the origins of the gluteal muscles. Likewise, a swing of the sacrum in dorsal direction, caused by the dorsal muscle group, is blocked by the pull of the muscles which extend the hip joint. Seen the other way round, the rigid pelvis is kept in balance against the lumbar vertebrae by the sacrotuberal ligament and the dorsal epaxial muscles. All muscle activities postulated here have been documented by EMG studies (e.g. Ritter et al. in dogs, Kienapfel et al. (2017/2018) in horses).

The *ischia* look like rearward-directed consoles, or short cantilevers, from which ischiocrural muscles take their origin. The postcoxal cantilevers are propped near the median plane by the lower ramus of the *pubes*. Kummer (1959a, b) has observed that the dorsal angulation of the ischia against the ilia in large mammals reduces the bending moments for the sake of compressive stresses. The results of the above analysis agree fully with those obtained from a (non-cursorial!) primate (end of Chap. 8.10, Fig. 8.51).

If seen *from in front*, the pelvis can be interpreted as ring-shaped structure, loaded by trunk weight, which is supported on either two hip joints or only one. The mode of stressing the pelvic girdle (like that of the shoulder region; see above) diverges in dependence from being supported symmetrically or asymmetrically: in the first case, the dorsal side (sacrum and the iliosacral joints) is compressed and the ventral (symphysis ossis pubis) side stretched. In the second load case, the dorsal side is stretched and the ventral compressed. In the absence of strong muscles which take over the tensile forces, the skeletal elements are connected by very thick and powerful masses of transverse fibres or by direct contacts between both sides.

7.8 Neck

Length of the neck does not seem to have much relevance for locomotion. Neither the big cursorial carnivores (wolves, cheetah) possess very long necks nor the young offspring of herbivores (foals, calves, fawns). Instead, the *long neck* of many herbivorous cursorials gives great reaching distance for collecting food without the necessity to move the inert mass of the trunk. This becomes most obvious in browsers, which collect food that grows at some height above ground (Fig. 7.38a). Because of their neck length, they can control and harvest a volume equal to a hemisphere which has its centre at the base of the neck. In the case of grazers, most superficial observers are tempted to put too much emphasis on the necessity to reach the ground. This is not wrong, but neck length also allows harvesting of a given area without changing the feeding station (as in Chap. 4.6). Any change costs energy for accelerating or decelerating body mass and therefore is avoided. The number of changes is kept low by a long neck (Fig. 7.38b; see also Preuschoft et al. 2011). Unpublished data (Schlunk 2010) show that especially the large grazers, horses and cattle tend to either maintain a slow, continuous speed during feeding or to stay at the same place until everything of value is consumed, before changing leisurely to a new feeding station. As is illustrated in Fig. 7.38c, the length of the neck also allows to take in sprouts or twigs of a promising plant besides the animal's track without changing locomotor speed.

7.9 Special Load Cases Among "Large Mammals"

In view of the notorious diversity of mammals and their modes of living (as mentioned in Chap. 7.1), which is mentioned in previous chapters (Sub-chaps. 3.3.5, 7.2 and 7.3), it seems pointless to expand on "alternative" body shapes and solutions of the related mechanical problems. In fact the latter are poorly known in

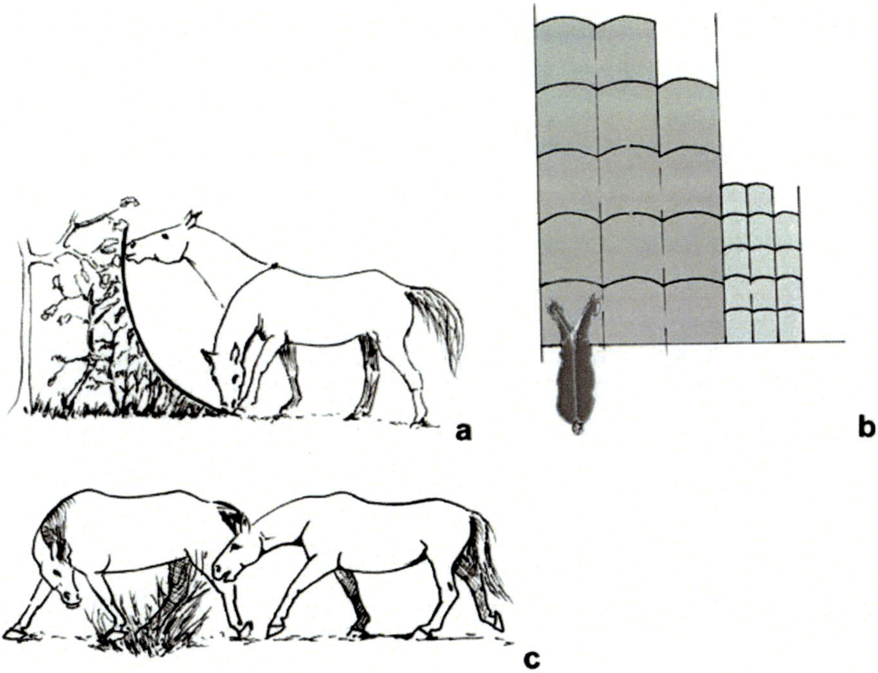

Fig. 7.38 Selective value of a long neck. (**a**) Without moving the mass of the trunk plus extremities, the long head and neck permit harvesting of a considerable volume, as can be seen in side view. (**b**) Seen from on top, the long-necked animal can control and harvest the grey area. According to our observations, this area is limited on both sides by the animal's avoiding strong lateral flexure of the neck. A long neck evidently opens access to a greater feeding volume (or feeding area, if low-growing grass is consumed). If the respective feeding station is changed by walking slowly forward, a much larger area (feeding volume) can be exploited than a shorter neck (light grey) would allow. (**c**) The long neck also permits a continuously walking horse to take in more of an attractive plant without stopping progression than a shorter one, because its head can remain longer at the same spot while the trunk mass moves forward

too many cases. Instead of "alternatives", some special load cases among the "large mammals" justify being considered at least in short form.

Many species among the cursorial mammals are characterised by horns (Bovidae), antlers (Cervidae) or other outgrows of the frontal, nasal or parietal bones (Giraffidae). They are often summarized as "*pecora*". (This term separates them from Camelidae, and the inclusion of Tragulidae is a matter of debate.) Their weapons are used primarily in intraspecific conflicts, rarely in defence against predators. In both cases, the use of weapons causes predictable stresses of the head and neck but has also consequences for the whole animal. These will be detailed in an example:

The two male *gazelles* in Fig. 7.39 press their armoured heads against each other – a behaviour typical for all pecora, bovids and cervids and modified also for

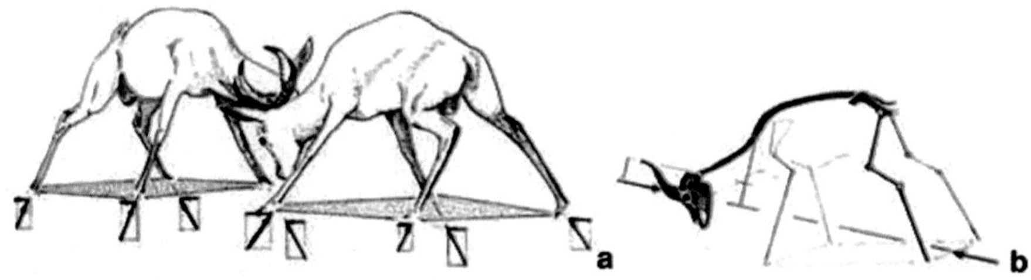

Fig. 7.39 Springbok males fighting. (**a**) The limbs are widely spread to enlarge the area of support (stippled). GRFs are estimated under the conditions that the slender metapodials are placed in the direction of the GRFs and that the sum of external forces in each plane must be zero. Vertical (*y*-) components on all fours add up to total body weight; horizontal (*z*-) components of the left side balance those of the right side. The distribution of body weight on fore- and hindlimbs can be chosen by the animal; in the present example, a ratio of 12/10 is assumed. (**b**) Schematical drawing with emphasis on skull and vertebral column, showing lever arms (l_o and l_n). Because hindlimbs are pushing forward, the trunk and neck are strongly compressed between the pelvis and atlanto-occipital joint. The external force in *x*-direction produced by the hindlimbs is identical to the force (F_k). It is transmitted through the axial skeleton, at the joints balanced by the dorsal musculature and finally transmitted to the horns of the opponent. This force has relatively short lever arms (l_o) at the atlanto-occipital joint (AOJ), where it can be balanced easily. Very long, however, are the lever arms of the external force at the base of the neck (l_n). Drawn from a photo published by Zachos, 2012

giraffes.[27] The external force exerted by the opponent is more or less parallel to the ground and combines with body weight to a resultant which must be balanced over the area of support. As can be seen, the latter is enlarged by placing the hindlimbs rearward and laterally. In the *x*-direction of the fighting partners, the resisting stance moments are very great. A deviation from this longitudinal *x*-direction seems to be a common tactic at least in bovids and cervids, against which a laterally sprawled position of all four feet enlarges the area of support to permit GRFs of different directions without tilting.

Inside the bodies of both partners, external force is transmitted to the head (Fig. 7.39b). It exposes the body to extreme bending, which is resisted by the retractor muscles of the hindlimb, the dorsal musculature of trunk and neck as well as the vertebral column. At the *basis of the neck*, the lever arm of the opponent's force is quite long; the deeper the head, the longer. Its impressive moments are countered by the pull of the strong nuchal muscles (mainly m. splenius, cleido-occipitalis, anterior part of m. trapezius) and by the strongly developed longissimus group of the neck muscles ("bull's neck"), which take their origins from the thoracal vertebrae and the withers and the trapezius from the shoulder blade. The spinal

[27] In sheep and goats, the external force is increased by ramming after an approach run or by falling down from a bipedal position in which the forelimbs are lifted from the ground. These contacts are obviously confined to short intervals.

The external force from the opponent is applied to the horns and then transmitted to the frontal bone. It is possible that a material-dictated damping of the impact takes place in the horns or antlers, but this has never been documented.

processes of the cranial thoracal vertebrae are extremely long in many large even-toed mammals, especially bovids, so providing outstandingly long lever arms. At the *atlanto-occipital joint*, however, external force has a rather short lever arm. The nuchal plane of the skull is large in all pecora, like in suids, and clearly larger than in horses, so that more muscle fibres can insert into the dorsal part of the occiput, and the most dorsal fibres have obviously long lever arms.

Cursorial mammals *without frontal weapons* also use their necks in intraspecific combat: camels as well as equids try to push the necks of their contrahents downward by placing their own neck across that of their partners. This means that body weight of the animal on top is transferred from the frontlimbs to the caudal or medium part of the neck. Pushing downward is mainly the role of the mm. scaleni. In the animal below, once more the dorsal neck muscles are responsible for counteracting the additional load.

Elephants are not counted among the "cursorial mammals" but possess traits characteristic of cursors: a compact body stem and long, rather slender limbs, mostly kept extended. Their body shape permits covering very long distances without exhausting their energy budgets. They are not "fast" when compared to other quadrupeds, and their most common gait is walking with duty factors of more than 50%. If in a hurry, elephants shorten their stance phases to less than 50% of cycle duration and increase step lengths by intercalating a phase of aerial suspension, in just one-limb pair. The resulting gait is similar to the "amble" of horses; first both hindlimbs and then both frontlimbs are without ground contact (Christian et al. 1999; Hutchinson et al. 2006). The first authors hold that pendulous up-and-down movements of the heavy head may contribute to storing energy, because elastic upward movements of the head help to accelerate the forelimbs. The investigation of the ground mechanics underneath the footprints (Schanz et al. 2013) yielded valuable information about the roll-off and ground-foot forces but could not be extended to high-speed locomotion. A really outstanding trait of Proboscidea is that they have transformed their upper lip and nose into a prehensile organ (Martin 2003; Martin and Niemitz 2002), which plays the same biological role as the long neck of herbivorous cursorial mammals (Chap. 7.8) by increasing reaching distance, especially in feeding.

7.10 Essential Results

Nearly all terrestrial mammals (except some specialists, like ant-eaters, pangolins, armadillos, ardvaark) are characterised by relatively heavy heads, with a masticatory apparatus that allows comminution of food within the oral cavity. Necks are sometimes short, sometimes long, as are the tails, which in terrestrial forms normally do not contain much weight. Locomotion takes place on limbs, which are moved in parasagittal planes. The construction of shoulder and pelvic girdles in mammals deviates widely from the arrangement in reptiles and birds.

Not alone the variation in body size suggests a separation of *small* from large mammals. The former usually have cylindrical, lengthy, flexible trunks, short necks and short frontlimbs, while the hindlimbs can be very long. The mobility of the body stem is particularly marked in the sagittal x/y-plane, and crouched postures are maintained through long periods. "Small" mammals mostly attain the lifestyle of "accelerators" or "escapers" being able to rush away in explosive escape – and also in sudden attacks. They prefer locomotion in an asymmetrical gait (bound or half-bound), which is not or rarely used by other tetrapods. It includes flexing and extending the trunk, especially in short-legged forms, and phases of aerial floating. Small mammal locomotion culminates in bipedal hopping, which is connected with special morphology. Outstanding examples are the kangaroos – some of them, however, being too large and heavy for our theoretical expectations. In contrast, hares and their relatives do not hop, although they are in a size order in which they should do so, and possess long hindlimbs. The reason seems to be the absence of a long tail, which could counteract the nose-down rotation produced at the onset of the aerial suspension. For this purpose, they need their frontlimbs.

"*Large*" mammals are characterised by a massive, rather stiff trunk and very slim extremities, often of considerable length. Necks may be short or very long. Many large mammals are endurant "wanderers" over long distances and at the same time rapid runners. Outstanding and well-known examples among these "cursorial mammals" are horses. Length of the limbs is provided by reduction in the number of bony elements and elongation of the distal segments, parallel to, but anatomically different from, the situation in birds. Long limbs are useful for reaching considerable locomotor speeds in the energetically not demanding symmetrical gaits over very long distances. Large mammals frequently use a symmetrical gait, mostly striding (walk and its derivatives), but also the trot (mostly in diagonal, but also in lateral sequence as pace trot in a strict sense), which includes phases of aerial suspension. The walk is largely confined to movements of alone the limbs, driven by the laws of the pendulum; the trot and the pace trot make also use of elastic structures: tendons or muscle-tendon complexes. The newly "invented", asymmetrical gait is modified in "large" mammals and allows reaching of very high speeds. The contribution made by the trunk to progression varies between the smaller (dorsomobile) and the larger (dorsostabile) forms. The long free limbs of "cursorial mammals" confront, however, the animal with difficulties in lying down and getting up.

The deep-reaching difference between "small" and "large mammals" is largely due to the mechanics of locomotion. An essential role plays the discrepancy between mass (a third power function of volume) and muscular strength (a second power function of cross-sectional area). A sharp borderline between small and large mammals does not exist; the mobility of the trunk and the length and use made of long extremities follow different necessities.

Tracks are documents of the spatial arrangement of supports and of the construction of autopodia. A regular, fairly dense arrangement of footprints along a track is advantageous for supporting the weight of the animal and for avoiding intervals in time and space without support. A relation to gaits exists, but complications are often underestimated; many attributions in the literature are erroneous.

The considerable mass of the body stem in large mammals is supported to a larger extent by the fore- than hindlimbs. The absence of a reliable, energy-saving bony connection between body axis and the frontlimbs is compensated by a strong development of the rib cage. The individual ribs are subject to different stress patterns. Examples are given and shapes are related to stresses.

The inertia of the great body mass of several large mammals makes mobile prehensile organs advantageous, by which contacts can be made without moving the entire body mass, so saving energy. The sensory organs, jaws and lips are part of the head, which is carried and can be moved freely on a long, mobile neck. Since several representatives of the large mammals (pecora) have weapons on their heads, their necks are exposed to extremely high stresses.

The need to have a prehensile organ is solved in an alternative, completely different way by the trunks of elephants. This group also possess traits of cursorials, without belonging to that highly specialised group. An alternative way to develop a prehensile organ in the autopodium of the anterior limb is realised among several rodents and carnivores and in a particularly elaborate fashion among primates – which will be dealt with in Chap. 8.

Chapter 8
Primates: The Group Including Humans

Keywords Orbits · Danger of arboreal life · Injuries · Prehensile autopodia · Transmitted forces · Friction · Gaits on continuous substrates · Interrupted substrates · Bridging · Leaping · Brachiation · Stressing of body segments · Big size · Sexual dimorphism · Climbing

8.1 What Are Primates? Which Characteristics Do (and Did) They Have in Common?

8.1.1 Eye Sockets in Tarsiers

Primates are a diverse group of mammals, difficult to describe in few, short words. They vary in size between *50 grammes (Microcebus, Cebuella) and 150 kg (male Gorilla).*[1]

The morphological traits of dentitions, closure of orbits and variations of organ systems as well as embryological details, which are used often in zoology, are not easily visible in the living animal. The best way to recognise a living primate are the prehensile terminal parts of their limbs, hands and feet (Figs. 8.1 and 8.2). Among mammals, most primates, especially the monkeys and apes, stand out by their long and very mobile extremities (Schultz 1933, 1936). In the majority of species five digital rays are slender, often very long, and can be abducted from each other to a varying degree and be extended and flexed strongly. The dorsal sides of the terminal phalanges are commonly equipped with flat nails. Exceptions from this rule are the small-sized New World Callitrichidae, which possess claws or clawlike nails following Sprankel, and a number of non-simian primates which possess on the 2nd – rarely also on the 3rd – digit "toilet claws". The volar surfaces of the terminal phalanges and of the palms and soles are equipped with thick, compliant pads, and the whole or at least a large part of these pads are covered by "friction skin". This is characterised by patterns of ridges ("dermatoglyphics") and the absence of hairs. Friction skin is mechanically tough and

[1]More about body weight is given in Chapters 8.11, 8.12 and 8.13.

H. Preuschoft, *Understanding Body Shapes of Animals*,
https://doi.org/10.1007/978-3-030-27668-3_8

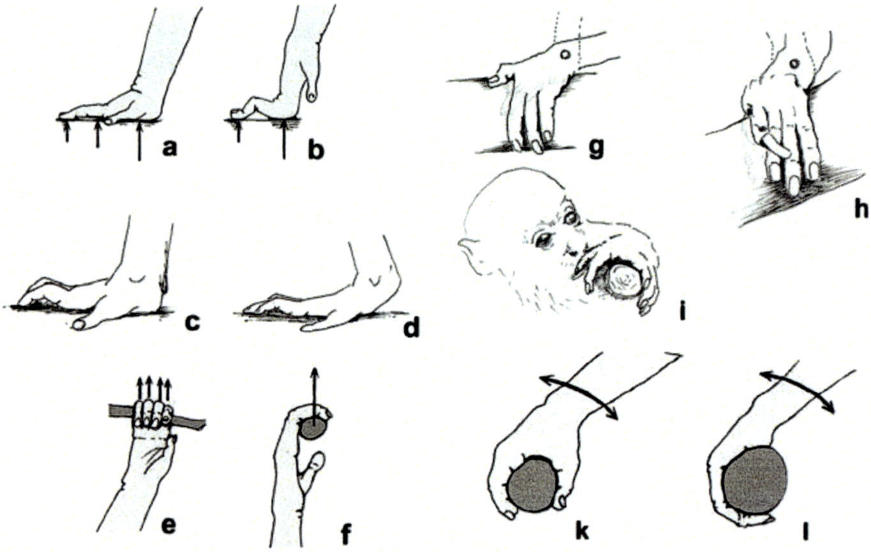

Fig. 8.1 Variability among primates of hand and foot use in locomotion. (**a**) Palmigrade hand of an arboreal monkey and (**b**) digitigrade hand of a terrestrial monkey in side view, both transmitting compressive forces. The arrows indicate the distribution of GRF at midstance. (**c**) Plantigrade foot of a chimpanzee and (**d**) semiplantigrade foot of a vervet monkey transmitting compressive GRF (see also the roll-off in a Japanese macaque, Fig. 8.4f–k). In several species, individuals change between palmigrade and digitigrade hand postures, where the first is preferred on narrow and compliant substrates, the latter on firm ground. (**e** and **f**) Transmission of tensile forces through a hook hand in palmar and in ulnar view. (**g**) "Power forceps" in the hand and (**h**) in the foot of a lorisid, both seen from the left side. The other, not visible branch of the forceps is formed by the very strong pollex or hallux. Once the forceps grip is closed, the hand or foot remains at its place and the limb (zeugo- and stylopodium) rotates about the carpal or distal tarsal joints, while the forearm and lower leg make progress from the continuous to the dotted outline. (**i**) "Schizodactylous" hand of a NW monkey (*Alouatta*), in which rays I and II are separated from III, IV, and V for gripping. Like in all NW primates, the thumb is not opposed to the "lateral" fingers. This hand posture reduces the distance of roll-off in walking along a branch (extending from carpus to the tips of the metacarpals II and III) and allows exerting SRFs on both sides (see below). (**k**) Transmission of torsional moments through hands or feet requires the existence of two opposing branches (in most cases the first ray against the parallel "lateral digits" II–V, as in (**g**), (**h**) and (**k**), but in thumbless *Ateles* or *Colobus*, the (long) lateral digits act against the carpal ball (or in the above-described schizodactylous condition rays I + II against III, IV and V). The small circles at the wrist or ankle in (**g**) and (**h**) indicate the pivot of rotation during progressing above hand- or foothold. (Modified from Preuschoft, 2004)

contains numerous receptors for palpation and accelerations (Textbooks of Histology; Biegert 1961; see below, Chapters 8.3 and 8.4).

Objects, especially food objects, are normally seized by the *hands*, which carry them to close inspection to the eyes, nose, lips or teeth.

Tails among primates are extremely variable, ranging from complete absence (apes, humans Nakatsukasa et al. 2003b), through very short (several macaque species, lorsids, *indris*) to long and elegant (other macaque species, relatives of

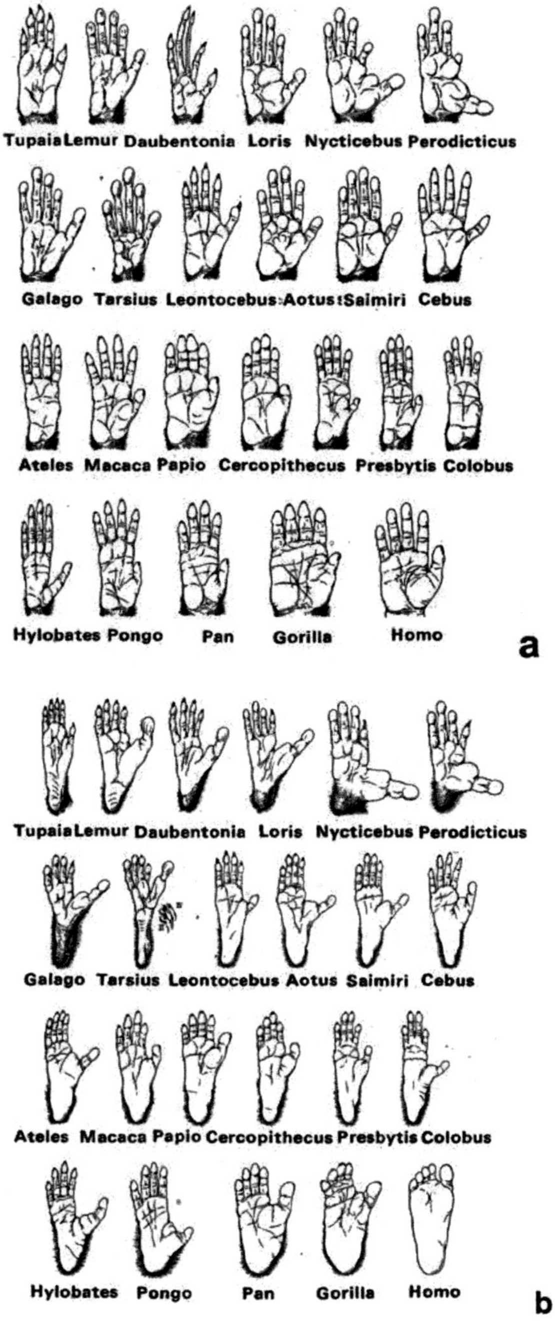

Fig. 8.2 (**a**) Volar aspects of prehensile hands and (**b**) plantar views of prehensile feet, all drawn to the same size. Note the pads and the short, flat nails or long claws as well as the sometimes hairy heels. (From Schultz 1956 and Biegert 1961)

guenons, leaf monkeys, tarsiers, galagos, sifakas) and finally a true prehensile organ (Atelinae and Alouattinae from the New World; for more details, see also below Chapters 8.3, 8.4.

A separation into two *major subgroups* is useful and easy to recognise: strepsirrhines (pointed face with a protruding nose, upper lip tethered to the upper jaw, moist rhinarium (surface around the nostrils)) and simians (rounded heads, with a voluminous brain capsule, normally short jaws, hairy and mobile upper lip, dry nasal apertures). The second group comprises the monkeys and apes, aside from the repeatedly mentioned, but aberrant *tarsiers*. Among the simian group, the New World monkeys, in which the nostrils are widely apart, occupy two size classes: the small marmosets and tamarins (Callitrichidae) and the larger Cebidae, which include the prehensile-tailed howler, woolly and spider monkeys. In the Old World monkeys and the apes, the nostrils are separated by a narrow septum. The monkey group is split, according to their feeding habits and specialisation of their digestive tracts, into leaf monkeys (Colobinae) and the morphologically guenon-like but diverse Cercopithecidae. The small gibbons (Hylobatidae) and the impressively large African (gorilla, chimpanzees) and Asian (orangutan) apes, as well as humans, are classified into the "superfamily" Hominoidea.

The *fossil history* of the primates takes its origin in the lower Eocene with the lemur-like Adapidae (New World, Notharctinae; Europe, Cercamoniinae, among them *Europolemur*; Asia, no documents). These forms flourished in the Eocene, but most of them vanished in the Oligocene. An exception were the recent Lemuridae, which appeared in the upper Eocene of Africa. The Adapidae were replaced by early representatives of the simians, nearly at the same time in the Oligocene of South America and of the Old World (Godinot 2014). A well-dated monkey has been found in the lowest Miocene of Panama (Bloch et al. 2016). Tarsiers can be traced back into the middle Eocene. Still debated is the first appearance of simian primates. According to Franzen et al. (2009, 2012) and Gingerich (2012), *Darwinius* from the early Eocene must be seen as a representative of this advanced group. In the Miocene and Pliocene, Hominoidea were more abundant among the Old World primates than Cercopithecoidea (Godinot 2014). The well-documented specialisations of the locomotor apparatus in Hominoidea (see below and Chap. 9) seem to have evolved after the Miocene (Nakatsukasa et al. 2000, 2003a).

The dominating *sensory organ* of primates are the eyes. This corresponds to the preferredly diurnal activity of the animals. Both eyes are placed side by side and orientated forward. Their fields of vision overlap to a large extent and therefore allow stereoscopic vision. The evolutionary value of this ability for animals living on a frequently interrupted, three-dimensional environment like the branches, twigs and foliage of trees is evident and has been described often. Less often was the disadvantage mentioned that perception of objects – or movements – behind the head is excluded. Primates also possess the ability to perceive colours – whereby a weakness in discriminating red and green exists in many "lower" primates, even in monkeys. The brains, and their bony capsules, are quite large in comparison to other mammals.

In those forms which have adopted a nocturnal lifestyle, the eyes are much enlarged. In all monkeys and apes, the eyes are surrounded by the *bony orbit*, the morphology of which has attracted much attention. In the order of primates, the closure of the orbits against the temporal fossa is seen as a taxonomically relevant character. In the genus *Tarsius* (Fig. 8.3), an arrangement of bones is found, which is interpreted as transitional between ("primitive") strepsirrhines and ("advanced") haplorrhines. Rosenberger and Preuschoft (2012) have taken into consideration the mechanical constraints posed by the huge eyeballs of tarsiers. As a means to perceive even smallest amounts of light, the eyes in these animals became extremely large and therefore heavy. At the same time, *Tarsius* evolved to be an extremely gifted leaper (see below, Chap. 8.8). Leaping implies very high accelerations, at takeoff as well as in landing. The big eyes therefore exert huge forces, which the skull must be able to sustain.

Accelerations of 8–12 times earth gravity have been measured under lab conditions in another leaper of similar size, *Galago moholi* (Günther 1989), and may be even greater. In tarsiers, each of the huge eyeballs has about 25% of brain mass. At rest, the force exerted by the eyeballs on their socket is

$$F_g = \text{mass} * \text{gravity}, \tag{8.1}$$

and directed toward the earth surface. During takeoff or landing, the inertial resistance against being moved (or braked), however, is

$$\text{mass} * \text{acceleration}$$

and acts against the direction of movement, either as compressive force from the eyeballs against the orbital wall or as a tensile force luxating the eyeballs out of their bony sockets. To keep the eyeballs in place, tough strands of connective tissue ("periorbital ligament") surround the eyeballs and tie them into the orbit (Fig. 8.3). As Niemitz (1984) has observed, the lids are closed during the landing of a tarsier. The author assumed that this behaviour is a means to protect the eyes but it may as well be an efficient means to keep them in their positions within the orbits (Fig. 8.3c, e) against mass inertia. The (large) inertial force combines with (relatively small) gravitational force to the resultants F_r as shown in Fig. 8.3.

Primates live in very diverse *habitats*, ranging from rocky cliffs (gelada, capuchin monkey, macaques), over open, even dry savannah (baboons, patas) or woodland savannah (baboons, vervets, patas), open forests without continuous canopy (chimps, gorillas, hulman, macaques, New World monkeys, especially marmosets, as well as bushbabies, catta and sifaka) to dense rain forests (orangutan, gibbons, leaf monkeys, prehensile-tailed monkeys, many lemurs, lorisids). Many primate genera are strictly specialised for living in particular biotopes and therefore vulnerable by habitat destruction (orangutan, gibbons, most leaf monkeys, many strepsirrhines). Other forms (most macaques, hamadryas and other baboons, vervet monkeys, to some extent chimpanzees) are generalists, and this allows them even to invade human habitats or at least close neighbourhood. All their habitats have in common substrates for locomotion in *three dimensions*. Trees as well as rocky cliffs are

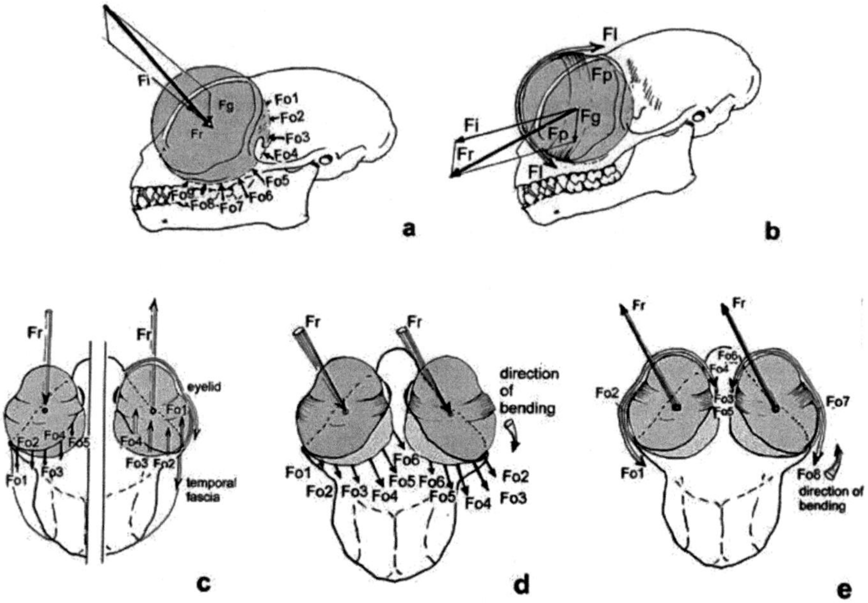

Fig. 8.3 The huge eyes in the genus *Tarsius* are subject to mechanical conditions which require morphological "adaptations". The "evolutionary early" separation of the orbits from temporal fossa by outgrows of several elements of the skull is a mechanically relevant trait. (**a**) Lateral view during takeoff. The eyeballs are pressed by inertial forces (F_i) into the bony orbit. (**b**) In landing, the inertial force is directed away from the skull, and therefore tension-resistant structures are required in the form of connective tissue (lig. periorbitale) and the (closed!) lids (Fl). The exact direction of the inertial forces is determined by the position of the head and the direction of acceleration or deceleration. These factors vary within limits. Our analysis is based on the motion analyses published by Niemitz, and Peters and Preuschoft, both in 1984. (**c**) Top view of a *Tarsius* skull cut along its median plane into two halves. The resultants F_r are drawn without their components in z- or y-directions. On the left side, the situation during takeoff is shown. The force exerted by inertia, pressing the eyeball into its socket, is split into the forces $F_{o1} - F_{o5}$, which are distributed over a large area, including the rear wall of the orbit. On the right side, the inertial force exerted by braking pulls the eyeball away from the orbit, to which it is fixed by the lig. periorbitale and the eyelid. Again F_r is split into the partial forces $F_{o1} - F_{o5}$, which tend to swing the zygomatic bone forward. This element is kept in place by the temporal fascia and the m. masseter, which exert considerable force components toward the occiput. (**d**) and (**e**) Top views during takeoff (**d**) and during landing (**e**). In both cases, the inclination of the resultant forces F_r is indicated. A deviation from the longitudinal axis of the head is assumed to exemplify the possible variation of reaction forces against the orbit (**d**) or out of the orbit (**e**). In (**d**) the partial forces $F_{o1} - F_{o6}$ assume different values, because they act on different, asymmetric parts of the socket. Inertia of the eyeballs exerts a rotating moment on the head, which must be counterbalanced by neck muscles. In **e**, the pulling force evoked by inertia during deceleration requires tough connective tissue (F_{o2} and F_{o7}), as well as activity of the eyelids (F_{o1} and F_{o4}) on left, F_{o6} and F_{o8} on right side. These tensile forces are transmitted to the zygomatic bones and rotate the head to the left. This movement must also be balanced by neck muscles. (From Rosenberger and Preuschoft 2012)

characterised by pronounced inclination of substrates. Due to the geography of rocky outcrops, or the shapes of branches and twigs, sometimes only very small areas of support are available for carrying body weight. In other cases (stems), the thickness of stems or inclined boughs and their slippery surface make it difficult to find a firm hold. Interruptions and empty interspaces are ubiquitous. The trees do not alone offer safe resort or protection; they also offer resources. Most of the resources are to be found in the periphery of trees: young leaves, buds, flowers, fruit – and arthropods which also are after these nutritious parts.

8.2 How Dangerous Is Living in the Three Dimensions of Trees, High Above Ground? Frequency of Accidents/Injuries

Living in trees, or on steep cliffs, is dangerous, since it implies the *danger of falling down*, be it because of loosening the grip or of breaking the substrate. Empirically, about 33% (= 1/3!) of the gibbon and orangutan skeletons stored in scientific collections show the traces of bone fractures, chimpanzees and gorillas about 20%, monkeys between 28% (*Nasalis*) and 12% (*Leontocebus*, as reported by Schultz 1956; Buikstra 1975; Lovell 1987). On top of the list of injuries seem to range baboons, among which 80% of the individuals show such lesions (Bramblett 1967). Niemitz (personal communication) found fractures in about 13% of the tarsiers he investigated.[2]

Falling down from a tree onto the ground or a deeper bough entails more acute dangers than the fall of an animal running on the ground (Preuschoft 1990b). The speed (v) of a falling body is

$$v_s = (2 * G * h)^{1/2} \tag{8.2}$$

where G is earth acceleration or gravity and h is the height of the free fall. The speeds set out in Tables 8.1 and 8.2 are impressively high, near the limits of running speeds. The body contains, according to Eq. 1.6, an impulse (P) which is proportional to mass and height above ground.

$$P = m * v = m * (2Gh)^{1/2} \tag{8.3}$$

At the end of the fall, the body is stopped by a braking force (F), which is the greater, the more rapid the braking takes place.

[2]It should be noted that the hurt animals have survived their accidents long enough for healing – and this without any sort of intervention by surgeons and specialists taking care!

Table 8.1 Forces occurring in a monkey or other mammal of 12 kg falling down from a tree. $v_s = (2Gh)^{1/2}$; $p = m * (2Gh)^{1/2}$; $F = (2Gh)^{1/2} * t^{-1}$. (From Preuschoft, 1990b)

Height *h* [m]	Speed *v* [m/s]	Impulse *p* [kgm/s]	Force *F* at 0.5 s [N]	Force *F* at 0.7 s [N]	Force *F* at 0.9 s [N]	Force *F* at 1.5 s [N]
5.00	9.90	118.85	2331.94	1665.64	1295.51	777.34
10.00	14.01	168.09	3297.82	2355.58	1832.12	1099.21
15.00	17.16	205.86	4038.97	2885.02	2243.94	1346.32
20.00	19.81	237.71	4663.71	3331.28	2591.02	1554.59

Table 8.2 Forces occurring in a monkey or other mammals of 12 kg, which falls down on a level surface while running at given speeds. Height of CoM above ground assumed to be 0.35 m. $v_r = (v_h^2 + v_s^2)^{1/2}$; $F = m * v_r * t^{-1}$; $p = m * v_r$

Running speed v_h [m/s]	Resulting speed v_r [m/s]	Impulse *p* [kgm/s]	Force *F* at 0.5 s [N]	Force *F* at 0.7 s [N]	Force *F* at 0.9 s [N]	Force *F* at 1.5 s [N]
5.00	5.65	67.74	1329.06	949.31	738.40	443.02
10.00	10.34	124.06	2433.96	1738.53	1352.21	811.29
15.00	15.23	182.72	3584.97	2560.70	1991.72	1195.05
20.00	20.17	242.05	4749.02	3392.20	2638.40	1583.04

$$F = P/t = m * (2 * G * h)^{½} \tag{8.4}$$

where t is the time interval available for braking. In the case of a primate falling down from a tree, the time span t is very short, since neither the ground nor a strong branch gives way compliantly. If the fall ends in a mass of thin branches, or foliage, deceleration is much slower, so that the downfall leads to a happy end. The mobility of body segments can also be used to slow down the deceleration. Our own measurements on humans in controlled landings showed deceleration intervals of 0.4 to 1.3 seconds, while the impact forces were 5–9 times body weight (Preuschoft 1985). There is no obvious difference in deceleration intervals between landings on the feet or on other parts of the body. If, however, an animal running on +/− level ground tumbles and falls, it is decelerated more slowly. According to video films in public TV, it takes several seconds, until the impulse contained in a falling antelope or a race horse is used up and the body comes to a complete halt. Calculation of the acting force F (as in Tables 8.1 and 8.2) must include the height of the CoM above the ground (evoking the vertical speed v_s) which must be added to the velocity of running (v_h)

$$F = mv_r/t, \text{ where } v_r = \left(v_h^2 + v_s^2\right)^{½} \tag{8.5}$$

To keep the risk low, and to reach their preferred resources, primates have to cope with the limited strength of branches and peripheral twigs and with three dimensions (see below). For getting along with such difficult supports, they possess autopodia with special properties: the above-mentioned *prehensile hands* as well as *prehensile feet.* These are (see above) the most characteristic traits of the order. Payne et al. (2006) emphasised the length of muscle fibres in apes, which is in contrast to humans. The fascicles stretch along most of muscle length, while tendons are short. This arrangement provides maximal lift; the possibility to move the segments actively over a wide angle of excursion. This obviously is of advantage "for locomotion in a three-dimensionally complex arboreal environment". Humans, by contrast, have relatively short muscle fascicles and long tendons – this will be dealt with in Chap. 9. An interesting point is that gibbons show several morphological details similar to humans – a fact possibly due to the outstanding length of their extremities.

8.3 How Are Prehensile Feet or Hands Used? Morphological Variation, and Which Forces Are Acting?

Prehensile autopodia allow to expose extremities to *tensile*, rather than the usual compressive forces (Pouydebat et al. 2006). Tensile forces acting on extremities means reducing the compressive stresses in the skeletal elements. The splitting of *primate autopodia* into five rays (as can be seen in Figs. 8.1 and 8.2) is commonly taken without further consideration simply as retaining a relic from "primitive" forerunners (see above, Chap. 5.4). This notion is, from the viewpoint of descriptive morphology, completely correct. The statement, however, should by no means block the question, whether there exist mechanical reasons for maintaining the five rays. In fact, such reasons can be found: at the toes and fingers, very high forces are exchanged between the animal and its substrate. Deficiencies of force transmission or local weakness of the substrate may lead to loosening the firm hold – which is dangerous for an arborealist (see previous Chap. 8.2). The safety of a grip is improved because of statistical reasons if several points are recruited for force transmission – especially if these can be adapted independently to the locally existing conditions, which means if each of the digits possesses its own musculature. The ensuing differentiation of the musculature in the autopodia of primates is an argument against their "primitive" nature. It should be kept in mind that the number of five is not always maintained! (See also below Chap. 8.7.)

The *feet of primates* (Figs. 8.1 and 8.2b) vary considerably, though within characteristic limits. Lengths of the toes and loading the heel may differ among the genera or families. In nearly all species, the powerful hallux or first toe with its metatarsal can be abducted from or adducted to the "lateral" toes. This does not hold true for humans, the only primate without a prehensile foot. In orangutans the powerful hallux is very short, especially if compared to the long lateral digits. In many individuals the terminal phalanx or the long flexor tendon of the first ray are lacking. In many strepsirrhines, especially lorisids, the second ray is more or less reduced (this will be dealt with in the next paragraph). The heel in some genera is suited, because of its breadth and soft fatty cushion, to carry a great share of substrate reaction force GRF. In other forms, especially strepsirrhines and tarsiers, the narrow and long heel is covered by hairy skin, and normally not in contact with the substrate (see below). In most, especially the highly arboreal primates, the toe flexors are extremely strong. Since they pass behind the talus, they exert a considerable plantar flexing moment at the ankle joint. An additional plantar flexion by a share of GRF is avoided, if the heel does not at all get into contact with the substrate. In monkeys, sometimes a resting position can be observed with relaxed toe flexors and the heel on the ground, as well as a semiplantigrade position (Figs. 8.1d and 8.4f-l) during movement in which the digital flexors are pressing the toes against the substrate, from which the heel is lifted. This lifting keeps down (see above) the plantar flexing moment around the ankle joint and increases at the same time the "functional length" of the whole hindlimb. Because of the length of the primate feet, the "rolling off" during the stance phase becomes obvious.

Primate hands vary mostly in the proportions of digital "rays", including metacarpals and phalanges (Figs. 8.1 and 8.2), in contrast to the free "digits", which consist alone of phalanges. *Length* of the hands in many primates seems to be a very specific trait, although it is not easy to pin down its characteristics numerically. Because of the size variation among primates, hand proportions must be expressed in relation to other measurements, which represent body size. This sometimes leads to not satisfying results: in relation to arm length, for example, the hands vary in simian primates between 32% (*Leontocebus)* and 24% (*Papio*), where apes are between 25% (*Hylobates*, *Gorilla*) and 29% (*Pan*, *Pongo*). Human hands exhibit inobvious relative hand lengths of 24.3% (Jouffroy et al. 1993). Many strepsirrhines show higher values, but lorisines have low ones. In relation to trunk length, hand size becomes more obvious: humans with 37% and orangutan with 63% (Schultz 1930). The ratio between the lengths of the carpus and the metacarpals in humans does not deviate clearly from other primates. The lengths of digital rays, however, show considerable, lifestyle-dependent variation. The longest among the lateral fingers, III, is in humans short if compared to apes but longer than in some monkeys. In comparison to the third, digits II and IV are of similar length, a few percent shorter than digit III. This ratio between digits is termed "*mesaxonic*" and judged by Jouffroy et al. (1993) to be "basic mammalian". If the digits III and IV have the same length, the pattern is called "*paraxonic*". Hands of this type mostly have a rather short digit II. This holds true for many strictly arboreal NW and OW monkeys, among them *Ateles* and *Colobus*, in which digit I, the thumb, often is completely absent. If digit IV is longer than III, the pattern is called "*ectaxonic*". It exists in most strepsirrhines. In the lorisids, it is combined with great length of the thumb, which forms, together with the fourth ray, the typical "power forceps". As detailed below, the second ray is short in lorisids.

In most species, the *first ray (the thumb)* can be abducted from and adducted to the lateral fingers. The relative length of the thumb is often between 40% and 56% of digit III. In lorisids, and in humans it reaches 66%. In lorisids the long first ray is strongly developed and can be widely separated. Their second ray is reduced, and the longest finger is digit IV (see the above mentioned, "power forceps"). It should be noted that the metacarpal III is longer than that of the fourth digital ray and the obvious length of ray IV is determined by the free fingers (Jouffroy and Lessertisseur 1979; Jouffroy et al. 1993). The so-realised "*ectaxonic*" configuration implies the advantage of allowing maximal span of the grip (Fig. 8.2b and below). Its disadvantage is that the roll-off during walking is difficult to perform by a (long) plantigrade hand without loosening the grip. Only if the "power forceps" formed between pollex and toe IV is very close to the wrist, the forelimb can swing like an inverted pendulum about the carpal joint after closing the forceps around a twig. The dimension of the prehensile organ (sensu stricto) in x-direction of locomotion should be and is indeed very short, if ray II, between pollex and the longest and strongest lateral finger, is moved out of the way. This is accomplished in bushbabies (Galagidae) by hyperextension of the basal and flexion of the middle joint. In lorises (Fig. 8.1a), *ray II is reduced*, being very short and weak.[3] This usage of the hand as power forceps requires in walking along a substrate, especially in the first half of the stance phase, a pronounced abduction of the hand to the ulnar side (see below; Nieschalk and Demes 1993). At the same time, the finger flexors must exert very great forces for securing the grip.

The *1st ray*, or thumb of the hand, varies not alone in length, strength and the degree of abduction; it can be completely *absent*, in the not closely related genera *Ateles* (spider monkey, from the New World) and *Colobus* (guereza, from Africa). In the apes, it is not really reduced in comparison to trunk length, but its development lags behind that of the elongated lateral digits, in comparison to which it appears to be short (Fig. 8.1a). These thumbless hands form *hooks* (see below), in which all lateral fingers, especially rays III and IV, are working side by side. Fingers III and IV have the same length ("*paraxonic*" configuration). Most New World monkeys possess five-rayed hands, the thumbs of which cannot be widely separated from the other fingers and are orientated parallel to them. In grasping objects, some of them moderately adduct the thumb. This, however, is far from a true opposition of the thumb as in some Old World monkeys. When marching along a substrate, all fingers are placed parallel. Especially the members of one family (cebids) have a tendency to place digits I and II on one side of the branch and digits III, IV and V on the other, assuming the so-called schizodactylous position with a cleft between digits II and III (Daver et al. 2012).

The hands (and also the feet) may be placed with the full length of the autopodium on the substrate and are rolled off (Figs. 8.1 and 8.4) in most primates from carpus or tarsus (especially the heel), respectively, to the balls/pads beneath the heads of the metatarsals or to the tips of the digits. This is called *palmigrade posture* of the hand or *plantigrade posture* of the foot (see Fig. 8.1a–c). If only the distal part of the foot makes contact with the substrate (Fig. 8.4f–l) and the heel is permanently clear from the ground, this is called a *semiplantigrade* posture. This is often combined with the *digitigrade* posture of the hands, in which substrate contacts

[3]A very similar power forceps is developed in chameleons. In these forms, however, two digital rays form the medial/radial branch of the forceps, while two or three rays form the lateral/ulnar.

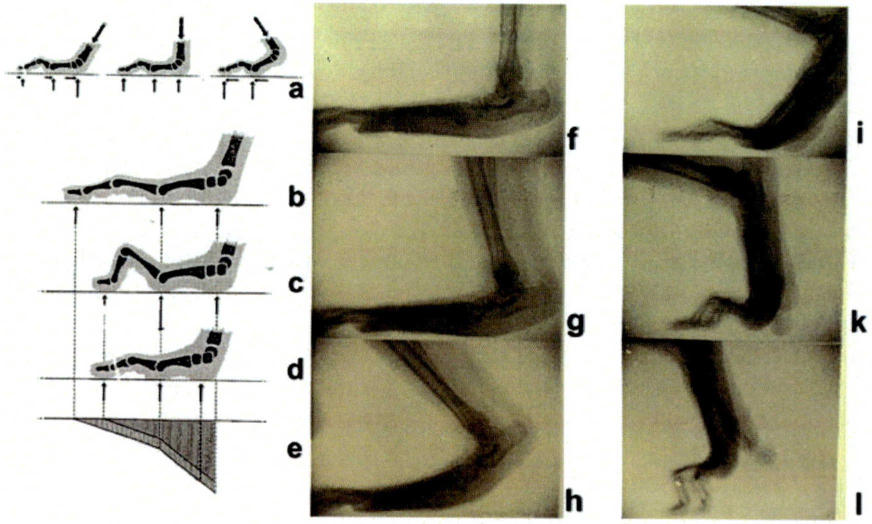

Fig. 8.4 Palmigrade and plantigrade autopodia under compressive loads. (**a**) Rolling off of a palmigrade monkey's hand. From left to right: shortly after touchdown, midstance, shortly before liftoff. The share of body weight on this limb is indicated by a downward pointing arrow; parts of the GRF in vertical and in horizontal direction are indicated by arrows pointing upward or horizontally. The external forces acting against the toes grow during roll-off, which means from left to right. (**b**) Long fingers extended during midstance, (**c**) long fingers flexed, (**d**) short fingers extended. (**e**) Modified from Preuschoft, Bending moments along the autopodium (hatched), drawn for long fingers and shortened fingers. Flexion of fingers and morphological shortening lead to reduced bending moments. (**f–l**) A series of x-ray photos of a foot in *Macaca fuscata* during roll-off from touchdown (**f**) to liftoff (**l**). Like in other OW monkeys, the heel does not touch the ground, so establishing a semiplantigrade position. The tarsometatarsal joint is actively extended during the second half of the stance phase (**i, k, l**), while the toes are functionally (not morphologically) shortened by hyperextension in the basal and plantar flexion in the middle joint (like in **c**). Compare Fig. 8.15e

are made alone by the balls/pads covering the capitula metatarsalium and the terminal joints and phalanges of digits. This is applied by primarily ground-living monkeys (*Macaca species*, *Papio*, *Cercopithecus aethiops*, *Erythrocebus patas*), in which the rolling off takes place from the balls beneath the metacarpophalangeal joints to the tips of the fingers. So the metacarpus increases the functional length of the limb (Fig. 8.1a, b). The more important walking on flat ground becomes to a species, the greater the share of GRF on that ray (III) which points in the direction of locomotion. Of course this is only possible, if the parts making contact with the environment are strong enough to sustain the acting loads (=external forces) without damage. This means a prevalence and strengthening of ray III, provoking what is called mesaxonic arrangement (Jouffroy et al. 1993).[4]

[4]It should be remembered that this way of loading leads in terrestrial mammals to the perissodactylous feet of odd-toed mammals, while the even-toed mammals follow the "paraxonic" pattern (see Chap. 7).

In the palmi- and plantigrade postures, the carpal or tarsal joints are exposed to considerable (dorsiflexing) moments. These are partly balanced by the powerful toe flexors and by the carpal or tarsal flexors. In the forelimb, especially the m. flexor carpi ulnaris inserts into a very long lever, namely, the os pisiforme, which is reminiscent of the heel bone in the hindlimb. The advantage of semiplantigrade or digitigrade positions is the elongation of the "functional limb length", which allows great step lengths (Fig. 7.15 in Chap. 7). The disadvantage is the reduction of the "prehensile organ" in a strict sense. If this is needed, for example, in climbing on trees, the digitigrade posture can be replaced by a palmigrade use of the hand (e.g. in vervets or macaques).

In spite of all characteristics which hands and feet have in common, a tendency toward *division of labour* between anterior and posterior limb is visible in all primates. It can be observed in locomotion, where it has been pinned down as "front steering, rear driving" in the sense of Kimura et al. (1979), which is in contrast to the "front steering, front driving" in cursorial mammals. In primates, the hindlimbs carry a greater share of body weight and produce more propulsion than the frontlimbs (see also Demes et al. 1994). This has to do with their short necks, and their hindfeet being placed far forward below their CoM, and allows their elevation on the hindlimbs, or their sitting for long periods on the soles of their feet and their tuber ischiadicum, which is covered among monkeys by nothing than horny ischial callosities. This relieving the frontlimbs from locomotor obligations makes them free for other biological roles, for example, the acquisition of food. While strepsirrhines sometimes take in their food directly by their pointed snout, they make use in other cases of their hands for seizing objects and leading them to the mouth. In monkeys and apes, this is the dominant, nearly exclusive technique. Every object of interest for the primates is taken up by the hands and carried for optical, olfactory, tactile inspection to the eyes, nose, lips or jaws – or for being taken in as food. The jaws can be short, so that the adductor muscles have long lever arms in relation to the bite forces. This either saves force (with the effect of reduced head weight) or gives the animals remarkably high bite forces. The existence of prehensile hands on long, mobile arms as grasping organs makes pointed and elongate jaws superfluous.

Prehensility of autopodia entails *great weight* of both segments, the zygo- and autopodia, because the differentiated muscles are heavy and the skeletal parts need to be strong enough to sustain muscle and external forces. In fact, thanks to the efforts of M. Günther, Preuschoft et al. (1998) have found the relative weights of these segments (segment weight/total body weight) in primates about twice as much as in non-primates. This influences the choice of gaits (see below), namely, walk and canter, while trot is extremely rare.

8.3.1 *Qualities of the Transmitted Forces*

8.3.2 *Morphology of Dermatoglyphic Ridges, Pads and Nails*

8.3.3 *Any Connections Between Them?*

With their special properties, prehensile hands and feet favour the grasping of substrates (branches, twigs, rocks) of their environment and allow getting hold of tiny particles. From a mechanical viewpoint, all this can be traced back to the ability of primate autopodia to exert not alone the ubiquitous *compressive forces* against an object (Figs. 8.1 and 8.4) but also *tensile forces* (Figs. 8.1 and 8.5) and even *torsional moments* (Fig. 8.6). Transmission of compressive forces does not demand special configurations (for examples see above, Chap. 7.5 and Fig. 8.1). Transmission of tensile forces is possible and often realised by simple hooks. Many animals for this purpose make use of claws (e.g. lizards, birds, sloths, bears, among primates the small New World callitrichids). Most primates form their hands or feet into hooks (Figs. 8.1e, f and 8.5). The often-discussed *hook hands* in apes, the African

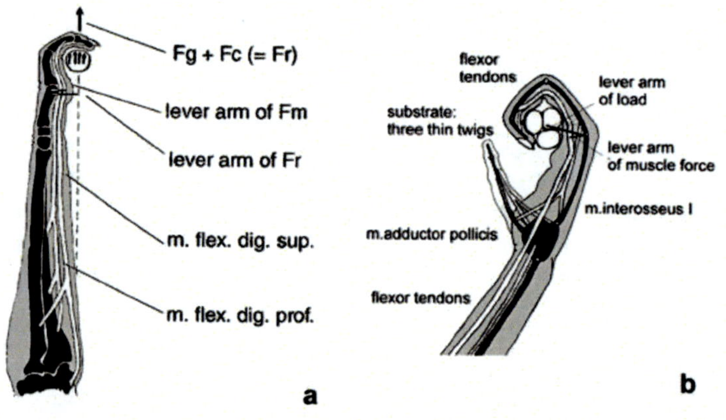

Fig. 8.5 Hook-shaped hands. (**a**) The rotating moments at the joints of the forearm are smallest, if the carpal joints, carpometacarpal junctions, metacarpophalangeal and perhaps even the interphalangeal joints are kept close to the prolonged external tensile force (= SRF, $F_g + F_c$). This gives the hand the typical hook position in which sometimes the middle joint, but always the distal interphalangeal joint, is flexed to form the hook. The longer the fingers, the wider the hook but also the bigger the rotational moments which must be balanced by the digital flexor muscles. The lever arm of the tendons is always shorter (1/3 to ½) than that of the external force (F_r), which consists of body weight (F_g) plus central force (F_c). Using the hand as a hook is only possible if sufficient muscle force can be provided, and any strengthening of the flexor muscles is connected with increased weight of the arms. (**b**) Special case of a hook hand, in which not the crooked shape but friction is responsible for the handhold. The three thin twigs are grasped so forcefully that friction between them and the fingers is strong enough to carry the animal. The intrinsic muscles of the pollex are not active and the thumb is not required for grasping hold and therefore can be reduced. (From Preuschoft et al. 2016)

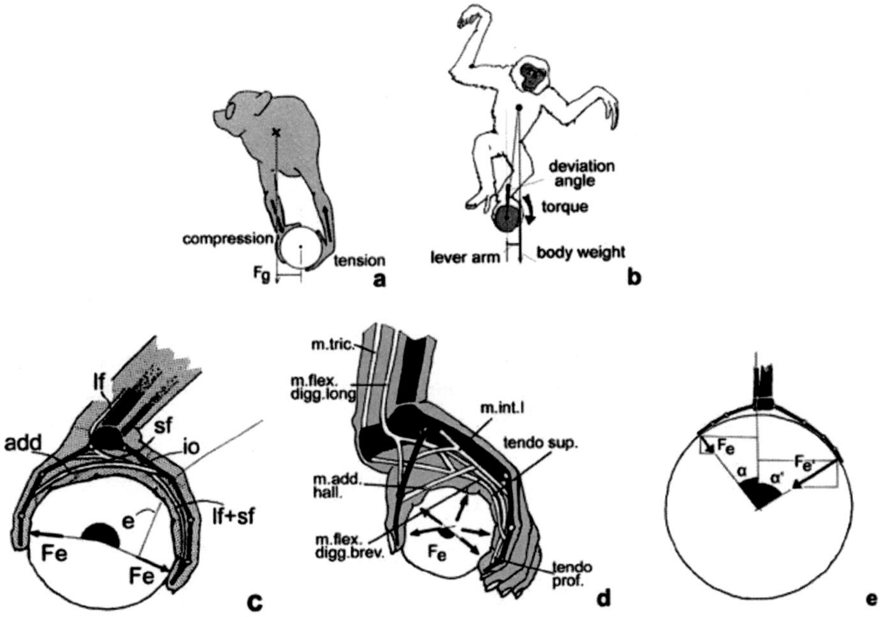

Fig. 8.6 Torsion and friction between a primate and its support. On a thin, narrow twig, the CoM may act lateral to the area of support on the substrate (**a, b**). Body weight (F_g) assumes a lever arm (l), which yields a tilting moment (as in Eq. 2.1). To prevent falling down, this tilting must be counterbalanced, either by two limbs, one exerting tensile and the other compressive force (black arrows, **a**), or by only one prehensile autopodium (**b**). Prehensile hands (**c**) and prehensile feet (**d**) possess intrinsic muscles which can produce compressive forces (F_e) between autopodium and substrate without exerting influence on the carpal or ankle joints. The longer the digits are, the longer the lever arm of the reaction force which acts from the substrate on the digit (**c, d**). (**d**) shows the direction of all reaction forces, which act between branch and foot, while (**c**) shows only those force components, which act on the fingertips. (**e**) The central angle in the sense of Cartmill (1974) is drawn for a hand grasping a thick branch. The forces (F_e) exerted by the terminal phalanges are indicated by black arrows. The forces have components which act against each other, their size being proportional to the sine of the central angles α and α'. These components increase friction. Because one of them is greater, no equilibrium exists, but it can be established by actions of the forearm muscles and shifting the SRF of this limb. The forces F_e also have components in the direction of body weight, proportional to the cosine of the central angle. If the reaction forces are large enough, they make the hand slide away from the branch, as antagonists of weight force, so making a safe grip impossible. If the central angles $\alpha + \alpha'$ are >180°, a minor force component is exerted in proximal direction against carpus (**c**) or tarsus (**d**). This strengthens the safety of the grip. Parts from Preuschoft et al. (2016). In (**c**), abbreviations are used as follows: *add*, adductors of the thumb or hallux; *Fe*, the force exerted by the digits or reaction forces of the substrate; *io*, m. interosseus I; lf, long (= deep) digital flexor muscle; sf, short (=superficial) digital flexor muscle

leaf monkeys and the spider monkeys of the New World are easy to analyse. The four lateral fingers are placed side by side, so that they brace each other in the radioulnar plane. In several genera (siamangs, gorillas), webs are extending distally up to the middle joints. As can be seen from Figs. 8.1 and 8.5, the joints of all fingers are in about the same flexed position. This open "hook" is a perfect means for transmitting tensile forces and to loosen the grip quickly. The longer the fingers are, the greater the hook, by which thick branches can be surrounded, or several thin twigs can be bundled to make use of their combined tensile strength (Fig. 8.5b). The force arm lengths of the flexor tendons at the basal and middle joints in African apes and gibbons are about one third of the load arm lengths (Fig. 8.5a). Therefore the tendon force must be about three times the external load supported by the arm – and this may well be a multiple of the weight of the whole animal! Closing the grip by opposing the thumb would limit the rotation of the hand on the branch and perhaps make loosening the handhold more difficult. Since the thumb is not essential for forming the hook, it can be reduced or even be absent. It should be noted that humans also form a "hook hand" when loads are carried. The only difference between a human carrying a bag or a suitcase and an ape hanging down from a branch is the inverted direction of the external load and the hand in space. If the "hook" posture of the hand is maintained in quadrupedal walking, the dorsal surfaces of the intermediate phalanges may be placed on the substrate, so realising the all-too-often-discussed "*knuckle walking*" (see below).

For transmission of *torsional moments* (Fig. 8.6), there exist two ways: first, by fitting forms (*"Formschluss" in German*) and, second, by "pure" friction (*"reine Reibung" in German*) or adhesion between the hand or foot and the substrate. An example for the former is the hook around a sturdy branch or the inobvious interdentation between the soft, deformable pads or cushions and the uneven surface of a branch; an example of "pure" friction is the well-known "sticking" or adhesion of a moist hand on a smooth surface. Both forms of friction are difficult to separate, and primates make use of both. The role played by friction in non-human primates becomes evident, if an ape is dangling from a bundle of twigs or a vertical, smooth rope (similar to Fig. 8.5b).

If friction coefficients and surfaces are given, and assumed to be constant, the resistance against friction or the ability to adhere depends on the *compressive force between animal and substrate*. This relationship is, however, not linear; under very high pressure, a viscoelastic process comes into action, which leads to much reduced friction. In order to increase the compressive forces, there are two possibilities: a basic principle inherent to all prehensile hands and feet is exerting compression from both sides on the substrate (Fig. 8.6), with digital ray I from one side and rays II–V from the other. The sum of these forces is zero. The efficiency of such a grip depends upon width of the span between the 1st and the longest lateral ray, in most cases this is IV (see above, "ectaxonic shape" of the autopodium). Determining factor of grip safety is the "central angle" (Cartmill 1974, 1979, 1985). The relevant force components are proportional to the sine of the central angle $\alpha + \alpha'$. The force components countering body weight are proportional to the cosine of the central angle. Only if the central angle in the schematic prehensile hand of Fig. 8.6 is wide, the opposed (horizontal) force components (F_{e1} and F_{e2}) are great enough to cause sufficient friction. A narrow central angle (thick branch or stem, Fig. 8.6c) makes the vertical force components so large that the hand is pressed upward, away from the substrate, and slips on the surface. A very wide central angle (= thin branch or stem) leads to force components directed against the carpus or tarsus, where the external load (F_a) or weight force is applied. The total forces so become larger than necessary for sustaining body weight, but the grip is safe.

Of course the long tubular bones of primate extremities are particularly prone to fractures, because the noxious forces, which can be applied to the body in an accident, can act on long lever arms. Shortness of the limbs increases safety, and therefore there must be good reasons for the evolution of long limb segments. In the absence of such reasons, short length is to be expected (as noted above).

8.4 Any Connections Between Friction, Dermatoglyphic Ridges, Pads, and Nails?

Friction in a strict sense depends upon *moisture* of the surfaces, *structure* of substrate and skin and *compressive force* between substrate and skin.

As is generally known, dry fingers slip easier on surfaces than moist ones. Too much liquid, however, again leads to slippery fingers. In all primates the skin on palms and soles is hairless, compliant but very tough and covered by dermal ridges, which form characteristic patterns. At least since Whipple (1904), this skin structure is related to friction. This "friction" skin is equipped with a great number of sweat glands, which keep the surface moist. There exist still other, additional means to moisten the grasping surfaces: bushbabies, for example, which come from their sleeping nests, have dry and warm hands and feet, which easily slip over surfaces. Before starting their nightly activities, bushbabies "wash" their autopodia with limited amounts of urine. The hands and feet become cold and stick very tightly to different materials.

Strictly speaking, there is a difference between static friction and slide friction, the former being in the foreground. Material-dependent properties of the substrate are summarised as "friction coefficient" and are dictated by the environment – while the properties of the structure of the palms and soles of the respective animal seem to be developed in order to fit to the pre-existing properties of substrates.

According to Niemitz (1977, 1990), the epidermis is tied by connective tissue to the corium, the fibres of which are so orientated that greater forces can be sustained perpendicular to the ridges (see also Nieschalk and Klauer 1989). Niemitz noted in some of the pads a concordance of ridges with the directions of local stresses. The characteristic arrangement of papillae of the corium in double lines along the dermal ridges may be another hint to a mechanical function of the dermatoglyphics. Schulz (1995) has observed that collagenous fibres (type I) are concentrated below the trenches between the dermal ridges and run parallel to them, while other fibres, running at right angles, connect neighbouring ridges (Fig. 8.8d). To make visible the deformation of the plantar pads, a compliant silicon model of the fingertip was equipped with regularly spaced markers, which are dislocated under load. The dislocations are shown in Fig. 8.8h. They correspond to the fibres crossing the

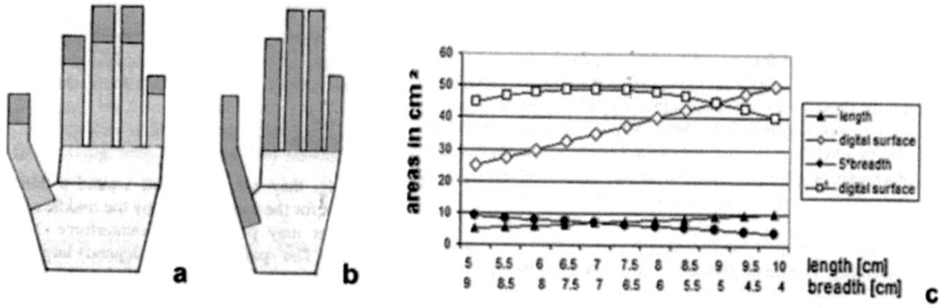

Fig. 8.7 Length and width of hands. (**a**) The size of the contact areas of the digits is given before (light grey) and after (dark grey) elongation of the fingers. The length increase of the digits (**a**) entails an enlargement of the contact area. Friction is reduced, because the compressive force between digits and substrate is more dispersed. If the breadth of the entire autopodium becomes smaller with length increase (**b**), the contact area remains small and therefore the compression between autopodium and substrate is on a higher level. (**c**) Mutual dependence of length and breadth (black symbols) and resulting digital surfaces (white symbols) – if breadth is either maintained (diamonds) or reduced (squares). (From Preuschoft et al. 2016)

ridges, and these can be expected to run at right angles to the strains. When following individual ridges, the distances between the markers under load become greater by some 4%, especially if a shearing component is superimposed. This elongation can be countered by the collagenous fibres parallel to the ridges. In the majority of load cases, the shearing component will deviate from the longitudinal axis in distal-radial direction (see above, vertical climbing as well as tool using), so favouring "loop" patterns, which are open to the ulnar side.

Niemitz (1977) and Cartmill (1979) have checked the *friction of the complete soles* on diverse substrates. They concluded that the friction skin leads to increased adhesion, but this does not answer the question why the ridges form the well-known patterns. The influence of the dermatoglyphic ridges has been investigated by Buck and Bär (1993). Any load can be applied either in the direction parallel to ridges or perpendicular to them. For their experiments the authors have checked the most distal part of the terminal segments, where ridges are arranged from ulnar to radial in z-direction (Fig. 8.8a) and at right angles to the longitudinal axis of the digit in x-direction (black arrows). The authors have made human experimental subjects slide their fingertips along selected surfaces: smooth, polished stone, wood, glass and plastic, in the direction of ridges and at right angles. They found in the direction across ridges the friction resistance to be dependent upon compressive force and 20% greater than parallel to the ridges, if stone or wood was used as a substrate. These materials indeed form the substrates on which primates move under natural conditions. The very hard and very smooth surfaces of hard plastic and glass did not show any difference. The previously mentioned, very high compressive forces have even lead to reduced friction, what is due to viscoelastic laws. The results obtained give reason to understand the function of ridge patterns: we may expect that ridges are so formed that many of them cross the most frequent direction of SRFs. At least on the fingertips of humans, this expectation seems to be fulfilled: the most frequent

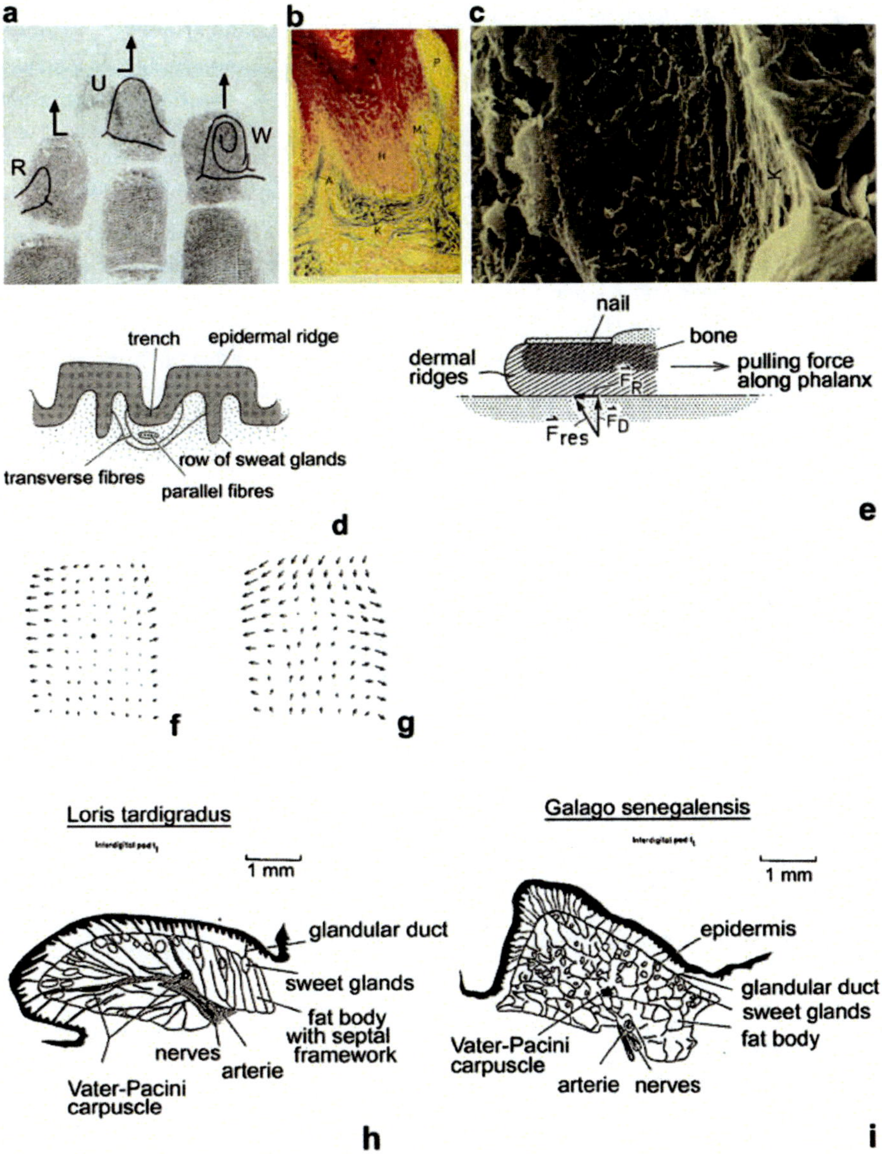

Fig. 8.8 Friction skin. (**a**) Tips of human 2nd to 4th fingers from a right hand with common dermatoglyphic patterns. Most frequent is the "ulnar open loop" as shown on digit 3, less frequent is the "whorl" on digit 4, and rare is the "radial open loop" on digit 2. All are suited to sustain distally directed force components (black arrows – see also **e**), the whorl more or less in line with the finger, in the loops deviating toward the "closed" side. (**b**) Stained cross section through the trench separating two dermal ridges. Collagenous fibres run parallel and at right angles to the trenches between ridges. (**c**) Bundle of collagenous fibres on right side and parallel to the trench. (**d**) Interpretative scheme of fibre arrangement. (**e**) Model of a terminal segment pressed against the

loop patterns are open to the ulnar side – in agreement with the predominant direction of external forces which run distally and in the majority of load cases have a component from the ulnar to the radial side of the hands. These conclusions are in full agreement with the findings of Schulz (described above). It should be noted that the digital ridge patterns are inherited and species-specific. See also Niemitz (1990).

Underneath the ridge-bearing skin, there exist at the tips of fingers and toes, and at defined places of the palms and soles, thick, soft and deformable *pads*, consisting of connective tissue and sweat glands; sometimes (not in the digital pads of lorisids and tarsiers! Nieschalk and Klauer 1989) lobuli of fat cells are wrapped in tough and dense connective tissue (Fig. 8.8h, i). In smaller forms, these pads are particularly obvious and prominent; in the large primates, they are levelled in with the other parts of the autopodia. When forces are transmitted through a hand or a foot, these pads are in contact with and pressed onto the surface of the seized object. The biological role of these pads is clinging to minor roughnesses of seized objects, since their pliant structure adapts itself to the form of the substrates (Figs. 8.1k, l, 8.6c, d and 8.8e-g). Their mechanical function is distributing the external forces on the skeletal elements. Their microscopic structure is determined by sensory organs and sweat glands dispersed in the epidermis and dermis (textbooks of anatomy or histology) and by masses of fibres of connective tissue which attach the skin to the skeletal parts (Nieschalk and Klauer 1989). The same authors found "small ligaments parallel to the epidermal ridges". Niemitz (1990) related the dominating direction of fibres to the prevailing external forces (compare also the dermatoglyphic ridges above).

Preliminary, not yet worked out approaches by Witzel and Preuschoft (2008) to the stress flow within the fingertips seem to indicate a functional connection between the soft cushions and *flat nails*. In Fig. 8.9 the greatest stresses are within, or very close to the skeletal element. Compressive forces occur plantar to the phalanges, at least in the two investigated load cases (concentric, perpendicular to the palm and excentric that is superimposed by a component parallel to the palm). The greatest tensile forces occur between the skeletal element and the nail, although no force is applied directly to the latter. The pad seems to be squeezed under compression to both sides of the bone (see also the silicon model of Schulz, Fig. 8.8 e-g), and the deformation of the pad finds resistance in the nail. If, however, the nail is loaded directly, tensile stresses between the phalanx and the nail are quite high, and the nail is being compressed (see also above, Sub-chap. 7.2.1). Equipping the

Fig. 8.8 (continued) substrate and pulled proximally. The resultant of both (F_{res}) acts obliquely on the pad. (**f**) Strains in a deformable model of a fingertip under centric compression. Shown are the dislocations of marks on a silicon model after increasing the load. (**g**) Strains in the fingertip under compression plus longitudinal tension. Like in (**f**), ridges can be expected to be perpendicular to strains. (**h** and **i**) Sagittal sections through an interdigital pad in slender loris (**h**) and bushbaby (**i**). The arrangement of fibres differs, although the animals are members of the same family. The hydraulic elements are sweat glands, rather than fat. (**a** From Rodewald and Zankl 1981; **b–g** from M. Schulz unpublished 1993; **h, i** from Nieschalk and Klauer 1989)

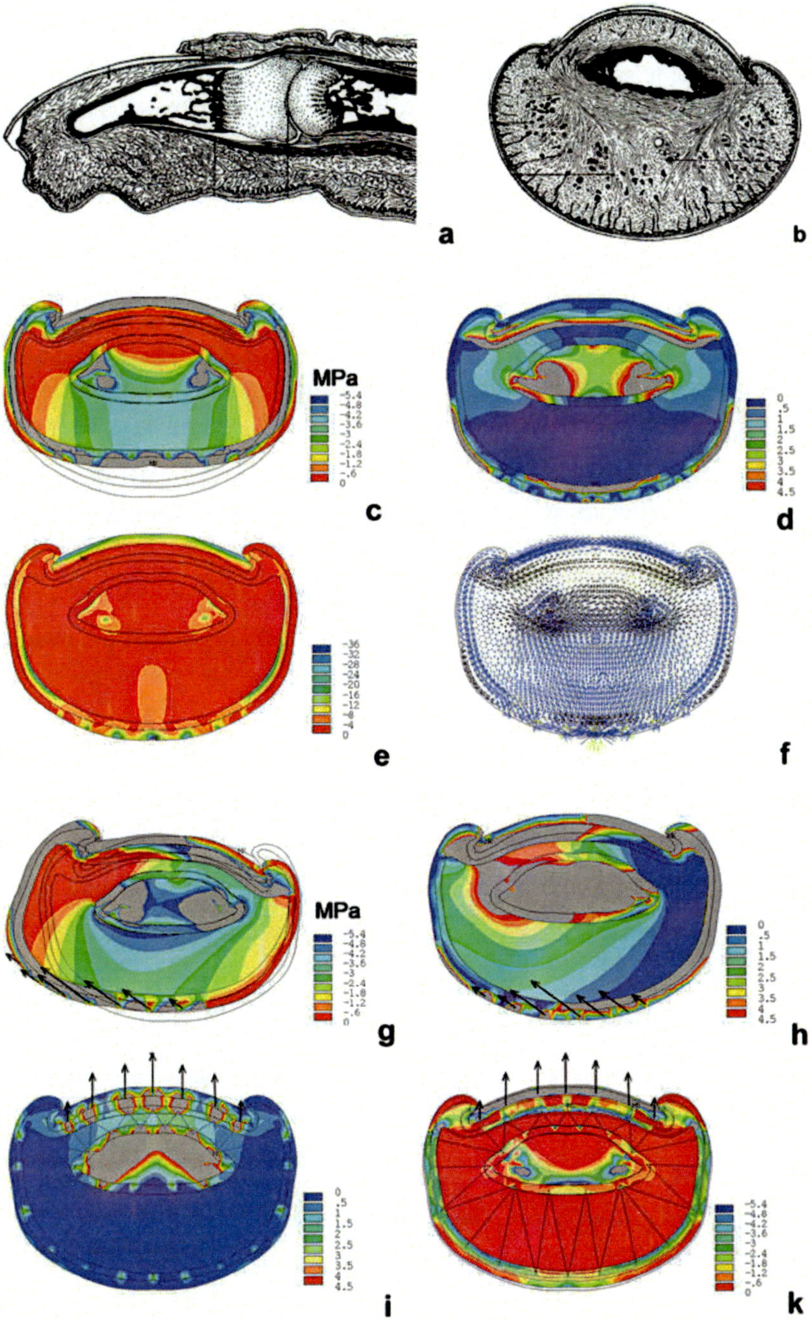

Fig. 8.9 Fibre arrangement in human fingertips. **(a, b)** Longitudinal and transverse sections through terminal segments (From Bucher-Wartenberg 1989). The compression-resistant hydraulic elements are fat lobuli and sweat glands. The masses of fibres indicate the directions in which tensile

two-dimensional FE model with tension-resistant fibres in rather unspecific radial directios, similar to that observed in sections (**a**), reduces the total stresses considerably. A causal relationship between pads, nails and connective tissue at present is not much more than a working hypothesis. The mechanical stresses during a variety of behaviours and their possible relationship to nails and fibre arrangement within the pads also need to be studied further.

The toes are pressed actively against the substrate by contraction of digital flexors and adductors (Fig. 8.6d, e). The strongest flexor muscles (mm. flexores digitorum pollicis, hallucis, communis longus) are located on the zeugopodium, and their tendons in many primates are common to all or several digits (Preuschoft 1961, 1963a, b; Landsmeer 1993). These tendons cross the carpal or tarsal joint(s). That means, the long digital flexors have a rotating moment at the joint between autopodium and zeugopodium (in Fig. 8.6d the wrist, in Fig. 8.6e the ankle joint). This moment may be useful for maintaining the external equilibrium or be an irritating factor. A disturbing influence of the long flexor tendons can be excluded by using only the *intrinsic muscles of the hand or foot*, which can produce independent rotational moments at the digital joints (and therefore increased frictional resistance), while they have nothing to do with the carpal or tarsal joints: mm. abductor pollicis, hallucis, flexor brevis of first and fifth ray, interossei, adductores pollicis et hallucis, and flexor brevis digitorum – if not originating from the long flexor tendon (as described in Preuschoft (1961, 1965); more details see below). Because the compression between autopodia and substrates (and therefore also the frictional resistance) can be chosen freely, primates can transmit torsional moments between animal and substrate, which allows moving on very narrow substrates (Fig. 8.6), or suspend themselves on very thin twigs (Fig. 8.5b). Prehensile autopodia of comparable potentials exist in other animals as well (chameleon, parrots, kusu, koala, rodents), but they are especially typical of and very obvious in primates.

Very long digits are a means to span a wide central angle. Length of digits, however, enlarges the contact area between autopodium and substrate. The effect is

---

Fig. 8.9 (continued) forces can be sustained. (**c–e**) Two-dimensional FESA models show the magnitudes of compressive (**c**, from red to green, light blue) and tensile (**d**, from dark blue to green and yellow) stresses. (**f**) shows the direction of stresses, either compression (>—<) or tension (<-->). The dominating compressive stresses are greatest between volar surface and bone, as well as besides and within the nail. Tensile stresses occur at both sides of the phalanx, running toward the margins of the nail. Both are in agreement with stress trajectories (**f**), if absolute sizes are taken into consideration. (**g, h**) Under oblique loading (arrows indicating the resultant SRF), tensile stresses become greater and asymmetrical. (**g**) Compressive, (**h**) tensile stresses. If the direction of external loads is reversed, the stress pattern would also be reversed. (**k**) External forces are applied to the nail (arrows directed upwards); in addition, mechanically "stiff" fibres (black lines), reduce all above shown stresses and resulting deformation to nearly zero. (**i**) To sustain external forces that are applied to the nail, not to the pad (arrows), the former must be tied to the phalanx by the densely packed fibres on its palmar side, while the pad on the fingertip receives no force. (From Witzel and Preuschoft 2008) reduced compression between the hand/foot and the seized branch and subsequently reduced friction. A way out of this conflict is narrow shape of the autopodia, at least of digits (Fig. 8.7). Of course, the narrowness of digits finds its limits in the strength necessary to sustain the high forces that occur in gripping. This the more so, since length of digits gives the external substrate reaction forces against the fingertips a very long lever arm and therefore a great rotational moment. This must be counterbalanced by a very strong musculature and high energy expenditure.

$$F_{\text{ext}} * h_{\text{ext}} = F_{\text{m}} * h_{\text{m}} \tag{8.6}$$

The strong intrinsic and flexor musculature of the digits, in combination with their lengths, make the zeugopodia and especially the autopodia of primates about two times heavier than the respective segments in other mammals (Preuschoft et al. 1998). This has consequences for the gaits used by primates (see below).

In many (preferably arboreal) primates, the fourth ray is longer than the others are, because the fourth can reach the greatest distance from the first ray (see above, "ectaxonic" configuration). If the interspace between the first and the second ray is widely open, this allows grasping of particularly thick substrates. The essential factor seems to be not so much the width spanned, than the area opened between the 2nd and the abducted 1st ray (Preuschoft et al. 2016). This is characteristic for gibbons and sifakas, in which the m. adductor pollicis/hallucis takes its origin not from the shaft of the second metapodial but from its basis, so forming the often observed deep cleft between the first and the second ray. Nearly the same effect is realised, if – like in the heavy lorisids *Nycticebus* and *Perodictius* – the 2nd digit becomes very short and unable to contribute high forces (Figs. 8.1 and 8.2).

If alone compressive forces are transmitted between the animal and its substrate, long digits are not advantageous (see above, Fig. 8.4). Instead, the animal saves energy, if the toes or fingers are shortened, be it morphologically (reduced length) or functionally by hyperextension of the basal and flexion of the middle joints. The latter is similar to the autopodia in carnivores and can be observed often in primates (e.g. in human toes!). As Figs. 8.2 and 8.4 show, the lever arms of the external loads are shortened, so that muscle force (= energy) can be saved.

8.5 How Do Primates Move Around in Their Habitat? Modes of Locomotion, Gaits on Continuous Surfaces and the So-Called Climbing

The diversity of primate movements have induced several attempts to classify "locomotor categories". Quite influential has been Napier (1960, 1961); Napier and Napier (1967). These authors, however, emphasised the lack of precision and the danger of misclassifications. The same observation was made by Oxnard (1979a, b). In my opinion, all attempts to classify primates into "locomotor categories" have not contributed to understanding morphology, and therefore I prefer a different approach.

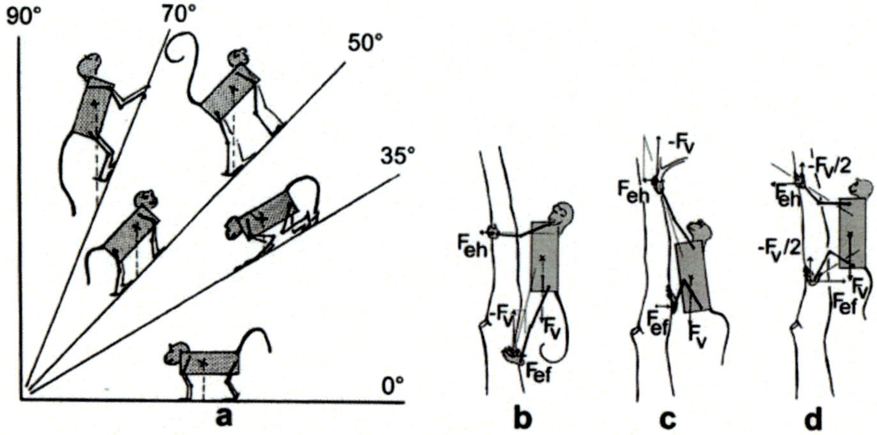

Fig. 8.10 "Climbing" on inclined substrates. (**a**) Inclination of substrates exerts influence on the distribution of body weight on front- and hindlimbs. CoM is indicated by a cross (x), and the gravity vertical by a broken line. If the head points downward, the frontlimbs receive a greater share of body weight (35°); if the head is kept up, the hindlimbs have to bear more of body weight than frontlimbs (45° and more). Length of the limbs emphasises the shifting of weight toward hindlimbs, as shown in 50° inclination. If the gravity vertical passes behind the hindfeet (70° and more), body weight exerts a moment rotating the animal rearward. This must be countered by tensile forces transmitted through the frontlimbs. (**b–d**) The often discussed "vertical clinging and climbing". (**b**) Vertical substrates pose conditions similar to those on steeply inclined substrates (as in **a**). If a substrate is +/– perpendicular, body weight may be supported alone by the hindlimbs. Since the moment of body weight rotates the trunk dorsally, away from the stem, it must be counterbalanced by tensile forces exerted by the frontlimbs. The longer the hindlimbs, the smaller the tensile force on the upper limbs. Extension of the hip and knee joints saves energy. This climbing technique can be observed in lemurs and sifakas. Humans, when climbing on trees, have to cope with the same conditions (see Fig. 8.11). (**c**) Body weight may also be carried alone by the anterior limbs. Since the gravity vertical passes dorsal to the hands, a moment rotates the trunk ventrally. In most cases, the hindlimbs are used to push the body clear from the stem. Expenditure of energy is reduced, if the forelimbs are long and the hindlimbs are short and strongly flexed in hip and knee joints (like in apes). If, however, body weight is distributed on fore- and hindlimbs (**d**, as can be observed in most monkeys), the horizontal components of the external force are the greater, the farther the CoM is kept away from the stem. This leads to greater total SRFs and increases friction. (From Preuschoft, 2002)

At a first, superficial glance, the most characteristic locomotor propensity of primates seems to be "climbing" in their arboreal environment. A close-up view (Arms 1998; Arms et al. 2002; Preuschoft 2002; Morbeck 1977, 1979; Rose 1973a, b, 1979), however, shows that primates whenever possible prefer quadrupedal walking on level substrates (for the gaits used on +/– level surfaces see below). These substrates (sufficiently strong branches, thin twigs, stems or cliffs) are often steeply inclined (see also Niemitz 1990). Their inclination determines the transmitting of forces as well as load distribution on front- and hindlimbs, but not the gaits used (Fig. 8.10).

Under consideration of the above observations, the locomotor mode "climbing" was defined by Preuschoft (2002) as "locomotion on steeply inclined or even vertical substrates, or by making use of a series of hand- and footholds at various levels. It implies a gain or loss of height above ground. The gravity vector may be outside the area of support, which entails the need to transmit rotating moments, thus, tension as well as compression has to be applied from the animal's body on the substrate". This definition includes all variants of "below-branch" locomotion. Phases of rest imply the same necessities as locomotion. Obviously, friction is a very important factor for the transmission of forces between animal and substrate.

As long as inclination of the substrate is moderate, both autopodia are in safe contact with the underlying substrate, where *compressive forces* are exchanged (Fig. 8.10a). The gravity vertical comes closer to the lower contacts, and therefore body weight is shifted toward that pair of limbs which is below. This the more so, if limbs are extended, as illustrated in 50° inclination of Fig. 8.10a. Since most primates tend to keep their heads up, even if climbing downward, in most cases the hindlimbs receive a greater share of body weight. In several studies, Kimura (1991, 1992, 1995, 2002, 2010 in Kyoto, Abstract Volume) has shown that indeed those primate species which spend more time on trees than on ground have stronger and longer hindlimbs. The more ground-dwelling baboons, macaques and vervets tend to a reinforcement of the frontlimbs and so come closer to the load distribution in cursorial mammals, which carry more load on the front than on the hindlimbs. Kimura's results are in agreement with those obtained by Strasser (1992) and Jungers et al. (1998). The interpretation of the proportions as functional necessity is confirmed by the same difference in non-primate mammals (Kimura 1995). There is, however, one notable exception: the howler monkey, *Alouatta*, frequently climbs downward head first according to Youlatos and Gasc (1994) and so puts more load on the frontlimbs than other monkeys (Fig. 8.10a, 30°).

If the substrate is steeper (>60°), the gravity vertical passes behind the hindfeet, and the frontlimbs are under *tensile force*. Precondition is of course that the hands are able to transmit tension (Fig. 8.10a) and the feet are able to transmit compressive force (Fig. 8.10 middle) on the respective substrates, so establishing a "division of labour" as noted above (see also below). The fixation on the substrate is largely due to "*friction*" in a wide sense. A special case of climbing on inclined substrates is "vertical climbing" (Napier and Walker 1967; Cartmill 1974; Jungers 1979, Fig. 8.10b–d). This has received in the primatological literature a somewhat exaggerated importance.[5] No matter whether directed up- or downward, climbing on a stem can be performed in at least two ways: first by safe *anchoring of the hindfeet* to the stem (Fig. 8.10c). The hindlimbs then are under compressive force. Because the

[5]In the form of "vertical clinging and leaping", it has been treated like a distinct mode of living and even as an important stage of evolution. In fact there are primate species which spend much time clinging to +/– vertical substrates – as well as primates which leap frequently. It is misleading to look at "vertical climbing" as special (and primitive) mode of life. It instead is simply a technique to reach the interesting and promising canopy. The stem is just that part of the tree which connects the (for primates) relevant periphery of trees with the ground. Too many authors seem to have neglected that not alone upward but also downward climbing plays a role: if someone has climbed upward, he must come down again! Admittedly, upward requires the input of mechanical work (force * distance), but climbing downward must be controlled as well, in order to avoid free falling.

CoM must be kept at distance from the support and the gravity vertical is dorsal to the footholds and so has a rotating moment about the connection between both feet as a pivot, the body tends to rotate away from the supporting stem. The tensile force which keeps it in its place is provided in non-human primates by the hands and arms. Long hindlimbs are an advantage, because they permit the horizontal force components at the hands as well as the feet to be small (Fig. 8.10c). This mode of using vertical substrates is often used by lemurs and sifakas, which possess the long and strong hindlimbs typical of leapers. The same holds true for humans, in which the hindlimbs are elongated to fulfil different functions (see below and Chap. 9). The second possibility for vertical climbing is *suspending the body's weight mainly by the anterior limbs* (Fig. 8.10b). The frontlimb then is under tensile force, and the elbow and shoulder joints are extended. Because the gravity vertical is at distance from the handholds, a rotating moment tends to swing the body against the stem. In most cases, the compressive force necessary to keep the body in its place is provided by the hindlimbs. The moment is small and the required force low, if the belly is close to the support and the hindlimbs are short, or at least strongly flexed and abducted (Preuschoft 1961). Hip, knee and ankle joints are close to the compressive force, and therefore not too much muscle force is necessary. Again the length of the supporting limb contributes to keeping the horizontal force components small. The resultant of the great vertical and the smaller horizontal component is shorter than in the case of shorter frontlimbs, and so the energetic costs are kept low in long-armed forms (like lesser and large apes).[6]

Of course, transitions between the described extreme examples are common. All primates may well distribute their body weigh on both pairs of limbs. In monkeys, the combination of hind- and frontlimb support seems to be most usual (Fig. 8.10d). By removing the CoM from the stem, they may increase the horizontal force components of the substrate reaction force and so adapt the pressure between palms, soles and substrate to produce the necessary friction. Hirasaki et al. (1992) have shown how spider monkeys and Japanese macaques modify their state of equilibrium in accordance with length and strength of limbs and with the surface qualities of the substrates.

Humans who ascend (or descend) trees or masts must solve the same problems (Fig. 8.11). The strongest limbs, the legs, are used for supporting body weight. Precondition for this technique is a reliable and firm hold of the feet. Whereas this is not a problem for the grasping feet of a lemur, humans need in the absence of prehensile feet a technical device: either climbing irons with spikes, or a sling, which presses the feet from both sides against the stem, or extreme abduction and flexion of the hips, flexion of the knees and supination of the feet, which are placed close to the hip joint. The latter techniques produce, without overstressing the adductor muscles, high pressure of the feet against the stem, which evokes sufficient "*friction*" in a

[6]An exception is Tarsius. According to Niemitz, 1984, tarsiers, when resting on their prefered vertical stems, flex their hindlimbs strongly, and extend their frontlimbs moderately upwards, indicating tensile SRFs. The small animal so is prepared for the next explosive leap.

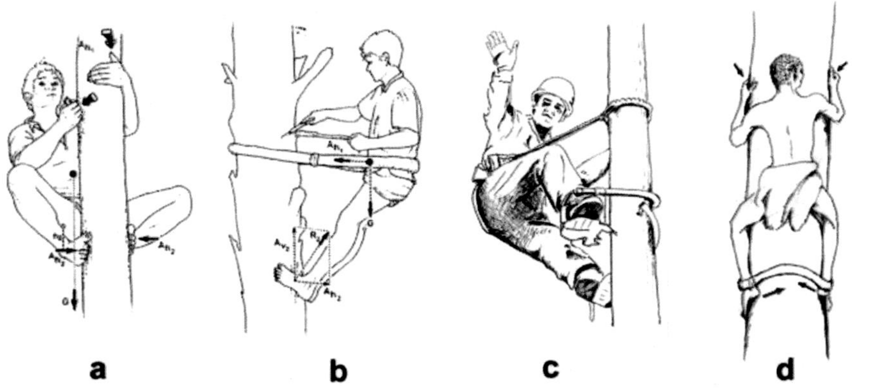

Fig. 8.11 Humans ascending or descending tree stems. (**a**) Native travel guide in Brazil, collecting a sloth, without using any device. (**b**) Palm wine toddler in Central India, using a belt to relieve the hands. (**c**) Workman in Germany ascending the mast of an electric cable, using a belt and climbing devices to replace prehensile feet. (**d**) Pygmy in the Congo basin ascending a tall tree, the feet tied together in order to exert sufficient pressure for friction. In all cases the feet carry most of body weight. (**a** and **b** from Preuschoft et al. 1992)

wide sense. Keeping balance by the hands is possible only for limited time, and hands are anyway used for other purposes. Therefore, often a belt is used – which has no other function than taking over the tensile force component in an comfortable way. Hands are free for working.

On ground and other wide substrates (boughs, branches – depending on size), apes and monkeys prefer *walking* (Fig. 8.12a, b) on four, rarely on two, limbs (gibbons). Among strepsirrhines, walking is the clearly preferred gait in lorisids and often used in lemurs and in the larger *Galago* species – at least in difficult conditions. Galagos, sportive lemurs and hapalemurs make use of the same gaits as other small mammals: aside from walking, hopping, half-bounding and bounding have been observed (see also below, p. 17). *Propithecus* likes bipedal hopping very much when on ground.

The walk is defined as a striding locomotion, in which the stance phases of each limb are longer than swing phases (= duty factor > 50%). The laws governing the walk have been described in Sub-Chapters 7.5.3 and 7.5.5: each limb is either loaded during the support phase, in which it behaves like an inverted pendulum (Fig. 8.12a), or swinging forward without load like a pendulum, the pivot of which is moving continuously forward (Fig. 8.12b). Monkeys diverge from other mammals of about the same size by their rather long step lengths. They reach, in spite of their low frequency, considerable marching speeds. Squirrel monkeys, as an example of a small monkey, march with 0.4–0.9 m/s speed in walking and tamarins with 0.4–0.7 m/s. If their speed is very high, phases of areal suspension of either the hind- or frontlimbs can be observed. This is comparable to the toelt or running walk of cursorial mammals (see above). Unfortunately, this gait has been termed "trot" by many authors – which is misleading. Walking of monkeys is also characterised by

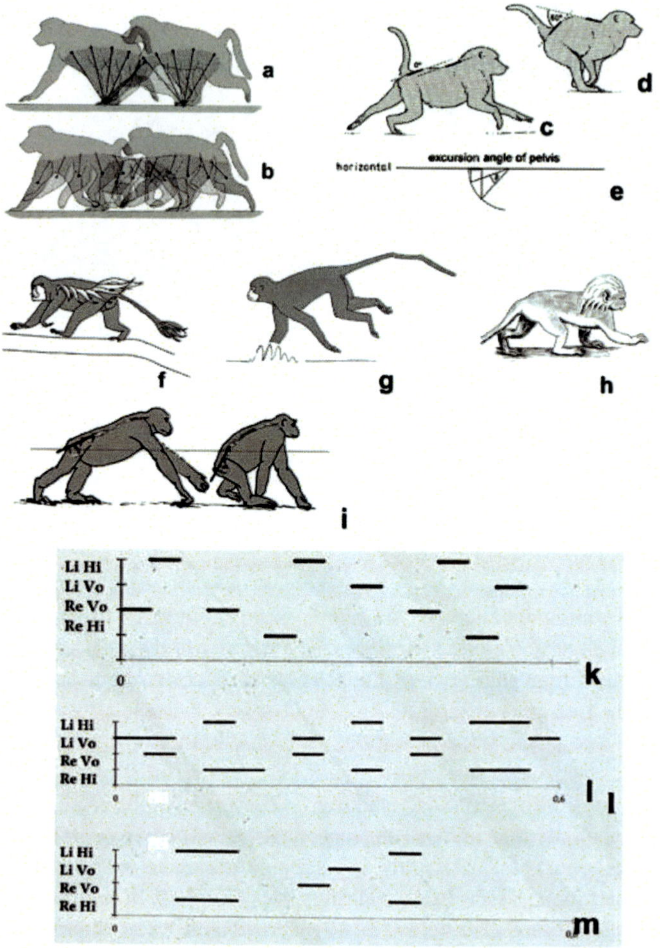

Fig. 8.12 Locomotion of primates on level substrates. (**a, b**) Most frequent is the walk, shown here in a baboon. (**a**) Stance phases of front- and hindlimbs representing inverted pendula, above which a part of body mass is swung forward. (**b**) Swing phases, in which the front- and hindlimbs are swinging forward like a pendulum with its pivot moving continuously forward. (**c** and **d**) As faster gait, primates prefer the gallop or bound while avoiding the trot. Baboon in two phases of the (rotary) gallop, with extension (**c**) and flexion (**d**) of the trunk axis, especially the pelvis. (**e**) Superimposing both extremes shows the contribution of trunk flexion to cycle length. (Modified from Preuschoft 2004) (**f**) Black and white guereza, (**g**) vervet monkey galloping (both taken from Rose 1979). (**h**) Tamarin (*Saguinus oedipus*). (**i**) Chimpanzee in two phases of (lateral) gallop. (**k**) Footfall sequence of a galloping tamarin (*Saguinus oedipus*) in comparison with a bounding (**l**) and a half-bounding (**m**) conspecific. (From Preuschoft 2004). A bounding *Cheirogaleus* is shown in Fig. 7.7a–d

pronounced up-and-down movements of the shoulder and the hip region – just because of their great excursion angles. The most often seen *diagonal footfall sequence* (M. Rose 1979) may be connected with the monkeys using rather thin and pliant substrates. The branches are under less torsion, if an extremity is placed on either side, and the animal gets additional safety because of placing the extremities on both sides. Diagonal footfall is also favoured by their greater shoulder and thorax width (see above, Chaps. 4.3 and 5.7). It also seems to be connected with a position of the CoM in the posterior half of the body, whereas the position of the CoM in the anterior half of the body is tied to lateral footfall sequence (Kimura et al. 1979; Kimura 1985; Rollinson and Martin 1981). Some of the preferably ground-living monkeys (vervet monkeys, Vilenski 1989; Vilenski and Moore 1992; baboons, Gambaryan 1974) change easily to the lateral footfall sequence if moving on ground. This may be due to the fact that they are not compelled to take care of the weakness of pliant substrates. Walking is rather safe and comfortable, but only possible on continuous substrates (ground, branches), not in a less dense canopy and not when moving at right angles to the branches or twigs. If these substrates are separated by varying interspaces, a continuous rhythm cannot be maintained. If dense twigs are available, but too thin to carry the weight of, e.g. a slow loris (*Nycticebus*), the animal just turns upside down to suspend itself and continues its walking locomotion as before (Jouffroy and Petter 1990; Ishida et al. 1990). In walking, step lengths of front- and hindlimbs must be the same, which favours limbs of equal lengths. If there are differences in limb length, like in the African apes, they are compensated by different excursion angles of front- and hindlimbs (Kimura, personal communication).

8.5.1 *Fast Gaits on Continuous Substrates*

In spite of their taking long steps (see above) primates – like other small mammals – often do not reach walking speeds which permit competition with larger competitors. Therefore, they must use more speedy gaits, even if these consume more energy than walking.

The true *trot* can be seen rarely in adult monkeys; youngsters try but use it less when growing up (Nakano 1996). Obviously, the strong musculature of their prehensile hands and feet makes the autopodia and zeugopodia so heavy (Preuschoft et al. see above), that pendulum lengths of the forward swinging limb pendula are so great that high step frequencies become energy-consuming. In addition, the synchronised footfalls of hands and feet in the trot exert high forces on the substrate. These forces may exceed the strength of substrates or initiate an undue swinging of the substrate.

The fastest gait available is a gallop (Fig. 8.12c–i). This was detailedly described and related to their respective modes of living by M. Rose (1973a, b, 1979) in East African *Colobus guereza*, vervet monkeys and baboons and also by Morbeck (1975, 1977). Aside from (mostly transverse, Fig. 8.12c) gallop, black and white monkeys, weighing 5–10 kg, also use – rarely – the bound. Galloping can also be observed frequently in macaques and in the larger baboons (5–20 kg, Fig. 8.12b).

Dorsomobility is pronounced in all monkeys. For the larger chimpanzees, galloping is the usual fast gait, and in siamangs (own observations) the characteristic footfall sequence can also be observed, while bipedal running is more frequent and probably faster. Squirrel monkeys (0,7–1.1 kg, *Saimiri sciureus*) also prefer a true gallop, and the really small tamarins from the New World (0.2–0.3 kg, *Saguinus oedipus*) use the gallop as often as bound and half-bound, in both gaits flexing and extending the back (Fig. 8.12i; Arms 1998; Arms et al. 2002). Both forms change from the walk into the asymmetric gait at about 0.7 (tamarins) or 0,9 m/s (squirrel monkeys). Among strepsirrhines, the smaller a species is, the more often it seems to use the bound. A good example is *Cheirogaleus medius* (see above Sub-Chapter 7.3.2). This makes the gallop in tamarins more remarkable. *Propithecus* and *Galago* frequently hop when on ground, the larger sifakas with pronounced "lead" of one side, and disrupted footfall, the smaller bushbabies with symmetrical movements.

8.6 What About Moving in an Arboreal Habitat with Interrupted Supports? Particular Modes of Arboreal Locomotion

An arboreal habitat is not only composed of a three-dimensional meshwork of branches and twigs; characteristic are also interspaces between these structures. For overcoming these gaps, primates have developed rather special behaviours: bridging, leaping, suspensory behaviour and "brachiation".

8.7 Bridging

Slow climbers, that is, larger *Galago* species, lorisines, leaf monkeys, siamangs, chimpanzees and especially orangutan, use *bridging* (Fig. 8.13) often. The animal anchors itself by three, at least two limbs and stretches one forelimb until a new, distant support is reached. Once this distant handhold is reached, an additional autopodium is placed on it and loaded by a growing share of body weight, carried by the first, then also the second forelimb and finally also by the hindlimbs. Fixation at the beginning of the manoeuver implies very great moments, because the body's CoM is kept distant from the support and so has a long load arm. Bending moments are demanding for the substrates and energy-consuming for the animal. In many cases, alone both hindlimbs are used to provide compressive and tensile forces as far apart as possible (Fig. 8.13). Alternatively, the new handhold can be pulled toward the ape. Its flexing brings it close enough to the active animal to change its support. Especially orangutans are perfect in bending pliant stems to the limits of their strength and using their elastic recoil as a means for transportation over considerable distances. Essential

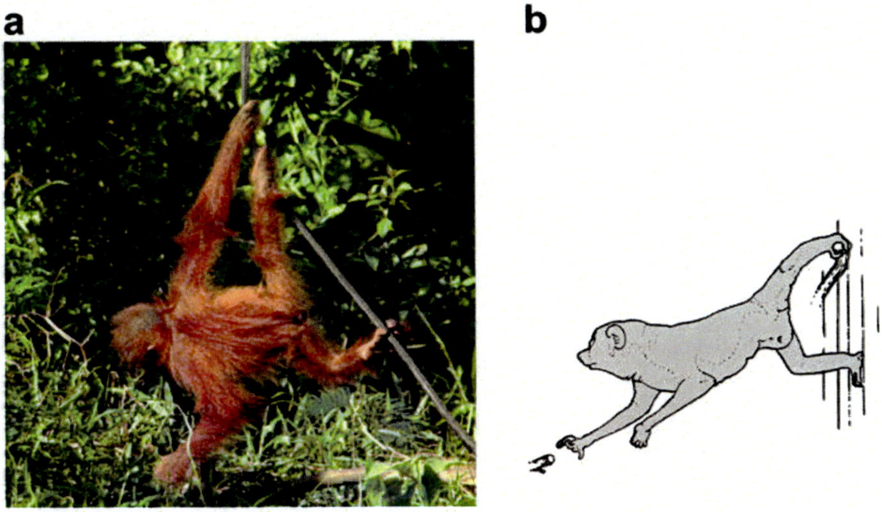

Fig. 8.13 Bridging. (**a**) *Pongo pygmaeus*. In addition to the abducted right foot, the right hand takes over tensile forces resulting from body weight. The left foot exerts compressive force against the rope, so that the CoM can be shifted to the left side. (**b**) *Loris tardigradus*. The bending moments are balanced alone by the hindlimbs. (**a**) by courtesy of S. Preuschoft; (**b**) from Schulze and Meier, 1995

for these behaviours is the possession of grasping hands and feet. These do also exist in, e.g. sloths, and therefore bridging is also a characteristic of their locomotion.

8.8 Leaping

Another means to overcome discontinuities of substrates is acyclic *leaping* (Fig. 8.14, Günther et al. 1991; Hall-Craggs, 1965, 1977). Most strepsirhines (with the exception of lorisines!), tarsiers and nearly all monkeys are fairly good jumpers. The physical laws relevant for this mode of locomotion are easier to understand in acyclic leaping than in the cyclic repetition of bipedal hopping. The latter does not play such an important role for arboreal animals – exactly because of the discontinuous substrates which are so unpredictably interrupted by narrow or broad gaps between twigs and branches. In fact the specialised leapers tend, if on flat ground, to take refuge to bipedal hopping, rather than walking. The latter gait (walking) can be seen (as noted above) in larger *Galago* species and lemurs; the smaller *Galago* species normally use, when hopping, both their hindlimbs side by side, while the larger sifakas (Propithecus) place their hindlimbs on the ground one after the other, so that one of the hindlimbs leads, the other trails. The effect of this time shift between footfalls is an elongation of ground contacts and, as a consequence, reduction of GRF and better control – like in the case of gallop if compared to bound or half-bound (see above Chap. 7.3).

The most important trait of *specialised leapers*, the *long hindlimbs*, has been explained above as means to accelerate the body during long intervals (Chap. 5.10).

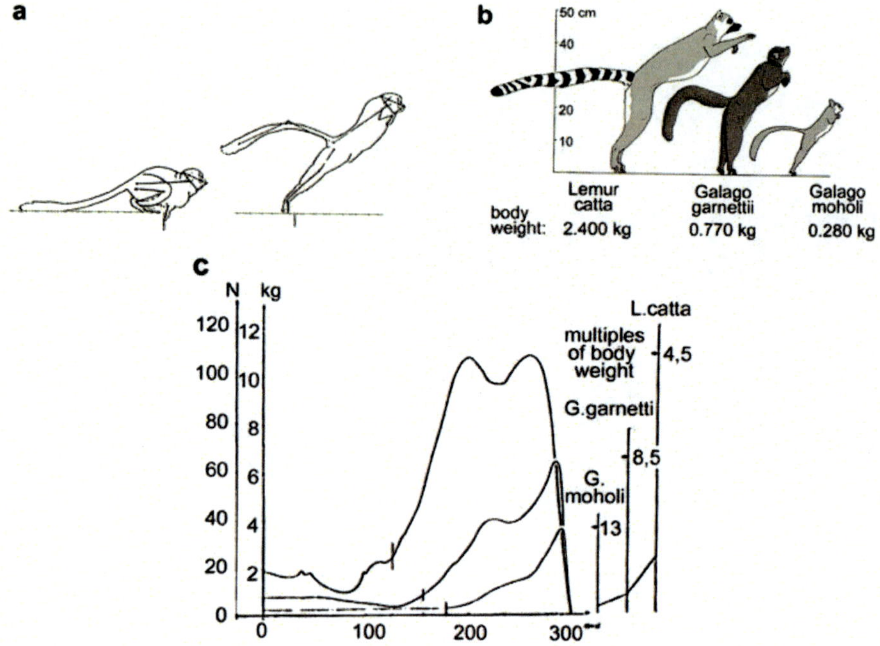

Fig. 8.14 Leaping. (**a**) Bushbaby at the beginning of takeoff and just before losing substrate contact. The initially strongly flexed hindlimbs are fully extended before the ground contact is lost. (**b**) Three strepsirrhines of different sizes and masses at the end of takeoff. All animals reached the same height of about 2 m in the experiments. Note the different postures of the arms: the heavier the leaper, the more are the arms swung forward (see also Fig. 8.20). (**c**) GRFs at takeoff in the same three strepsirrhines. Horizontal axis: time in milliseconds (ms). On the vertical axis on the left side is shown the maximal force in N and on the right side the force in multiples of body weight. The ring-tailed catta crouches before taking off, lowering its CoM, and prestretching the extensor muscles. During the entire takeoff phase, it exerts the same high force against the force plate. This is in contrast to the smaller bushbabies, especially *G. moholi*, which exert the maximal force only during the final stages, when the tarsal joints are extended. (From Günther et al. 1991)

Among primates, elongate hindlimbs are found in tarsiers and in small strepsirrhines (bushbabies) but also in medium-sized sifakas (Propithecus) and indris. Fairly long are also the hindlimbs in many lemurids, and in tamarins and marmosets, while they are less obvious in monkeys – maybe because these medium-sized or large forms also possess relatively long forelimbs.

Forelimb length provides – if moderate – advantages in walking and long reaching distance (compare above further advantages of forelimb length will be discussed below). The opposite trait, reduction of forelimb length (Galago), entails a shifting of the CoM in caudal direction and so facilitates maintenance of trunk posture against the accelerations during takeoff (see below). In landing, the frontlimbs are not recruited by the most specialised leapers, which break the impulse of the flying body by the hind- rather than forelimbs (Peters and Preuschoft 1984).

Accelerations for single leaps are produced by contraction of muscles (while in hopping a part of the invested energy can be recovered by elastic stretching and recoil as in running and in asymmetric gaits; see Sub-Chapter 7.5.6). The

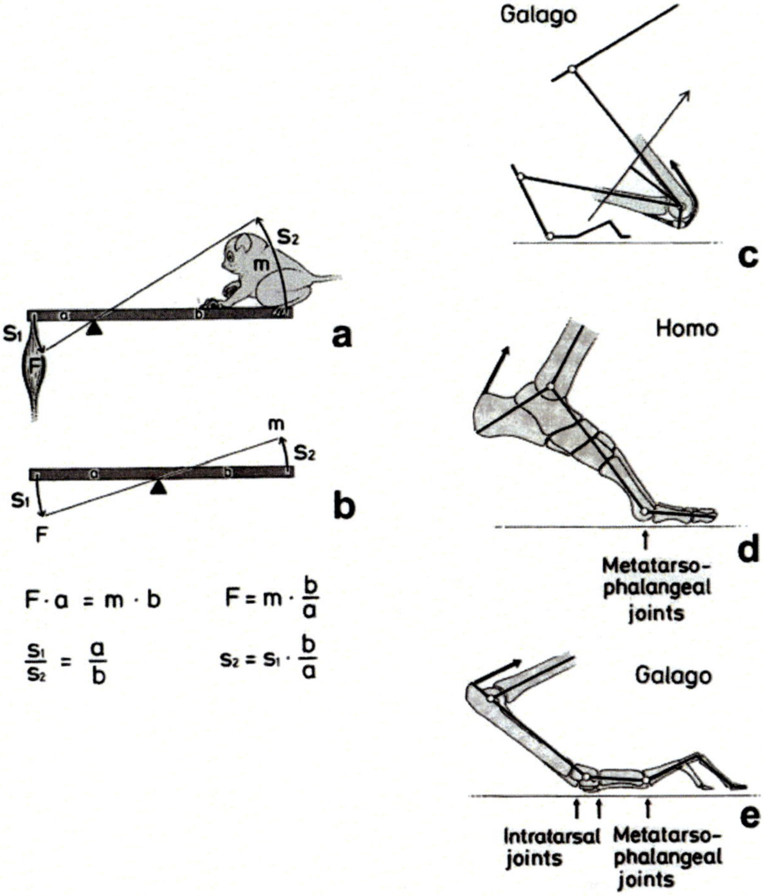

Fig. 8.15 Ratio force arm: load arm. (**a**) A long load arm (on the right side of the pivot) combined with a short force arm (on the left side of the pivot) leads, if the speed of a muscle contraction (= shortening per time unit) is constant, to higher peripheral velocity (distance s_z covered per time unit) of the moved lever. (**b**) By contrast, a short load arm in combination with a long force arm saves force (= energy) but yields less peripheral speed. This variant is most clearly realised in the heavy gorillas. (From Demes and Günther 1989). (**c**) Unfavourable ratio between power arm and load arm in the knee joint of a bushbaby, taking off from a crouched posture. Only small animals possess sufficient force to overcome these asymmetries between power arm and load arm, because of their notorious force reserves (Chap. 1). (**d**) Rigid foot in *Homo*, while the heel is lifted from the ground. (**d**) Intratarsal (or mid-tarsal) joint between the elongated collum calcanei and the long os naviculare on one side and metatarsus and phalanges on the other in *Galago*. The mid-tarsal joint permits a rolling off of the prehensile foot and is probably balanced by elastic ligaments. Compare Figs. 8.4f–l and 8.16. (From Günther 1989)

acceleration over time is proportional to the forces exerted by the animal against its substrate (Fig. 8.14). An important factor for acceleration is the *ratio between force arm and load arm*, as illustrated in Fig. 8.15a. During a rapid muscular twitch, a short force arm, e.g. the tuber calcaneum is moved over a given distance with a high

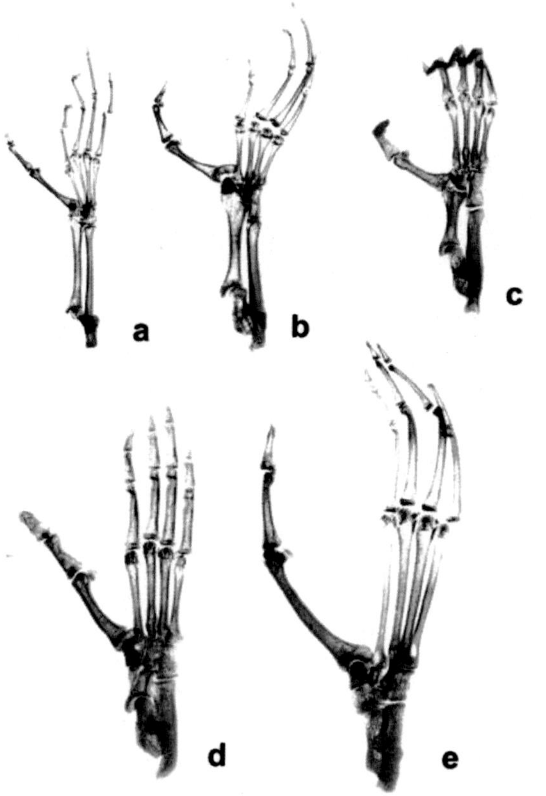

Fig. 8.16 Foot skeletons of leapers. **(a)** *Tarsius*, **(b)** *Galago*, **(c)** *Phaner*, **(d)** *Lepilemur*, **(e)** *Indri*. Note the elongation of the collum calcanei and the os naviculare, to a minor degree also the collum tali and the os cuboideum. All these elements contribute to length of the load arm, while the force arm, tuber calcanei, remains short in all of them. The parts of the foot which perform prehensile gripping are also but not the tarsals. Phalanges and metatarsals rather seem to be proportional to body size. (From Jouffroy and Lessertisseur 1979)

angular velocity (v_ω). The excursion of the long load arm, e.g. the anterior part of the foot, as shown in Fig. 8.16, is effected with the same angular but a much higher peripheral velocity (v_{peri}), because the toes are swung through a much longer distance during the same time interval.

$$\text{Peripheral velocity } (v_{peri}) = \text{angular velocity } (v_\omega) * \text{radius length } (l), \qquad (8.7)$$

where the latter (l) is the length of the foot segment distal to the ankle joint. This fact was emphasised first by A.H. Schultz (1963a, b). It is the reason why the heel bone (tuber calcanei) is short in specialised leapers (tarsiers and bushbabies, to a lesser extent also in *Phaner* and *Lepilemur*, Lemurinae and Indriidae), while the anterior part of the calcaneus (collum calcanei), the talus neck (collum tali) and especially the navicular are *elongated as the load arm* (Figs. 8.15 and 8.16). This morphological specialisation is unusual; other primates, that is, New World tamarins and marmosets (Schultz 1963a, b), and most other mammals (kangaroos, hooved animals, carnivores and rodents) have elongate metapodials. In these mammals, not the tarsus but the metatarsals undergo special morphological changes, which, however, lead to the same mechanical effect, namely, a long load arm between ankle joint and most distal ground contact.

That most primates follow an evolutionary pathway different from that in other mammals seems to be due to the typical primate characteristic, the *possession of a prehensile foot*. The "prehensile organ" or gripping forceps in a strict sense consists of the distal tarsals, metatarsals and digits. Although its constituents are different, its functional properties are similar to those in the feet of arboreal birds (see Chap. 6.6). A mobile intertarsal joint is developed between the grasping organ, that is the gripping forceps and its shaft. This gripping apparatus is maintained, probably because of their arboreal lifestyle, and shifted into a distal position by the elongation of the navicular and the calcaneus neck. This attempt to explain primate foot morphology, however, cannot be applied to marmosets. The situation in prosimians is also in some conflict with Fischer's, Blickahn's and Witte's generalisation of an additional hindlimb segment.

Exactly the opposite specialisation is realised in the *feet of the gorilla*, best visible in comparison to the chimpanzee: the tuber calcanei is long, and the part of the foot anterior to the ankle joint (the load arm) is short (see below, Figs. 8.15b, 8.37). The heavy gorillas do not jump (in contrast to the less heavy chimps) and need less force of the m. triceps in all other gaits.

A disadvantage of the elongated foot segments is derived from their length-related weight (Fig. 8.17) and the required mass of musculature. Producing the great forces to overcome the enormous rotational moments of the distal foot requires strong muscles. These muscle forces are provided by the digital flexors (mm. flexores digitorum longus et brevis, which at the same time secure the foothold) and by the mm. tibialis posterior and triceps. These powerful muscles (and perhaps elastic ligaments on the plantar side of the intertarsal joint) accelerate

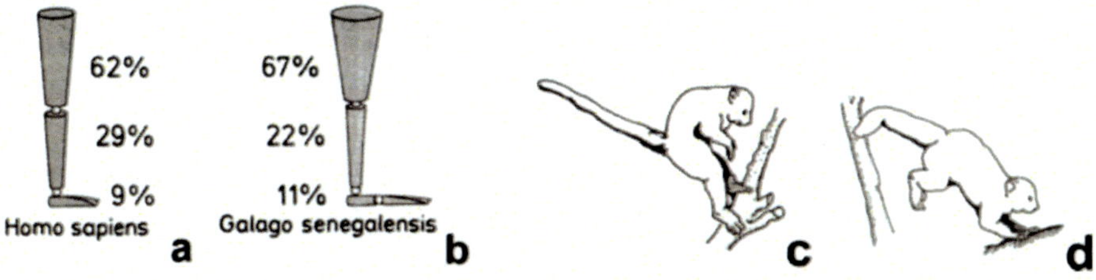

Fig. 8.17 Distribution of mass on the hindlimb, modes of landing. (**a**) Human, as example of a walker and (**b**) the specialised leaper *Galago senegalensis*. Proportions are similar, but not identical! (From Günther et al. 1992). (**c**) Landing on hindlimbs after a broad jump, (**d**) landing on frontlimbs after a short jump. Note the tail in both examples. *Hapalemur griseus*. (Redrawn from Peters and Preuschoft 1984)

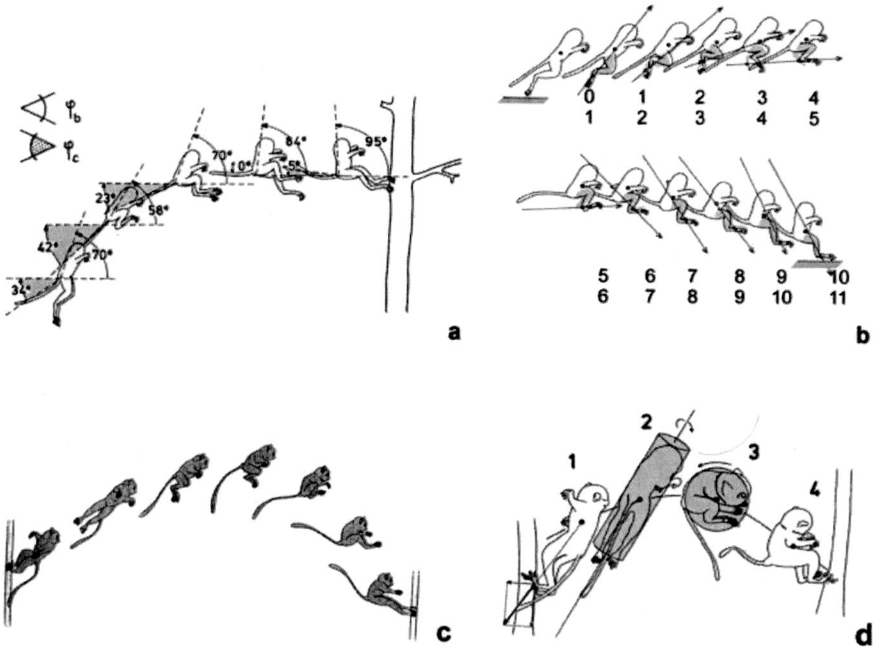

Fig. 8.18 Leaping tarsiers. (**a**) Jumping upward from ground to a stem; (**b**) hopping, on top the first six and below the following six stages (From Niemitz 1984); (**c**) leaping backward from one stem to another. The CoMs of the hindlimb segments are always kept on or close to the trajectory of the body's CoM. (**d**) Rotation about two axes and the body shapes which facilitate these rotations. In all drawings, the horizontal distance between stages does not represent the covered distance. The hindlimbs are swung forward during aerial suspension, and that means without the possibility to exert forces on the airborne body (From Peters and Preuschoft 1984). In all cases, the flexion of the tail initiates rotation about a horizontal (z-)axis. The landing is prepared by adducting the limbs and flexing the trunk ventrally, so that a ball shape is approached in stage 3. It offers but small mass moments of inertia, and therefore the dorsiflexion of the tail rotates the body about a z-axis to swing the hindlimbs forward. Extension again is made in such a way that segment-CoMs are moved on the trajectory of the whole body's CoM. The impulse for longitudinal rotation in leaping backward is effected by the hindlimbs during takeoff. The +/– cylindrical outline in stage 2 facilitates rotation about the x-axis, because the mass moment of inertia is low. Rotations can be slowed down by spreading the extremities and so increasing the mass moments of inertia

the body during the last stages of takeoff (Günther 1985; "sauteurs du pied", Figs. 8.14 and 8.16). The muscles make, however, the zeugopodia and feet heavy. This is not favourable for swinging the hindlimbs forward during free flight (as shown in Fig. 8.18). As a consequence, the larger leapers (*Propithecus* and *Indris*) do not have such an elongate foot (Fig. 8.16) but accelerate by extending their hip and knee joints ("sauteurs de cuisse").

Characteristic differences in the *kinematics of leaping* have been worked out by Günther (1985; Günther et al. 1992) on the basis of high-speed video recordings and force plates. While three strepsirrhines (*Galago moholi*, 0.250 kg; *Galago garnetti*,

0.770 kg; and *Lemur catta*, 2.400 kg) took off from squatting postures, macaques (*M. fuscata*, 3.600 kg) and humans (experimental subject: 82 kg) preferred the takeoff with a (very short) approach run that provided some acceleration in the horizontal before giving up ground contact. Small and large forms reached in the experiments the same heights. Small strepsirrhines take off from a deeply crouched posture, so that the full length of the hindlimb is used (Figs. 8.14a and 8.18d). Larger indriids (Demes and Günther, 1989), macaques and other monkeys flex their hindlimbs only to a limited extent. They avoid the stage, in which the knees are far in front, the hip joints far behind of the limb's functional axis, so giving the external GRF the maximal rotating moments (Fig. 8.15c). Because of their size-dependent smaller ratio of muscle force/body mass (see above, Chap. 1.4, Fig. 1.6), the gain of velocity would at least be reduced (if not interrupted) during this stage. Interestingly, the actual acceleration distance in the larger species is not shorter than in small leapers (about 15 cm in all investigated primates). GRF increases slowly in small *Galago moholi*, reaching its maximum shortly before liftoff when ground contact is lost (Fig. 8.14c). The same holds true for the body's speed. In the larger *Lemur catta*, the GRF increases more rapidly and remains on the maximal level until toe-off, so that a maximum impulse (p) is exerted. The velocity of the animal increases continuously throughout the whole takeoff phase. The medium-sized *G. garnetti* is in both parameters intermediate. After recalculation of GRF from Newton into multiples of body weight (Fig. 8.14c) (scales on right vertical axis), *G. moholi* produces 13 times body weight at takeoff, *L. catta* only 4,5 times body weight and *G. garnetti* an intermediate value of 8,5 times body weight.

In the x-direction of movement, the *acceleration of the body* is transmitted from the hindlimbs to the trunk in the hip joints, and the resultant GRF makes the trunk, head and forelimbs rotate downward, unless the dorsal epaxial musculature counteracts (Fig. 8.18, and see above Chaps. 5.10 and 7.4). Preuschoft and Fritz (1977) have calculated for a leaping tarsier that an increase of pressure within the abdomen (Fig. 8.19; by the muscles of the body wall and the pelvic flour as well as the diaphragm) by not more than 10% counterbalances all ventrally bending moments. The small compressive force acting in the abdominal cavity has a long lever arm at the intervertebral joints and therefore leads to a complete relaxation of the dorsal muscles. According to everyday observation (jumping horses, weightlifters), the contraction of the hypaxial muscles is a common means to relieve the epaxial musculature from stabilising the trunk. A reduction of frontlimb mass and length also has the consequence that the vertebral column is relieved, but is possible only, if the anterior limbs are not required for landing. Peters and Preuschoft (1984) have observed that less good leapers among the strepsirrhines (*Lemur*, *Hapalemur*) after short leaps put the frontlimbs down first (Fig. 8.17c, d) and so produce the impulse necessary to stop forward rotation and to move the hindlimbs forward.

Quite typical for the best leapers is the *landing* on the (long!) hindlimbs, to brake the flying mass slowly, with but moderate expenditure of energy. This permits making use of the long limb springs not alone at takeoff but also at landing. For doing so, however, bipedal leapers as well as hoppers must solve the problem how to bring their long and heavy hindlimbs from their initial position trailing behind the

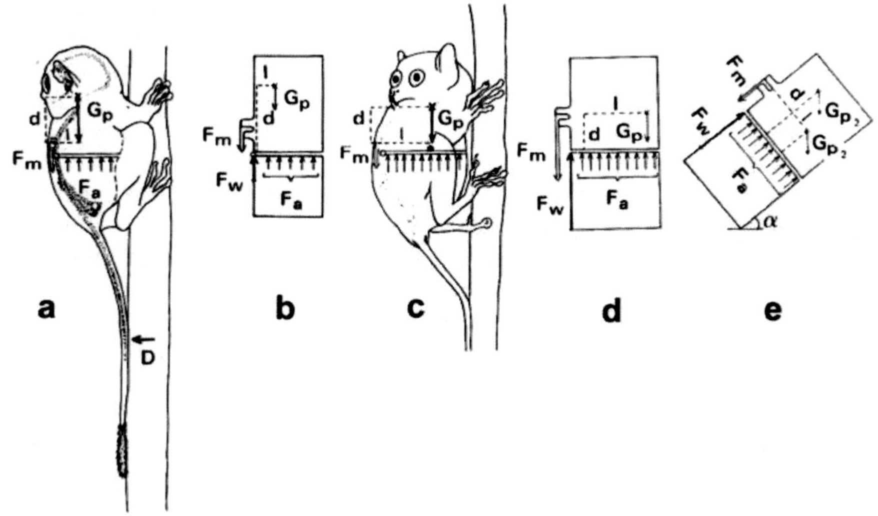

Fig. 8.19 Effect of intra-abdominal pressure according to (Preuschoft et al. 19179). *Tarsius* clinging to a stem in typical position which can be maintained for hours (**a** and **c**). The rearward rotating moment of body weight (see Fig. 8.10) can be countered in two ways: by the hands or the tail. For the sake of simplicity, it is assumed that the arms do not exert a tensile force, and the animal carries its body weight alone by its feet while balancing the backward rotating torque by pressing its tail against the stem. (**b**, **d** and **e**) The weight of the head, neck, arms and cranial half of the trunk (G_p) must be balanced at the intervertebral joints – of which only one is shown. During takeoff, the inertia of the masses cranial to the joint is multiplied by the observed acceleration, so that G_p reaches values of about ½ of total body weight times 13. The comparison of (**a**) and (**c**) emphasises the influence of the exact position of the head on the position of the CoM. While weight acts always vertically (**a–d**), the inertial forces depend upon the direction of the leap (**e**). Inertia as well as weight can be counterbalanced alone by the pull of the dorsal muscles (F_m). In such a case, all mentioned forces compress the vertebrae with the force F_w. Alternatively, a contraction of the ventral abdominal muscles leads to a compression of the intestines. This compression of the intestines exerts a force (F_a) which is opposed to the moments flexing the trunk. Although F_a is distributed over a large area, it can be simplified as one force in the middle of the cross section. This concentrated force has a long lever arm. If intra-abdominal pressure (F_a) is raised by 10%, the resulting moment is large enough to counterbalance the compressive force between the vertebrae completely; they are instead pulled apart

trunk (where they have been used for acceleration) into a position in front of the trunk, in order to break the impulse of the flying body (Fig. 8.18) or, when hopping, to prepare the next hop. All this must take place during free flight, while no external forces act on the body.

From a physical point of view, two traits are helpful: 1st minimising mass moments of inertia, which resist being rotated, and 2nd moving another part of the body, for example, the arms or the tail, into the opposite direction. Regarding point 1, Peters and Preuschoft (1984) have shown how tarsiers flex the hindlimbs during the flight phase so that the CoMs of all segments are adducted along the flight trajectory of the body's CoM. So flexion of the hindlimbs is performed during flight without evoking rotational moments (Fig. 8.18b, c). After flexion of the hindlimbs, the body stem is flexed as well, so that the mass moments of inertia are minimised and do not resist strongly a rotation of the hindlimbs forward, the more so since extension takes place again with the segmental CoMs on the trajectory of flight. The distal segments of the leapers consist, like those of birds and cursorial mammals (see above), of nothing than bones for sustaining compression, tendons and ligaments to control the joints and skin to cover the deeper structures. The heavy muscles, which determine the positions of the joints, are shifted proximally and concentrated on the thigh or the proximal half of the lower leg. The physical conditions in walking, running, hopping and leaping are nearly the same: minimising mass moments of inertia in spite of maximal limb lengths (see also Fig. 8.17a, b). Therefore it is quite difficult to draw conclusions about one or the other locomotor modes alone on the basis of morphology (see above).

The head downward rotation is stopped, and the head upward rotation is initiated in small strepsirrhines and in *Tarsius* by a dorsiflexion of the long tail (Figs. 8.17 and 8.18b). A support by air resistance applied to the tail is not required in tarsiers (Peters and Preuschoft 1984), but it is possible. In fact the tails of bushbabies, *Ptilocercus* and, among rodents, squirrels and jerboas are equipped with long, often stiff hairs. The not hairy but heavy tails of kangaroos are also swung upward during the airborne phase of hopping. In primates of greater mass (sifakas (Demes et al. 1991, 1992), macaques, apes, humans (according to Demes and Günther 1989; Preuschoft 1990b), also vervets), a tail cannot provide a sufficient rotating moment – unless it was 4 m long or quite heavy. Larger primates therefore use retraction of their arms to rotate their hindlimbs forward (Fig. 8.20d).

The anteverted position of the arms is attained in the initial phase of takeoff, where their arms are swung forcefully forward-upward (Fig. 8.20a–c). The emphasis put on this foreswing increases with body mass (compare Fig. 8.20a, b). It leads to the effect that a part of body mass (e.g. humans, 9%; chimpanzee, 20%) has a considerable velocity before extension of the hindlimbs takes place (Fig. 8.20c), so that the latter either must exert less force or the jump carries farther than without the pre-acceleration of a part of body mass. The anteverted frontlimbs stabilise the position of the body while airborne (because of their mass inertia), especially when not kept exactly in line with the trunk axis (Dunbar 1994). Once anteverted, the forelimbs can be swung ventrally downward in order to rotate the hindlimbs forward (Abb 8.20d; from Demes and Günther 1989).

A tricky manoeuvre is the *rearward takeoff* from clinging to a stem. This is most common in *Tarsius*, since these animals spend more time clinging to vertical supports than other species do (Fig. 8.18c, d; Peters and Preuschoft 1984). The behaviour is not restricted to tarsiers; it can also be seen regularly in *Propithecus* and sometimes in *Galago*. Dunbar (1994) has documented and analysed a rearward-directed takeoff in *Macaca*. In all cases, the impulse for rotation about the longitudinal (x-) axis can be created during takeoff. To facilitate rotation, the animal's body assumes a cylindric form, in which mass moments of inertia are minimal (Fig. 8.18d). For landing, the animal orientates its ventral side toward its aim, to make use of its limbs for braking. Therefore the rotation must be stopped, while the

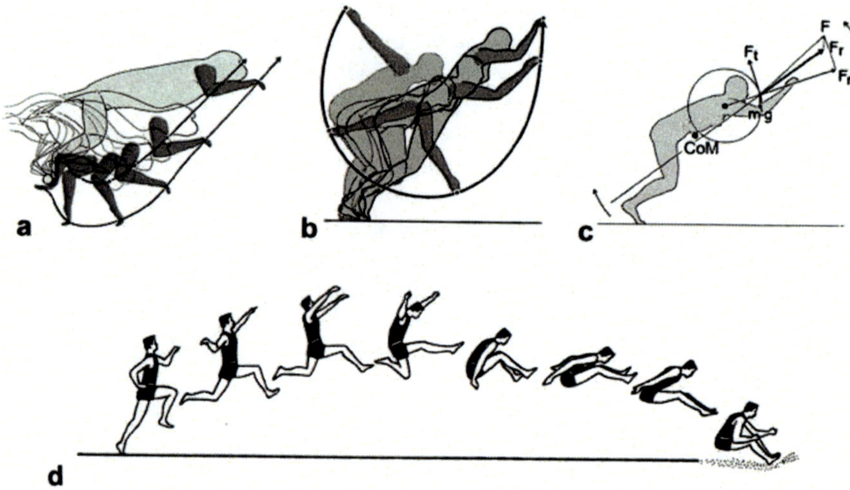

Fig. 8.20 Movement of the arms at takeoff. (**a**) Vervet monkey, (**b** and **c**) humans. The trajectories of the elbow and hand are indicated in (**a**, **b** and **c**). While the arms begin their swing forward, the hindlimbs are flexed; this "crouching" movement prestretches the extensor muscles and prolongs the acceleration distance. (**c**) Lifting the arms at the beginning of takeoff pre-accelerates a part of body mass and so facilitates acceleration of the CoM by exertion of the force F_r. If the arms are heavy, as in chimpanzees, this effect is particularly important. (**d**) Takeoff and landing of a broad jumper. In the first half, especially during takeoff, the arms are elevated and help in accelerating the body mass; in the second half, the arms are moved forward-downward and so help to rotate the hindlimbs forward into the landing position by producing the required counter-impulse. (From Demes and Günther, 1989, and Demes, 1991). See also Fig. 8.14b

animal is airborne, and no external forces can act on it. A swing of the tail in a direction opposite to rotation (Fig. 8.18c, d) is a means to slow down or stop the rotation. All these movements are superimposed on what happens in side view (see above and Fig. 8.18a, b).

In contrast to small leapers (*Tarsius* or *G. moholi*, see above), monkeys and indriids spread out their extremities in x- and/or z-directions (Fig. 8.21b) while airborne in wide, mostly downward directed jumps. This position increases the mass moment of inertia of the body (*J*), because

$$J = m_{\mathrm{i}} * l_{\mathrm{i}}^2 \tag{8.8}$$

and so stabilises its position in space without input of energy (Dunbar 1994).

Several strepsirrhines (*Indris*, *Propithecus*, *Avahi*, *Galago*) possess a loose skinfold between the upper arm and trunk. This fact has been taken by Demes et al. (1991) as the reason to investigate the possibility that such a fold can be used in the sense of a parachute for gliding (Fig. 8.21a). The velocities of takeoff are clearly greater in the heavier forms than in the light-weight bushbabies (10–8 m/s versus 7–5 m/s). The authors have calculated the aerodynamic forces acting on the ventral

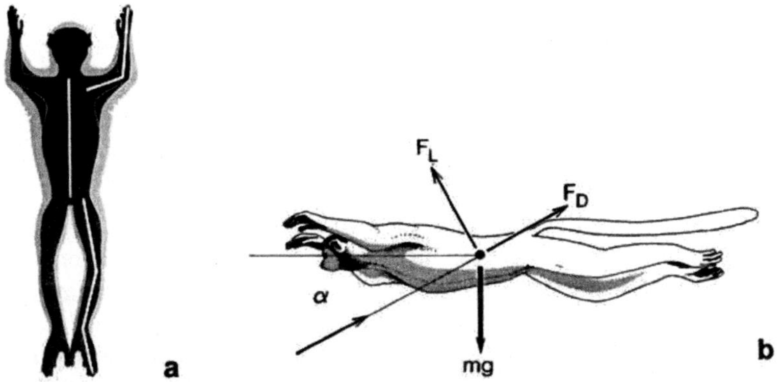

Fig. 8.21 Alternative behaviour while airborne. A relatively heavy leaper, the sifaka extends its extremities during free flight. This posture increases its mass moment of inertia and so stabilises its position in space. At the same time, it exposes an increased area to air resistance. (**a**) Schematic illustration of the skinfolds in *Propithecus* running from upper arm to trunk (black, limited by white line) with fur (light grey). (**b**) Its speed forward in x-direction is provided during takeoff; m_g is the weight force, which accelerates the body downward. The resulting air resistance therefore acts in the direction of arrow F_D, the angle α being the "sliding angle" between the horizontal and movement in space. F_D has a decelerating "drag" component in x-direction and a carrying "lift" component F_L perpendicular to it. (From Demes et al. 1991)

surface of the animals with and without fur. The ratio area/body mass is greater in the small galagos and smaller in the larger indriids. In the small bushbabies, the surfaces exposed to air resistance are large enough to give the animal an advantage of 5% leaping distance. In the larger sifakas, the high velocity exerts a major influence, because the square of velocity is part of the formulae which describe the effects of drag (D) and lift (L):

$$K_L = c_L * A * (\rho/2) * v_o^2/m * G \text{ and } K_D = c_D * A * (\rho/2) * v_o^2/m * G \quad (8.9)$$

$$c_L = K_l * m * G/(\rho/2) * A * v_o^2, \quad (8.10)$$

and

$$c_D = K_D * m * G/(\rho/2) * A * v_o^2 \quad (8.11)$$

where K are dimensionless numbers that contain all variables which influence the aerodynamics of the flying body; A is the area exposed to streaming air; ρ is the density of air; c is the coefficients of either lift or drag (which have been derived from the result!); v_o is the velocity at takeoff; m is mass; and G is earth acceleration.

8.9 Suspensory Behaviour, Arm Swinging and "Brachiation"

For relatively large animals, access to nutritious parts of the tree in the outer foliage is possible (see also above, Chap. 8.1), if the animal suspends itself below twigs or branches (Fig. 8.22). This is the simple reason, why "*suspensory behaviour*" has been observed in many primate species (Fig. 8.23). The technique is optimised in lorisinae, which often and during long periods are found below their substrates (Jouffroy and Petter 1990; Ishida et al. 1990). The essential property for suspension is the ability to transmit tensile forces from the substrate to the animal. This is accomplished in primates by prehensile hands and feet and – if present – a prehensile tail.

In suspended postures, the direction of the major external (substrate reaction) force acting on the extremity is opposite to that prevailing in walking. Limbs are under tension, rather than compression. In the joints between autopodia, zeugopodia, stylopodia and the trunk, the extending, not the flexing, moments of body weight must be balanced, of course by other muscles. The activity of these muscles has been

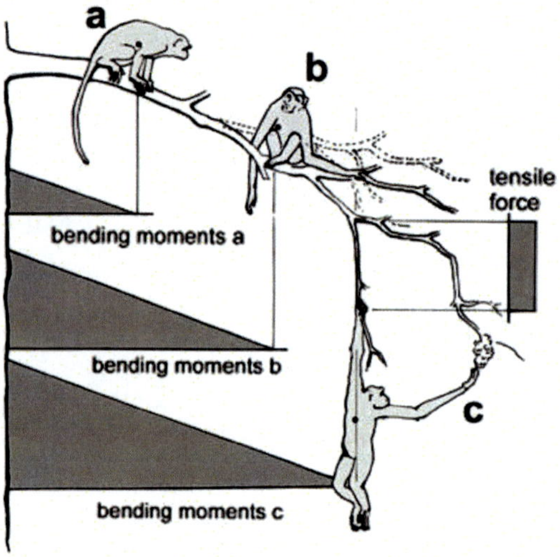

Fig. 8.22 Access to the periphery of a tree depends on the strength of branches and twigs. Bending moments and tensile stresses to which the branch is exposed by the weight of the primate are grey. Thick branches close to the stem are accessible by walking (**a**), even if the branches are inclined upward or downward by the weight of the (in **b** sitting) primate. The original position of the branch is stippled. The major problem of this primate is keeping its CoM perpendicularly above the narrow support. Transmission of torsional stress, like in Fig. 8.6, is very helpful but costs energy. Thin twigs offer not much bending, but considerable tensile strength. Therefore, even the thinnest twigs can be reached, if the primate is able to suspend itself from the twigs (**c**). For this purpose, prehensile organs (hands, feet, tails) are useful, in spite of their energy requirements, since maintenance of external equilibrium is no longer a problem. (From Preuschoft et al. 2016)

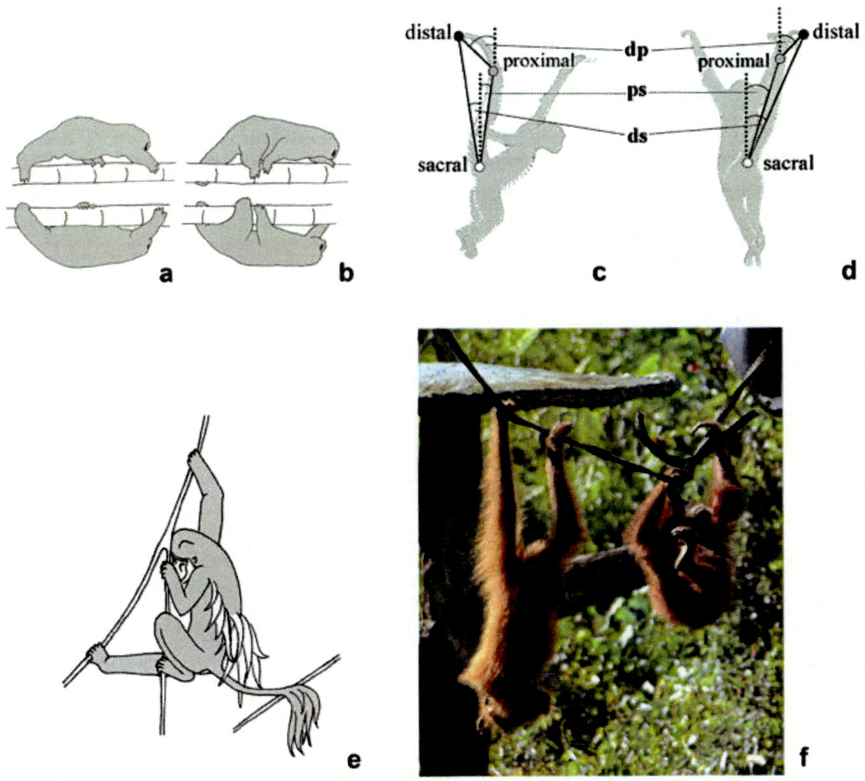

Fig. 8.23 Examples of suspended postures and locomotion. (**a**, **b**) Slow loris, *Nycticebus coucang*. Changing its position above/below the substrate is possible without effort, like in all lorsinae (From Jouffroy and Petter 1990). (**c**, **d**) *Ateles* suspended on a stretched rope in side view (**c**) and from behind (**d**). The seemingly regular change from one hand to the other becomes secure by the intermittent action of the tail, but an energy-saving, rhythmic pendulous swinging like in gibbons is not realised. Anyway, the movements of the tail are optimised to least energy requirement (From F. Martin 2003). (**e**) Black and white guereza, *Colobus*, during feeding in suspension (From M. Rose 197). (**f**) Orangutan youngsters from Borneo (Courtesy S. Preuschoft). Note the "hooks" formed by hands and feet

checked in the slow loris, *Nycticebus*, by EMG (Jouffroy and Stern 1990). Footfall sequence usually is the same as in (diagonal) walking.[7]

Several New World monkeys frequently move suspended by their arms (*Ateles* 45%, *Lagothrix* 23–38% of their time, Martin 2003). This behaviour is often supplemented by a *prehensile tail* (Fig. 8.23, see also above). The most efficient

[7]In the example of the likewise suspended sloths, another method of anchoring body weight with the aid of claws as hooks is highly developed. In addition, Nyakatura and Andrada (2012) have investigated the question whether its locomotion can also be interpreted as a combination of pendulous swinging of all four limbs. The answer of the authors was NO.

tails have a naked, hairless field on the ventral sides near the tail tip, covered by dermatoglyphic skin, with ridges mainly in transverse (z-) direction. The powerful and highly differentiated musculature allows precisely controlled movements in the intervertebral joints. The vertebrae short and equipped with marked apophyses, especially in the most proximal and the most distal third of the tail (Ankel 1962). Both traits permit precise movements. Suspension alone by the tail can be seen often in spider monkeys as well as in woolly monkeys (*Ateles* and *Lagothrix*). The third genus equipped with a prehensile tail, the howler monkeys (*Alouatta*), brachiates rarely (Youlatos and Gasc, 1994). Instead its most preferred locomotor mode is, similar to most other South American monkeys (Cebinae) or cercopithecines, walking on top of branches (Youlatos and Gasc 1994), while the tail acts as an assurance of the position in the tree. Less precise and less efficient are the movements of a "rolling tail", which can be seen in several capuchin monkeys (*Cebus spec.*). No hairless field is developed, musculature is less strong, and the individual vertebrae are longer. Therefore, the tail is less compliant than a prehensile tail in a strict sense. Nevertheless, it is rolled around an appropriate substrate and so helps to secure the external equilibrium in a challenging environment. Similar, though less expressed, behaviour can been seen occasionally in Old World monkeys.

As mentioned above, many New World and Old World monkeys possess long arms, and among all simians, *forelimb length scales positively* with body mass (Jungers 1984, 1985). Most obvious and often documented is the length of the arms in apes (Table 8.3); among the large apes, it is further emphasised by the shortness of hindlimbs. In the lesser apes, arm length is exceptional, but the hindlimbs are also quite long (Jungers 1984; Tuttle 1972; see also muscular anatomy as emphasised by Payne et al. 2006). Overestimating of the actually realised mode of locomotion and connecting them with postulated, not observed behaviour has lead early (Gregory 1916) to classify the locomotor mode of apes as *brachiation*. This attribution has since persisted since a long time and directed many studies astray on misleading pathways. In fact, bimanual suspension and locomotion, that is, brachiation in a strict sense, is not the outstanding capability of the large apes. According to a study of Hollihn (1984), even the smallest of them, the bonobos hardly stand more than 8 pendulous swings in sequence. This shortcoming can be confirmed on the basis of mechanics (see below).

In fact, arboreal, rather heavy animals have more than one good reason to develop long arms. Especially in collecting food, a long *reaching distance* is an advantage. As mentioned above, all higher primates tend to seizing objects by their hands to approach them to the sense organs at the head for close inspection or for being ingested. This makes long jaws unnecessary, and because lever arms of the mandibular adductors are short in relation to the load arms, either the bite forces can be higher than in long-snouted forms or the same bite forces can be produced by less developed and therefore less heavy muscles (see also below in Chap. 9). Figure 8.24 shows that everything what the primate can reach from its position is within a hemisphere, the radius of which is total arm length, the centre being the thoracoscapular junction. This is reminiscent to the conditions described for long-necked sauropods in Chap. 4 and for long-necked herbivorous mammals in Chap. 7. Long

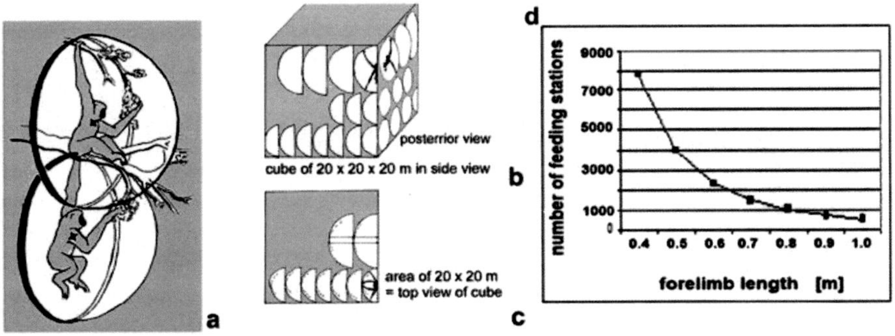

Fig. 8.24 Advantages of long arms in feeding. (**a**) Reaching distance without change of location ("feeding stations", similar to Grand 1972). (**b**) For harvesting or controlling a given volume of vegetation, long arms with long reaching distance require fewer changes of the feeding station than shorter arms would do. If projected into a plane with only two dimensions (**c**), it becomes visible that shorter arms would need a greater number of the potentially dangerous changes; (**d**) calculated number of necessary changes of body position as a function of arm length. (From Preuschoft et al. 2016)

arms save the energy necessary for frequent changes of the position of the primate in space (=feeding stations) and increase safety, because the animal can stay for longer time on a substrate which does not fail under its weight. This point gains importance, when the frequency of falling down from a tree and subsequent lesions are taken into consideration (see above, Chap. 8.2).

The most obvious use made of long arms is for locomotion. Well-known and clearly understood is the relationship between the enormously long arms and *arm swinging in hylobatids* (Fig. 8.25; Jungers and Stern 1981, 1984; Preuschoft and Demes 1984, 1985). Gibbons grasp hold of a "superstratum", commonly by using the hand as a hook, transmitting tensile forces (Fig. 8.5a). The joints of the arm (carpal, elbow, shoulder) are more or less extended. The active muscles have been identified by Kummer (1970), who also emphasised that only moderate muscle force is required. Jungers and Stern (1981, 1984) and Larson and Stern (2009) have documented in a series of EMG studies the activity of muscles experimentally. There were no contradictions between theory and experiment but differences between *Hylobates* and *Ateles*, who is equipped with a prehensile tail. The characteristic locomotion of hylobatids consists of sequences of pendulous swings, where the head, trunk, hindlimbs and free arm are the "mass" of the pendulum, the arm on which the animal is suspended, the "chord" and the grasping hand the pivot (Preuschoft and Demes 1984). The alternating contacts of the hands may overlap in time (duty factor more than 50% of cycle duration, like in the walk of other mammals (Fig. 8.25a)) or may be separated by a phase of aerial suspension (Fig. 8.25a, two cycles on left, like in the trot of tetrapods or in "running" of birds).

As is known, the swinging of a *pendulum* does not consume much energy, because the mass of the pendulum conserves potential energy, which is transformed

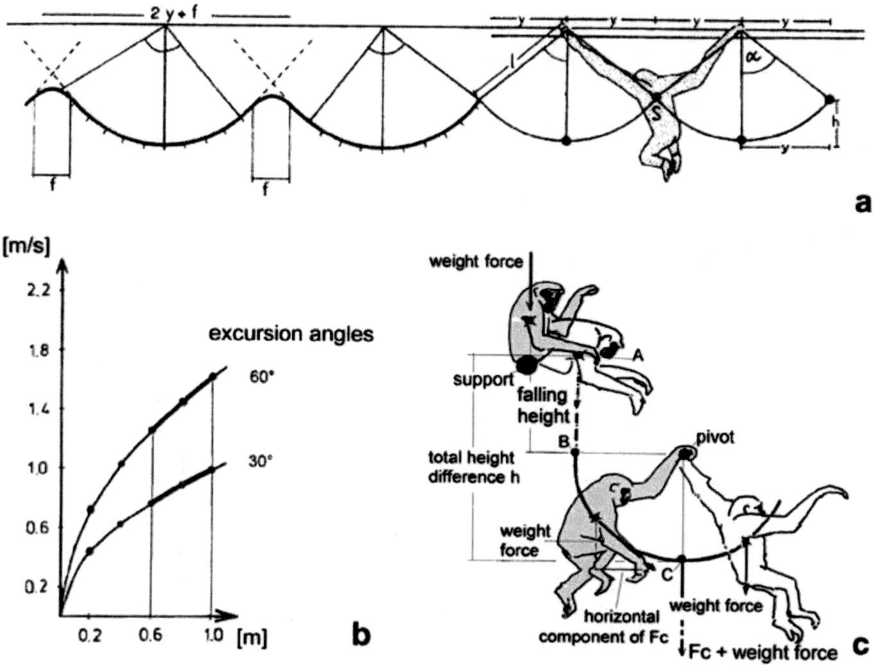

Fig. 8.25 Arm-swinging locomotion, interpreted as a pendulum. (**a**) Two cycles on right: Gibbon in continuous contact with substrate. Cycle length here is called *y*, lift is *h*. (**a**), two cycles on left: If arm swinging includes a phase of free flight ("ricocheting"), the CoM (here "*S*") follows the trajectory indicated by the fat line. The airborne phase is a means to cut off those stages in which least progression takes place and lift is greatest; it thereby increases cycle length ($y + f$). (**b**) Speed of arm swinging depends upon arm length and excursion angle. The realistic angles of 30° and 60° are shown. (**c**) Rapid escape of a gibbon sitting on top of a branch. The longer its arms are, the greater will be the total height difference *h* and therefore the kinetic energy contained in its body. This energy is transformed into horizontal speed. See text below. (From Preuschoft et al. 2016)

into kinetic energy during the downswing, and due to the inertia of mass, the upswing leads again to a gain of potential energy (Fig. 8.25a). Only as compensation for losses by "internal friction", energy must be put into a pendulous system. Precondition for the interpretation of a system as "mathematical pendulum" is a mass great enough in relation to the mass of the chord. To check this point, Preuschoft and Demes (1984) have proposed the following argument: the laws of the mathematical pendulum predict that half of the pendulous swing ($T/2$) needs about 1 second, if the length of the chord is the same as in a hylobatid arm. In fact, 0.8–1 seconds have been observed in relaxedly moving gibbons. By contrast, in a "physical pendulum", in which the mass of the arm is considered, the much more complex calculation yields a pendulous period T about 5% longer. In view of this small difference, the simpler approach was preferred. Later studies (e.g. d`Aout and Vereeke, 2011; Bertram and Chang 2001 – see below) have controlled this view, and added much precision, but not changed the overall result.

The *speed of locomotion* in an arm-swinging gibbon depends from the peripheral velocity of the pendulum (Fig. 8.25a). This of course becomes greater with increasing length of the chord. From this, the characteristic morphological traits of a hylobatid arm can be derived: length and limited weight. Considerable muscle force is required for keeping the fingers flexed to the hook and for making sure that any upswing brings the body back to the same height as at the beginning of the downswing, just for *compensating the losses* due to internal friction. The lifting of the CoM can be accomplished in several ways. Preuschoft and Demes (1984) have found that the forceful whipping down of the free arm (Jungers and Stern, 1984) leads to a shifting of the CoM by 5–10 cm away from the pivot at the deepest point of its trajectory. During the following upswing, the free arm is moved upward and forward. Since the arm mass in a gibbon is 10–12% of total body weight, the CoM wanders back toward the pivot: the pendulum chord becomes shorter and its peripheral velocity greater.

Bertram and Chang (2001) have modelled the gibbon (in both gaits) as a combination of *two cylindrical bodies* or segments: light-weight suspending arm and the heavier complex segment consisting of free arm, both hindlimbs, trunk and head. Pivot between the segments is the shoulder joint. During "continuous contact", the heavy segment swings in the same frequency as the arm. In "ricocheting", however, the heavy "trunk" segment lags behind the arm movement and reaches its deepest point in the middle of the swing. During the upswing, the swing of the heavy segment overtakes the long arm and so shortens the chord of the whole system. Therefore, the CoM of the whole system is farthest away from the common pivot in the middle of the swing, so that least energy is required.

The physical conditions of long forelimbs are used by gibbons also for braking after jumps or falling downward and for a quick start (Fig. 8.25c). Imagine a gibbon sitting on top of a tree, when an eagle attacks: it will jump or let itself fall downward and escape into the dense foliage. Its potential energy is

$$E_{pot} = m * G * h \tag{8.12}$$

If this can be transformed into kinetic energy

$$E_{kin} = ½\, m * v^2, \tag{8.13}$$

$$\text{both are equal}: ½\, m * v^2 = m * G * h \tag{8.14}$$

This formula can be dissolved

$$v_{max}{}^2 = 2 * G * h, \ \text{or} \ v_{max} = \sqrt{(2 * G * h)} \tag{8.15}$$

The greater the total falling height ($h + l$), the faster the locomotion during escape. The change of direction from vertical to horizontal requires the input of centripetal force (F_c). This can be the smaller, and acting for a longer time interval, if the radius (l) of curvature is longer.

$$F_c = m * v_{max}^2/l \tag{8.16}$$

This can also be written in the form

$$F_c = 2 * m * G * l/l = 2 * m * G \tag{8.17}$$
$$\text{or } F_c = 2 * F_w,$$

whereby the weight force (F_w) must be added. Because

$$v_{max}^2 = 2 * G * (h + l) \text{ or } v_{max} = \sqrt{2} * G * (h + l), \tag{8.18}$$

the centripetal force F_c can also be written as a function of maximal speed:

$$F_c = 2 * F_w * (1 + h/l) \tag{8.19}$$

Great falling height and a slow redirection are possible if the animal possesses long arms which make the radius (l) of redirection long – clear advantages for long limb length.

The critical point of the whole system, where the maximal force comes into action, is the pivot of the *supporting hook hand*. The centripetal force F_c in Fig. 8.25c) can easily reach three or five times body weight (Preuschoft et al. 2016). This requires considerable force to be provided by the digital flexors (see above, hook hand), which are not alone strong but also heavy. On the other side, the mass of the arm is limited, because a thin chord (= low muscular mass) is the precondition for realising the favourable properties of a mathematical pendulum. The force provided by the muscles cannot be increased without losing the properties of a mathematical pendulum. This is one simple reason why the body size of arm-swinging brachiators is limited – which is confirmed by Hollihn's (1984) observations.

Another *limitation for length* increase of the arms is posed by the rotating moments in the joints (Figs. 8.26a and 8.27a). During movements, more than in static situations, the mass moments of inertia grow by the square of length. As a consequence of further elongation of the segments, and of the relative weight of the arm, a more than proportional increase of muscle mass is required for balancing the joints. Exactly this increase of muscle force is again connected with weight increase, which must be avoided. As can be seen from Fig. 8.26b, the available muscle forces (FM) are larger than the forces required at maximum speed (=excursion angle 90°).

Finally, long arms influence the external equilibrium of a primate, in sitting on a branch as well as in suspension from a twig (Fig. 8.27). Reaching out for an aim (food) produces a rotating moment in the sitting gibbon (Fig. 8.24a or Fig. 8.27a) which depends upon arm weight (F_a) and arm length (h_a). To remain seated, the animal relies on the inversely rotating moment of the weight of the reduced body F_{wr} multiplied by the lever arm h_{wr}. The contact closest to the extended arm acts as pivot in this situation, namely, the left foot. The moment $M_a = F_a$ times h_a cannot become

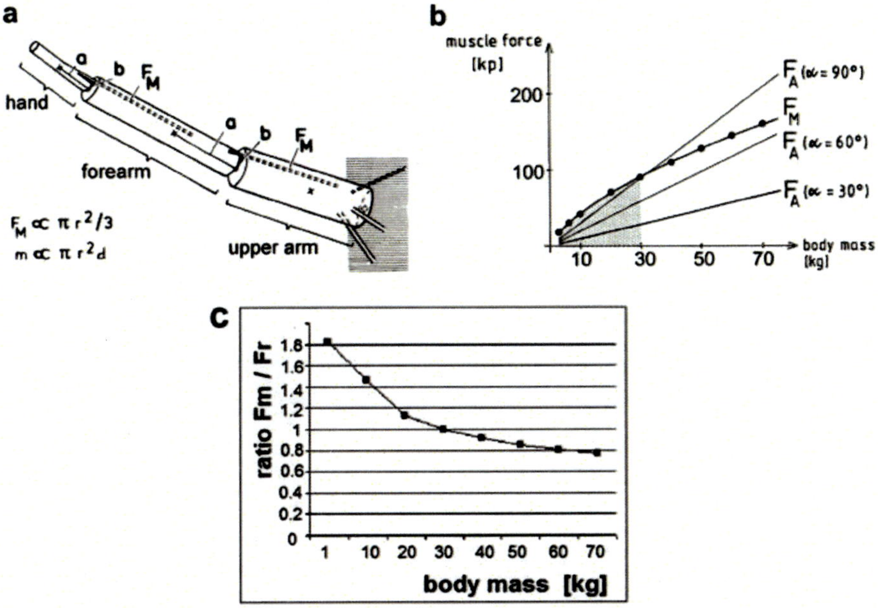

Fig. 8.26 Limitations set to arm length. (**a**) Length of arm segments alone increases weight, and with increased weight, additional muscle force is required for balancing the joints. Therefore, length has an influence on the required muscle mass, and that means on arm weight. As shown in Fig. 8.5a, the finger flexors in the hook posture of the hand counterbalance the weight of the body in static situations, as well as the inertial forces during locomotion. The latter grow linearly with speed, and speed depends on excursion angles, as shown in (**b**) for 30°, 60°, 90°.This means that muscle forces (F_A) necessary for keeping the intermediate and terminal phalanges in equilibrium against the external tensile force also grow linearly. However, body mass is proportional to a third power function, while the force F_m which is really available in the finger flexors grows slower, following a second power function (see Chap. 1). The muscle force (F_M, dots) an ape can dispose of therefore grows slower than the linear functions. In gibbons, the greatest force reserves are available if masses vary between <5 kg ("small" gibbons) and > 10 kg (siamang): If body weight is more than 30 kg (large apes), the available force lags behind what would be necessary. (**c**) shows the ratio between available force (F_m) and external force (F_r) which extends the finger joints. Evidently, smaller forms are better in suspending their weight from a hook hand. A clearcut borderline does not exist, but if body mass is about 15 kg or above, the ratio F_m/F_r decreases less rapidly, (**a** and **b** from Preuschoft and Demes 1984, **c** from Preuschoft et al. 2016)

greater without the animal losing external balance – unless the animal does neither lift the right arm, nor grasp hold of the twig beneath its right hand, nor change the pivot by placing its left foot toward the active arm. All these options, however, are connected with the disadvantage of additional expenditure of energy! Likewise, the gibbon hanging down from a branch in Fig. 8.27b is rotated about its handhold at the branch as a pivot. External equilibrium exists, if

$$F_a * h_a = h_{wr} * F_{wr} \tag{8.20}$$

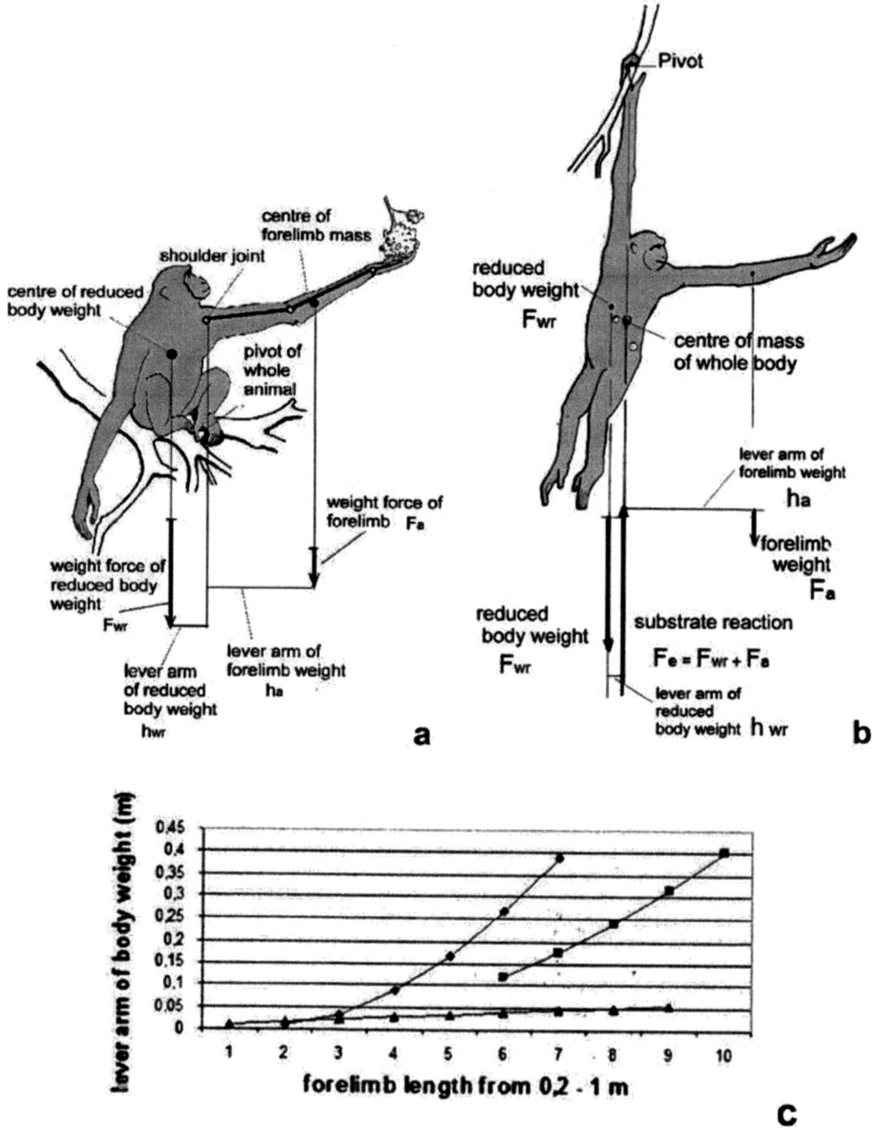

Fig. 8.27 Limitation of arm length set by the external equilibrium. (**a**) Gibbon sitting on top of a branch. (**b**) Gibbon suspended below a branch. The outreaching arm shifts the CoM toward the target. External equilibrium is defined by the balance of rotating moments of the extended arm (F_a) and of body weight reduced by the active arm (F_{wr}). Any movement to maintain balance against a further elongated arm costs additional muscle force (= energy). (**c**) Lever arms of body weight, reduced by the stretched-out forelimb, plotted against forelimb length. Triangles: forelimbs are in a constant relationship to body weight. Diamonds: body weight kept constantly at 5 kg (like in gibbons). Squares: body weight 10 kg, forelimb length more than 0.60 m (like in siamangs). (From Preuschoft et al. 2016)

Again the gibbon has the option to balance itself, for example, by flexing the hip joints ventrally, but this costs additional energy! In both cases, a limitation of arm length is the least energy-consuming means to solve the mechanical problems. An evolutionary trade-off between the advantages and disadvantages of arm length is to be expected.

Quite interesting in this context is the fact that the absolute length of the cranial extremity is in the size order of 0,5–0,8 m among all hominoids, regardless of their overall body size (Schultz 1956).

8.10 Why These Characteristic Shapes of Skeletal Elements and No Others? Functional Morphology of Anatomical Details

Essential for the evolution of skeletal elements are the mechanical stresses which act on the respective element: the skeleton needs to be as light as possible (i.e. in other words, it must be constructed by a minimum of material), and it must be strong enough to sustain the stresses occurring during everyday life. The analysis of mechanical stresses in skeletal segments nowhere is so far developed as in primates (e.g. Patel and Maiolino, 2013). This of course is due to the close similarity of other primates with humans. Detailed knowledge of the predictable mechanical loads is most relevant for successful therapeutic interventions in orthopaedics and surgery, and the problems on how to conceive therapy have stimulated research. In all primates, the function and construction of fingers, toes and metapodials is identical.

The overall shape of primate hands and feet with five rays and 2-3-3-3-3 segments of the fingers and toes is commonly seen as an old mammalian heritage, but there are also arguments for a functional explanation (see above, Chap. 8.3).

The uniform arrangement of tendons on the digits has been mentioned above in Chap. 5.4. Among primates, the more "primitive" and rather small strepsirrhines and tarsiers normally have common *flexor and extensor tendons*, which split in the palm or sole into the strands for the digits, while the more "evolved" (e.g. Takahashi et al. 2002) and mostly heavier monkeys and apes show a variety of separations of the muscle bellies from which the digital tendons arise. The most marked separation in catarrhines is that of the thumb, in correspondence with the mobility of this ray (see also in Chap. 9.2.2).

On the flexor side of the autopodia, every mammal is confronted with the necessity to keep balance in the three joints. This is possible with length-adaptable tendons for each segment (Fig. 8.28a). Due to this necessity, the *arrangement of tendons* is so uniform among mammals (Preuschoft 1987; Landsmeer 1993). It determines the way of loading of all digital rays: on the volar side of the digit, the supported loads commonly are much greater than those acting against the dorsal or lateral sides (see below). If loads are directed against the dorsal side of the digit, the necessities are different (Fig. 8.28b), because the tendons follow the convexity: flexing the fingers to form a fist would require a length increase of the extensor

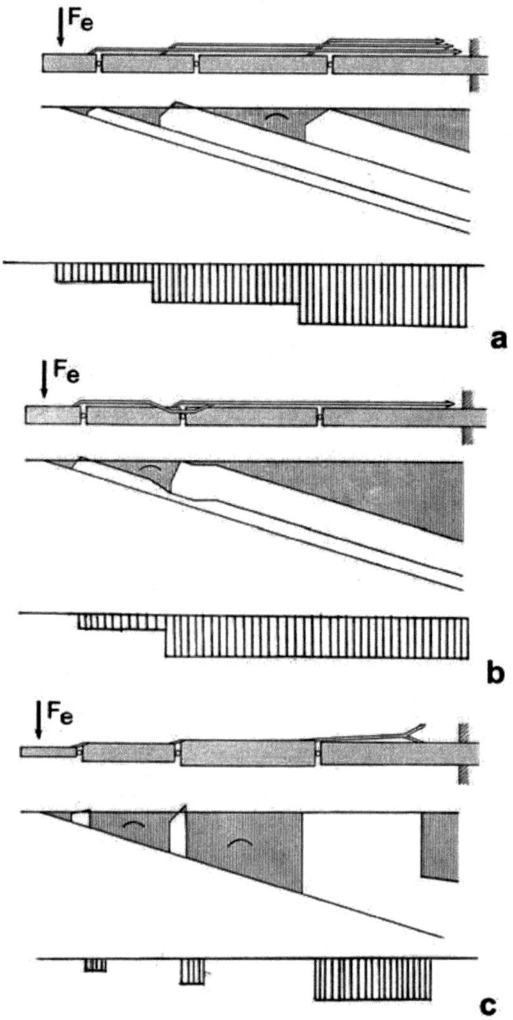

Fig. 8.28 Stress distribution along the "lateral" digits. Instead of an isolated digital ray, a cantilever is assumed, on which an external load is acting. These forces (F_e) are arbitrarily assumed to be of the same size – in contrast to the realistic loading that acts on digits. (**a**) The load F_e is acting against the palmar side of the hand, which is turned upside down. F_e evokes bending moments (white triangular areas in the diagrams). Because the cantilever (= digital ray) is interrupted by joints, each of the joints must be kept in balance by a tendon (double lines, from left to right that of the deep flexor, superficial flexor and interosseus muscles). The arrangement of tendons obviously allows independent flexion of all joints. Less evident is their effect on the stresses. The grey triangles show the bending moments, which remain when the ties (= muscles) are tensed. The exerted tensile forces cause compression (vertically hatched rectangles) but reduce the bending moments at the most distal joints to zero. Along the phalanges, a remainder of bending moments persists (grey areas), which must be sustained by the bony elements. (**b**) External load applied to the dorsal side of the nail. The distal joint is balanced by the force exerted by those strands of the extensor tendon which pass lateral to the middle interphalangeal joint. These fascicles cross the joint axis and therefore do

tendon to 127% of its length in extension. The action of the extensor tendon can be maintained, if alone a median slip of tendon fibres inserts into the base of the middle interphalangeal joint, while the fibres destined for the distal joint pass on the radial and the ulnar sides of the middle joint. So the sequence of movements is guaranteed: flexion takes place first in the middle joint. Not before this is to some degree flexed, a flexion in the distal joint becomes possible and extension takes place first in the distal joint, provided that the intermediate joint is also extended (Landsmeer 1949; Haines 1951; experimental confirmation by Sarrafian et al. 1970). The *sequence of flexing* or extending actions is defined by the relative lengths of the slips and does not require neural coordination or control. This guarantees even in most rapid and poorly controlled grasping a first contact with the volar pads on the fingertips and makes a first contact by the dorsal sides of the nails improbable. External forces applied to a digit in radioulnar or tibiofibular (z-) directions entail rather high bending moments (Fig. 8.28c) to be sustained by the skeletal elements, because there exist no ties to protect the metacarpal or the phalanges from being bent.

The shapes of the *terminal phalanges* vary considerably among primates, to a large part intraindividually (Fig. 8.29; Godinot and Beard 1993; and in v. Koenigswald et al. 2011, 2012). This variation is a good reason to focus on these seemingly unconspicuous, small elements. The first digit in the majority of forms has a flat and broad terminal phalanx with a flat nail, while the terminal phalanges of lateral digits may be blunt, with thick, rugose tips, or slender and elongate, with flat or vaulted nails or even claws. Many primates, especially strepsirrhines, possess nails, which are strongly curved in longitudinal as well as in transverse direction and exceed the soft parts of the fingertips. These nails can be compared with the reinforcements of the most highly loaded parts, claws or hooves, in other mammals. Among the non-simian primates, one or two digits have long and pointed so-called toilet claws. The shapes of the nails are commonly seen as specialisations to their biological roles. Their shapes, however, depend upon the stresses which they sustain (see below). In contrast to the above described terminal phalanges, the toilet claws are not exposed to extreme forces which decide about life or death.

The *external substrate reaction forces* (F_e) acting against the fingertips in the dorsovolar plane are the highest among the loads which occur in everyday life

Fig. 8.28 (continued) not have a lever arm at the middle joint. The intermediate joint is balanced instead by the median fascicle of the extensor tendon (see also text below and Fig. 8.30b). At the proximal joint, the full tendon force of both fascicles comes into action, and in addition, the lever arms are longer (not illustrated). **(c)** A load in the radio-ulnar plane applied to the terminal phalanx is balanced at the distal and middle joints by nothing except the collateral ligaments. The proximal joint, however, is controlled by the interosseus muscles, which extend along the length of the metapodium. The bending moments at the joints are reduced to zero, while the shafts of the phalanges are exposed to much greater bending moments than in side view. The diameters of the phalanges become proximally greater in accordance with the proximally increasing compressive stresses. The metapodial, by contrast, does not suffer from high bending, because it is protected by the interosseus muscles. If the bending moments exceed the strength of this muscle, a movement takes place in the radio-ulnar plane. In addition, the four metapodials brace each other when loaded in the radio-ulnar plane. (From Preuschoft 1987)

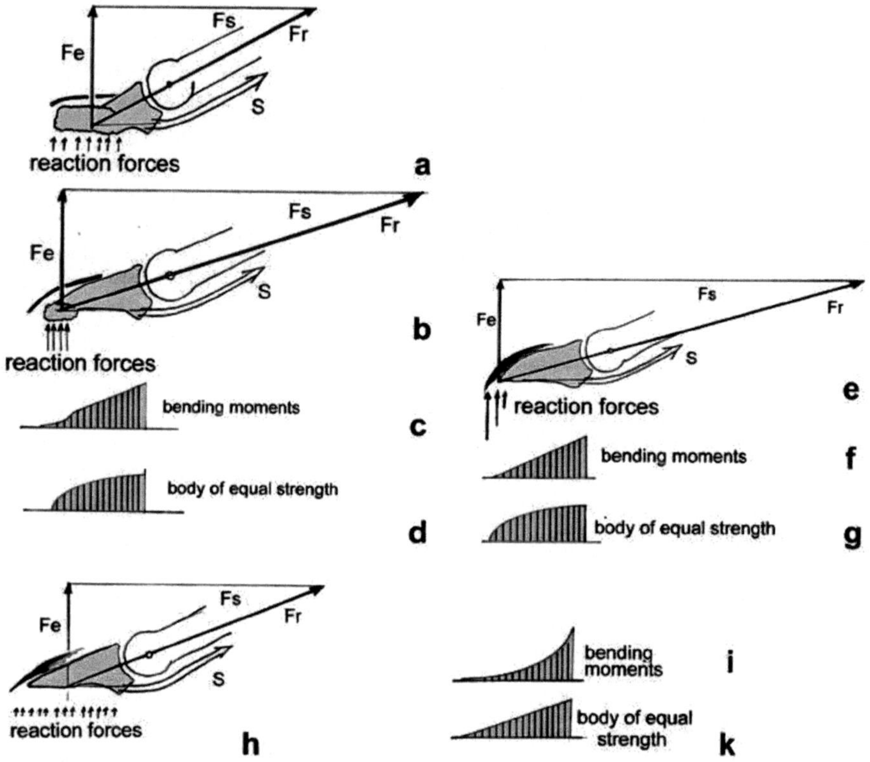

Fig. 8.29 Stressing of terminal phalanges in lateral view. (**a**) Substrate reaction forces (black vertical arrows) are distributed by the soft pad on a large surface to the apical tuft = tuberositas unguicularis. The resultant SRF F_e combines with the pull of the tendon (S) to a joint resultant F_r, which is inclined against the axis of the phalanx. Equilibrium is established only if F_r passes through the rotational axis of the joint. If so, no motion takes place. (**b**) SRFs habitually concentrated on the distal part of the terminal phalanx (**b, c, d**) and the rigid nail (**e, f, g**) require high tendon forces and lead to a resultant F_r, which is in line with the terminal phalanx. (**c, d, f, g**) Bending moments and body of equal strength against bending stresses. (**e**) Because of the protruding nail, the external force is totally concentrated on the tip of the digit. It has a still longer lever arm than in (**b**), so that a greater tendon force (F_s) must be provided. The joint force (F_r) does not deviate from the phalanx's long axis, and the joint is not tilted upward. The size of the joint force requires a large joint surface. Because the tendon inserts at a tubercle near the proximal joint, the terminal phalanx in this case is exposed to bending stresses (**f**), which require a considerable strength, provided by the parabolic outline of the element (**g**). (**h**) A toilet claw meets external forces which may be applied to the whole length of the element. They combine to a total reaction force, which has a shorter lever arm than that shown in (**e**). The tendon force as well as the joint resultant may be small. The bending moments (**i**) increase more than linearly; the body of equal strength (**k**) has a triangular outline. (Terminal phalanges are taken from v. Koenigswald et al. 2011; Godinot and Beard 1993)

(as has been mentioned above). They can easily exceed body weight, when the body is accelerated. Depending upon the form and strength of the nails, the stresses in terminal phalanges are different. As shown in Fig. 8.29a, a soft pad at the tip distributes F_e on the terminal phalanx (see also Fig. 8.8). At the apical tuft (= tuberositas unguicularis), the forces exerted from externally are balanced by the rotating moments of the flexor tendons. The joint force (F_r), resulting from external load and pull of the flexor tendon (F_t), is somewhat inclined against the shaft of the terminal phalanx, so that a dorsally tilted joint surface is at right angles to the force (Preuschoft 1969a, b, 1970a, b, 1971). The chosen example is typical for the first digits in all primate autopodia and for the lateral digits possessing flat nails, like in Hominoidea. A highly vaulted and protruding nail leads to a concentration of external forces on the tips of the digits (Fig. 8.29b, e). This requires very high tendon forces, the linearly increasing bending moments along the phalanx are greater than in Fig. 8.29a, and the joint resultant (F_r) is less inclined. The strength of an element exposed to bending by a load concentrated on the tip depends from the square of height, multiplied by its breadth (Eq. 2.3 and Fig. 2.4; see Chap. 2.1; Fig. 8.29c, d, f, g). A body of equal strength, in which stresses are on the same height, therefore, is characterised by a convex profile which follows, if breadth is constant, a square root function (Fig. 8.29d, g), because the height of the cross section must grow proximally by $\sqrt{h}$.

This form is characteristic of the *lateral digits of most monkeys*, and lemurs, but there exist many transitions between this variant and the one described above. They also exist in fossil hominoids (Preuschoft 1973a, b). If the nails are still higher vaulted and longer, and approach true claws, like in callitrichids, the external forces are concentrated alone on the tips of these claws. This arrangement comes close to the situation as was described above for cats (Fig. 7.3, in Sub-Chapter 7.2.1). The stresses are sustained alone by the horny claws, from which the forces are transmitted by connective tissue to the bones. Because of this unknown force flow, the bony terminal phalanges must not follow exactly the theoretically derived shapes.

A third variant are the often-discussed *toilet claws* of strepsirrhines and *tarsiers* (see above). Used alone for self-grooming, these are clearly weaker and less strong than the terminal phalanges used for locomotion. In addition, toilet claws are typically exposed to external forces, which act against their whole length. If a toilet claw is used like a comb, hairs will resist its being moved over its entire length, so that bending moments increase more than linearly. The partial forces sum up to a resultant, which acts at about half the length of the segment. Its lever arm is rather short, so that no powerful flexor tendon is required. The body of equal strength is characterised by straight, triangular outlines. A good example for such a claw shows Figs. 8.10, 8.19c, 8.29h (*Tarsius*) in v. Koenigswald et al. (2011).

Aside from the dependence of claw shape from their biological roles (being inserted into the bark of trees, or into the skin of prey animals, or just protecting the toe tips (see above in Sub-chap. 7.2.1)), there is, however, also a distinct mechanical influence exerted by the insertion of the flexor tendon. If it is restricted to the basis of the terminal phalanx, for example, on a tubercle, the bending moments grow up to the proximal joint surface (Fig. 8.29b, e). Such an insertion is favoured by pronounced excursion angles of the joint. If, by contrast, the insertion of the flexor tendon extends distally, it acts like a tie along the volar surface, which reduces the bending moments in the proximal part of the phalanx (Fig. 8.29a). This arrangement of fibres is associated with limited excursions of the joint. It can be observed in hominids and has been related to the asymmetry of the insertion in the 1st ray of humans (Preuschoft 1970a, b). The breadth of the phalanges does not contribute much to the strength of the element like height, because Eq. 2.3, where ΔA is determined by breadth. Nevertheless, the substrate reaction force may well contain a component in a medio-lateral (= radioulnar or tibiofibular) direction, simply because substrates are often inclined. The lateral digits are placed side by side in the direction of gravitational force (Figs. 8.13 and 8.19). These components are balanced at the joints by tight collateral ligaments and require considerable bending strength of all, not alone the terminal phalanges, as shown in Fig. 8.28c. In fact, the ***structural strength of phalanges*** in the medio-lateral plane is usually greater than in dorsovolar direction and particularly pronounced in the pollex and hallux, which are not braced and protected by their neighbours (see below, Fig. 8.34). This breadth of the terminal phalanx also entails a broad joint surface, expanded in medio-lateral direction.

The *middle phalanges* (of the foot as well as the hand; see Preuschoft 1969a, b, 1970a, b, 1973a, b) receive the joint force (F_{re}), which deviates dorsally from the longitudinal axis of the segment, so compressing the element and pushing it in dorsal direction (Fig. 8.30). Its extending moment at the middle joint is countered by the

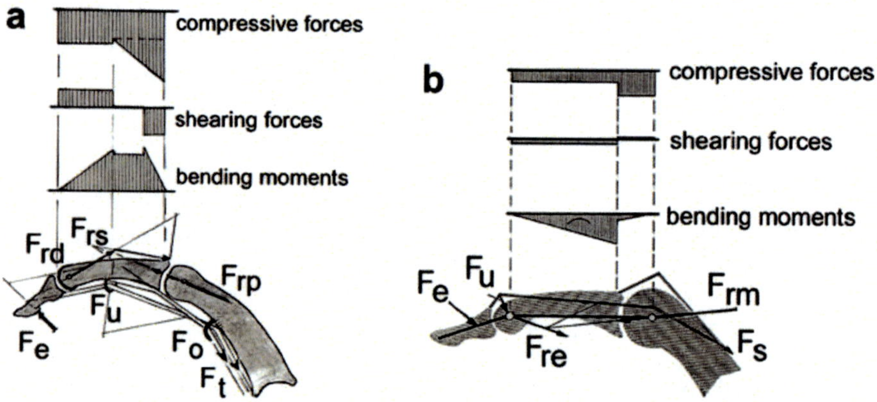

Fig. 8.30 Stresses in a middle phalanx. (**a**) In flexed digits, the intermediate phalanx is pushed dorsally by the joint force F_{rd}, which evokes bending moments (shown in the lowermost diagram). The joint surface is typically tilted in volar direction and extends more on the volar than on the dorsal surface of the phalanx. The force F_{rd} is opposed, though not necessarily balanced, by the force F_u, evoked by redirection of the deep flexor tendon (F_t). It is transmitted (and distributed along the shaft of the phalanx) by the fibrae anulares of the tendon sheath (black sling). Equilibrium in the middle joint is established by the superficial tendon, which exerts the force F_o. The insertion of this tendon is extended along most of the volar side of the shaft, so that it not only rotates the middle phalanx in volar direction but at the same time also acts as a tie, keeping the volarly convex bending moments low. It compresses the bony phalanx between joint surface and insertion (diagram on top). The joint force F_{rp} in the interphalangeal joint passes through the middle joint's axis. (From Preuschoft 2018). (**b**) Stresses occurring in the middle phalanx if a load is applied to the dorsal side of the nail. The extensor tendon keeps balance in the proximal interphalangeal joint by the insertion of its median slips, in the distal joint by the lateral strands, which have no lever arm at the intermediate joint see also Fig. 8.28b and text below

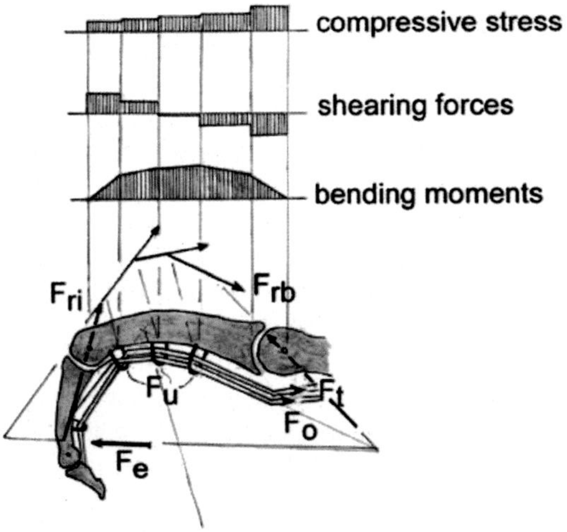

Fig. 8.31 Stresses in a proximal phalanx. The most dominating external force F_e acts against the flexed middle phalanx. The middle joint is kept in equilibrium by the flexor tendons, which exert the forces F_o and F_t. The resultant of these three forces F_{ri} acts against the head of the basal phalanx and tends to "extend" the basal joint dorsally. This action is gradually countered by the forces created by redirecting the pull of the tendons (F_u) and transmitted to the element by the fibrae anulares of the tendon sheath (black slings). In the basal joint, the resultant force F_{rb} passes through the joint axis. A curvature of the basal phalanx approaches the bony element to the acting forces and therefore is a means to reduce bending stresses. This is divergent from the middle phalanx, where the bending moments are kept low by the "tie" formed by the insertion of the superficial flexor tendon (as shown in Fig. 8.30). If the basal joint is not in equilibrium, the mm. interossei can provide rotational moments, or external SRFs may be applied in addition. The compressive forces increase proximally. (From Preuschoft 2018)

redirectional force F_u, which is evoked by the change of direction of the deep flexor tendon. The redirectional force is transmitted to the phalanx by the fibrae anulares of the tendon sheath. To keep the intermediate joint in balance (that means to direct the joint force F_{rm} toward the axis of the joint), the action of the superficial flexor tendon F_o is also required, for example, in the highly loaded hook position of the hand. The insertion of this tendon is distributed over a large part of shaft length, and so the inserting fibres act like a tie to relieve the bone from dorsoconcave bending. At the same time, the inserting superficial tendon exerts proximally growing compressive forces on the phalanx, which require a proximal increase of the cross-sectional area.

In a typical highly loaded position (like in the hook hand), the joint force F_{ri} of the middle joint deviates dorsally from the *proximal phalanx* (Fig. 8.31). This segment is kept in its place mainly by the redirectional force F_u, which has small compressive components in the direction of the phalanx. This force is transmitted on the skeleton by the fibrae anulares of the flexor sheath, which ties both flexor tendons to the bone. These fibres are distributed over a considerable part of shaft length. The proximal phalanx is more curved than the middle phalanx. In the case of the African apes, only the distal fourth of phalanx length is curved; the proximal ¾ are nearly straight

(Susman 1979). In orangutans and gibbons, however, the curvature comprises the entire length of the proximal phalanx. Unpublished data collected by Creel, Susman, Stern and Preuschoft (1979) from phalanges of apes indicate the dependence of these curvatures alone from the lengths of the phalanges, not from variants of hand use in the investigated species. Anyway, these characteristic curvatures of the phalanges lead to the effect that the local resultants of all acting forces fall in line with the bone's axis. Consequently, the skeletal element is under compression, while bending is kept on a low level (Fig. 8.31b; more details in Preuschoft 1969a, b, 1970a, b, 1973a, b). Therefore the tubular bones assume a circular cross section and do not need long diameters for providing great bending resistance. In addition to the flexor tendons, the *mm. interossei* exert a rotating moment at the basal joints, which is commonly underestimated. Because of their marked pinnation, these muscles exert surprisingly great forces, which allow, for example, a flexion of the basal joint, combined with extension of the intermediate and the terminal joints.

External forces which act against the *dorsal sides of the digits* are always much smaller than those occurring in grasping. This has to do with use made of the autopodia and the lack of force-distributing pads on the dorsal surfaces. At the base of the intermediate phalanx inserts alone the median slip of the extensor tendon. If this is acting against an external force, bending moments occur along the shaft (Fig. 8.30c). Because of the limited force, the stresses probably do not reach values, which lead to an optimised shape. In the typical "knuckle-walking" position (Fig. 8.34a,c), the volar sides of the terminal and middle phalanges are in firm contact (Susman 1979), the former often also braced against the capitulae metacarpalium. At the basal joint, this complex is kept in balance by the flexor tendons, not by the extensors.

Similar to the proximal phalanges, the *metapodials* are kept in their positions against high loads acting at the digits, palms or soles mainly by redirectional forces of the flexor tendons (Fig. 8.32; Preuschoft 1969a, b, 1970a, b, 1971, 1973a, b). In grasping, the external forces against the digits lead to resultants (F_{rg} or F_{rcap} in Fig. 8.32) that flex the metapodials dorsally, and the flexor tendons are forced to change their direction by their tendon sheaths. Although both tendons have clearly longer lever arms in the metapodium than along the digits, the tendon sheath compels the tendon to follow any deviation from the straight line, exerting at all curves a redirectional force, the most important illustrated in the "hook hand" (Fig. 8.32a, b). If a "hook" position is assumed, the carpal or tarsal joints are often not exposed to bending moments, since the external reaction force from the substrate has no lever arm. With the input of additional energy, this position can be modified by activity of the carpal or tarsal flexors or extensors. The insertions of these muscles usually can be traced on the basis of the metapodials (Fig. 8.32c, d). In plantigrade or semiplantigrade positions of a hand or a foot, additional external forces against the heads of the metapodials come into action, which also contribute to dorsiflexion (this is particularly important for the human foot; see below in Chap. 9.2). The mm. interossei are located more on the plantar or volar side of the metapodia than on the dorsal, so that they exert considerable moments bending the shafts of the metapodia dorsally convex and so reduce its overall stress by acting as a tie.

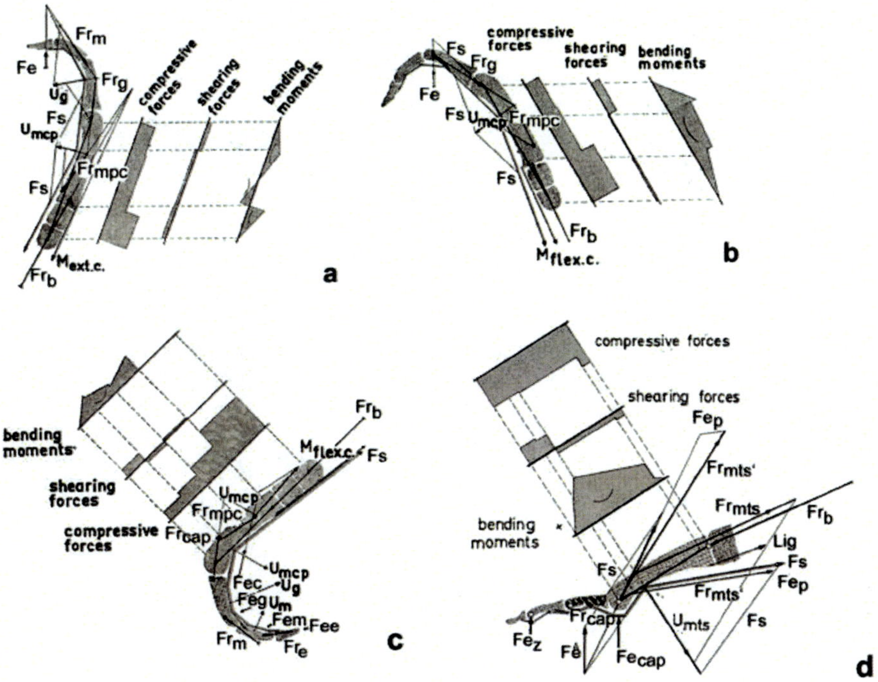

Fig. 8.32 Stresses in metapodials. (**a**) Hook hand with carpal joint close, but somewhat volar to the prolongation of the external force that acts against the terminal/middle phalanges. The joint resultant in the metacarpophalangeal joint is directed dorsally and its moment countered by the redirectional force of the digital flexors, as in Fig. 8.31. A contraction of the carpal extensors is to be expected. (**b**) Hook hand with wrist joint dorsal to the SRF. To keep the wrist joint in balance, the carpal flexors must be active. (**c**) Metatarsal of a grasping foot, which also supports a part of weight (F_e). Parts of the total external force act against the free digits and against the ball beneath capitula of the metapodium, the basis of which is fixed by ligaments. (**d**) First metatarsal with extended toe during roll-off. The acting GRF is divided into a part directed against the toe and a part against the ball of the foot (Fez and Fecap, their resultant being Fe). The difference between hands and feet in primates is that the latter more often are kept at right angles to the external loads and therefore subject to stronger bending moments. The joint at the metatarsal's base is balanced by mm. tibialis posterior, peroneus brevis and peroneus longus or the tarsal ligaments. Their proximal insertion causes bending moments along the metatarsal's shaft, in spite of the redirectional force U_{mts} produced by the toe flexors. Note that the redirectional force evoked by the flexor tendons in these drawings are called U, not F!. (From Preuschoft 1970a)

The most marked difference between metapodials and phalanges is the more rounded, less broad *cross section* of the former. Because of this shape, they are not suited to resist the same stresses in medio-lateral direction as the digits are (Fig. 8.33). In fact, such forces do not play the same role as in the free digits (Figs. 8.28, 8.34), because of two reasons: first, the metapodials brace each other, so that the total force component in the medio-lateral plane is distributed over four

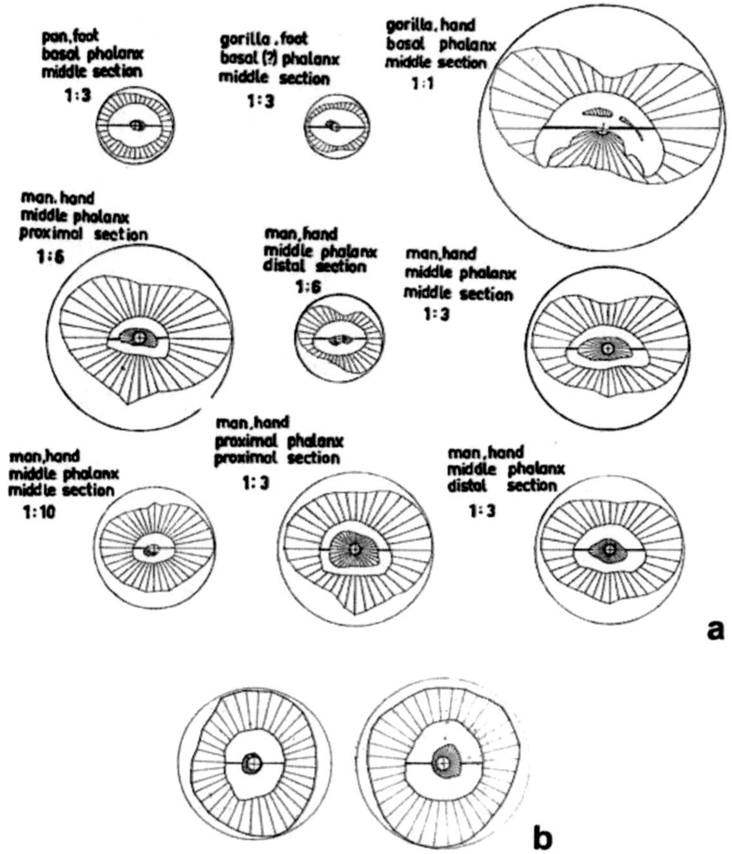

Fig. 8.33 Cross sections through skeletal elements: intermediate and basal phalanges (**a**), metacarpals (**b**) and metatarsals (**c**). The planes of bending are rotated about the cross section's centres, and the area moments of inertia (I) are drawn for every tenth plane. For better visibility, the longest of them surrounds each cross section. The pivotal axes of normal, "physiological" flexion/extension are indicated by horizontal, heavy lines. The isolated digits commonly have high shape-dictated strength in the mediolateral plane – but this does not necessarily hold true in the hook hands of apes. Like the metapodials, they are not exposed to strong mediolateral bending, because the individual digital rays brace each other. The same condition exists in the (lateral) toes of humans. Enlargement factors are added for each cross section, because the varying size does not allow presenting the values drawn to the same scale

skeletal elements, and second the mm. interossei, take over the tensile component of bending as active ties.[8]

[8]In strictly arboreal primates (orangutan and gibbons), which spend a large part of their life suspended from thin twigs, Susman (1979) has observed a tendency to strengthen the fourth digital ray (approaching the "ectaxonic" hand shape of strepsirrhines, as noted above) and rather narrow cross sections of the proximal phalanges in comparison to African apes. Both details are easy to explain: the external substrate reaction forces in medio-lateral direction act first of all against the

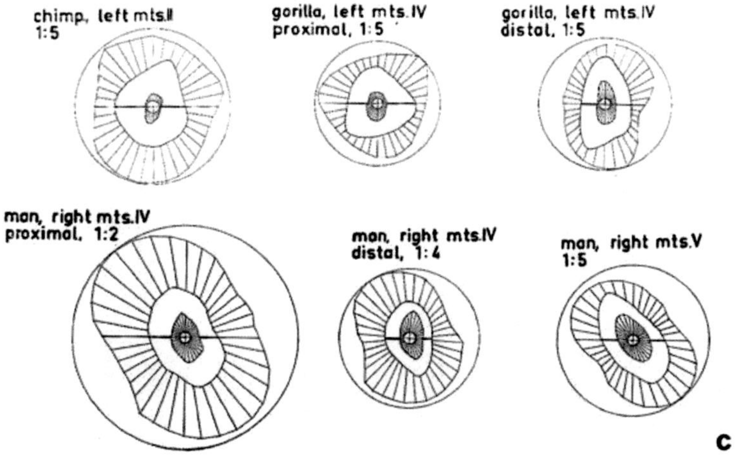

Fig. 8.33 (continued)

Nevertheless, metatarsals often show oval cross sections (Fig. 8.33c), indicating frequent load components in medio-lateral direction superimposed on those in the dorsoplantar plane. Especially the fifth metatarsal is exposed, because of its lateral position and its fixation by the m. abductor metatarsalis V, to bending stress components in the medio-lateral plane. The same holds true for the first metatarsal, due to its separation from the lateral rays and the strong mm. abductor hallucis and extensor et flexor hallucis. These muscles insert into the basis of the element, but the m. adductor pollicis inserts slightly proximal to its capitulum, where external loads are applied. These muscles provoke (or reduce) bending moments along the shaft in a medio-lateral plane, which superimpose those acting in the dorsoplantar plane. The pattern of shape-dictated strength depends closely upon the use made of the foot and will be taken up in Chap. 9, Fig 9.2.

Experiments (Preuschoft 1963a, b) have shown that about one half of the juvenile (!) chimpanzees and gorillas can extend their fingers only, if the wrist joint is flexed, and with extension of the wrist, the fingers assume a flexed posture. This shortness of the flexor tendons makes a hyperextension of the fingers impossible when the hand is

fifth and fourth ray (Fig. 8.27b), while the third and second receive only smaller partial forces. The fingers are usually held side by side, parallel to each other. That finds its morphological correspondence among siamangs in the well-known, distally extended webs between the proximal digits. This position also contributes to a bracing of the individual fingers by their neighbours.

An obvious speciality are the sickle-shaped cross sections of basal phalanges in gorillas and siamangs. They offer great resistance to bending, like the oval cross sections of most other primates, but the highest strength can be observed not in the dorso-palmar but rather in laterally or medially inclined planes. In the dorso-palmar plane, the external substrate reaction forces seem to be so constant, that bending moments are excluded completely by the curvature of the elements, and only moderate shape-dictated strength is required.

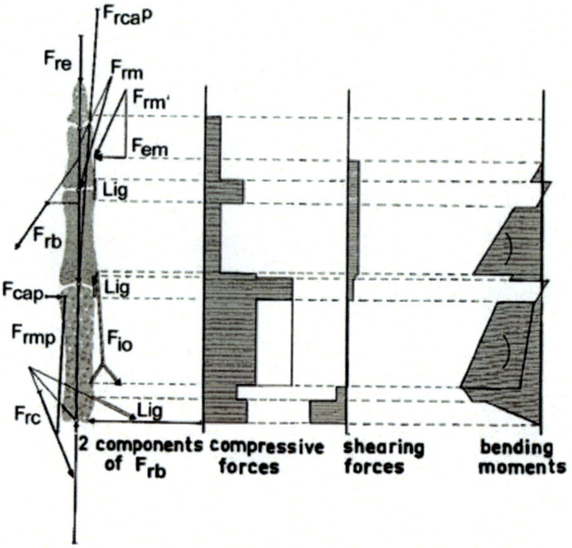

Fig. 8.34 Loading of a right-hand digit in the radio-ulnar plane. See also Fig. 8.28c. In the present illustration, an external force, which acts against the volar side of the terminal phalanx like in Fig. 8.28a, is assumed, but not visible in this projection. It must be balanced by the digital flexors, which exert a force along the digital ray (F_{re}). The (visible) resultant of both forces (F_{rm}) compresses the middle phalanx. In addition to this longitudinal force, a radially directed force component (F_{em}) is applied to the middle phalanx. A giving way at the middle joint is excluded because of the collateral ligament (*Lig*), which brings the bending moments down to zero but compresses the phalanges close to the joint. The resultant F_{rb} (of F_{re} and F_{em}) deviates from the basal phalanx, causing compressive and high bending stresses along this element. In the absence of protecting tendons (= ties) along the digit, the bony phalanges are exposed to the bending moments and must offer resistance. At the basal joint, however, the collateral ligaments (*Lig*) are slack and permit excursions. The joints are controlled by the mm. interossei (F_{io}), which, when active, at the same time act as ties for the metapodials, reducing bending moments further (hatched area, minus the outlined, also hatched quadrangle). In addition, the metapodium can be braced by its neighbour, which provides a force in ulnar direction (F_{cap}). At its proximal basis, the metapodium is fixed by the strong ligaments of the carpus or tarsus, respectively

rolled off on the ground in palmigrade or semipalmigrade posture. The African apes therefore place the dorsal sides of their intermediate phalanges on the ground, performing the often-discussed "*knuckle walking*" (Tuttle 1970). The forearm and the carpal joint are kept roughly in line with the metacarpus, so that the muscles crossing the wrist joint are used just for keeping the joint resultant in an optimal direction or controlling this joint. Gorillas tend to turn the back of their hands into the x-direction, while chimps more often turn their hands with the dorsum laterally in z-direction. Orangutans, when on the ground confronted with the same problem, mostly flex their fingers to a fist and transmit most of the GRF through the carpus.

The theoretically possible hand positions in *knuckle walking* are shown in Fig. 8.35. The left drawing (a) shows what can be observed rarely while the right

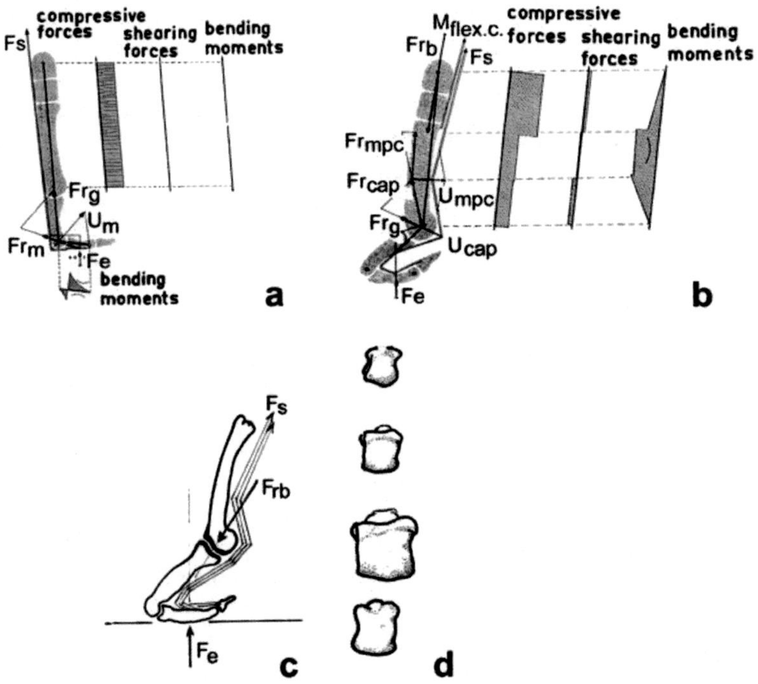

Fig. 8.35 Knuckle walking. (**a**) If the basal joint is extended and the GRF passes plantar to the interphalangeal joint, a contraction of the digital extensor is necessary and leads to similar bending stresses along the intermediate phalanx as shown in Fig. 8.30b. (**b**) Typical position in a knuckle walking ape, according to Preuschoft (1973a, b): terminal and middle phalanges are propped against the basal phalanx and the capitulum metacarpalis. The GRF (F_e) is applied evenly along the middle phalanx (impressions of equal depth in soft ground); the total external force exerts a dorsiflexing moment in the metacarpophalangeal joint, which must be balanced either by the force of the flexor tendons or by direct contact between the soft pads on the free digit and the balls of the hand. Redirectional forces evoked by the pull of the tendons are called *U*, not F. (**c**) Loads acting on a third digit of a chimpanzee, walking on the knuckles according to Susman (1979). Equilibrium is established like in (**b**). The joint force presses the phalanx against the joint surface, which is dorsally expanded by the "knuckle-walk-facet" on the metacarpal head (fat lines). (**d**) Distal views on the joint surfaces of os metacarpale III of hominoids. From top downward: *H. sapiens*, *Pan troglodytes*, *G. gorilla*, *Pongo pygmaeus*. Note the great breadth on the dorsal sides in the African apes. (**c** and **d** Redrawn and modified from Susman 1979)

(b) shows the most common posture. Because in Fig. 8.35b,c the fingertips are braced by the subdermal pad covering the capitulae metacarpalium, no extensor activity is necessary, and not bending occurs in the intermediate phalanges, while the flexor muscles balance the basal joint.

In the literature, the "*knuckle walking*" leads its own life, largely independent from its real significance. It is often discussed and overestimated as a presumed evolutionary stage on the way of our ancestors to bipedality (Tuttle 1975). There are reasons to bring it down and to reduce its imagined importance to the individual solution of a mechanical problem faced by heavyweight apes which have – and need – outstanding abilities in suspending themselves by their hook hands.

According to Kimura (1985, and personal communication), chimpanzees carry not more than about 20% or 1/5 of their body weight on the frontlimbs. For gorillas, the share may be somewhat greater, though no precise data exist. This is in line with the distribution of total body weight primarily on the posterior, rather than the anterior pair of limbs, which also indicates that the forces between the substrate and fingers are in the size order of 1/5 of body weight. A hook hand, by contrast, regularly carries the full body weight – sometimes, if accelerations play a role (see above, gibbons), even a multiple! The flexed middle phalanges carry the load at about half of their length, so that the lever arms of external load and tendon force are about 1/3 (see above, gibbons). The tendon force therefore is something like three times the external force – or more – than body weight. Along the basal phalanx or the metacarpal, a force of carefully estimated three or four times body weight comes into action. This seems to be a good reason to expect that morphology is specialised for the necessities of gripping, not of knuckle walking.

While the occurring stresses in knuckle walking are moderate and must not lead to morphological specialisations, the skin on the dorsum of the intermediate segments is hairless and callous, but no typical, clear-cut pads are developed. Skeletal traits have been postulated, among which only the observation of Tuttle (1967, 1969a, b) is realistic. He found at the dorsal margins of the metacarpal heads in gorillas an expansion of the facet, which give the basal joint of the proximal phalanx an enlarged area for transmitting joint forces in dorsiflexed position (Fig. 8.35d). This is in line with the great breadth of the dorsal diameters of the metacarpal joint surfaces, especially in African apes, while these areas are more restricted in the monkey-like hands (Preuschoft 1973a, b) – simply because the very high joint forces occur in grasping with flexed joints because of the forceful contraction of the flexor muscles.

Not convincing are in my opinion "adaptations" of the carpals, as postulated by Richmond and Strait (2000) and Richmond (2001). My doubts are based on the fact that up to now no really complete and mechanically correct analysis of the carpus in primates exists (Koebke 1993, and also below). As long as no well-founded knowledge exists about the mechanical conditions in the carpus, all imaginations about "adaptations" rest on assumptions, which are in the best case correlations between morphology of the carpals and locomotor modes, that is, between a behaviour and a morphological trait (Fisk 1970; Sennwald 1987; Taleisnik 1976, 1988). Likewise, the search for specialisation for "knuckle walking" in the hand skeletons of australopithecines and their predecessors is not very promising (see also in Chap. 9.2).

At its volar side, the carpus forms the "*carpal tunnel*" between eminentia radialis and ulnaris, which guides the tendons of the strong digital flexors (Fig. 8.36c). The eminentiae are formed by the hamulus ossis hamati and the os pisiforme. The height of the eminentiae gives the mm. flexores carpi radialis et ulnaris long lever arms, which saves force. Of particular importance seems to be the height of the pisiforme in the semiplantigrade, preferably ground-living monkeys.

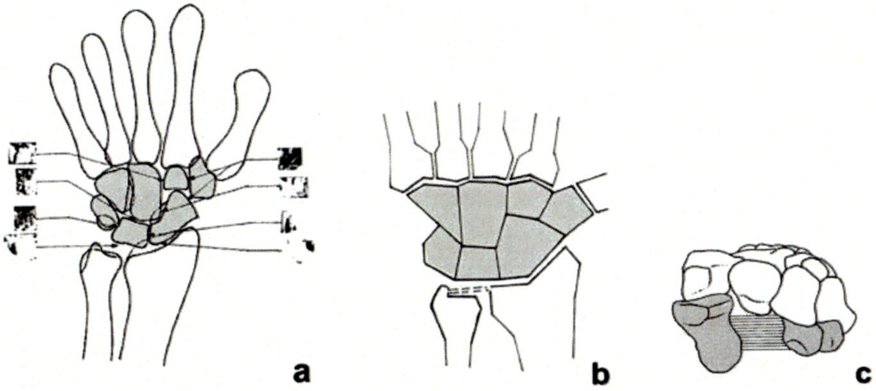

Fig. 8.36 Approaches to force flows within the carpus. (**a**) Redrawn x-ray of a left human wrist, comprising metacarpals, carpal elements, ulna and radius. The locations of pressure gauges are indicated. (**b**) Scheme of the left carpus, on which the calculations are based. The analysed load cases are set out in Table 8.3. (**c**) Carpal tunnel (hatched) of a left hand in *Cebus* spec., seen from proximal. The eminentia radialis (grey) is formed by the hamulus of the os hamatum, the eminentia ulnaris (grey) by the os pisiforme. Both are connected by the lig. transversum carpi, which covers the carpal tunnel. (**a** and **b** from Abker and Mahl 1993, **c** from Preuschoft et al. 1993)

The *placement* and *orientation of the individual carpalia* to each other change with positions of the wrist and loading of the hand. No conclusive rules could yet be identified (Jouffroy and Medina 2002; Jouffroy, personal communication). This the more so, since not even the number of carpal bones is constant. The interpretation of Koebke (1993) of longitudinally orientated pillars formed by the carpals comes probably closer to the mechanical conditions than the description commonly presented in anatomical textbooks, which discriminate a distal and a proximal transverse row.

An attempt to analyse the mechanics of the *complex system of the carpus* at least in one species (*Homo sapiens*) has been made by Abker and Mahl (1993) in their doctoral thesis. The authors started from the repeatedly mentioned fact that really high stresses occur when great loads are carried by the hand. Digital flexors or extensors, as well as carpal extensors or flexors, must be active and compress the *carpus*, often by pulling the metacarpals proximally, but their exact directions vary with the point(s) where loads are applied. The flexor tendons exert forces of about three to four times the external load. In addition, the carpal flexors and extensors must be activated, to keep the wrist in the chosen position. Within the carpus, these compressive forces are broken down into components, the sum of which must be zero.

In view of the immense *variability of hand postures*, some of them were selected which are of importance in work physiology. These are hand in zero position, hand abducted by 15° in radial direction and hand abducted in ulnar direction by 10° and by 20°. In the dorsoplantar plane alone a dorsiflexion of 25° was investigated, which is considered in work physiology as particularly important. The force application at

Table 8.3 Forces assumed by Abker and Mahl (1993) for the investigated load cases. Because the carpals change their relative positions in dependence upon the position of the wrist and hand, nine load cases have been selected as examples for a close-up analysis

Example A	10 N	Axially on each metacarpal
Example B	20 N	Axially on metacarpals I and II
Example C	50 N	Total, distributed on all metacarpals as follows: I, 20N; II, 12N; III, 7N; IV, 6N; V, 5N
Example D	50 N	Axially on metacarpal I
Example E	50 N	Axially on metacarpal II
Example F	50 N	Axially on metacarpal III
Example G	50 N	Axially on metacarpal IV
Example H	50 N	Axially on metacarpal V
Example I	12.5 N	Axially on each lateral metacarpal: II, III, IV and V

the metacarpus depends on the equilibrium at the finger joints, and the values considered in the calculations are listed in Table 8.3.

The *directions of the compressive forces* were (first) calculated in a simplified model (Table 8.3) and (second) measured between the neighbouring carpal elements with "Fuji-Druckmessfolien". Eight localities were selected under consideration of accessibility without cutting ligaments. These localities (Fig. 8.36a) corresponded to some of the intercarpal joints of the model used for calculations. So the measurements can be taken as experimental tests of the calculations. The calculated and the measured compressive forces did in fact not deviate by more than 3–16%, in most cases 10–15%, and so seem to confirm each other. The only exception from this convergence is the joint between trapezium, trapezoideum and scaphoid with 35% divergence. (Comparable data exist only for the contacts in the "proximal row", between radius and carpus or ulna and carpus (Tencer et al. 1988; Viegas et al. 1987; Viegas et al. 1989), which do not in principle deviate from those presented here.)

The *force flows* in the various positions of the hand vary considerably; the measured or calculated forces deviate clearly among the load cases listed in Table 8.3. The force pattern was determined by the digital ray to which the external load was applied; the force flow could be traced from the bases of the loaded metacarpal through the proximal adjacent carpals to the forearm bones. The force values increased proximally at the insertions of the carpal flexor and extensor muscles. In the investigated humans, the highest values were observed between radius and carpus, the contacts to the ulna did not exceed ¼ of these forces.

Components of the total compressive *forces acting at the wrist* parallel to the joint surfaces are equivalent to shearing, and these are countered by reinforcements of the ligamentous apparatus, which is known from descriptive anatomy. The mechanical analysis could explain those ligaments in considerable detail. The force components perpendicular to the joint surfaces correspond to the relative density and the direction of trabeculae of the spongy bone, as well as the reinforcements of subchondral bone plates. The magnitudes of compressive forces are not obviously correlated with the sizes of the contact areas between the carpal bones, since there are areas in reserve for changes of joint positions.

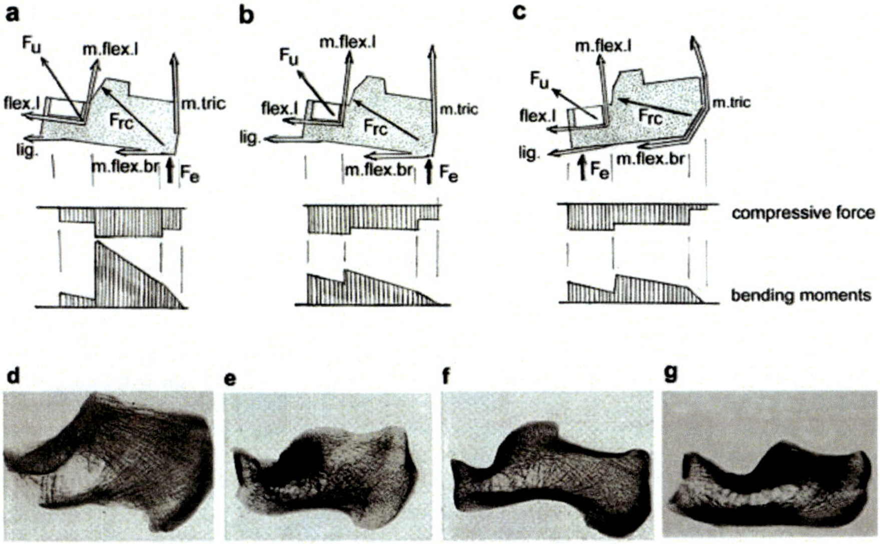

Fig. 8.37 Forces acting, and stresses evoked in the heel bones of humans (**a**), chimpanzees (**b**) and an OW monkey (**c**). The compressive forces and bending moments (hatched diagrams) are not drawn to scale. (**b**) Forces acting on the heel bone in a long-toed ape, or NW cebid deviate from the human situation (**a**), since the digital flexors replace partly the function of the m. triceps, which consequently exerts less force. External force and force of m. triceps vary as well as those of the mm. flexores digitorum, although the greatest share is probably produced by the m. triceps. Consequently, the resultants of all forces applied to the heel can vary within limits, but are directed in all cases mainly against the facies talaris posterior. To a lesser degree, compression is transmitted through the facies cuboidea. The greater the force of the m. triceps, or the external SRF, the steeper their resultant and the greater the bending moments in the corpus calcanei. (**c**) In cercopithecoid monkeys (*Papio* as an example), no SRF acts against the heel, which is not in contact with the substrate. The tendon of the m. triceps often passes around the rear border of the calcaneus, as "m. plantaris longus", and only a part of its fibres insert into the tuber calcanei. (**d–g**) The pattern of spongy trabeculae inside the heel bones becomes visible on x-ray images of thick slices through the bones. It varies between *Homo* (**d**), *Pan* (**e**), *Gorilla* (**f**). While similar in chimps and humans, the element in gorillas (see above) indicates more bending strength, required because of its greater length. (**g**) The pattern in *Papio* differs, like that of other cercopithecoids. Its stress trajectories surround not one but two "singular points" (**d–f**). (From Preuschoft 1970a, b). See also Fig. 9.5 in Chap. 9 for the talus

The results of this study neither justify the conventional interpretation of the carpus as a combination of two transverse rows, nor did it confirm the new model of three columns completely. The assumptions in the literature (Richmond and Strait 2000; Richmond 2001 as above; Lewis et al. 1970; Lewis 1969, 1983) about the "function" of particular carpal bones are not in agreement with what has been found by Abker and Mahl (1993).

The mechanical conditions within the *tarsus* are better known than the mechanics of the carpus (see Preuschoft 1969a, b, 1970a, b, 1971). The biggest and easiest to understand part of the tarsus is the *calcaneus*, which exhibits some morphological variation between the families or subfamilies (Fig. 8.37c, d). As mentioned above repeatedly, it offers a lever for the rather powerful m. triceps surae. In cebids and Hominoidea, the posterior part of the calcaneus receives also an often high external force. All this is clearly visible in side view. At its anterior (subtalar) part, the element is anchored to the neighbouring tarsals by strong ligaments (including aponeurosis plantaris) and by muscles taking their origins from the heel bone (m. plantaris, part of the flexor digitorum brevis). The tensile forces exerted by these ligaments evoke reaction forces which are sketched in Fig. 8.37a, b. Its being loaded as a lever leads to eccentric (bending) stresses inside the corpus calcanei and the corresponding arrangement of trabeculae (Fig. 8.37d–g). This, however, differs between hominoids and cercopithecoids (which do not apply external loads on the heel, see above). The density of trabeculae indicates the magnitude of local stresses – as shown in diagrams a and b. The greatest density can be observed close to the posterior subtalar joint. The orientation of his facet, however, deviates from the sagittal plane through the foot.

No tendon inserts directly into the *talus*, and the extensor tendons of the digits and of the tibialis anterior are tied by a ligamentous sling to the talus and so exert force on it. Likewise, the strong flexor tendons passing the talus on the posterior side are also forced to change their directions and so exert redirectional forces on the talus. At the *posterior margin* of the talus, and at the *sustentaculum tali*, these redirectional forces are applied to the skeletal elements (Fig. 8.38k), push the talus toward the toes and support the sustentaculum. Because of the strength of the flexor tendons in non-human primates, the latter can be expected to be broad. If cut in transverse direction (y/z-plane), the sustentaculum exhibits a trabecular pattern typical for spongy bone under bending loads. The posterior subtalar joint surface of the heelbone is more inclined in the non-human primates than in humans, which lack the transverse force components of tendons acting on the abducted first toe.

The *joint closest to the ground*, about which the *hindlimb rotates* in locomotion, varies: plantigrade species, like *Homo*, *Gorilla*, *Pan*, *Hylobates* and Cebidae, put the entire sole, including the heel, on the substrate, and the limb rotates about the tibio-talar joint as its pivot. The semiplantigrade cercopithecoidea do not possess a plantar process at their tuber calcanei (Fig. 8.37g). At least during movement, the heel is kept clear from the ground and does not contribute to transmission of weight force. The pivot of the inverted limb pendulum is a combined movement of the tibio-tarsal and the intertarsal joints, as shown in Fig. 8.4f–k. In relaxed standing, however, the heel can be put on the substrate and so allow a relaxation of the mm. flexores digitorum and triceps. Tarsiers and most strepsirrhines lift the heel permanently from the substrate and make contact alone by their distal tarsals and the metatarsus. This can be seen from the fur-covered skin on the heel, which is typical of the best leapers (galagos, tarsiers) and lorisinae. The joint closest to the ground is between the posterior and the anterior tarsal bones. Callitrichidae rotate their hindlimbs about a pivot in the heads of the metatarsals (Arms 1998; Arms et al. 2002; direct evidence from x-rays presented by Voges 1999). The advantage connected with lifting the heel from the ground is a gain of "functional limb length", which, for example, increases walking speed (see above). Shifting the contact between animal and substrate in distal direction leads to an elongation of the load arm and therefore to a faster movement of the most distal parts because peripheral speed depends on

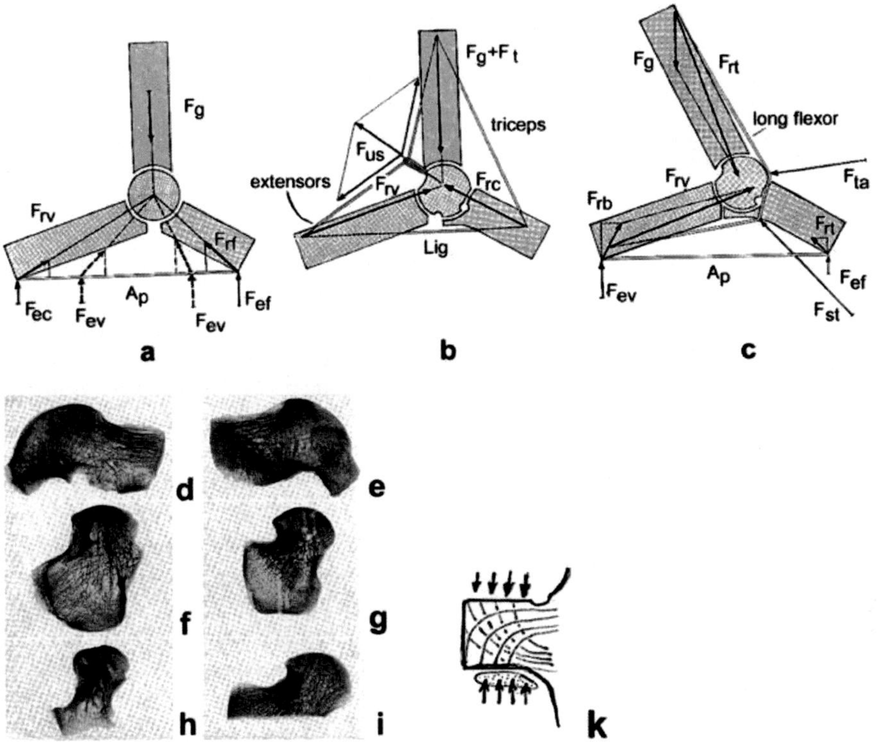

Fig. 8.38 The talus is kept suspended between the anterior and the posterior parts of the foot skeleton and the lower leg. (**a**) *Fg* from above presses the heel with force F_{ef} and the anterior part of the foot with force F_{ec} against the ground. Both parts are tied by a plantar ligament, or m. flexor digg. brevis, or m. quadratus plantae (A_p), and therefore the resultant forces F_{rv} and F_{rf} are created, which balance F_g. If the GRF does not act against the most distal elements, but is distributed over the length of the sole, represented by the broken lines F_{ev}, the resultant forces assume a steeper direction and keep A_p low. In (**b**), the pull of the m. triceps, or an external force ($F_g + F_t$), acting against the heel rotates the forefoot downward. The tendon of the pedal extensor keeps the anterior part of the foot in balance. Where it changes its direction (sling around extensor), it exerts the force F_{us} which keeps the talus in place. In (**c**), the toe flexors (*flex.long.*) contribute to equilibrium. The tendons change their direction at the rear margin of the talus (creating F_{ta}) and at the sustentaculum tali (creating F_{st}). In all load cases, the talus receives forces from three sides, which sum up to zero. (**d–k**) The pattern of spongy trabeculae reflects the force application and force propagation within the element, trabeculae being concentrated at the joint surfaces, from where they distribute. (**d, e**) Human talus in side view: (**f**) human, (**g**) and (**i**) chimpanzee, (**h**) baboon, all parallel to dorsum pedis. The direction of the collum tarsi depends upon the abduction of the hallux (see also Fig. 9.6). (**k**) Human sustentaculum tali cut in a transverse y/z-plane, to show the redirectional force exerted by the digital flexor tendons from below or the joint force acting in the anterior subtalar joint from above. The arcuate trabeculae indicate stresses caused by the redirection of the flexor tendons (upward) and the joint forces (downward) which vary in size. (From Preuschoft 1970a, b)

angular velocity times radius (as detailed above under leaping). It is, however, only possible, if the rotating moment of the toe flexors combined with the m. triceps at the ankle joint is great enough to counterbalance the load of the external force. The strength of the digital flexors may also be the reason behind lifting the heel from the ground: the rotational moment of these tendons is so strong that it must be counterbalanced by the opposed moment of body weight, if ground contacts of the heel are eliminated.

The most important excursions of the *ankle joint* take place between trochlea tali and distal joint of the tibia and are visible in side view (Fig. 8.4f–k). Both malleoli form something comparable to a "fork", which guides dorsiflexion and plantar flexion of the foot. On the malleolus tibialis is a distal extension of the joint surface, which in monkeys ends up in a medially facing, cup-shaped mould on the collum tali for the protruding malleolus. This arrangement permits transmission of forces along the tibia. Likewise, on the lateral side provides the fibula mainly a brace against force components in medio-lateral direction, parallel to the trochlea, and to a lesser degree in the direction of the fibula. The trochlea tali in the great apes is much more elevated at its lateral than on its medial margin. This has been related by Morton (several widely known papers since 1922 and his book 1948) to the (in fact occasional) loading of only the lateral margin of the foot in chimpanzees. By contrast, Preuschoft (1970a, b, 1971) connected this inclination of the trochlea tali to the abducted first ray (see also below, Chap. 9.2).

The *navicular*, the two lateral *cuneiform bones* and the *cuboid* of apes and monkeys are essentially stressed by compression along the longitudinal axis of the foot. These longitudinal forces are mainly caused by the resultant of GRFs against the balls and toes and the digital flexors, sometimes also by ligaments (like the lig. plantare longum in *Homo*). The extremely elongated navicular and the equally long neck of the calcaneus in specialised leapers have never been the subject of mechanical analyses. The most *medial cuneiforme* bears a more or less cylindrical joint facet, which permits movements of the first metacarpal about a nearly dorsoplantar axis.

The *zeugopodium* in primates consists of two parallel skeletal rods. In the hindlimb, *tibia* and *fibula* are (see above, Chapter 5.5) arranged side by side, more or less in the frontal plane (= z-direction) of the animal. The contact with the femur is established alone by the robust tibia (Fig. 8.39); the slender fibula, however, takes part in forming the distal articular contact with the foot, where both malleoli form a "fork" (*"Malleolengabel"*; see above), which guides the trochlea tali. Both long bones are connected by the *interosseous membrane*, most fibres of which descent from the tibia to the fibula, but there are also fibres in the opposite direction. The fibres reduce the laterally concave bending along the tibia and evoke laterally concave bending of the fibula. At the same time, the oblique fibres transfer a part of the compressive force from the fibula back to the tibia. If the tibia is slender and slightly curved with the concave side toward the fibula (as, e.g. in the apes), the share of load sustained by the fibula becomes greater and the element stronger. The long distance between the lower leg bones gives the segment a high strength against bending (Eq. 2.3 and elsewhere).

As long as *radius and ulna* are loaded side by side in a supinated posture (Fig. 8.40), the situation in the forearm is similar to that in the lower leg. A

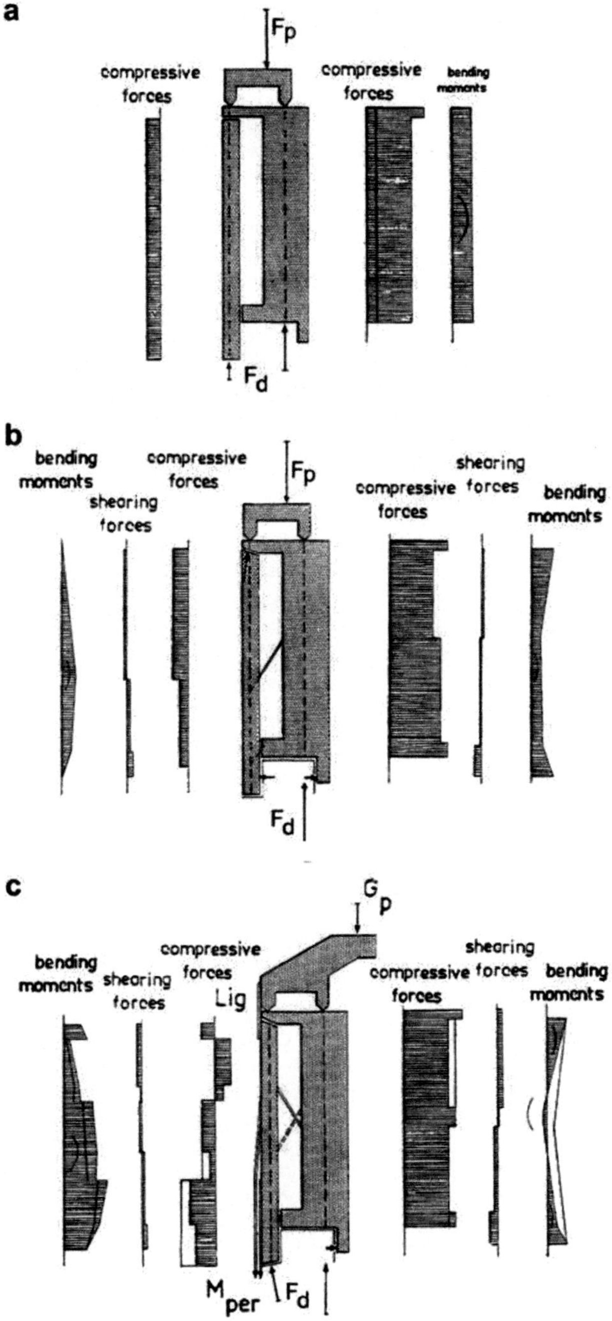

Fig. 8.39 Mechanical conditions in the (right) tibia and fibula seen from in front, with joint forces and stress diagrams. (**a**) The compressive force F_p (= weight) is applied through the femoral condyles to the proximal joint surface of the tibia and evokes a reaction force (F_d) against the distal

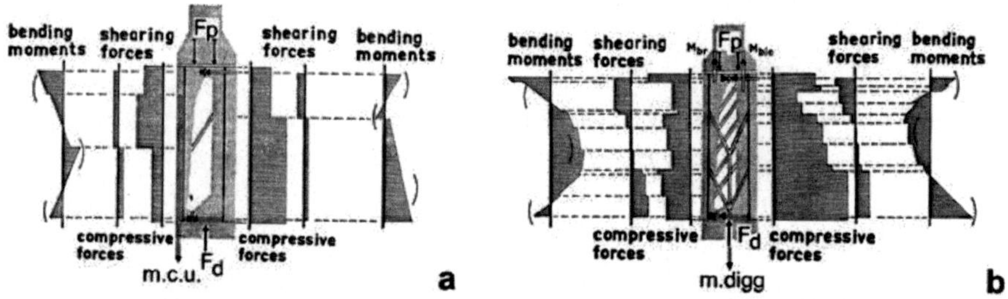

Fig. 8.40 Stressing of radius and ulna side by side in a supinated position. (**a**) A compressive load (F_d, regardless where it comes from, see text) is applied from the carpus to the radius and its reaction force at the elbow (F_p) from the humerus to both forearm bones. Because of the eccentric application of forces, both elements are subject to bending moments. In the middle of the system, a considerable share of the total compression is passed from the radius to the ulna by oblique fibres of the interosseous membrane, of which only one is shown. The bending moments along radius and ulna are reduced, or even inverted because of the interosseous fibres. The carpal flexor or extensor on the ulnar side (m.c.u.) is introduced to show the influence of muscular activity. (**b**) The distribution of oblique fibres along the whole length of the forearm bones smoothens the diagram of bending moments. Among the forearm muscles, the finger flexors (*m. digg.*) are the strongest. They take their origins from both forearm bones. If muscles are contracted, they increase the compressive stresses between origin and carpal joint and bend radius and ulna convex to each other. A means to keep these bending moments low is the often observed inverse, outward curvature of the skeletal elements

compressive force acts in longitudinal direction between the carpus and the elbow. This force in most cases will be evoked as a reaction to the pull of the strong digital and carpal muscles; sometimes, however, it may stem from carrying body weight – as, for example, in a knuckle-walking chimpanzee. Between ulna and carpus exists a compliant discus articularis, which undergoes deformation rather than transmitting force. Consequently, in the "higher" primates, the carpal joint force acts primarily on the radius, while the ulna does receive a small share, if any (see below). The oblique fibres of the interosseous membrane transmit gradually a considerable part of the

---

Fig. 8.39 (continued) joint surface. Depending upon the elasticity of the bone, a part of the load is transmitted on the fibula (indicated on the left side), while the tibia is relieved, but subject to some bending moments, because of the eccentricity of force and counterforce. Minimal elastic deformation of the tibia transfers a (moderate) share of total compression to the fibula downward to the joint surface between fibular malleolus and talus. The arcuate lines on or beside the diagrams indicate the direction of bending. (**b**) The interosseous membrane excludes a slipping of the fibula away from the tibia. Instead of the expanded membrane, just one fibre is introduced. This oblique fibre transfers a part of the compressive stresses from the fibula back to the tibia, In addition, the tension within the fibres causes bending moments in the fibula and reduces the bending moments along the tibia, provided that a horizontal force component (horizontal and inclined arrows) can be resisted. (**c**) If the knee is abducted and the foot adducted (as is common among the large apes), the mm. peronei must keep the tibio-talar joint in balance. In this case, fibres of the membrana interossea are tightened, which have the direction opposite to those described above. Bending of the tibia is near zero; the fibula, however, is exposed to greater bending moments than before and to higher "normal" forces. In its proximal portion, tensile forces may occur, while muscles taking their origins from the fibula (mm. peronei, soleus) may act as a stress-reducing tie and take over the tensile force from the fibula. (From Preuschoft 1970a) compressive force from the radius to the ulna, so that both forearm bones are pressed against the humerus. On the radial side, this compressive force is the only one which acts between radius and humerus. Therefore, the joint surface is at right angles to the shaft. In contrast, the joint force between ulna and humerus may well assume different directions (e.g. Fig. 8.44h, i). In accordance with the compressive forces illustrated in Fig. 8.40a, the distal half of the radius and the proximal half of the ulna possess the larger cross-sectional areas to take over the proximally decreasing or increasing compressive forces.

In many lorisids and in fossil primates from the Eocene (*Europolemur kelleri*, Franzen 2000) and Oligocene (Simons 1972; Fleagle 1983; Preuschoft 1974; Preuschoft et al. 1993), the *ulna is much stronger developed* than in "higher" primates. Preuschoft et al. (1993) have connected this fact with the ulnar abduction of the hand in walking along thin twigs (Figs. 8.41a and 8.42a). In this frequently used locomotion, the finger flexors must exert great forces for grasping handholds. For these modes

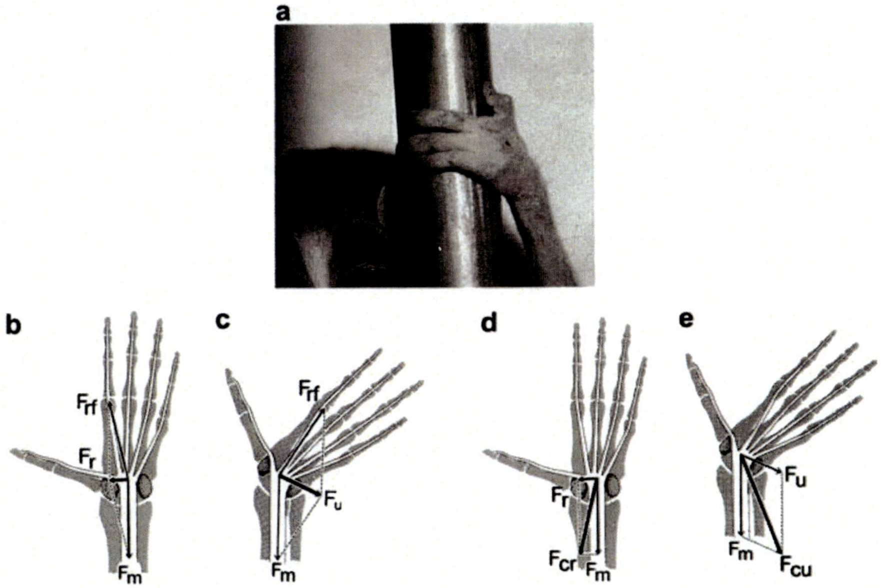

Fig. 8.41 Forces acting on the ulna. (**a**) *Loris tardigradus*; frequently assumed abduction of the hand in ulnar direction during vertical climbing along a branch. (**b**) Human, mesaxonic hand seen from its volar side: the forces of the common flexor tendon proximal to the carpus (F_m) and the resultant of the tendons flexing the digits (F_{rf}) include a blunt angle; therefore a (small) redirectional force (F_r) acts against the eminentia radialis. (**c**) If the hand is abducted in ulnar direction, the angle between F_{rf} and F_m opens toward the ulnar side; a (rather large) redirectional force (F_u) acts against the eminentia ulnaris of the carpal tunnel (see Fig. 8.36d). (**d, e**) In the joint between the proximal carpals and the forearm bones, the pull of the flexor tendon (F_m) and the redirectional forces (F_r) and (F_u) from (**b**) and (**c**) come to action. These forces combine to the resultants F_{cr} in the mesaxonic or entaxonic hand and F_{cu} in the abducted, ectaxonic hand. In the latter, the ulna receives a greater joint force than the radius. (From Preuschoft et al. 1993)

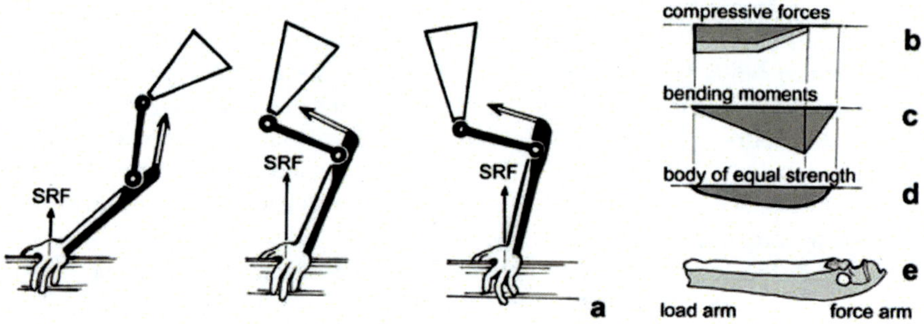

Fig. 8.42 (**a**) Three stages of the stance phase in *Saimiri* as an example for loading the ulna, beginning with touchdown (left) to liftoff (right). The element is bent like a beam on two supports, by the SRF (black arrows) and the pull of the m. triceps brachii (double line/white arrows). (**e**) As an example of several "primitive" primates, the fossil forearm bone in the Eocene adapiform *Europolemur* from Messel is shown. (**b**) Diagrams of compressive forces, (**c**) bending moments along the ulna if loaded as in (**a**). The bending moments require a body of equal strength (**d**), which resembles the fossil ulna. (From Preuschoft and Franzen 2012)

of climbing, the proportions of "ectaxonic" hands are advantageous, since the elongated 4th digit yields the widest span for gripping with the "power forceps". The muscle and tendon forces act along the forearm. The distribution of tendons on the fingers is unimportant, as long as each of the lateral digits receives a deep and a superficial flexor tendon. The individual carpal bones can be neglected and the whole carpus treated as a unit. Because of the ulnar abduction of the hand, the tendons change their direction and so exert a redirectional force against the eminentia carpi ulnaris (Fig. 8.42c). This force is transmitted through the carpus on the distal part of the ulna (Fig. 8.42e). The same holds true for climbing "below branch" in suspension. The ulna so becomes the most highly loaded element of the forearm. It is indeed stronger developed in extaxonic strepsirrhines, than in the mesaxonic hands of Old World monkeys and apes. Primates less prone to moving along thin branches orientate their metacarpus commonly in line with their forearm. During contraction of the finger flexors in a "mesaxonic" position (like in apes and humans), no redirectional force is exerted in ulnar direction (Fig. 8.42b, d). If the flexor tendons of the 1st and 2nd digit are particularly strong ("entaxonic" hand position), a force may occur in radial direction, pressing against the eminentia radialis carpi, that is, the os hamatum. This leads to a shift of the reaction force toward the radius, leaving the ulna less stressed. The joint force between the carpus and the radius deviates radially from the forearm and so gives reason for an inclination of the distal joint surface against the longitudinal axis of the radius.

The *femur* in most primates is rather straight; the round, ball-shaped caput femoris has a long collum femoris. It resembles the shapes known from most other mammals, and in fact it is moved preferably in a parasagittal plane (Fig. 8.43c). In orangutans, the femoral neck is connected in a wide angle with the femoral shaft, similar to the "coxa valga" in humans. The ligamentum capitis femoris is poorly developed, slender and long in gorillas and missing in most orangutans. Correspondingly,

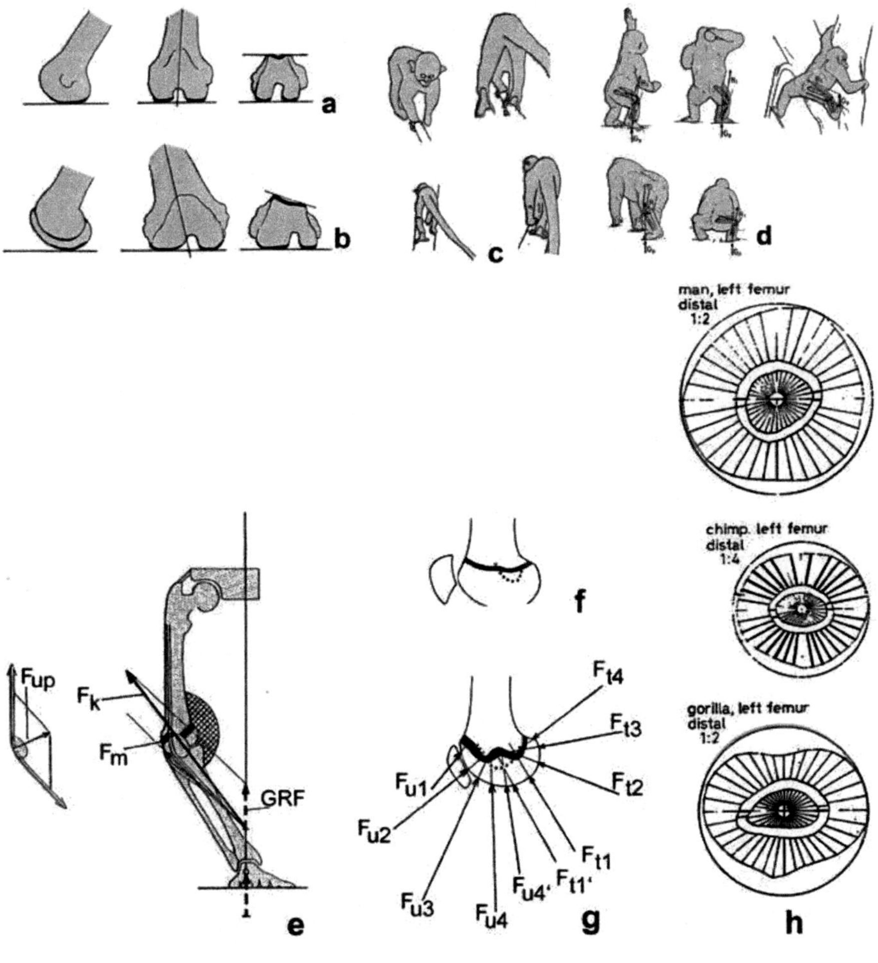

Fig. 8.43 Knee joints in primates. Left femoral condyles of a baboon (**a**) and female gorilla (**b**) as examples for monkeys and apes; seen in lateral (left), anterior (middle) and distal view (right). Facies patellaris is emphasised by heavy line. (**c**) Squirrel monkey as an example of a quadrupedal monkey, to show the limb and especially knee movements in a parasagittal plane. Note also the "schizodactylous" hand in anterior view. (**d**) Juvenile and adult gorillas with abducted hindlimbs. The abduction of the knee is usually combined with adduction of the foot. (**e**) The knee joint in apes is balanced in the frontal y/z-plane mainly by a contraction of the m. vastus lateralis. The pull of the ligamentum patellae and the GRF combine to the resultant F_k, which compresses the medial condyle more than the lateral. Note the position of the trochlea tali, as discussed in Chap. 9.2. The joint forces in the knee joint (F_u and F_t) of a gorilla (**g**) vary much more than in a human (**f**). Consequently, epiphyseal sutures are more folded. (**h**) Cross sections through the distal third or fourth of femoral shafts in African apes possess their greatest strength in z-directions, in agreement with load case (**e**). (From Preuschoft and Tardieu 1996)

there is no rugosity on the vertex of the femoral head. All traits seem to be related to the free mobility of the hip joint (Fig. 8.43d), but a really satisfying analysis has never been presented. While both distal condyles in strepsirrhines and monkeys have about the same size, they are asymmetric in (especially the African) apes (Fig. 8.43a, b). The medial condyle has a longer radius of curvature than the lateral. Its joint surface is greater, though it does not in all cases project farther distally. The condylo-diaphyseal angle in anterior view is about 90° or less; the cross sections through the shafts have longer medio-lateral than anterior-posterior diameters (Fig. 8.43h). In distal view, the medial margin of the facies patellaris is obviously higher than the lateral (Fig. 8.43a, b). In orangutans, this asymmetry is less pronounced; in other primates (monkeys, tarsiers and strepsirrhines), it is missing or at least not obvious (as noted above, and see below in Chap. 9.2). Preuschoft (1970a, b, 1971) has these from ancient times known facts explained by the varus position of the abducted knee joint: in anterior view, the femora of apes in critical, load-bearing phase of climbing and walking are abducted (Fig. 8.43d). The most extensive excursion of the femur against the trunk is possible in abduction (Preuschoft 1961, 1963a, b), where the thigh movements are not restricted by the belly. This is best to perform, if the *ilium*, from which the hip flexors and hip abductors take their origins, is expanded laterally. The foot, however, must be placed as close to the median plane of the animal as possible, in order to minimise the danger of breaking the substrate and to avoid spending too much energy. Preuschoft and Tardieu (1996; see also Tardieux and Preuschoft, 1996) have repeated the analysis. They have estimated the direction of the joint resultants in various positions of the knee and related them to facets of the epiphyseal plates. These were found to offer a multitude of surfaces perpendicular to force components which occur during locomotion (Fig. 8.43f, g).

The knee in the African apes is exposed in anterior view to medial flexion by the GRF and balanced by the lateral head of the m. vastus and the ligamentum patellae (Fig. 8.43e). The produced force changes its direction in the patella, which therefore is pushed in medial direction. This force component combines with the redirectional force in side view and presses the patella against the medial border of the facies patella. It can only be sustained, if the medial margin is elevated and the patellar facies inclined. The medial condyle receives more of the joint force than the lateral, and the shaft of the femur needs – and possesses (see above) – increased bending strength in the medio-lateral plane (Fig. 8.43b, g). In the lower leg, the resulting bending moments are sustained because of the great distance between fibula and tibia. (Remember Eq. 2.3: strength increases by the second power of distance between the resisting rods and see above.)

In monkeys walking frequently on the ground, the most proximal part of the *humerus* shows a curvature ("retroflexion") of the shaft axis (Fig. 8.44b) and a "retroversion" as well as a distal extension of the joint surface on the caput ((Fig. 8.44b–d); Krause 2008; Scherf et al. 2009; Scherf, 2013). The humerus of hominoids, however, is straight, in some apes slightly curved concavely on the anterior (flexor) side (Fig. 8.44e). The joint surface occupies a large part of the globular humeral heads. In comparison to the large joint surface on the spherical caput humeri, the glenoid surface of the scapula is always small (Fig. 8.44f).

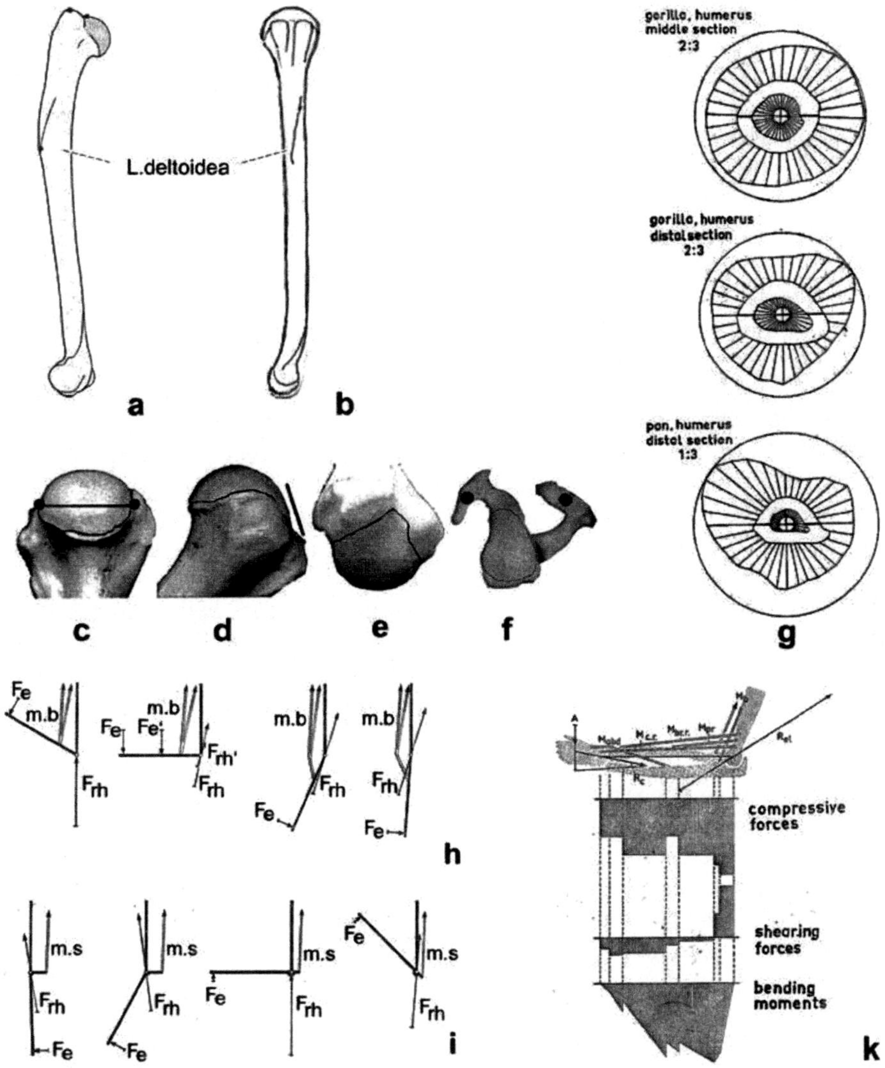

Fig. 8.44 Humerus in primates. (**a**) Lateral view of a more terrestrial OW monkey, showing the retroflexion of the shaft proximal to the linea deltoidea, the retroversion of the caput and the anteversion of the distal joints. (**b**) Humerus of an ape. In some individuals, especially gorillas, the whole shaft is curved concave on the side of the flexors. (**c**) Caput humeri of an OW monkey, seen from caudal side (fat line showing breadth); (**d**) viewed from lateral side, the projecting tuberculum provides a long lever arm (fat line) for the m. supra spinam; (**e**) cranial view. In the apes, the glenoid joint facet on the head is directed more medially than in quadrupedal monkeys, exhibiting the so-called torsion. (**f**) Facies glenoidalis, processus coracoideus (dot) pointing to the left, and acromion (dot) pointing to the right. (From Krause 2008). (**g**) Cross sections through a humerus at midshaft (top) and at two thirds of length in gorilla and chimpanzee. (**h, i**) Joint forces in the elbow depend upon position and external force, of which examples are shown. If the external

Obviously, the joint surface on the caput humeri provides a reserve, which allows force transmission in different positions of the scapula, while the glenoid facies of the shoulder blade possesses the area necessary for distributing the joint force without doing damage to the joint cartilage. In quadrupedal posture, the resultant of all forces acting in the shoulder joint deviates dorsally from the greatest length of the humerus.

In the *elbow joint*, however, the resultants of all forces acting in the elbow joint deviate from the long axis toward the flexor side (Figs. 8.44h). The humerus shaft in about two thirds of its length has a circular cross section (Fig. 8.44g), which allows withstanding forces that deviate in any direction from the bone's axis. In its distal third, its cross section becomes broad in radioulnar direction, with a flat dorsal aspect (Fig. 8.44g). This shape is connected with the very tough and strong collateral ligaments of the elbow joint, which exclude any movement in the radioulnar plane and transfer tensile forces resulting from bending in the radioulnar plane. Since most external forces act against the hand, they have long lever arms and therefore considerable moments. The two bony elements are at distance from each other and so can transmit these moments across the elbow joint.[9] In the proximal half of the upper arm, the strong shoulder and elbow muscles act as ties of the humerus and take over the tension produced by bending the segment.

The swinging of the GRF like an inverted pendulum has been documented in various monkey species experimentally by D. Schmitt (1998, 2003). It implies the rolling off of the autopodium on the substrate and is characteristic of quadrupedal locomotion in general (see above). The illustrated positions of the limb segments (Fig. 8.45) are preferred by "more terrestrial" Old World monkeys (Larson and Stern 2007; D. Schmitt 2003a, b) and differ slightly from those in "more arboreal" monkeys – as also presented by D. Schmitt (1998, 2003). In all phases of walking, the elbow must be kept in balance by the m. triceps brachii.

Typical for the walk of monkeys are their *long steps*. In contrast to other mammals, they put *the hand* on the substrate *clearly in front of the eyes* (M. Schmidt 2008; M. Schmidt et al. 2002). At midstance, and before liftoff, the GRF passes behind the shoulder joint, and this must be kept in equilibrium by the

---

Fig. 8.44 (continued) force (F_e) extends the elbow joint (**h**), and the elbow flexors (*m.b*) are active, joint forces deviate from the humerus axis toward the flexor side, requiring an anteversion of the joint surface. If, however, the external force (F_e) flexes the elbow (**i**), and the m. triceps (m.s) is activated, joint resultants often assume a direction along the shaft and require a direction of the joint surface at more or less right angles to the shaft. (**k**) Flexors of the elbow joint and muscles of the forearm, which reduce bending of the forearm bones. (From Preuschoft 1973a, b, 1975)

---

[9]The mechanical situation is similar to that in finger joints (see above) where also deformations in one plane are excluded by passive structures – which therefore need to resist high bending moments.

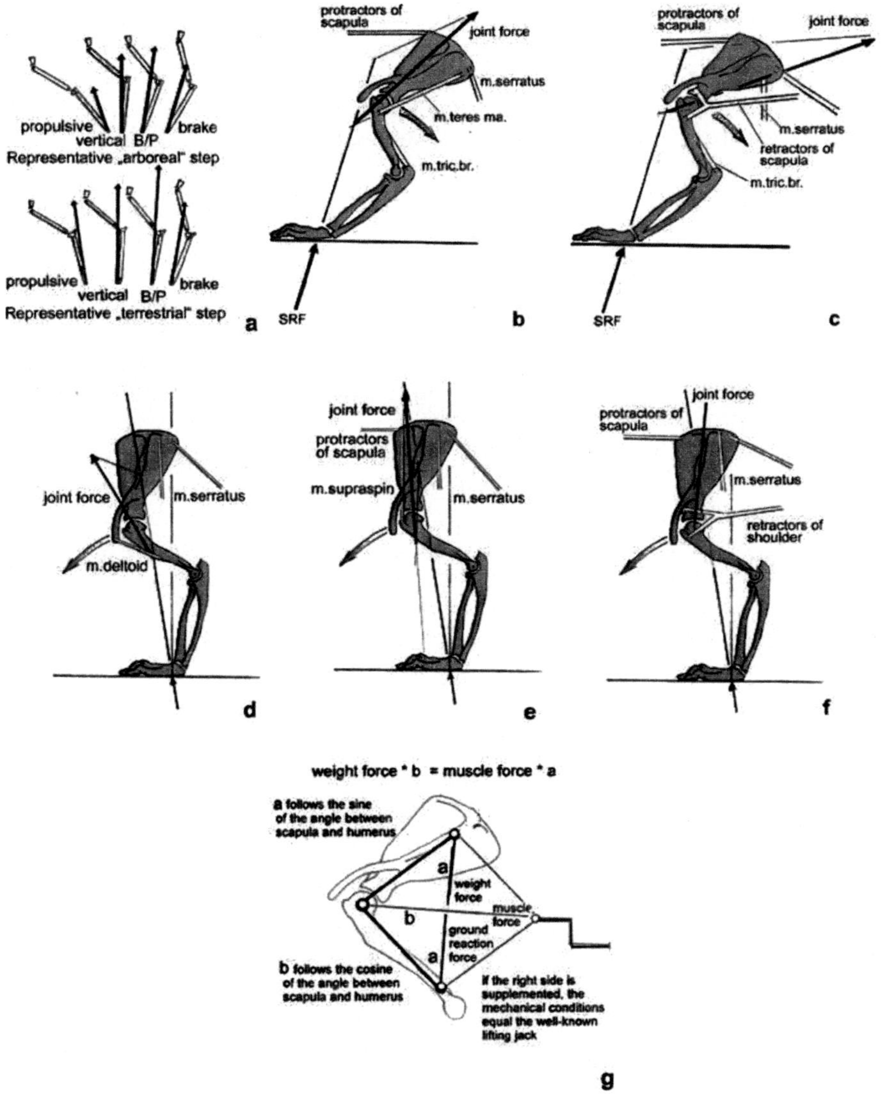

Fig. 8.45 Forelimb of a walking monkey and activity of shoulder muscles. (**a**) shows substrate reaction forces in "more arboreal" and "more terrestrial" monkeys according to D. Schmitt (1998). (**b–e**) Positions in x-ray recordings of the NW monkey *Saimiri*, obtained by M. Schmidt (2001, 2002). The data of M. Schmidt deviate by their greater excursion angles, GRFs passing at first substrate contact in front of the shoulder joint, rather than behind, and liftoff before the GRF being behind the elbow joint. The results obtained by both authors seem to encompass variations of the locomotor repertoire which can be chosen by the animals. The grey, bowed arrow near the shoulder indicates a tendency of the entire limb to swing rearward (**b, c**) or forward (**d–f**) about the wrist, if the elbow joint is fixed. This tendency can be blocked without input of energy by the (not illustrated) clavicula and the underlying ligamentum subclavium, which assume different directions at the beginning of the stance phase (**b, c**), its lateral part is directed rearward, at midstance mainly

mm. deltoideus and supraspinatus or the retractors of the shoulder (see below). In all cases (Fig. 8.45b–d), the joint resultant is at an angle against the humerus shaft. In the shoulder joint exist several possibilities, of which two are investigated here.

Immediately after touchdown, the GRF can be in front of the *shoulder*. Passive movements in the glenoid joint are excluded by the m. teres humeri or by the retractors of the shoulder, primarily mm. latissimus and pectoralis profundus. Both muscles insert into the humerus and at the shoulder blade and act together as the "lifting jack" (which is commonly used in motorcars and well-known Fig. 8.45g). They pull the shoulder joint rearward and so effect an extension. Because these muscles pull in caudal direction, a protractor of the shoulder blade must also be active to exclude a shifting of the scapula on the thorax. Protractors are mm. rhomboideus and levator scapulae, or the cranial part of the m. trapezius. Their action is necessary in addition to the weight-bearing m. serratus lateralis. The compression-resistant *clavicula* points rearward at touchdown, shortly before liftoff and forward toward the cranium, so bracing the shoulder in the respective directions.

The resultant force acting across the *glenoid joint* assumes variable directions against the humeral head. Thanks to its globular shape, the head offers to each resultant a surface at right angles (Fig. 8.45). Walking on the ground leads primarily to resultants in a parasagittal plane, but not so much in z-direction. This fits to the dorsal extension of the narrower, not so globular joint surfaces on the humerus in the more terrestrial monkeys. In the more arboreal primates, the resultant joint force is less one-sided, so that the joint surface must be and in fact is more spherical (Krause 2008). The insertion of the m. deltoideus into the humerus shaft distal to the head has the effect that the joint force acts along the (retroflexed) proximal shaft. The above-mentioned retroflexion places the proximal shaft into the same direction as the most relevant joint force (Fig. 8.45d).

Fig. 8.45 (continued) medially and in the later phases (**e, f**) forward). (**b**) If the hand is placed far in front of the glenoid joint (as is characteristic of primates!), activity of either the m. teres major or the m. latissimus is necessary to keep the joint in balance. Both muscles swing the joint resultant (fat arrow) toward the dorsal side of the humerus (see also below). For keeping the scapula in its place on the thorax, muscle forces are necessary: mm. trapezius or rhomboideus pulling cranially and caudal fibres of the m. serratus pulling rearward. In later phases of roll-off, either the deltoid muscle is active (**d**), or the m. supraspinatus (**e**), or the m. latissimus (**f**). This retractor of the shoulder exerts (like in **c** or **f**) force at the scapula as well as at the humerus; its function is that of a "knee lever", as is commonly known from a lifting jack (**g**) in motor cars. In cases (**d–f**), the joint force is directed to the supraspinous, cranial part of the scapula and deviates rearward from the longitudinal axis of the humerus, in agreement with the retroverted humeral joint surface. In (**d**) the curvature (retroflexion) of the humerus shaft contributes to saving muscle force by elongating the force arm, and in (**e**) a high tuberculum majus increases force arm length of m. supra spinam. Great length of the scapula and an open angle between the scapula and humerus obviously are of advantage. (**g**) shows a humerus with scapula, superimposed by a lifting jack instead of the m. latissimus/pectoralis. (From Preuschoft et al. 2003)

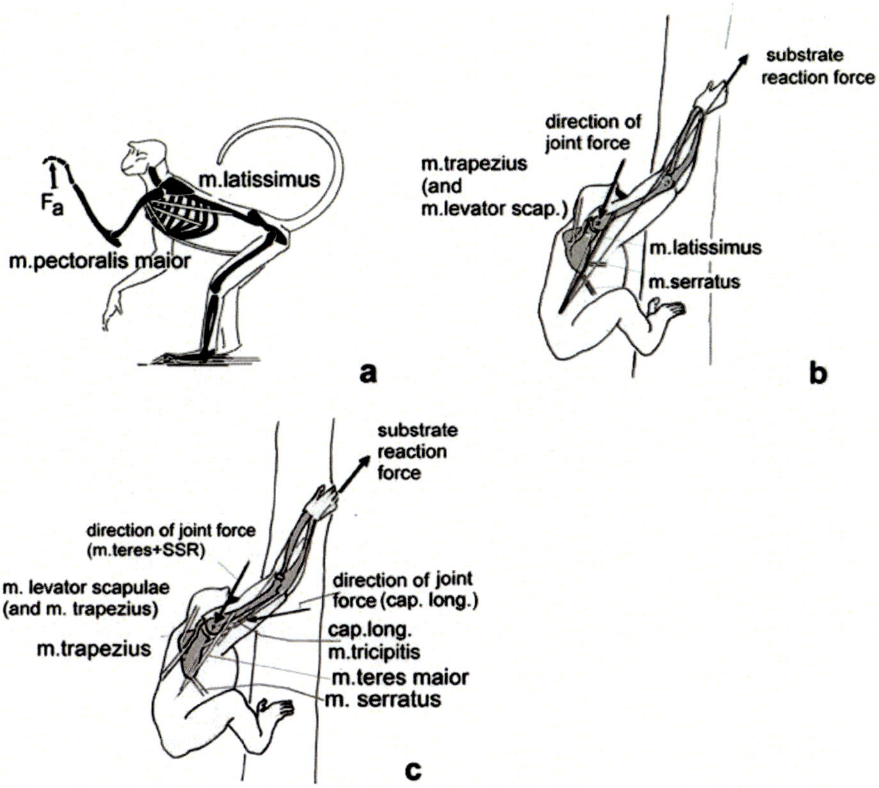

Fig. 8.46 Primate forelimbs in "elevated" positions. (**a**) Hanuman langur, grasping a branch higher than its shoulder. The "retractors", mm. pectorales and latissimus, in this case lift the body in relation to the fixed handhold, where the external force F_a is applied. Equilibrium in the joint can be established by the additional activation of the m. teres major. (**b, c**) In the "vertical climbing" chimpanzee, the same conditions must be fulfilled as in (**a**). Several muscles may contribute to keeping balance in the shoulder complex. In (**b**) the m. latissimus is assumed to be active. Since it spans (like the m. pectoralis major) not only the shoulder but also the scapulo-thoracal junction, the shoulder blade must be fixed on the thorax. This is accomplished by the protractors (m. levator scapulae, cranial part of m. trapezius, m. rhomboideus), and a rotation of the scapula with its caudal angle in dorsal direction under the influence of the joint resultant is inhibited by the m. serratus. In (**c**), the glenoid joint is assumed to be balanced by the mm. teres major and the caput longum m. tricipitis. The joint resultant has a similar direction as in (**b**), along the axillary margin of the scapula, but to keep the latter in its place on the thorax, the caudal part of the m. trapezius must be activated in addition to the muscles active in (**b**). If the m. pectoralis is recruited, a dislocation of the scapula in ventral direction is blocked by the clavicle. (From Preuschoft et al. 2003)

For arboreal animals, it is of advantage, often necessary (as in suspensory behaviour or vertical clinging to stems), to grasp a handhold in front of, or on top of, their head (Fig. 8.46a, b) and far lateral to the shoulder (Fig. 8.48). Because of the length of their forelimbs, the external force is applied in side view in front of and higher than the shoulder joint. Elevation and abduction of the arms do not, or rarely,

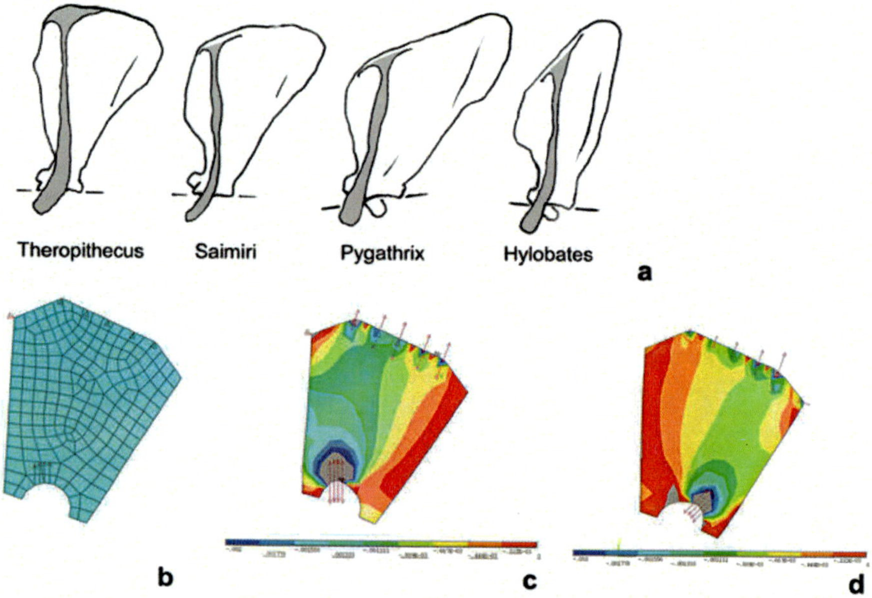

Fig. 8.47 Shapes of the scapula in primates. (**a**) Shoulder blades of four selected locomotor types, drawn to about the same length. The animals have, from left to right, more terrestrial locomotor modes (*Theropithecus*) and increasingly more arboreal and suspensory habits (*Saimiri*, *Pygathrix*, *Hylobates*). (From M. Schmidt 2001). Note the decreasing length of the spina scapulae and the increasing length of the (reinforced) margo axillaris. Both lead to a narrower angle between margo axillaris and margo vertebralis scapulae. The plane of the glenoid fossa is indicated by a nearly horizontal line and the spina marked in grey. (**b**) Meshwork of a FES analysis and acting forces (tiny arrows near the joint and the vertebral margin). (**c**) Two-dimensional model of a shoulder blade loaded like in the ground-walking monkey in Fig. 8.45; (**d**) two-dimensional model of a shoulder blade loaded like in the climbing ape in Fig. 8.46b, c. Highest stresses grey, dark blue, light blue; low or absent stresses yellow or red. (From Preuschoft et al. 2003)

occur in other, especially large mammals. Elevation and abduction require contractions of the *shoulder muscles* (which alone span the glenoid joint) as well as *shoulder girdle muscles* (which pass the shoulder blade and insert into the trunk; see Figs. 8.45 and 8.46). These muscle activities are not only postulated theoretically but have been documented experimentally by EMG (e.g. Larson and Stern 2007). They exert influence on the shape of the scapula, the ribs and the whole thorax – as will be shown below.

The shapes of the *shoulder blade* in monkeys and apes have been described repeatedly since Ashton and Oxnard (1961); examples can be seen from Fig. 8.47a. While the scapulae of cursorial mammals resemble a triangle with two equal sides, and the spina scapulae runs along the greatest "length", the higher primates possess a "short" scapula, in which the cranial margin is short and the axillar and vertebral margins are long. So the fossa supra spinam is much smaller than the fossa infra spinam; the caudal angle is acute and the cranial angle blunt. The orientation of the glenoid surface varies as can be seen from Fig. 8.47e. All primates possess a *clavicula*. The *thorax* of most primates is broad (in z-direction) and not deep (in y-direction), although there exists a divergence between monkeys and apes. Why these traits, in addition to the shoulder blade, vary in common is an important question, which has been investigated by Preuschoft et al. (2010).

Early authors, like Schultz (1933, 1930) (or even Mollison 1911), had noticed that the thorax in apes is not so narrow and deep as in monkeys and other mammals of the same size. The same characteristic holds true for humans, but Schmid (1983, 1991) has emphasised that the broad, not deep, *funnel-shaped thorax* (Fig. 8.50) of the apes clearly differs from the vaisted, *barrel-shaped rib cage* of humans (see Chapter 9: Human Evolution). Recently, Kagaya et al. (2008a, b) broke these observations down to the curvature of individual ribs.

All mentioned characteristics can be explained by the same mechanical analysis, aiming at the girdle skeleton of the forelimb: the *joint forces* illustrated in Figs. 8.45 and 8.46 act not alone on the *humerus* but also on the *shoulder blade*, at least up to the insertions of the shoulder muscles in a strict sense. According to the FES analyses of schematised scapulae during walking (Figs. 8.45, 8.47) or climbing with hands above the head (Figs. 8.46, 8.47), the high compressive stresses are confined to the neighbourhood of the glenoid surface, from where they distribute widely over a large area. In walking on all fours, the joint resultant, and that means high compressive forces (green) are close to the cranial margin, in front of the spina scapulae. The form of the scapulae with long cranial margins fits in the more terrestrial monkeys (as well as those of the cursorial mammals) to this stress distribution (Fig. 8.47a, Theropithecus) and provides large areas for insertion and great length of fibres. The glenoid surface is nearly perpendicular to the spina scapulae. Typical for stresses evoked in climbing with anteverted humeri and hands lifted above the head is a direction of the joint force close to the axillary margin of the scapula, caudal to the spina (Fig. 8.47a, *Pygythrix or Hylobates*). The glenoid surface in these forms is nearly perpendicular to the spina and to the axillary margin (Fig. 8.47a). Especially the apes, which often suspend themselves by the forward or laterally lifted arms, possess an obviously reinforced axillary margin of the shoulder blade, where strong muscle contractions are to be expected.

Seen from the front, the fixation of the trunk to the anterior limb and therefore the strength required from the ribs in the more terrestrial monkeys (baboon, macaques) does not essentially deviate from the situation in cursorial mammals. Therefore, morphological characters resembling those in these mammals can be expected and indeed exist (Fig. 8.49a, c). Placing the hand lateral to the shoulder joint, however, requires contractions of the mm. pectoralis profundus et superficialis, as well as latissimus (Fig. 8.48a). These muscle activities have been proved by EMG recordings of (Larson and Stern 2007; Stern et al. 1980). They lead to joint resultants, which are directed from lateral to medial, and therefore push the shoulder blade medially. The latter is fixed to the thorax on the ventral side by the clavicle and on the dorsal side by the m. serratus lateralis. A marked *curvature of the ribs* gives the mm. serratus, latissimus and pectoralis a medio-lateral direction, and therefore the

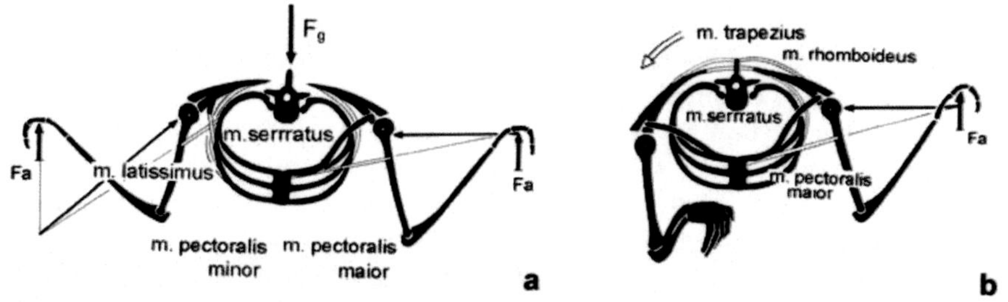

Fig. 8.48 Loading of primate frontlimbs in an abducted posture, like in Fig. 8.46a, viewed from in front. (**a**) Symmetrical support by both arms. The external force F_a has the same magnitude on both sides, but on the left side of the reader, the shoulder joint is balanced by the mm. latissimus dorsi, on the right by the m. pectoralis superficialis. Realistic may be a combination of both variants. In both cases, the joint resultant is determined by its medial component. The shoulder blade is kept in its position on the thorax on the left side by the m. pectoralis profundus and the m. serratus lateralis, on the right side by the same muscle and the clavicle (which is drawn only on this side). Again a combination of both variants may be realistic. (**b**) Asymmetrical support by only one forelimb. Alone the m. pectoralis superficialis is recruited for keeping balance. The joint resultant pushes the scapula in medial direction, and its giving way is prevented by the m. serratus lateralis. A twisting of the shoulder complex against the supporting arm under the influence of its weight is excluded by the mm. rhomboideus and trapezius. For maintaining balance in the trunk, the oblique muscles of the body wall are responsible, as in quadrupeds. The adducting mm. latissimus, serratus and pectorales superfic. et prof. exert forces with considerable components in the frontal plane (z-direction) at their origins from the anterior ribs. These slender rods can sustain the forces only if they are orientated in the direction of the muscular forces. This is equivalent to the well-known pronounced curvature of the ribs, which form the broad but not deep thorax – in contrast to the straight anterior ribs in cursorial mammals. (From Preuschoft et al. 2010)

reactive compressive forces within the ribs act along the rib axis, so that no bending moments occur along these slender rods.

Interestingly, very similar internal forces occur, if the hand grasps hold of a substrate below the trunk but exerts, besides weight force, an additional force component in medial direction (Fig. 8.49b, d). D. Schmitt (2003a, b) has shown experimentally that the SRF during walking along a pole contains such a component. The latter is necessary, because the CoM must be kept above a support, the breadth of which is narrower than shoulder breadth of the animal. This is in accordance with our considerations of the footfall sequence (see above). The total SRF is laterally inclined, according to D. Schmitt (Fig. 8.49b) and balanced by contractions of the mm. pectoralis major et minor as well as latissimus. These muscles redirect the acting force so that it meets the glenoid surface (Fig. 8.49d). Equivalent activities of these muscles have been documented by Larson and Stern (2007). The joint force is received by the scapula optimally, if it deviates in frontal view from the vertical (= sagittal plane) and approaches the horizontal (= frontal) plane of the trunk.

In both load cases, the marked curvature of the ribs places the bony rods in a direction parallel to the pull of mm. pectoralis, latissimus and serratus (Figs. 8.48 and 8.50c, d) if a handhold is seized above the head or lateral to the shoulder joint, as

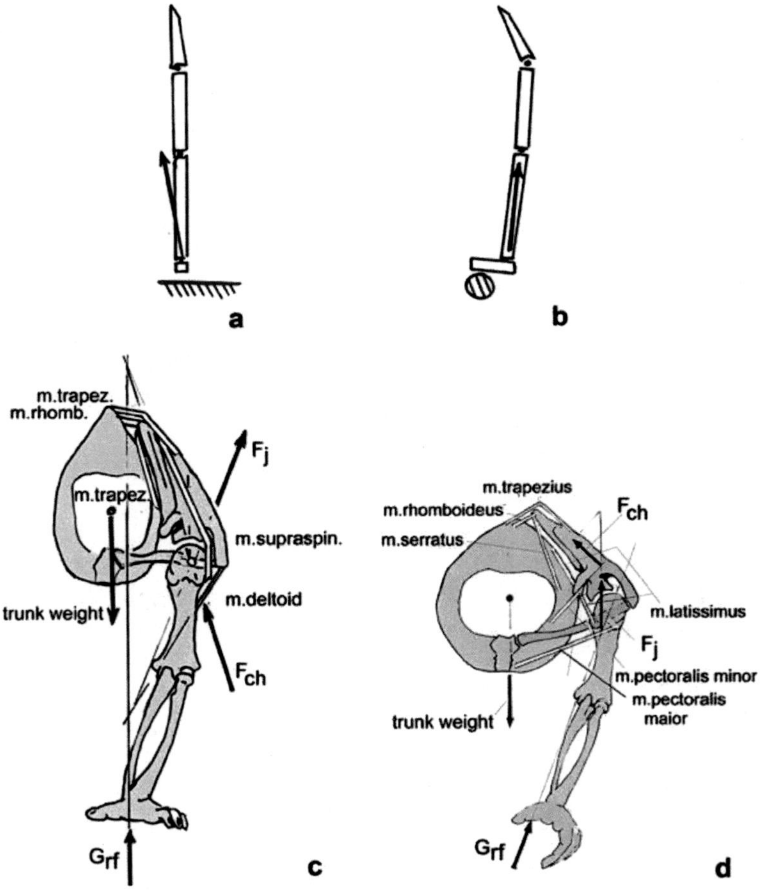

Fig. 8.49 Forces acting in the shoulder joint, shoulder blade and thorax, in anterior view. **(a, b)** External substrate reaction forces according to experimental data obtained by D. Schmitt 2003a, b. The more terrestrial monkey walking on flat ground **(a)** produces a GRF (arrow) which passes medial to the glenoid joint, halfway to the median plane of the animal. More arboreal monkeys walking along a pole **(b)** produce a laterally inclined SRF, which passes along the forearm and humerus, crossing the glenoid joint. **(c)** The typical load case in more terrestrial monkeys is similar to the situation in cursorial mammals (Sub-Chapter 7.7.4). The shoulder joint is balanced by the mm. deltoideus and/or supraspinatus, both producing an upward directed joint resultant (F_j). Both muscles exert a redirectional force component (not drawn), which redirects the force between humeral head and glenoid surface at the scapula medially. This force is countered partly by the clavicle. The trunk is hung up by the m. serratus lateralis and secured in asymmetrical support by the mm. rhomboideus and trapezius. These muscle forces plus GRF combine to the force F_{ch}, which acts along the scapula. **(d)** Characteristic load case in a more arboreal monkey walking along a pole. The laterally directed SRF is countered by the mm. pectoralis major, minor and latissimus; the joint resultant F_j has a more medial direction than in **(c)**. By the pull of the mm. serratus, rhomboideus and trapezius, which keep the trunk in its place, like in **(c)**, F_j is redirected to the force $F_{ch,}$ which runs along the shoulder blade. Consequently, the latter finds its appropriate position between the sagittal and the frontal plane. In addition, it is also propped by the clavicle

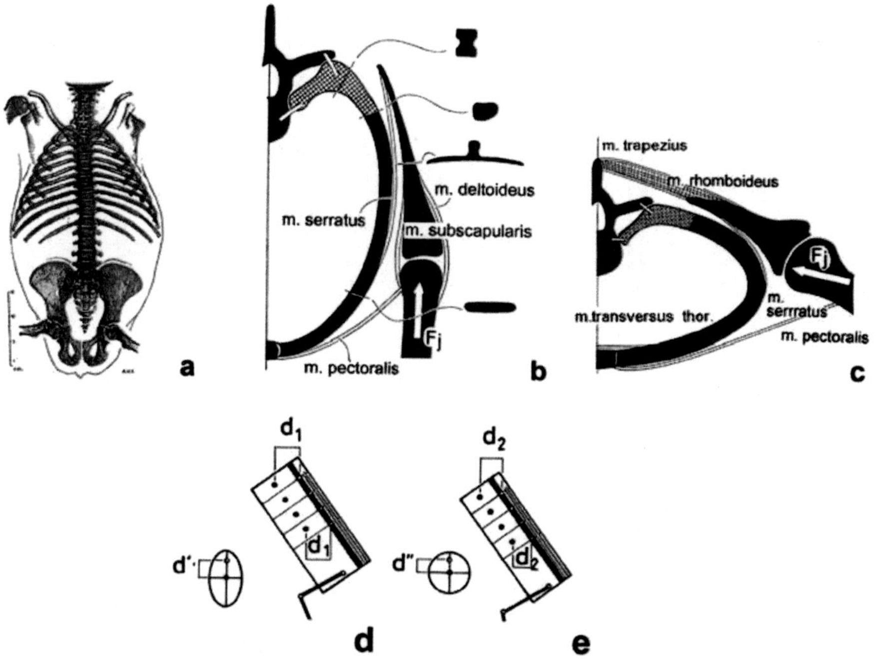

Fig. 8.50 Divergent shapes of the thorax and curvature of ribs. (**a**) In most primates, especially apes, in this case a (juvenile) orangutan, the trunk is short. Note the funnel-shaped thorax, the cranially swung clavicles and the pelvis. (Redrawn from an x-ray published by Nakatsukasa et al. 2004). (**b**) Stressing of ribs in a quadrupedal cursorial mammal, like a baboon. The joint force of the glenoid joint is directed dorsally (white arrow) and pushes the scapula in dorsal direction. The rib cage is connected to the scapula by the mm. pectoralis and serratus lateralis. Most of the rib is under compression (black); its dorsalmost part in addition is bent (stippled) – as is the scapula. The profiles of the ribs and of the scapula (thanks to the spina) are suited to sustain the occurring stresses, as becomes evident by the profiles (set out to the right). (**c**) Stressing of the ribs in a primate climbing with laterally abducted or elevated arm. The joint force is directed medially (white arrow). The pronounced curvature of the rib leads to the effect that again compression prevails in the bony rod, except in the dorsalmost part – which is bent in the same way as (**b**). In contrast to the deep trunk of ground-dwellers (**d**), the less deep, more rounded rib cage of arborealists (**e**) makes upright postures of the trunk less energy demanding, because the lever arms of the body segments (*d*1 or *d*2, respectively) are shorter in (**e**) than in (**d**). This facilitates not only the (rare) upright standing but also the (frequent) upright sitting. (From Preuschoft et al. 2003). ***See also Figs.*** 9.14*c* ***and*** 9.17*b, c*

described above. Because of the coincidence of ribs with the compressive reaction force to the pull of muscles, the slim bony elements are not exposed to being bent and so can provide sufficient strength against the only acting force, namely, compression.

The flow of forces so explains the broad, not deep *trunk* (Figs. 8.49 and 8.50) which is characteristic in particular for the apes, animals, often suspeded by their forward or laterally lifted arms. Still the characteristic "funnel shape" of their rib cage with narrow cranial and increasing width toward the abdomen meets the requirements of quadrupedal body postures (see above, Sub-chap. 7.7.4). This trait gets importance in comparison with humans (see also Sub-chap. 9.2.1).

The clavicle is exposed to compressive (Fig. 8.48 and 8.49c) and, in special cases, also to tensile forces (Fig. 8.49d). If so, the lig. subclavium and the m. subclavius are structures suited to take over such tensile forces.

In most (higher) primates, the trunk is short, and the *vertebral column* comprises 7 cervical, 12 or 13 thoracal, rib-bearing, 4 to 5 lumbar and 3 or more sacral vertebrae. In about 1/3 of gorillas exist only three lumbar vertebrae. The axial skeleton is supplemented by the long ilium. In most primates, it is fixed firmly to the sacrum by three and in apes and humans by five sacral vertebrae. The long distance between the caudal limits of the thorax and the pubic symphysis provides much space for the often obvious expansion of the belly, which contains, especially in leaf-eaters, a voluminous digestive tract.

The girdle skeleton of the hindlimb, the *pelvis* (Fig. 8.51a, b), appears to be a part of the trunk – where the interpretation of the pelvis as a bowl, carrying the intestines (as derived from the human condition), is not really imposing in the non-human primates nor in other quadrupeds. As mentioned above (Chaps. 4.3, 6.4 and 7.7.6), the *anterior view* of the pelvis has been compared to a ring. Its part ventral to the hip joints, on both sides of the pubic symphysis in symmetrical support on both hindlimbs, is exposed to tensile force. The part dorsal to the hip joints, consisting of the ilium and sacrum of both sides, is exposed to compression if supported symmetrically by both hindlimbs. By contrast, during asymmetrical support on only one limb, the ventral part is under compression but the dorsal under tension in one-limb support. On the ventral as well as on the dorsal side exist nearly immobile junctions in the form of the pubic symphysis and the auricular surfaces connecting the ilium and sacrum. Both connections are large enough to permit transmission of compressive force and tied firmly together by very tough fibres of connective tissue. Yet unclear is why the ventral and the dorsal joints persist without being ossified, while they do not visibly move.

The pelves of strepsirrhines, ceboids and cercopithecoids do not exhibit much variation. The *ilium* appears cranial to the hip joint as a straight rod and caudal to the hip joint the *ischia* as cantilevers braced in the middle by the *pubes* (Fig. 8.51e–g and below). In side view, the ilia are a bit inclined against the lumbar spine but nevertheless seem to prolong the axial skeleton. Their cross sections mostly resemble an L. In the apes, the very broad ilia resemble plates, orientated nearly in the frontal plane. Only few static analyses have yet been attempted. In most of the body postures investigated by Preuschoft (1999), the feet are placed more or less cranial to the hip joint. Therefore hip extending muscles must be active, to keep equilibrium against the share of weight force carried by the hindlimb. No matter which muscles are activated, they combine with weight force F_g to a joint resultant F_h, which passes from the hip joint along the ilium and crosses the iliosacral junction (Fig. 8.51f–h). The ilia therefore are under compressive force, not under bending moments. Likewise, no rotational moments do occur in the iliosacral junction. Both joints, the hip and the iliosacral junction, are in equilibrium as long as the resultant maintains its

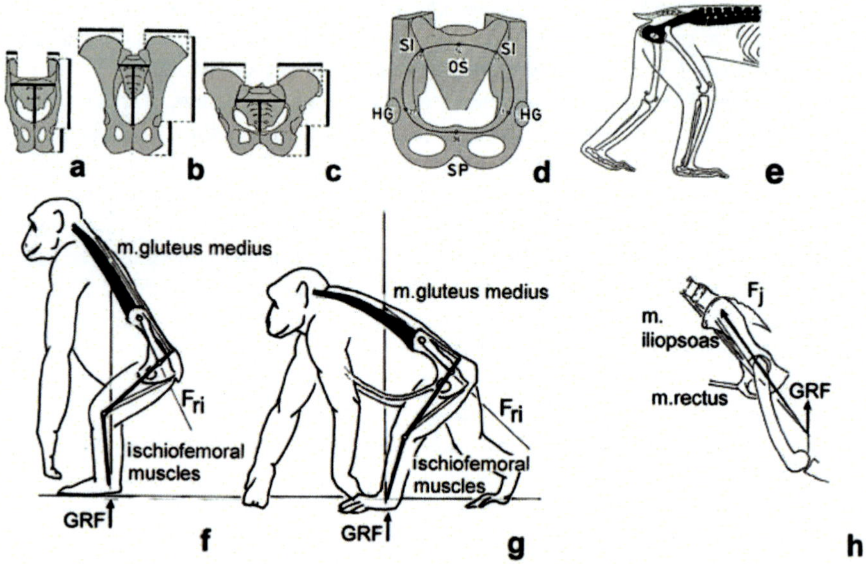

Fig. 8.51 Pelvis of non-human primates. (**a**) the pelves of most primates are similar. Only the apes diverge by the width of their iliac blades (**b**). Hominids deviate from both (**c**). Characteristic distances are emphasised. The traditional interpretation of the pelvis is the comparison with a bony ring (**d**), which can be loaded either symmetrically or asymmetrically. (**e**) X-ray photo of a Japanese macaque, *Macaca fuscata*, in side view. This is one of the bipedally trained individuals; therefore it has a lumbar lordosis. Note the posterior cantilevers formed by the ischia, both of which braced by the pubes at the midline. (**f, g**) Chimpanzee in bipedal (**f**) and quadrupedal (**g**) body posture. In both, the hip joint is balanced by the ischiofemoral muscles in combination with the m. gluteus medius. The latter muscle also provides tensile forces, necessary for keeping balance in the forwardly inclined trunk (**f**). If the foot is placed far behind the hip joint, the flexors of the hip joint become active (**h**). Only the strongest of them (m. iliopsoas) is shown. The joint force F_i again passes along the ilium and crosses the iliosacral joint, so that moments do not occur

direction. Only one reinforcement exists in the form of a strong strand of connective tissue between the sacrum and the ischium. No matter whether this has the form of a "sacroischial ligament" (as in primates or carnivores) or of a "ligamentum latum" (as in ungulates), it excludes any widening of the angle between the ilio-ischiadic axis and the sacrum, even if the muscles of the gluteus group are not active. As a whole, the statical situation in primates is similar to that in cursorial mammals (see above, Chap. 7) in which the iliosacral junction is often surprisingly small, restricted on one or two vertebrae.

As noted above, and shown in Fig. 8.51a, the *ilia* in the *apes* are expanded in the medio-lateral (z-) direction. The greatest amount of bone substance so is located parallel to the dorsal contour. This arrangement combines high compressive strength with low bending strength in side view and a large area for muscle insertions. On the dorsal surface, it is the m. gluteus medius, which occupies the ilium, on its ventral side the m. iliacus. The former sustains, aside from extending the hip joint, the tensile component of the bending stresses which occur in the often assumed semi-erect trunk postures exactly where they reach their highest values (Fig. 8.51f), close to the dorsal contour of the lower trunk. The obvious breadth of the ilia provides a large area for insertions of a great number of muscle fibres. The length of the ilia entails great length of muscle fibres of both, mm. gluteus and iliacus, and that permits wide active excursions of the femur in the hip joint. There may well exist mechanical conditions similar to those in the frontally orientated scapula, but this line of thinking has never been worked out.

Very strong extensors of the hip joint take their origins from the *ischia*. The complex of both ischia and pubes provides a bony cantilever for the ischiofemoral and the hamstring muscles. A bending of the ischia is reduced, or eliminated, by a bracing of the ischia on the ventral pubic bones and by the strongly developed lig. sacrotuberalia, which act as strong ties on the dorsal side. Some of the extensors of the hip joint, the so-called mm. adductors, originate from the pubis and contribute only to extension as long as the femur is in an anteverted position.

A detailed and really convincing study of the pelvis, with the only adequate FE method, is still a demand for the future.

The *tail among primates* varies in length, and number of segments, often within closely related groups. In most strepsirrhines and tarsiers, the tail is long, but *Indris* and the lorisines have only short appendages, or no tail at all (*Loris tardigradus*). New World monkeys in most cases have long tails, though *Cacajao* does not, and in the same group, the tails of Alouattinae and Atelinae are outstanding prehensile organs (see above, Chap. 8.3). Cercopithecoids usually possess long tails (as indicated by their name!), but some of the baboons (drills and mandrills) have a very short tail, and among the macaques are species (*M. fascicularis*) in which the tail has more than the length of head and trunk and others (*M. sylvanus*, *M. nigra*) which are completely tailless. In between are *M. nemestrina*, or *M. fuscata*, which possess short, rudimentary tails. The Hominoidea do not at all possess a tail. The most differentiated forms of caudal vertebrae can be found in forms which also possess a strong and diversified musculature, namely, the New World monkeys with prehensile tails (see above, Chap. 8.3; Ankel 1962).

8.11 Body Sizes of Primates, Facts and General Conditions

Most of the strictly arboreal primates are below 10 kg, but there are exceptions: siamang (ca. 12 kg), male *Nasalis* (20 kg, while females are about 10 kg) and the large apes, which vary from about 30 kg (*Pan paniscus*) to 80 kg (male *Pongo*). The still heavier male *Gorilla* with up to 150 kg are not counted among the strictly arboreal animals. A tendency toward a more terrestrial lifestyle is common among macaques and especially baboons, which weigh between 10 kg (females) and 27 kg (males), but also in the smaller patas (6–11 kg) and vervet monkeys (3,6–5,4 kg). The phylogenetically oldest fossil forms from the Eocene have had small to medium sizes: 0.5 kg (*Europolemur koenigswaldi*, *Darwinius massillae*) to 2.2 kg

(*Europolemur kelleri*, according to Franzen 1987; Franzen et al. 2009). From the Eocene in North America, *Notharctus*, *Cantius* and *Tetonius* are known who had sizes between 0.1 kg and 5 kg. The Oligocene and Miocene apes (*Aegyptopithecus*) were much larger (Fleagle 1983; Bown et al. 1982). During the Miocene, the radiation of apes has led to forms that reached the sizes of *Gorilla* or even exceeded this species (*Gigantopithecus*) in the late Miocene and the Pliocene. The predecessors of Cercopithecoidea are poorly known but seem to have been smaller. The group reached the biggest size in the baboon-like, probably terrestrial *Theropithecus oswaldi* from the East African Plio-Pleistocene with estimated 80–100 kg.

8.11.1 (How) Is Overall Body Size Limited in Primates?

A general *limitation for body size* of primates is given by their arboreal mode of living. Because of the limited strength of branches (see above, bridging, Fig. 8.13), arboreal animals cannot be extremely heavy. It seems that male gorillas (about 150 kg) are close to the absolute borders of arboreality. According to Schaller (1963), the big male mountain gorillas commonly sleep on the ground. This was not confirmed for other gorillas at other places, which also built sleeping nests. Male orangutans (ca. 80 kg), however, are as strictly arboreal as are females and youngsters. The next largest tree-dwellers seem to be *Nasalis* males with about 20 kg. The powerful-looking drill and mandrill males probably exceed this size but are, like most baboons, anyway prone to terrestriality. In the very big *Theropithecus oswaldi* (who is only known as fossil, see above), fore- and hindlimbs of nearly the same length and markedly short digits seem to prove ground-living habits. Of the still larger *Gigantopithecus*, not more than indeed gigantic teeth and lower jaws are known, so that there is no realistic basis for any conclusions about lifestyle of these biggest of all primates. There are ideas flying around connecting them to the legendary Yeti. But all these imaginations are preformed by living apes or humans and lack any scientific relevance.

An additional limitation of body size is the above-detailed (Chapter 1.4) lagging of available muscle force behind mass inertia. In the really existing primates, this seems not to play an essential role as a general borderline for body size.

8.11.2 Why are Some Primates Sexually Dimorphic?

Aside from the above-mentioned *size variation among primate species*, there is a remarkable tendency for developing sexual dimorphism. While in several primates the females are as heavy as males (*Hylobates*), or maybe even larger (*Lemur catta*, Jolly 1966), in most species the males are heavier. A median value is where females have between >50 and 80% of the weight of males (e.g. in humans (80%), also chimpanzees (70%), macaques, colobines, New World monkeys), but in several *Papio* species or in *Nasalis*, also in *Pongo* and *Gorilla*, (Schaller, 1963) the females weigh less than 50%, sometimes 40% of the males. Aside from the divergence in weight, nearly all male primates have longer and broader, also more projecting canine teeth than the females. This is reason enough to search for causes.

The first attempts to understand the *sexual dimorphismus* have been made on the field of *biological roles*. There being no deep-reaching divergences between the lifestyle of the sexes (biotopes, climate, nutrition), the interest of researchers shifted toward relations between species-specific social structures and mating systems. Indeed a loose connection seems to exist (Clutton-Brock, van Schaik et al.): some of the most dimorphic species are living in harem-like groups (gelada, hamadryas, drill, mandrill, patas, *Nasalis*, *Gorilla*). These forms are not markedly territorial. In contrast, species of *Hylobates*, *Saguinus* and *Callithrix* live in pairs, where both sexes take part in defending the borders of their territories. Between these extremes, several macaques like wanderoo, *M. silenus*, vervets and other guenons, as well as colobines, form rather strictly separated one-male groups without being expressly dimorphic. Savannah baboons, *M. fuscata*, *M. mulatta* and chimpanzees live in multi-male groups, where the baboons show a marked and the others a not-so-obvious sexual dimorphism.

An important argument for smaller size in females was proposed by Crook 1966: Distribution of body mass on several, smaller individuals – which keeps nutritional requirements of the individual low and increases the potential number of births. This model implies that the strong males take over defending a territory and maintaining group structure. These roles, however, have not been verified in all cases.

Although all this is somewhat confusing, there are clear *mechanical advantages* connected with body size and therefore with sexual dimorphism: in physical agonistic encounters, most primates (but not the gibbons!) push, wrestle or cling. All, inclusive of the small apes, kick, blow and bite. The principle advantages of body size for all these aggressive behaviours have been discussed above in Chapter 1.4 "Body Size" and play the same role for "sexual dimorphism". The most essential factor is that the impact, which one body exerts against the other in collisions, depends on its impulse (*P*), which is proportional to its mass and the square of speed (see Eqs. 1.5, 1.6, 1.7, and 1.8).

In *a direct impact* of two bodies (first attack in wrestling, pushing, kicking, blowing), the force transmitted (*F*) grows with the negative acceleration (= deceleration, $-a$) of both partners, offensor (o) and defensor (d), and mass (*m*).

$$F_o = m_o * -a_o \text{ and } F_d = m_d * -a_d \tag{8.21}$$

A difference between both values results in a movement toward that partner having the smaller product of mass times acceleration. The interdependence of factors is illustrated in Fig. 8.52. Instead of force, the kinetic energy (mass times

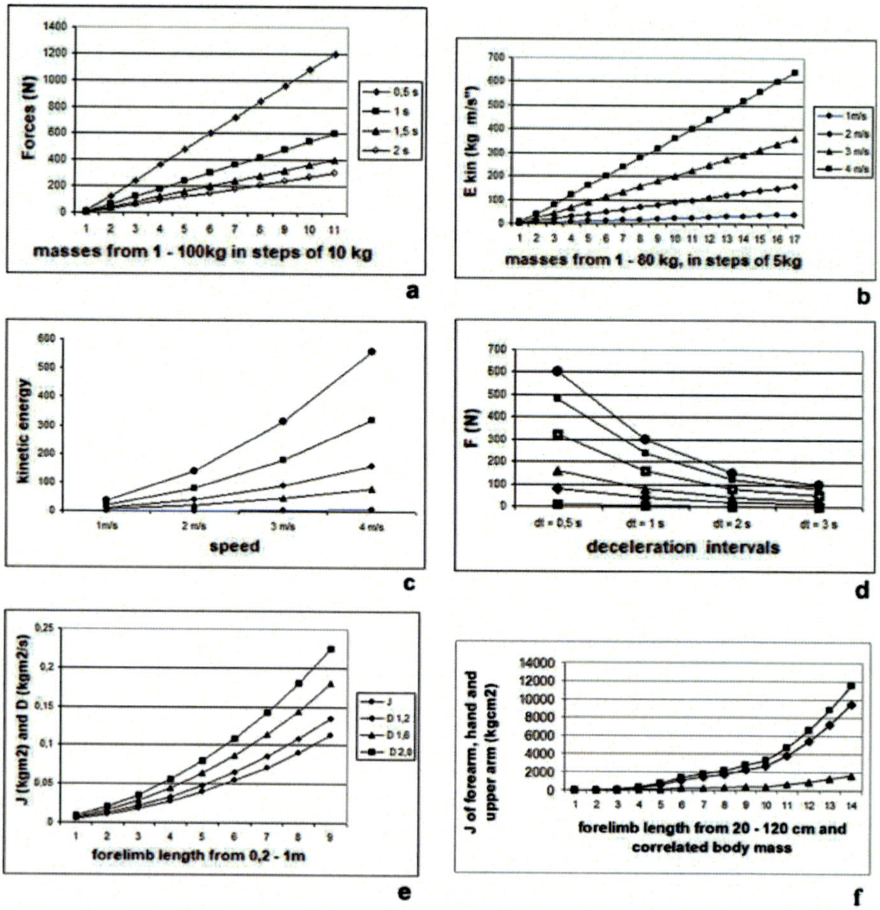

Fig. 8.52 Effects of body size in agonistic encounters. Forces exerted at impact (**a**) as well as the kinetic energy contained in a body (**b**) depend on body mass, if speed is constant. The four graphs in (**a**) show the variance of impact times, in (**b**) the influence of speed on kinetic energy. In (**c**), the kinetic energy is plotted against speed for primates of different masses: talapoin (both sexes, diamonds), baboon and proboscis monkey (females, triangles; males, dots); orangutan (females, squares; males, rosettes). The steps from one curve to the next at the same speed show the effect of size. (**d**) The forces at impact decrease with contact time or duration of force exchange during deceleration (= braking). Drawn for primates of varying masses: talapoin monkey (both sexes, black squares), baboon and proboscis monkey female (diamonds) and male (triangles); chimpanzee female (large squares) and chimpanzee male and human female (small squares); human male (dots). Note the differences between females and males. (**e**) Mass moment of inertia of forelimbs (J, dots) and rotational impulse (D) plotted against forelimb lengths. The graphs are drawn for varying braking intervals (1.2 s, diamonds; 1.6 s, triangles; 2.0 s, squares). (**f**) In more detail, mass moments of inertia of the forelimb segments: forearm (squares, indeed larger than the following!), hand (diamonds), upper arm (triangles)

square of velocity) can be compared (Fig. 8.52b). As a consequence of Fig. 8.52a, the heavier animal compels the lighter one to give way. In spite of this, a smaller animal can surmount a bigger one, if its speed during the attack is greater (Fig. 8.52c) and is retarded at impact more rapidly (Fig. 8.52b, d), so that the deceleration (−a) is greater. In other words, a smaller animal weighing 10 or 25 kg may well bring down a bigger one that weighs 20 or 45 kg – provided that the smaller one moves two times or three times faster than the bigger primate.

The degree of deceleration, or the size of the transmitted force, is difficult to estimate, because neither the own movements of the attacked individual are known in advance nor the place where the impact takes place (Fig. 8.52b, d). The same situation exists as in a fall by accident (see above). Everybody knows from own experience the difference between an impact at the dorsal side of the shoulder region (where compliant musculature distributes the force over a large area) and, for instance, an impact at the anterior margin of the shinbone where the hard bone is covered by nothing except skin. Gibbons and chimpanzees frequently make use of their heels in agonistic contacts. This allows the transmission of strong impulses on a small surface without damping.

The same laws hold true, if the two bodies are not pushing against each other, but if one grips hold of the other and pulls, be it by its hands or by its teeth.

In *blowing*, primates take advantage of the proportions of their arms: the mass of their forearms is rather heavy (because of the strong digital muscles, as detailed above) and the peripheral velocity of the blowing hand is great because of arm length. Especially in the long-armed apes, the latter leads to particular impulse or angular momentum (D), since peripheral velocity (v_u) of the blowing hand is great

$$v_u = \text{angular velocity} * \text{“radius”}, \tag{8.22}$$

where “radius” is the distance between the blowing hand and the shoulder joint.

The impulse (D) inherent to the arm also grows with its mass, the square of arm length and angular velocity (ω):

$$D = \Sigma\left(m_i * l_i^{\,2}\right) * \omega \tag{8.23}$$

That means arm length is responsible for (peripheral) velocity of the blowing hand. Their respective influences are illustrated in Fig. 8.52e. At impact, the hand is decelerated. The degree of deceleration is again, as above, difficult to estimate in generalised form, because it depends upon the hit body part and on the own movements of the fighters in relation to each other: the hand of the attacker moves in addition to its own movement, while the target may minimise the impulse by moving away or increase it by moving toward the attacker.

Against biting or being bitten, *not mass but volume* is an efficient means of defence. With increasing diameters, the parts of the victim between the dangerous teeth of the attacker become smaller. As Fig. 8.53a shows, the upper and lower teeth exert at a given gape their full force from both sides of the bitten object if the latter is

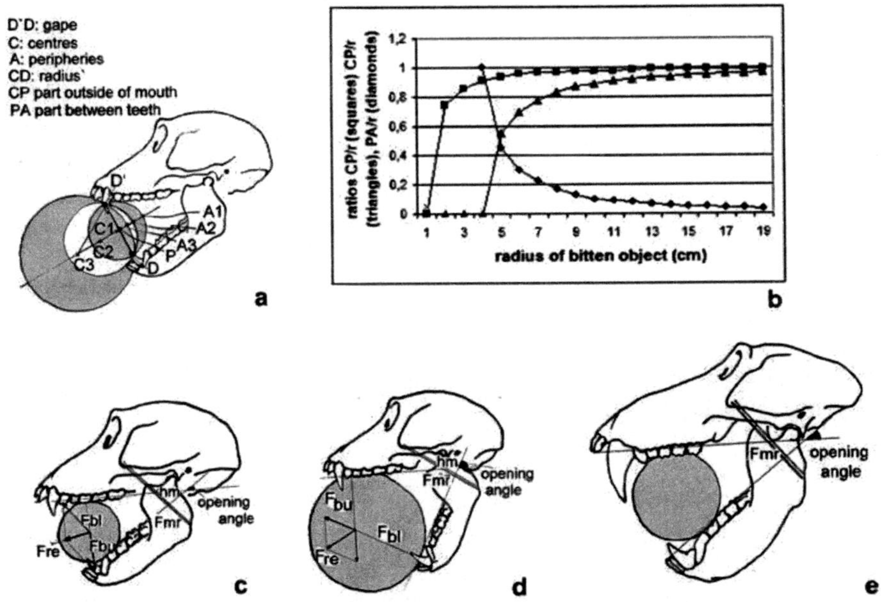

Fig. 8.53 Biting, gape and volume of bitten victim. (**a**) The female mandrill with jaws opened to an arbitrarily chosen gape can seize objects of limited diameters (alternating white and grey circular areas). If diameters are too long, the bigger objects do not fit between the teeth. The bigger the object (or the longer its diameter), the smaller the share of the object between the front teeth of the monkey in comparison to its total cross section. (**b**) The numerical relationship between the parts of the bitten object between the teeth and its radius. The parts of the bitten object inside the mouth between the teeth (triangles) or outside the mouth (diamonds for the gape shown, squares for a reduced gape) are plotted against the radius of the bitten object. The larger a bitten object (= the longer its diameter), the smaller is its part exposed to bite forces between the teeth of the attacker (**c**, **d**). One means to seize bigger objects is increasing the opening angle of the jaws (**d**). This, however, reduces the lever arms of the adductor muscles (double line in **c** and **d**) and therefore bite force. It also increases the (not useful) force F_{re}. Another means to make biting on thicker objects more efficient is elongation of the jaws, like in the male mandrill (**e**). Even if the opening angle is the same as in the female (**c**), elongation of the jaws allows seizing of bigger objects (i.e. a bigger enemy) but requires more force of the muscles and more strength of the bony jaws, both being equivalent to greater weight. (**c**, **d**) The resultant of bite forces exerted by the upper and lower jaws, which squeezes the object out of the mouth (F_{re}), becomes greater at wider opening angles (proportional to the cosine of this angle). In (**d**), the wider gape leads to a reduction of the effective bite forces at right angles to the tooth rows. F_{re} also presses the object against the cutting edges of the upper canines and into the canine/first premolar edge. This is best visible in (**e**); in (**c**) and (**d**) the canines are drawn longer than realised in most mandrill females, to emphasise their effect against the force F_{re}. (Skulls redrawn from Schultz 1971)

slender, that means its cross-sectional area (circle filled with grey) is small and its radius (r) is short. The thicker the bitten object (= the longer its radius, white and grey circles), the smaller becomes the part between the teeth of the attacker. The part between the teeth (PA) can be expressed by the ratio between the part between the teeth (CP) and the length of the radius (r).

$$r^2 = CP^2 + (CC'/2)^2; \text{ and } CP = \sqrt{r^2 - (CC'/2)^2} \qquad (8.24)$$

If the bitten object (victim) is very big, not more than a fold of its skin is left between the teeth.

On the side of the attacker, there are three means to manage the problem evoked by volume, all of which connected with disadvantages:

1. Directing the bite to the extremities, rather than the more voluminous trunk of the victim. This is what has been observed in intraspecific fights between primates.
2. A wide gape (= opening angle between upper and lower jaws). The obvious disadvantages being shorter lever arms of muscles and greater force component pushing the bitten object out of the mouth.
3. Elongation of the jaws (disadvantages: longer jaws require stronger musculature and stronger jaws). Both result in increasing weight of the head! (Fig. 8.53)

8.12 Essential Results

Because of their similarity with humans, primates as a whole have attracted intensive research and so are, in spite of their variability, a rather well-known group of mammals. A number of rules that have general validity (Chapters 1.4, 2.2, 5.7, 6.5, 7.3, 7.4) have been identified first in primates.

All specialisations of non-human primates are connected with living in trees, beginning from the limited body size and prehensility of hands and feet, through mobility of the long limbs and the dominance and orientation of the eyes to the use made of the autopodia of the forelimbs (hands) as organs to collect and sometimes to diminute food. A variety of unusual locomotor modes has been developed, which permit moving on interrupted substrates like the branches of trees: bridging, leaping and arm swinging. Despite the characteristic specialisations, living in trees is dangerous and entails a surprisingly great number of serious lesions. Nevertheless, several primates have a strong tendency to spend much of their time on ground. But these forms still are very gifted climbers and possess characteristics typical of the whole order.

According to fossil documents, one of the descendants of the primate order is the ancestor of pre-humans and early and modern humans (Chap. 9). These forms adopted a way of (bipedal) locomotion on ground, which does not exist in any other animal. The physical conditions, however, are completely fulfilled.

The shapes of skeletal elements in primates are better analysed than in any other group: phalanges, metapodials, carpus/tarsus, zeugopodia, stylopodia and shoulder girdle are known in more detail than in other animals. This should, however, not conceal the fact that neither the shoulder girdle, nor the pelvis, nor the carpus has been subject to a really complete and convincing analysis.

Chapter 9
Evolution of Hominids

Keywords Pre-humans · Early humans · Human foot · Lower leg · Knee joint · Length of hindlimb · Pelvis · Lumbar lordosis · Trunk shape · Kinetics of bipedality · Human hand · Morphometry of skull · Mechanics of human skull

9.1 What Do We Know About Early Humans and Prehumans?

Among the primates, one group stands out by a huge brain, short dental arcade without prominent canines, habitually upright body posture, absence of prehensile characteristics in the foot and instead an integrated big toe and bipedal locomotion: the *Hominidae*. Descriptions of human evolution are fairly numerous, in the present context I just would like to mention the very precise and rich books of Klein, 1999 and of Henke and Rothe1994, as well as Steitz, 1993 and Young, 2013. Together with the mentioned traits, most members of this group show less obvious characteristics: vertically orientated, chisel-shaped front teeth (including the canines), the development of a long thumb, strongly reinforced second and third finger ("entaxonic" hand in the terminology of Jouffroy et al. 1993), reduction of the lateral toes for the sake of the big first toe, length and slender shape of the trunk, in comparison to the hindlimbs "weakly" developed forelimbs, deep position of the shoulder and constriction of the caudal opening of the rib cage, particular shape of the pelvis.

In a first, rough approximation, two subfamilies can be discriminated (Figs. 9.1 and 9.2): the *Homininae*, existing recently and fossilised ("early humans"), and the *Australopithecinae*, known only as fossil "prehumans". The group mentioned first contains with *Homo sapiens* the most investigated and best known living being. The latter group seems to be more variable than the former. This may be simply a consequence of its long existence, about twice as long as that of *Homo*.

In the fossil representatives of both subfamilies, most parts of the skeleton are known, as well as body proportions. Habitats and mode of living are derived from accompanying fauna and flora, admittedly not completely, but in essential. Their

H. Preuschoft, *Understanding Body Shapes of Animals*,
https://doi.org/10.1007/978-3-030-27668-3_9

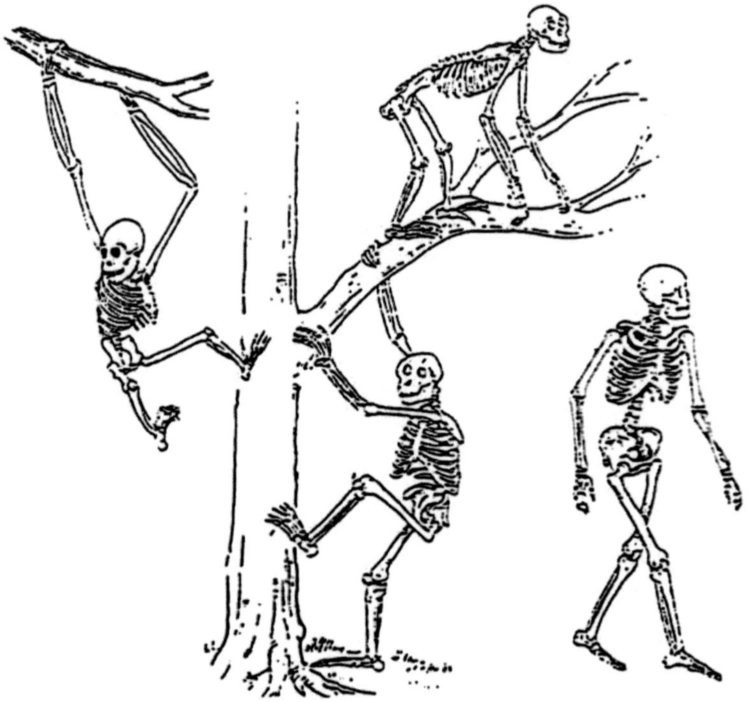

Fig. 9.1 Locomotor modes among hominoids, commonly seen as stages of human evolution. On top: two examples of arboreal locomotion. *Left* suspended from branches (orangutan) and *right* arboreal quadrupedalism on a branch (chimpanzee). The descent from trees seems to have taken place gradually in short steps (*Ardipithecus, Australopithecus*). *Homo* is a purely terrestrial, rather specialised cursorial mammal. (From Preuschoft (2004))

environments were not the tropical rain forests but rather its marginal areas, woodland savannah, stuck with groups of trees and gallery forests (like those landscapes, which today exist widely spread in East and South Africa). Vegetation reached from woodland with continuous canopy to "wooded steppe" with isolated trees, separated by wide areas of grassy or uncovered ground (McHenry 1991). Water obviously was amply available to the extent that most fossils have been recovered from sites which have been lakeshores or riverbanks (Niemitz 2004, 2010) (see also below, Chap. 9.5).

Aside from their basically similar body shapes and their varying size, the skulls of early human forerunners differ markedly from those in *H. sapiens* and have undergone remarkable changes during evolution. This process continues even in recent times and has given reason for taxonomic separations. This point will be treated below in Chap. 9.7.

The body shape of the genus *Homo* (Fig. 9.2, which will be analysed first) can be traced back to at least 1.5, more probable 2 MY. The very complete skeleton of the youth of Nariokotome (Walker and Leakey 1993; Ruff and Walker, 1993)

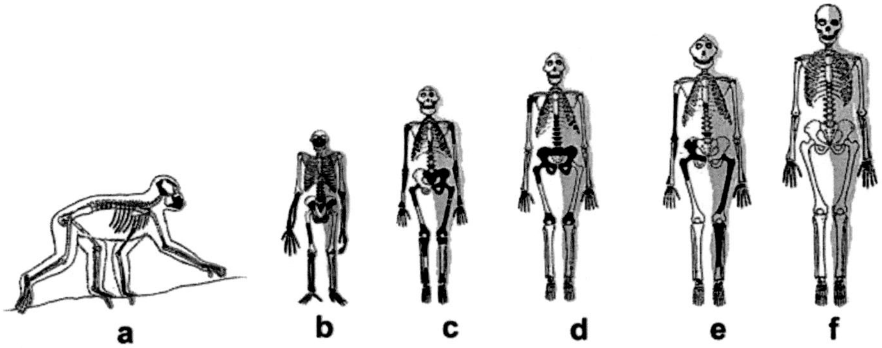

Fig. 9.2 Proportions of extinct and extant constructional types of hominines: black are preserved parts of the originally described skeleton; the supplements from other finds or reconstructed elements are indicated by empty contour lines. Grey shades on the right sides in (**b**–**f**) are estimated body outlines including soft parts. The skulls are better documented than the postcranial skeleton (see Chap. 9.6), but shown here as outlines only. Same scale in all cases except (**a**), which is the earliest well-known representative of the Hominoidea, *Proconsul* from the East African Miocene. (From Walker et al., 1993). (**b**) *Ardipithecus ramidus*. (Redrawn from White et al. 2009). (**c**) *Australopithecus afarensis*, (**d**) *Australopithecus africanus*, (**e**) *Australopithecus (Paranthropus) robustus*, (**f**) *Homo erectus*; all from Preuschoft, 2021 Note the increasing hindlimb, decreasing frontlimb lengths, the funnel-like ribcage in (**c**–**e**) and the gradual integration of the hallux into the plate-like foot. The (deep) position of the shoulder is a human-specific trait, which may have been present in the fossils or not (see text)

documents the body proportions typical for humans.[1] The site of Dmanisi, Republic of Georgia, seems to be with 1.8.–2.0 million years several millennia older than Nariokotome. Aside from a series of surprisingly variable skulls, with brain volumes of 600–800 cm^3, and a very massive jaw apparatus, postcranial material has been described. It suggests statures of 145–166 cm, body weights between 40 and 50 kg, and proportions typical of *Homo*. Stone tools of an early Olduvan type are associated with the bones (Bosinski, Lordkipanidze & Weidemann, 1995; Jöris 2008; Garcia, 2004; Lordkipanidze et al. 2007). This converges with Olduvan assembleges found with H. erectus in Africa (Clarke et al. 1994).

Under consideration of the older, but less complete finds, we may assume that the human body proportions exist since roughly 2 MY, that is since the Early Pleistocene. The importance of the doubtlessly humanlike femur fragment of "*Orrorin tugenensis*" (Pickford 2001; Senut et al. 2001), aged ca. 5 MY, is difficult to integrate into our knowledge, since fossil postcranial material of comparative age is lacking completely. Recently, the complex from the *Rising Star Cave* not far from Johannisburg, South Africa, became widely known from the press as "*Homo naledi*". Up to now, the material is not dated, although it seems to be relatively

[1]This is in contrast to earlier descriptions (such as Ruff and Walker 1993), but one of the excavators and authors of the first detailed description (A. Walker, at the Senckenberg-Conference in 2003) has identified a neglected fragment as a missing thoracic vertebra, which supplements the trunk length to a body height of nearly 160 cm. Since his age, estimated from the dentition, was 11–12 years, this corresponds to the height of the adult of more than 180 cm.

young. As far as can be derived today, it represents a small and rather early form of *Homo*, although recently published reconstructions (Berger et al. 2015) of the postcranial skeleton claim existence of combinations (funnel-shaped rib cage, pelvis), which are not found together in other fossils of this type. Nearly all parts of the skeleton are present, often from more than one individual. The best preserved parts are the nearly complete hand and foot skeletons. The hand has (aside from the surprisingly massive terminal phalanx of the pollex) the proportions of a human hand. Likewise, the foot skeleton resembles very much that of a modern human in all traits which can be interpreted. Only in fragments preserved is the pelvis, the sacrum is missing, so that a reconstruction is not reliable. The tibia is similar to that of a modern human, but the femur has a long neck and rather wide shaft-neck angle as is characteristic for *Australopithecus*.

Long time before the appearance of the genus *Homo*, since the Early Pleistocene, existed *prehumans* (Australopithecinae) in at least two morphs (probably genera: *Australopithecus* and *Paranthropus*; Fig. 9.2). Reliably dated fossils reach back to 3.3 MY (Hadar, Sterkfontein: "Little foot"; Clarke 1998, 1999, 2008; Clarke & Tobias 1995; Partridge 1986). Others may belong to the Early Pliocene *Ardipithecus* with an age of 4.6 MY is older than the Australopithecines s.s. are. According to the reconstructions of White et al. and Lovejoy et al. both 2009, the proportions of this creature are closer to modern apes than those of the later forms. Aside from their seeming strength, the pronounced curvature of the forearm bones, is a strong argument to assume extremetly strong flexor muscles, as shown in Fig. 8.40b, where the outward curvature of ulna and radius yield resistance against bending by the active muscles. The general impression of forelimb-dominated climbing is confirmed by its well-documented, short-thumbed hand, which looks remarkable powerful and the also well-documented foot, which possesses a strong hallux, that clearly could be abducted from the fairly long "lateral" toes. The calcaneum is not completely preserved, the pelvis is strongly deformed, and resembles that of an australopithecine. Representatives of the subfamily, especially the robust variant *Paranthropus*, survived until about 1.5 MY, being contemporaneous with *Homo*. Both the above-mentioned, prehuman genera differ from *Homo* by their brain volumes, jaws and dentitions and more so by their total body shapes and their smaller size (120 cm or 150 cm, respectively), where the larger genus *Paranthropus* is within the range of variation of *Homo*. That these forms have been bipeds has been claimed early (Le Gros Clark 1959, according to Clarke 2008). This assumption was based on the pelvis, although nobody has made a serious attempt to work out in detail the mechanically relevant traits which characterise the pelvis of a bipedal hominoid. The distal condyles of the femur are of different height and document a genu valgum, which indeed is of advantage for standing and walking on the hindlimbs (see Preuschoft (1971); also Fig. 9.10). On the basis of a biomechanical analysis, the skeleton of a foot (later known as OH 8) and lower leg bones (OH 35) has been identified by Preuschoft (1971), as that of a bipedalist with relatively weak toe flexors and a relatively rigid passive apparatus (see also below, Fig. 9.8). The final proof for bipedal locomotion of prehumans was provided by the famous foot tracks from Laetoli in East Africa, 1974 (Fig. 9.3; Leakey and Hay, 1979 as well as Leakey and Harris, 1987). The tracks of several bipedally walking individuals are

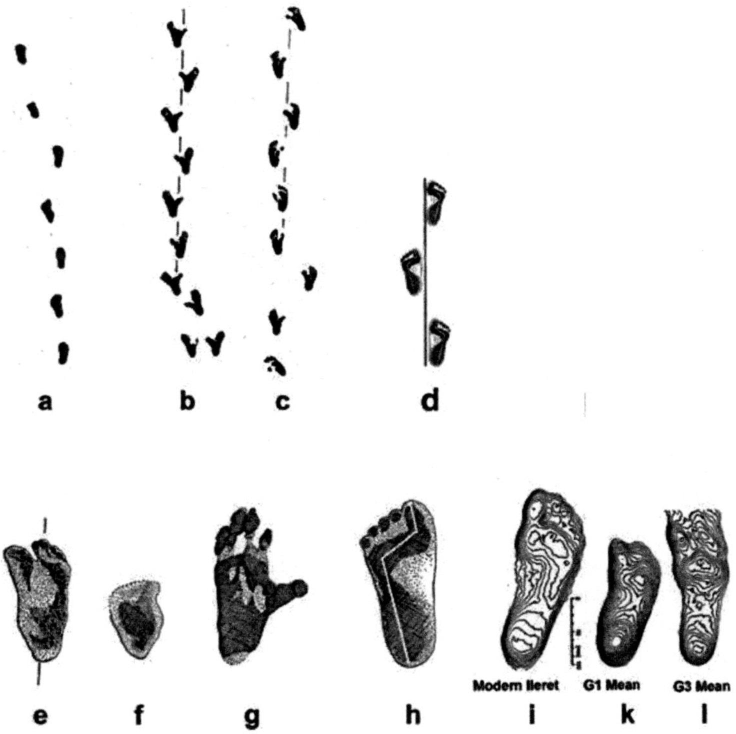

Fig. 9.3 (**a**) Foot tracks from the newly discovered site S (L8) from Laetoli, Northern Tanzania, (**b**) a juvenile gorilla, (**c**) a juvenile chimpanzee. (From Preuschoft 1963a, b). (**d**) Adult human. (**e, f**) Individual (right) footprints from Laetoli, (**e**) from site G1, redrawn from Deloison (1991); (**f**) from site A, redrawn from Day (1991). (**g**) Individual (left) footprint of a macaque, as found in wet snow, as an example of cercopithecoid monkeys. Note the long toes, the impression of the hallux and the absence of a heel, which is not in contact with ground. (**h**) Individual (left) footprint of a human, with trajectory of total GRF. In (**e–h**), the deep impressions are cross-hatched, shallow impressions stippled. (**i**) Optical laser scan image of a footprint of modern *H. sapiens*, added for comparison. (**k, l**) Footprints of the Laetoli series (averaged optical laser scan images of G1, $n = 11$, and G3, $n = 5$). All have in common the deep impression of the heel, a low longitudinal vault, deep impression of the adducted hallux, toes slightly longer than in humans and toes directed somewhat medially. ((**i, k, l**) From Bennett et al. 2016)

3.5–3.7 MYs old, slightly older than the skeletal remains from Hadar. On the same stratum have been found skull fragments, which are similar to *A. afarensis*. New footprints (Site S) have been detected at the same locality, and described by Masao et al. (2016). Several later studies unveiled additional morphological traits characteristic for bipedal locomotion (e.g. Latimer and Lovejoy 1989; Stern and Susman, 1983; Susman 1988, 1989 on new material from Swartkrans; Masao et al. 2016). Aside from this, there were always arguments, which relate details of the fingers and toes to the abilities of the australopithecines as climbers. The relatively long and curved pedal phalanges of several finds indicate at least better grasping abilities than in hominines. The very robust and massive skeleton of the hand has been taken as indicating usage of the anterior limb for locomotion (Susman 1979; Susman and Stern 1991; Larson 2015) or for tool using, as Susman 1989, argues. The best arguments for their climbing propensities are the powerful anterior and short, relatively weak posterior extremities. The partial foot skeleton "Little foot", for which Clarke and Tobias (1995) emphasise the abductable first toe, points into the same direction.[2] The assumption of an abductable first toe, however, is in contrast to Susman's (1989) observation of flat joint surfaces between os cuneiforme mediale and metatarsale I, which indicate adducted posture and immobility. Recently published parts of a foot skeleton from Ethiopia (Haile Selassie et al. 2010 and 2012) show clearly a cylindrical joint surface at the basis of the first metatarsal. The other metatarsals from the same complex, however, are quite long and slender, looking much more like those of apes than of those of humans. It should be considered to count them among non-human, rather than human hominids – in spite of the attribution proposed by the authors.

A find which became known under the name *Australopithecus sediba* (Berger et al. 2010; Williams et al. 2013) is hardly 2 MY old. It comprises an 11–13-year-old adolescent and a fully grown female, with an estimated stature of 130 cm. Preserved is a complete upper extremity with shoulder blade and hand and fragments of the pelvis. Reconstructions of the latter seem to be more similar to the human pelvis than in other australopithecines.

All this taken together, australopithecines and other pre-humans were probably better climbers (Fig. 9.1), but less endurant and less fast bipedalists than hominines. With Klein (1999, on pp. 210 ff.) and also Tattersall ("While we were not alone 2000"), we should consider the existence of different locomotor modes in early phases of human evolution as possible. On the basis of a very comprehensive collection of data and current ideas, but without own contributions, Brandt (1995) tried to demonstrate that the group discussed here as "prehumans" are a "highly specialised" taxon of primates, without being part of the human evolution.

Most primitive *stone ("pebble") tools* reach back to not more than 2.6 MY ago (Semaw et al. 1997, former estimates from Olduvai, Bed 1, indicate ages of more than 2.0 MY, Fig. 9.29). The famous "handaxes" and flakes of the "Acheulian", which seem to have been made by *H. erectus*, are about 1.75 MY old (from Konso, Southern Ethiopia, Beyene et al. 2013). While being very informative for later evolutionary stages of the genus *Homo*, stone tools do not yield so important information for the earliest stages – aside from the notable fact that tools have been produced, Susman (e.g. 1989) maintains that *Paranthropus* hands exhibit traits typical of tool using.

[2]The ability to abduct the first toe for gripping does not seem to be a decisive argument against bipedal locomotion. The (very variable!) footprints of bipedally walking chimpanzees often show the hallux closely adducted to the medially deflected lateral toes. The hallux may bear a large share of GRF or not. In this respect, the footprints of chimps, with their clearly mobile hallux, look similar to fossil footprints from Laetoli. The variable weight distribution on the toes in chimps is in clear contrast to gorillas, which abduct and load their big toes in walking on the ground in bipedal as well as in quadrupedal posture (Preuschoft 1963a, b). Jungers (2009) noted "that climbing and terrestriality are not mutually exclusive".

4–5 MY ago, which means in the Early Pliocene, the traces of *prehumans* (*Ardipithecus* (*4, 6 MY*), *Sahelanthropus tchadensis*, *Graecopithecus* (*6–7 MY*)), get lost among the *remains of large apes*, of which most, however, are clearly older and come from the Late (*Sahelanthropus*, *Graecopithecus)* or Middle Miocene (*Sivapithecus*, *Ramapithecus*, aside from those listed below). One of the most characteristic traits of Homininae are the chisel-shape of the canines, which do not jut out over the tooth rows – in contrast to the sectorial, projecting canines of recent apes and their fossil kin. This trait, combined with a parabolic rather than u-shaped dental arcade (Fig. 9.37 or Fig. 9.39), however, appears gradually in the course of evolution – as does relative brain size. Both do not yield a clear-cut divergence. Overall body shape has been estimated for *Ardipithecus* (Lovejoy, 2009; Lovejoy et al. 2009; Simpson et al. 2019) and is known for *Dryopithecus*, *Oreopithecus*, *Pliopithecus* and *Nacholapithecus*. The first mentioned genera resemble in many details recent large apes, where, however, the special evolutionary acquisitions of African (gorilla, chimpanzee, bonobo) and Asian apes (orang-utan, gibbons) cannot be discerned. The latter genera (*Pliopithecus* and *Nacholapithecus*) seem to be less forelimb-dominated and more similar to quadrupeds. Still older than these forms is *Proconsul* from the Lower Miocene, the body proportions of which match those of arboreal quadrupedalists among the New World or Old World monkeys – although without a tail.

Because of its huge size, the Pleistocene *Gigantopithecus* from Southern China always again attracts interest. Since the genus is known alone by its teeth, three very massive lower jaws and a palate, all considerations are more or less speculative. A lower jaw is also the only document of a slightly smaller species from the Upper Miocene/Early Pliocene of the Sivalik Hills (E. Simons, 1972).

9.2 Which Morphological Specialisations Have Been Developed in *Homo* Compared to Those of His Closest Relatives? How Do Fossil Hominids Fit to Humans?

In the preceding chapter, several characteristics have been made apparent by comparisons between non-human and human primates. This is partly due to the lack of interest for understanding the peculiarities of various species. Instead of analysing the morphology of a particular animal, alone the differences diverging from humans were perceived and described in detail. The reader, if interested, therefore should look up items like tarsals, especially talus and calcaneus, lower leg bones and femur, pelvis, etc. in Chap. 8.

While the variety of hominines was greater in the geological past, most traits and their functions have been worked out for modern humans, mostly by comparisons with and in contrast to our closest living relatives, the apes. This procedure is also chosen for the following text.

The most dominant traits of the hominids are *upright body posture* and *bipedal locomotion* (Figs. 9.1 and 9.4; Witte et al. 2003). Both require particular

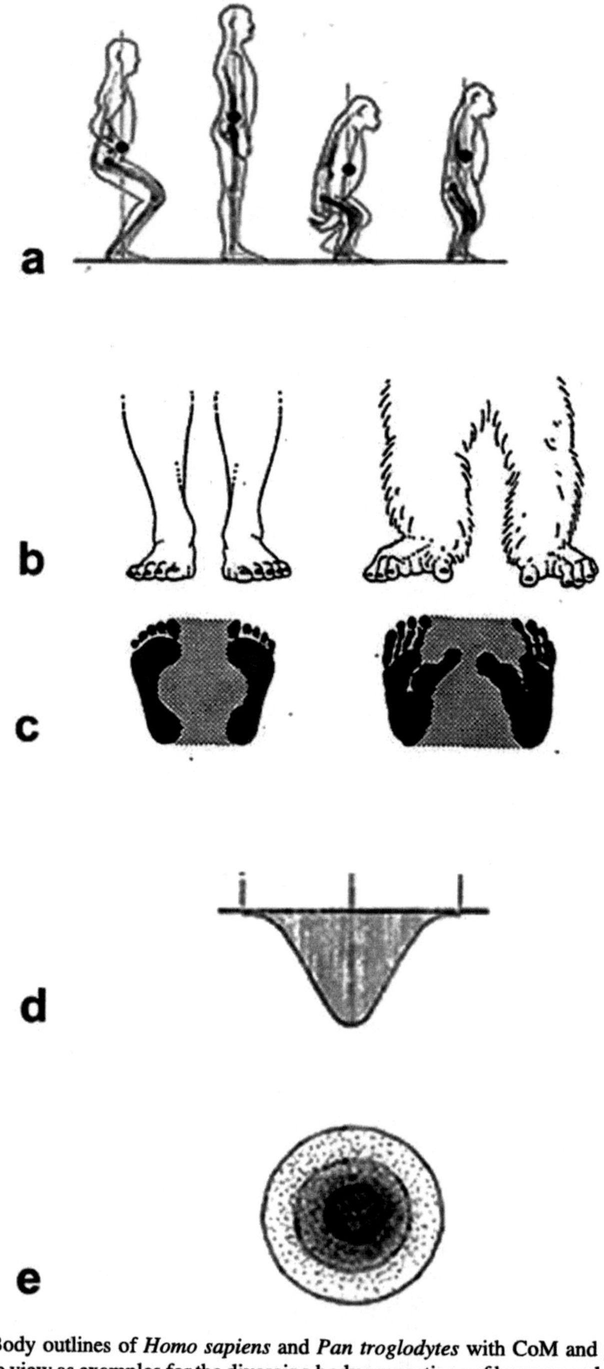

Fig. 9.4 (**a**) Body outlines of *Homo sapiens* and *Pan troglodytes* with CoM and gravity vertical (=GRF) in side view as examples for the diverging body proportions of humans and apes, which are specialised for different modes of living. In upright standing, the GRF passes close to the axes of the

specialisations of body shape and will be treated one after the other. Aside from standing and walking, modern humans tend, like all primates, to squat or sit when resting, eating, investigating objects closely. In all body postures, their forelimbs are not loaded statically and completely free from locomotor tasks.

The primary mechanical problem of bipedal posture is *keeping the CoM* perpendicularly above the area of support (Fig. 9.4a, as emphasised repeatedly by Kummer 1959a, b, 1965). This is in bipeds much smaller than in quadrupeds. Since parts of the body, especially arms and head, are moved independently, the position of the CoM is by no means constant. Alone the circulating mass of blood causes measurable deviations. The gravity vertical oscillates during two-limb support, and because of statistical reasons, it can be expected that the gravity vertical hits the ground most often near the centroid of an elliptical area following a Gauss distribution. This is shown in Fig. 9.4c,e in top view (*y*-direction) as well as in anterior view (*x*-direction, Fig. 9.4d). A quite economic way to enlarge the area of support is directing the toes laterally and the heels medially, so that the short lever (heel) receives a greater share of the load than the long lever (metatarsus and toes). The area of support can be chosen to be large by a great distance between both feet (Fig. 9.4b) or small, for example, in standing on one hindlimb. The plantigrade foot of a primate then can be imagined as a plate, on which the leg is balanced by stays and shrouds like the mast of a sailing boat (Fig. 9.5a). During locomotion, the resultant of the gravity vertical and its ground-parallel components (GRF; see also Chaps. 5.4, 5.6 and 8.3) hits the ground on each foot first at the heel (heel strike), wanders close to or along the lateral border towards the toes (roll-off) and is finally shifted from the lateral fifth towards the medial first metatarsal ball and the big toe (push-off; Fig. 9.19). External forces (GRFs) which are applied close to the heel may be greater than at midstance but have a short lever arm and so need less muscle force for being balanced at the ankle joint than GRFs applied to the anterior part of the foot or confined to the toes. In static situations, the external forces (GRFs) applied directly to the heel may assume the value of body weight, and in the heelstrike more than 2 G. The muscles (mm. tibialis anterior, extensores digitorum) which counterbalance the ankle joints, insert into the

Fig. 9.4 (continued) joints, so that no energy-consuming muscle contractions are necessary for keeping the joints balanced. If the CoM is lowered, the gravity vertical in humans passes far behind the forward-displaced knee joint, as a consequence of the long hindlimbs, characteristic for humans. Because of the shortness of the human foot, the GRF cannot be shifted forward to shorten the load arm at the knee joint and so requires strong activity of the knee extensors. This is not the case in the short-legged and long-toed chimpanzee, or gorilla, which therefore do not need the same powerful m. quadriceps femoris as is typical for humans. (**b**) Anterior views of human (left) and chimpanzee (right) hindlimbs. (**c**) Area of support (grey) is determined by the lengths of the feet and their breadth plus distance between them. The share of body weight supported by each foot depends upon the point where the gravity vertical hits the ground. (**e**) The statistical variation of the gravity vertical, shown along the *y*-axis, and in top view from above (**f**), to show the variation in *x*- and *z*-directions. Twofold and threefold standard deviations are indicated by increasingly dark shades of grey. The lateral abduction of the anterior part of the foot (in humans often characterised as "valgus-position") expands the diameter of the area of support into a less frequently and less highly loaded area (see also Fig. 9.10)

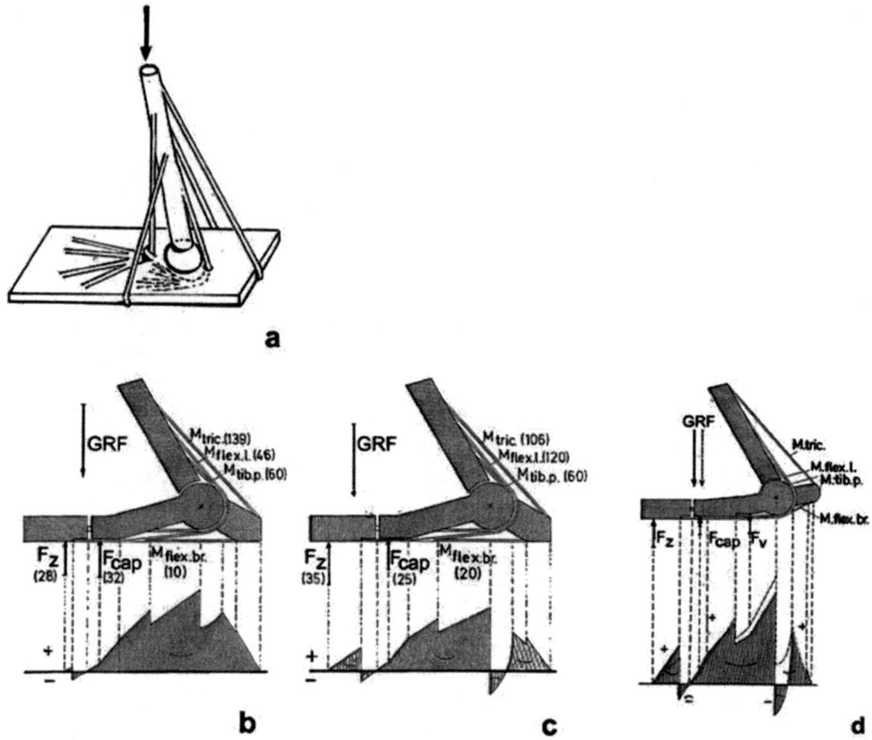

Fig. 9.5 (**a**) The human foot provides a considerable area of support for keeping balance, like a base plate. The lower leg, drawn as a single rod, carries the load (GRF, downward directed arrow) and transmits body weight to the plate. It is kept upright on top of the plate by muscles, and reaction forces concentrate at the margins of the plate. The bending moments along the foot become best visible in side view: In (**b–d**) the plate is projected into the *x*/*y*-plane and looks like a beam, which is bent (diagrams on bottom). At the joints, bending must be zero. The heel in (**b**, **c**) is close to the ground, but does not carry weight. The total GRF is distributed evenly on the toes and the balls (F_z and F_{cap}). The basal toe joint is counterbalanced by the m. flexor digitorum brevis, which takes its origin from the heel, and the m. flexor digitorum longus, originating from the lower leg. The latter muscle exerts a rotating moment at the ankle joint, and both together reduce the bending moments along the foot (hatched area). Close to the ankle, where bending moments reach their greatest values, m. tibialis posterior (on the medial side of the foot) and mm. peronei brevis et longus (on the lateral side) act as additional ties which reduce bending moments further. The human foot in (**b**) is characterised by short toes and thus needs less force in the toe flexors. Balance at the ankle joint is completed by the m. triceps surae, which bends the heel strongly upward. (**c**) In the foot of an ape, the toes are longer and can take over a bigger share of GRF than in humans. As a consequence, the toe flexors exert more force to keep the toes in balance, and this leads to lesser bending moments along the passive structures (hatched area, same scale as in (**b**)) of the foot, while the m. triceps is recruited to a lesser degree. The smaller bending moments require a less rigid construction of the passive apparatus, ligaments and bones. In both variants, the shifting of a part of the total GRF on the heel allows the m. triceps to exert less force. (From Preuschoft 1970b). (**d**) For comparison, the foot of a cercopithecoid monkey lifts the heel from the ground, so that no external force comes into action. Two variants of loading are superimposed. A moderate proximal shift of the GRF (broken arrow in contrast to the original solid arrow) puts less weight on the toes and therefore requires less force of the digital flexor. This entails greater bending moments along the foot (unhatched areas), especially since an additional ground contact at the anterior tarsals (Av) is recruited. The m. flexor longus exerts less but the m. triceps greater force. The scale of bending moments is twice that of (**b** and **c**). (From Preuschoft 1970a)

os cuneiforme I, or into the toes, evoking bending moments similar to, but smaller than those to be dealt with below. The load case at heelstrike is often overestimated, whereas the load case shortly before pushoff is often overlooked.

When the whole GRF acts against the toes and the balls, the *m.* triceps must exert much force for balancing the ankle joint. The necessary force F_t is determined as in Eq. 2.5 by the equilibrium of moments, dissolved like in Eq. 8.16.

$$F_t = \mathrm{GRF} * l_v / l_c \tag{9.1}$$

where l_c is length of the heel, and l_v the length of the forefoot.

Total GRF is normally less than 2 G, but can become higher, according to our measurements up to 4 G in high jumpers. Therefore, concentration of GRF on the anterior part of the foot requires plantar flexing muscle forces that are greater than external forces and entail the greatest bending moments along the foot (Fig. 9.5 b, c d).

The *human foot* is highly specialised to only one mechanical function in full contrast to the hand and to the versatile use made of the foot in apes or monkeys (see Chap. 8). The human foot can perform its function, that is transmission of compressive forces between the moving organism and the ground, without restrictions if covered by shoes, provided these also permit minimal movements of the toes as a whole and minor deviations from pure compression, namely, force components parallel to the ground by frictional resistance. As has been shown above (Chaps. 5, 6 and 7), totally different constructions of the feet are possible and work perfectly – maybe even more efficient. But the human foot was evolved from the prehensile foot typical of all primates and had to optimise its shape on this pre-existing basis. The same principle, namely, that evolution took its starting point from the needs posited by the arboreal lifestyle, holds true for overall body shape.

The really obvious specialisations of the human foot are the non-abductable, but very strong hallux, the very weak and short "lateral" toes, the talus and the calcaneus. The characteristics of the foot skeleton in *Gorilla* as an example for African apes and *Homo* are compared in Fig. 9.6a–d. The joint surfaces of the calcaneus are shown from above, in Fig. 9.6e, f in an oblique view from medial behind. The resultant of the tendon and muscle forces which are projected into the plane of the drawing is clearly more medially directed in the ape, because of its abducted hallux. The direction of the subtalar joints must be expected at right angles with this resultant, and the direction of the caput and collum tali is also in agreement. All these details at the same time prevent rotational moments about the talus in the plane of the dorsum pedis.

The *muscular anatomy* in human feet and those of African apes, described in terms of origins, insertions and course of muscles and their tendons, is very similar (Preuschoft 1961, 1970a, b, shown schematically in Fig. 9.6). There exist, however, characteristic *quantitative differences* between the muscles in the human hindlimb and those in the apes. As noted above, muscles are the weakest element in the locomotor apparatus, and therefore this material must be present in large amounts – which also means that muscles contribute the greatest share to the weight of body

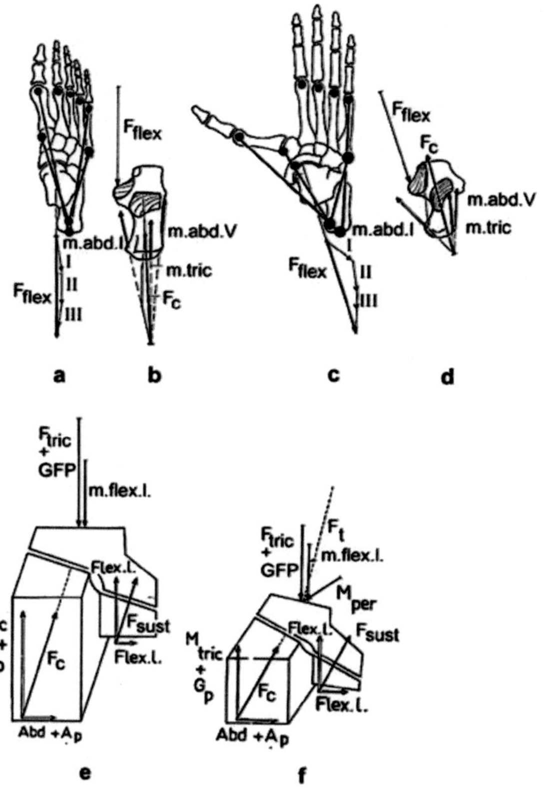

Fig. 9.6 (**a**) Human foot with adducted hallux and short toes, (**b**) human calcaneus, both viewed from above. (**c**) For comparison, gorilla foot with abducted hallux, and long toes, and its calcaneus (**d**). The forces exerted by the short toe flexors (heavy lines between the black dots marking the insertions) are combined at the plantar surface of the tarsus, where they form a common tendon. The direction of all forces acting at the heel bone (F_c, F_{flex}) depends from the directions assumed by the loaded digits. The same holds true for the other intrinsic muscles of the planta pedis. The resultants of the redirected tendons in the human foot (F_c) are stronger and closer to the foot's longitudinal axis than in the ape, where they deviate medially. The surfaces of the subtalar joints are orientated at right angles to these resultants, the neck of the talus parallel. (**e**) (*Homo*) and (**f**) (*Gorilla*) schematic views from medial-behind on the calcaneus (with sustentaculum tali – see Fig. 8.38) and talus. Because of the permanent adduction of the big toe, the plantar muscles in humans (**e**) have lesser components in medial direction, parallel to the sole. The toe flexors and the plantar aponeurosis in apes (*m. abd* + *Ap*) as well as the long digital flexor (*m. flex l, or Flex. l.*) commonly are stronger and more medially directed in the apes, while in humans the parasagittal force components acting at the heel, m. triceps surae (*m. tric*) and *GRF* are greater. The resultant of all forces acting against the trochlea tali is inclined in the apes with their abductable big toe, but closer to the vertical in humans (from Preuschoft 1970a)

segments. While this agrees with common experience and general observations, precise data are rare. According to Preuschoft (1963b, Fig. 9.7), the intrinsic foot muscles of humans weigh 2.5% of the total hindlimb musculature, in a gorilla about 4.5%. The toe-flexing and toe-extending musculature in humans amounts to 5.9%, in

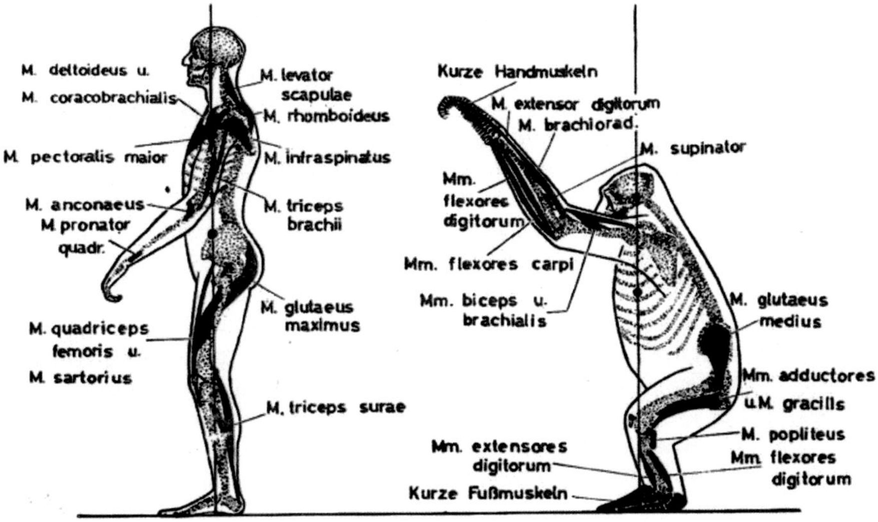

Fig. 9.7 Quantitative relations between the weights of muscles in humans and gorillas as examples for apes. Skeleton stippled, muscles which are relatively heavier than in the compared species are black. CoM and gravity vertical (=GRF) are indicated. Note in the present context especially the hindlimb muscles. (From Preuschoft 1963b)

the gorilla 9.5% and in orang-utan 11.6% of total hindlimb muscles. This part seems to become smaller with ontogenetic growth. The share of the m. triceps surae in humans is 13.3% but in gorillas 7.6% of hindlimb muscles. In contrast to the toe muscles, this share seems to become greater with ontogenetic growth. The repeatedly noted shift of muscle mass into proximal direction can be observed best in the thigh: the m. gluteus superficialis (=maximus) in humans originates more proximal than in gorillas and weighs 12.1%. The more distally displaced m. gluteus superficialis (=ischiofemoralis) has only 9.5% of total hindlimb muscle mass. The human adductor group weighs 14.4%, and its m. adductor maximus becomes transformed distally into a (less heavy) tendon, while the adductor group of the gorilla occupies a share of 21.8% of total hindlimb musculature and has a fleshy insertion. What does not fit into this scheme is the m. quadriceps femoris in humans, the share of which is 24.8%, in contrast to the same muscle in the gorilla with 17.2%. This inflation obviously is due to the forwardly displaced position of the knee joint, which leads to a long load arm of GRF, the enormous rotating moments of which must be counteracted by muscles.

The *toes of the prehensile feet* in apes as well as in monkeys (Fig. 9.5) are long and therefore need markedly stronger toe-flexing muscles than humans with their short toes, in which particularly the intermediate and the terminal segments are reduced. While the long toes of non-human primates can be and often are loaded separately, this is not the case in humans, where the lateral toes brace each other against load components in the medio-lateral plane – similar to the metatarsals (see

Figs. 8.35 and 9.6). That is the reason why the basal phalanges II–V in humans are not exposed to high bending moments in this plane and possess circular cross sections, rather than the oval profiles of phalanges in non-human primates (Fig. 8.35). The intermediate and terminal phalanges of the big toe, however, are exposed to force components parallel to the substrate in transverse (z-) direction and possess adequate shapes. The m. flexor digitorum superficialis, which in humans takes its origin alone from the heel bone, in all monkeys originates to a greater or lesser degree from the tendon of the m. flexor digitorum profundus. Therefore its tensile force (and the rotating moment in the ankle joint) adds up to the latter, and the m. triceps surae is less strongly developed than in humans (see above, last section). Because of the reduced rotating moment of the m. triceps, the heel bone of apes and monkeys does not need the special stays on its plantar side, formed in humans by the entire m. flexor brevis digitorum, the m. plantaris (missing in nearly all apes) and the obviously strong aponeurosis plantaris, especially its pars fibularis (Fig. 8.37).

A *simple, two-dimensional mechanical model* of the primate foot can be derived from the model in Fig. 9.5a by projection into side view (Fig. 9.6b–d). The ground plate is replaced by a beam, and in an uniaxial joint, a rod (lower leg) is balanced upright. The load (GRF) is attached to this rod. This arrangement allows the shifting of the load between heel and toes (Preuschoft 1969a, b, 1970a, b, 1971) like in rolling off. If the (low) partial forces along the sole are neglected, the highest ground reaction forces are applied at the heel and at the tips of the toes. The rotating moments at the ankle joint are countered by a tension-resistant structure in place of the m. triceps surae. The moments at the ankle change during roll-off, while the GRF wanders forward from the heel to the toes. The beam (=foot) is exposed to bending moments which grow from the heel rapidly to the ankle joint and less steeply from the toes to the same joint. The introduction of a joint at the base of the toe requires a tie on the plantar side for keeping the joint in balance. The longer the toe, the stronger must be the tie. If this tie is prolonged rearward to the heel, it reduces the total bending moments along the whole beam. The arrangement of the flexor tendons, which exists in primate feet, allows adapting of the tie function to variable parts of the foot: between metatarso-phalangeal joints and either the posterior border of the talus (mm. flexores digitorum longi), or the tuber calcanei (mm. quadratus plantae, flexor digitorum brevis, aponeurosis plantaris), or be confined to the heel alone (like the stays of the heel bone shown in Figs. 8.37 and 9.5). The toe flexors also exert considerable redirectional forces against the posterior margin of the talus and against the plantar side of the sustentaculum tali, which is broad in the apes but narrow in humans. This force component is passed on to the talus neck through the articulatio talo-calcanea media, in response to the high or low forces transmitted (Figs. 8.37 and 8.38). It should be noted that this joint surface is, as far as can be seen from the illustrations in Deloison (1991), rather small in the two calcanei from Hadar (AL 333 8, less so in the more damaged AL 333 55), indicating shorter toes than in the apes but longer digits than in humans. The toe flexors also possess a rotating moment at the ankle joint, and this relieves the m. triceps surae: the stronger the toe flexors, the less force must be provided by the m. triceps and vice versa. The strong tensile force applied by the m. triceps surae requires more strength

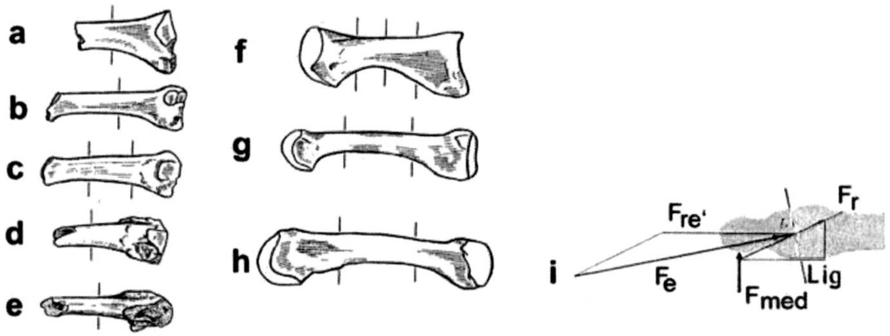

Fig. 9.8 The prehuman foot skeleton OH 8 is ascribed either to *Australopithecus* or to "*H. habilis*". (**a–e**) are the remains of the metatarsals in lateral view, from first on top to fifth on bottom. For comparison, a human os metatarsale I is depicted (**f**), and a human os metatarsale III (**g**). As example for apes, the os metatarsale III of a chimpanzee is shown (**h**). The obvious height of the bases in the fossil and recent hominin metatarsal indicate more bending strength of the foot than the short diameters in the prehensile foot of apes. (**i**) The also preserved right terminal phalanx of the fossil hallux in plantar view, with force components in longitudinal direction ($F_{re'}$) and from medial to lateral (F_{med}), the latter balanced alone by the ligaments. The "valgus" orientation of the interdigital joint and the insertion of the flexor tendon (not illustrated) are suited to the one-sided loading typical of humans, regardless of their wearing shoes or not (Barnett 1962) Drawing taken from Preuschoft, 1971

of the corpus calcanei than the application of ground reaction force does. (This point is neglected often, e.g. in Latimer and Lovejoy (1989).)

The same model has been used by Preuschoft (1971) for a biomechanical analysis of the foot skeleton later named OH 8 from Olduvai Bed I (Davies et al. 1964; Day and Wood 1968; Day 1985, Fig. 9.8). Unfortunately, the corpus calcanei and the toes are not preserved. The distal parts of tibia and fibula (OH 35) seem to belong to the same individual; the femur is documented at the same site and at Hadar (Ethiopia) by several, proximal and distal fragments. All these traits fit together, indicating humanlike bipedality. However, the attribution of these postcranial remains to hominid taxa: australopithecine or very early humans (hominines, the so-called *Homo habilis*) are still a matter of debate; associations of skull or dentitions, which allow taxonomic identification, with essential parts of the postcranial skeleton, are in most cases not beyond doubt. What we know from several sites (Hadar, Olduvai) are skeletons with the proportions of australopithecines, and pedal remains, which look fairly humanlike (Johanson and Edey 1981; Hartwig-Scherer and Martin 1991; Clarke and Tobias 1995)

The lengths and positions of the digits, as well as load distribution on the sole, are documented and can be derived from the known footprints (Figs. 9.3 and 9.8a–h). The proximally increasing bending strength of the metatarsals (visible from the diameters in side view, Figs. 9.5 and 9.8f, h) as well as the position of the tibio-talar joint surface in anterior view (Fig. 9.9d) make this foot and lower leg well suited for sustaining the stresses which occur in bipedal locomotion (though not to the extent of modern humans, as emphasised by Klein (1999), based on Day (1985)).

The proximal phalanges unearthed at Hadar (AL 333, *115*) are strongly curved, flattened in dorsoplantar direction and shorter than in modern apes but longer than in present-day humans (Deloison 1991). This yields evidence about the strong development of the long digital flexors and the functional independence of the individual toes. In these traits, australopithecine feet deviated from those in humans, without being developed to the extent as in apes. No convincing arguments have been found for an abduction of the big toe. By contrast, the tarsometatarsal joint is flat and does not seem to permit a position of the hallucal metatarsale other than close and nearly parallel to metatarsale II. All these conclusions were confirmed later by the famous foot tracks from Laetoli (see above). The clear characterisation of these footprints by Deloison (1991) has been confirmed by the scans of Kullmer et al. (2003). A detailed comparison of the fossil tracks with those of living bipedalists was made by Tuttle et al. (1991). These authors, like others, emphasise the similarity of the fossil tracks to humans. Day (1991) in a way complemented this study by revising a number of fossil foot tracks of *Homo* from all over the world. With respect to individual footprints, the conclusions drawn by Tuttle et al. 1991 as well as by Day (1991), are in conflict to Deloison (1991) in the same volume. In fact the fossil footprints seem to be so variable that arguments can be derived in favour of any conclusion.

Pedal phalanges from other fossil sites, but similar chronological age (or younger, like Susman 1989), are stronger and longer than in humans; their cross sections are not round, but oval (like in those of fingers); and the most proximal elements are often strongly curved. All show obvious marks at the insertions of the superficial flexor tendon, and the anular parts of the tendon sheaths, and often also at the insertions of the collateral ligaments. The static situation is detailed above in Chap. 8.10. By these traits, the pedal phalanges of prehumans are suited to sustain stresses which occur during grasping, with independent actions of the different rays. Such stresses seem to occur in climbing on trees or cliffs and are probably more similar to the use made of the foot by apes than by extant humans and fossil representatives of *Homo*. Many authors neglect this fact or drown it in repetitions of asserting the similarity with the human condition. Morphological similarity indeed is an important observation, and comparisons of the use made of morphological traits as well, but both do not give satisfactory information about the mechanically relevant situation, which after all governs the development of shape!

The tibiotarsal joint, or "upper" *ankle joint* of anatomy textbooks, has in extant humans in side, as well as in anterior, view not the same orientation relative to the tibia as in the African apes (Preuschoft 1971). In *side view*, the tibial joint surface shows in humans an angle that opens forward; in the apes, however, it is orientated more or less at right angles to the tibia (Fig. 9.9c). Both variants make sense under consideration of the position of the GRF to the knee joint and the resultant of all forces acting across the joint in both hominoids. Because of the length of the human hindlimb, the intersection between GRF and line of action of the m. triceps surae is well behind the knee joint (Figs. 9.4a and 9.9a, c). The tibio-talar joint is, under the action of body weight and pull of the muscle, in balance only, if the joint surface of the tibia is tilted forward. In the apes, by contrast, the knee joint is kept closer to the GRF, because of the shortness of limb bones as much as because of the length of the toes, which induce a forward shifting of the GRF. Therefore, the intercept between

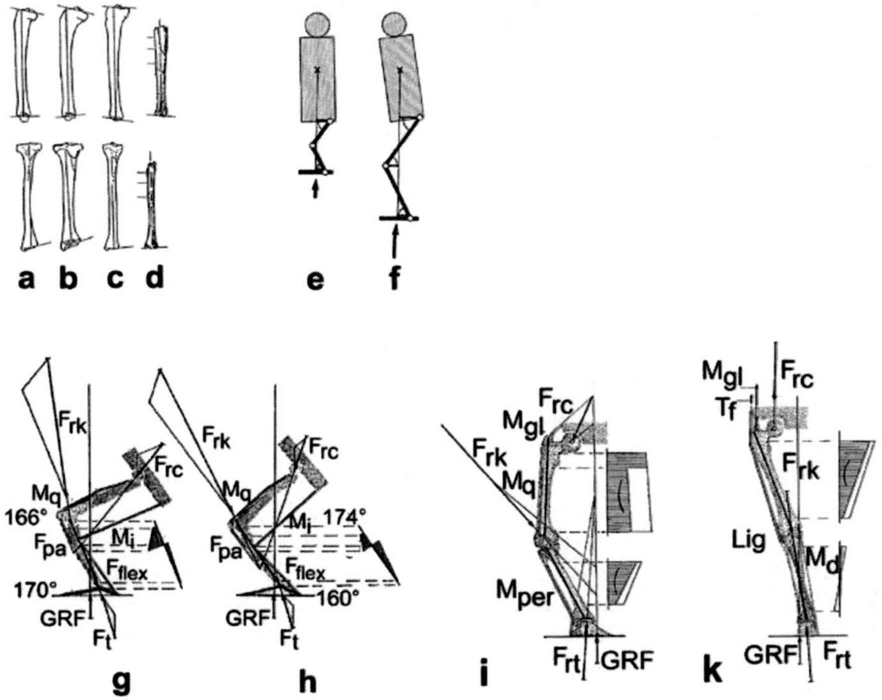

Fig. 9.9 (**a–d**) Top row: side view. Bottom row: anterior view of tibiae. (**a**) Chimpanzee, (**b**) gorilla, (**c**) human, (**d**) fossil tibia OH 62, all showing the different inclinations of the talar and femoral joint surfaces. (**e**) Short-legged and (**f**) long-legged hominoid standing with the same knee angle and location of GRF. In the long-legged form, the knee joint is invariably farther in front of the GRF. From Preuschoft (1971). (**g, h**) Consequences of (**e**) and (**f**): Loading the lower leg in lateral view differs. The resultants of all forces acting on the joints of the tibia (F_t and F_{rk}) have different directions in relation to the tibia's long axis. If the GRF passes closely behind the knee joint, F_t is more or less in line with the tibia, and therefore the most advantageous talar surface is at right angles. In contrast, the knee resultant F_k deviates from the longitudinal axis of the tibia and so a retroverted joint surface to receive F_{rk} is of advantage – and present in long-footed and short-limbed African apes. If however, GRF (like in **f** and **h**) passes far behind the knee joint and the talar joint resultant F_t forms a blunt angle with the tibia, so making a forwardly tilted joint surface advantageous, while F_{rk} is nearly in line with the tibia, so that a joint surface at right angles is optimal to reduce sliding – like in short-footed and long-limbed humans. (**i–k**) Two variants of loading the lower leg in anterior view. In (**i**), the GRF is shifted toward the medial margin of the foot (like in climbing apes, when the foot is grasping hold) and the flexor and the peroneus muscles exert more force. In (**k**) the GRF is in the middle of a narrow foot (*Homo*). The tibia in anterior aspect needs in forms with a narrow foot a talar joint surface more or less at right angles to the long axis, while in apes with their broad, prehensile foot the trochlear surface of the talus needs a high lateral margin and the tibia a marked tilting of the distal joint surface. Note the low bending moments in lateral view (**g, h**) and the high bending moments in the anterior view on the femora and the lower leg of apes (**i**). Both require, and possess, diverging strength properties, which are in accordance with the diverging morphology. (Modified from Preuschoft 1971). The resulting forces in the knee joints put more compression on the medial than on the lateral condyle in the apes. Compare the forces acting on the patella in the same positions in Fig. 9.10

GRF and the line of action of the muscles is far in front of the ankle, and the resultant falls in line with the tibia shaft or even deviates rearward. A tilting of the distal joint facet forward is not appropriate. Seen from in front, the high lateral margin of the trochlea tali, as well as a marked inclination of the distal joint facet of the tibia, is obvious in apes, while the ankle joint in humans is at right angles to the tibia and does not exhibit this inclination (Fig. 9.9b). Morton (1924) and Elftman and Manter (1948) have drawn a connection to load-carrying on the lateral side of the chimpanzee foot in combination with extreme abduction of the knee. This generally accepted explanation, however, neglects the influence of musculature, especially if load is carried not on the lateral but on the medial side of the foot. Preuschoft (1963a, b, 1979, 1971) has analysed the feet of apes in climbing, and the gorilla foot in walking on flat ground, where the hallux regularly transmits a part of body weight. In both load cases, the mm. peronei must be and indeed are activated to balance the tarsus against the medially shifted SRF or GRF. The resultant of both SRF and muscle force is very strong and inclined medially, so that it acts at right angles against the laterally elevated trochlea tali (Fig. 9.9d).

Non-human primates usually assume in bipedal standing a *flexed position of the knee joint* (the “flexed hip-flexed knee posture” of Crompton and his group See also Okada 1985). If the CoM is kept above or closely behind the knee joint (Demes et al. 2014), its extensor, the m. quadriceps femoris, must not exert much force. If the gravity vertical is shifted in front of the knee joint, the activity of the extensor is not at all required and can be replaced by that of knee flexing ischiocrural muscles. These hip extensors are activated anyway for balancing the hip, and they possess long lever arms at both joints. Such a position is, of course, easier to assume if femur and tibia are short (see also figures above). More detailed studies of the GRFs and the pathways of the CoM during bipedal locomotion have shown that at least in gibbons (Yamazaki 1992) as well as Japanese macaques (Ogihara et al. 2014), probably in all non-human primates (Ogihara et al. 2007), mechanical rules are used which differ from those in humans (see below).

Hindlimb length, in combination with the above-mentioned short foot, keeps the GRF clearly behind the axis of the knee joint (see also above). This leads inevitably to long load arms and outstanding rotational moments, which explain several characteristics of humans: they are counterbalanced mainly by the very strong m. quadriceps femoris (see Figs. 9.4a, 9.7, 9.9e-h); the great forces in the knee joint require very large contact surfaces (the knee is said to be the largest joint of the human body), and the above-mentioned strengths of the m. triceps and of the corpus calcanei are also consequences of the typical human proportions (the reason for them is dealt with below in Chapter 9.3) Strength properties of femora and tibiae are in agreement with loads (Drapeau and Streeter, 2006).

The expenditure of energy, however, for keeping their body upright and their CoM above the support area, is limited in humans. Instead they are able to stand or walk for long time intervals. This is largely due to the extended hip and knee joints, which are kept close to the prolongation of the GRF (=gravity vertical), so that rotating moments in these joints are kept low.

The often mentioned X-position of the *human knee in frontal* view is due to a slightly shorter length of the lateral condyle in comparison to the medial. This leads, if the femur is placed upright on a table, to the famous laterally open condylo-diaphyseal angle (Lovejoy, 1975, 2003; Lovejoy and Heiple, 1970; Lovejoy et al. 1973). This configuration approaches the joint in the middle of the human hindlimb to the line of action of body weight (=gravity vertical or GRF; Fig. 9.10) and to the carrying skeletal structure, so that bending moments are reduced. This implies the advantage of less strength and less weight, and distributes the joint force evenly on both femoral condyles (=areas of equal size) but requires slightly greater length of the medial than of the lateral condyle or, in other words, a laterally open condylo-diaphyseal angle. Although this trait is well-known and often discussed in the context of human evolution, it is not at all restricted to humans. Exactly the same mechanical principle is realised in cattle as well as other bovids in the carpal joints of their forelimbs (Fig. 9.10m), and the hindlimbs of many cursorial mammals show in relaxed postures a shorter distance between the hock joints than between the hooves and the knee joints (Fig. 9.10n). This "valgus position" approaches – as described above – the joint in the middle of the lengthy free limb (carpal and hock joints, respectively) to the GRF, reducing the bending moments in the y/z-plane and relieving the collateral ligaments. Especially in the hindlimbs, the hooves are often directed outward, so that the area of support is not narrowed. In addition, the m. triceps and the Achilles tendon get a component in the transverse plane. Exactly the same is the mechanical effect of the characteristic posture of humans with the toes turned slightly outward (see above, Figs. 9.3d and 9.4c).

In humans, both *femoral condyles* are of about the same size; alone the relatively larger area covered by the lateral meniscus seems to indicate somewhat higher loads on the medial than the lateral condyle. If both condyles of a femur are placed on the table and the joint is viewed from distal, the lateral margin of the *patellar facies* is higher than the medial (Fig. 9.10d). This elevation offers a surface at right angles to the force by which the patella is pressed against the femoral condyle: the pull of the m. quadriceps is nearly parallel to the femur and the pull of the ligamentum patellae parallel to the tibia. The change of direction between the two segments causes a redirectional force which presses the patella on the condyles. To prevent the patella from sliding laterally on the facies patellaris, the latter needs an inclination. The situation in the African apes, especially the gorilla,[3] is clearly different: the condylo-diaphyseal angle (see above) is open to the medial side, and their abducted knees are kept in balance not so much by the collateral ligaments, than by the m. quadriceps and the ligamentum patellae. The redirectional force presses the patella strongly against the medial rather than the lateral condyle (Fig. 9.10), and the medial instead

[3]Crompton et al. (2003); Thorpe and Crompton (2005) have observed that orang-utans frequently and for longer intervals carry most of their body weight on the hindlimbs while keeping balance with the aid of their elevated arms. In this "assisted bipedal locomotion", the hindlimbs are exposed to considerable forces, which are more similar to those in humans than to those in African apes. Indeed the asymmetry of the knee condyles is less than in chimpanzees and in gorillas. The same holds true for *Oreopithecus*, which seems to have used a similar locomotor mode like orang-utans (Moya-Sola and Köhler 1996; Rook et al.)

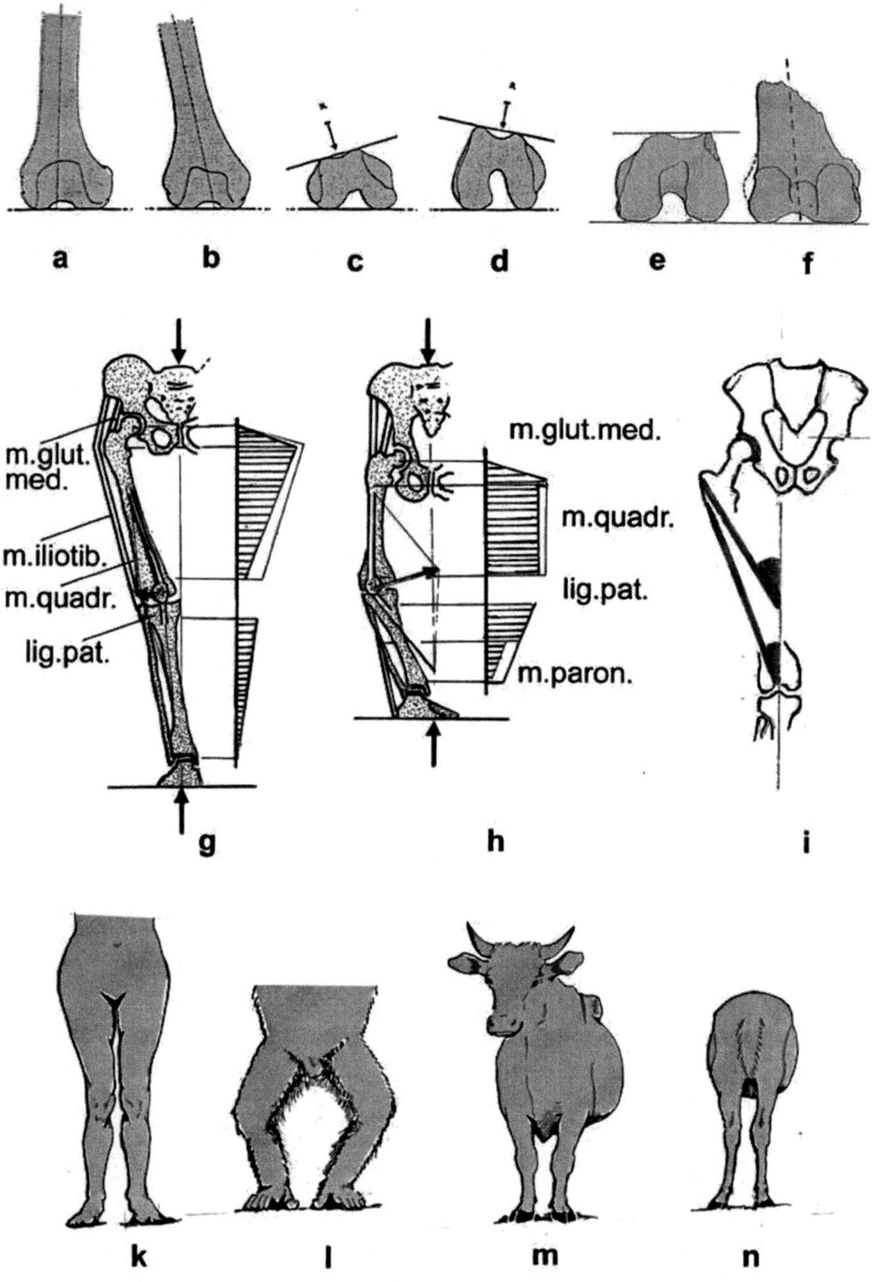

Fig. 9.10 (**a–d**) Right femora of an African ape (**a**, **c**) and a modern human (**b**, **d**), (**e**, **f**) distal femur of *Australopithecus* according to Le Gros Clarke (1967) as just one example. In (**a**, **b** and **f**) the femur is placed upright on a table, so that the condylo-diaphyseal angle becomes visible; in

of the lateral margin of the facies patellaris is elevated to exclude an uncontrolled luxation.

In many *fossil fragments* of prehuman forerunners, the relatively wide angle between the femur shaft and neck (=*collo-diaphyseal angle*) is documented, as well as the length of the neck and its anterior-posterior flattened profile. These traits often are summarised under "coxa valga". A partial analogy to this configuration (wide collo-diaphyseal angle) can be observed in orang-utan and is combined with the absence of a lig. capitis femoris (see above). The mechanical value or the consequences of this combination of traits have never been pinned down precisely, and therefore all conclusions drawn on this basis should be handled with care. The "coxa valga" is seen in medical science among modern humans as a deviation from the "normal" configuration.

The *human pelvis* (Fig. 9.12) is markedly different from that of apes, and therefore comparisons seem to be a promising approach to identifying the divergent modes of posture and locomotion. Unfortunately, up to now a really satisfying analysis (with the adequate method of FES) of *pelvic morphology* in apes, prehumans and recent humans, under inclusion of the femora and the lumbar spine, is still missing.

This is more annoying, since the pelvis of australopithecines has commonly been taken as a proof for bipedality in this group. In fact, all these repeated assertions are based on the simple fact that its shape is morphologically closer to that of humans than of apes, not on mechanical necessities or "adaptations". Its shape is – as long as

Fig. 9.10 (continued) (**c**, **d** and **e**), the femora are resting on the table with the posterior margins of their condyles in contact, to show the elevation of the medial (ape) vs. lateral lip of the facies patellaris, which prevents a slipping of the patella. (Compare Fig. 8.43g). (**g**, **k**) The human hindlimb showing the characteristic "valgus"- (*X*-) position, in contrast to the "varus" position of apes (**h**, **i**). The marked "varus"- (*O*-) position of the knee joint in frontal view leads to higher load on the medial than on the lateral condyle. The patella in (**g**, **h**) is exposed to a redirectional force, which presses it against the condyles. In (**g**) this force F_p is directed laterally like in humans and requires an elevated lateral lip of the facies patellaris. In (**h**), the force F_p is directed medially, like in the apes. It entails an elevation of the medial, rather than the lateral, lip of the facies patellaris. The increased load on the medial side is distributed over the large surface of the greater medial condyle, while the lateral meniscus covers a greater part of the joint surface and can assume the shape of a discus articularis (Preuschoft 1961). The postures entail different bending moments (hatched diagrams on right side) along the thighs and lower legs, which are in accordance with the profiles of the respective bones. Muscular activities visible in side view - like in Fig. 9.9. - are not illustrated. In (**i**) the pelvic breadth is the same as in (**g**, **h**), in order to show that a short femur needs more pronounced inclination to approach the knee joint to the vertical than a longer one. The condylo-diaphyseal angle must be expected to be and in fact is sometimes more marked in the short-legged australopithecines than in long-legged *Homo*. The valgus-position of the human knee joint (**k**) is in contrast to the varus-position of the knee of apes (**l**), but conforms to the joint in the middle of the free extremity in cursorial mammals (**m**, **n**), no matter whether the joint belongs to the front- (**m**, cattle) or hindlimb ((**n**) *Hippotragus*, where any other even-toed ungulate could also serve as an example). The X-position relieves the collateral ligaments of the knee, carpal or hock joint from being permanently tensed and the lower leg or metapodia from being exposed to bending (see also Fig. 9.4b) only well-preserved, not the reconstructed specimens are considered – surprisingly uniform throughout the group (Figs. 9.11 and 9.12).[4]

The existing approaches to a functional analysis of the pelvis take their starting point from a comparison of the pelvis with a ring (as in Pauwels 1965, 1968). Lenz (1993) has calculated the energy required for deformation in the case of a human and a chimpanzee pelvis under typical loads (Fig. 9.11). For sustaining the "deformation energy", a certain amount of bone material is required. This amount of bone is shown on the left half of the ape and the human pelvis. As can be seen, this approach yields a rough approximation but does neither explain the expanded outline of the ilium nor the three-dimensional curvature of the ischium-pubis complex.

If the pelvic region is viewed from the side (Figs. 8.51 and 9.12), at least some aspects become visible and give reason to create a hypotheses based on statics: *in the apes*, the weight force of the body stem (=gravity vertical) as well as the GRF usually tends to flex the hip joint ventrally. This holds true for bipedal as well as quadrupedal postures (Fig. 8.51).

The situation *in humans*, which spend a lot of time *standing upright*, is different. The weight of the trunk, neck and head in relaxed standing may well act behind the hip joint (Fig. 9.12h). Ventral muscles of the trunk wall keep the latter in balance, and the hip flexors are required for counterbalancing the hip joint. If muscle activity is low, or switched off, the ligamentum inguinale may be tensed (although no empirical confirmation seems to exist). This would be in line with the fact that it is in humans much stronger developed than in apes (Preuschoft 1961, 1970a, b), where this load case does not exist or only rarely exists.

In the often assumed ventrally *inclined trunk postures*, the vertebral column (like in the apes) is balanced by the dorsal muscle group, called erector spinae, which takes its origin (partly) from the iliac crest and a greater part from the sacrum. This is located at the height of the hip joint, not – as in the apes – at the height of the iliac neck. Since most of the hip extending muscles also originate from the ischium-pubis complex, tensile forces must be expected, which are sustained by the sacro-ischial ligaments (Fig. 9.12i) (which are stronger in humans than in apes and monkeys) and by shifting the origin of the m. gluteus maximus towards the sacrum. The expansion of the facies auricularis offers much strength against bending in the lowermost segment of the trunk. The part of the *ilium between the hip and the sacroiliac joint* (=iliac neck) is short and rotated towards the sagittal plane. Both traits obviously provide greater bending strength than exists in the ilium of an ape (Figs. 8.51 and 9.12). The arrangement of trabeculae in this area in humans shows the arcuate pattern as is typical of spongy bone under eccentric load.

The pelvic region is not alone the girdle of the posterior extremity but also, together with the lumbar vertebral column, the posterior *part of the trunk*. A simple,

[4]We have combined casts of the well-preserved left half of Lucy's pelvis with the right ilium Sts 14 to a complete, three-dimensional and symmetrical pelvis. In contrast, the reconstruction of the pelvis from the Rising Star Cave deviates from those of australopithecines and seems to be closer to that in humans. In view of the fragmentary state of the ilia and especially of the sacrum of so-called *Homo naledi*, it seems risky to draw precocial conclusions.

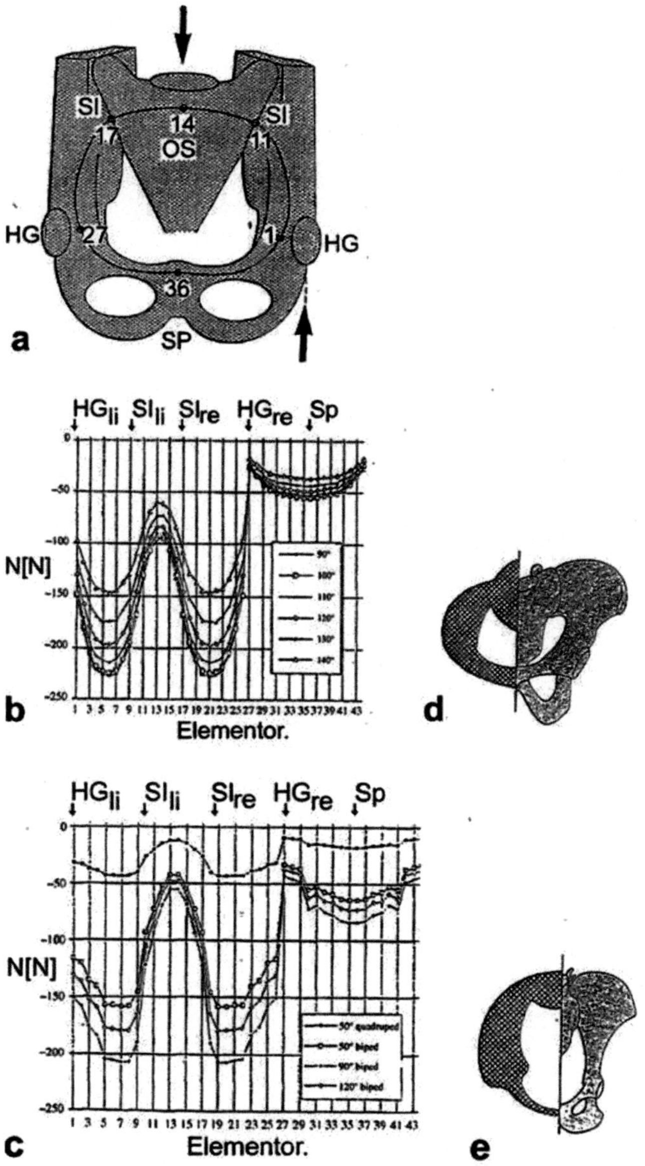

Fig. 9.11 (**a**) Generalised model of a primate pelvis in anterior view after Leutenegger (1987). Along the fat line, calculations have been made by Lenz (1993) for 43 points, most critical points indicated and named in the drawing. HG hip joint, SI sacro-ilium joint, OS os sacrum. Stresses are calculated for common inclinations of the pelvis in side view about an axis through both hip joints. The angle λ describes the inclination of the pelvic entrance against the horizontal. In humans it is in standing about 120°, in bipedal chimps around 90° and in quadrupedal posture around 50°. The greatest stresses occur in asymmetrical standing on only one (the left) hindlimb. (**b, d**) To demonstrate the calculated values, the "ring" is cut open at the left hip joint and unfolded from left to right into a plane. The amount of bone required for sustaining the loads at points 1–43 is illustrated on the left half of a human (**d**) and a chimpanzee pelvis (**e**); the real shape of the respective pelvis is shown on the right, projected into the plane of the ring model. The muscle forces are determined by the equilibrium in the hip joint, and curves are drawn for various inclinations

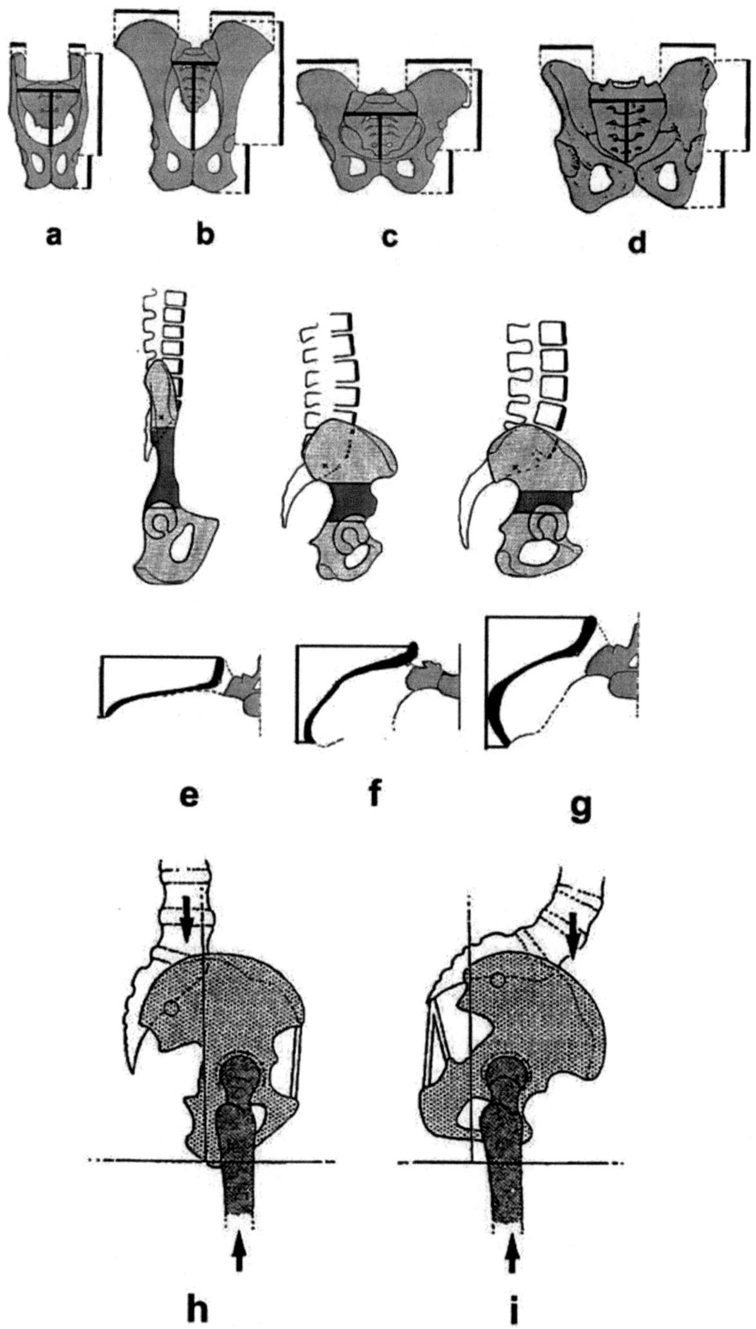

Fig. 9.12 (**a**) Pelvis of *Papio*, (**b**) *Pan,* (**c**) *Australopithecus* (combined *Sts14* and *A. afarensis*) and (**d**) *Homo sapiens* with characteristic dimensions, all drawn to about the same size. (**e–g**) In side views of the os coxae and on horizontal sections through the ilium,

but informative approach to the stresses in the trunk is illustrated in (Fig. 9.13: as in Preuschoft et al. 1988; see also Nakatsukasa et al. 1995) the head, neck, upper limb and trunk are simplified as a heavy column, or beam, balanced in the hip joint by ischiofemoral muscles (Fig. 9.12a, b). Obviously, the beam is tensed on its dorsal and compressed at its ventral side, as a consequence of the bending moments, which are evoked by the weights of the trunk segments. If, however, a cranial part of the body stem is dorsiflexed and elevated into a vertical posture, this part is no longer under bending moments, and those in the still inclined part become smaller. A totally upright trunk is only compressed by the weights of its segments and not bent. The farther rearward the "kink" in the beam, the more pronounced this effect. As illustrated in Figs. 9.12c and 9.13c, h, the "kink" of the body axis in humans is shifted so far caudally that the pelvis gets involved: the short iliac neck and the rearward displacement of the iliosacral junction contribute to a "lordosising" of the ilium. The iliac neck between the acetabulum and facies auricularis – or spina iliaca superior – is shortened and rearwardly curved, so that a deep incision of its posterior (=dorsal) contour occurs. This dorsiflexion of the ilium leads to the position of the iliosacral junction dorsal to the hip joint (rather than cranial to it, as in apes) and to the pronounced sciatic notch (incisura ischiadica) specific of humans. This, however, entails very high bending stresses in the (short) iliac neck. The necessary strength is provided by the breadth of the most highly loaded part of the ilium and its being swung towards the sagittal plane. The trabeculae of the spongy substance in this area are arranged in the arcuate pattern typical for bones under eccentric loads. The shortness and the swinging of the iliac blade can be seen more clearly in australopithecines than in apes, but both traits are less obvious in *Homo* (Fig. 9.12).

The characteristic stresses in the inclined trunk result from the weight forces of the body segments, which split into a compressive component parallel to the stem

Fig. 9.12 (continued) at the height of the greatest breadth of the sacrum, dorsal sides directed upward, the increasing rotation of the iliac blade toward the sagittal plane becomes visible (from left to right: *Pan*, *Australopithecus*, *Homo*). In each case the length of the projection into the lateral (vertical) and the frontal (horizontal) plane is indicated. The iliac "neck" is hatched, the ventral contours of the lumbar vertebrae are marked to emphasise the lordotic curvatures, and cross sections are black. *Australopithecus* and *Homo* provide in the *x*/*z*-plane greater amounts of bone material than *Pan*. This accumulation of bone material in the lateral aspect of the ilium increases the relevant diameter and so provides much strength in the hominids. In addition, the "neck" of the ilium becomes shorter in these forms. In homininae the pelvis is integrated into the lordotic curvature of the trunk and becomes itself "*lordosised*" as in Fig. 9.13h. The equilibrium in the hip joint must be controlled in all cases by muscle activity. This can be established by the same muscles as in other primates, provided that the postcoxal part retains its position – which is the case (see Fig. 9.13e-h). (**h**) Human pelvis in relaxed standing with CoM displaced rearward. Weight force is transmitted through the axial skeleton indicated by the downward directed black arrow at the vertebral column. Tensile force between pubis and spina iliaca anterior superior (sustained by the ligamentum inguinale) must be expected (see also Fig. 9.13). (**i**) Human pelvis in semi-erect posture with weight force (black arrow downward) acting in front of the hip joint. Tensile forces are to be expected in the ligamenta sacrotuberosum and sacrospinosum. This situation differs from that in apes, as shown in Fig. 8.51

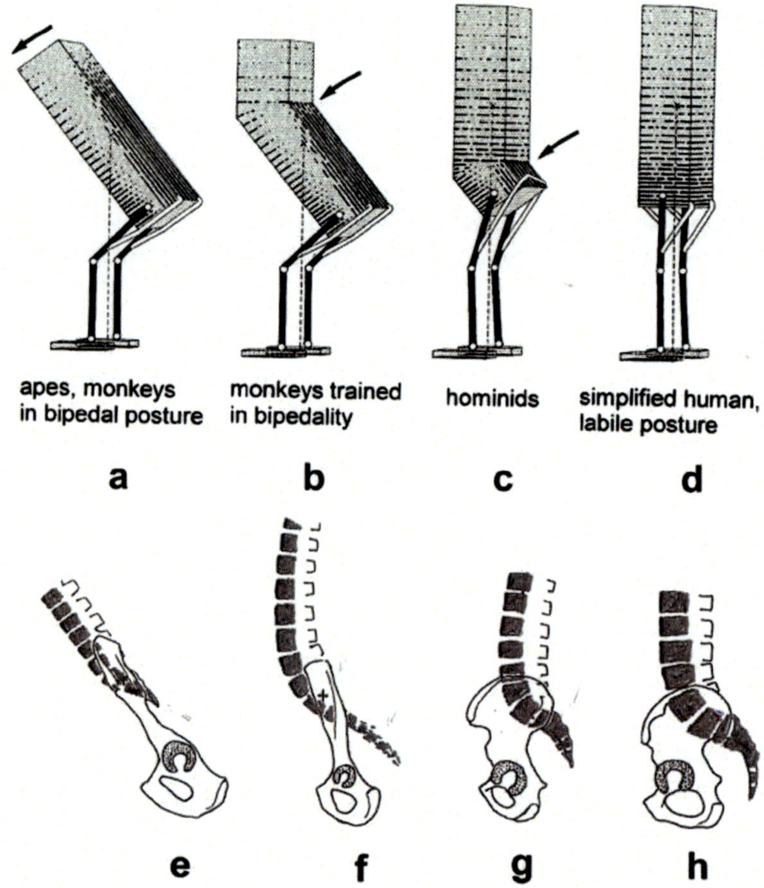

Fig. 9.13 Statics of body stem in bipedal postures. (**a**) If kept semi-upright, the body stem is loaded by compression and bending moments (see also Fig. 8.50). On the ventral side occur compressive (transverse hatching) and on the dorsal side tensile forces (longitudinal hatching). The latter are produced by the dorsal muscles and increase caudally. (**b**) If a cranial part of the body stem is lifted into a fully upright posture, this part is free from bending and exposed alone to compression (see also (**c**)). At the same time, the bending moments in the body stem caudal to the kink become smaller, until (**d**) bending becomes zero in completely upright posture, so that no muscle action is required, except for controlling the hip joint. The kink in the columns, or body stem, is functionally equivalent to the lumbar lordosis (**f, c, h**), which does not exist in normal/uninfluenced quadrupedal monkeys and apes (**e**), but is acquired (**f**) in monkeys trained to walk bipedally and in gorillas which have adopted bipedality (not illustrated). It existed in Australopithecines (**g**) and is well-known in humans (**h**). Most of the "human" lordosis is located between L_5 and S_1, in the "promontorium". This keeps the post-articular ischia and pubes in their position to the muscles, so that the coordination of muscles must not be completely changed. The more caudal the lordosis is located, the less muscle force is necessary for keeping the trunk in balance. The closer the kink of the beam is to the hip joint, the smaller are the bending moments, and the more "efficient" becomes this upright trunk posture. A totally upright trunk (**d**) is only compressed by the weights of its segments. This arrangement, however, is not realised (modified and partly redrawn from Preuschoft et al. 1988)

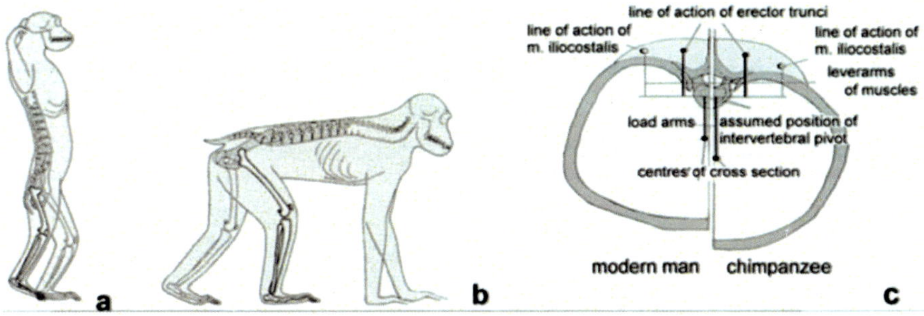

Fig. 9.14 (**a**) If subject to a strict training to walk bipedally, Japanese monkeys (and long-tailed macaques) develop within about 2 years a lordotic curvature comparable to that in humans. This curvature relieves the dorsal muscles. It is due to wedge-like intervertebral discs and persists in the quadrupedal posture (**b**) which the monkeys prefer in their free time. (**c**) Cross sections through the rib cage along costae V, VI or VII, on the left side of the drawing human and on the right side chimpanzee. In the thoracal section of humans, the 5th and adjacent ribs are dorsally curved, so that the lever arms of the muscles are longer in humans (left side) than in apes (right). The human vertebral column can be imagined as pushed ventrally into the thorax, so that the dorsal muscles have long lever arms, especially the iliocostalis, which is seen in human anatomy as a part of the "erector trunci" system, but compare Chap. 4. (Redrawn from Preuschoft et al. 1988)

axis (evoking compression in the vertebral column) and a component at right angles, which evokes bending moments (Fig. 8.50 d,e). These moments do not alone depend upon the segment masses but also upon the lengths of their load arms in relation to vertebral column and hip joint. If *the cross section* is narrow and deep (Fig. 9.14c; like in cursorial mammals, see Sub-chap. 7.7.4, 7.7.5, as well as Fig. 8.50), the CoMs of the segments have rather long lever arms (d_1) and therefore require much force to keep the intervertebral joints in balance. In an isodiametric trunk, the CoMs of the segments have short lever arms at the intervertebral joints (d_2). For a semierect posture, a shallow trunk is of advantage, as well as low-weight anterior limbs. Primates, however, cannot afford light-weight forelimbs, because of their arboreal habits – in contrast to, for example, kangaroos, which possess reduced and slender forelimbs. In addition to influencing the load arms, the marked curvature of *human ribs* shifts the thoracic vertebrae as a supporting column in ventral direction (Fig. 9.14c) and the muscles dorsally, giving them long lever arms, while the load arms of the segmental weights are shortened.

The advantageous lordotic curvature of the trunk axis is primarily *produced* by the vertebral column. The *lumbar lordosis* of *Homo* is commonly known (Fig. 9.13f, h) but often perceived alone as a means to prolong the lever arm of the dorsal muscles while shortening the load arms of segment weights. The most marked lordotic curvature in humans is far caudal, between the first sacral and the last lumbar vertebra in form of the wedge-shaped intervertebral disc, called promontorium. Surprisingly, the obvious advantage of the lumbar lordosis, the elongation of the force arms, decreases with forward inclination of the trunk (Fig. 9.15). The curvature

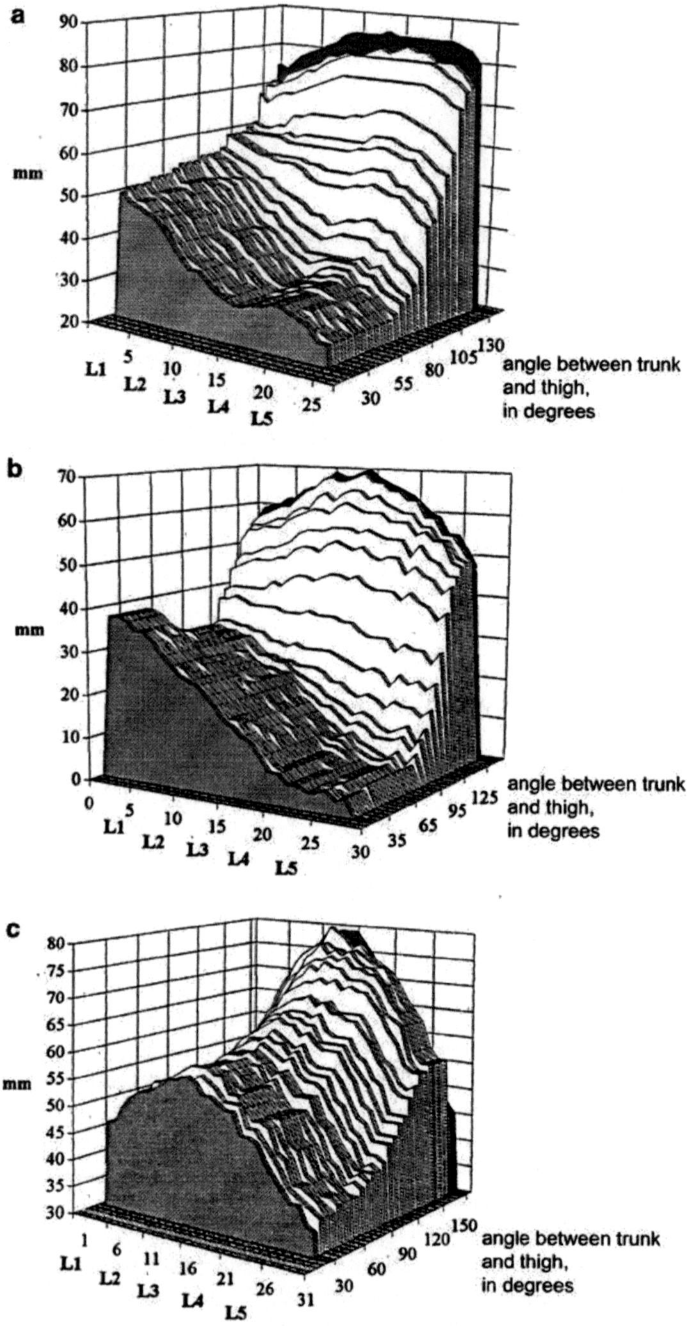

Fig. 9.15 Influence of gradual flexion of the hip joints on the curvature of the back. The longitudinal axis of the trunk (=contour of back) is tipped ventrally by 90°, and the contours of the back are arrayed behind and to the right for intervals of 5° hip flexion. Horizontal axis: on the

of the human lumbar spine usually disappears with flexion of the trunk (Jaeger 2003; Busching 1998; Pakusa 1998); alone the lowermost part between L_5 and S_1 is permanent. The lack of mobility in the lowermost part of the lumbar spine and the pelvis seems to confirm the above idea of "lordosising" the pelvis. Forward inclination of the trunk, or anteversion of the legs, is inevitably connected with flexion in the hip joints, and movements of the pelvis can provoke flexion of the vertebral column (Fig. 9.15; Busching 1998; Pakusa 1998).

Often neglected is the fact that the human lumbar lordosis exists only in combination with extension of the hip joint. As has been worked out by Pakusa (1998), the elastic resistance of inactive muscles, and at their limits the forces exerted by the gradually tensed ligaments, tilts the pelvis with its iliosacral junction rearward, as soon as the upper thigh is anteverted. This pelvic tilt is independent from active muscle contractions and can be observed in passively moved experimental persons. The tilting of the pelvis leads to a movement of the lumbar spine. According to Busching (1998) see also Preuschoft et al. 2012b), the lumbar lordosis disappears completely in 2/3 of all young humans if the hip joints are flexed by 30°; in 1/3 of the subjects, the hip flexion must exceed about 80°, before the lordosis is replaced by a straight line or a kyphosis. In more pronounced flexion, the lumbar spine is completely kyphotic throughout.

Abitbol (1989) has found that human children acquire their lumbar lordosis and the promontorium at the time they learn to walk. Children who do not start walking at the end of their first and beginning of their second year do not develop the lordosis. Doubtlessly the curvature of the body axis depends upon function.

In non-human primates, the "kink" in the trunk axis cannot be moved farther rearward (caudally), because of their long, straight ilium, so being confined to the lumbar section. This becomes evident in the case of Japanese macaques (*M. fuscata*), trained to walk bipedally. During months of training, the monkeys develop a lordosis of the lumbar spine, which is often more pronounced than in humans (Fig. 9.14a, b; Preuschoft et al. 1989). It is maintained even in quadrupedal posture and comes from brought forth by a wedge-like shape of the intervertebral discs, not by deformation of the vertebral bodies themselves. Their musculature, bones and insertions of tendons into the skeletal parts are developed accordingly (Nakatsukasa et al. 1995). The results confirm the conclusion drawn above that the dorsal curvature of the trunk axis depends on function. This conclusion is further confirmed by observations of several, male as well as female gorillas, which often assume volontarily a bipedal locomotion and show a marked lumbar lordosis.

Fig. 9.15 (continued) left side, levels of the segments L_1–L_5, measured in mm; on the right side, the angles between thigh and trunk axis. Vertical axis: the contours of the back. The grey area on the right side of the graph makes the change obvious. In anteverted positions of the thigh (=flexion of hip joints), as assumed in all kinds of sitting, the lordosis is replaced by a kyphotic curvature. (From Busching 1998). The lordosis disappears in about 32% of the subjects at hip flexions of 60° or less (**c**), in 68% at hip flexions of about 80° (**a, b**)

Australopithecines seem to have possessed a lordosis. Robinson (1972) found it in the fairly complete vertebral column of StS 15. A really satisfying, complete analysis of the pelvis, including the proximal femur and the vertebral column by using the FE method, remains a desideratum.

A similar lordosis exists in the neck of humans: as in the lumbar section, the neck muscles are quite distant from the cervical spine. This is emphasised by the shift of the foramen magnum from the caudal to the ventral side of the skull. The prominence of the occiput gives the uppermost neck muscles (between the skull and C3 long lever arms (Jaeger 2003)). The load arms of segment weights are short in humans (see also below, under skull).

Aside from its static function, the curvatures of the human vertebral column are often interpreted as a means for *elastic damping* the force impulses along the body axis (Fig. 9.16a). According to Fig. 9.16b, c, an elastic resilience in vertical direction is excluded by the height of the vertebral bodies. Alone a deformation of the intervertebral discs allows an up and down movement (Fig. 9.16d, e). The liquid nuclei are not compressible, but the surrounding fibres of the annulus fibrosus may change the angles between cranial and caudal branch (Fig. 9.16f, g) and stretch, so that the diameters and circumference of the whole element become greater. All this, however, has turned out to be pure theory. The results obtained by Recknagel and Witte (1996; also Witte 1996), who applied high negative accelerations to five standing male humans, document not the expected forward, but in contrast a dorsal tilting of the pelvis under these high loads. To further verify the situation, we have performed an additional in vivo experiment. Force impact was applied in drivers of motorbikes and in horseback riders directly to the ischia, not through the hindlimbs. The transmission of these forces was measured by accelerometers fixed by straps around the trunk at L4 and on top of the head. The forces applied to the pelvis arrived in full size at the head (Fig. 9.16h–m). The observed reduction of the accelerations at the lumbar area seems to be due to a lagging of the soft tissues between spinal processes and skin surface. This lagging of soft tissues behind skeletal elements has been investigated in much detail by Recknagel (1995).

The short, voluminous trunk of the apes implies a low mass moment of inertia (which means energy-saving mobility) and safety (in accidents the dangerous lever arms of unpredictably applied external forces are always short). *Homo*, however, is characterised by a rather long, slender trunk, with a heavy head on a long neck. The seemingly misfitting length of the human trunk repeatedly gave reason to inquire whether or not human body shape is a misconstruction, still "not fully adapted" to the existing conditions of life. Figure 9.17 illustrates the problem: inclination of the body stem indeed leads to very great bending moments, which consume muscle force and require strength of the vertebral column to resist compression. These static disadvantages are illustrated for various inclinations and length/diameter ratios in Fig. 9.17b, and their consequences for the mass moment of inertia are shown in Fig. 9.17c.

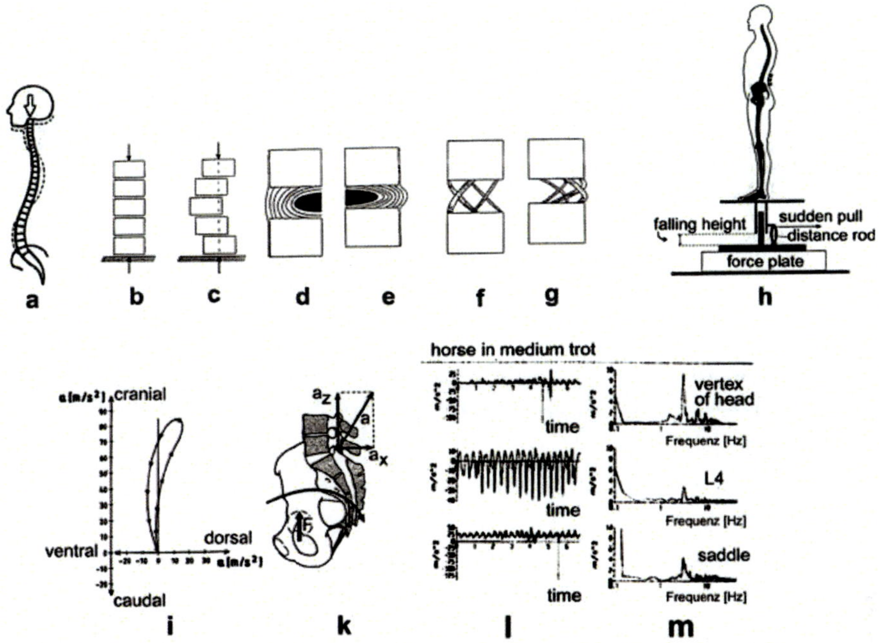

Fig. 9.16 The vertebral spine is in fact not elastic in longitudinal direction, as supposed in (**a**). The curved, lordosis-like arrangement of the five vertebral bodies in (**c**) does not allow a reduction of the total height of a straight column (**b**); therefore no "damping" effect is possible. The only means to reduce the height of the vertebral column is deformation of the intervertebral discs, as shown on a schematic longitudinal section (**d**, **e**). The oblique, annular fibres indeed permit an extension of the horizontal diameter, because of their helical arrangement (**f**, **g**). The experimental setup used by (Recknagel and Witte, 1996) to measure sudden force applications along the vertebral column (**h**). Five healthy young men were standing upright on a plate, designed to fall 30 or 50 mm after removal of a "distance rod". The elastic deformation of the vertebral column as imagined in (**a**) entails a forward tilting of the crista iliaca. In reality observed has been the opposite: a tilt rearward (*k*), with retarded recovery (**i**). Likewise, in humans mounted on a horse (**l**), or sitting on the saddle of a motor bike, the upward directed accelerations (**l**) in *y*-direction applied to the ischium reach the vertex in full size (**m**) and not delayed in time. This becomes visible best in the analysis of frequencies (**m**)

In the years between 1959 and 1966, Nachemson and Morrris, 1964 have measured, in famous (though for the patients quite risky!) experiments, the compressive forces in the nuclei pulposi of the intervertebral discs of human lumbar segments. Trunk flexions of up to 20° were considered. The experimental subjects varied in stature, weight and constitution. Jaeger (2003) has measured dimensions and lever arms from CT data of one individual and calculated the compression between vertebral bodies systematically for inclinations of 30°, 45°, 60° and 90° (Fig. 9.18a). In order to compare the calculated values with those measured experimentally, Jäger has recalculated his values on the body sizes of the experimental

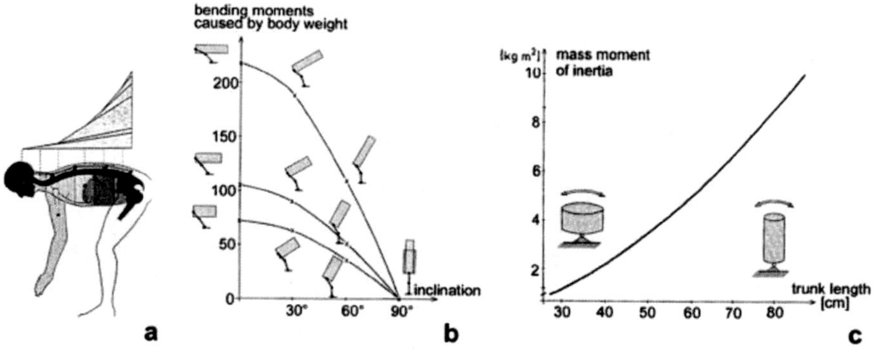

Fig. 9.17 (**a**) Human subject with trunk flexed forward, like in lifting a load or doing work on or close to the ground. The weights of all segments and of the frontlimbs act on lever arms which decrease in length from head to pelvis, so producing bending moments. These moments are drawn one on top of the next, adding up to a more than linear increase. (**b**) In horizontal posture, (left side) bending moments are greatest; in fully upright posture (on right side), bending at the sacrum/S_1 is zero, regardless of trunk shape. A slender, elongate trunk (as present in humans) evokes greater bending moments than a shorter and stouter trunk (like that in apes). The graph is drawn for three examples, in which the ratio of length to diameter varies, while volume (proportional to weight) is constant. (**c**) The mass moment of inertia grows faster than linearly with stem length. The proportions of a human body stem in contrast to those of an ape yield greater resistance against being flexed or extended/inclined forward or rearward (rearranged from Witte et al. 1991 and Jaeger, 2003)

subjects of Nachemson and Morris (1964) or Nachemson (1966), respectively. The pre-tension of the nuclei has also been considered. The obtained results of the low-risk theoretical study were in fair agreement with the high-risk experimental data (Fig. 9.18c).

The above attempts to understand human morphology leave us alone with leaves us alone a number of *unsolved questions*:

- Why are human hindlimbs so much longer than in our closest relatives?
- Why is the human trunk so long and slender – in contrast to the obvious advantages of a stout trunk?
- Why are human frontlimbs not shorter? Since it would cost less energy and be safer to have short, rather than long, limbs?

These (and other) questions cannot be (and have not been) answered by application of statics. It took much time, until we found that the human body proportions are evolutionary responses to the needs of locomotion, while the shapes can be understood on the basis of statics in much detail.

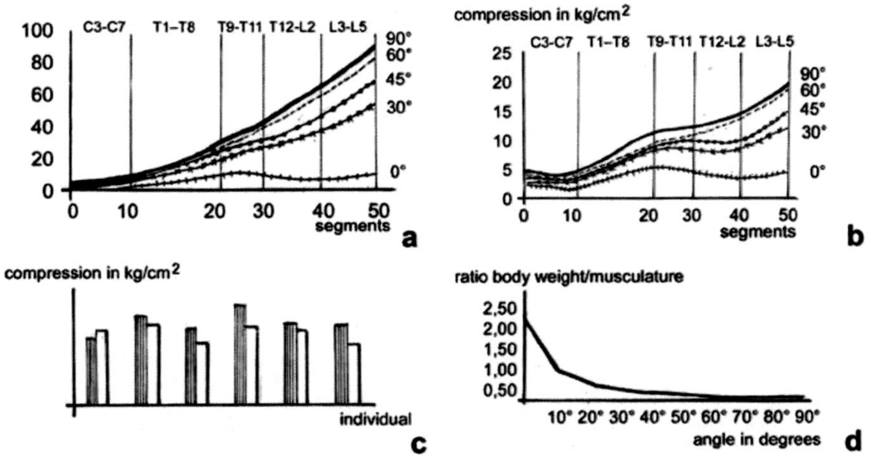

Fig. 9.18 (**a**) Bending moments calculated by Jaeger (2003) along the vertebral column in Nm for various inclinations of the trunk. (**b**) Calculated compression along the vertebral column in 10 N/cm^2. (**c**) Comparison of the measurements obtained experimentally by Nachemson (hatched columns) between vertebrae L_3 and L_4 with the size-corrected calculated values (white columns) in 0° of flexion (=upright posture). At least for moderate inclinations of the body stem, the values were similar. (**d**) In upright posture, the greatest share of compressive force along the vertebral spine comes from weight, whereas increasing inclinations of the trunk (from left to right) entail stronger muscle contractions and less influence of weight. The graph shows the ratio between the compression derived from bending and from weight force

9.3 Bipedal Locomotion and Its Influence on Body Shape

9.3.1 Why Are Human Hindlimbs So Long? Why Is the Human Trunk So Thin and Elongated?

On continuous substrates, the mechanical function of the foot leads to *footprints* and the sequence of them, *tracks* (Fig. 9.7). In contrast to the individual footprints, which yield information about the shape of the sole and foot construction (see above), tracks document an essential part of locomotor behaviour, namely, the loads which act on the foot during stance phase (GRFs) and the sequence of limb movements. In walking humans, the resultant of the gravity vertical and its ground-parallel components (GRF, see above in Chaps. 6.7 and 8) hits the ground first at the heel (heel strike), wanders close to or along the lateral border towards the toes (roll-off) and is finally shifted from the lateral fifth towards the medial first metatarsal ball and the big toe (push-off). The narrower the excursion angles, the greater the vertical component bearing body weight in relation to the braking and reaccelerating components in *x*-direction. External forces (GRFs) which are applied close to the heel have a short lever arm and need less muscle force for being balanced at the ankle joint than GRFs applied to the anterior part of the foot or confined to the toes. The medial margin of the foot, between collum calcanei and capitulum metatarsalis primum, makes no or only slight ground contact. In the last stage of the support phase (push-off), the entire load is on the toes, especially the big toe; force components from lateral to medial (z-direction) come into action and influence the shapes of the metatarsal bones and phalanges (see below).[5]

A comparison between humans and African apes, which spend much time on the ground, seems promising. The first *footprints of chimpanzees* have been published by Morton (1924, also 1948, see also Elftman & Manter 1935). His generalisation became widely known and is still taken for granted: that "apes" carry load alone on the lateral margin of the foot and place their slightly flexed toes with their fibular surface on the ground. Preuschoft (Preuschoft 1963a, b, 1970a, b, 1973a, b) has extended the comparison on other apes, especially gorillas (Fig. 9.3), and found clear differences between genera. The foot tracks of chimpanzees are more variable than those of the bigger and heavier gorilla (see also Kullmer et al. 2003) and of humans. Tuttle (1987) and Tuttle, Webb & Baksh (1991) have systematically investigated foot tracks of apes and included for comparing bipedally walking bears. Clarke (2003) also described footprints of chimpanzees. Chimps indeed tend, even if the variability of their footprints is thoroughly considered (Fig. 9.3c), to put most load on the lateral border of their foot and to place their lateral toes more or less flexed with their fibular sides on the ground. The big toe may be either placed loosely parallel besides the lateral toes or abducted from them, carrying considerable load. This resembles the typical (and less variable) foot posture of gorillas, which put much load on their heels, the (short) lateral toes abducted in lateral and the big toe abducted in medial direction. It carries much of body weight, and its posture implies a transverse arch of the foot skeleton. Orang-utans often flex their (long!) toes and place alone the lateral margin of the foot on the ground, often together with the dorsal sides of the toes. Most researchers, however, have not classified the orang-utan as an "adapted" bipedalist and therefore not taken this ape into consideration – until the results of Thorpe and Crompton 2005 became known (see above, Chap. 8). Very precise information about GRFs in bonobos and in gibbons has been provided by Vereecke et al. 2003, 2005.

Among the footprints unearthed at Laetoli, those of site G (Fig. 9.3e, k, l) have received most attention. The track at site A (Fig. 9.3f) shows less details and seems to indicate an extremely lax foot musculature, so that only the most loaded parts of the sole were imprinted. Anyway, the individual footprints are enigmatic (Deloison 1991). This, however, should not be taken as a reason to exclude them from consideration. The depths of these imprints do not much differ from those of

[5]The frequent and clinically relevant "valgus" position of the big toe (=abduction of the intermediate and terminal phalanges against its metatarsal) is often attributed to wearing poorly suited, ill-fitting shoes. However, it occurs also in humans who never had worn shoes. Typical human footprints are known from various fossil horizons, and the oldest known at present come from Ileret and Koobi Fora in Kenya (Day 1991; Bennett and Morse 2014) about 1.5 MY old, those from outside Africa Happisburgh (UK) and Roccamonfina (Italy; Bennett and Morse 2014) being less than 1 MY or 0.75 MY old.

site G, especially if a considerable "laxity" in the sense of Deloison is taken into consideration. Orang-utans and chimpanzees also sometimes leave tracks behind, which show complete relaxation of the toe-flexing muscles and therefore no imprints of the toes. In view of the differences between the living apes, it seems doubtful/ difficult to maintain the traditional and often repeated imaginations of the evolution of the human foot from a chimpanzee-like foot. Most authors do not even attempt to propose precise causes for an observed morphology and just postulate the existence of "adaptations" to "bipedal locomotion", where the only evidence are morphological similarities to humans, often wrapped in classifications like "autoapomorph" or "plesiomorph".

Bipedal locomotion is performed (as mentioned in Chap. 6) in *different gaits*: either by walking with duty factors of >50% of stride duration or by running, the latter being the equivalent of trotting in quadrupeds: duty factors are <50% of stride duration, greater step lengths due to intercalated phases of aerial suspension make it faster, and recovery of energy by making use of elastic propensities of muscle-tendon complexes seems probable. Hopping is possible, but rarely used by humans, since it requires too much muscle force, and a gait similar to hopping with elongated ground contacts by placing the feet on ground one after the other between phases of aerial suspension (an equivalent for "asymmetric", gallop-like gait, as described in Chap. 8 for indriids) can also be seen in playing children and in adults occasionally.

The most frequently used gait in humans is *the striding walk* (Fig. 9.19). *Endurant* that means little energy consuming, but sufficiently fast locomotion, is

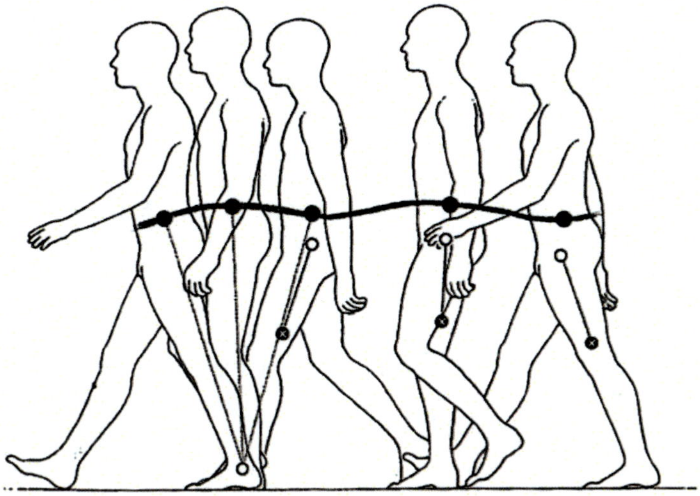

Fig. 9.19 Locomotor cycle of a walking human. The CoM follows an up and down pathway, which has its greatest height above ground at midstance. During the stance phase, first the right and then the left leg carries the body like an inverted pendulum. In the swing phase, the left leg performs a pendulum-like swing forward. Its pendulum length is determined by the distance between hip joint and the leg's CoM (black dot; white dots are the respective pivots). Each arm performs a pendulous swing in counterphase to the legs (modified from Preuschoft and Witte, 1991)

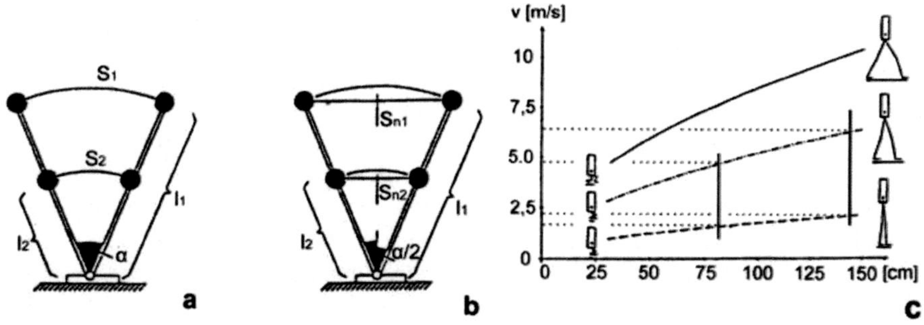

Fig. 9.20 Step length is determined by limb length. (**a**) In each stance phase, the limb covers the same angle (α), but the gain of distance parallel to the ground (s) is greater, if the leg is longer ($s_1 > s_2$). The same dependence of the distance (s) holds true, if the excursion angle α is divided into the excursion in front of and behind the vertical (**b**), because the distance $s/2 = \text{sine } \alpha * 1$. The inverted pendulum is not limited by own frequency, so that only its length is of relevance for walking speed if angular speed is constant. (**c**) The frequency of cycles, however, is determinded by the pendulous foreswing. The net locomotor speed v therefore follows a root function. For three excursion angles, the walking speed at minimal expenditure of energy is calculated on the basis of pendulum laws and plotted against limb lengths on the horizontal axis (Witte et al. 1991)

vital if dispersed resources are the basis of living, like fruit, tubers, water and game. In fact, humans are among the most endurant cursors among mammals. 30–40 km per day are not highly demanding distances for a modern man, a value which must be compared to bands of baboons, which normally move less than 12 km per day (Sue Altmann, pers.comm.). The biomechanical rules, which are obeyed in bipedal walking, have been mentioned above (in Chaps. 6 and 7). The basic interplay between suspended pendulum during foreswing and interested pendulum during stance (Cavagna et al. 1977; Mochon and McMahon 1980/1981) and the mechanical advantages and shapes (Witte et al. 1991; Preuschoft and Witte 1991, 1993; Preuschoft et al. 1992) have been found and analysed first in the example of *Homo sapiens*. As mentioned above and illustrated in Fig. 9.20, limb length is an efficient means to increase step length. The height to which the body's CoM is lifted can be and actually is modified by flexion of the knee joint. By doing so, the virtual pivot of the "inverted pendulum" is located below the ground surface, and the "pendulum length" (l in Fig. 9.20 and s in Fig. 9.21) becomes greater, so that step length is increased. In fact this knee flexion during midstance takes place and is in principle agreement with the reduction of GRF at midstance in walking humans.

Walking speed (v) is determined by:

$$v = \text{step length} * \text{step frequency}, \quad \text{or} \tag{9.2}$$

$$v = \text{stride length} * \text{stride frequency} \tag{9.3}$$

In order to maximise step length, it is essential to put the swing limb down in a nearly extended posture. This leads to a first and heavy ground contact by the posterior part of the foot, the often discussed "heel strike". Making a first ground contact with the forefoot is possible but rarely performed over long distances – simply because they do not conform with least energy consumption under the conditions given by foot morphology.

In *continuous contact locomotion* (=walk), the excursions of the limbs at the end of swing are limited by avoiding too great rotating moments of the GRF at the hip joint. This sets limits to step length, but the angle at which the excursion is stopped depends upon limb length. This is illustrated in Chaps. 5, 6, 7.3 for quadrupeds. Rather short step lengths can be seen in many foot tracks, and shortness of steps obviously has to do with saving muscle force by limiting the excursion angles. If the excursion angles are constant, the length of hindlimbs yields, according to Fig. 9.20a, b, greater step length and speed without input of energy. In view of their short step lengths, the Laetoli footprints are commonly interpreted as documents of walking, rather than running. Greater step lengths without undue angular excursions of the limb at the beginning and end of the stance phase can be reached by a phase of aerial suspension that is a duty factor of less than 50% of stride length (=trotting).

Frequency of pendulum swings, the other determinant of speed,

$$T = 2\pi\sqrt{(l/g)}, \tag{9.4}$$

depends on the shape of the swinging limb, which influences the position of the limb's CoM (Fig. 9.21a–d). The shorter the pendulum, the higher the frequency and

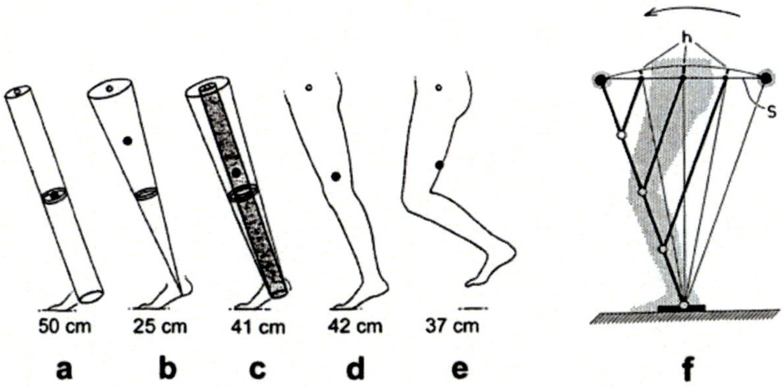

Fig. 9.21 Frequency of steps is determined by length of the limb pendulum. (**a–e**) The position of the leg's CoM varies with shapes of the limb, or more exactly, with distribution of masses: In a cylindrical column, the CoM is at 50% of length (**a**); in a cone-shaped leg, the CoM is at 24% of length (**b**). This shape, however, is not realistic, because of the insufficient compressive strength of the most distal section. In reality, the load-carrying skeleton plus skin occupies a cylinder of about ½ of the diameter along the entire length, from hip to foot (like in (**a**)). Only the musculature (ca ½ of the leg mass) can be concentrated proximally, like in a cone wrapped around the skeleton. Consequently, in (**c**) the CoM is at 41% of total limb length. This agrees nicely with the empirically determined position of the CoM of an extended human leg at 42% of length. By flexing the knee joint, the CoM is commonly moved to 37% of total length. (**f**) The position of the knee joint in the middle of the leg yields most lift (*h*), with given "work" (force ∗ lift) provided by the extensor muscles. The shorter the relative thigh length, the less lift of the hip. Relative shortness of the lower leg leads to the same result – as long as the excursion angle is the same (From Witte et al. 1991)

the faster the locomotion. As has been explained in Chap. 6, one of the means for reducing length of the limb pendulum is the concentration of mass on the pendulous limb to its proximal section, so that the CoM is shifted upward. The column formed by the skeletal elements needs nearly the same cross-sectional area along the entire limb, for sustaining weight force and accelerations. The arrangement of muscles is, within limits, variable, since the transmission of force can be secured by the much lighter tendons, instead of muscles. In fact, the upper thigh contains more musculature than the lower leg. In addition, the muscular bellies in the human hindlimb do not – in contrast to the apes – extend over the whole distance between origin and insertion but are confined to the proximal third or half. This is particularly obvious in the human m. triceps surae, which distal to the calf is transformed into the slender Achilles tendon. This muscle is (see above) strongly developed in *Homo*, while the digital flexors are (perhaps in the interest of mass distribution!) so weak that they can balance and move only short toes.[6] The same effect, shortening of the limb pendulum, is brought forth by flexion of the knee joint at midswing (Figs. 9.18 and 9.21e).

The position of the knee joint in the middle of the column representing the limb requires least "work" (=force $*$ lift) of the muscles (Fig. 9.21f).

The advantage of great limb length also favours high body stature. Individuals with hindlimb lengths of about 1 m ("functional hindlimb length", extrapolated from Flügel et al. (1986), corresponding to about 2,17 m stature) have an advantage of 20% over those with legs of only 0,5 m (corresponding to a stature of 1,08 m).

Of course, great length of the limbs implies disadvantages. With any elongation, an increase of mass is connected, and the resistance against being moved; the mass moment of inertia (J) grows by the square of limb length, because

$$J = l_i^2 * m_i \tag{9.5}$$

In flexed postures of the joints, increased length leads to long load arms (which are proportional to the cosine of the flexion angle between the segments, see Figs. 9.4a and 9.9e, f). So the control of long limbs as well as their moving requires increasing muscle force, and muscle force is proportional to muscle weight. (This has been quantified for the frontlimbs of brachiating apes in Chap. 8.) In *Homo*, each hindlimb occupies about 15% of total body mass (see also the comparison between ape and human muscle weights in Fig. 9.6).

In walking, the hindlimbs are decelerated before touchdown at the end of each foreswing and accelerated after each liftoff. The muscles, which extend or flex the hip joint, have their origins at the pelvis (mm. glutei, hamstrings and adductors and m. iliopsoas plus m. rectus femoris, respectively) and lumbar vertebrae (m. psoas). Therefore, they exert the same force as on the leg also on the trunk (Fig. 9.22) and cause *nicking movements* of the body stem in x-direction, which are the more

[6]The more common explanation emphasises the reduction of toes as a first step, because they are not longer needed for grasping, and as a consequence the subsequent reduction of the digital muscles.

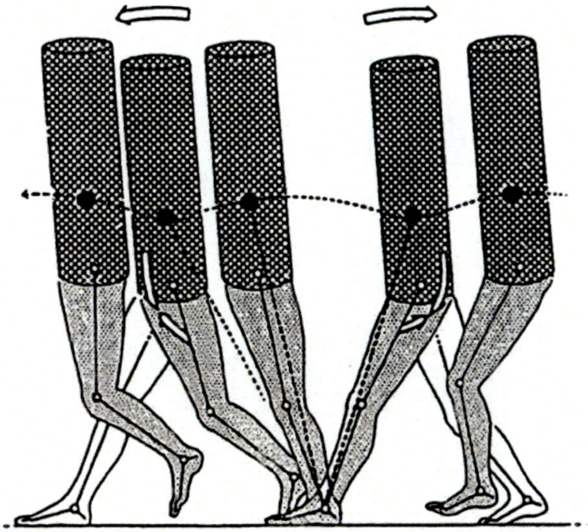

Fig. 9.22 Selective advantage of trunk length. The strong and correspondingly heavy hindlimbs are decelerated against the end of the foreswing (2nd from right) and reaccelerated against the end of the stance phase (4th from right). The muscles provoking anteversion take their origins from the trunk and are shown as bowed arrows, representing mm. glutei, adductores and iliopsoas, respectively. They cause fore and aft swings of the body stem about the hip joints. These "nicking" movements are suppressed without input of energy, if the body's mass moment of inertia is great enough. Here we have a convincing reason for the statically inadvantageous length and slender shape of the body stem. (From Witte et al. 1991)

pronounced, the longer and the heavier hindlimbs. A counterbalancing of these fore and aft movements is energetically cheapest by high mass moments of inertia of the body stem. An elongate and slender trunk shape offers this possibility (Fig. 9.24). Here we have the reason why our trunk shape is in contrast to the requirements of statics: humans are constructed for locomotion, not for sustaining static loads. If the mass moments of inertia are calculated in detail, a heavy head with an overproportionally heavy brain on a long neck turns out to provide a particularly great contribution!

The morphological traits optimal for walking are very similar to those required for running, hopping or leaping. This similarity, as well as the conclusions possible about lifestyle, will be discussed below.

9.3.2 *Human Forelimbs and Twisting of Trunk*

As shown in Figs. 9.19 and 9.24 the *forelimbs* are swinging fore and aft in counterphase to the hindlimbs. The contralateral arm is swung in phase with the

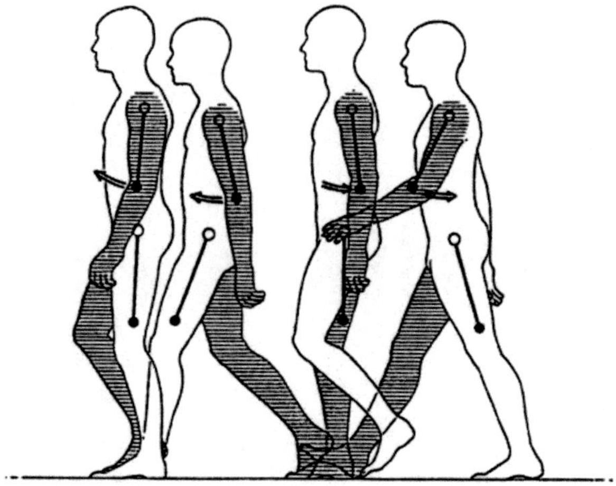

Fig. 9.23 Contribution of arm swings. The left arm (hatched) swings like a pendulum in phase with the contralateral leg (also hatched) rearward, the right arm, however, in synchrony with the left leg forward. Acceleration of the arm forward takes place at the same time as the deceleration of the forward swinging leg on the same side. The braking of the arm swing is synchronous with the beginning of the leg swinging forward. Pendulum length from shoulder to the arm's CoM is similar to that of the hindlimb pendulum (because of its mass distribution), and therefore their own frequencies are not very different (From Witte et al. 1991)

opposed leg. Mass distribution on the arms deviates from that on legs, so that the CoM of the relaxedly extended arm is distal to the elbow joint (where it is proximal to the knee). This doubtlessly has to do with the weak development of the upper arm and the strong development of the forearm, both in contrast to the hindlimb. The pendulum lengths of anterior and posterior limbs are close to each other, and their frequencies deviate moderately.[7] In any event, the forces acting between the pendulous arms and the trunk contribute to suppressing the nicking stem movements.

If *viewed from on top*, the mass inertia of the hindlimbs rotates the body stem about a vertical (y-)axis rearward at the onset of the foreswing and forward at its end (Fig. 9.24a). The rotating moments are countered by two factors:

1st The phase-reverted foreswing of the arms (Figs. 9.23 and 9.24) is opposed to a rotation caused by the mass inertia of the legs. The mass of the arms is with 4.5% of total body mass smaller than that of the legs (15%), but the lever arms of the legs (1/2 of biarticular pelvic breadth) are only one half of those of the arms (1/2 of shoulder breadth).

[7]Here again, speculations are possible whether the mass distribution on the arms is a consequence of bipedal locomotion or of using the hands in manipulation. The last mentioned possibility, however, is in contrast to the fact that apes have still stronger (and heavier!) forearm and hand muscles than humans.

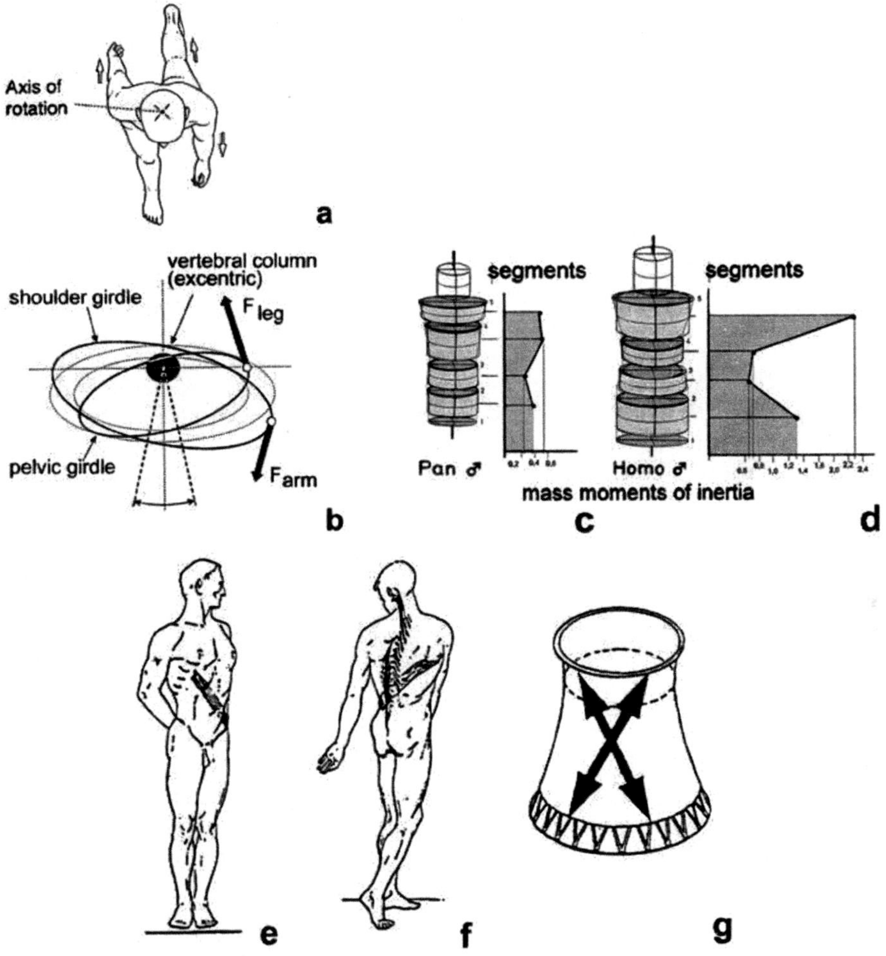

Fig. 9.24 Torsion of the trunk during walking. **(a)** Walking human seen from on top. The white arrows indicate the direction of mass inertia. Inertial forces lead to a rotation of the shoulder girdle forward and at the same time to a rotation of the pelvic girdle rearward. **(b)** The phase shift between shoulder and hip region entails a torsion of the trunk. The shorter load arm lengths of the heavy hindlimbs and the greater load arm lengths of the light forelimbs approach the rotating moments of both extremities. The eccentric position of the lumbar spine should be noted. **(c)** The trunk plus head of a chimpanzee has about the same circular outline at each level and therefore similar mass moments of inertia **(d)**. Humans **(d)** by contrast have pelvic and especially shoulder girdles, which are broad in frontal direction and shallow in sagittal direction. Their oval outlines make the mass moments of inertia great (**d** in contrast to **c**) and so provide considerable stability against rotations without input of energy. **(e, f)** The oblique ventral muscles probably store energy by elastic stretching and recoil. The tension-resistant elements form, when twisted, in the wall of a hollow cylinder a "waiste" – as can be seen in a cooling tower **(g)**. ((**a, b, g**) From Witte et al. 1991; (**e, f**) from Benninghoff 1938)

2nd In humans, the breadth of pelvis and shoulders is great in comparison to the situation in the apes. This alone increases the mass moment of inertia, which resist rotation about a vertical axis (Fig. 9.24a, c, d). The shoulder breadth in humans is attained by the simple means of lowering the height of the shoulder (=swing of the clavicle in caudal direction about the sternoclavicular joint). The characteristic muscle activity in carrying of loads by the hands has been documented Fritz et al. 2014.

As can be seen from Fig. 9.24a, in each cycle, two rotations in opposite directions take place: one of the pelvis under the influence of leg mass and the other of the shoulder under the influence of arm mass. This leads to a twisting of the lumbar section. The tension of the stretched muscles leads to a "vaste" of the trunk, which is visible in the skeleton: Schmid (1983, 1991) has worked out that the well-known "funnel"-shaped rib cage of the apes can be discriminated from the "barrel"-shaped rib cage typical of *Homo* (Fig. 9.2, see also M. Kagaya, who investigated in a number of primates the shapes of individual ribs in relation to the shape of the rib cage). All this is not in conflict with the rotational capabilities in chimpanzees, observed by (Thompson et al. 2015). *A. afarensis* possesses, according to Schmid, the same funnel-shaped rib cage as the apes. The same seems to hold true for *A. sediba* (Schmid et al. 2013). This means that *Australopithecus* lacks a human-specific specialisation which saves (in *Homo*) energy during endurant walking by elastic stretching and recoil of the oblique abdominal muscles.

9.4 Human hands

9.4.1 Is There a Decisive Difference in Hand Morphology and Hand Use Between Humans and Other Hominoids? Problem of "Precision Handling"

The organs, by which all primates make first contact to objects in their environment, are the *hands*. This requires a close-up view on prehensile and gripping behaviour. The principle, so to say qualitative ability of primate hands for gripping rather small objects, has never been denied. The interest on the hands of different species is rather focussed on the quantitative question of how precise the grip can be. In the case of extant forms, this is just a matter of direct observation; in the case of extinct forms, however, this is a matter of evaluating shapes and guessing. The divergent proportions of primate hands are described above, in Chap. 8.3.

While in most primates the relative *length of the thumb* (see above) varies around one half of digit III, it reaches in humans 66%. In comparison to our nearest biological relatives, the thumb in humans appears to be long and strongly developed. If the ratio between thumb and trunk length is considered, the seemingly long thumb of the human hand is not longer than in the apes, but in these, the lateral fingers are elongated, so that the first-glance-impression appears different (total hand length in

Homo 37%, 40–53% of ventral trunk length in large apes and 59% in gibbons; as noted above; Schultz 1971). Aside from the thumb, the second ray in the human hand is quite strong and its metacarpal longer than the third, although the entire digital ray II is always shorter than III, which also seems to be the strongest (Fig. 9.25a). In comparison to rays I, II and III, the fourth and especially the fifth look rudimentary. His impression is emphasised by their reduced active mobility, which is due to their musculature: *digits II and V* are extended and dorsiflexed by their own fibre bundles of the m. extensor digitorum communis, independent from the remainder of this muscle. In addition, a separate m. extensor indicis proprius exists, comparable to the mm. extensores pollicis brevis et longus, which give the thumb independent mobility. Flexion of the fingers is accomplished by the mm. flexores digitorum superficialis and profundus. While the deep tendons in about one half of humans originate from a common tendon in the palm, so that no independent flexion can be made, the other half possesses a completely separated deep flexor muscle and tendon, allowing independent flexion.[8] The superficial m. flexor digiti II superficialis is always independent from the other parts of the flexor muscle.

The finding gains importance if the situation in reptiles (Landsmeer 1993) is taken into consideration, as well as that in Ceboidea and Cercopithecoidea (collected from earlier literature by Preuschoft (1963a, b; see also Jouffroy, 1962, 1971) where common extensor and flexor muscles are present, which fan out on the dorsum and vola manus and form strands which distribute the muscle force on all fingers.

Only in the apes are the muscles separated, which flex the individual fingers. Not alone in "primitive" animals but also in small mammals the fused flexor tendons originate from common muscle bellies and do not split before reaching the palm (Landsmeer, e.g. 1993, and older; Preuschoft 1963a, b). According to Maier's (1971) detailed description of the hand musculature of the gelada, there is no obvious difference from other monkeys, aside from a more proximal split of the digital flexor tendon. The exchange of fibre bundles between the parts, however, makes an independent function impossible. Since this large form makes its living from "small object feeding" (Crook and Aldrich-Blake, 1968) the near absence of marked morphological specialisations is remarkable.

The *thumb in humans* can not alone be abducted from the "lateral" metacarpals but can also be moved away from the palmar plane. Doing so, it is turned some degrees about its longitudinal axis, so that its finger pad faces against those of the other fingers. Exactly this is, strictly speaking, the *opposition* of the thumb. It is best

[8] This fact can be verified by a simple experiment, which should be performed on both hands: the fingers of one's own left and then right hand are extended completely. Now alone the terminal phalanges of isolated fingers are flexed actively, without movements of the other fingers, and kept in flexed position against an external load. This is possible in about one half of all Europeans, US-Americans and Japanese (own observation). Independent flexion of the middle joint, however, is no problem, because this movement is performed by the m. flexor digitorum superficialis, which inserts at the middle phalanges and dispatches own bundles to each ray. If the lateral fingers are allowed to move as well, the experiment does not yield clear results.

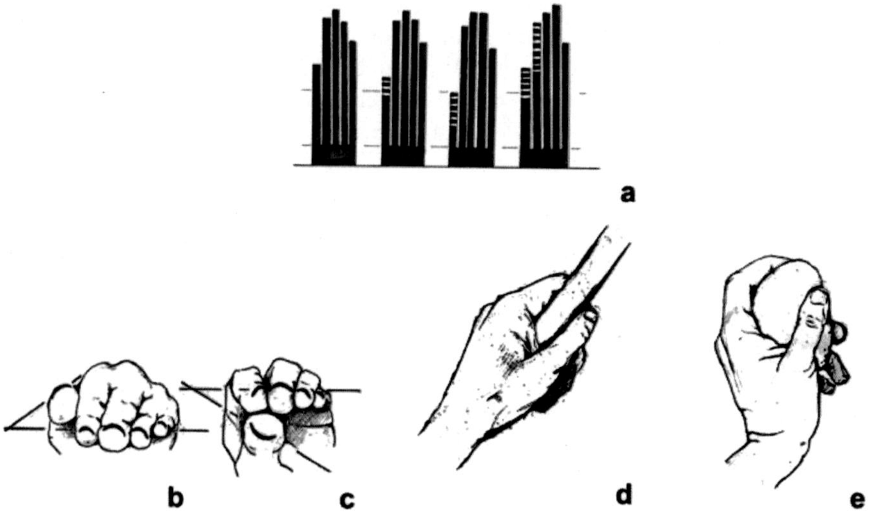

Fig. 9.25 Relations between and use made of human fingers and metacarpals. (**a**) Proportions of human hands (left) in comparison to other primates. The horizontal lines separate carpus, metacarpus and the most variable segments, the phalanges. The banded digits vary in length in monkeys: (second), and third the paraxonic hand in short-thumbed OW and NW monkeys, *Colobus* and *Ateles*; the fourth is the ectaxonic hand in most strepsirrhines (compare Chap. 8). (Data taken from Jouffroy et al. (1993, pp. 150 ff)). (**b**) Opposition (in a narrow sense) of the human thumb. Note the position of the finger nails in opposition (**b**) and reposition (**c**) of the first digit. (**d**) Human hand gripping the handle of a hammer. Like most tools for blowing, it is held in prolongation of the forearm, primarily by the digital rays I and II, where its position is under optical control. (**e**) Throwing a stone. Gripping between the radial digits I, II and III elongates the radius of the throwing movement by 8–13 cm and permits, at the end of the throwing movement, an additional acceleration of the projectile by plantar flexion of the wrist joints (compare the terminal phase of javelin throwing in Fig. 9.28e)

visible, if the position of the nail is observed, which includes a blunt angle with the lateral nails as long as the thumb is kept parallel to the other fingers; in opposition, however, the nails include an acute angle (Fig. 9.25b, c). This rotation about the longitudinal axis of the first ray cannot be seen in all primates and only sometimes in their feet. In many species, especially apes, and in the feet alone, an adduction/abduction about one axis of the carpometacarpal (or tarsometatarsal) joint takes place (Preuschoft 1961, 1963a, b, for gorillas and Preuschoft 1973a, b, for chimpanzees). The equipment of the hand with intrinsic muscles does not differ much in humans, apes (Sokoloff 1972) and many monkeys (Maier 1971). A really comprehensive investigation of the separated first digit and its musculature is missing, as well as a uniform terminology, especially of muscles. Therefore all relating statements have a touch of vagueness and lack of precision.

Many arguments aiming at the outstanding evolutionary position of humankind have been based on the "opposition" of the thumb and its length in relation to the lateral fingers. The most influential has been during the last decades the postulate of a "*typically human* precision grip" (Napier, 1961; Napier and Napier, 1967). Many

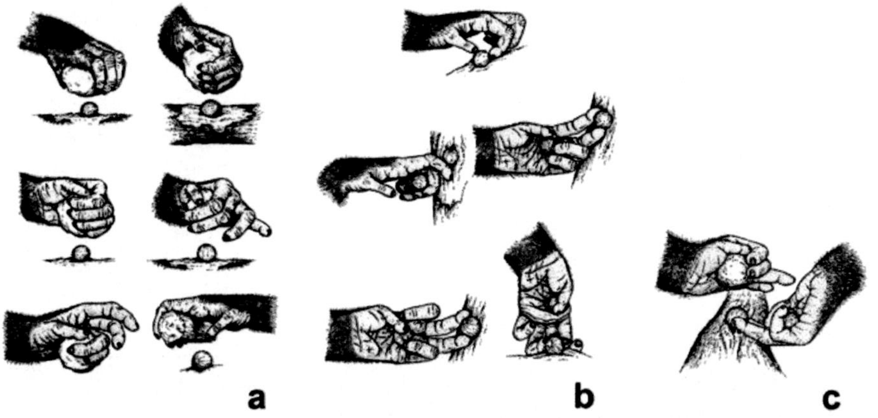

Fig. 9.26 Examples of gripping in wild-living chimpanzees. (**a**) Hands holding tools; (**b**) hands holding objects in place, (**c**) coordinated interaction of both hands. (From Boesch and Boesch 1993)

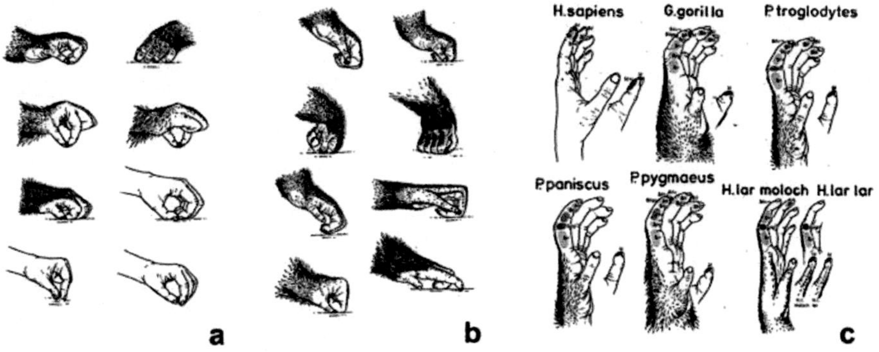

Fig. 9.27 (**a**) "Precision grip" in apes. (**b**) The typical human variant between the tips of the two medial fingers, which also occurs in apes. (**c**) Frequently used contact areas in precision gripping. The darker the area, the more frequent their use. Observations and experiments documented by Christel (1993a). (**d**) Chimpanzee holding a "hammer". (**e**) Hands fixing nuts to be cracked by a "hammer". (**f**) Both hands of the same animal cooperating: one fixes the nut, the other uses the tool

conclusions derived from this concept suffer from the fact that the abilities of non-human primates have been classified à priori, without experimental verification, as inferior to those of humans. The observations of Boesch and Boesch in the wild (Fig. 9.26), as well as Christel in experiments (both 1993) Fig. 9.27a, b, have unfortunately not found the attention they deserved. Christel (1993a, b); Christel and Fragazy, (2000) has also identified those parts of the fingers and the thumbs which make contact during grasping small objects (Fig. 9.27c). In particular the apes often make use of contact areas, which clearly deviate from the tip-to-tip contacts preferred by *Homo*, but obviously depend upon the proportions of the hands, especially the shortness of thumb in relation to length of fingers. These proportions (as shown above) can be connected convincingly with locomotion. In later experiments, Christel et al. 1998 have presented very small objects to her experimental subjects and counted the number of objects seized during a time unit, to measure the "efficiency" of a gripping technique. The values observed were not worse than those found in humans. The influence of volume of the seized object, and changes of the grip during lifetime have been studied by (Pouydebat et al. 2009, 2011).

"*Precision handling*", by the way, is primarily a problem in the *large hands of humans, apes* and some big monkeys. Small objects which raise the interest of a primate (like thin twigs or stalks, with diameters of 1 mm or less, seeds, insects, ticks, thorns) are of course easier to grasp by a hand which is 2.5 cm long and 1.5 cm broad, than by a forepaw of ten times these dimensions. In the latter case, special abilities are required are required and in fact present, as shown in Fig. 9.29.

Not very promising is the search for specialisation like "*knuckle-walking*" in the hand skeletons of australopithecines – simply because KW is a means to use a hand which is already specialised to form a hook for quadrupedal locomotion on ground. The hook hand is an aberrant specialisation of the apes and some other primates (see above, Chap. 8), while the australopithecines have been bipeds, and there was no need to use their hook hand as a balancing additional support on the frontlimbs. The ape-specific elongation of the metacarpus does not seem to be very pronounced in australopithecines (Clarke 2003, see also the hand of *A. sediba*). Several authors have emphasised the strength of the very robust phalanges and the curvature of the basal segments (Susman 1979; Susman et al. 1984). As has been detailed above (Chap. 8), the curvature is a means to keep the bending stresses in the phalanges low, if flexor tendons are very strong.

9.4.2 Are There Mechanical Reasons Which Favour Human Hand Proportions? In View of the Human Mode of Living?

In all, locomotory as well as idiomotoric movements of a primate, the radial margin with thumb and 2nd, "index" finger are within the visual field of the eyes (Figs. 9.25d,e and 9.26). Hand movements so are executed under *optic control* The ulnar margin, however, is only in some cases visible. While this is sufficient for mass movements, precise manipulation doubtlessly profits if performed under visual control by rays I to III. Among the non-locomotor movements, which are executed with great force, clubbing, hammering/pounding and throwing stand out. The use of "hammers" for opening nuts by chimpanzees has been described and documented by Boesch and Boesch (1993, see also Günther and Boesch 1993). The use of heavy stones by capuchin monkeys for the same purpose has been observed by Spinozzi et al. (2004) and Visalberghi et al. 1989 and 1995. Clubbing by chimpanzees in agonistic contexts has been documented by Kortlandt in 1967,

1968. Doubtlessly such movements occurred early in human evolution, be it in offensive context (in attacking a potential prey or enemy) or in a defensive (in defence against a predator or competitor).

Repeatedly, Marzke (1997, 2013), Marzke and Shackley (1986) has investigated the forces occurring in tool-making or working. A great number of movements performed in *working* (be it preparation of or working with stone tools, primitive agriculture or highly specialised industrial work) consist in essential is striking with a tool from above downward, whereby the gravitational force acting on the tool is used as additional enforcement. Simple every day experience shows that a tool for striking downward should be lengthy and have the greatest part of its mass at its far, distal end like in a club, hammer or axe.[9]

The mechanical basis of this experience has already been mentioned above: the impact exerted on an object (or prey, or enemy) is proportional to mass and speed of the *club, hammer or axe*. The tools are following a circular path about the pivot. As pivots function the wrist, elbow (hammering) or shoulder or even scapulothoracic joints, the handles of the tool elongate the distance between pivot and head of the tool ("radius"). The longer the radius, the greater the speed of and the impact exerted by the tool, if the angular velocity is the same (what is probable if the same effort is put into each of the movements):

$$v_{\mathrm{per}} = \omega * 1 \tag{9.6}$$

The angular momentum (D) inherent to the tool is

$$D = m * l^2 * \omega, \tag{9.7}$$

and grows with the square of peripheral velocity. The force (F) exerted is proportional to mass and deceleration,

$$F = m * -a \tag{9.8}$$

and the negative acceleration depends first on the peripheral velocity of the tool and second on the stiffness of the object which is hit. The longer the handle, the greater the peripheral speed – and the bigger the mass of the tool's head, the heavier the blow.

The muscles moving the digits, wrist, elbow, shoulder and shoulder girdle, which are recruited for these movements, are the same as those needed for carrying body weight in forelimb-suspended locomotion in the branches of a tree. Especially in those cases, in which the full length of the arm is used, and the movement has its pivot in the scapulothoracic joint, the activated muscles are quite similar to those in primates climbing with the hands above the head (see above, Chap. 8). Therefore,

[9]According to Boesch and Boesch (1993), it seems that wild-living chimpanzees, which are using tools for cracking nuts, do not or not in all cases make use of this experience.

the static and quasi-static situation in the *shoulder joint* of working, clubbing or throwing humans is similar to that in climbing apes. As a consequence, the scapula must be characterised by a strongly reinforced and long axillary and a short cranial margin. This expectation indeed is fulfilled by recent humans and the few fossil scapulae which are preserved. Larson (2012) has raised explicitly the relevant question “did australopithecines climb trees?”, searching an answer based on scapular morphology. Aside from her discussion of growth processes, she found traits indicating overhead use of the forelimb: the angle between axillar margin and glenoid surface nearly rectangular and the direction of the spina scapulae in adult australopithecines to be intermediate between those found in apes and in humans. The tricky problem is that the same conditions prevail in climbing, working and clubbing (=using tools or weapons), so that morphology hardly permits unambiguous conclusions from the scapula on either “locomotion” or any sort of “tool use”.

Both factors, length and mass of the tool or weapon, find their limits in decreasing handiness of the handled object when its becomes too long or too heavy. A lengthy tool can easier be directed in prolongation of the metacarpals, or the forearm, if its longitudinal axis exceeds the radial margin of the hand, not the ulnar (Fig. 9.25d, e). As noted above, it is held primarily by the rays I, II and III, while rays IV and V together with the hypothenar do not more than providing a prop. The grip can be controlled visually. If such sorts of tool use or tool making play a role for survival, the development of thumb, second and third ray on the expense of rays IV and V, is favoured. This can explain on the level of biological roles the “*entaxonic*” proportions of the human hand. A particular version of tool using may have been of outstanding importance: throwing.

For throwing, the conditions detailed above for blowing also hold true – provided that the usual technique throwing overhead (“Schlagwurf”) is applied (Fig. 9.28), which favours precisely aimed actions. The same muscles of the arm plus shoulder girdle are active as in clubbing or striking. The *mechanics of throwing* have been the object of several studies, all of which concentrated on the standardised form of *javelin throwing* – in which all efforts are concentrated on reaching great width, while the projectile is a constant. Baumeister and Wolniak (1986, unpublished) have been the first, and Baumeister (1991) expanded this approach (*see also* Ogihara et al. (2012). Roach et al. (2013) and Lieberman et al. (2008, 2015) have used similar considerations as we did and have included very detailed information about muscle morphology of humans and chimpanzees. The width of a throw depends – as in leaping – upon take-off speed (of the projectile at losing the contact between hand and javelin) and on the take-off angle. Simply because of empirical reasons, the latter is usually chosen at about 35°.

A difference between blowing and throwing is that in the latter movements of the entire body are *conducted* to increase maximal take-off velocity: run-up, “Stemmschritt”, translation of the upper part of the body stem, rotation of the shoulder girdle about a vertical axis and swinging forward of the throwing arm, finally flexing the wrist joint (Fig. 9.28a,e). The latter already has a considerable velocity, before the rotation and forward swing of the arm are initiated. The impulse (P_1) of the body segments below the waist (having the mass m_l in Fig. 9.28b,c) is

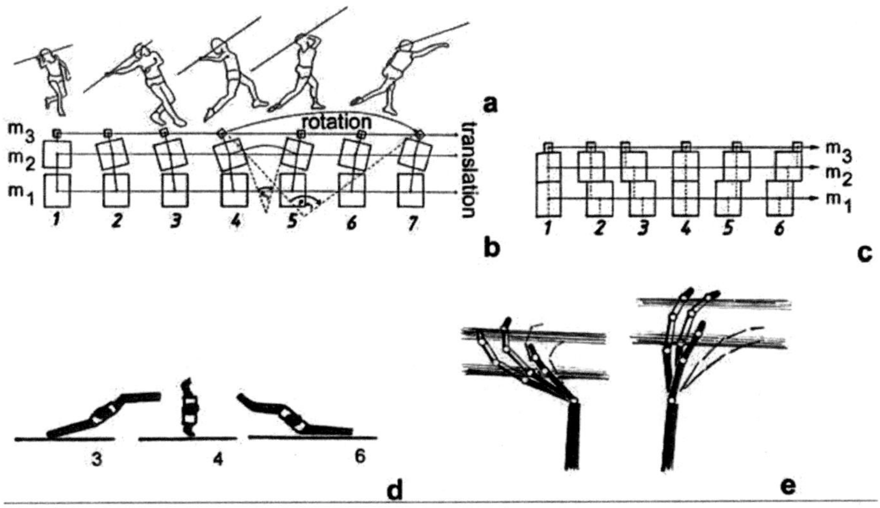

Fig. 9.28 Throwing. (**a**) Phases of javelin throwing in side view. (**b**) Relative movements of partial masses against each other shown as hyperextensions and flexions, (**c**) the same dislocations shown as translatory movements. The "whiplash movement" gives the smaller, more distal masses (m_1 body below waist in the lumbar region, m_2 trunk and free arm above waist, m_3 throwing arm) gradually increasing speeds. (**d**) Schematic top view of a thrower, showing three phases of rotating the shoulder girdle about a vertical axis and the movement of the arms, while the javelin does hardly change its direction. (**e**) Two phases of hand movements in the final stage of throwing a spear, or stone from dorsiflexion (right) to extension (left). The wrist joints are on the same height in both drawings. Pollex: heavy lines. Digits II and III are covered by the spear and therefore indicated by double lines. Digits IV and V are not involved and presented symbolically as thin lines. The slapping of the hand in the final stage gives the projectile additional speed a-d from Baumeister, 1991

transmitted completely to next upper segments above the waist (having the mass m_2) and the impulse P of the major part of the body (masses $m_1 + m_2$) to the arm (with mass m_3) and finally to the projectile. The so-called whiplash effect describes in short that the body parts contain gradually smaller masses, and therefore the transmitted impulse yields gradually greater velocities. If

$$m_1 > m_2 > m_3, \quad \text{and} \quad P = m * v_0, \tag{9.9}$$

$$\text{then is } \frac{P_3}{m_3} = v_3 > \frac{P_2}{v_2} = v_2 > \frac{P_1}{m_1} = v_1 \tag{9.10}$$

Indeed, the speeds measured by Baumeister are from ground upward 7:22:28 m/s, and this is reciproke to the masses (4:3.3:1). The impulse (P) produced at the highest point of the hand's trajectory is transmitted to the object thrown. As long as the projectile has constant mass, the peripheral velocity of the hand at the time of losing the contact with the projectile decides alone on the width of the throw or the penetrating force exerted at hitting the aim. All previous authors have neglected the final movement of the throwing hand. As is shown in Fig. 9.28e, it gives the projectile additional speed before contact with the hand is lost. Because the impulse contained within a body is constant and does not change by itself, the greatest speed is reached, if the mass of the arm is small in comparison to the mass of the thrower. This implies that *Homo* is better in throwing than australopithecines have been and apes are – whereby the enormous muscle force of the latter has a positive effect on the width and penetration power of their throws, but not on aiming accuracy (see below).

Trunk rotation is favoured by the above-described waist of the trunk, which is specific for humans, in spite of recent estimates of thorax shape in newly found australopithecines (see above). The greater the *breadth of the shoulders*, the greater the peripheral velocity. The swing of the throwing hand takes place not alone in the sagittal plane (*x*-direction) shown in Fig. 9.28a but also about a vertical axis in *z*-direction in a horizontal plane (Fig. 9.28d). Its pivot is the shoulder joint, and its effective "radius" – or *lever length* – is 0.55–0.6 m ("Armlänge ohne Hand" after Flügel et al. 1986). Hand flexion at the wrist joint takes place in the sagittal plane; the forearm rotates from supination through neutral into pronounced pronation. In order to increase length of the lever, which is the distance between the pivot and the projectile, especially the length of the forearm is useful. Just on the basis of experience, throwers shift the contact with their projectile as far away from the wrist joint as possible. (This can also be seen in shot-putting, where the weight of the projectile sets limits to the length of the lever arm: all trained shot-putters make use of the length of the metacarpus!). The javelin is guided, like a stone used for throwing, at three points, between the thumb and the index and middle finger. The relevant lengths contributing to "lever length" are the distances between radiocarpal joint and the terminal phalanx of the thumb, the middle joint of the second and third digits. By this "entaxonic" grip, the lever is elongated to 0.08–0.11 m, average 0.09 m, which is 15% more than would be possible if a "power grip" would be applied. Because the kinetic energy transmitted to the projectile is proportional to the square of lever length, the thrower gets an advantage of 28% more speed ($l^2 = 3600$ instead of 2600 in the classical "power grip" with all 4 fingers). This may well be the evolutionary advantage for the development of the thumb and index finger, in short of an entaxonic hand, which differs from that in apes and provides an elongated lever. Indeed the ratio between the lengths of metacarpals II to metacarpal III is between 85 and less than 100% among monkeys, 101 (chimp) to 109 (gibbon), but 112 in humans (Jouffroy et al. 1993). According to published photos of the hand of *Australopithecus sediba*, the second metacarpal exceeded the third in thickness as well as in length. H. erectus from Nariokotome does not include a hand.

Living African apes (chimpanzees according to Kortlandt 1967) and gorillas (according to Perret, 1997; pers. observation in the Zoo at Krefeld) prefer clearly the skidding technique, in which the *projectile is accelerated by a hanging arm*. This allows maintaining the quadrupedal posture. The technique is efficient because of the enormous force available in the arms of these animals. Such a technique is also used in sports (discus throwing and hammer throw). Greatest widths are reached, if the mass of the projectile is great and the radius (=lever) of the rotation elongated (like in hammer throwing). In addition, for throwing a discus or hammer, a very long acceleration distance is allowed, by accumulating speed over one or more rotations of the whole body. Although the rotation impulse (*D*) transmitted to a projectile is quite high, and the widths of throws are great, hitting an aim seems to be difficult when using this technique.

9.5 Are Mechanically Based Conclusions Possible Regarding the Lifestyle of Prehumans? Beyond the Common Correlations on the Level of "Biological Roles"?

Although we are, regarding the locomotor modes of the fossils, confined to drawing conclusions, these are quite reliable, because they are based alone on mechanical strength of parts and physical facts. This does not necessarily yield the desired information about the biological role of a given morphological detail (as, for instance, the abducted big toe; see above).

While morphology of hands and (to some extent also) feet unveils but gradual differences between apes and modern man, the *proportions of the limbs* in fossil *Homo*, *Australopithecus* and *Paranthropus* indicate a dichotomy of the evolutionary lines (Figs. 9.1, 9.2).

Reliable arguments can be derived alone from fossils which comprise remains of the skull and relevant parts of the upper as well as lower extremities. Finds of this type are rare: *Ardipithecus* (White et al. 2009), *A. afarensis* (AL 288, Johannson et al. "Lucy") Johanson and White, 1979 and "Little foot" (Clarke 1998 and 1999) are the oldest; younger seems to be OH 8, OH 62 attributed to *H. habilis* (Hartwig-Scherer and Martin 1991), *A. sediba* (DeSilva et al. 2013) and the so-called *H. naledi* from the Rising Star Cave (Berger et al. 2010). Nariokotome (*Homo erectus*, Walker and Leakey 1993) and the Neanderthalians and their kin fulfil this condition – more or less.

The proportions of the older prehumans are characterised by long, impressively strong arms and "weak", rather short, hindlimbs and so deviate, according to various authors, from the well-known human proportions: long, very muscular lower, but relatively short, weak upper limbs. Prehuman arms resmble somewhat those of modern apes, see also Venkataramana et al. those of the early humans are very similar to modern man and demonstrate specialisation for endurant, and moderately fast locomotion on flat ground.

Concerning the anterior extremity, the problem is that widespread forms of arboreal locomotion and typical movements, performed under high loads (working, blowing, throwing), lead to the same stress patterns. Therefore it is difficult to draw conclusion from shapes (although they are closely connected with stress patterns) to behaviours.

9.5.1 Walking or Running?

The morphological traits optimal for walking are very similar to those required for running, hopping or leaping.

Günther et al. (1992) have discussed the "preadaptation" of leapers to permanent bipedality, which has in fact been considered in older literature. The authors have found differences between the autopodia of leapers and walkers, which are advantageous for only one of the locomotor modes and therefore are in conflict with the assumption of a "preadaptation". According to Yamazaki (1985, 1992); Yamazaki et al. (1983), Kimura et al. 1983; Yamasaki et al. 1997, the bipedal locomotion of apes is more similar to running than to walking (see Fig. 9.29). Its most important traits are a duty factor of 50% or less of the whole cycle and a period of aerial suspension, at least a lifting of the CoM when the support limb is changed.

Liebermann, as well as Carrier (Symp. in Paris, TV; Bramble and Lieberman 2004), emphasise the importance of running, especially hunting big game for the evolution of human body shape (see above). We think that walking is an every day locomotion which any healthy individual can maintain over distances of 30 or 40 km per day, while running is always an exception, which can be performed only by selected and trained individuals. The finding of Carrier et al. (2011) should be kept in mind that "humans are 60–70% more economical when they walk than when they run". There exists a long-standing divergence of ideas between my group (we argue on the basis of the ubiquitous and extremely endurant walking) and Liebermann, Bramble and Carrier (whose arguments are based on an assumed preference of early humans for running – which is accessible only for selected - always male - individuals). The tricky point is that a decision between the diverging opinions up to now is not possible from the viewpoint of biomechanics and functional morphology. The mechanical demands raised by both gaits, which have given reason for body shape, are so similar that a decision based on morphology is not possible. All traits claimed for one gait are also necessary for the other. As an energy-conserving spring mechanism (Witte et al. 2002), the flattening and restoring of the vault of the longitudinal arch of the foot have been proposed by Carrier, but a passive spring mechanism has been postulated by earlier researchers without the hypothesis of running. My doubts against this interpretation are based on the too short lift, such a spring would provide. In addition, an old x-ray study of Puff and Rosemeyer (1970), showed that the longitudinal arch becomes higher during carrying load in walking by a reflex contraction of the intrinsic foot musculature.

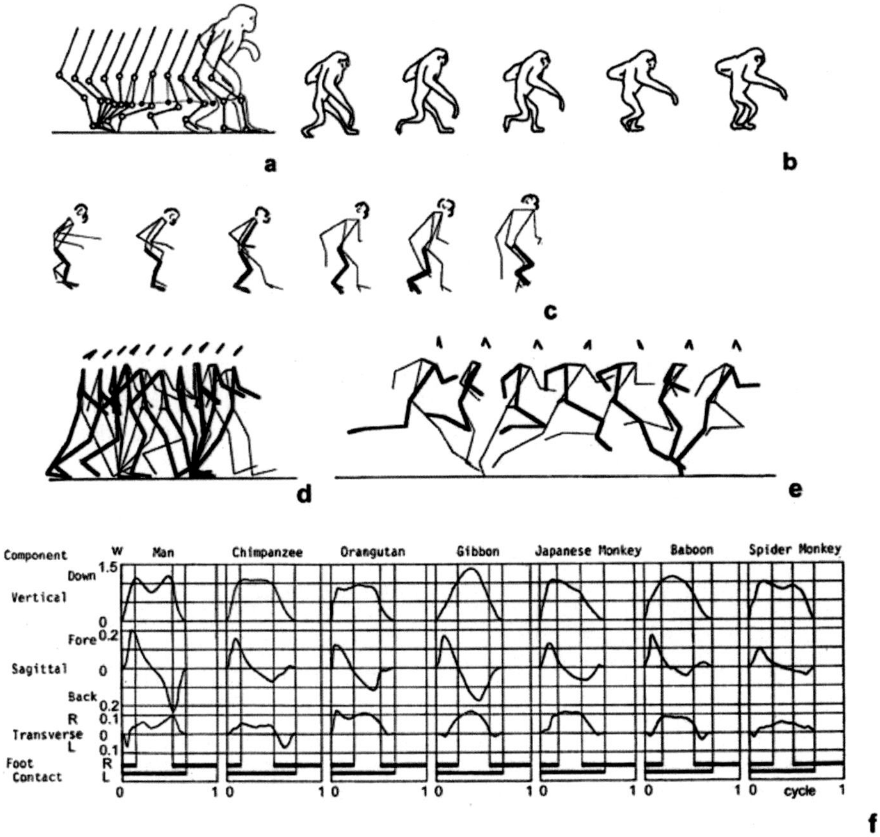

Fig. 9.29 (**a**) Bipedal locomotion in apes and monkeys (a gibbon is shown as example) usually has characteristics of human running: bent knee, bent hip postures, often a phase of aerial suspension ((**b**), second phase). The CoM reaches its highest position during swing phase (**a**), and spring mechanisms may be used (Ishida et al. 1984); (**c**) bipedal gibbon according to Schmid and Piaget (1994). These authors have put most emphasis on trunk rotation and its diverging from the human pattern. Right hindlimb emphasised by fat line. (**d, e**) For comparison, human walking (**d**) and running cycles (**e**), also from Schmid and Piaget (1994). (**f**) GRFs of humans, bipedal apes and monkeys. Note the varying scales of the forces. (From Kimura 1985; see also Li et al. 1996). For walking humans, the two-humped vertical components of GRF in y-direction are characteristic, which also may occur in walking cursorials (see Fig. 7.15b). The indentation becomes deeper with increasing locomotor speed (Alexander, 1976). According to Yamasaki (1985), it seems to represent a centrifugal force, which lifts the body mass during midstance. The GRFs of both feet overlap, so that the greatest force is transmitted to the body during two-limb support (a fact emphasised by Kimura 1985). The vertical GRFs in bipedal apes are similar to those usual in quadrupedal mammals, as shown in Figs. 7.6 and 7.22. On bottom the force components in x-direction

9.5.2 *Use made of the hands*

Not very promising is the search for specialisation like "knuckle-walking" in the hand skeletons of australopithecines – simply because KW is a means to use a hand (which is specialised to form a hook) for quadrupedal locomotion on ground. The hook hand is an aberrant specialisation of the apes and some other primates (see above, Chap. 8), while the australopithecines have been bipeds, and there was no need to use their hook hand as a balancing additional support on the frontlimbs.

The ape-specific elongation of the metacarpus does not seem to be very pronounced in australopithecines (Clarke 2003, see also *A. sediba* hand). Several authors have emphasised the strength of the very robust phalanges and the curvature of the basal segments (Susman 1979; Susman et al. 1984). As has been detailed above (Chap. 8), the curvature is a means to keep the bending stresses in the phalanges low, if flexor tendons are very strong. The metric characteristics of the human hand make sense if activities like throwing, clubbing, hammering are accepted as essential factors.

9.6 Characteristics of the Hominoid Skull Shape

The shapes of the skull to be discussed in the following subchapter also do not give much help for understanding the reasons for morphological changes, in view of the mosaic nature of quite divergent lifestyles. For example, the chewing apparatus may well have developed in a totally different, independent direction as the feet have.

While the postcranial skeleton, especially that of the hindlimb, seems to have evolved in a rather straightforward way to the shape represented in the genus *Homo*, the shapes of skulls show more variation than body proportions or postcranial skeletal morphology. Since the dentitions make the most intimate contact with parts of the environment in acquisition and processing of food, and the braincase contains the outstanding human characteristic, namely, the huge brain (see below), the jaw apparatus and the skull cap promise a wealth of information. However, skull characteristics must by no means be connected with shape characteristics of the postcranial body. An additional, often encountered problem is that no convincing connections exist between dental characteristics, jaw shapes, shape of the braincase and postcranial morphology.

9.6.1 *Morphology of Skulls*

Whoever is dealing with skull shapes can hardly avoid discussions about *taxonomy* (Figs. 9.30, 9.31, and 9.32). The great majority of ideas concerning human evolution was and still is based on skull and dental morphology. This is fully justified, in view of the amount of details, which have undergone changes during evolution and

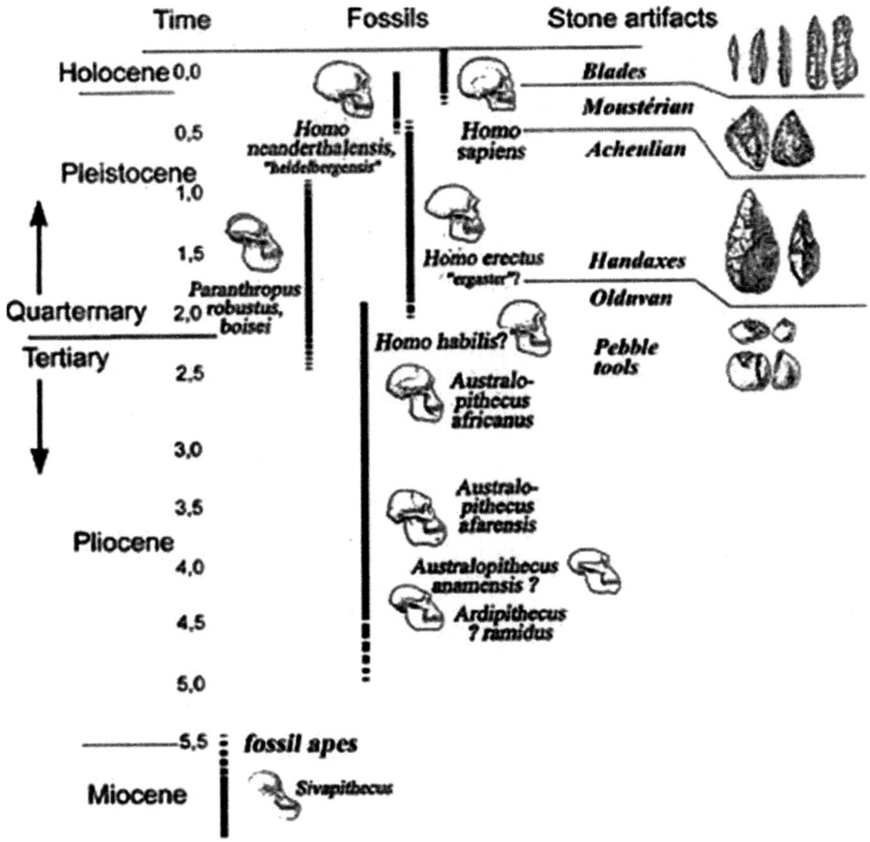

Fig. 9.30 Fossil hominids through geologic times. On the far left are the limits of the "periods", followed by "epochs" and absolute chronology in millions of years (MY). In the middle are fossil prehumans and early humans characterised by well-preserved or reliably reconstructed skulls, and their existence of heavy vertical lines. These lines are interrupted, where the existence of a form is not documented beyond doubt. Similarity to modern humans increases from left to right. For later evolutionary stages, the most characteristic stone tools are shown, but in a diverging scale. (Modified from Klein, 1999)

because the chances for preservation of skulls, jaws and especially teeth during fossilisation are relatively good, due to their compact shapes, smallness and chemically resistant materials. Alone by dental and skull characteristics, a multitude of taxa has been discriminated ("morphospecies"). In view of the long time intervals which have elapsed between embedding of the not too numerous fossils ("chronospecies") and in view of the huge geographical distances between them, it is quite possible that many of the postulated "taxa" are real (My background is close to that of Peters, 1970; Peters and Gutmann, 1971). However, there are very good reasons for doubt: first, the erection of a new taxon is tempting, since it is often

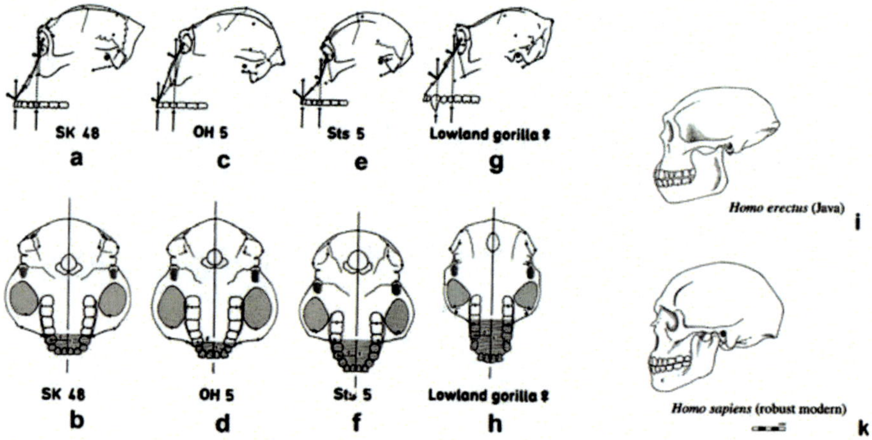

Fig. 9.31 **(a–d)** Selection of hominoid skulls in lateral (upper row) and basal view (lower row). The outlines are based on our Cartesian coordinates of cranial landmarks (dots), in **(g, h)** on mean values of female gorillas. Projection of the upper jaw or tooth row as possible spots of load application beyond the roots of the zygomatic arch as most anterior muscle insertions are indicated in the upper row as vertical lines and in the lower row by dark grey shading. (From Preuschoft et al. 1989). Since *H. erectus* (from Java, **(i)**) and a robust specimen of *H. sapiens* **(k)** are not contained in **(a–h)**, illustrations from Klein (1999) are added

considered to be a highlight in a scientific carrier. Second, in most cases fossils are remains of only one or very few individuals and therefore do not give any idea about the individual variation within a breeding group the most convincing definition of "species". (Sudhaus and Rehfeld, 1992; Storch and Welsch, 1992)

In order to find a measure for judging how large a *distance between related forms* of hominoids can be to justify discrimination between (the already defined) subfamilies, genera and species, Creel and Preuschoft (1971), see also Preuschoft (1989c) have drawn morphometrical comparisons between hominoid skulls. For widely used and commonly known anatomical landmarks (Fig. 9.31, Table 9.1; Martin 1928; Mollison 1911), three coordinates were measured in a space, defined by the Frankfurt horizontal and a line connecting both ear openings. The data, 91 independent variables, were size-corrected and averaged in the case of recent forms and treated as the centroids of ellipsoids in the case of fossils. The values were compared by multivariate statistical procedures. The resulting distribution along discriminant axes was plotted in 2, 3 or even 4 dimensions, as shown in Figs. 9.32, 9.33 and 9.34. As can be seen, the differences between the commonly accepted genera – large apes, australopithecines, early *Homo* and modern humans – are great and do hardly overlap. More complicate, however, is the discrimination between clusters considered to be "species". The geographical groups of *H. sapiens* (Figs. 9.32 and 9.33) are closely packed; their deviation from Neanderthalians (which existed up to about 30,000 years in Western Europe, beginning at about 100,000 years; Fig. 9.33b) on at least two canonical axes is visible. The dichotomy

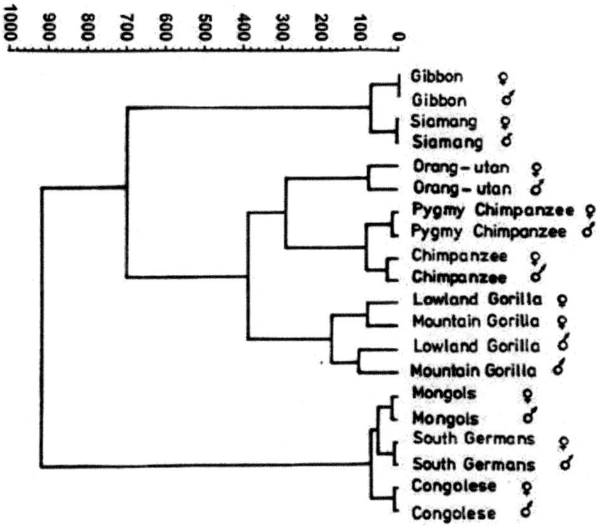

a

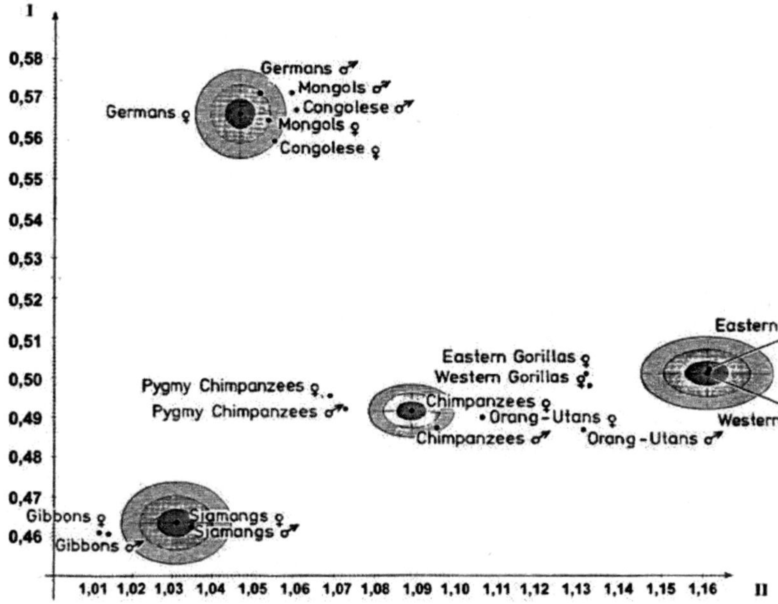

b

Fig. 9.32 (**a**) Dendrogram based on D^2 distances between the groups under investigation. (**b**) Two-dimensional plot of the results of a multiple discriminant analysis based on 91 independent variables (Cartesian coordinates) of skull landmarks in living hominoids. The centroids of the groups are shown by dots. In some examples, the variation is indicated by ellipses (dark grey 1 sd, middle grey 2 sd, light grey 3 sd). Note the huge sexual dimorphism in orangutan and gorillas, as well as the clear separation of gibbons and of recent humans from large apes (From Preuschoft, 1989c)

Table 9.1 List of the Hominoidea included into Creel's and Preuzschofts morphometrical comparison of skull shapes

Recent		Fossil	
1	Lowland gorilla ♀	18	Le Moustier
2	Lowland gorilla ♂	19	Swartkrans
3	Gibbon ♀	20	Skhul 5
4	Gibbon ♂	21	La Quina
5	Orang-utan ♀	22	La Ferrassie
6	Orang-utan ♂	23	Steinheim
7	Bonobo ♀	24	La Chapelle
8	Bonobo ♂	25	Gibraltar
9	Mountain gorilla ♀	26	Lautsch
10	Mountain gorilla ♂	27	Predmost ♂
11	Siamang	28	Cro magnon
12	Chimpanzee ♀	29	Oberkassel ♂
13	Chimpanzee ♂	30	Brno 3
14	Modern human, Congo ♀	31	Combe Capelle
15	Modern human, Congo ♂	32	Zinjanthropus (OH 5)
16	Modern human, South Germany ♀	33	KNMR 406
17	Modern human, South Germany ♂	34	Rhodesia
18	Modern human, Mongolia (♀)	37	Mt. Circeo
19	Modern human, Mongolia (♂)	38	Pithecanthropus robustus
		39	Choukoutien reconstr.
		40	Sterkfontein 5
		41	Predmost ♀
		42	Oberkassel ♀

The recent groups are represented by >20 fully grown individuals of each sex. The fossils are casts

of *H. sapiens* and *H. neanderthalensis* has been the object of fierce debates, reflected in varying names given to the respective fossils (e.g. *archaic sapiens*).[10]

[10]Morphologically, sufficiently preserved skulls can be attributed with a probability of far more than 10,000 to one of the groups (Creel and Preuschoft 1971; Preuschoft 1989c). Especially the Neanderthalians from Western Europe were with a stature of 165 cm, often smaller but probably heavier than the morphologically modern humans. The joints are broader; shafts of long bones are more curved (Neanderthal, Spy, La Chapelle aux Saints, La Ferrassie). In contrast to the Western European representatives, some of the Neanderthalians from the Near East, that is from Skhul Cave in Mount Carmel, Kebara (Israel), Shanidar (Iraq) and el-Amud (Israel) were as tall as in "morphologically modern man". *Homo floresiensis* from Indonesia (100,000 or 60,000 years old) seems to have been smaller than modern humans, but deviating details of his proportions have not been described. Enigmatic are these fossils because they resemble more the next oldest group: *H. erectus*. If only fragments or isolated teeth are preserved, morphology fails and alone molecular biology can yield results. The first information about DNA sequences in fossil humans, produced by the MPI group around Pääbö in Leipzig, emphasised the divergence between both forms. Later publications of the same group, however, concentrate on the content of "Neanderthalian genes" in recent populations of *H. sapiens*. This implies a sort of hybridisation. Under the assumption of constant rates of mutations (molecular clock), this has been estimated to have taken place something like 200,000 years back. From this time, however, no "Neanderthalers" in a strict sense are known. That all forms have common ancestors has never been questioned and implies the (observed) existence of shared genes.

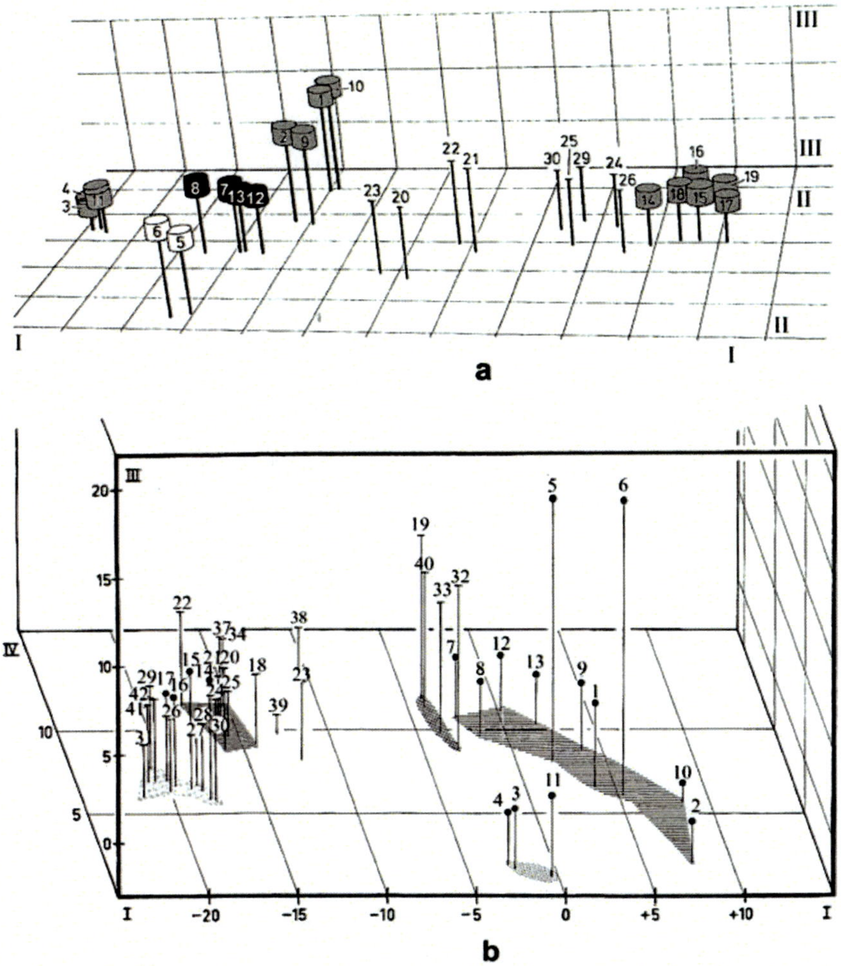

Fig. 9.33 (**a**) Distribution of the average hominoids on four dimensions. Discriminant axes I (from right to left), II (from near to far side of the diagram) and III (height of column above the plane). On discriminant axis IV, only the orangutans (white, 5 and 6) and the chimpanzees (black, 7 and 8, including 12 and 13, bonobos) deviate clearly and in opposite directions from the other forms. Both species of *Gorilla* (1, 2, 10, 9) and the orangutans (5, 6) stand out by their enormous sexual dimorphism. Geographically separated populations of modern humans (14–19) occupy the right limits of the total variation. The distances from each other are so large that they represent a caveat against attribution of moderate divergences to taxonomic differences. Aside from the group means (cylinders), fossils (horizontal crossbars) are included: 20–23 Australopithecines, 24–30 early *Homo*. (**b**) Three-dimensional plot of the results of a multiple discriminant analysis based on 91 variables (as in Fig. 9.32a, b). Discriminant axes I (inverse to (**a**)), III and IV are shown. Dots mark the centroids of groups; crossbars mark individual fossils. The position of several groups (from left to right: *H. sapiens*, *H. neanderthalensis*, australopithecines, hylobatids, large apes) on discriminant axes I and IV are emphasised by hatching. Note the clear difference between modern humans and neanderthalians. (From Preuschoft, 1989)

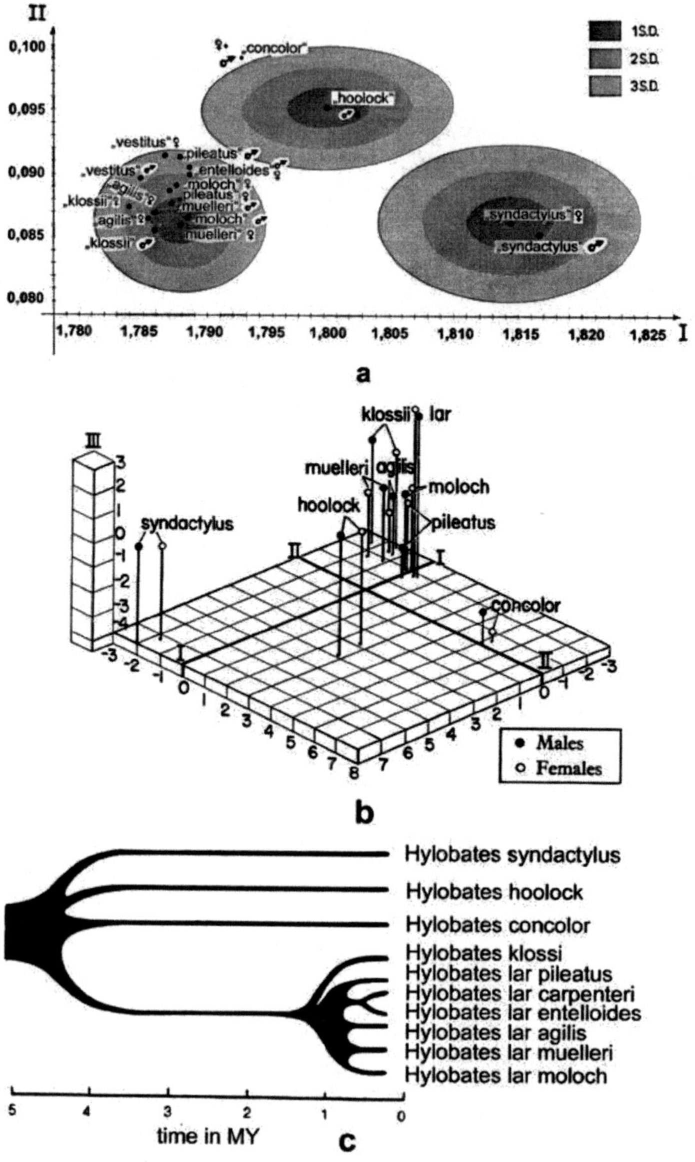

Fig. 9.34 (**a**) Position of male and female samples of hylobatid species on discriminant axes I and II. The concentric ellipses centred on the female "muelleri", "hoolock" and "syndactylus" should enclose 63% (dark grey) and 99.9% (light grey), respectively, of the individuals belonging to each of these groups. Note the divergence of siamangs (no overlap), "hoolock", "concolor" (considerable overlap) in contrast to the closeness of the "lar"-group (much overlap and high probability to misclassify individuals). (From Creel and Preuschoft 1976). (**b**) Three-dimensional plot of hylobatids along discriminant axes I, II and III, which contain 46% of total variation contained in 90 size-adjusted cranial and dental variables. (From Creel and Preuschoft 1984). (**c**) Cladogram of hylobatidae with estimated chronology of splitting (based on the "molecular clock") as a result of the phenetic analysis. Unfortunately, we were unable to differentiate the crested gibbons ("concolor"-group) because of the lack of clear information about origins of the skulls we had at hand. (Taken from Preuschoft 1988)

The common ancestors of both reach back to nearly 2 million years (Fig. 9.30). They have been found in Northern Africa (Moroc, Algery), Europe (Steinheim, Mauer), Middle Asia (Dmanisi, Georgia), East Asia (Choukoutien, China, Trinil, Modjokerto in Java; Flores, both in Indonesia) and South Africa (Sterkfontein, *Telanthropus*, also *H. naledi*). Since they existed during a long time, and were distributed over a large part of the Old World, variation is not surprising, although all have been named *H. erectus*. Their variation has been emphasised already by Creel and Preuschoft (1971), Preuschoft (1989c) (see Fig. 9.33b). Individual variation within a fossil group has become more than evident by the finds from the site of Dmanisi (Georgia, Lordkipanidze et al. 2013). Of course it is possible that several taxa are hidden within this variation. This is pre-empted by frequently postulated names – like *H. habilis*, *H. ergaster*, *H. rudolfensis*, *H. heidelbergensis* and *H. floresiensis* – which all are without a solid and detailed definition.

While the large apes have restricted geographical distributions, alone the small apes live in widely spread habitats of South East Asia and are split into a considerable number of groups, commonly considered to be species. Even among the most distant of these groups, hybrids are known (siamang and lar gibbon, "siabon"), natural groups in their native habitat usually establish hybrid populations where the habitats are in touch. Schilling and Preuschoft (1984) have compared a particular behaviour, namely, sonagrams of the songs of several of these subgroups and hybrids. Obviously, the songs are inherited, and the knack was that hybrids of the first generation were exactly in the middle between their parents; backcrosses of the second generation occupied positions exactly between the hybrids and one of the original groups. For investigating this in all details, Creel and Preuschoft (1976, 1984) concentrated their efforts on the small apes and their morphological relationship. The authors started off from the concept of the species as a natural, reproductive community. As in the example of modern humans, minor differences between groups are considered to be nothing but local populations (see above). The number of *crested gibbons* ("concolor" group) we could measure in European and US-American collections was large enough to separate them from other gibbons, but not sufficient to proof differences within this group, and clear information about origins of the skulls often was not available. The separation between populations seems to be similar to that in the "lar" group. Figure 9.34 shows the results: gibbon clusters are very similar to each other, and not more than four groups (siamang, crested gibbons, hoolock and lar gibbons) can be discriminated reliably. Most of the others overlap morphologically to a large extent. The similarity can best be demonstrated by the probability to misplace individuals into a group considered to be a different taxon.[11] In contrast to the probability to misplace

[11]Our morphological comparisons include, aside from metrics, also pelage colours and non-metric characters; details for our interpretation are given in Creel and Preuschoft (1984). Anyway, our conclusions are in contrast to those drawn by molecular geneticists, who claim generic status for the four groups mentioned above (e.g. Roos (2016). Of course it is largely a matter of taste which degree of divergence is considered to define a "genus" or a "species". Nevertheless, before postulating deep-reaching changes of commonly accepted taxonomy, the ultimate sense of grouping animals (namely, providing a general overview and being understood by non-specialists) as well as the situations in other taxa should be taken into consideration.

H. sapiens or *H. neanderthalensis* of >1:10,000, the value relevant for lar gibbons is about 1:4.

The morphometric investigation of skull shape has not been continued, except in the case of fossils, where the comparison of distances between the common anatomical landmarks turned out to be easier to obtain and to yield nearly the same results as our coordinates. Discussion at present is dominated by molecular biological, "genomic" traits which are more expensive to determine but easier to handle by multivariate techniques – largely the same as developed in the old days of morphology. Most molecular biologists generally take any minor difference between two groups as the argument to split them into separate species, even genera (e.g. *Nomascus* instead of *H. concolor*).

9.6.2 Brains

The most important trait of all hominids, which follows the same general tendency over all evolutionary stages, is the *enlargement of the brain*. Modern chimpanzees have brain volumes of 380–420 ccm and gorillas up to 500 ccm. Australopithecines with their estimated body masses of 30–60 kg have brains of about 500 ccm; the earliest humanlike skulls, so-called *H. habilis*, vary between 600 and 900 ccm, while *H. erectus* sensu stricto between 800 and 1200 ccm. Recent *H. sapiens* has usually brain sizes between 1300 and 1500 ccm, Neanderthalians often had more than that.

The evolutionary advantages of *huge brains* and the connected abilities are obvious and do not need to be explained in detail. Nearly all parts of the brain are involved, the exception being the bulbus olfactorius, and most pronounced is the inflation of the neocortex. The work of Stephan (1967, 1972, 1981); Stephan et al. 1981; see also Maier 2017) contains the best, quantitative information. He has sectioned in series the brains of a great number of primates. The volumes of functionally defined parts of the brains were compared.

"Die Abhängigkeit des untersuchten Gehirnabschnittes vom Körpergewicht wird für basale Insektivoren ermittelt und im doppelt logarithmischen System als Regressionsgerade dargestellt. An dieser lässt sich für jede beliebige Körpergrösse ermitteln, wie gross die Struktur bei einem typischen basalen Insectivor ist (Basalgrösse). Die bei einer progressiven Art wirklich erreichte Strukturgrösse wird mit der für das gleiche Körpergewicht gültigen Basalgrösse verglichen. Der Quotient zwischen beiden Werten (*Progressionsindex*) drückt zahlenmässig aus, um wie viel mal so gross die Struktur bei der progressiven Art im Vergleich zu einem basalen Insectivoren gleicher Körpergrösse ist" cited from Starck (1981).

The "index of progression" of the neocortex is an outstanding measure of "evolutionary height", at least in primates. Some representative values are set out in Table 9.2.

Although the enlargement of the braincase is the most obvious trait which can be pursued through hominid evolution, *other traits* are likewise of importance: reduction of canines, reduction of the masticatory apparatus, shortening of the tooth rows with

Table 9.2 Index of progression of the neocortex in primates, after Stephan (1972), cited in Starck (1981) and Maier (2017)

Cheirogalinae	9–12
Lemurini	17–23
Tarsius	21–32
Callitrichidae	26–29
Cebidae	>40–60
Ateles	45–55
Colobidae and Cercopithecidae	>30–45
Gorilla	>30
Pan	<60
Homo	156

diminishing of prognathism, formation of a chin, shifting the foramen magnum forward in rostral direction, ventral flexion of the skull base, elevation of the nasal roof and formation of the steep forehead. For most of these traits, there exist inexpensive, more or less detailed "explanations" on the level of "biological roles". As examples, the shortening of the tooth rows and reduction of prognathism have been connected with consumption of soft, perhaps cooked, rich nutrition (Wrangham et al. 1999, Niemitz 2004, 2010). The reduction of the originally prominent canines was connected with giving up of biting in agonistic contexts, which made these weapons superfluous, shifting of the foramen magnum with increasing frequency of upright trunk posture and the improved balancing of the head on top of the cervical spine. These assumed connections may well be correct, but the explanations are not satisfying, because no convincing relation between trait and its role is given. Instead, we have tried to pin down the mechanical advantages of some selected cranial traits.

9.7 Are There Mechanical Reasons for Hominoid Head Shape?

The first condition for any *mechanical approach* is recognizing the way in which skeletal elements are strongly loaded in every day life. In the case of the skull, first the weight of the head, second its accelerations and third the acquisition and intake of food must be taken into consideration. Point 1 (Fig. 9.35) requires a concentration on the skull base and knowledge of the epaxial muscles; point 2 needs consideration of the partial masses, including the brain as well as epaxial and the hypaxial muscles. It has yet rarely been considered, but see below. Point 3 (see below and Fig. 9.37) means focussing on the jaws and the parts between origins and insertions of the masticatory muscles. All muscle insertions are spread over large areas and therefore more difficult to handle than reaction forces in the joints, which result from weight, mass inertia and muscular forces.

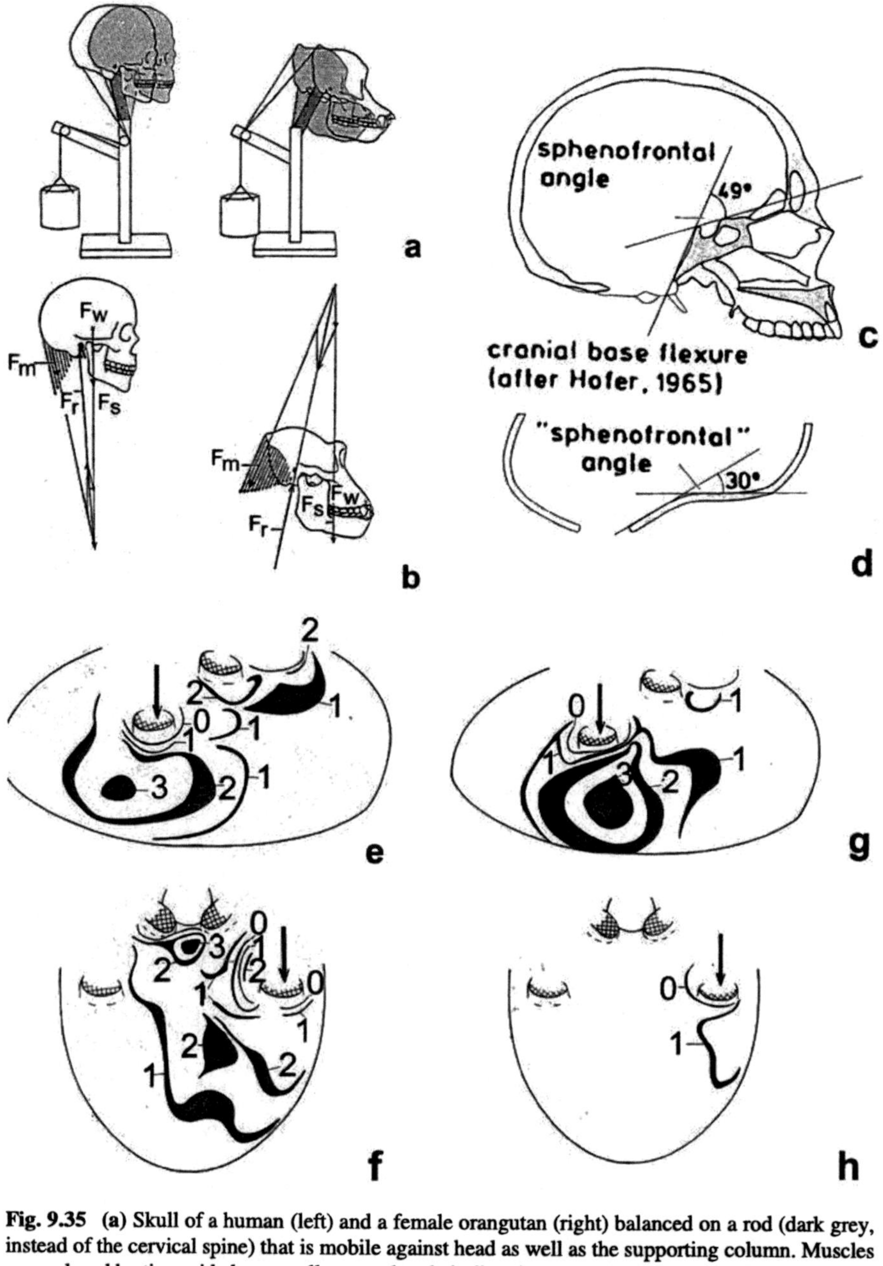

Fig. 9.35 **(a)** Skull of a human (left) and a female orangutan (right) balanced on a rod (dark grey, instead of the cervical spine) that is mobile against head as well as the supporting column. Muscles are replaced by ties guided over pulleys, so that their direction at the skull-neck (=atlanto-occipital) joint is similar to the pull of the nuchal muscles. The force exerted (no matter which of them is active) is condensed to its line of action. This allows drawing of the parallelograms of the forces ((**b**), muscles and weight of head) and the resultant (F_r) crossing the atlanto-occipital joint. F_r falls in line with the above-mentioned grey brace of the head. Note the diverging direction of the resultant F_r, which favours the well-known forward (in humans) or rearward inclination (apes) of the foramen magnum. An excess of muscle force leads to dorsiflexion of the neck. **(c)** Basal kyphosis of a human skull on a median- sagittal section. **(d)** Section through the model (a "shell", turned upside down) which has served for photoelastic experiments. **(e–f)** On the inverted model act vertical forces, concentrated either on an occipital condyle (**e**, **g**) or one of the mandibular joints (**f**, **h**). Both evoke roughly similar stress patterns. (From Demes 1985a, b)

Rarely preserved in unearthed skeletons, and not sufficiently investigated, are remains of *the skull base*. An exception is the H. erectus- skull OH 9, which was described in detail by Maier and Nkini (1985).

As observed first by Schultz in the 1930s, in the apes and in humans the heavy head is carried on the cervical spine in different ways. According to Demes (1985a), *head weight is balanced* by neck muscles. The resultant of weight and muscle force (F_r) determines the direction of the cervical vertebral column. This is illustrated in the experiment of Fig. 9.35a by the mobile rod simulating the vertebral column. In humans, a lordotic curvature of the cervical spine is of advantage, since it provides long lever arms for the nuchal muscles. In the apes, the long spinal processes, in combination with the forwardly displaced position of the head, give the nuchal muscles also long lever arms, so that moderate forces are required. The head position shown for apes also fits well to the semi-pronograde posture of the body axis, as shown, for example, in Fig. 8.52f. A fully upright position of the body axis, which is typical for humans (see Figs. 9.12 and 9.21), however, would lift the jaws, especially the tooth rows, into a rather uncommon, nearly vertical position. The way out is usually seen in a shifting of the foramen magnum forward (Fig. 9.35b). This also can be seen as a rotation of the face downward about the atlanto-occipital joint. The foramen magnum is clearly inclined dorsally in the apes and slightly inclined ventrally in humans. If the resultant F_r is orientated at right angles to the foramen, no potentially dangerous shearing forces acting on the brain stem or medulla oblongata do occur. While muscle forces are distributed over a large part of the squama occipitalis, the joint force is concentrated at the occipital condyles.

This obvious and easy-to-observe shift of the occipital condyles and the foramen magnum is commonly seen in relation to the enormous brain size in humans. This factor may well be of relevance. Both traits are connected, however, with a flipping downward of the facial skeleton, which brings the tooth rows back into their usual horizontal, or downward tipping position. This flip becomes visible – and measurable – in the "*kyphosis*" of the skull base (Fig. 9.35c), be it basal or prebasial in the sense of Hofer (1965).

Regardless of its direction, the resultants act against the occipital condyles, a constituent of the *skull base*. The other forces acting against the base of the skull are the joint reactions exerted by the mandibular condyles alternatingly at either side when biting or chewing. This concentrated force application allows an experimental approach by placing the skull base upside down on a support and concentrate weights (=loads) on appropriate points of its basis. It can be expected that stresses distribute like in a shell (Fig. 9.35c) over large areas. Demes (1981, 1982, 1984, 1985a) tried to make them visible by covering a model with a photoelastic coating. This method is not longer used. It is described in Demes (1985a, b) and also in Preuschoft (1970a, b); for more details see Kummer (1959a, b). Demes found similar stresses, no matter whether they are caused by forces applied to the mandibular joints or to the occipital condyles (Fig. 9.35c). The stresses spread preferably towards the squama temporalis and the occiput, where they meet the insertions of mandibular and nuchal muscles and close the circles of forces.[12] Bone material in these areas is particularly dense and strong. On the rostral and occipital flanks of the os petrosum, stresses are lower than elsewhere or zero. The passages (foramina) of nerves and vessels which connect the brain with the periphery make use of these low-stress areas. The often discussed basal kyphosis (shown in Fig. 9.35b) entails a reduction of the stresses in the anterior part of the "shell".

An approach to head shape through feeding has been selected and dealt with in detail already in preceding parts of this book, especially Chap. 2. Our own (e.g. Preuschoft et al. 1983, 1986a, b) studies on hominoid head shapes as well as those of other authors started off from the general characteristics of primate skulls: large braincase; occipitally closed eye capsules; nearly complete dental arcade on a non-kinetic, rigid skull; and the only mobile element being the mandible. The skull (including the mandible) has been described in all details very carefully, and the fluctuations of its shape during ontogenetic development, growth and evolution have been documented in many studies (see Chapter 9.6). Allometric aspects were in the foreground (e.g. Schmid and Stratil 1986). Allometric studies, although being descriptive by nature, often include a good deal of functional morphology (Demes 1991; Demes et al. 1986; Demes & Creel 1988). Precondition for the creation of testable hypotheses is an overview, if not thorough knowledge of the biomechanical conditions prevailing in the parts under study. Questions must be raised concerning areas on which insertions of muscles or bite (chewing) forces are distributed, and estimates must be included concerning the directions, lever arms and magnitudes of forces.

The first parts of the body which make contact with food are the teeth, and some points mentioned in Chap. 2 should be recalled here (Fig. 9.36).

On the teeth act forces which have, in addition to their "normal" component perpendicular to the tooth rows, also a component in medio-lateral direction. This leaves traces on the surface of the teeth, especially on the cones of molars. These wear facets on upper and lower cheek teeth have been analysed in detail by R. Kay (Kay and Hiiemae 1974; Kay and Hylander 1978) and independently by W. Maier (1977a, b, 1978, 1980) and related to the chewing movements of the lower against the upper jaw.

Specialisations of tooth morphology and of jaw shape can be expected to be the first of all consequences of changes of feeding habits. A change of the latter – no matter whether enforced by a change of the environment or by a change of lifestyle – away from consuming rather soft, fleshy fruits, sprouts, marrow to mechanically

[12]In the large apes, as well as in carnivores, the nuchal muscles are in direct contact with the posterior part of the m. temporalis, the fibres of which form a sort of continuous tie along the occipital and parietal parts of the skull cap (see in Chap. 2: Sub-Chapter 2.2.3). In many, especially large forms (*Panthera*, male *Gorilla*, male *Pongo*), a bony plate or occipital crest is developed between both muscles. It allows both muscles to act separately, although in many cases (shaking the head) the tensile forces of the neck muscles are transmitted directly on the skull roof. In some, more gracile australopithecines and in members of the genus *Homo*, it remains questionable, whether the circles of forces are indeed closed via the skull roof.

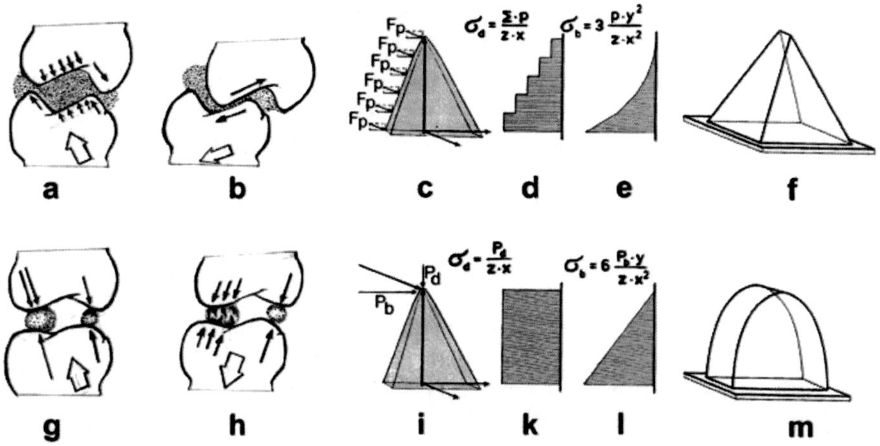

Fig. 9.36 Cusps of molars and their relation to ingested food. (**a, b**) Soft, but tough material is formed to a bolus, which fits into the interspace between the cusps of molars. Leaves, sprouts and many fruits contain tough fibres, but no really hard components. For diminution of these fibres, the cusps need long flanks (**a, b**), between which friction is exerted. High cusps are considered to be typical for leaf eaters. The reaction forces from the bolus are distributed over the height of the cusp (**c**). Compressive forces within the cusps increase toward the roots (**d**); bending moments (**e**) increase parabolically. The optimal shape for sustaining the evoked stresses in such a cusp has the outline of a pyramid (**f**), and the most frequent traces of usage are "scratches". (**g, h**) Hard particles or shells lead to a concentration of the bite force on the cusp (**i**). Compressive stress ((**k**), here named P instead of F) has the same size along cusp height, and bending moments (**l**) increase linearly toward the bases. These stresses can only be sustained if cusps are low and blunt. The optimal shape is that of a dome with parabolic outlines (**m**), and the traces of wear are "pits". (After Preuschoft et al. 1989)

more resistant, harder and smaller fruits, seeds, kernels or shell-protected invertebrates (see above, Niemitz 2010; Shabel 2005) favours specialisations of *tooth morphology*, because teeth are the first part of the body in immediate contact with its environment.

Two traits are of relevance:

1. The thicker the enamel layer (like in australopithecines, compared to apes, (C. Jolly 1970a, b); Spears and Crompton 1994 and later), the more resistant the tooth. The immediate traces left by food particles under high forces on the enamel layer may have two forms: pits or scratches. During several years many attempts have been made to derive the sort of food from the occurrence of these traces (e.g. Grine, 1981), but it seems that pits as well as scratches are temporary phenomena which appear and disappear with the food consumed most recently.
2. The shape of cusps. The relation between tooth shape and stresses has been detailed in Chap. 2, with emphasis on the pointed, sectorial teeth of lower vertebrates. Each cusp of a primate molar is exposed to loads similar to those acting on the more slender fang teeth of predators. The basic rules are the same in

both cases, although cusps are much lower than fang teeth. The partial loads normal to the crown are distributed over a large area and grow along the flanks of the cusps, so that the compressive stresses exerted against food substance increase towards their bases. The transverse force components also are distributed, and the bending near the tip of a cusp is minor, at its basis, however, quite large (Fig. 9.36). The outlines of an optimised cusp are determined by the square root of the bending moments (see above, Figs. 2.19 and 2.23 in Chap. 2), that is, the slopes are linear. These traits indeed can be observed in species, which live largely on leaves (e.g. gorillas or leaf monkeys). Crushing of hard shells, or the cases of seeds or nuts, by contrast, leads to loads concentrated on small areas, mostly the tips of cusps. The stresses evoked by such loads are of the same height from tip to base, and the bending moments grow linearly (Fig. 9.36). The optimised profile resembles a dome (vide above, in Fig. 2.4, 2.5, 2.8). Hard-object feeders like orang-utans therefore can be expected to have low and blunt cusps. In order to increase the number of individual small objects, e.g. seeds which are broken in one chewing stroke, molar surfaces must be rather large (above, Fig. 2.11, 2.17). This, however, leads to a decrease of bite pressure – unless the musculature is further strengthened. As further means to increase molar strength, enamel wrinkles are useful (see above, Chap. 2). They also reduce the slipping of chewed objects away from the areas where most pressure is exerted. All traits mentioned here are present in the *Paranthropus*-like fossils, to some degree also in *Australopithecus* (e.g. *A. afarensis*). The remarkable strength of the australopithecine jaw apparatus has often been mentioned.

The most promising part of the skull for a mechanical analysis seems to be the *lower jaw*, because rather well-defined (bite) forces as well as the mandibular joint forces act concentrated, while the muscle insertions are confined to only a part of the element. Like in all animals equipped with jaws, the stresses evoked by biting must be sustained alone by the skeletal element, which obviously does not fulfil additional "functions" (see Sub-Chapter 2.2.1, Figs. 2.25, 2.26; Demes et al. 1984.)[13]

The way of loading a mandible in higher primates is the same as in all animals yet investigated.[14] As mentioned above (in Chap. 2, Figs. 2.16, 2.26), it is a continuous skeletal element of its own, separated from the skull. While the lower jaws of monkeys, apes, prehumans and early human species show the expected, curved

[13]These factors have given reason for studying the jaws of the shark *Heterodontus* (Nobiling 1977; see above, Chap. 2). Likewise, the first of Hylander's exquisite experimental studies (1975, 1979, 1980, 1984, 1986) on the masticatory apparatus were concentrated on the mandible.

[14]K. Hiiemae (see above, and in Chap. 2, Figs. 2.15, 2.16) has recorded the mandibular movements during biting and chewing in a large number of animals on x-ray movies. She found the same pattern in reptiles and mammals, consisting of an "opening stroke" to increase the distance between lower and upper teeth, an upward directed "closing stroke" to approach the lower to the upper teeth and a "power stroke" which guides the lower teeth upward medially into full occlusion ("phase 1") and beyond ("phase 2"). While opening and closing do not require much effort, the power stroke is executed with great force, and during this phase, the bolus of ingested materials is being severely compressed.

and forwardly inclined dental rows in the symphyseal region between the canines of both sides, alone *H. sapiens* seems to stand out in its prominent "chin" (Schwartz and Tattersall, 2000). The search for a mechanical explanation of this particular shape requires a detailed understanding of the mandibular shape. The attempt to understand the human peculiarity therefore leads to a mechanical analysis of the entire element. The calculations of Wolff (1982, 1984), as well as Demes et al. (1984), have used the horizontally curved beam in Fig. 9.37 as a simple model, which will be described here. Witzel has used with the FESA and FESS much more fine-grained methods. Unfortunately, these results have never been published in the detail they deserve. A very short summary is contained in Witzel et al. (2011).

If projected into a frontal plane, the mandible appears as a beam, hung up at both sides by muscles (Fig. 9.37a). Acting are muscle forces (F_{mb} and F_{mn}), joint forces (F_{tb} and F_{tn}) and chewing or biting forces (F_z), the latter applied to only one side. From muscle forces, the joint forces are subtracted, where F_{mn} is near zero, according to experiments. Bending moments (M) and this mean stresses depend, if the magnitude of the load (F_z) is given, from the ratio a/b, that is, the distance between the joint on the bite side and the bite point. $a + b = l$, so that

$$M = \frac{F_z * a * b}{l}, \tag{9.11}$$

which can be transformed into

$$F_z = \frac{M * l}{a * b}$$

F_z is greatest, if a and b are of the same size: the closer the teeth are to the sagittal plane, the greater use can be made of the force available on the balancing side. This is advantageous for biting on the front teeth and a very good reason to approach the tooth rows – at least the molars – of both sides to each other. This trait can be observed in apes, prehumans and many herbivores, as well as in the heterodont shark (Chap. 2).

If viewed from the side (Fig. 9.37b, c), the lower jaw can be seen as a cantilever. Its length (in the direction of the occlusal plane) determines the bending moments. These differ between balancing (left) and biting side (right), because upward directed forces are transferred from the balancing to the biting side at the most anterior (mesial) end of each ramus. This leads to high torsion in the symphyseal region, between both canines, because the lever arms of the acting forces are long (Fig. 9.37b, c). Torsional moments are best visible in occlusal view on the tooth rows (Fig. 9.37d, e, f). Since the tooth rows diverge from the body of the mandible, torsional moments may also be evoked along both halves of the mandible (Figs. 9.37e, f and 9.38). This raises the question, why teeth diverge (in different directions!) from the perpendicular to the occlusal plane. An answer may be found in the grinding medio-lateral movements during the power stroke. Aside from the bite forces exerted perpendicular to the tooth row (see above, Fig. 9.37b, c), force

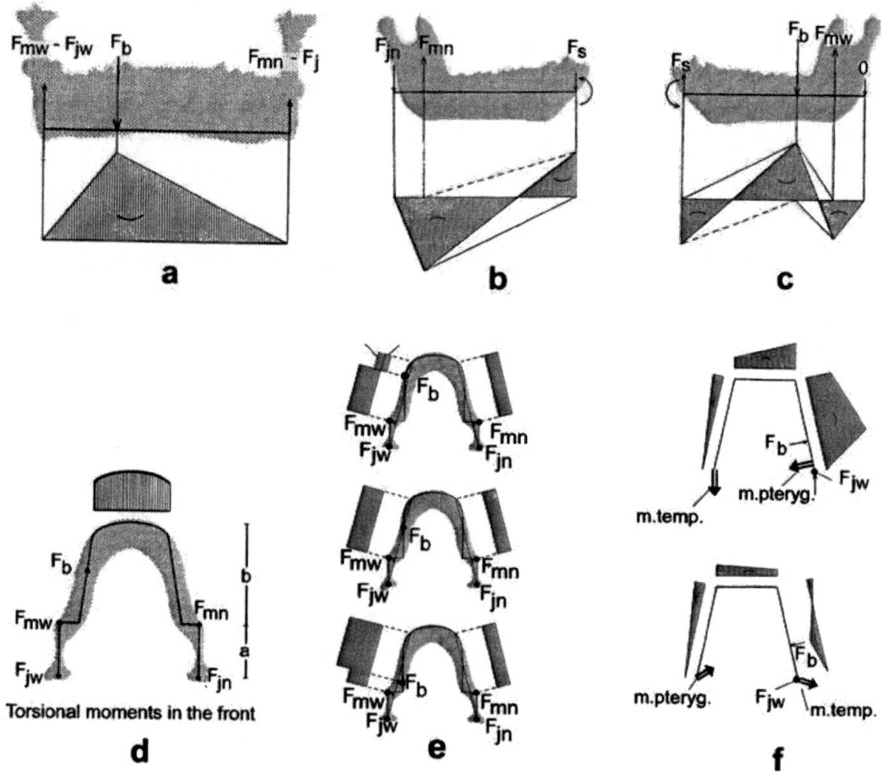

Fig. 9.37 Simple model of a hominoid mandible as a curved beam. Grey: mandible; hatched: moment areas, not drawn to the same scale; continuous lines: models, simplified for calculations. (**a**) Anterior view. The chewing force (F_b) is applied to a right premolar and produced by muscle forces F_{mw} and F_{mn}. The muscle force (F_{mn}) on the left side becomes greater, if the bite point is closer to the midline; this also allows exerting a greater total bite force. (**b**) Lateral view on the balancing side and (**c**) on the biting side, the latter with bite force (F_b). The patterns of bending moments are different on both sides and vary with the (changing) position of the bite point. (**d**) Torsional moments in the most anterior (symphyseal) region between the canines of both sides in occlusal view. The longer the tooth rows, the more pronounced is torsion. (**e**) Occlusal view with bite points on the left canine (*top*), anterior left molar (*middle*) and last molar (*bottom*). The shaded areas represent the corpus mandibulae; the continuous arcades symbolise the tooth rows. The diagrams show the torsional moments caused by the eccentric positions of the teeth along the tooth rows between canine and last molar. The obvious and well-known difference between humans and large apes (gorilla as an example) in the position of the canines leads to different bending moments in the anterior region of the mandibles. (**f**) During power stroke, the medio-lateral movements of the mandible and the resulting bite forces cause bending in the occlusal plane. The centre of rotation is the joint on the biting side. *On top* The transverse component of bite force acts in buccal direction, like during phase I; on *bottom,* in lingual direction like during phase II. Muscle forces are *on top*: the more or less horizontal posterior fibres of m. temporalis (F_{mn}), and on the biting side, medial components of the pterygoid muscles (F_{ma}). *On bottom*, F_{mn} stands for medial components of the pterygoid muscles and F_{ma} for fibres in the middle of the m. temporalis. (Modified from Demes et al. 1984)

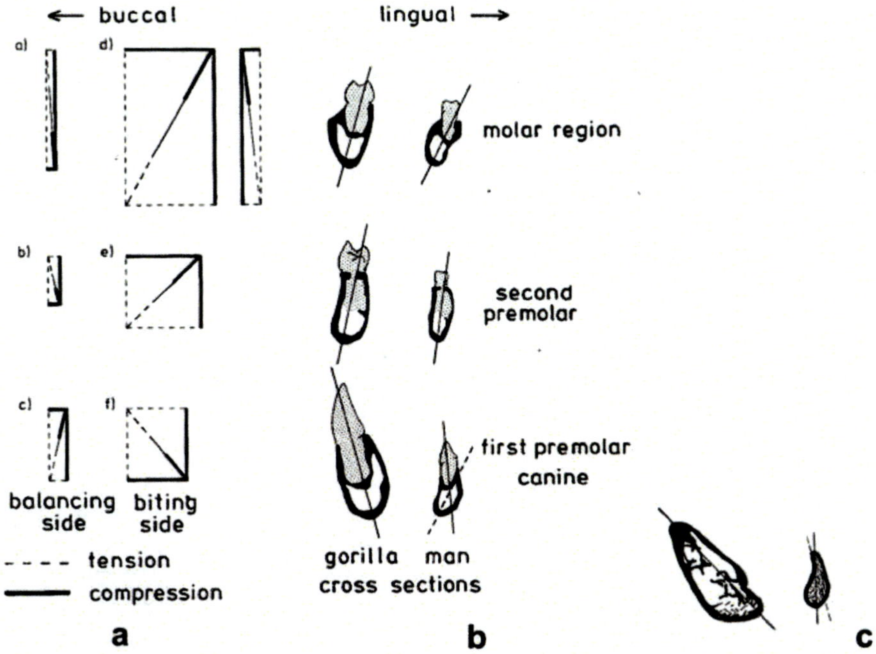

Fig. 9.38 (**a**) Bending moments in lateral and occlusal view are superimposed to a "resultant bending plane". The values at the biting side are higher than those at the balancing side. (**b**) Sections through the mandibles in a gorilla and a human indeed show the greatest bending strength in the same planes. (**c**) The middle of the mandible, the "symphyseal region" in non-human primates as well as early humans is inclined outward, but in modern humans a "chin" is developed. In apes, the inclination is in accordance with greatest bending moments. In the anterior parts of the mandibles of modern humans, there exist two planes that offer particular bending resistance, indicated by continuous (stronger) and broken lines (less strong). The concentration of cortical bone material along the peripheries of the sections indicates in addition considerable torsional strength. Torsional moments reach their greatest values in the symphyseal region of apes – due to the long lever arms provided by their long tooth rows (see Fig. 9.37b). Great strength against torsion is provided by long diameters, and this seems to explain the ape-specific "inflated" profile in this region. (From Demes et al. 1984)

components in the occlusal plane (Fig. 9.37f) are exerted during phases I and II of the power stroke. The bending moments evoked by both components along the bowed cantilever are combined in Fig. 9.38a). Comparison of the resultant bending planes with the directions of greatest bending strength shows a coincidence. While the situation along the lateral tooth rows is similar in apes and humans, the symphyseal region is different: modern humans show two planes of nearly the same maximal strength; the fossil species of *Homo*, like Neanderthalians, do not. The first plane is inclined in labial direction (continuous line in Fig. 9.38c, like in a gorilla), and the second in lingual direction. The latter situation implies a protruding "chin". Decisive for the development of a "chin" seems to be the ratio between bending moments in anterior and lateral view, and this depends from the ratio between bicondylar width and length of the tooth row (or length in anterior view/lateral view). With shortening of the corpus mandibulae, the bending moments in side view become smaller, while the influence of breadth in anterior view remains the same. The morphological specialisation must follow this preponderance of bending in frontal view and assume in modern humans the rearward inclination. Several authors have observed that tooth rows in *H. neanderthalensis* are hardly longer than in *H. sapiens*, but their jaw length is greater, because of the toothless "trigonum retromolare" behind the third molar.

Much more complex than the lower is the *upper jaw*, especially because of its connection with the cases of sense organs and brain capsule. Indeed one of the most puzzling parts of our morphology is the *hominid midface*. Before a new analysis is described, some attempts to explain details by using a global approach should be kept in mind.

The steep arrangement of the lower jaw, upper jaw, midface and forehead, one on top of the other in humans, deviates strongly from all other hominoids and is unique, although it obviously was attained by gradual changes from an ape-like face to the human condition. The very detailed experimental data about stresses in the skull, as were obtained by Hylander (on primates) and his group or by S. Herring (preferred experimental animal: pig), can probably not be applied directly to humans because they have been obtained in more long-snouted mammals. Explicit attempts to understand human skull shape can be found dispersedly in the anthropological and anatomical literature, like in Toldt (1914), whose three-pillar theory was also incorporated in Benninghoff's textbook, 2nd edition, 1938).[15]

The two most recent species of *Homo* show a similar basic arrangement but also several enigmatic differences: *H. neanderthalensis* exhibits a more pronounced prognathism, lack of canine fossa and marked torus supraorbitalis. The explanation of these differences challenges anthropologists since a long time ago. Rak (1986) has proposed an interpretation of the facial skull of Neanderthalians as a cylinder in the frontal plane (Fig. 9.39a). This emphasises the use made of the anterior (incisor and canine) teeth and leads to the assumption that premaxilla, anterior maxilla plus its nasal processes, zygomatic bone and nasal roof are under compression, while tensile stresses in the postmolar, pterygoid parts of the head are taken over by muscles. This interpretation does not explain the nasal opening. Demes (1987) put forward the idea

[15]This theory will again be taken up below. A revision and summary of the literature are given in Hofer (1965); more recent are Endo (1966, 1969, 1970, 1973), Endo and Adachi (1988), Endo and Fukushima, 1973, Rak (1986) and Demes (1988). Seipel (1948) and Tappen (e.g. 1958, 1967) tried to find evidence about the stress distribution in the facial skull and neurocranium with the split-line technique. Split lines, however, represent just the most superficial layer of the bone and do not give much information about the stresses within the bony elements. (The latter become visible in FES analyses; see below). Endo focussed mainly on possible functions of the torus supraorbitalis, which was seen as a beam, bent by the compressive force along the maxillary processes of the zygomatic bones and the nasal roof, while tensile forces acting downward at both sides through the zygomatic bone and the masseter.

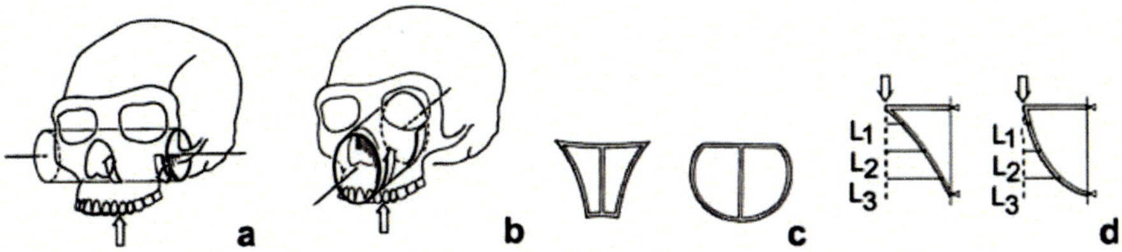

Fig. 9.39 (**a**) Facial skull of a neanderthalian, interpreted by Rak (1986) as a cylinder, the axis of which is parallel to the frontal plane. The arrows indicate bite forces acting on the incisor and canine teeth, which cause compression in the "anterior pillar" (premaxilla, nasal roof and zygomatic bones), while the mm. masseter and temporalis pull the posterior wall of assumed cylinder downward. (**b**) Interpretation of the same facial skull by Demes (1987) as a cylinder, the axis of which is in the sagittal plane. The same essential bite forces (arrows) are applied alternating to the canines and premolars of only one side. Bite forces lead to compression in the anterior pillar of the bite side and to tension in the mm. masseter and temporalis of the balancing side. (**c**, **d**) Scheme of sections through the walls of the maxillary cavity, without (**c**) or with (**d**) the concavity called "fossa canina" in *H. sapiens* or *H. neanderthalensis*, respectively. (**d**) Bite forces, applied to the tooth rows, have longer lever arms if a fossa canina exists (**e**). If the profile is curved laterally, the lever arms are shorter (**f**), and therefore less material is needed to provide the same strength. (From Demes 1987)

that the facial skull should be compared with a cylinder in sagittal direction (Fig. 9.39). Both authors start off from a broad dental arcade and the observation that anterior teeth are well developed and show strong abrasion. Both holds true for Neanderthalians. The most relevant bite forces are assumed to act against the canines and premolars. Because of the width of the tooth row, lever arms for torsion are long, and compressive forces flow alternately along the left or right maxilla to the anterior root of the jugal arch, where tensile forces can be applied alternately to the right or left side. Torsional stresses are very low or do not exist at all near the centre of the loaded structure (cylinder), and therefore this model explains the nasal cavity. Because torsional strength grows with the square of the distance of small parts of the cross-sectional area from the torsional axis, this model also offers an explanation for the convex lateral walls of the sinus maxillaris (absence of a fossa canina), which is so obvious in many Neanderthalian skulls.

A rather comprehensive (though insufficient regarding methods) picture of the *evolutionary changes* from Miocene apes through australopithecines to modern humans has been sketched by Preuschoft (1989b). Starting point of evolution towards modern humans seems to have been the facial skeleton and braincase of an ape-like Miocene forerunner. As a representative of this early stage, the skull of a female chimp is drawn in Fig. 9.40. The (even moderate) length of the jaws, and their prognathism, favours using the jaws as primary contacts with objects of interest as well as using the anterior teeth as weapons in agonistic contexts. It entails, however, long load arms of bite forces. To keep bite forces on a high level, a very massive (and heavy) musculature is necessary. During the power stroke of mastication, force components normal to the occlusal plane are created and components in the occlusal plane, the latter at right angles to the row of teeth behind the canines (Figs. 2.20, 2.26, Table 2.1 in Chap. 2, and also Fig. 9.37 and 9.38). The curvature of the dental

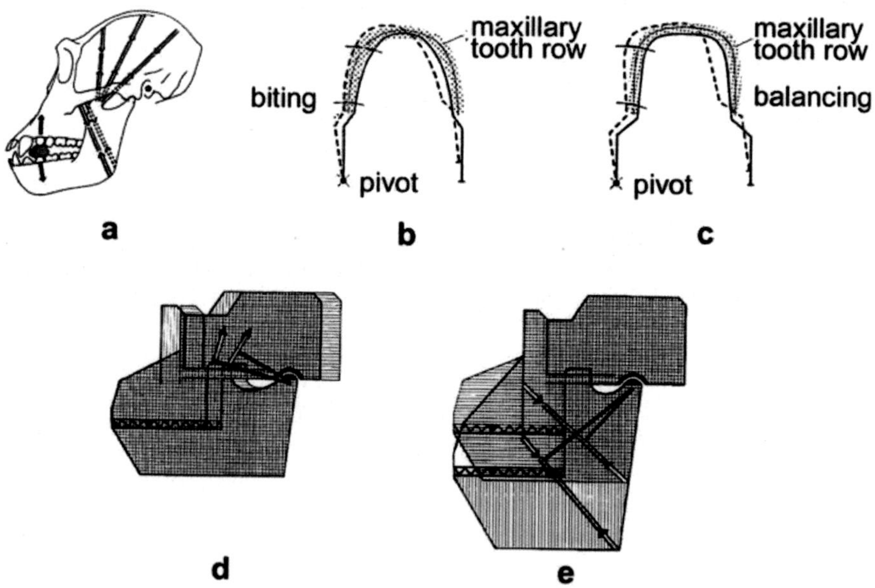

Fig. 9.40 (**a**) Forces acting in side view on the jaw apparatus of an ape, for example, a female chimp. Mm. temporalis, masseter and pterygoideus internus adduct the lower jaw; objects between the teeth ("bolus") press the upper jaw upward and the lower jaw downward. (**b**) Occlusal views on the jaws of a human and an ape (**c**) during the power stroke. The lower jaw (broken lines) is rotated about the temporo-mandibular joint on the biting side and approached eccentrically to the upper (stippled), until full occlusion is reached (continuous lines) and beyond in phase II. The highest forces occur during the power stroke. They must be sustained and therefore are the decisive factor for the shape of the dental arcade (Wolf 1982, 1984). (**d**) Shortening of the jaws (=shifting forward the origins of muscles (arrows double lines) on the braincase in relation to the jaws) gives the m. temporalis a longer lever arm at the TMJ (continuous lines), while the lever arm of the bite force decreases. (**e**) Increasing the height of the midface (=upper jaw) gives the mm. masseter and pterygoideus medialis a longer lever arm. This transformation yields more bite force, but does not save energy, because muscle fibres become longer. (From Preuschoft et al. 1989)

arcades (parabolic in humans, Figs. 9.37, 9.40; U-shaped in apes, Figs. 9.37, 9.40c) may have to do with the size of the canines but is not important in the present context. The bite forces can be increased (without increasing muscle mass) by shortening of the dental arcades and by giving the midface great height (Fig. 9.40d, e).

A breakthrough for the functional interpretation of skull shape was attained by application of the *FE technique* by Witzel in several steps – as noted above in Chap. 2 where the concepts FESA and FESS are explained (Witzel and Preuschoft 1999, 2002, 2005; Preuschoft and Witzel 2002, 2004a, b, 2005; Preuschoft et al. 2001, 2002, 2004). If FESS is used, the flow of (primarily compressive) forces within an amply chosen "Bauraum" can be made visible. If a new model is created, consisting of nothing than the stressed parts of the Bauraum, the new, "reduced"

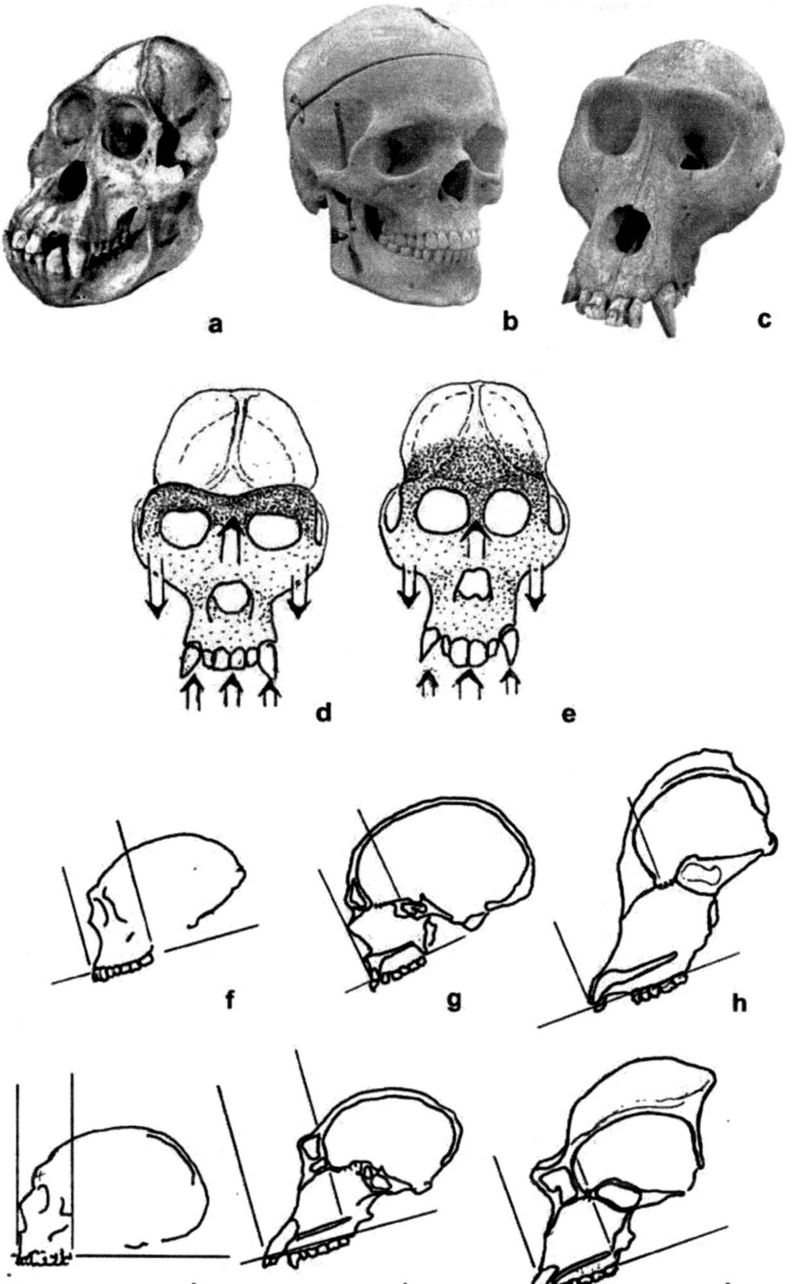

Fig. 9.41 Skulls of (**a**) an adult male orangutan, (**b**) adult male human and (**c**) a sub-adult male gorilla. Note, aside from the prognathism in the apes, the divergence of the foreheads. (**d**) Adult male gorilla and (**e**) adult male orangutan seen from in front and above, with schematic illustrations

model exhibits reasonable similarity to the real skull. Further reiterations of this process lead to more condensed stress flows within the "reduced" model and to more precise similarity of the stress-bearing model to the real skull.

First of all, the question must be raised and answered, which forces act on the skull. The next most important precondition is the position and direction of the bite forces in relation to the braincase, from which the muscles take their origins.[16] Other preconditions are the position of the orbits and the insertions of the masticatory muscles. These traits can be made plausible by their biological roles (e.g. orbits) or phylogeny (like the divergent arrangement of masticatory muscles in reptiles vs. mammals). By contrast, the nasal, and maxillary cavities, as well as the palatal vault, must not be given à priori but are direct consequences of stress flows. Skull cavities just represent empty spaces not occupied by the stresses.

In comparative morphology, aside from the human chin (see above), especially the prominence of the *human nasal roof* demands an explanation. Schmittbuhl et al. (1996) observed that modern human skulls possess, in contrast to prehumans, but similar to several species of simians (guenons, gibbons), a low midface, in which the upper margin of nasal aperture is higher than the lower margins of the orbits (Fig. 9.42a). Starting off from this observation, Witzel and Preuschoft (1999) have investigated the stress flow in a rectangular plate, in which the orbital and nasal openings are cut out. The bite forces exerted from below by the mandibular teeth flow on both sides upwards in an anteriorly concave curvature, towards the forehead (Fig. 9.42b). This corresponds to the "fronto-nasal pillar" as described in Benninghoff's textbook and to the anteriorly hollow profile of the midface, which

---

Fig. 9.41 (continued) of the major forces (arrows) exerted on the anterior teeth, running through the "nasal pillar" and caused mainly by the m. masseter. The supranasal and supraorbital region of the frontal is exposed to bending moments as documented below. The impressive supraorbital torus in gorillas (stippled) has thin walls which surround a spacious frontal sinus (**l**). This hollow, box-shaped beam provides high resistance against being bent. Orangutans (**a**, **e**, **h**) commonly lack a torus, as well as frontal sinuses. Their forehead – like that in humans – can be compared with a plate, which offers great resistance against being bent in the plane of its breadth. (**f–l**) Bite forces can be applied along the whole length of the tooth rows, between the medial incisors and the last molars. Their greatest components are more or less perpendicular to the occlusal plane. In humans as well as orangutans (**g**, **h**), the forehead is in line with the chewing forces acting against the molars, but in chimpanzees, gorillas and forerunners of modern humans (**f**, **i**, **k**, **l**), the frontal bone is not in the plane of the molar forces, but instead in front of the braincase. All this can be seen in connection with or even as a consequence of the "basal" or "prebasial kyphosis" of the human, of the "aerorhynchic" position of the jaw apparatus in relation to the braincase of *Pongo* skulls. See above and Fig. 9.36

---

[16]The bite forces habitually act against varying parts of the tooth rows and can assume varying directions, while equilibrium is established by various muscles. This makes the investigation of many load cases necessary. The observed stress flows are summed up. The parts of the skull model are shaped to sustain the greatest among all occurring loads.

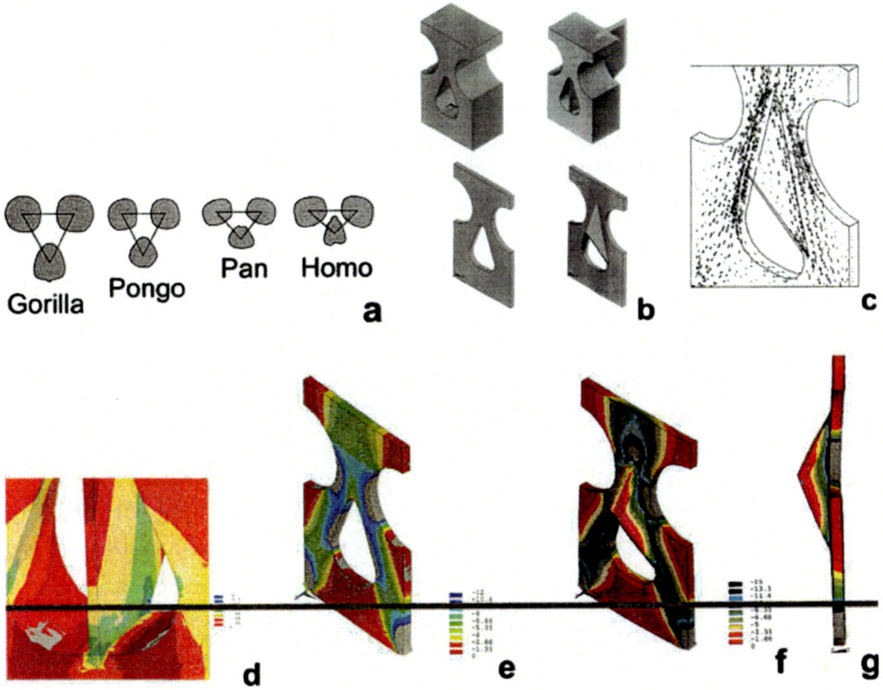

Fig. 9.42 (**a**) Orbitae and apertura piriformis in several hominoids, according to Schmittbuhl et al. (1996). (**b**) Simple approximations to the midface in form of rectangular solids, which are investigated with the aid of the FES method as in Witzel and Preuschoft, 1999. Forces are applied from above at the occipital margin and the counterforce from below near the "rostral" margin. The former represent the reaction forces from the forehead; the latter represent the bite forces on incisors. (**c**) Shows the trajectories of stresses in the thin plate in (**b**). The density of trajectories (=arrows) indicates the magnitude of stresses (compare (**f**)). In (**d–g**), the stresses are coded by colours, where red is very low, yellow or green moderate and blue and grey very high. (**d**) Detail of the model in (**b**), top right, with nasal septum. Stresses (yellow, green) occur in the lateral walls of the nasal cavity (=medial wall of sinus maxillaris), in the vomer and the palate. At the anterior margin, at the location of the spina nasalis, they assume the highest values (light blue). (**e**) Stress distribution in a thin, vertical plate. Supporting forces from below are applied to the posterior teeth. Because the upper limitation of the nasal cavity is between the orbits, stresses become very high (grey). If the same plate is equipped with a prominent nasal roof ((**e**), best visible in side view, (**g**)), the stresses become smaller. Nevertheless, the regions presented in green, blue and even grey consist in the native skull of massive and dense cortical bone, while the green and yellow regions in the skull consist of thin bony lamellae

is common among hominoids, but could never be explained (Fig. 9.42c). Between the force applications on the molars of both sides exist also stresses. Their location corresponds to the palatal plate. If a nasal septum is introduced, the stresses show a direction from occipital-top to rostral-bottom, as does the bony vomer. The stresses in this humanlike model reach their highest values in the region of the spina nasalis (Fig. 9.42d), a structure characteristic of humans but not of other primates (see below). The lateral walls of the nasal cavity – or medial walls of the sinus maxillaris – are also stressed. Between the nose and the orbits, the stress flows are bundled (Fig. 9.42g), and stresses reach high values. This becomes evident, if the model consists of a thin, rather than a thick, plate (Fig. 9.42e). A means to distribute stresses in this area over a large cross-sectional area is broadening the lateral parts of the nasal roof and to approach it to the sagittal plane: that means elevating the nasal roof above the level of the midface (Fig. 9.42f). This increasing the amount of load-carrying material, bone, by elevation of the nasal roof reduces the stresses (force/area) on both sides and at the anterior margin of the nasal cavity. This explains the prominence of the human nose and leads to a concentration of stress in the middle of the forehead. A close-up view on several non-human primates shows that this arrangement of the nasal roof for keeping low the stresses is in fact not a speciality of humans but occurs in the same form in several low-faced primates, like guenons and gibbons. In these forms, however, the trait is much less obvious, because the prognathic upper jaw is more prominent and camouflages the prominence of the nose.

The above-described, very simple model can be made more similar to a human midface by giving it the form of a *semi-cylinder* with a vertical axis (Witzel and Preuschoft 2002). This model provides an ample volume as "Bauraum" for the skull parts between spina nasalis and the pterygoid processes in sagittal direction and between the left to the right jugal arch in frontal direction (Fig. 9.43a). The "Bauraum" is compressed by bite forces acting from below upward, and the counterforce exerted by the muscles pulling the skull roof (m. temporalis) and the zygomatic arches (m. masseter) downward. The m. temporalis has its origin outside the margins of the model, but the exerted force is transmitted through the frontal, temporal and pterygoid bones into the upper and posterior parts of the Bauraum, so closing the circle of forces. The application of the reaction force from the skull roof at the forehead is visible on the upper side of the model and the force components acting against the temporal bone at its rear border. The orbits and nasal opening are given á priori. All tooth forces are assumed to be acting at the same time against the entire row of teeth. The relative sizes of these tooth forces depend on the geometry of the mandible (Wolff 1982, 1984). So we created a synchronous superposition of the naturally occurring stress patterns, evoked subsequently by biting and chewing. In reality, however, simultaneous biting on all teeth leads to low forces. Therefore, bite forces are always concentrated on one or two upper teeth, because muscle forces do not allow exerting the maximal bite forces on all teeth at the same time. To control the influence of unilateral biting, we applied in one of our FE models the realistic chewing force to few molar teeth of one side alone. No essential divergence from the stress pattern found in simultaneous loading could be identified.

The thin bony jugal arch is kept in its place by the fascia temporalis. This strong aponeurosis (pulling upward), and the masseter (pulling downward), combine to a resultant which presses the jugal arch medially (Witzel et al. 2004; see also Chap. 2, and below), a direction in which the slender and fragile-looking element is very strong. The pattern of stresses found in the loaded model coincided in principle with the ossified parts of the midface (=upper jaw in a wide sense, including the jugal

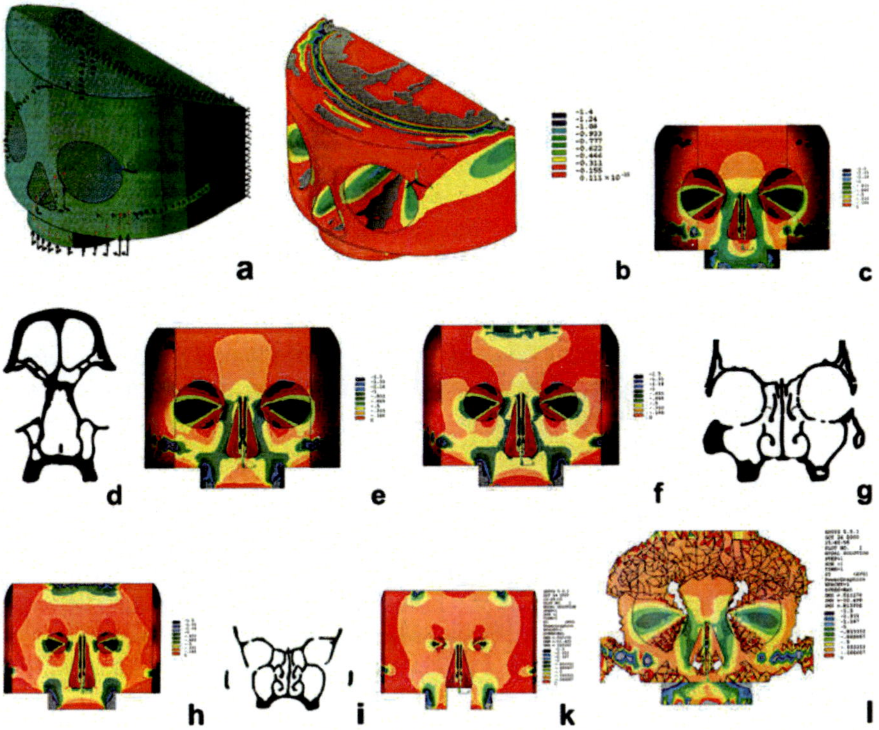

Fig. 9.43 Human midface as a mechanically created form. (**a**) The model used in Fig. 9.39 is enlarged to include the jugal arches and has, like a semi-cylinder, a bowed anterior margin. It is transparent to make visible the applications of bite forces and the insertions of muscles (tiny arrows). (**b**) Preliminary model without the forces acting at the jugal arch. The forces evoke stresses which are coded as colours. Note that compressive stresses have a negative sign. The most highly stressed regions are hidden deeply below the surface. (**c**) Frontal section at the height of the lower forehead through the loaded model. (**d**) A human skull cut by a saw at about the same height, but slightly more occipitally than in (**c**). Both are rostral to the lateral margins of the orbitae. The alveolar process, palate, the lateral walls of the nasal cavity and the upper margins of the orbitae, as well as the forehead, are highly stressed, and the bony structures coincide with major stress flows. (**e, f**) Sections through the model at the height of the maxilla and the zygomatic bone. (**g**) Saw cut through zygomatic bone (left of reader) or zygomatic arch (right). Again the alveolar process, palate and lateral wall of the nasal cavity as well as the nasal septum are under high stress and correspondingly ossified. Bone material (black) and stressed areas (yellow, green, blue) are obviously similar, the lateral wall of the orbits is nearly closed. The major difference from (**c** and **d**) is the lateral wall of the maxillary sinus, which connects the tooth row with the zygomatic bone. The maxillary sinus itself turns out to be a not-stressed region which does not contribute to the strength of the upper jaw. The orbitae are nearly completely surrounded by stresses – and by bone. (**h**) Frontal section through the loaded model at the height of the 3rd molar. Note the not stressed frontal sinus above the orbits (red) and the high stresses at the upper forehead. (**i**) Saw cut through a human skull at the height of the 2nd molar. (**k**) Frontal section through the model at the rear margin of the maxillary sinus/pterygoid process, upper margin behind the frontal bone. In (**l**) all not stressed parts are eliminated, and the slices placed one behind the other. So appears a loaded three-dimensional model of the human face, "reduced" to those parts which are recruited by stress flows as the summary of the illustrations (**c**–**k**). It resembles a human midface. (From Witzel and Preuschoft 2002)

arches and the walls of nasal capsula, septum nasi, orbitae, walls of the maxillary sinus and pterygoid region). The similarity to the midfacial skeleton becomes still more obvious, if in a second step a model is created, which is reduced to the stress-bearing parts of the primary model (Fig. 9.43l).

Our results support the old *three-pillar theory* of Toldt (1914). Bite forces on the incisors and canines are transmitted along both sides of the nasal opening and nasal cavity by premaxilla, maxillary process and os nasalis to the frontal bone. The Figs. 9.43c, e, f and h show a stress flow in the forehead, which begins in the middle at the "glabella", expands upward and laterally joins the highly stressed bearings at the anterior and lateral margins of the model and indicates reasons for the special configuration in modern human skulls. Bite forces applied to the premolars flow through the maxilla, zygomatic bone, anterior root of the jugal arch and the lateral margin of the orbitae to the braincase (as far as the circle of forces is not closed through the masseter!) and chewing forces on the molars flow along the occipital wall of the sinus maxillaris and the ala pterygoidea to the anterior skull base. The FES technique proves that, although the latter bite forces are quite high, the spreading of the stresses within the "shell" allows surprisingly thin walls. Due to its suspension by the fascia temporalis, the (in transverse, z-direction) thin jugal arch is not exposed to outstandingly high bending moments by the m. masseter in y-direction. Instead, it is loaded by the resultant in medial (z-) direction, derived from forces transmitted by fascia temporalis and created by the m. masseter.

The special stressing in the jugal arch in the example of *Macaca fascicularis* has been the object of a very elaborate study by Curtis et al. (2011). The authors confirm the interpretation of Witzel et al. (2004) about the role of the temporal fascia and explain details of the sutura zygomatico-temporalis from its function as a sliding joint.

The model in Fig. 9.43 depicts the situation in the short-jawed human skull. Its tooth row is arranged perpendicularly below the anterior braincase, so that the bite forces, acting more or less at right angles to the tooth rows, flow upwards, directly to the forehead. In the apes, however, as well as in early *Homo* and prehumans, the jaws project forward from the braincase (Fig. 9.41c) being "prognathic" (see below). Because of the frequently described kyphotic angle between clivus and anterior skull base (see also above, Demes' and Demes et al.'s investigation of the stresses in the skull base), the entire face assumes a position flexed ventrally in relation to the braincase (Fig. 9.41). In humans, a line perpendicular to the tooth row through the incisors passes near the forehead, and a line perpendicular to the tooth row through the posterior molars passes occlusal to the forehead. This situation differs from that in apes, where the first line passes far anterior to the forehead and the second close to the internal wall of the braincase. Hofer (1965) has emphasised a commonly underestimated divergence of *Pongo* from the African apes: the Asian apes lack the basal (or prebasal) kyphosis; thus their face with the anterior part of the tooth row is lifted against the clivus, and the basal angle becomes small. This "aerorhynchic" position of the tooth rows in *Pongo* (Figs. 9.35c; 9.44g, h) leads to the consequence that all biting and chewing forces are more or less parallel to and only somewhat in front of and frequently behind the most anterior part of the braincase, the forehead.

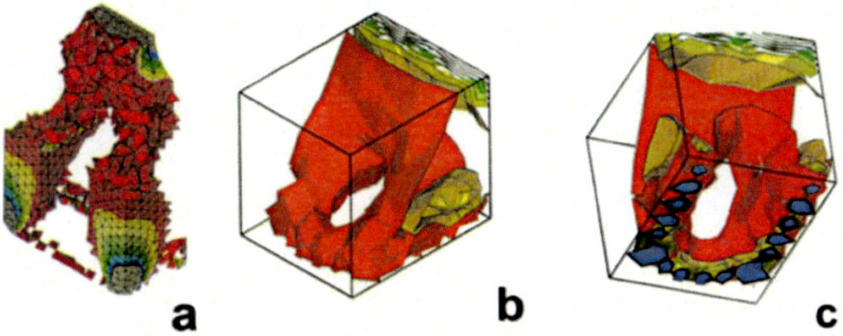

Fig. 9.44 (**a**) Nasal opening and dental arcade. Oblique view from left top on the "reduced" model of a prognathic upper jaw, which remains after elimination of all not stressed finite elements, which are visible as small polygons. Note the concave profile and the horizontal flow of stresses between both sides in the direction of the palate. (**b**) Another, more advanced, and smoothened FESS model of the snout, rostral to the orbitae, same view as (**a**). This shape is typical for prognathic forms. (**c**) The same "reduced" model, seen from below. Bite forces have been applied to the teeth which are marked in blue. All models consist of only those finite elements which contribute to sustaining stresses evoked by potential bite forces. Common characteristics are a nasal cavity, an open passage from nasal to oral cavity and a more or less concave profile. Note the high stresses at the upper margins (forehead) and at the zygomatic bones. (From Preuschoft and Witzel, 2004)

This is similar to the skull of *H. sapiens* as described above. The strength of a *platelike forehead* against being bent is remarkably high (Fig. 9.41a, e, h, 9.44). In the skulls of *Gorilla*, *Pan*, australopithecines and early *Homo* species, however, the bite forces exerted on the teeth pass, because of the basal kyphosis, rostrally to the braincase, exerting considerable torsional moments between the braincase and the midface (Fig. 9.41k, l). The profile of the torus supraorbitalis (=walls of the sinus frontalis) is suited to sustain these torsional moments. Prehumans and early humans have positions between the extremes represented by *Gorilla* and *Pongo*.

As mentioned above, the skulls of the *apes and of all ancient forerunners* of our species are called "prognathic", because their jaws protrude rostrally in front of the orbits. Their anterior profile is slightly concave. The similarity between our FESS results and the human skull encouraged us to continue the FES synthesis, under inclusion of the prognathic skulls of apes and fossil hominids. As a first step, alone the protruding snout has been modelled. The remarkable results are that the profile is slightly concave, as in nearly all hominoids (Fig. 9.44), and that a nasal cavity opens between the maxillary "pillars". The finite elements in this region are not recruited as stress-bearing parts. Very high stresses are visible below the surface of the forehead and at the anterior root of the jugal arch. The lack of a palate in the more advanced model (Fig. 9.44b, c) and the open passage between the nasal and oral cavity seem to be due to our neglecting the components of the bite forces in the occlusal plane.

This having in mind, a new model was created, which allows the application of bite forces in front of the braincase (Fig. 9.46). The latter were given á priori, as well as the nasal cavity – which corresponds to a space not recruited for sustaining of stresses (see above). The unstressed region in fact is higher than the cavity left open

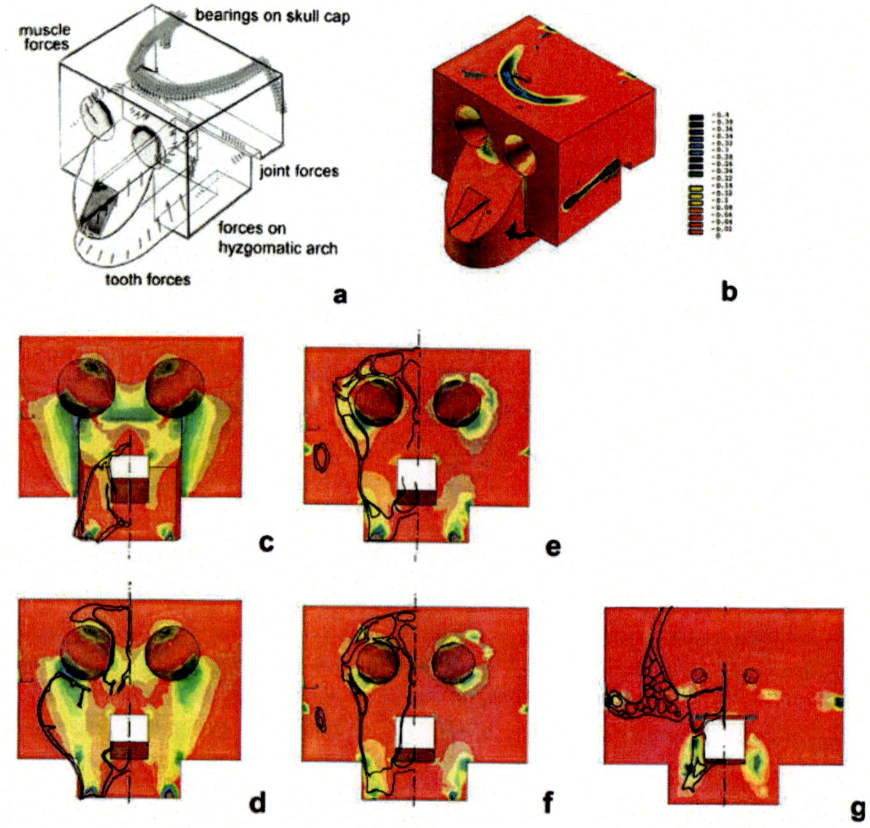

Fig. 9.45 **(a)** The "Bauraum" provided for an unspecific hominoid skull with the forces acting on it (arrows). Aside from its prognathic jaw, the model includes the skull base and the mandibular joints. Bite forces from below are opposed by bearings at the forehead and at the sides. Orbits and nasal aperture are given (although the latter is not necessary, because it is an unstressed region; see above). **(b)** External aspect of the loaded model. **(c, d, e)** Sections through the loaded model. **(c)** Behind the canines; at the anterior premolars. **(d)** At first molar. **(e)** At 2nd molar, both cutting through the torus supraorbitalis, **(f)** between the 2nd and 3rd molar. **(g)** At the occipital end of the tooth row, through the pterygoid flanges. On the left side of each image, the outline of a CT section through the skull of a (subadult male) gorilla is superimposed. The stress flows are essentially similar to the distribution of bone material. (From Preuschoft and Witzel 2004)

in the model (Fig. 9.44c). Near the median plane of the midface, a high, narrow region remains unstressed, no matter which upper tooth is loaded. This becomes most obvious, if bite forces are applied to only one side. As noticed above, the maxillary sinus, lateral to the nasal cavity and below the orbits (see Fig. 9.45c, e, f), is also free of stress. Just to make the model more similar to a real skull, we extended the available Bauraum in occipital directions to include the mandibular joint (TMJ). The bite forces were applied not simply at right angles to the occlusal plane but slightly inclined to take into account the transverse and longitudinal components of

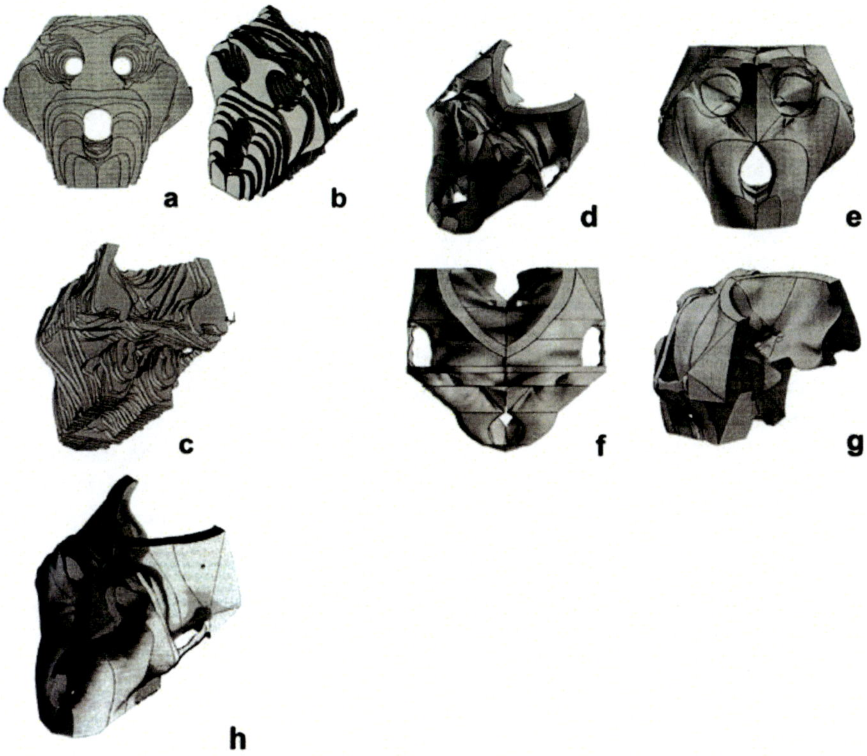

Fig. 9.46 (**a–c**) Slices cut through the "reduced" model of Fig. 9.45, consisting of those finite elements which carry stresses above an arbitrarily chosen value, while all not stressed finite elements have been eliminated. After aligning the sections and smoothing (**d–h**), a three-dimensional incarnation of the stress flows resembles closely the gorilla skull which has been used for defining the exact positions of the external forces. (**a**) Anterior view, (**b**) seen from left top, (**c**) viewed from left caudal, (**d**) viewed from left top, (**e**) anterior view, (**f**) top view, (**g**) viewed from left behind, (**h**) from top left. (Modified from Preuschoft and Witzel 2004)

the bite forces. The degree of inclination was derived from the direction of the tooth roots in the gorilla skull.

For comparing the stress-bearing parts of the model with the real skull, we cut the skull of a subadult, male gorilla into thick slices. The cuts through the skull are copied on the left side of the sections through the model. As can be seen from Fig. 9.45, there is a remarkable coincidence between the distribution of bone material and the stress flows. The coincidence becomes even more pronounced, if the bite forcesare applied to only one side. Stress flows end with the last molar, as does the posterior wall of the maxillary sinus, and the pterygoid processes, which are the most occipital braces of the upper jaw (Fig. 9.45g).

After elimination of the not stressed finite elements, a new, "reduced" model was created, which consisted alone of the stress-bearing parts (Fig. 9.46a). This was exposed to the same loads as before and again the unstressed FEs were eliminated.

After this first reiteration, the remaining model was cut into thick slices, and these slices were aligned along a longitudinal axis (Fig. 9.46a, b). After passage through a smoothing programme, the reduced model had the shape illustrated in Fig. 9.46d-h. The similarity of this model with the native skull of a gorilla, which had been used for locating the bite forces in relation to the braincase (Fig. 9.41c), is evident. In addition to its general similarity, it includes surprising details, like the blown-up anterior root of the jugal arch, the supraorbital torus, vault of the palate, separation of the pterygoid processes, joint surface in the temporal, skull base, mass of the os petrosum and anterior cerebral groove (Fig. 9.46b, c).

Among excavated or at fossil sites collected human remains, the fragments of the skull cap (roof and lateral walls, occiput) are the most common, after teeth and fragments of the mandible. This is probably due to their structure, which resists mechanical as well as chemical influences. While the latter seems to be connected with the very dense bone material, the former is due to the sandwich structure of inner and outer tabula, connected by a layer of spongy bone. The remarkable thickness (and strength) of the *hominine skull cap* is the more surprising, since the ape-like and *Australopithecus*-like skulls, from which a much stronger musculature originates, are clearly less thick and do not show the marked sandwich structure (Meyer, 1990). This seems at a first glance to contradict the hypothesis that skeletal parts are shaped by mechanical influences and that the highest among the every day forces determine shape. The undoubtedly high bite forces obviously are sustained by surprisingly thin lamellae of the bone, thanks to the shell characteristics of the brain capsule. The thick-walled brain capsule of hominines seems to be in conflict with this notion.

In an attempt to solve this problem (Witzel 1994, 2011; Witzel and Hofmann 1993; short version in Witzel et al. (2011)) have developed the following line of thinking: the voluminous brain has the same specific weight as liquor cerebrospinalis, in which it is suspended. Both are subject to accelerations as soon as the head is moved. At the inner surface of the braincase, where the entire mass of the brain is in contact with the walls, these accelerations can lead to considerable forces. Falx cerebri and tentorium cerebri subdivide the mass in the interior and therefore are tensed by the accelerated masses. The tensile forces are transmitted from the mentioned structures to the connective tissue of the dura mater and finally the bony walls of the braincase. At the points where connective tissue is anchored into the bone, only those partial forces are applied which are caused by parts of the mass. Especially for the falx cerebri, the directions of stresses have been calculated by Witzel, and these match the visible directions of fibre bundles.

The forces evoked by acceleration of the head grow with the size of brain and liquor mass. For the sake of simplicity, the braincase can be assumed to be a sphere. Because

$$\text{Force} = \text{mass} * \text{acceleration}, \tag{9.12}$$

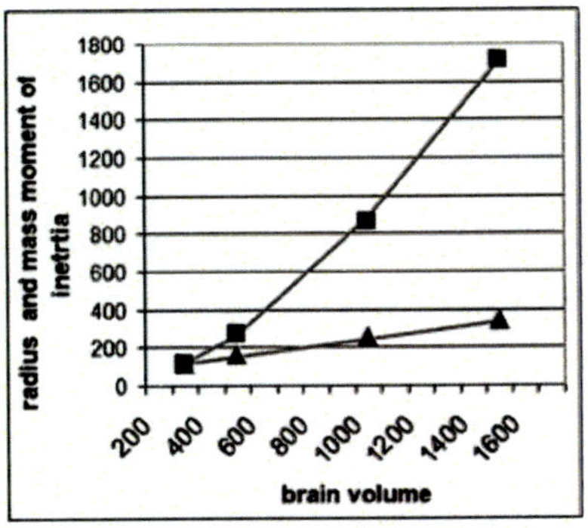

Fig. 9.47 With increasing volume of the braincase (on *X*-axis), the diameters (triangles on *Y*-axis) grow linear; the mass moments of inertia of mass units (squares on *Y*-axis), however, grow more rapidly, following a parabolic function. See text

the force evoked by the same acceleration inside the braincase of a human with 1500 ccm brain volume is five to three times larger than inside the braincase of an ape with a volume of 300–500 ccm. In addition, because growing brain mass implies an increase of volume, the lever arms (r) of masses grow with the cube root of volume. The latter is

$$\text{Volume} = 4\pi/3 * r^3, \quad (9.13)$$

and the radius r

$$r = 3.\sqrt{4\pi/3 * \text{volume}} \quad (9.14)$$

Mass moment of inertia (J) of brain and liquor depends on the square of the radius (r^2). From Fig. 9.47 can be seen that the mass moment of inertia, which must be overcome to move the head in a human with a brain volume of 1500 ccm, is six times larger than in an ape with a brain volume of 300 ccm. The large human brain exerts six times the force exerted by the brain of an ape. This larger force acts on a braincase with a longer span – which therefore must be considerably stronger. All these factors have been taken into consideration for a FES synthesis of a typical Neanderthalian skull (Figs. 9.48 and 9.49). The forces derived from mass inertia and transmitted through the membranous structures, falx cerebri and tentorium cerebelli dictate the overall shape. Its low vault and length depend upon the preconditions given á priori in the dimensions of the skull base and the muscle insertion at the occiput. Not completely in line with the realistic skull is the region of the browridges. The model does not explain the divergent shapes of the braincase in Neanderthalians and anatomically modern humans!

The face of the typical Neanderthalian in Figs. 9.48 and 9.49 has been created in a FESS like that of a recent human in Figs. 9.43: the laterally convex contour of the upper jaw below the anterior root of the jugal arch is, in contrast to modern humans, similar to the gorilla skull (see above) (Figs. 9.41c, 9.46b, e, h).

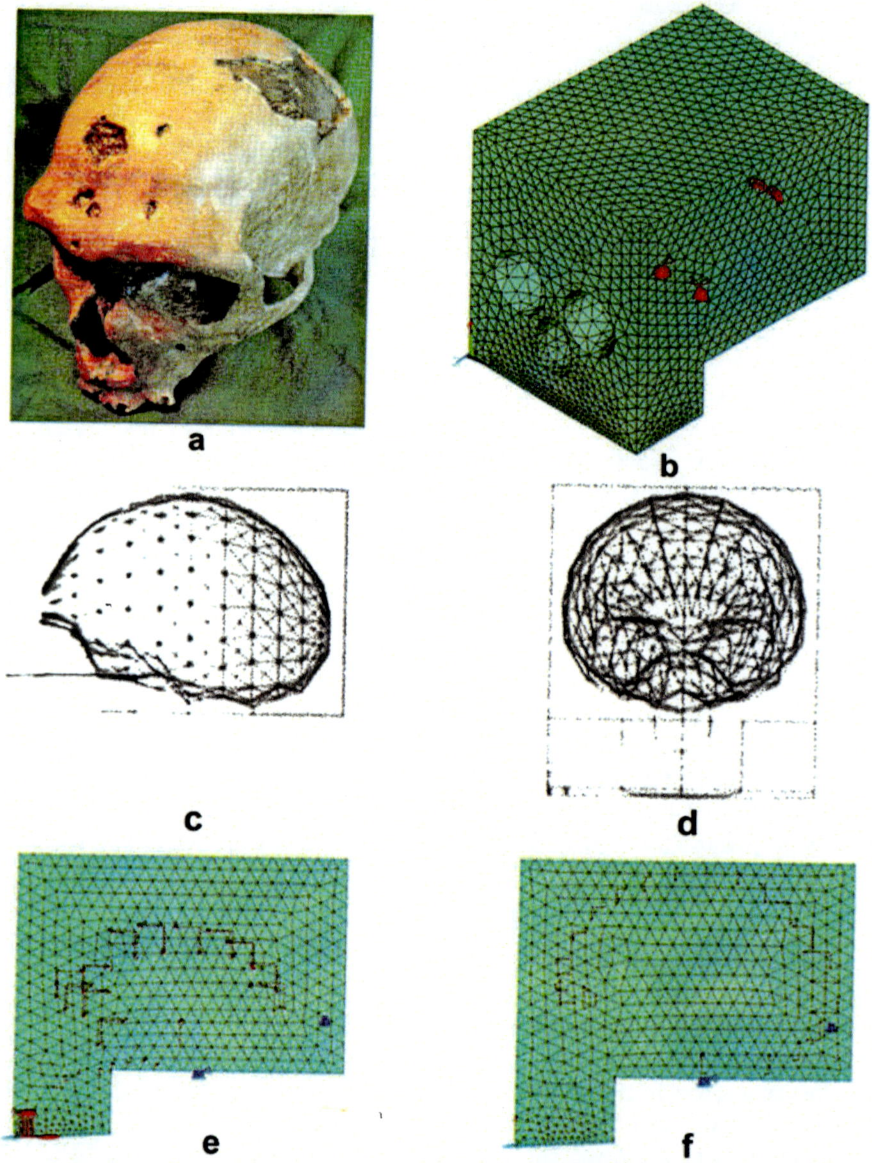

Fig. 9.48 FES synthesis of a neanderthal skull, step 1. (a) The skull of the famous neanderthalian from La Chapelle aux Saints. (b) "Bauraum" for a FES synthesis of a skull with similar proportions, position of the tooth row to the braincase, position of the occipital condyles, and muscles. In contrast to those treated above, this model includes the braincase (c, d). The red dots in (b) and the red arrows in (e) as well as the blue dots in (e, f) indicate the forces applied to the temporal bone and occiput (tension, produced by mm. temporalis, masseter and pterygoideus medialis) as well as the occipital condyles (compression by the joint resultant) and the anterior teeth. In the interior of the skull vault, tensile forces exerted by the falx cerebri and the tentorium cerebelli are transmitted through the dura mater (f). These forces are caused by accelerations of the brain mass. (From Witzel et al. 2011)

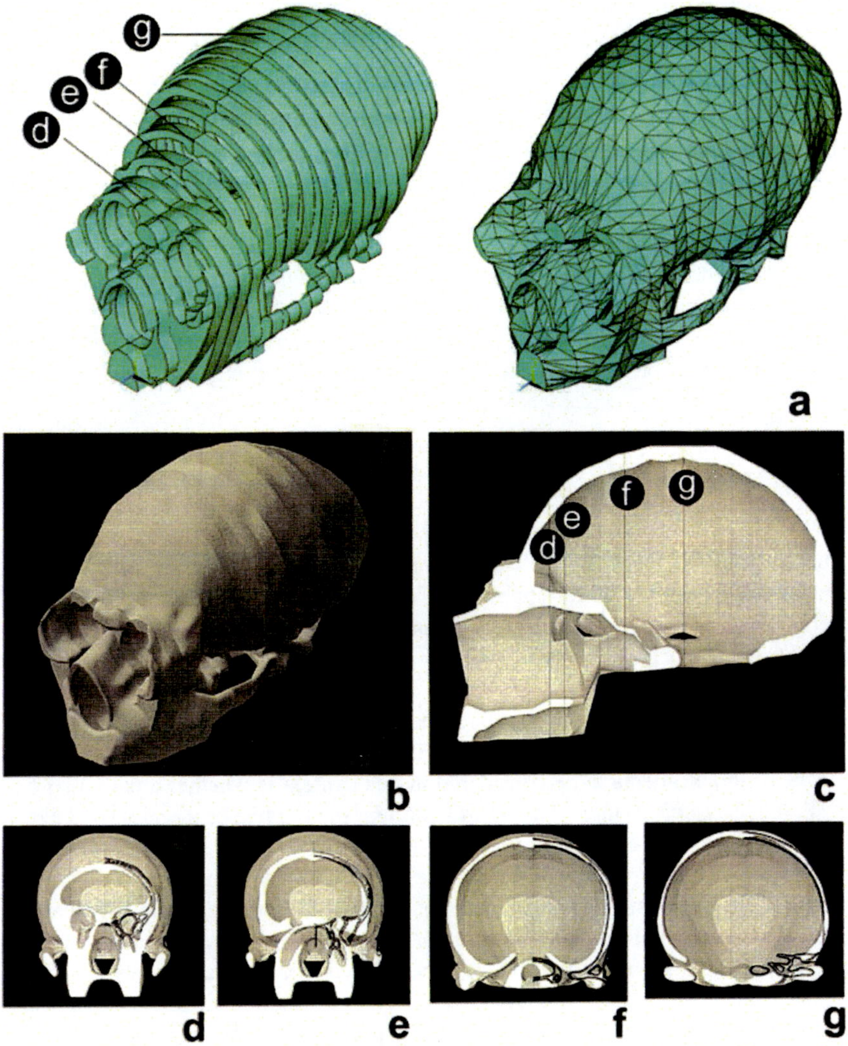

Fig. 9.49 FES synthesis of a neanderthalian skull, step 2. (**a, left**) Sections through the reduced model of a neanderthalian skull lined up from rostral to occipital. Only those finite elements are considered which carry stresses above an arbitrarily chosen threshold. The letters within the black cirdles indicate the cross sections shown below. (**a, right**) The reduced model with meshwork for the 2nd re-iteration of the synthesis. (**b**) Smoothed model, seen from right top. (**c**) Section along the median sagittal plane through the reduced model. The letters **d-g** indicate, like in (a) left, the cross sections **d, e, f,** and **g** in the row on bottom, showing frontal sections through the same reduced model, superimposed on the left side (of the model) with CT images of a skull. Note the similarity of both. (From Witzel et al. 2011)

9.8 Can Ultimate Reasons Be Pinned Down for "Hominisation" of Body Shape and Skull Configuration?

The above-described results, especially those obtained by the FESS, explain the immediate, "proximate" reasons for particular shape characteristics, first those related to bipedality and second those related to the typical "human" shape of the head. In the current discussion, aspects for both parts are densely interwoven, but I will make an attempt to keep them separate. Fossil finds sometimes include post-cranial material as well as remains of the cranium, and both do not always indicate the same direction. Mosaic evolution is to be expected, and therefore separate explanations must be found for each body part.

9.8.1 Acquisition of Bipedality

Although we are, regarding the locomotor modes of the fossils, confined to drawing conclusions, these are quite reliable, because they are based alone on mechanical strength of parts and physical facts. This does not necessarily yield the desired information about the biological role of a given morphology (as, for instance, the abducted big toe; see above).

The shapes of the skull, as discussed below in a special subchapter, do not give any help, in view of the mosaic nature of quite divergent lifestyles.

One of the problems to be considered in this context is whether a particular trait exists in australopithecines or alone in Hominines and whether the proposed explanation holds true for both subfamilies.

An often raised question is WHY the precursors of humans ultimately have "decided" to walk on two, rather than four limbs – like the monkeys (macaques, patas, baboons) which switched from more arboreal to a more terrestrial lifestyle. In other words, the "biological role" of bipedality is yet not definitively identified. Preuschoft (2004) worked out that bipedality is for a "higher" primate *by no means a difficult task to perform*, in view of the group-specific hindlimb dominance, length of the limbs in relation to trunk length, thoracic shape and rib curvature, position of the CoM and the upright position of the body stem in space, with head up and arms moving overhead and all around. Aside from the "division of labour" between forelimbs and hindlimbs, carrying of body weight is in many, if not all primates, the major task of the hindlimbs – in contrast to the situation in "real quadrupeds" (special data provided by Kimura, 2010). The specialisations of the genus *Homo* in comparison to australopithecines can completely be explained as optimisations of body shape to bipedal, energy-saving locomotion over long distances. Load-carrying abilities were also enhanced from *Australopithecus* to *Homo* (Crompton et al. 2003; Wang and Crompton 2004). The former, however, were also bipedalists. Still the question is open, whether there are ultimate reasons which pushed the earliest representatives of the evolutionary line, the australopithecines, into the direction of bipedality, instead of falling back on the quadrupedal behaviour of the monkeys (as, e.g. baboons, djelada, patas, vervets) and African apes. All attempts in this direction necessarily imply leaving the firm ground of mechanics and taking refuge to more *speculative* approaches. This, however, is absolutely necessary, because any novel solution starts off from a theoretically conceived idea, which is later weighed in its relation to facts (Kant, Popper 1968). Speculation about the reasons for the existence of characters or of widely spread preferences in our ancestry is important and should never be restricted or suppressed as "not scientific"!

A considerable array of ideas are around and will be looked through in the following: There is agreement of opinions concerning the increasingly *cooler and dryer climate* during the Pliocene, which has favoured the expansion of *open grassland* and increased the distances between trees or groups of trees, often near rivers or lakeshores. In these patches of trees or forests, the precursors had found their food resources and refuge. Retention of climbing abilities, as in australopithecines (Larson, 2012), makes much sense and obviously implies additional advantages. Of course, the forerunners of our species were endangered as prey of the contemporaneous raptors. Passing through the growing grassy interspaces – or even exploitation of its resources – became an increasingly urgent problem. It became necessary to cover the distances without expending much energy and sufficiently fast. This evolutionary tendency is doubtlessly the cause for hominines in the narrow sense ("*savannah theory*"). Finding or capturing prey in the open grassland or scavenging seems a reasonable idea for improving the diet. This, however, meant competition with carnivores – aside from the danger of themselves falling prey. The ability to use stones, at a later stage spears thrown at the enemy, would offer a means to defend themselves and at the same time to capture swift prey. Indeed, several characteristics of the human hand imply clear advantages of their owners. General body shape is in accordance with this idea, but whether these advantages can be attributed to prehumans or alone to *Homo*, is at present a matter of taste.

Present-day humans are extremely adaptable in their choosing an *environment*. It should be noted that they avoid dense rain forest – or destroy it, wherever possible. In contrast, landscapes formed by humans, and not dictated by the technical requirements of agriculture, always are characterised by open grassland, shadow-providing isolated trees or groups of bushes and trees, rocks and standing or flowing water. Such "gardens", parks" and "paradises" have been produced in many cultures, from Europe to the Far East, while existing without human influence in East and South Africa. That such landscapes are considered as "beautiful" or "attractive" could be an expression of universal desires of human beings. Humans like climbing on rocks or trees without fear, that is, they are not afraid of the third dimension but make use of it with pleasure.[17] The great affinity of humans to shallow water has been emphasised by Niemitz (2000, 2002b, 2004, 2010, 2011). Superficial support seems to be given to his opinion by the taphonomic condition in Nariokotome (imbedded in shallow water) and by the fact that the preserved foot tracks are in

[17]To evaluate this statement, human behaviour should be compared with the very unhappy state of dogs or most hooved cursorial animals (horses!) on grids or scales without closed back walls.

moist underground (but where else?). Niemitz proposes well-founded ideas for the biological role of bipedality and skull shape. Most important is for Niemitz the collecting of aquatic invertebrates as a source of animal proteins for the inflating brain. If searching food in shallow water is an essential factor, assuming and maintaining bipedal posture keep most of the body dry and not exposed to heat loss by becoming wet and the hands free for gathering and capturing food animals.

Niemitz's publications also provide an argument *why* bipedality was chosen instead of quadrupedalism. Such an argument is missing in other lines of thinking (see below) but necessary (Preuschoft 2004).

Wheeler (1984, 1985, also 1991) proposed a different explanation why bipedality has been acquired. He also starts off from living in open, sun-blasted grassland, where one of the most urgent problems is getting rid of superfluous body heat. In the absence of a reflecting haircoat, and of a brain-cooling blood circulation from the nasal cavity to the brain, upright body posture for one minimises the area of body surface which is exposed to sunshine during the hottest hours of the day. Thermoregulation, that is, cooling, is further enhanced by reduction of the hair cover or pelage, and the remaining sparse hair tracts are arranged to favour and guide an air stream upward along the body to improve thermoregulation by cutaneous evaporation. The hair tracts, especially of the lower body half, facilitate cooling by opposing the direction of natural convection rising up the body of a bipedal hominid. The human hair-tracts are suited to maximise airflow over the wetter skin surface. They are not in agreement with the superficial currents in swimming and also deviate from the hair tracts in apes. Whether this explanation can be applied for hominines or australopithecines remains open.

9.8.2 Why Have Humans Acquired Their "Entaxonic" Hand Proportions?

The question must be answered, why hominoidae have developed their long thumb and the versatile second finger, proportions of the hand, which deviate somewhat (Jouffroy et al. 1993) from the forms realised in other hominoids. Many, if not most authors, try to solve the problem by simply connecting human hand shape and the "precision grip" said to be characteristic for humans. These attempts neglect two facts: first that "precision gripping" is not really a specific of humans (as has been discussed above) and second that morphological specialisations must inevitably be able to sustain the greatest among the usually occurring stresses, if the danger of a bone fracture because of overstressing is to be excluded. Reasons for "adaptations" must be sought among the regularly occurring high loads. The majority of precision grips are performed against low resistance so that a morphological specialisation cannot be expected. It may be argued that in the absence of loads on the frontlimbs in human locomotion, the threshold for morphological specialisations is lowered, so that perhaps the production of stone tools leaves traces in morphology. Actually,

Susman has described traits of the hand bones from which he connects with tool use. Unfortunately, nobody has yet attempted a detailed analysis, on which load cases such an specialisation can be based. This is the more surprising, because work sciences should have opened access to such approaches.

The approach to hand shape, as presented above, lends support to one of the *existing hypotheses about the origin of modern humans of the genus Homo.* Always again, researchers are worried about the question how a relatively small being without special weapons and without the ability to escape predators by swift running or rapid and acrobatic climbing on trees could survive in a savannah habitat which is also the home of plenty of large and strong predators. The question becomes more acute, if one imagines that this being had defended prey against the predators or taken it over from them. A very efficient means against predators are blows with clubs or still better: throws with one of the ubiquitous stones – not to speak of spears. These tools keep the thrower outside the reaching distance of the predator, which perhaps does not even realise from where the sudden pain comes. The literature to this point (Calvin 1983; Fifer 1987; also Kirschmann 1999) has been revised by Dunsworth et al. (2003). It is unnecessary to emphasise again that aimed throwing requires high quality of neuromuscular control, which can be provided only by a well-developed brain. If the ability to well-aimed throwing in prehumans and early humans was present, a real chance existed in controversies with large predators – as well as obtaining swift prey like many medium-sized or small animals, which could by no other means be reached. All this seem to be very good reasons to develop morphological specialisations for throwing, and a considerable number of such detailed specialisations (which discriminate humans from prehumans and/or apes) are proposed above. The use made of throwing may have occupied the anterior limb to the extent that it loses any importance for locomotion.

Several attempts have been made to derive from *human quadrupedal locomotion,* (especially *"creeping"* in young children) information about the origin of bipedality (e.g. Burnside 1927; Sparrow 1987; Niemitz 2002a, b). The results show an impressive inefficiency and variability of human quadrupedalism. The support phases are long, frequency is low. The footfall sequences are sometimes lateral, sometimes diagonal. The repeated referring to "knuckle-walking" contributes to confusion. As a whole, analysing quadrupedal locomotion in humans has not provided much insight into the reasons why bipedality has been acquired. "Bipedal posture seems to have been assumed very soon after abandoning some sort of suspensory posturing" (Tuttle 1969b; Sparrow 1987). This would be in line with recent information such as Moya Sola on *Oreopithecus* or Thorpe and Crompton (2004) on orang-utans.

9.8.3 Skull Shape and Dentition

The ultimate reasons for the evolutionary changes in *the form of the skull* are usually searched in the inflation of the telencephalon, and a change of *feeding habits* (see above), both in combination with a loss of importance of canines as a weapon. These explanations of the most marked traits (large or very large braincase, reduced prognathism, change of tooth shapes) on the level of "biological role" may well be correct. The ability for stereoscopic vision, which determines the position of the eyes, seems plausible (for all higher primates, not alone for hominines) on the same level. However, the details of skull shapes (nasal cavity, maxillary sinus and anterior cranial groove) are in surprising agreement with the results of our FES synthesis; they are obviously the consequence of mechanical necessities. Seemingly minor variations lead to changes in the force flow, which manifest themselves in marked morphological differences (prominent chin, browridge/torus supraorbitalis, inclination of forehead, fossa canina) which have been the subject of far-reaching, but poorly founded conclusions in the past.

Modern humans are *omnivorous*. Aside from a multitude of plants, where fruit, seeds, roots and tubers play a prominent role, animal matter is commonly consumed, be it insects, crustaceans, mussels or fish, birds and mammals. This seems to fit with the shapes of teeth. The *dentitions of australopithecines*, however, are characterised by reduced front teeth, large and "molarised" premolars and often impressively large molars. The shapes of the cones fit better to hard-object feeding (hard shells or seeds) than to grinding grasses and sedges (which has been proposed on the basis of C-isotopes). They seem to exhibit the "human" traits in exaggerated form. The *nutrition of our ancestors* is a favourite object of fierce debate, in which romantic imaginations (our ancestors were heroic hunters of big game animals) or ethically motivated beliefs (our ancestors lived in harmony with the surrounding nature as gatherers of plant material) are receiving more emphasis than appropriate for a discipline aiming at understanding causation. The role attributed to hunting rapidly moving and dangerous mammals depends largely on personal preferences of the author. Intermingled is this load of beliefs with prejudices about sex-specific roles (males hunted, while females at home took care of their offspring, perhaps gathering ubiquitous and non-dangerous food objects). In a rather arbitrary way, analogies are selected from the highly diversified social life of monkeys, apes or "primitive" human populations, which per se are relics in remote environments.

Well-founded is the widespread notion that the *large brains* of hominines (see above) could have developed only by providing animal proteins and essential fatty acids. This requirement can be fulfilled by heroic big game hunting but also by scavenging. Both imply conflicts and competition with the large predators: cats, hyaena, bears and wolves. The physiological necessities, however, can also be fulfilled by the not so hero-like gathering of small invertebrates (insects, bivalves, crustaceans and possibly amphibians), as has been proposed by Niemitz (2002a, b, 2004, 2010). His ideas found a strong confirmation by the independently obtained results of Shabel (2005), who has observed the same distribution of C14-isotopes in the dentitions of South African *Paranthropus*, clawless otter (*Aonyx*) and marsh mongoose (*Atilax*), both living on aquatic invertebrates. During the last years, however, a considerable number of new studies on stable carbon isotopes (e.g. Cerling et al. 2011; Lee-Thorp et al. 2012; Van der Merwe et al. 2003) have been published, which are summarised by Sponheimer et al. 2013: "These studies suggested that australopiths ate ^{13}C-enriched plants, such as C_4 grasses or sedges, or consumed animals that had eaten these plants"... "such as grass seeds or sedge underground storage organs, were added to an otherwise C_3 diet" ..."... we have increasing confidence in the pattern of hominin dietary evolution from a carbon isotopic perspective. Before 4 Ma, most hominins avoided the abundant (grass- dominated) C_4 (or CAM) resources that we know were available to them. By about 3.5 Ma, C_4 (or CAM) plants (that is grass seeds – or spades and stalks – and/or sedge underground storage organs) had become, in addition to a more fruit-orientated diet, a frequent component of an expanded hominin dietary repertoire. After this time, the overall range of African hominin isotope space changed little, although some taxa (e.g. the extremely robust *P. boisei*) followed isotopically narrow paths, whereas others (e.g., *Homo* and *P. robustus*) remained in mixed C_3/C_4 space". The conclusions of the authors are drawn under consideration of tooth wear and dentognathic morphology. This information does not contradict Shabel's results. They can be understood as confirmation of the more general ideas about the environment, in which hominisation took place.

An original idea has been proposed by Wrangham et al. (1999): the early control of fire has opened access for prehumans to otherwise not digestible food resources by cooking tubers. In a cooked state, these materials do not require an extremely strong chewing apparatus and no devices for tearing objects apart by highly developed front teeth. I feel tempted to relate the dental characteristics of prehumans to their consuming mainly vegetable matter and hard-shelled invertebrates rather than "cooked" material and meat.

9.9 Essential Results

Starting off from typical and widely spread primate characteristics, the postcranial locomotor apparatus of recent humans is strictly specialised to fulfilling the static conditions of bipedal posture, and to permit endurant bipedal locomotion. This holds true for the rigid feet, long and extended zeugo- and stylopodium, pelvis, upright and elongate trunk – even the forelimbs and the head are adjusted to walking, or running. While some traits can be explained by statics, some of the most puzzling characteristics (length of hindlimbs, length and profiles of trunk) turn out to be advantageous only if kinetics are considered.

Human forelimbs are lightly built and rather weak. They are suited to saving energy in walking, and to act as prehensile organs for first contacts with items of the environment. Their proportions, especially the emphasis on the most radial digital rays I, II and II differ from the arrangements in other primates. Their strength, differentiation and length seem to be connected with the fact, that these elements are under permanent optical control, and therefore suited for movements required in working, using tools and throwing.

The fossil forerunners of humans show a combination of several not so highly developed specialisations to their documented bipedality and traits necessary for efficient climbing. The australopithecine hindlimbs, pelvis and trunk yield the same information concerning their locomotor habits: suited for bipedality, but not so much specialised as in Homo, while many traits exist which make “climbing” easier, safer, and more efficient than in our species. All foot tracks younger than 2 MY, that is since the appearance of the genus Homo, show the same characteristics as those imprinted by modern humans. This is in full agreement with documented foot morphology.

The best arguments for climbing propensities of pre-humans are the powerful anterior and their short, relatively weak posterior extremities. Australopithecines were probably better climbers (Fig. 9.1), but less endurant and less fast bipedalists than hominines. With Klein (1999, on pp. 210 ff.) and Tattersall (“While we were not alone”) we should consider the existence of different locomotor modes in early phases of human evolution as possible.

The head is the envelop of a very large brain, balanced on top of the vertebral column. While these characters can be derived from the configuration in all primates, the arrangement of the anterior teeth is different: chisel-shaped “incisors” in a strict sense, the canines not prominent. The entire jaw apparatus and the whole midface is bent sharply downward or kinked ventrally, a fact usually identified and named “basal kyphosis”. This falls in line with the development of the brain. Details of human skull shape have been traced down to the distribution of mechanical stresses identified by means of the FES-method, and possible arguments for biological roles of human characteristics of skull shape are discussed.

Chapter 10
Open Questions

Keywords Mutations · Selection · Causal Morphogenesis · Wolff's Law · Antlers · Bivalves · Sea Urchins · Tails

In spite of their fragmentary nature, the information contained in the preceding Chapters 1–9 permit some generally valid conclusions. These are detailed in the concluding "Essentials" at the end of the chapters and must not be repeated here. A considerable number of problems, however, are not solved, or the solutions proposed here are not really satisfying. These are discussed in the following sections.

10.1 Is Shape Inherited or Ad Hoc Developed as a Response of Tissue to Stress?

A precondition of all which has been developed in the foregoing chapters is what has been postulated often under the concept "**mosaic evolution**": Evolutionary changes do not necessarily comprise the whole body, but act on isolated, defined traits[1]. It may well be that one condition influences several morphological traits at the same time (as in optimising the early human body to bipedal walking, see Chap. 9). On the other hand, seemingly contradictory optimisations may well show up in the same organism (as optimisations for bipedality as well as climbing in trees in orangutans, and *Oreopithecus*, Chap. 8, or *Australopithecus,* Chap. 9).

The preceding text leaves open – as could be expected – a number of questions. The most critical of these questions seems to be how the optimised shapes of skeletal elements and connective tissue structures are acquired. There exist obviously two options: First, either the morphological "traits" are inherited in all details or, second, morphological traits may appear as direct response of the tissues to locally acting mechanical stresses.

[1] Stern & Susman (1991) have opposed a "total morphological pattern" to a "magic trait".

H. Preuschoft, *Understanding Body Shapes of Animals*,
https://doi.org/10.1007/978-3-030-27668-3_10

Ad 1. The best known and generally accepted way to the acquisition of (new) shapes is the interplay of **undirected mutations and subsequent selection**: All morphological traits are caused and controlled in detail by genes. Undirected at random occurring mutations may equip offspring with altered morphological traits. No doubt that evolution must start off from existing and, therefore, "functioning" organisms (as emphasised by the "Frankfurt school", e.g. Gutmann, or Grasshoff) and can only change already existing "traits". With these new characteristics, the organism can either occupy pre-existing "ecologic niches" or even create new ones. Only those morphological traits will persist, which fulfil clearly defined functions, necessary for the animal and implying advantages. Identification of these advantages, as attempted in the foregoing text, allows in some cases a quantification of the advantages. If this is possible, the "evolutionary pressure" acting in a special direction can be quantified as well. Examples are given in the text.

As noted in Chap. 1, most biologists will decidedly tend to prefer this option. A problem connected with this theory is the enormous number of genes necessary for isolated, often minimal, morphological or behavioural traits which must be expected.

Ad 2. The reaction of connective and supporting tissue to stress is widely known as "Wolff's law". Pauwels' (1965, 1980) investigations concerning "**causal morphogenesis**" have confirmed Wolff's law and documented the direct dependence of connective and supporting tissues (= bones and cartilage) upon exactly defined mechanical conditions. On this tissue reaction rest fracture healing and the success of accident surgery as well as orthopaedics (e.g. Amtmann 1968, Amtmann and Doden, 1981), and also the processes of normal growth (as the acquisition of the lumbar and sacral lordosis (Abitbol 1987, 1989). A changed function (e.g. more frequent and more extended walking, burrowing) leads immediately to an adequate change of morphology. This process is impressively copied by the FES-syntheses of the skull (Chaps. 2 and 9), which imply a direct dependence of skull shape upon mechanical conditions, if a framework of preconditions (size of brain, position of eyes, position of tooth rows) is given.

As noted in Chap. 1, medical people will give this second option more sympathy, because their daily experience comprises healing processes which doubtlessly follow the mechanical conditions. The possible (mechanical) advantages of traits and the (mechanical) conditions under which shapes may be developed are the focus of the present text.

A rule, saying that skeletal (and connective tissue) structures develop alone under the influence of mechanical stresses, however, is in opposition to several facts: The process of functional adaptation of skeletal shape to mechanical stress can only take place if structures to be stressed are already existing. In the experiments on **embryonic chicken** (Christ, H.-J. Jacob, M. Jacob, B. Brand-Saberi, e.g. 1993), the first "Anlage" of extremity bones and joints connecting them occurred without mechanical influence – while muscular activity later on became clearly responsible for the normal development. If muscle activity, which means mechanical loading, was excluded, the "anlagen" were resorbed. Very similar results were obtained in mammals. Furthermore, it depends on the existence of muscles. Their development is only in one way determined by stress: although their shapes and places within the

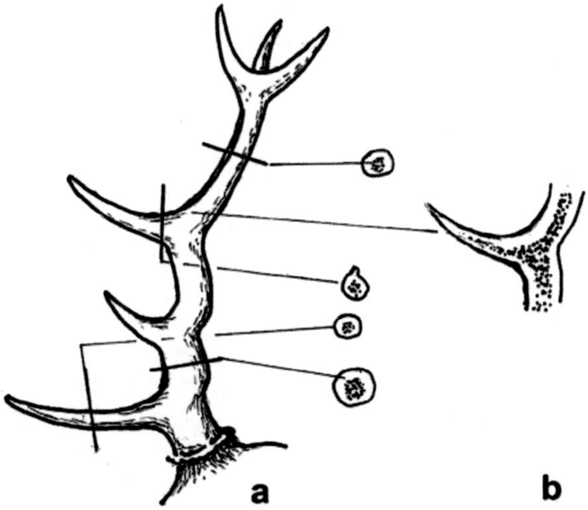

Fig. 10.1 European red deer. (**a**) left side antler, cut as indicated to show the arrangement of compact and the spongy bone substance. (**b**) For the middle spike (third from below), a longitudinal section is shown. Note the slightly broader compact bone along the concavity at the base of the spike. Bone structure in fallow deer is very similar. Redrawn after Rumbler, 1914

organism are invariable, they may become stronger by being heavily loaded and may become weaker if not stressed.

The classical arguments against causal morphogenesis are the **antlers of cervids**. These outgrows develop without any direct influence of mechanical stresses and have species-specific shapes, for example, in moose, red deer, European roe deer or the medium-sized American cervids (Ceacero 2016). They assume their function not before they are completely developed. Their internal structure seems reasonable from a mechanical viewpoint (Fig. 10.1a): Close to the surface, the stems and all branches have a sort of cortex, consisting of compact bone, while the interior is filled with rather dense spongy substance. This arrangement is resistant against compressive and tensile forces as well as against bending moments, and so allows sustaining high loads in any direction, applied to any place – as may be the case in fighting. In the concavity between deviating spikes diverging from stems, sometimes a reinforcement can be seen (Fig. 10.1b, after Rhumbler 1914). The "*bizarre structures*" (Gould 1974) of several dinosaurs (Padian and Horner 2010) seem to fall into the same category. Likewise, the **teeth** develop their shapes in complete absence of external forces in pockets below the surface. Not before they have assumed their nearly complete shape they are exposed to typical – and very high – loads. The **jaws** as well have assumed their basic shapes before being exposed to their physiological loads – at least in Eutheria. In newborn marsupials, however, a mechanical stressing takes place very early in extrauterine development (see Chap. 2). Their mechanical needs are completely fulfilled by the observed shapes.

Finally, causal morphogenesis cannot at all explain the functionally so important **lengths** of the segments or their proportions – even if some influence of stresses on longitudinal growth has been reported. After all, there does not exist any way except assuming an influence of internal (genetic) factors for the development of shapes.

10.2 Is Pauwel's (Wolff's) Law Also Valid in Invertebrates? The Case of Bivalves

The highly advantageous ability to heal fractures of skeletal elements and to close soft-part defects by producing scars of connective tissue rests on and is controlled by the reaction of bone and fibres to compressive or tensile stresses (causal morphogenesis). From this, we may expect a sort of responsiveness of tissues not alone in vertebrates, but as well among invertebrates. To check this expectation, we need to know first which loads are acting during everyday life on the hard substances, second whether these structures have "optimised" shapes, which allow sustaining these loads with a minimum of material, and finally, whether shapes change with changing loads. In my opinion, the following four approaches have particular importance for answering one or more of these questions.

Currey & Taylor (1974; see also Wainwright, Biggs, Currey & Gosline 1976) found in bivalves a considerable variety of shell structures. While all other bivalves lead a sedentary life, Pectenidae can actively swim by contracting their (single) adductor muscle and expelling a jet of water near the hinge (Alexander 1967). When closing the valves, clear-cut forces are exerted on them. Valves are made of the most resistant material ("nacre") which gives most strength in combination with lowest weight. This is in contrast to the shells of other, sessile forms, which possess thick shells with minor mechanical strength (extremes are oysters). This seems to indicate some degree of "adaptation" to the respective modes of life. On the other hand, the pectenid shells are either vaulted or flat, while being stressed the same way in both cases. The profile of the shells all around has nearly the same thickness (=diameters). Both characteristics are arguments against a functional "adaptation" of shells. Preuschoft, Fritz & Krämer (1975) investigated the stress patterns in the shells of *Pecten* with the aid of enlarged models covered by a brittle lacquer ("stresscoat technique"). If an object covered by this stresscoat, it will break the cracks orientated at right angles to the tensile forces. If the object is covered in a loaded state, and unloaded after drying, the crack will appear at right angles to the compressive stresses. The obtained results were verified at selected spots by strain gages. Instead of the pulling force of the adductor muscle on the inner side, a compressive force was applied to the outer side of the shell (Fig. 10.2a). The closing movement is exerted first against water resistance (distributed over the entire inner surface), finally concentrated against the margins of the opposite valve. Regardless of its being loaded against evenly distributed water resistance, or against the free margins of the shell, the highest stresses were always tensile, running like a girdle parallel to the free margin. Somewhat lower are the bending stresses radiating from the insertion of the adductor muscle and stretching the inner while compressing the outer side of each valve. The vaulted and the flat shells were subject to the same stress patterns. The existing shape characteristics do not at all correspond to the pattern of stress: Pectenids have shells with a wave-like profile, the waves of which radiate from the umbo and become higher and broader towards the periphery (Fig. 10.2a). This of course makes them sensible against radial rupturing under the influence of tensile forces. There is neither an enforcement of the shells close to the

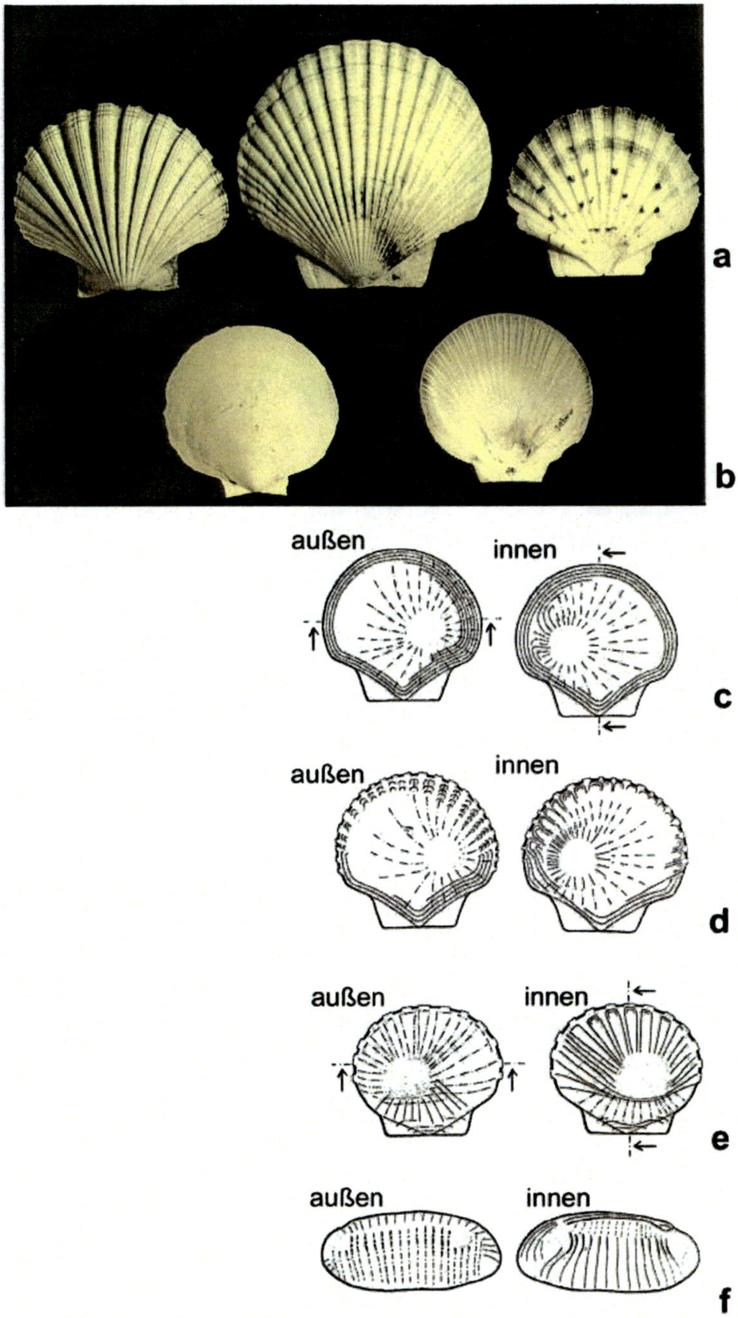

Fig. 10.2 Shells of bivalves. (**a**) crenulated shells of *Pecten maximus*. Two on left are vaulted, the right one is plane; (**b**) smooth shells of *Lentipecten*, left from outside, right from inside. (**c–f**) Stress flows in the valves. The objects are resting on their free margins, and the applied forces simulate the pull of the adductor muscle(s). Results of stresscoat experiments. Continuous lines: tensile stresses, interrupted lines: compressive stresses. On left side is shown the external, on right side the internal surface. (**c**) Smooth shell of *Lentipecten*, (**d**) the vaulted shell of *Pecten*, (**e**) the plane shell of *Pecten*, (**f**) shells of *Elliptio*, which represents the common form of mussels with two adductor muscles. Rearranged from Preuschoft et al. 1975

insertion of the adductor muscle, nor does the vaulting correspond in any way to the stresses. Instead, like in all bivalves, the profile has its centre in the umbo. No growth-dependent changes of shape can be observed, except a moderate increase of shell thickness. For comparison, we have also investigated the stress patterns in the oval valves of *Elliptio* (Fig. 10.2f), which are closed (as in most mussels) by two adductor muscles. Here again the highest (tensile) stresses occurred close to the free margin and bending between the free margin and the hinge (Fig. 10.2f). Again there exists no thickening of the shells, nor any curvature of the shell's profile suited to reduce the bending stresses evoked by the muscles.

This is in clear contrast to the skeletal structures in vertebrates.

10.3 Validity of Wolff's (Pauwel's) Law in Invertebrates: The Case of Sea Urchins

In "regular" sea-urchins, the chewing apparatus, the so-called "lantern of Aristoteles", is formed by five teeth, which are moved one against the other by strong muscles. The interior structure of these teeth (Märkel 1969; Märkel et al. 1977) is comparable to the (much later analysed) structure of gnawing teeth as in beavers (see above, Chap. 2). The external shape of the regular sea-urchins, which live in an aquatic environment on the ground (Fig. 10.3a), corresponds well to that assumed by a liquid heavier than its surrounding and filled into a tension-resistant sac (Fig. 10.3b, Seilacher 1991; Otto 1977, oral comm.).

Irregular sea urchins, the "sand-dollars", however, live subterraneously, buried in sediment. According to Seilacher (1979), young individuals have uninterrupted, circular outlines, but if they grow older and bigger, their external margins develop incisions which are cut out deeper and deeper with growing size. In many cases, there appear on all sides closed openings (Fig. 10.3f, k). U. Witzel investigated their shape and its variation mechanically in an unfortunately still unpublished study. He has started from the assumption that the essential way of loading the external skeleton of these more or less horizontally buried animals is compression by the overlying sediment. Its weight exerts compressive forces, distributed evenly over the dorsal and ventral surfaces. The most endangered part, needing the strongest resistance against being compressed, is close to the free margin. Reinforcement in this girth is provided by pillars (Fig. 10.3c, d, e). The compressive stresses lead to bending of the dorsal as well as the ventral plate, which becomes greater with width of span between the reinforced margins. The resistance against breakdown under the load becomes smaller with increasing diameter. Opening of indentations at the periphery or of holes (lunules), the side walls of which connect dorsal and ventral surfaces, reduces the span of the area on which sediment weight is applied. Its most important effect is providing additional connections between the dorsal and ventral sides, which reduce the span of the loaded surface and insert additional props.

The obtained results can be seen as indicating the ability of an invertebrate to adjust shape to mechanical stresses (Fig. 10.3g, h).

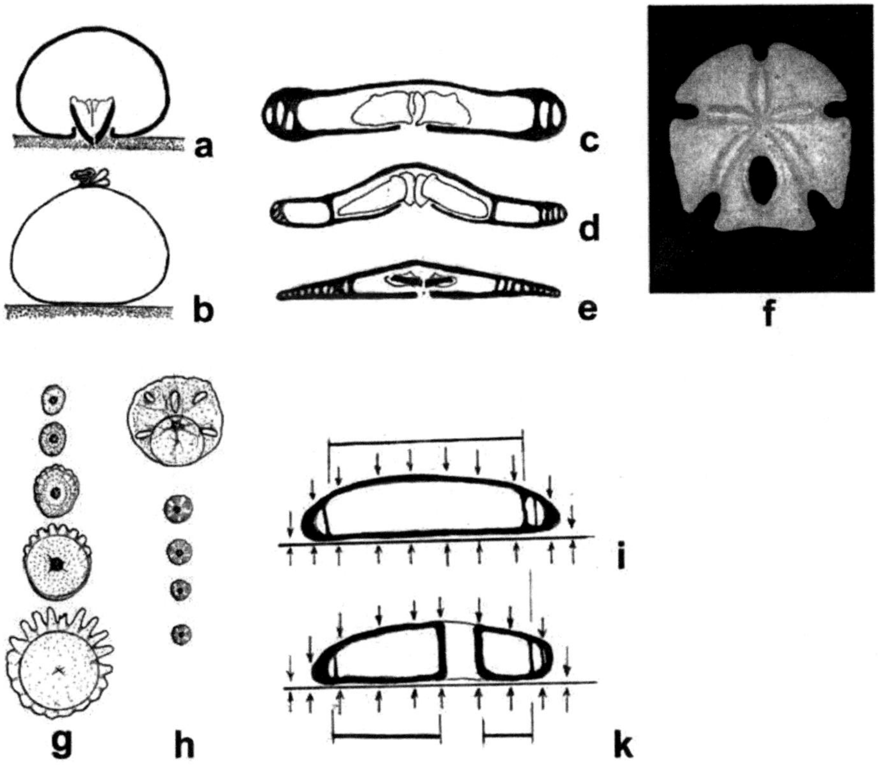

Fig. 10.3 (**a**) The external shape of an echinoid test is similar to that of a rubber ballon filled with liquid plaster (**b**). (**c–e**) Median sections through sand dollars, which live buried in the sediment. They are permanently exposed to the weight of the overlaying material. Near the margins, the kinking of the margin is excluded by an assemblage of pillars as reinforcements. These pillars become more frequent with increasing size. (**g, h**) Growth stages of sand dollars. During growth, many forms develop "lunules" (**g**), or indentations of the free margin (**h**). Both morphological changes reduce the span covered by the dorsal and the ventral plate. Redrawn after Seilacher 1979. (**f**) An irregular sea urchin in dorsal view, showing indentations as well as a "lunule". (**i**) Schematic cross section through a sand-dollar to show the distributed load (= weight of sediment). An opening, no matter whether an indentation or a "lunule" (**k**), shortens the diameters of the area without supports between upper and lower surface

10.4 Homology of Tails

Most animals, water-living as well as land-living, possess a tail. In mammals, this can be a short, in many cases not actively moved, appendix (seals, some rodents, martens, cursorial animals) but as well an important organ of propulsion (whales, dolphins, sea-cows), or a means to maintain balance during locomotion (birds, kangaroos, leapers among rodents, primates and several tree-living marsupials). A tail may also serve as a signal for communication (birds, monkeys, lemurs, dog-like carnivores, some even-toed cursorials, hippo) or for comfort (martens, herpestidae, dogs, cats) or for protection (pangolins, armadillos). It can also be reduced to a skeletal rudiment which is externally not visible (hominoidea). The variability of its development even among close relatives (macaques, baboons – see above, Chaps. 7 and 8) is obvious.

In reptiles and amphibians, the tail can be missing completely (anurans) or partially (turtles, several fossil synapsids, the mammal-like reptiles), but may also be an important device for propulsion (crocodiles, lizards, newts – when in water) or at least an aid for locomotion (land- and tree-living lizards, large terrestrial reptiles of the past). In fishes, the tail is commonly the dominating instrument for propulsion, rarely for fixation at firm structures (sea-horse, hippocampus), but in rare cases, it may also be missing (sunfish, Mola).

The innervation of the tail in mammals is provided by collector nerves in all four quadrants of the tail's cross section. They have their origins in a plexus caudalis the roots of which come from the lower lumbar segments. This innervation is in contrast to the tails of reptiles, amphibians and fishes. According to handbooks of comparative anatomy (Guibé, 1970, and others), in these forms the spinal cord continues until the end the canalis vertebralis and detaches in each segment of dorsal and ventral roots, which combine to the segmental nerves, that reach the segmented musculature and the skin above them on the shortest possible ways.

In comparative anatomy, the innervation of the skin and especially of the musculature is taken as the decisive criterion for homology of the respective structures. If this is taken for granted, the tails of mammals are not homologous to the plesiomorph tails of "lower" vertebrates, but instead must be considered as newly developed, "apomorphic" structures.

10.5 Further Research Required

Within the foregoing text, there have been touched a number of points which would profit from additional research. In several cases, the problems can be solved by application of the FES technique.

- The fixation of nails (and claws) and their influence on stress distribution in 3D within the terminal segments of digits and the digital pads (Chaps. 7 and 8).
- Pelvis including trunk and thigh. The three-dimensional shape of the (not alone mammalian, but also the avian and reptilian) pelvis is yet poorly understood, the existing approximations are insufficient or too crude (Chaps. 7 and 8). To understand the situation completely, an expansion of the investigated part on the lower lumbar region and on the upper thigh cannot be left aside. As soon as the latter is included:
- The special problem of the femoral neck and its relation to bipedality must be given attention.
- In spite of the impressive FES synthesis of the hominoid skulls provided by U. Witzel (Chap. 9), further studies seem extremely promising, for example to

verify my hypothesis concerning the frontal region in skulls with or without heavy-looking browridges, or the influence of prominent canines vice their integration into the row of incisor teeth. In addition, the question should be addressed whether the relative skull length (which was taken as a starting point in the synthesis of the neanderthalian skull) has an influence on the configuration of the facial skeleton – what may well lead to the facial profile of modern humans.

- Stress flow within the (primate, especially human) foot, with particular emphasis on the tarsals during roll-off (Chaps. 8 and 9), and its relation to the directions of articulating surfaces.
- Stress flows within the carpus under special consideration of variable load cases (Chaps. 8 and 9). This point includes identification of those regularly occurring load cases in which the stresses assume great values, and the relation of the joint surfaces between carpals as well as the arrangement of the trabeculae within them.

Other problems can be approached by less demanding methods, but need a great deal of imaginative and innovative power to find solutions:

- The mechanical advantages and disadvantages of both adducted stylopodia (=sprawled limbs of "lower tetrapods") and extremities moving in parasagittal planes (in birds and mammals) need to be worked out under consideration of the "high walk" of reptiles more detailedly than before. The ideas presented in Chap. 7 end up with (feeble) attempts to explain the divergence on the basis of "biological roles" or of "a different evolutionary line". Because the parasagittal movements have been acquired in two classes of outstandingly mobile animals (birds and mammals) independently, there should exist a really stringent mechanical reason.
- Closely related to that question, it seems worthwhile to reconsider in every detail the coincidence of several evolutionary changes among Permian-Triassic representatives of the mammal-like reptiles: The often discussed changes towards synapsid skull shape may well be connected with the mammal-like arrangement of the jaw musculature and the heterodont dentition. At about the same time appeared the forward-directed ilium and the strong development of the ischium-pubis complex with a foramen obturatum. This may well be related to a re-arrangement of the hip-extending muscles; the corresponding replacement of the reptilian caudofemoral muscle seems to be documented by the much smaller diameters of the proximal caudal vertebrae and a shortening of the tail. Regarding the shoulder girdle, the reduction of the ventral plate formed by the elements named coracoid, procoracoid, sternum, interclavicula etc. seems to be paralleled by the development of a scapular spine. These changes in the girdle skeletons may well be indicators for the preference of parasagittal limb movements. This idea is not new, see Romer 1945 (p. 289), but should be reconsidered from mechanical viewpoints. Both complexes, the skull as well as the limb movements, seem to represent a sort of evolutionary alternative to the modes preferred by reptiles.

- Dermatoglyphic patterns in relation to the external (frictional) forces applied in representative phases of locomotion (Chap. 8).
- All details rotating around the so-called "opposition" of the thumb. First of all, an overview about variability of muscles among all primates should be attempted, together with a uniform terminology; second, an unambiguous system for defining the movements is necessary; third a comparison of the use made of the thumbs in performing various, species-specific tasks.
- Functional morphology of morphological traits in invertebrates. A general approach towards "hydraulic constructions" (coelenterates, molluscs and tunicates) has been proposed by the Senckenberg Group or Frankfurt School. Considerable work has been done on arthropods, especially on their locomotion (Full, 1989; Blickahn and Full, 1993), but also on the way how their joints function. The results obtained on arthropods are not at all considered in the present book. As far as I can see, no ideas have been persued about the existence of "functional adaptation".

This book would never have been written without the strong influence of my academic teachers (and a number of other persons!).

I owe thanks to Peter Kramp, who introduced me to his idea of academia, where nothing mattered apart from thinking with precision; Dietrich Starck, who showed me the marvels of embryology and comparative anatomy; Benno Kummer, who guided me to functional anatomy and encouraged my first steps on this field. Helmuth Hofer has given considerable support to my way into primatology. Frank Westphal and Adolf Seilacher led me towards paleontology and helped in a critical phase of my career. Ulrich Witzel introduced me into the enormous value of finite element systems which is essential in analysing complex spatial structures, and often improved my attempts to solve problems with the aid of mathematics. My most successful co-workers and co-authors were (in alphabetical order) Andreas Christian, Norman Creel, Brigitte Demes, Martin Fritz, Michael Günther, Stefan Recknagel, Hartmut Witte. They all kept a watchful eye on my tendency to consider things as self-evident and insisted on really exact step-by-step thinking. Stephan Recknagel designed and used ingenious measuring devices. Without their input, most of the ideas contained in this book would never have been worked out. The same kind of influence, though more restricted on particular problems, was exerted from the many students, doctorands, and co-authors who have worked in and with my group. Most of them show up in the List of References.

Particular thanks are due to my family. First to my wife Lotte, who sustained me for more than 6 decades, and raised but moderate objections against my travelling to other places – be it for meetings or for research work. Second to my daughters, who loved me and took motivating interest in my work – at least most of the time – and third to many quadrupedal companions, hooved, pawed or beaked, who confronted me with ever new questions.

References

Abitbol, M.: Evolution of the sacrum in hominoids. Am. J. Phys. Anthropol. **74**, 65–81 (1987)

Abitbol, M.: Sacral curvature and supine posture. Am. J. Phys. Anthropol. **80**(3), 379–389 (1989)

Abker, U.M., Mahl, M.-A.: Untersuchungen zur Druck- and Kraftverteilung in der Handwurzel des Menschen. Doctoral Thesis Med. Fakultät, RU Bochum (1993)

Adachi, K., Nishizawa, S., Endo, B.: The trajectory of the point of application of the resultant force of body mass at different walking speeds. Statistical analysis of human walking. Folia Primatol. **66**, 160–180 (1996)

Adoutte, A., Balavoine, G., Lartillot, N., de Rosa, R.: Animal evolution: the end of the intermediate taxa? Trends Genet. **15**(3), 104–108 (1999)

Ahlberg, P.E., Clack, J.A.: A firm step from water to land. Nature. **440**, 747–749 (2006)

Alexander, R.M.N.: Functional Design in Fishes. Hutchinson University Library, ISBN 0-09-104751-X (1967)

Alexander, R.M.N.: The orientation of muscle fibres in the myomeres of fishes. Journal of the Marine Biological Association U.K. **49**, 263–290 (1969)

Alexander, R.M.N.: Size and Shape, (Institute of Biology Studies in Biology). Edward Arnold, London (1971)

Alexander, R.M.N.: The Chordates. Cambridge University Press, Cambridge, UK (1975)

Alexander, R.: Mechanics of bipedal locomotion. In: Davies, P.S. (ed.) Perspectives in Experimental Biology, pp. 493–504. Pergamon Press, Oxford, UK (1976)

Alexander, R.M.N.: Locomotion of Animals, (Tertiary Level Biology). Springer (1985)

Alexander, R.M.N.: Elastic Mechanisms in Animal Movement. Cambridge University Press, Cambridge, UK (1988)

Alexander, R.M.N.: Principles of Animal Locomotion. Princeton University Press, Princeton, NJ (2006)

Alexander, R., Mc, N., Vernon, A.: The mechanics of hopping by kangaroos. J. Zool. **177**, 265–303 (1975)

Alexander, R., Mc, N., Dimery, N.J., Ker, R.F.: Elastic structures in the back and their role in galloping in some mammals. J. Zool. **201**, 135–152 (1985)

Almécija, S., Moyà-Solà, S., Alba, D.M.: Early origin for human-like precision grasping: A comparative study of pollical distal phalanges in fossil hominins. PLoS One. **5**, e11727 (2010)

Almécija, S., Tallman, M., Alba, D.M., Pina, M., Moyà-Solà, S., Jungers, W.L.: The femur of *Orrorin tugenensis* exhibits morphometric affinities with both Miocene apes and later hominins. Nat. Commun. **4**, 2888 (2013). https://doi.org/10.1038/ncomms3888

Almécija, S., Smaers, J.B., Jungers, W.L.: The evolution of human and ape hand proportions. Nat. Commun. **6**, 7717 (2015)

H. Preuschoft, *Understanding Body Shapes of Animals*,
https://doi.org/10.1007/978-3-030-27668-3

Amtmann, E.: Mechanical Stress, Functional Adaptation and the Variation Structure of the Human Femur Diaphysis. Ergebnisse der Anatomie and Entwicklungsgeschichte vol. 44 (3). Springer, Heidelberg (1971)

Amtmann, E.: Biomechanical interpretation of form and structure of bones: role of genetics and function in growth and genetics. In: Morbeck, M.E., Preuschoft, H., Gomberg, N. (eds.) Environment, Behavior, and Morphology: Dynamical interactions in Primates, pp. 347–366. G. Fischer-Verlag, New York (1979)

Amtmann, E., Doden, E.: Anpassung der Knochenstruktur an mechanische Beanspruchung. Z. Morphol. Anthropol. **72**, 1–21 (1981)

Ando, K., Fujiwara, S.: Farewell to life on land – thoracic strength as a new indicator to determine paleoecology in secondary aquatic mammals. J. Anat. **2016**, 1–10 (2016). https://doi.org/10.1111/joa.12518

Andrada, E., Haase, D., Sutedjy, Y., Nyakatura, J.A., Kilbourne, B.M., Denzler, J., Fischer, M.S., Blickhan, R.: Mixed gaits in small avian terrestrial locomotion. Nat. Sci. Rep. **5**, 136636 (2015). https://doi.org/10.1038/srep13636

Ankel, F.: Vergleichende Untersuchungen über die Skelettmorphologie des Greifschwanzes südamerikanischer Affen (Platyrrhina). Z. Morphol. Ökologie der Tiere. **52**, 131–170 (1962)

Anonymous: Tiger claw. KfK Nachrichten (1995)

Anthoney, J.: Le neuraxe des reptiles. In: Grassé (ed.) Traité de Zoologie, Tome XIV, Fasc. II, pp. 202–332. Masson, Paris (1970)

Arms, A.: Lokomotionsverhalten und Fortbewegungsweisen kleiner Primatenarten (*Saguinus oedipus* und *Saimiri boliviensis peruvianus*). Masters Thesis, Fac. Biology, Göttingen (1998)

Arms, A., Voges, D., Fischer, M.S., Preuschoft, H.: Arboreal locomotion in small New World monkeys. Z. Morphol. Anthropol. **83**, 243–263 (2002)

Ashton, E.H., Oxnard, C.E.: Functional adaptation of the primate shoulder girdle. Proc. Zool. Soc. London. (1963)

Baier, D.B., Gatesy, S.M., Jenkins, F.M.: A critical ligamentous mechanism in the evolution of avian flight. Nature. **445**, 307–310 (2007)

Barnett, C.H.: The normal orientation of the human hallux and the effect of footwear. J. Anat. **96**, 489–494 (1962)

Baumeister, E.T.: Belastungen und Traumatologie des menschlichen Bewegungsapparates beim Speerwerfen. Doctoral thesis, Med. Fakultät RU Bochum (1991)

Baumeister, E., Wolniak, H.J.: Functionelle und antomische Grundlagen der Entstehung von Ellbogenbeschwerden bei Speerwerfern. Masters Thesis, Sports Sciences Ruhr-Universität Bochum

Bennett, M.R., Morse, S.A.: Human Footprints: Fossilised Locomotion? Springer, Heidelberg/London (2014). https://doi.org/10.1007/978-3-319-08572-2

Bennett, M.R., Reynolds, S.C., Morse, S.A., Budka, M.: Laetoli's lost tracks: 3D generated mean shape and missing footprints. Sci. Rep. **6**, 21916 (2016). https://doi.org/10.1038/srep21916

Benninghoff, A.: Spaltlinien am Knochen. Anat. Anz. **60**, 189–206 (1925)

Benninghoff, A.: Lehrbuch der Anatomie Des Menschen, vol. 1, 2nd edn. Lehmann, München (1938)

Berger, L.R., de Ruiter, D.J., Churchill, S.E., Schmid, P., Carlson, K.J., Dirks, P.H., Kibii, J.M.: *Australopithecus sediba*: a new species of *Homo*-like australopith from South Africa. Science. **328**, 195–204 (2010). https://doi.org/10.1126/science.1184944; pmid: 20378811

Berger, L.R., Hawks, J., de Ruiter, D.J., Churchill, S.E., Schmid, P., Delezene, L.K., Kivell, T.L., Garvin, H.M., Williams, S.A., DeSilva, J.M., Skinner, M.M., Musiba, C.M., Cameron, N., Holliday, T.W., Harcourt-Smith, W., Ackermann, R.R., Bastir, M., Bogin, B., Bolter, D., Brophy, J., Cofran, Z.D., Congdon, K.A., Deane, A.S., Dembo, M., Drapeau, M., Elliott, M. C., Feuerriegel, E.M., Garcia-Martinez, D., Green, D.J., Gurtov, A., Irish, J.D., Kruger, A., Laird, M.F., Marchi, D., Meyer, M.R., Nalla, S., Negash, E.W., Orr, C.M., Radovcic, D., Schroeder, L., Scott, J.E., Throckmorton, Z., Tocheri, M.W., VanSickle, C., Walker, C.S.,

Wei, P., Zipfel, B.: *Homo naledi*, a new species of the genus *Homo* from the Dinaledi chamber, South Africa. eLife. **4**, e09560 (2015). https://doi.org/10.7554/eLife.09560
Berillon, G., Marchal, F.: Nos encetres ont-ils modifié le climat? Pour la Science (Èdition francaise de Scientific American). 330, pp. 75–83 (2005)
Berman, D.S., Henrici, A.C., Kissel, R.A., Sumida, S.S., Martens, T.H.: A new diadectid (diadectomorpha), *Orobates pabsti*, from the Early Permian of central Germany. Bull. Carnegie Mus. Nat. Hist. **35**, 1–36 (2004)
Bertram, J.E.A., Chang, Y.H.: Mechanical energy oscillations of two brachiation gaits: measurements and simulation. Am. J. Phys. Anthropol. **115**, 319–326 (2001)
Beyene, Y., Katoh, S., WoldeGabriel, G., Hart, W.K., Uto, K., Sudo, M., Kondo, M., Hyodo, M., Renne, P.R., Suwa, G., Asfaw, B.: The characteristics and chronology of the earliest Acheulean at Konso, Ethiopia. Proc. Natl. Acad. Sci. USA. **110**(5), 1584–1591 (2013)
Biancardi, C.M., Minetti, A.E.: Biomechanical determinants of transverse and rotary gallop in cursorial mammals. The Journal of Experimental Biology 215, 4144–4156 (2012)
Biegert, J.: Volarhaut der Hände und Füsse. Primatologia II, part 1/3. Karger-Verlag, Basel (1961)
Biewener, A.A.: Biomechanics of mammalian terrestrial locomotion. Science. **250**, 1097–1103 (1990)
Biewener, A.A.: Muscle-tendon stresses and elastic energy storage during locomotion in the horse. Comp. Biochem. Physiol. B. **120**, 73–87 (1998)
Bilo, D., Nachtigall, W.: Biophysics of bird flight: questions and results. In: Nachtigall, W. (ed.) Physiology of Movement – Biomechanics, pp. 217–234. G. Fischer-Verlag, Stuttgart (1977)
Blickhan, R.: Running and hopping. In: Wieser, W., Gnaiger, E. (eds.) Energy Transformation in Cells and Animals, pp. 183–190. Thieme, Stuttgart (1989a)
Blickhan, R.: The spring-mass model for running and hopping. J. Biomech. **22**(11–12), 1217–1227 (1989b)
Blickhan, R.: Flow pattern in the vicinity of fish during undulatory locomotion. Abstracts of the first World Congress of Biomechanics, San Diego, p.116
Blickhan, R.: Biomechanik der Axialen Aquatischen und der Pedalen Terrestrischen Lokomotion. Habilschrift an der Universität des Saarlandes Homberg (Saar), 1992. Shacker-Verlag, Aachen (2007)
Blickhan, R., Full, R.J.: Mechanical work in terrestrial locomotion. In: Biewener, A. (ed.) Biomechanics: Structures and Systems: A Practical Approach. Oxford University Press, Oxford, UK (1992)
Blickhan, R., Full, R.J.: Similarity in multilegged locomotion: bouncing like a monopode. J. Comp. Physiol. A. **173**, 509–517 (1993)
Blickhan R., Nachtigall, W.: New aspects of the energetics of fish locomotion. In: Rayner, J. and Bone, Q (eds): Mechanics and Physiology of Animals, Plymouth, pp. 23 ff. (1991)
Blickhan, R., Bleckmann, H., Breithaupt, T., Tautz, J.: The spacial distribution, time course, and spectral energy content of flow fields caused by swimming trouts. In: Erber, J., Menze, R., Pflüger, H.-J., Todt, D. (eds.) Proc.of the 2nd International Congress of Neuroethology, p. 81. Thieme, Stuttgart (1989)
Blickhan, R., Braun, K., Nachtigall, W.: Biomechanik de Fischlokomotion. Verh. Deutsch. Zool. Ges. **83**, 629 (1990)
Bloch, J.I., Woodruff, E.D., Wood, A.R., Rincon, A.F., Harrington, A.F., Morgan, G.S., Foster, D. A., Montes, C., Jaramillo, C.A., Jud, N.A., Jones, D.S., MacFadden, B.J.: First North American fossil monkey and early Miocene tropical biotic interchange. Nature. **533**, 243–246 (2016). https://doi.org/10.1038/nature17415
Bock, W.J.: The avian skeletomuscular system. In: Farner, D.S., King, J.R., Parkes, K.C. (eds.) Avian Biology, pp. 119–257. Academic Press, New York (1974)
Bock, W.J.: Cranial kinesis revisited. Zoolog. Anz. **238**, 27–39 (1999)
Bock, W.: Dual causality and the autonomy of biology. Acta Biotheor. **65**, 63–79 (2017). https://doi.org/10.1007/s10441-016-9303-2

Bock, W., von Wahlert, G.: Adaptation and the form-function complex. Evolution. **19**, 269–299 (1965)

Boeker, H.: Vergleichend-biologische Anatomie der Wirbeltiere. G.Fischer, Jena (1935-7)

Boesch, C., Boesch, H.: Different hand postures for pounding nuts with natural hammers by wild chimpanzees. In: Preuschoft, H., Chivers, D.J. (eds.) Hands of Primates, pp. 31–43. Springer, Wien (1993)

Bone, Q., Marshall, N.B.: Biologie der Fische. G. Fischer, Stuttgart (1985)

Bosinski, G., Lordkipanidze, D., Weidemann, K.: Der altpaläolithische Fundplatz Dmanisi (Georgien, Kaukasus). Jahrbuch des Römisch-Germanischen Zentralmuseums. Bonn. **42** (1995), S.21–S203 (1995). ISSN0076-2741

Bown, T.M., Kraus, M.J., Wing, S.L., Fleagle, J.G., Tiffney, B.H., Simons, E.L., Vondra, C.F.: The Fayum primate forest revisited. J. Human Evol. **11**, 603–632 (1982)

Boyde, A.: Enamel structure and cavity margins. Operat. Dent. **1**, 13–28 (1976)

Boyer: Abhandlungen über Knochenkrankheiten. 2. Bd. Würzburg (1819)

Bragulla, H.H., Homberger, D.G.: Structure and functions of keratin proteins in simple, stratified, keratinized and cornified epithelia. J. Anat. **214**, 516–559 (2009)

Bramble, D.M., Carrier, D.R.: Running and breathing in mammals. Science. **219**, 251–256 (1983)

Bramble, D.M., Lieberman, D.E.: Endurance running and the evolution of *Homo*. Nature. **432**, 345–352 (2004)

Bramblett, C.A.: Pathology of the Darajani baboon. Am. J. Phys. Anthropol. **26**, 331–340 (1967)

Brandes, G.: Buschi, Vom Orang-Säugling zu Backenwülster. Quelle & Meyer, Leipzig (1939)

Brandes, R., Lang, F., Schmidt, R.F. (eds.): Physiologie Des Menschen. Thieme, Stuttgart (2019)

Brandt, M.: Der Ursprung des Aufrechten Ganges. Zur Fortbewegung der Plio-pleistozänen Hominiden. Hänssler-Verlag, Neuhausen-Stuttgart (1995)

Brain, J., Reif, W.: A new terminology of aquatic propulsion in vertebrates. Neues Jahrbuch der Geologie und Paläontologie, Abhandlung **164**, 162–167 (1982)

Bronder, S.: Kinematik und Strömungsanalyse beim Schwimmen der Vipernatter (*Natrix maura*). Diplomarbeit Univ. d, Saarlandes (1988)

Bruna, M., Blickhan, R., Nachtigall, W.: Die Haut des Fisches als Sehne: Festigkeit der Haut von Forellen. Verh. Deutsch. Zoolo. Ges. **82**, 258 (1989)

Bucher, O., Wartenberg, H.: Cytologie, Histologie und mikroskopische Anatomie des Menschen. Hans Huber, Bern, 11. Aufl. (1989)

Buchholtz, E.A.: Implications of vertebral morphology for locomotor evolution in early Cetacea. In: Thewissen, J.G.M. (ed.) The Emergence of Whales, pp. 325–351. Plenum Press, New York/London (1998)

Buchholtz, E.A.: Vertebral osteology and swimming style in living and fossil whales (Order: Cetacea). J. Zool. **253**, 175–190 (2001a)

Buchholtz, E.A.: Swimming styles in Jurassic ichthyosaurs. J. Vertebr. Paleontol. **21**, 61–73 (2001b)

Buck, C., Bär, H.: Investigations on the biomechnical significance of dermatoglyphics. In: Preuschoft, H., Chivers, D.J. (eds.) Hands of Primates, vol. 284, pp. 285–306. Springer, Wien (1993)

Bühler, P, 1972: Sandwichstrukturen der Schädelkapseln verschiedener Vögel. – Zum Leichtbauprinzip bei Organismen. = Sandwich Structures in the skull capsules of various birds. – The principle of lightweigth structures in organisms. IL 4 Mitt. Inst. Leichte Flächentragwerke 4: 40–50

Bühler, P.: Functional anatomy of the avian jaw apparatus. In: King, A.S., McLelland, J. (eds.) Form and Function in Birds, vol. 2, pp. 439–468. Acad Press, New York (1981)

Buikstra, J.A.: Healed fractures in *M. mulatta*: age, sex and symmetry. Folia Primatol. **23**, 140–148 (1975)

Burnside, L.H.: Coordination in the locomotion of infants. Genet. Psychological Monographs **2**, 284–372 (1927)

Busching, M.: Eine experimentelle Studie zur Formveränderung der Lendenwirbelsäule und der Beckenkippung in der Sagittalebene in Abhängigkeit von der Oberschenkelwinkelung. Doctoral Thesis Medical Faculty, Ruhr-Universität Bochum (1998)
Calvin, W.H.: A stone's throw and its lauch window: timing precision and ist implications for language and hominid brains. J. Theor. Biol. **104**, 121–135 (1983)
Cardini, A.: Craniofacial allometry is a rule in evolutionary radiations of placentals. Evol. Biol. **46** (3), 239–248 (2019)
Carrier, D.R.: Activity of the hypaxial muscles during walking in the lizard *Iguana iguana*. J. Exp. Biol. **152**, 453–470 (1990)
Carrier, D.: Action of the hypaxial muscles during walking and swimming in the salamander *Dicamptodon ensatus*. J. Exp. Biol. **180**, 75–83 (1993)
Carrier, D.R.: Function of the intercostal muscles in trotting dogs: ventilation or locomotion? J. Exp. Biol. **199**, 1455–1465 (1996)
Carrier, D. R., Heglund, N. C., Earls, K. D.: Variable gearing during locomotion in the human musculoskeletal system. Science, New Series, vol. 265, No. 5172, pp. 651–653. (1994)
Carrier, D.R., Deban, S.M., Fidschbein, T.: Locomotor function of the pectoral girdle "muscular sling" in trotting dogs. J. Exp. Biol. **209**, 2224–2237 (2008)
Carrier, D.R., Anders, C., Schilling, N.: The musculoskeletal system of humans is not tuned to maximise the economy of locomotion. PNAS. **108**(46), 18631–18636 (2011). https://doi.org/10.1073/pnas.1105277108
Carter, D.R., Beaupré, G.S.: Skeletal Function and Form. Cambridge University Press, Cambridge, UK (2001)
Cartmill, M.: Pads and claws in arboreal locomotion. In: Jenkins, F.A. (ed.) Primate Locomotion, pp. 45–83. Academic Press, New York (1974)
Cartmill, M.: The volar skinn of primates: its frictional characteristics and their functional significance. Am. J. Phys. Anthropol. **50**, 497–510 (1979)
Cartmill, M.: Climbing. In: Hildebrand, M., et al. (eds.) Functional Vertebrate Morphology, pp. 73–88. Harvard Univ. Press, Cambridge, MA (1985)
Cassini, G.H., Vizcaíno, S.F.: Modeling the mammalian jaw mechanism: a comprehensive synthesis. J. Mamm. Evol. **22**(1), 121–122 (2015)
Cavagna, G.A., Heglund, N.C., Taylor, C.R.: Mechanical work in locomotion: two basic mechanisms in minimizing energy expenditure. Am. J. Phys. Anthropol. **233**, 233–261 (1977)
Ceacero, F.: Long or heavy? Physiological constraints in the evolution of antlers. J. Mamm. Evol. **23**(2), 209–216 (2016)
Cerling, T.E., Mbua, W., Jirera, F.M., Manthi, F.K., Grine, F.E., Leakey, M.G., Sponheimer, M., Uno, K.T.: Diet of *Paranthropus boisei* in the early Pleistocene of East Africa. PNAS. **108**(23), 9337–9341 (2011). https://doi.org/10.1073/pnas.1104627108
Chatterjee, S.: The Rise of Birds. 225 Millions Years of Evolution, 2nd edn. Johns Hopkins University Press, Baltimore (2017)
Christ, B., Jacob, H.J.: Morphogenesis, Musterbildung und Zell-Migration als Teilprozesse der Extremitäten-entwicklung bei Amniotenembryonen. Eine Synopsis experimenteller Befunde. Verh. Anat. Ges. **80**, 115–126 (1986)
Christ, B., Jacob, H.J., Brand-Saberi, B., Grim, M.: On the development of the human hand. In: Preuschoft, H., Chivers, D.J. (eds.) Hands of Primates, pp. 405–421. Springer, Wien (1993)
Christel, M.: Grasping techniques and hand preferences in hominoidea. In: Preuschoft, H., Chivers, D.J. (eds.) Hands of Primates, pp. 91–108. Springer, Wien (1993a)
Christel, M.: Greiftechniken und Handpräferenzen verschiedener catarrhiner Primaten beim Aufnehmen kleiner Objekte. Biol. Thesis Freie Universität Berlin (1993b)
Christel, M.I., Fragaszy, D.: Manual function in *Cebus apella*. Digital mobility, preshaping, and endurance in repetitive grasping. Int. J. Primatol. **21**(4), 697–719 (2000)
Christel, M.I., Kitzel, S.T., Niemitz, C.: How precisely do bonobos (Pan paniscus) grasp small objects? Int. J. Primatol. **19**, 165–194 (1998)

Christian, A.: Zur Biomechanik der Lokomotion vierfüssiger Reptilien (besonders der Squamata). Courier Forsch.-Inst. Senckenberg 180, Frankfurt (1995)

Christian, A.: Zur Biomechanik der Fortbewegung von *Leptictidium* (Mammalia Proteutheria). Courier Forsch-Inst. Senckenberg. **216**, 1–18 (1999)

Christian, A.: Scaling effects in monitor lizards. In: Horn, H.G., Boehme, W., Krebs, U. (eds.) Advances in Monitor Research III, Mertensiella (Supplement to Salamandra), pp. 1–9. Chimaera Buchhandelsgesellschaft, Frankfurt a.M (2007)

Christian, A., Dzemski, G.: Neck postures in sauropods. In: Klein, N., Remes, K., Gee, C.T., Sander, P.M. (eds.) Biology of the Sauropod Dinosaurs, pp. 251–260. Indiana Univ. Press, Bloomington (2011)

Christian, A., Garland Jr., J.: Scaling of limb proportions in monitor lizards (Squamata, Varanidae). J. Herpetol. **30**, 219–230 (1996)

Christian, A., Preuschoft, H.: Deducing the body posture of extinct large vertebrates from the shape of the vertebral column. Paléo. **39**(4), 801–812 (1996)

Christian, A., Horn, H.G., Preuschoft, H.: Biomechnical reasons for bipedalism in reptiles. Amphibia-Reptilia. **15**, 275–284 (1994a)

Christian, A., Horn, H.G., Preuschoft, H.: Bipedie bei rezenen Reptilien. Natur and Museum **124**, 45–57 (1994b)

Christian, A., Koberg, D., Preuschoft, H.: Shape of the pelvis and posture of the hindlimbs in *Plateosaurus*. Paläont. Z. **70**, 591–601 (1996)

Christian, A., Müller, R. H. G., Christian, G., Preuschoft, H.: Limb swinging in elephants and giraffes and implications for the reconstruction of limb movements and speed estimates in large Dinosaurs. Mitteilungen des Museums für Naturkunde, Berlin, Geowiss. Reihe **2**:81–90 (1999)

Clark, J.D., de Heinzelin, J., Schick, K.D., Hart, W.K., White, T.D., WoldeGabriel, G., Walter, R. C., Suwa, G., Asfaw, W., Vrb, E., Haile-Selassie, Y.: African *Homo erectus*, old radiometric ages and young Oldowan assemblages in the Awash valley, Ethiopia. Nature. **264**, 1907–1910 (1994)

Clarke, R.J.: First ever discovery of a well-preserved skull and associated skeleton of *Australopithecus*. South African J. Sci. **94**, 460–464 (1998)

Clarke, R.J.: Discovery of the complete arm and hand of the 3.3 million-year-old Australopithecus skeleton from Sterkfontein. South African Journal of Science **95**, 477–480 (1999)

Clarke, R.J.: Latest information on Sterkfontein's Australopithecus skeleton and a new look at *Australopithecus*. South African J. Sci. **104**(11 & 12), 443–449 (2008)

Clarke, R., Tobias, P.V.: Sterkfontein member 2 foot bones of the oldest South African hominid. Science. **269**, 521–524 (1995)

Clutton-Brock, T.H.: Primate social organisation and ecology. Nature (London). **250**, 539–542 (1974)

Clutton-Brock, J.: Horse Power. Harvard University Press, Cambridge (1992)

Conroy, G.C.: Primate Postcranial Remains from the Oligocene of Egypt. Contribution to Primatology 8. Karger-Verlag, Basel (1976)

Cope, E.D.: The Origin of the Fittest. Appleton, New York (1887)

Cope, E.D.: On the phylogeny of the vertebrata. Proc. Am. Phil. Soc. **30**, 278–281 (1892)

Creel, N., Preuschoft, H.: A canonical analysis of cranial dimensions. In: Proc. 3rd Internat. Congress of Primatology, Zürich 1970, Karger, Basel, vol. 1, pp. 36–43 (1971)

Creel, N., Preuschoft, H.: Cranial morphology of the lesser apes: a multivariate statistical study. In: Rumbaugh, D.M. (ed.) Gibbon & Siamang, vol. 4, pp. 219–303. Karger, Basel (1976)

Creel, N., Preuschoft, H.: Sytematics of the lesser apes: a quantitative taxonomic analysis of craniometric and other variables. In: Preuschoft, H., Chivers, D.J., Brockelman, W.Y., Creel, N. (eds.) The Lesser Apes, pp. 562–613. Edinburgh University Press (1984)

Crompton, A.W., Hiiemae, K.: Molar occlusion and mandibular movements during occlusion in the American opossum, *Didelphis marsupialis*. Zool. J. Linnaean Soc. **49**, 21–47 (1970)

Crompton, R.H., Li, Y., Thorpe, S.K., Wang, W.J., Savage, R., Oayne, R., Carey, T.C., Aerts, P., van Elsaker, I., Hofstetter, A., Günther, M.M., D'Aout, K., Declerq, D.: The biomechanical evolution of erect bipedality. Cour. Forchungsinst. Senck. **243**, 135–146 (2003)

Crook, J.H.: Gelada baboon herd structure and movement: a comparative report. Symp. Zool. Soc. Lond. **18**, 237–258 (1966)

Crook J.H., Aldrich-Blake, P.: Ecological and behavioral contrasts between sympatric ground-dwelling primates in Ethiopia. Folia primatologica **8**, 92–227 (1968)

Currey, J.D., Taylor, J.D.: Mechanical behaviour of mollusc shells. J. Zool. London. **173**, 395–406 (1974)

Curth, S., Fischer, M.S., Nyakatura, J.A.: Neoichnology of a belly-dragging lizard. - Analogies to early reptile locomotion? Ichnos **21**, 32–43 (2014)

Curtis, N., Witzel, U., Fitton, L., O'Higgins, P., Fagan, M.: The mechanical significance of the temporal fascia in *Macaca fascicularis*: an investigation using Finite Element Analysis. Anatomical Record. **294**, 1178–1190 (2011)

Cuvier, G.: Das Tierreich geordnet nach seiner Organisation als Grundlage der Naturgeschichte der Tiere und als Einleitung in die vergleichende Anatomie. Dt. Übersetzung A.V. Streubel, Brockhaus, Berlin 1846 (1831)

D'Août, K., Vereecke, E.E. (eds.): Primate Locomotion. Linking Field and Laboratory Research. Springer, New York (2011)

D'Aout, C., Aerts, P., de Clercq, D., de Mester, K., von Elsaker, L.: Segment and joint angles of hindlimb during bipedal and quadrupedal walking of the bonobo (*Pan paniscus*). Am. J. Phys. Anthropol. **119**, 37–51 (2002)

Daeschler, E., Shubin, N.H., Jenkins, F.A.: A Devonian tetrapod-like fish and the evolution of the tetrapod body plan. Nature. **440**, 757–763 (2006)

Dagg, A.I.: The locomotion of the camel (*Camelus dromedarius*). J. Zool. **174**, 67–78 (1974)

Darwin, C.: The Origin of Species by mens of natural selection. Murray, London Translated into German by D. Haek: Die Entstehung der Arten durch natürliche Zuchtwahl. Reclam. Leipzig (1859)

Daver, G., Berillon, G., Grimaud-Hervé, D.: Carpal kinematics in quadrupedal monkeys: towards a better understanding of wrist morphology and function. J. Anat. **220**(1), 42–56 (2012)

Davies, P.R., Day, M.H., Napier, J.R.: Hominid fossils from Bed 1, Olduvai Gorge, Tanganyika. Nature. **201**, 967–970 (1964)

Dawson, T.J.: Monotremes and Marsupials: the Other Mammals, pp. 1–87. Edward Arnold, London

Dawson, T.J., Webster, K.N.: Energetic characteristics of macropodid locomotion. In: Coulson, G. M., Eldridge, M. (eds.) Makropods: Kangaroos, Wallabies, and Rat Kangaroos, pp. 99–108. CSIRA Publishing, Collingwood (2010)

Day, M.H.: Hominid locomotion – from Taung to the Laetoli footprints. In: Tobias, P.V. (ed.) Hominid Evolution: Past Present and Future, pp. 115–127. Alan R. Liss, New York (1985)

Day, M.: Bipedalism and prehistoric footprints. In: Coppens, Y., Senut, B. (eds.) Origine(s) de la Bipédie chez les Hominidés, pp. 199–214. CNRS, Paris (1991)

Day, M.H., Wood, B.A.: Functional affinities of the Olduvai hominid 8 talus. Man. **3**, 440–455 (1968)

Day, S.W., Higham, T.E., Holzmnan, R., van Wassenbergh, S.: Morphology, kinematics, and dynamics: the mechanics of suction feeding in fishes. Integr. Comp. Biol. **55**(1), (2015). https://doi.org/10.1093/icb/icv032

DB-Mitteilungen: Mobil, (2001)

Deban, S.M., Olson, W.M.: Suction feeding in a tiny predatory tadpole. Nature. **204**, 41–42 (2002)

Debrunner, A.M.: Orthopädie, 2nd edn. Hans Huber, Bern (1985)

Deloison, Y.: Les Australopithèques, merchaient-ils comme nous? In: Coppens, Y., Senut, B. (eds.) Origine(s) de la Bipédie chez les Hominidés, pp. 177–186. CNRS, Paris (1991)

Demes, B.: Die Festigkeit von Primatenschädeln gegenüber mechanischen Beanspruchungen. Z. Morphol. Anthropol. **72**, 47–64 (1981)

Demes, B.: The resistance of primate skulls against mechanical stresses. J. Hum. Evol. **11**, 687–691 (1982)

Demes, B.: Mechanical stresses at the primate skull base caused by the temporomandibular joint force. In: Chivers, D.J., Wood, B.A., Bilsborough, A. (eds.) Food Acquisition and Processing in Primates, pp. 407–413. Plenum Press, London (1984)

Demes, B.: Biomechanics of the Primate Skull Base. Advances in Anatomy, Embryology and Cell Biology, vol. 94. Springer, Berlin (1985a)

Demes, B.: Biomechanics of the primate skull base. In: Duncker, H.R., Fleischer, D.G. (eds.) Functional Morphology in Vertebrates. Fortschritte der Zoologie, vol. 30, pp. 139–143 (1985b)

Demes, B.: Another look at an old face: biomechanics of the Neanderthal facial skeleton reconsidered. J. Human Evol. **16**, 297–303 (1987)

Demes, B.: Biomechanische Allometrie: Wie die Körpergrösse Fortbewegung und Körperform von Primaten bestimmt. Courier Forschungsinstitut Senckenberg 141 (Habilitations-Schrift, Medizinische Fakultät, Ruhr-Universität Bochum) (1991)

Demes, B., Creel, N.: Bite force, diet, and cranial morphology of fossil hominoids. J. Human Evol. **17**, 657–670 (1988)

Demes, B., Günther, M.M.: Wie die Körpermasse den Springstil von Halbaffen und deren Proportionen bestimmt. Z. Morphol. Anthropol. **77**, 209–225 (1989)

Demes, B., Preuschoft, H., Wolff, J.E.A.: Stress- strength relationschips in the mandible of hominoids. In: Chivers, D.J., Wood, B.A., Bilsborough, A. (eds.) Food Acquisition and Processing in Primates, pp. 369–390. Plenum Press, London (1984)

Demes, B., Creel, N., Preuschoft, H.: Functional significance of allometric trends in the hominoid masticatory apparatus. In: Else, J.G., Lee, P.C. (eds.) Primate Evolution, Cambridge University Press, Cambridge, UK, pp. 228–237 (1986)

Demes, B., Forchap, E., Herwig, E.: They seem to glide. Are there aerodynamic effects in leaping prosimian primates? Z. Morphol. Anthropol. **78**, 373–385 (1989)

Demes, B., Fleagle, J., Jungers, W.L.: Takeoff and landing forces of leaping strepsirrhine primates. Journal of Human Evolution **37(2)**, 249–292 (1991)

Demes, B., Jungers, W.L. Gross, T.S., Fleagle, J.: Kinetics of leaping primates: Influence of substrate orientation and compliance. Amercan Journal of Physical Anthropology **96(4)**, 419–420 (1992)

Demes, B., Larson, S.G., Stern, J.T., Jungers, W.L., Biknevicius, A.R., Schmitt, D.: The kinetics of primate quadrupedalism: "hindlimb drive" reconsidered. J. Hum. Evol. **26**, 353–374 (1994)

Demes, B., Thompson, N.E., O'Neill, M.C., Umberger, B.R.: Center of Mass mechanics of chimpanzee bipedal walking. Am. J. Phys. Anthropol. **156(3)**, 422–433 (2014). https://doi.org/10.1002/ajpa.22667

Dempster, W.T., George, R., Gaughran, L.: Properties of body segments based on size and weight. Am. J. Anatomy. **120**, 33–54 (1967)

DeSilva, J.M., Holt, K.G., Churchill, S.E., Carlson, K.J., Walker, C.S., Zipfel, B., Berger, L.R.: The lower limb and mechanics of walking in *Australopithecus sediba*. Science. **340**, 1232999-1–1232999-5 (2013)

Dimery, N.J., Alexander, R., Mc, N., Ker, R.F.: Elastic extension of leg tendons in the locomotion of horses (*E. caballus*). J. Zool. London. **210**, 415–425 (1986)

Doden, E.: The relationship between the function and the inner cortical structure of metacarpal and phalangeal bones. In: Preuschoft, H., Chivers, D.J. (eds.) Hands of Primates, pp. 271–284. Springer, Wien (1993)

Drapeau, M., Streeter, M.A.: Modeling and remodelling responses to normal loading in the human lower limb. Am. J. Phys. Anthropol. **129**, 403–409 (2006)

Dubois, E.: Paleontological investigations on Java (Translation of "Palaeontologische onderzoekingen op Java"). In: Meikle, W.E., Parker, S.T. (eds.) Naming our Ancestors, pp. 37–40. Waveland, Prospect Height (1892/1994)

Dunbar, R.I.M.: The social ecology of gelada baboons. In: Rubenstein, D.I., Wrangham, R.W. (eds.) Ecological Aspects of Social Evolution: Birds and Mammals, pp. 332–351. Princeton Univ. Press, Primceton (1986)

Dunbar, D.C.: The influence of segmental movements and design on whole-body rotations during the airborne phase of primate leaps. Forces and stresses as factors which influence adaptation and morphology in primates. In: Crompton, R.H., Preuschoft, H. (eds.) Zeitschrift f. Morphologie und Anthropologie, vol. 80, pp. 109–124 (1994)

Dunsworth, H.J., Challis, H., Walker, A.: Throwing and bipedalism: a new look at an old idea. Courier Forsch.-Inst. Senckenberg. **243**, 105–110 (2003)

Dupret, V., Sanchez, S., Goujet, D., Tafforeux, P., Ahlberg, P.E.: A primitive placoderm sheds light on the origin of the jawed vertebrate face. Nature. **507**, 500–503 (2014)

Dusek, J., Ehrlein, H.-J., von Engelhardt, W., Hörnicke, H.: Beziehungen zwischen Trittlänge, Trittfrequenz und Geschwindigkeit bei Pferden, Zeitung f. Tierzüchtung und Züchtungsbiologie **87**:177–188 (1970)

Elftman, H., Manter, J.T.: Chimpanzee and human feet in bipedal walking. Am. J. Phys. Antropol. **20**, 69–79 (1935)

Endo, B.: A biomechanical study of the human facial skeleton by means of strain-sensitive lacquer. Okajima Folia Anat. Japonica **42**, 205–217 (1966)

Endo, B.: Mechanical analysis of the form of the facial skeleton. In: 7th Congress International des Sciences Anthropologiques et Ethnologiques, vol. 2, pp. 346–353 (1967)

Endo, B.: Structure of the human facial skeleton from the viewpoint of statics. In: Proc. of the 19th Japan National Congress for Applied Mechanics, pp. 8–14 (1969)

Endo, B.: Analysis of stresses around the orbit due to masseter and temporalis muscles respectively. J. Anthropol. Soc. Nippon. **78**, 251–266 (1970)

Endo, B.: Stress analysis on the facial skeleton of *Gorilla* by means of the wire strain gage method. Primates. **14**, 37–45 (1973)

Endo, B., Adachi, K.: Biomechanical simulation study on the forms of the frontal bones of the recent human facial skeleton by using a two-dimensional frame model with stepwise variable cross-section members. Okajimas Folia Anat. Japonica **64**, 335–349 (1988)

Endo, B., Fukushima, Y.: Untersuchungen zur Beanspruchung des menschlichen Gesichtsschädels beim Kauen mittels der zweidimensionalen spannungsoptischen Methode. Okajimas Folia Anat. Japonica **49**, 413–432 (1973)

Falaturi, P.: Zwischen Huf und Erde, Teil 1. Pferdhuf: Die Belastungen beim Auffussen. Tierwelt 47, November 2000; Teil 2. Hufschutz verändert das Gangmuster. Tierwelt 48, Dez. 2000 (2000)

Falkingham (2014)

Farke, A.A., Wolff, E.D.S., Tanke, D.H.: Evidence of combat in triceratops. Plos One. **4**(1), e4252 (2009). https://doi.org/10.1371/journal.pone.0004252

Fasnacht, M., Hess, N., Frey, E., Weiser, H.P.: Finite-element analysis in vertebrate palaeontology. Senckenbergiana Lethaea. **82**, 195–206 (2002)

Ferrara, T.L., Clausen, P., Huber, D.R., McHenry, C.R., Peddemors, V., Wroe, S.: Mechanics of biting in great white and sandtiger sharks. J. Biomech. **44**, 430–435 (2011). https://doi.org/10.1016/j.jbiomech.2010.09.028

Fierstine, H.L., Walters, V.: Studies in locomotion and anatomy of scombroid fishes. Mem. South. Calif. Acad. Sci. **6**, 1–31 (1968)

Fife, M., Bailey, C., Lee, D., Carrier, D.R.: Function of the oblique hypaxial muscles in trotting dogs. J Exp. Biol. **204**, 2371–2381 (2001)

Fifer, F.C.: The adoption of bipedalism by the hominids: a new hypothesis. Hum. Evol. **2**, 135–147 (1987)

Fischer, M.S.: Die Locomotion von *Procavia capensis* (Mammalia, Hyracoidea). Ein Beitrag zur Evolution des Bewegungssystems der Säugetiere. Habilitationsschrift Fakultät Biologie, Universität Tübingen (1994a)

Fischer, M.S.: Crouched posture and high fulcrum, a principle in the locomotion of small mammals: The example of the rock hyrax (*Procavia capensis*) (Mammalia, Hyracoidea). J. Hum. Evol. **26**, 501–524 (1994b)

Fischer, M.S., Blickahn, R.: The tri-segmented limbs of therian mammals. J. Exp. Zool. **305A**, 935–952 (2006)

Fischer, M.S., Lilje, K.E.: Hunde in Bewegung. Franckh Kosmos Verlags GmbH, Stuttgart (2011)

Fisk, G.: Carpal instability and the fractured scaphoid. Ann. R. Coll. Surg. Engl. **46**, 63–76 (1970)

Fleagle, J.G.: Locomotor adaptations of Oligocene and Miocene hominoids and their phyletic implications. In: Ciochon, R.L., Corruccini, R.S. (eds.) New Interpretations of Ape and Human Ancestry, pp. 301–324. Plenum Press, New York (1983)

Flügel, B., Greil, H., Sommer, K.: Anthropologischer Atlas,Grundlagen und Daten. Edition Wötzel, Frankfurt a.M (1986)

Fortelius, M.: Ungulate cheek teeth: developmental, functional and evolutionary interrelations. Acta Zool. Fenn. **180**, 1–76 (1985)

Fortuny, J., Marcé-Nogué, J., Heiss, E., Sanchez, M., Gil, L., Galobart, À.: 3D bite modeling and feeding mechanics of the largest living amphibian, the chinese giant salamander *Andrias davidianus* (Amphibia:Urodela). PLoS. **10**(4), e0121885 (2015)

Franzen, J.L.: Ein neuer Primate aus dem Mitteleozan der Grube Messel (Deutschland, S-Hessen). Cour. Forschungsinst. Senck. **91**, 151–187 (1987)

Franzen, J.L.: *Europolemur kelleri* n. sp. von Messel und ein Nachtrag zu *Europolemur koenigswaldi* (Mammalia, Primates, Notharctidae, Cercamoniinae). Senckenb. Lethaea. **80**, 275–287 (2000). https://doi.org/10.1007/BF03043672

Franzen, J.L., Gingerich, P.D., Habersetzer, J., Hurum, J.H., von Koenigswald, W., Smith, B.H.: Complete primate skeleton from the middle Eocene of Messel in Germany: morphology and paleobiology. PLoS One. **4**(5), e5723 (2009). https://doi.org/10.1371/journal.pone.0005723

Franzen, J.L., Habersetzer, J., Schlosser-Sturm, E., Franzen, E.L.: Paleopathology and fate of Ida (*Darwinius masillae*, Primates, Mammalia). Palaeobiodiversity Palaeoenvironments. **92**(4), 567–572 (2012)

Fritz, M., Wagner, H., Preuschoft, H., Schultz, J.: Ermittlung der Muskelaktivität zur Stabilisierung des Schultergürtels beim Tragen von Lasten mit einem oder beiden Armen. Zbl Arbeitsmed. **64**, 26–34 (2014). https://doi.org/10.1007/s40664-013-0008-y

Frost, B.J.: The optokinetic basis of head bobbing in the pigeon. J. Exp. Biol. **74**, 187–195 (1978)

Fujiwara, S.I., Kuwazuru, O., Inuzuka, N.: Relationship between scapula position and structural strength of rib cage in quadruped animals. J. Morphol. **270**, 1084–1094 (2009)

Full, R.J.: Mechanics and energetics of terrestrial locomotion: from bipeds to polypeds. In: Wieser, W., Gnaiger, E. (eds.) Energy Transformation in Cells and Animals, pp. 175–182. Thieme, Stuttgart (1989)

Gaetano, L.C., Mocke, H., Abdala, F.: The postcranial anatomy of *Diademodon tetragonus* (Cynodontia, Cynognatha). J. Vertebr. Paleontol. (2018). https://doi.org/10.1080/02724634.2018.1451872

Gambaryan, P.P.: How mammals run. In: Anatomical Adaptations. John Wiley & Sons, New York (1974). (Halsted Press, John Wiley & Sons)

Garcia, T.: Cadres stratigraphique, magnétostratigraphique et géochronologique des hominidés fossiles du site de Dmanisi en Géorgie. Muséum National d'Histoire Naturelle, Paris (2004)

Gatesy, S.M.: Caudofemoral musculature and the evolution of theropod locomotion. Palaeobiology. **16**(2), 170–186 (1990)

Gauthier-Pilters, H., Dagg, A.I.: The Camel: its Evolution, Behavior, and Relation to Man. Universitäy of Chicago Press, Chicago (1981)

Gellmann, K., Bertram, J.: The equine nuchal ligament 1: structural and material properties. Vet. Comp. Orthop. Traumatol. **15**, 1–6 (2002)

Gemballa, S.: Vergleichend-anatomische Untersuchungen am Lokomotions-apparat der Actinopterygier. Phylogenetische Rekonstruktion und funktionelle Hypothesen. Doctoral Thesis Biological Faculty, Tübingen (1995)

Gersten, K.: Einführung in die Strömungsmechanik. Shaker-Verlag, Aachen (2003)

Gingerich, P.D.: Primates in the Eocene. Palaeobiodiversity Palaeoenvironments. **92**(4), 649–663 (2012)

Godinot, M.: Fossil record of the primates from the Paleocene to the Oligocene. In: Handbook of Paleoanthropology. Springer, Berlin/Heidelberg (2014). https://doi.org/10.1007/978-3-642-27800-6_68-1

Godinot, M., Beard, C.: A survey of fossil primate hands. In: Preuschoft, H., Chivers, D.J. (eds.) Hands of Primates, pp. 335–378. Springer, Wien/New York (1993)

Gomberg, N., Morbeck, M.E., Preuschoft, H.: Multidisciplinary research in the analysis of primate morphology and behaviour. In: Morbeck, M.E., Preuschoft, H., Gomberg, N. (eds.) Environment, Behavior, and Morphology: Dynamical Interactions in Primates, pp. 5–21. G. Fischer-Verlag, New York (1979)

Görke, O.: Beitrag zur funktionellen Gestaltung des Schädels bei den Anthropomorphen und Menschen durch Untersuchung mit Röntgenstrahlen. Arch Anthropol. **1**, 91–108 (1904)

Gössling, R.: Biologische Vorteile und mechanische Probleme der langen Schnauze am Beispiel Plateosaurus. Beitrag zur funktionellen Schädelmorphologie mit Hilfe der Finite Elemente Struktur-Synthese (FESS). Verlag Dr. Hut, München (2010)

Grand, T.I.: A mechanical interpretation of terminal branch feeding. J. Mammal. **53**, 198–201 (1972)

Grand, T.I.: Body weight: its relation to tissue composition, segment distribution, and motor function I. interspecific comparisons. Am. J. Phys. Anthropol. **47**, (1977).: 211–240 and 241–248

Grand, T.I.: Adaptation of tissue and limb segments to facilitate moving and feeding in arboreal folivores. In: Montgomery, G.G. (ed.) I: The Ecology of Arboreal Folivores, pp. 231–241. Smithonian Institution Press, Washington (1978)

Grand, T.I.: Motion economy within the canopy: four strategies for mobility. In: Rodman, P.B., Cant, J.G.H. (eds.) Adaptations for Foraging in Non Human Primates, pp. 54–72. Columbia Univ. Press, New York, NY (1984)

Grand, T.I.: The functional anatomy of body mass. In: Damuth, J., McFassen, B.J. (eds.) Body Size in Mammalian Palaeontology, pp. 39–47. Cambridge Univ. Press, Cambridge, UK (1990)

Grassé, P.P. (ed.) Traité de Zoologie, 16 vols., Masson, Paris (1970)

Grasshoff, M.: Kurze Geschichte der Evolutions theorien. Morphisto Wissenschaftsverlag, Frankfurt (2014)

Grasshoff, M., Gudo, M.: Die Evolution der Tiere. Senckenbergische Naturforschende Gesellschaft, (Schweizerbart'sche Verlagsbuchhandlung, Stuttgart) (2001)

Gray, J.: Animal Locomotion. Weidenfeld & Nicholson, London (1968)

Gregory, W.K.: Studies on the evolution of the primates. Bull. Am. Mus. Nat. Hist. **35**, 239–496 (1916). New York

Griffiths, R.I.: Shortening of muscle fibres during stretch of the active cat medial gastrocnemius muscle: the role of tendon compliance. J. Physiol. **436**(1), 219–236 (1991)

Grine, F.E.: Trophic differences between "gracile" and "robust" australopithecines: a scanning electron microscope analysis of occlusal events. S. Afr. J. Sci. **77**, 203–230 (1981)

Grote, K.-H., Feldhusen, J.: Dubbel – Taschenbuch für den Maschinenbau. Springer, Berlin (2007)

Grzimeks Tierleben V: Knochenfische, Amphibien. Kindler-Verlag, München (1970)

Grzimeks Enzyklopädie: Säugetiere, Bde. 1–5 Kindler-Verlag, München (1988)

Grzimeks Tierlieben vols. IV, V, VI, Kindler-Verlag, 1970 and 1972

Gudo, M., Grasshoff, K.: The origin and early evolution of chordates: the "Hydroskelett-Theorie" and new insights towards a metameric ancestor. Senckenb. Lethaea. **82**, 325–346 (2002)

Gudo, M., Homberger, D.G.: The functional morphology of the pectoral fin girdle in the spiny dogfish (*Squalus acanthias*): implications for the evolutionary history of the pectoral girdle of vertebrates. Senckenb. Lethaea. **82**(1), 241–252 (2002)

Guibé, J.: La musculature. In: Grassé, P.P. (ed.) Traité de Zoologie. Tome XIV, Fasc. II, pp. 701–812. Masson, Paris (1970)

Günther, M.M.: Funktionsmorphologische Untersuchungen zum Sprungverhalten an mehreren Halbaffenarten (*Galago moholi, Galago (Otolemur garnetti, Lemur catta)* Ph.D. Thesis, Freie Universität Berlin (1989)

Günther, M.M., Boesch, C.: Energetic cost of nut-cracking behaviour in wild chimpanzees. In: Preuschoft, H., Chivers, D.J. (eds.) Hands of Primates, pp. 109–129. Springer, Wien (1993)

Günther, M.M., Kumakura, H., Nakano, Y., Ohta, H.: High-speed motion analysis of leaping in small primates. In: Dewitz, W., Kaiser, I. (eds.) Bewegungabläufe,. Darstellung, Manipulation, Analyse, pp. 70–85. ZEAM, Freie Universität, Berlin (1989)

Günther, M.M., Ishida, H., Kumakura, H., Nakano, Y.: The jump as a fast mode of locomotion in arboreal and terrestrial biotopes. Z. Morphol. Anthropol. **78**, 341–372 (1991)

Günther, M.M., Preuschoft, H., Ishida, H., Nakano, Y.: Can prosimian-like leaping be considered a preadaptation to bipedal walking in hominids? In: Matano, S., Tuttle, R.H., Ischida, H., Goodman, M. (eds.) Topics in Primatology, Vol. 3; Evolutionary Biology, Reproductive Endocrinology, and Virology, pp. 153–165. Tokyo Univ. Press, Tokyo (1992)

Gutmann, W.F.: Die Hydroskelett-Theorie. Aufsätze und Reden der Senckenbergischen Naturforsch. Ges. **21**, 91 (1972)

Gutmann, W.F.: Die Evolution hydraulischer Konstruktionen. W.Kramer, Frankfurt a.M., 2. Auflage 1995 (1989)

Gutmann, W.F.: Konstruktionszwänge in der Evolution: Schwimmende Vierfüsser. Nat. Mus. **124** (6), 165–187 (1994)

Haeckel, E.: Generelle Morphologie der Organismen. 2 Volumes. Berlin (1866)

Haile-Selassie, Y., et al.: An early *Australopithecus afarensis* postcranium from Woranso-Mille, Ethiopia. Proc. Natl. Acad. Sci. U. S. A. **107**, 12121–12126 (2010)

Haile-Selassie, Y., et al.: A new hominin foot from Ethiopia shows multiple Pliocene bipedal adaptations. Nature. **483**, 565–570 (2012)

Haines, R.W.: The extensor apparatus in the hand. J. Anat. Lond. **85**, 257–259 (1951)

Hall, B.K.: Bones and Cartilage: Developmental and Evolutionary Skeletal Biology. Academic Press, Elsevier, San Diego/London (2005)

Hall, B.K.: The Neural Crest and Neural Crest Cells in Vertebrate Development and Evolution. Springer, New York (2009)

Hall-Craggs, E.C.B.: An analysis of the jump in the lesser galago (*Galago senegalensis*). J. Zool. London. **147**, 20–29 (1965)

Hall-Craggs, E.C.B.: Physiological and histochemical parameters in comparative locomotor studies. In: Martin, R.D., Doyle, G.A., Walker, A.C. (eds.) Prosimian Anatomy, Biochemistry and Evolution, pp. 830–845. Duckworth, London (1977)

Harder, W.: Anatomy of fishes. Parts I and II. Schweizerbart'sche Verlagsbuchhandlung (Nägele & Obermiller), Stuttgart (1975)

Hartmann, D. S.: Ecology and behaviour of the manatee (*Trichechus manatus*) in Florida. Special Publication No. 5 The American Society of mammalogists (1979)

Hartwig-Scherer, S., Martin, R.D.: Was "Lucy" more human than her child? Observations on early hominid postcranial skeletons. J. Human Evol. **21**, 439–449 (1991)

Heiss, E., Natchev, N., Gumpenberger, M., Weissenbacher, A., van Wassenbergh, S.: Biomechanics and hydrodynamics of prey capture in the Chinese giant salamander reveal a high-performance jaw-powered suction feeding mechanism. J. R. Soc. Interface. **10**, 20121028 (2013a). https://doi.org/10.1098/rsif.2012.1028

Heiss, E., Aerts, P., Van Wassenbergh, S.: Masters of change: seasonal plasticity in the prey-capture behavior of the Alpine newt *Ichthyosaura alpestris* (Salamandridae). J. Exp. Biol. **216**, 4426–4434 (2013b). https://doi.org/10.1242/jeb.091991

Heiss, E., Aerts, P., Van Wassenbergh, S.: Aquatic-terrestrial transitions of feeding systems in vertebrates: a mechanical perspective. J. Exp. Biol. **221**(8), jeb154427 (2018). https://doi.org/10.1242/jeb.154427

Henke, W., Rothe H.: Paläoanthropologie Springer-Verlag, Berlin (1994)

Hennig, W.: Phylogenetische Systematik. Parey, Berlin (1982)

Herring, S.W., Rafferty, K.L.: Cranial and facial sutures. Functional loading in relation to growth and morphology. In: Davidovitch, Z., Mar, J. (eds.) Biological Mechanisms of Tooth Eruption, Resorption and Replacement by Implants, pp. 269–276. Boston, Harvard Societny and Advanced Orthodontics (2000)

Herring, S.W., Rafferty, K.L., Liou, Z.J., Marshall, C.D.: Jaw muscles and the skull in mammals. The biomechanics of mastication. Comp. Biochem. Physiol. **131**, 207–219 (2001)

Hertel, H.: Struktur, Form, Bewegung. Krausskopf-Verlag, Mainz (1963)

Herzog, K.: Anatomie und Flugbiologie der Vögel. G.Fischer-Verlag, Stuttgart (1968)

Hiiemae, K.: Masticatory movements in primitive mammals. In: Anderson, D., Matthews, B. (eds.) Mastication, pp. 105–118. Wright, Bristol (1976)

Hiiemae, K.: Mammalian mastication: a review of the activity of the jaw muscles and the movements they produce in chewing. In: Butler, P.M., Joysey, K.A. (eds.) Development, Function and Evolution of teeth, pp. 359–398. Acad. Press, London (1978)

Hiiemae, K.: Functional aspects of primate jaw morphology. In: Chivers, D.J., Wood, B.A., Bilsbourough, A. (eds.) Food Acquisition and Processing in Primates, pp. 257–281. Plenum Press, London (1984)

Hildebrand, M.: Symmetrical gaits of horses. Science. **150**, 701–708 (1965)

Hildebrand, M.: Analysis of the symmetrical gaits of tetratpods. Folia Biotheoretic. **6**, 9–22 (1966)

Hildebrand, M., Goslow, G.: Vergleichende and funktionelle Anatomie der Wirbeltiere. Springer, Berlin/Heidelberg/New York/Tokyo (2005)

Hildebrand, M. Goslow, G.: Analysis of vertebrate structure. 5th ed., John Wiley & Sons, Inc., Hoboken, New Jersey (2001)

Hildebrand, M., Bramble, D.M., Liem, K.E., Wake, D.B.: Functional Vertebrate Morphology. Harvard Univ. Press, Cambridge (1985)

Hirasaki, E., Matano, S., Nakano, Y., Ishida, H.: Vertical climbing in *Ateles geoffroyi* and *Macaca fuscata* and ist comparative neurological background. In: Matano, S., Tuttle, R.H., Ischida, H., Goodman, M. (eds.) Topics in Primatology, Vol. 3; Evolutionary Biology, Reproductive Endocrinology, and Virology, pp. 167–176. Tokyo Univ. Press, Tokyo (1992)

Hochmuth, G.: Biomechanik Sportlicher Bewegungen. Limpert, Frankfurt a.m (1967)

Hofer, H.: Zur Kenntnis der Suspensionsformen des Kieferbogens und deren Zusammenhänge mit dem Bau und mit der Kinetik des Schädels bei den Knochenfischen. Zool. Jb. Anat. **69**, 321–404 (1945)

Hofer, H.: Die morphologische Analyse des Schädels des Menschen. In: Heberer, H. (ed.) Menschliche Abstammungslehre, pp. 145–226 (1965)

Hohn, B.: Walking with the shoulder of giants: Biomechanical conditions in the tetrapod shoulder girdle as a basis for sauropod shoulder construction. In: Klein, N., Remes, K., Gee, C.T., Sander, P.M. (eds.) Biology of the Sauropod Dinosaurs, pp. 182–196. Indiana Univ. Press, Bloomington (2011)

Hohn-Schulte, B.: Form and Function of the Shoulder Girdle in Sauropod Dinosaurs. A Biomechanical Investigation with the Aid of Finite Elements. Doctoral thesis (2010)

Hohn-Schulte, B., Preuschoft, H., Witzel, U., Distler-Hoffmann, C.: Biomechanics and functional preconditions for terrestroial lifestyle in basal tetrapods, with spezial consideration of *Tiktaalik roseae*. Hist. Biol. **2013**, 1–15 (2013). https://doi.org/10.1080/08912963.2012.755677

Hollihn, U.: Bimanual suspenspory behaviour: Morphology, selective advantage and phylogeny. In: Preuschoft, H., Chivers, D.J., Brockelmann, W., Creel, N. (eds.) The Lesser Apes. Evolutionary and Behavioural Biology, pp. 85–95. Edinburgh Univ. Press, Edinburgh (1984)

Homberger, D. G.: The Lingual Apparatus of the African Grey Parrot, *Psittacus erithacus* Linn, (Aves: Psittacidae): Description and Theoretical Mechanical Analysis. Washington, D.C., The American Ornithologists' Union. Ornithological Monographs, No. 39. Foster, M. S. and Johnston, D. W (1986)

Homberger, D.G.: The case of the cockatoo bill, horse hoof, rhinoceros horn, whale baleen, and Turkey beard: the integument as a model system to explore the concepts of homology and non-homology. In: Dutta, H.M., Datta Munshi, J.S. (eds.) Vertebrate Functional Morphology: Horizon of Research in the 21st Century, pp. 317–343. Science Publishers Inc., Enfield (2001)

Homberger, D.G., Brush, A.H.: Functional-morphological and biochemical correlations of the keratinized structures in the African Grey Parrot, *Psittacus erithacus* (Aves). Zoomorphology. **106**, 103–114 (1986)

Homberger, D. G., Barnett, H., Okunbakin, T., Ham, K., Matthews, K. L., Butler, L., Bragulla, H. H.: The microarchitecture of the cornified epidermal sheath of the cat claw. 8th Internat. Congress of Vertebrate Morphology, Paris (Abstract) (2007)

Homberger, D.G., Ham, K., Ogunbakin, T., Mtthews, K.L., Butler, L., Bragulla, H.H.: The structure of the cornified claw sheath in the domesticated cat (*Felis catus*): implications for the claw shedding mechanism and the evolution of cornified digital end organs. J. Anat. **214**, 620–643 (2009)

Hoyt, J.W.: Hydrodynamic drag reduction due to fish slimes. In: Wu, T.Y.T., Hoyt, D. F., Taylor, C.R. 1981: Gait and the energetics of locomotion in horses. Nature. **292**, 239–240 (1975)

Hummel, J., Clauss, M.: Megaherbivores as pacemakers of carnivore diversity and biomass: distributing or sinking trophic energy. Evol. Ecol. Res. **10**, 925–930 (2008)

Hummel, J., Clauss, M.: Sauropod feeding and digestive physiology. In: Klein, N., Remes, K., Gee, C.T., Sander, P.M. (eds.) Biology of the Sauropod Dinosaurs: Understanding the Life of Giants, pp. 11–33. Indiana University Press, Bloomington/Indianapolis (2011)

Hutchinson, J.R., Schwerda, D., Famini, D., Dale, R.H.I., Fischer, M., Kram, R.: The locomotor kinematics of African and Asian elephants: changes with speed and size. J. Exp. Biol. **209**, 3812–3827 (2006). https://doi.org/10.1242/jeb.02443. See also Ren et al

Hylander, W.L.: The human mandible: lever or link? Am. J. Phys. Anthropol. **43**, 227–242 (1975)

Hylander, W.L.: Mandibular function in *Galago crassicaudatus* and *Macaca fascicularis*; an in vivo approach to stress analysis of the mandible. J. Morphol. **159**, 253–296 (1979)

Hylander, W.L.: Loading patterns and jaw movement during the masticatory power stroke in macaques. Am. J. Phys. Anthropol. **52**, 239 (1980)

Hylander, W.L.: Stress and strain in the mandibular symphysis of primates: a test of competing hypotheses. Am. J. Phys. Anthropol. **64**, 1–46 (1984)

Hylander, W.L.: In vivo bone strain as an indicator of masticatory bite force in *Macaca fascicularis*. Archs. Oral Biol. **34**, 713–722 (1986)

Hylander, W.L.: Implications of in vivo experiments for intrpretating the functional significance of "robust" australopithevcine jaws. In: Grine, F.E. (ed.) Evolutinary History of the "Robust" Australopithecines, pp. 55–83. Aldine de Gruyter, New York (1988)

Hylander, W.L., Johnson, K.R.: The relationship between masserter force and electromyogram during mastication in the monkey Macaca fascicularis. Arch. Oral Biol. **34**, 713–722 (1989)

Hylander, W.L., Johnson, K.R.: Strain gradients in the craniofacial region of primates. In: Davidovitch, Z. (ed.) The Biological Mechanisms of Tooth Movement and Craniofacial Adaptation, pp. 559–569. Ohio State University College of Dentistry, Columbus (1992)

Hylander, W.L., Johnson, K.R.: Jaw muscle function and wishboning of the mandible during mastication in macaques and baboons. Am. J. Phys. Anthropol. **94**, 523–547 (1994)

Internet: oceanwide-expeditions.com: rockhopper penguin

Ishida, H., Kimura, T., Okada, M.: Patterns of bipedal walking in anthropoid primates. In: Kondo, S., Kawai, M., Ehara, A., Kawamura, S. (eds.) Proceedings from the Symposia of the Fifth Congress of the International Primatological Society, pp. 287–301. Japan Science Press, Tokyo (1975)

Ishida, H., Kumakura, H., Kondo, S.: Primate bipedalism and quadrupedalism. Comparative Electromyography. In: Kondo, S. (ed.) Primate Morphophysiology, Locomotor Analysis and Human Bipedalism, pp. 59–79. Univ. Tokyo Press (1985)

Ishida, H., Jouffroy, F.K., Nakano, Y.: Comparative dynamics of pronograde and upside down horizontal quadrupedalism in the slow loris (*Nycticebus coucang*). In: Jouffroy, F.K., Stack, H., Niemitz, C. (eds.) Gravity, Posture and Locomotion in Primates, pp. 209–220. Il Sedicesimo, Firenze (1990)

Ishida, H., Kunimatsu, Y., Takano, T., Nakatsukasa, M.: *Nacholapithecus* skeleton from the middle Miocene of Kenya. J. Human Evol. **46**, 67–101 (2004)

Isler, K.: Characteristics of vertical climbing in gibbons. Evol. Anthropol. **11**(suppl.1), 149–152 (2002)

Isler, K.: 3D-Kinematics of vertical clinging in hominoids. PhD Thesis Zürich (2003)

Jackowski, J., Witzel, U.: Kraftflusslinienanalyse axial belasteeter Implantatkörper in der proximalen Maxilla mit Hilfe der Finite-Element-Methode. Int. J. Oral Maxillofac. Implants

Jaeger, F.: Die Biomechanik der menschlichen Wirbelsäule in abhängigkeit vom Grad der Ventralflexion. Doctoral Thesis Medizinische Fakultät der RU Bochum (2003)

Janis, C.M.: Aspects of the Evolution of Herbivory in Ungulate Mammals. Harvard University, Doctoral thesis (1979)

Janis, C.M.: Prediction of primate diets from molar patterns. In: Chivers, D.J., Wood, B.A., Bilsborough, A. (eds.) Food Acquisition and Processing in Primates, pp. 331–340. Plenum Press, London (1984)

Janis, C.M., Buttril, K., Figueirido, B.: Locomotion in extinct giant kangaroos: were Sthenothurines hop-less monsters? PLoS One. **9**(10), e109888 (2014). www.plosone.org

Jenkins, F.: The postcranial skeleton of African cynodonts. Bull. Peabody Mus. Nat. Hist. **36**, 1–216 (1971)

Jenkins and Goslow, 1983

Jenkins, J., Thomason, J.J., Norman, D.B.: Primates and engineering principles: application to craniodental mechanismas in ancient terrestrial predators. Senckenb. Lethaea. **82**(1), 223–240 (2002)

Johanson, D.C., Edey, M.E.: Lucy, the Beginnings of Humankind. Simon and Schuster, New York (1981)

Johanson, D.C., White, T.D.: A systematic assessment of early African hominids. Science. **202**, 321–330 (1979)

Johanson, D.C., White, T.D., Coppens, Y.: A new species of the genus *Australopithecus* (Primates, Hominidae) from the Pliocene of Eastern Africa. Kirtlandia. **28**, 2–14 (1978)

Jolly, A.: Lemur Behaviour. The University of Chicago Press, Chicago/London (1966)

Jolly, C.J.: *Hadropithecus*, a lemuroid small-object-feeder. Man. **5**, 619–626 (1970a)

Jolly, C.J.: The seed-eaters: a new model of hominid differentiation based on a baboon analogy. Man. **5**, 5–26 (1970b)

Jones, K.: The axial skeleton: a missing puzzle piece in the evolution of cursorial locomotion in horses. 74 th SVP- Meeting, Berlin Abstract (2014)

Jones, K.: Evolutionary allometry of lumbar shape in Felidae and Bovidae. Biol. J. Linnaean Soc. (2015). ***published in the meantime?***

Jones, K.: Preliminary data on the effect of osseous anatomy on ex vivo joint mobility in the equine thoracolumbar region. Equine Vet. J. **48**(4), 502–508 (2016)

Jones, K.E., Holbrook, L.T.: The evolution of lateral accessory articulations in the lumbar region of perissodactyls. J. Vertebr. Paleontol. e1224892 (2016). https://doi.org/10.1080/02724634.2016.1224892

Jöris, O.: Der altpaläolithische Fundplatz Dmanisi (Georgien, Kaukasus). Monographien des Römisch-Germanischen Zentralmuseums. 56 farbige Abbildungen, 35 Taf., 8 Farbtaf. 3 Beilagen. Verlag Schnell und Steiner. ISBN: 3795421403 (2008)

Jouffroy, F.K.: La musculature des members chez les lémuriens de Madagascar. Mammalia. **2**, 1–326 (1962)

Jouffroy, F.K.: Musculature des members. In: Grassé, P.P. (ed.) Traité de Zoologie XVI, pp. 1–475. Masson, Paris (1971)

Jouffroy, F.K., Lessertisseur, J.: Relationships between limb morphology and locomotor adaptations among prosimians: an osteometric study. In: Morbeck, M.E., Preuschoft, H., Gomberg, N. (eds.) Environment, Behavior, and Morphology: Dynamical Interactions in Primates. G. Fischer-Verlag, New York (1979)

Jouffroy, F. K., Godinot, M., Nakano, Y.: Biometrical characteristics of primate hands. In Preuschoft & D. Chivers(eds.) Hands of primates. Springer, Wien, pp. 133–172 (1993)

Jouffroy, F.K., Medina, M.F.: Radio-ulnar deviation of the primate carpus: an y-ray study. Z. Morphol. Anthropol. **83**, 275–289 (2002)

Jouffroy, F.K., Petter, A.: Gravity-related kinematic changes in lorisine horizontal locomotion in relation to the position of the body. In: Jouffroy, F.K., Stack, H., Niemitz, C. (eds.) Gravity, Posture and Locomotion in Primates, pp. 199–208. Il Sedicesimo, Firenze (1990)

Jouffroy, F.K., Stern, J.: Telemetered EMG study of the antigravity versus propulsive actions of knee and elbow muscles in the slow loris (*Nycticebus coucang*). In: Jouffroy, F.K., Stack, H., Niemitz, C. (eds.) Gravity, Posture and Locomotion in Primates, pp. 221–236. Il Sedicesimo, Firenze (1990)

Jouffroy, F.K., Godinot, M., Nakano, Y.: Biometrical characteristics of primate hands. In: Preuschoft, H., Chivers, D.J. (eds.) Hands of Primates, pp. 133–171. Springer, Wien/New York (1993)

Jungers, W.L.: Interlimb proportions in humans and foaail hominines: variability and scaling. In: Grine, F.E., Fleagle, J.G., Leakey, R.E. (eds.) The First Humans- Origin and Early Evolution of the Genus Homo. Third Stony Brook Human Evolution Symposium and Workshop, pp. 83–100. Springer, Dordrecht (2009)

Jungers, W.L.: Climbing and terrestriality are not mutually exclusive (2011)

Jungers Jr., W.L.: Locomotion, limb proportions and skeletal allometry in lemurs and lorises. Folia Primatol. **32**, 8–28 (1979)

Jungers Jr., W.L.: Scaling of the hominoid locomotor skeleton with special reference to the lesser apes. In: Preuschoft, H., Chivers, D.J., Brockelmann, W., Creel, N. (eds.) The Lesser Apes. Evolutionary and Behavioural Biology, pp. 146–169. Edinburgh Univ. Press, Edinburgh (1984)

Jungers, W.L., Larson, S.G., Harcourt-Smith, W., Morwood, M.J., Sutikna, T., Rokhus Due Awe T. Djubiantonoe: Descriptions of the lower limb skeleton of Homo floresiensis. Journal of Human Evolution, **57**, 538–554

Jungers Jr., W.L.: Body size and scaling of limb proportions of primates. In: Jungers, W.L. (ed.) Size and Scaling in Primate Biology, pp. 345–381. Plenum Press, New York/London (1985)

Jungers, W.L., Stern, J.: Preliminary electromyographical analysis of brachiation in gibbon and spider monkey. Int. J. Primatol. **2**, 19–33 (1981)

Jungers, W.L., Stern, J.: Kinesiological aspects of brachiation in lar gibbons. In: Preuschoft, H., Chivers, D.J., Brockelmann, W., Creel, N. (eds.) The Lesser Apes. Evolutionary and Behavioural Biology, pp. 119–134. Edinburgh Univ. Press, Edinurgh (1984)

Kagaya, M., Ogihara, N., Nakatsukasa, M.: Morphological study of the anthropoid thoracic cage: scaling of thoracic width and an analysis of rib curvature. Primates. **49**, 89–99 (2008a)

Kagaya, M., Ogihara, N., Nakatsukasa, M.: Clavicular length of hominoids: a comparison with the thoracic cage width. IPS Post-Congress symposium on Comparative functional morphology in proimates. Durham, U.K. Abstract (2008b)

Kagaya, M., Ogihara, N., Nakatsukasa, M.: Morphological study of the anthropoid thoracic cage: rib orientation and its implication for orthograde postural behaviour. Oral presentation at the post-congress symposium on Comparative functional morphology in primates. Durham (2008c) (*may have appeared in 2010 or 2012) Where*?

Kälin, J., Knüsel, L.: Über die Lokomotion der Krokodilien. - Revue Suisse de Zoologie. Géneve. **51**, 389–393 (1944)

Kane, E.A., Marshal, C.D.: Comparative feeding kinematics and performance of odontocetes: belugas, Pacific white-sided dolphins and long-finned pilot whales. J. Exp. Biol. **212**, 3939–3950 (2009). https://doi.org/10.1242/jeb.009852

Karakostis, F.A., Harvati, K.: Auf die Finger geschaut. Senckenberg Wissenschaftsmagazin. **149** (1–3), 16–19 (2019)

Karakostis, F.A., Lorenzo, C.: Morphometric patterns among the 3 D surface of human hand enthesis. Am. J. Phys. Anthropol. **160**, 694–707 (2016)

Karakostis, F.A., Hotz, G., Scherf, H., Wahl, J., Harvati, K.: Occupational manual activity is reflected in the patterns among hand entheses. Am. J. Phys. Anthropol. **164**, 30–40 (2017)

Karakostis, F.A., Hotz, G., Tourloukis, V., Harvati, K.: Evidence for precision grasping in Neanderthal daily activities. Science Adv. **4**, eaat 2369 (2018)

Kaup, L.: Studien an den Hinterextremitäten bipeder Nagetiere. Diss. Kiel, Mathem.-Naturwiss, Fakultät (1973)

Kawano, S.M., Blob, R.W.: Terrestrial force production by the limbs of a semi-aquatic salamander provides insight into the evolution of terrestrial locomotor mechanics. Preprint May 2021. https://doi.org/10.1101/2021.05.01.442256

Kawano, S.M., Blob, R.W.: Propulsive forces of mudskipper fins and salamander limbs during terrestrial locomotion: implications for the invasion of land. Integr. Comp. Biol. **53**(2), 283–294 (2013)

Kay, R.F.: Functional adaptations of primate molar teeth. Am. J. Phys. Anthropol. **43**, 195–216 (1975)

Kay, R.F.: Molar structure and diet in extant cercopithecidae. In: Butler, P.N., Joysey, K.A. (eds.) Development, Function and Evolution of Teeth, pp. 309–339. Acad. Press, London (1978)

Kay, R.F., Hiiemae, K.: Jaw movement and tooth use in recent and fossil primates. Am. J. Phys. Anthropol. **40**, 227–256 (1974)

Kay, R.F., Hylander, W.L.: The dental structure of mammalian folivores with special reference to primates and Phylangeroid3w (Marsupialia). In: Motgomery, G.G. (ed.) The Ecology of Arboreal Folivores, pp. 173–191. Smithsonian Institution, Waschington DC (1978)

Kemp, T.S.: The primitive cynodont *Procynosuchus*: structure, function and evolution of the postcranial skeleton. Philos. Trans. R. Soc. **288**, 217–258 (1980a)

Kemp, T.S.: Aspects of the structure and functional anatomy of the middle Triassic cynodont *Luangwa*. J. Zool. **191**, 193–239 (1980b)

Kemp, T.S.: Mammal-Like Reptiles and the Origin of Mammals. Academic Press, London (1982)

Ker, R.F.: Elasticity of hand and forefoot tendons. In: Preuschoft, H., Chivers, D.J. (eds.) Hands of Primates, pp. 257–269. Springer, Wien (1993)

Key, A.J.M., Dunmore, C.J.: Use of hands and fingers in hominines probably. Am. J. Phys. Anthropol. (2015)

Kienapfel, K.: Die biomechanische Konstruktion des Körperstammes bei Pferden als Beispiel für quadrupede Tiere unter spezieller Berücksichtigung der während der Lokomotion auftretenden Spannungen aufgrund von EMG-Aufzeichnungen und Verhaltensbeobachtungen. Doctoral Thesis Fakultät für Biologie und Biotechnologie RU Bochum (2014)

Kienapfel, K.: The effect of three different head-neck-positions on the average EMG activity of thre important neck muscles in the horse. J. Anim. Physiol. Anim. Nutr. **99**, 132–138 (2015)

Kienapfel, K., Preuschoft, H.: Was bewirkt das Aufrollen des Pferdehalses? Einflüsse der Halsstellung auf die Dehnung der Weichteile. Pferdeheilkunde. **27**(4), 358–370 (2011)

Kienapfel, K., Preuschoft, H.: Lever arm lengths in the horse's atlanto-occipital joint and its consequences for mobility of the head (in prep)

Kienapfel, K., Laebe, S., Preuschoft, H.: Do tracks yield reliable information on gaits? Part 1: the case of horses. Fossil Record. **17**, 1–9, www.foss-rec.net/17/1/2014 (2014). https://doi.org/10.5194/fr-17-1-2014

Kienapfel, K., Preuschoft, H., Wulf, A., Wagner, H.: (electronic version in 2017): The biomechanical construction of the horse's body and activity patterns of three important muscles of the trunk in the walk, trot and canter. J. Anim. Physiol. Anim. Nutr. https://doi.org/10.1111/jpn.12840. Internal Article ID: 14721312 (2018)

Kimura, T.: Bipedal and quadrupedal walking of primates: comparative analysis. In: Kondo, S. (ed.) Primate Physiology, Locomotor Analyses and Human Bipedalism, pp. 81–104. Univ. Tokyo Press (1985)

Kimura, T.: Long and robust limb bones in primates. In: Ehara, A., Kimura, T., Takenaka, O., Iwamoto, M. (eds.) Primatology Today, pp. 495–498. Elsevier, Amsterdam (1991)

Kimura, T. Hindlimb dominance during primate high-speed locomotion. Primates **33**(4), 465–476 (1992)

Kimura, T.: Long bone characteristics of primates. Zeitschrift für Morphologie und Anthropologie, **80**, 265–280 (1995)

Kimura, T.: Primate limb bones and locomotor types in arboreal or terrestrial environments. Z. Morphol. Anthropol. **83**, 201–219 (2002)

Kimura, T.: Kyoto meeting, Abstract volume (2010)

Kimura, T., Okada, M., Ishida, H.: Kinesiological characteristics of primate walking: its significance in human walking. In: Morbeck, M.E., Preuschoft, H., Gomberg, N. (eds.) Environment, Behaviour and Morphology: Dynamic Interactions in Primates, pp. 297–311. G. Fischer Verlag, New York (1979)

Kimura, T., Okaka, M., Yamazaki, N., Ishida, H.: Speed of the bipedal gaits of man and non-human primates. Ann. Sci. Nat. Zool. Paris. **138**, 145–158 (1983)

Kirschmann, E.: Das Zeitalter der Werfer. E. Kirschmann, Hannover IDSBN 3-00-004783-2 (1999)

Klein, R.G.: The Human Carreer, 2nd edn. Univ. of Chicago Press, Chicago (1999)

Klein, R.G.: Stable carbon isotopes and human evolution. Stable carbon isotopes and human evolution. Proc. Natl. Acad. Sci. U S A. **110**(26), (2013).; 2013 Jun 25 PMC3696760

Klein, N., Remes, K., Gee, C.T., Sander, P.M.: Biology of the Sauropod Dinosaurs: Understanding the Life of Giants. Indiana University Press, Bloomington 331 pp. (2011)

Klima, M.: Development of shoulder girdle and sternum in mammals. In: Duncker, R., Fleischer, D.G. (eds.) Vertebrate Morphology, pp. 81–85. Fischer, Stuttgart (1985)

Klima, M.: Early Development of the Shoulder Girdle and Sternum in Marsupials (Mammalia: Metatheria) Advances in Anatomy, Embryology and Cell Biology, vol. 109. Springer, Berlin/New York (1987)

Koebke, J.: Functional morphology of the human carpus. In: Preuschoft, H., Chivers, D.J. (eds.) Hands of Primates, pp. 235–244. Springer (1993)

Köhler, M.: Skeleton and habitat of recent and fossile ruminants. Münchner Geowiss. Abhandlungen, Reihe A Geologie und Pläontologie 25:8–88 (1993)

König, K.E., Skewes, O., Helmreich, M., Böck, P.: Macroscopic and histological investigation of guanaco footpads (*Lama guanaco*, Müller 1776). J. Morphologxy, **276(3)**, 331–341 (2014)

Kortlandt, A.: Experimentation with chimpanzees in the wild. In: Starck, D., Schneider, R., Kuhn, H.-J. (eds.) Neue Ergebnisse der Primatologie/Progress in Primatology, pp. 208–224. G. Fischer-Verlag, Stuttgart (1967)

Kortlandt, A.: Handgebrauch bei freilebenden Schimpansen. In: Rensch, B. (ed.) Handgebrauch und Verständigung bei Affen und Frühmenschen, pp. 59–102. Hans Huber, Bern (1968)

Kostuj, T.: Längenänderungen der menschlichen Unterarmmuskeln in vivo. Doctoral Thesis, Fakultät für Medizin, Ruhr-Universität, Bochum (2002)

Kozma, E.E., Webb, N.M., Harcourt-Smith, W.E.H., Raichlen, D.A., D'Aout, K., Brown, M.H., Finestone, E., Ross, S.R., Aerts, P., Pontzer, H.: Hip extensor mechanics and the evolution of walking and climbing capabilities in humans, apes, and fossil hominins. Proc. Natl. Acad. Sci. (2018). https://doi.org/10.1073/pnas.1715120115

Krause, C.: Die Beweglichkeit des Schultergelenks arboricoler und terricoler quadrupeder Primaten. Doctoral Thesis, Biologisch-Pharmazeutische Fakultät Universität Jena (2008)

Kuhn, C., Frey, E.: Walking like caterpillars, flying like bats - pinniped locomotion. Palaeodiversity Palaeoenvironments. **92**, 197–210 (2012)

Kull, U.: Turgeszenz, Hydraulik, Information und das Maschinenkonzept in der Biologie. In: Maier, W., Zoglauer, T. (eds.) Technomorphe Organismuskonzepte. Modellübertragungen zwischen Biologie und Technik, pp. 199–212. Frommann –Holzbog, Stuttgart (1994)

Kullmer, O., Schrenk, F., Dörrhöfer, B.: High resolution 3D-image and analysis of ape, hominid and human footprints. In: Franzen, Köhler, Moya-Sola (eds.) Walking upright. Cour. Forschungsinst. Senck. **243**, 85–91 (2003)

Kümmell, S.: Die Digiti der Synapsida: Anatomie, Evolution und Konstruktionsmorphologie. Shaker-Verlag, Aachen (2009)

Kummer, B.: Bauprinzipien des Säugerskeletes. Thieme-Verlag, Stuttgart (1959a)

Kummer, B.: Biomechanik des Säugersekeletts. In: Helmke, J.-G., Lengerken, H.V., Starck, D. (eds.) Handbuch der Zoologie, Bd. 8, **6**(2):1–80, Berlin (1959b)

Kummer, B.: Das mechanische Problem der Aufrichtung auf die Hinterextremität im Hinblik auf die Evolution der Bipedie des Menschen. In: Heberer, H. (ed.) Menschliche Abstammungslehre, pp. 227–248. G. Fischer, Stuttgart (1965)

Kummer, B.: Die Beanspruchung des Armskeletts beim Hangeln. Ein Beitrag zum Brachiatorenproblem. Anthropol. Anz. **32**, 74–82 (1970)

Kummer, B.: Biomechanics of the mammalian skeleton, problems of static stress. In: Nachtigall, W. (ed.): Physiology of movement - Biomechanics. Bewegungsphysiologie - Biomechanik. G. Fischer Verlag, Stuttgart. Fortschr. Zool. **24**, 57–73 (1977)

Kuschel, C.H.: Vergleichende funktionell-morphologische Untersuchungen der Vorderextremitäten von Ohren- und Hundsrobben (Otariidae und Phocideae). Masters Thesis, Fakultät für Biologie, Ruhr-Universität, Bochum (1994)

Kuschel, T., Preuschoft, H.: More than wet feet- Whole body involved! On the interaction of propulsory habits and vertebral column in aquatic mammals. Centennary Meeting of the Paläontologische Gesellschaft, Berlin. Terra Nostra Abstract (2012)

Kuznetsov, A.N., Luchkina, O.S., Panyutina, A.A., Kryukova, N.V.: Observations on escape runs in wild European hare as a basis for the mechanical concept of extreme cornering with special inference of a role of the peculiar subclavian muscle. Mammalian Biology/Zeitschrift für Säugetierkunde. **84**, 61–72 (2017)

Landsmeer, J.M.F.: The anatomy of the dorsal aponeurosis of the human finger and its functional significance. Anat. Rec. **104**, 31–44 (1949)

Landsmeer, J.M.F.: Anatomical and functional investigation on the articulation of the human fingers. Act. Nat. **24**, 2511–2569 (1955)

Landsmeer, J.M.C.F.: The hand and hominisation. Acta. Morph. Neerl.-Scand. **25**, 83–93 (1987)

Landsmeer, J.M.F.: Evolution and the hand. In: Preuschoft, H., Chivers, D.J. (eds.) Hands of Primates, pp. 323–333. Springer, Wien/New York (1993)

Langdon, J.H., Bruckner, J., Baker, H.H.: Pedal mechanics in early hominids. In: Coppens, Y., Senut, B. (eds.) Origine(s) de la Bipedie chez les Hominidés, pp. 159–167. CNRS, Paris (1991)

Larson, S.G.: Evolutionary transformation of the hominin shoulder. Evol. Anthropol. **16**, 172–186 (2007)

Larson, S.: Did australopithecines climb trees? Science. **338**, 478 (2012). https://doi.org/10.1126/science.1230128

Larson, S.G.: Rotator cuff muscle size and the interpretation of scapular shape in primates. J. Hum. Evol. **80**, 96–106 (2015)

Larson, S.G., Stern, J.T.: Role of supraspinatus in the quadrupedal locomotion of vervets (*Cercopithecus aethiops*). Implications for interpretations of humeral morphology. Am. J. Phys. Antropol. **79**(3), 369–378 (1989)

Larson, S.G., Stern, J.T.: Humeral retractor EMG during quadrupedal walking in primates. J. Exp. Biol. **210**, 1204–1215 (2007)

Larson, S.G., Stern Jr., J.T.: Hip extensor EMG and forelimb/hind limb weight support asymmetry in primate quadrupeds. Am. J. Phys. Anthropol. **138**(3), 343–355 (2009). https://doi.org/10.1002/ajpa.20935

Larson, S.G., Schmitt, D., Lemelin, P., Hamrick, M.: Limb excursion during quadrupedal walking: how do primates compare to other mammals? J. Zool. Lond. **255**, 353–365 (2001)

Latimer, B.M., Lovejoy, C.O.: The calcaneus of *Australopithecus afarensis* and its implications for the evolution of bipedality. Am. J. Phys. Anthropol. **78**, 369–386 (1989)

Latimer, B.M., Lovejoy, C.O.: Hallucal tarsometatarsal joint in *Australopithecus afarensis*. Am. J. Phys. Anthropol. **82**(2), 125–113 (1990)

Latimer, B., Ohman, J.C., Lovejoy, C.O.: Talocrural joint in African hominoids: implications for *Australopithecus afarensis*. Am. J. Phys. Anthropol. **74**, 155–175 (1987)

Lauder, G.V.: Aquatic feeding in lower vertebrates. In: Hildebrand, M., Bramble, D.M., Liem, K.F., Wake, D.B. (eds.) Functional Vertebrate Morphology, pp. 210–229. Harvard University Press, Cambridge, UK (1985)

Lauder, G.V., Prendergast, T.: Kinematics of aquatic prey capture in the snapping turtle *Chelydra serpentina*. J. Exp. Biol. **164**, 55–78 (1992)

Lauder, G.V., Shaffer, H.B.: Functional design of the feeding mechanism in lower vertebrates: unidirectional and bidirectional flow systems in the tiger salamander. Zool. J. Linnean Soc. London. **88**, 277–290 (1986)

Lauder, G.V., Shaffer, H.B.: Design of feeding systems in aquatic vertebrates: major patterns and their evolutionary interpretations. In: Hanken, J., Hall, B.K. (eds.) The Skull, vol. 3, pp. 113–149. Univ. of Chicago Press, Chicago (1993)

Le Gros Clark, W.E.: The Antecedents of Man, 1st edn. Edinurgh Univ. Press, Edinburgh (1959)

Leakey, M.D., Harris, J.M. (eds.): Laetolil; a Pleistocene site in Northern Tansania, pp. 490–496. Oxford Science Publications Clarendon Press, Oxford (1987). Introduction to the hominid footprints by M.D. Leakey

Leakey, M.D., Hay, R.L.: Pliocene footprints in the Laetolil beds at Laetoli, northern Tanzania. Nature. **278**, 317–323 (1979). https://doi.org/10.1038/278317a0

Lee, D., Bertram, J., Todhunter, R.: Acceleration and balance in trotting dogs. J. Exp. Biol. **202**, 3565–3573 (1999)

Lee, Y.N., Lee, H.J., Kobayashi, Y., Paulina Carabajal, A., Barsbold, R., Fiorillo, A.: The First Crocodilian Trackway from the Upper Cretaceous Bayanshiree Formation of Mongolia. SVP-Meeting in Berlin, Abstract (2014)

Lee-Thorp, J.A., van der Merwe, N.J., Brain, C.K.: Diet of *Australopithecus robustus* at Swartkrans from stable carbon isotopic analysis. J. Human Evol. **27**, 361–372 (1994)

Lee-Thorp, J.A., Sponheimer, M., Passey, B.H., de Ruiter, D.J., Cerling, T.E.: Stable isotopes in fossil hominin tooth enamel suggest a fundamental dietary shift in the Pliocene. Philos. Trans. R. Soc. Lond. Ser. B Biol. Sci. **365**(1556), 3389–3396 (2010). https://doi.org/10.1098/rstb.2010.0059. Review. PMID: 20855312

Lee-Thorp, J., Likius, A., Makaye, H.T., Vignaud, P., Sponheimer, M., Brunet, M.: Isotopic evidence for an early shift to C_4 resources by Pliocene hominins in the Chad. Proc. Nat. Acad. Sci. US, (PNAS). **109**(50), 20369–20372 (2012)

Leinders, J.J.: On the osteology and function of the digits of some ruminants and their bearing on taxonomy. Z. Säugetierkd. **44**, 305–318 (1979)

Lemell, P., Lemell, C., Snelderwaard, P., Gumpenberger, M., Wocheslånder, R., Weisgram, J.: Feeding patterns of *Chelus fimbriatus* (Pleurodira: Chelidae). J. Exp. Biol. **205**, 1495–1506 (2002)

Lenz, S.: Zur Biomechanik des Beckens: Eine vergleichend-Funktionsmorphologische Studie. Doctoral Thesis Medizinische Fakultät RU Bochum (1993)

Leutenegger, W.: Neonatal brain size and neurocranial dimensions in Pliocene hominids: implications for obstetrics. J. Human Evolution **16**, 291–296 (1987)

Lewis, O.J.: The hominoid wrist joint. Am. J. Phys. Antropol. **30**, 251–268 (1969)

Lewis, O.J.: The evolutionary emergence of refinement of the mammalian pattern of foot architecture. J. Anat. **137**, 21–45 (1983)

Lewis, O.J., Hamshere, R.J., Bucknil, T.M.: The anatomy of the wrist joint. J. Anat. **106**, 539–558 (1970)

Li, Y., Crompton, R.H., Alexander, R.M., Günther, M.M., Wang, W.J.: Characteristics of ground reaction forces in normal and chimpanzee-like bipedal walking by humans. Folia Primatol. **66**, 13–159 (1996)

Lieberman, D.E., Bramble, D.M., Raichlen, D.A.: Integration of the head and forelimb in bipedal hominids. Oral presentations at the International Congress of Vertebrate Morphology-Congress in Paris, Abstract (2008)

Lieberman, D.E., et al.: Muscles used in throwing. (*Mit Schimpansen??*) (2015)

Lighthill, M.J.: Hydrodynamics of aquatic animal propulsion. Ann. Rev. Fluid Mech. **1**, 413–445 (1969)

Lighthill, M.J.: Aquatic animal propulsion. J. Fluid Mech. **44**, 265–301 (1970)

Lilje, K.E.: Artiodactyla. Eine morphometrische und kinematische Studie. Masters Thesis, Fakultät für Biologie und Pharmazie, Universität Jena (2000)

Lindner, D.L., Maretta, S.M., Pijanowski, G.L., Johnson, A.L. Smith, C.W.: Measurement of bite force in dogs: a pilot study. J. Vet. Dent. **12**(2), 49–52 (1995) PMID: 9693626

Lindsey, C.C.: Form, function and locomotory habits in fish. In: Hoar, W.S., Randall, D.J. (eds.) Fish Physiology, pp. 1–100. Acad. Press, London/New York (1978)

Lintner: Suction Feeding, vol. 2, p. 25 (2010)

Lintner, M., Weissenbacher, A., Heiss, E.: The oropharyngeal morphology in the semiaquatic giant Asian pond turtle, *Heosemys grandis*, and its evolutionary implications. PLoS One. **7**(9), e46344 (2012). pmid:23029486

Lochner, F.K., Milne, D.W., Grom, J.J.: In vivo and in vitro measurements of tendon strain in the horse. Am. J. Vet. Res. **41**, 1929–1937 (1980)

Löhr, R., Ungerechts, B.: Experimentelle Bestimmung der optimalen Fingerstellung beim Kraulschwimmen. Leistungsport. **4**, 312–315 (1976)

Loitsch, C.: Kinematische Untersuchung über den Galopp von Pferden (*Equus caballus*). Doctoral Thesis, Fakultät für Biologie, Ruhr-Universität Bochum (1993)

Lordkipanidze, D., et al.: Postcranial evidence from early *Homo* from Dmanisi, Georgia. Nature. **449**, 305–310 (2007). https://doi.org/10.1038/nature06134, ISSN0028-0836

Lordkipanidze, D., Ponce de León, M.S., Margvelashvili, A., Rak, Y., Rightmire, P., Vekua, A., Zollikofer, C.P.E.: A complete skull from Dmanisi, Georgia, and the evolutionary biology of Early *Homo*. Science. **342**, 326–341 (2013). https://doi.org/10.1126/science.1238484

Lovejoy, J.O.: Biomechanical perspectives on the lower limb of early hominids. In: Tuttle, R.H. (ed.) Primate Functional Morphology and Evolution, pp. 291–326. Mouton Publishers, The Hague (1975)

Lovejoy, O., Heiple, K.G.: A reconstruction of the femur of *Australopithecus*. Am. J. Phys. Anthropol. **32**, 33–40 (1970)

Lovejoy, O., Heiple, K.G., Burstein, A.H.: The gait of *Australopithecus*. Am. J. Phys. Anthropol. **38**, 757–779 (1973)

Lovejoy, O., et al.: Careful climbing in the Miocene. The forelimbs of *Ardipithecus* are primitive. Science **326**, 70e1 (2009). https://doi.org/10.1126/science.1175827

Lovejoy, O., Suwa, G., Spurlock, L., Asfaw, B., White, T.D.: The pelvis and femur of *Ardipithecus ramidus*: the emergence of upright walking. Science. **326**, 71e1–71e6 (2009a)

Lovejoy, C.O., Latimer, B., Suwa, G., Asfaw, B., White, T.D.: Combining prehension and propulsion: the foot of *Ardipithecus ramidus*. Science. **326**, 72e1–72e8 (2009b)

Lovell, N.C.: Skeletal pathology of pongids. Am. J. Phys. Anthropol. **72**, 227 (1987). Abstract

Lucas, P.W.: The dental-dietary adaptations of mammals. N. Jb. Geol. Paläont. Mh. **8**, 486–512 (1979)

Lucas, P.W., Luke, D.A.: Chewing it over: basic principles of food breakdown. In: Chivers, D.J., Wood, B.A., Bilsbourough, A. (eds.) Food Acquisition and Processing in Primates, pp. 283–301. Plenum Press, London (1984)

MacFadden, B.J.: Fossil Horses: Systematics, Paleobiology, and Evolution of the Family Equidae. Cambridge University Press, Cambridge (1992)

Maier, W.: Vergleichend- und funktionell-anatomische Untersuchungen an der Vorderextremität von *Theropithecus gelada* (Rüppell 1835) Abhandl. Senckenberg Naturforschende Ges. **527**, S.1–S.284 (1971)

Maier, W.: Die Evolution der bilophodonten Molaren der Cercopithecoidea. Z. Morphol. Anthropol. **68**, 25–56 (1977a)

Maier, W.: Die bilophodenten Molaren der Indriidae (Primates) – ein evolutionsmorphologischer Modellfall. Z. Morphol. Anthropol. **68**, 307–344 (1977b)

Maier, W.: Zur Evolution des Säugergebisses. Natur und Museum **108**, 288–300 (1978)

Maier, W.: Konstruktionsmorphologische Untersuchungen am Gebiss der rezenten Prosimiae (Primates). Abh. Senckenbergischen naturforschenden Ges. **538**, 1–158 (1980)

Maier, W.: The functional morphology of the dentition of the Tarsiidae. In: Niemitz, C. (ed.) Biology of Tarsiers. Fischer, Stuttgart (1984a)

Maier, W.: Tooth morphology and dietary specialisation. In: Chivers, D.J., Wood, B.A., Bilsbourough, A. (eds.) Food Acquisition and Processing in Primates, pp. 303–330. Plenum Press, London (1984b)

Maier, W.: Cranial morphology of the therian common ancestor, as suggested by the adaptation of neonate marsupials. In: Szalay, F.S., Novacek, M.J., McKenna, M.C. (eds.) Mammal Phylogeny, vol. 2, pp. 165–181. Springer, New York (1993)

Maier, W.: Die Erfolgsgeschichte der Säugetiere und des Menschen: Ein evolutionsbiologisches Szenario. In: Betz, O., Köhler, H.R. (eds.) Die Evolution des Lebendigen. Attempto-Verlag, Tübingen (2008)

Maier, W.: Die Entstehung tierischer Vielzeller und die Kambrische Explosion. In: Maier, W., Werenburg, I. (eds.) Schlüsselereignisse der organismischen Makroevolution, pp. 93–145. Scidinge Hall, Zürich (2014)

Maier, W.: Der Weg zum Menschen. Ausgewählte Schriften zur Evolutionsbiologie der Wirbeltiere. Scidinge Hall, Zürich (2017)

Maier, W., Nkini, A.: The phylogenetic position of Olduvai Hominid 9, especially as determined by its basicranial evidence. In: Delson, E., Tattersall, J. (eds.) Ancestors. The Hard Evidence, pp. 249–255, Alan Liss, New York (1985)

Maier, W., Schneck, G.: Konstruktionsmorphologische Untersuchungen am Gebiss der hominoiden Primaten. Z. Morphol. Anthropol. **72**, 127–169 (1981)

Manter, J.T.: The mechanism of swimming in the alligator. J. Exp. Zool. **83**(3), 345–358 (1940). Philadelphia

Märkel, K.: Morphologie der Seeigel-Zähne. II Die gekielten Zähne der Echinacea. Z. Morphol. Tiere. **66**, 1–50 (1969)

Märkel, K., Gorny, P., Abraham, K.: Microarchitecture of sea urchin teeth. In: Nachtigall, W. (ed.) Physiology of Movement – Biomechanics. Fortschritte der Zoologie, Bd. 24, pp. 103–114. G. Fischer, Stuttgart (1977)

Marshall, C.D., Kovacs, K.M., Lyderson, C.: Feeding kinematics, suction and hydraulic jetting capabilities in bearded seals (*Erignathus barbatus*). J. Exp. Biol. **211**, 699–708 (2008). https://doi.org/10.1242/jeb.034686

Martin, R.: Lehrbuch der Anthropologie. G. Fischer, Jena (1928)

Martin, F.: Organisationsprinzipien zielgerichteter Bewegungen flexibler Greiforgane. Doctoral Thesis, Fachbereich Biologie, Freie Universität Berlin (2003)

Martin, F., Niemitz, C.: 3D-kinematics of trunk movements in captive feeding African elephants (*Loxodonta africana*). 4th Internatl. Symp. of Physiology and Behaviour in Wild and Zoo Animals. Berlin. Advances in Ethology 37 Supplements to Ethology, 115 (2002)

Marzke, W.M.: Precision grips, hand morphology, and tools. Am. J. Phys. Anthropol. **102**, 91–110 (1997)

Marzke, M.W.: Tool making, hand morphology and fossil hominins. Phil. Trans. Royal Soc. London B Biol. Sci. **368**, (2013). https://doi.org/10.1098/rstb.2012.0414

Marzke, W.M., Shackley, M.S.: Hominid hand use in the Pliocene and Pleistocene. Evidence from experimental archaeology and comparative morphology. J. Hum. Evol. **15**, 439–460 (1986)

Masao, F.T., Ichumbaki, E.B., Cherin, M., Barili, A., Boschian, G., Iurino, D.A., Menconero, S., Moggi-Cecchi, J., Manzi, G.: New footprints from Laetoli (Tanzania) provide evidence for marked body size variation in early hominins. Elife. **5**, e19568 (2016). https://doi.org/10.7554/eLife.19568

Mayr, E.: Proximate and ultimate causations. Biol. Philos. **8**, 93–94 (1993)

Mayr, G.: Evolution der Vögel. In: Storch, V., Welsch, U., Wink, M. (eds.) Evolutionsbiologie, pp. 177–182. G. Fischer, Stuttgart (2013)

McElroy, E.J., Wilson, R., Biknevicius, A.R., Reilly, S.M.: A comparative study of single-leg ground reaction forces in running lizards. J. Exp. Biol. **217**, 735–742 (2014). https://doi.org/10.1242/jeb.095620

McHenry, H.: Analysis of the postcranium of early hominids. In: Coppens, Y., Senut, B. (eds.) Origine(s) de la Bipédie chez leas Hominidés, pp. 133–142. CNRS, Paris (1991)

Meers, M.B.: Maximum bite force and prey size of *Tyrannosaurus rex* and their relationship to the inference of feeding behaviour. Hist. Biol. **16**, 1–12 (2002)

Melville, H. Moby Dick or The Whale. Harpers, New York (1851). German Translation by W. Stüver, Verlag von Th. Knaur Nachf., München 2014 (1851)

Metzger, K.A., Daniel, W.J.T., Ross, C.F.: Comparison of beam theory and finite element analysis with in vivo bone strain data from the alligator cranium. Anat. Record. **283 A**, 331–348 (2005)

Meyer, W.: Vergleich von Wanddicken- und Krümmungsmassen des Hirnschädels von fossilen Hominidae und rezenten Hominoidea. Masters Thesis, Universität Göttingen (1990)

Mickoleit, G.: Phylogenetische Systematik der Wirbeltiere. Verlag Dr. Pfeil, München (2005)

Mochon, S., McMahon, T.A.: Ballistic walking. J. Biomech. **13**, 49–57 (1980)

Mochon, S., McMahon, T.A.: Ballistic walking – an improved model. Math. Biosci. **52**, 241–260 (1981)

Mollison, T.: Die Körperproportionen der Primaten. Gegenbaurs Morphol. Jahrb. **42**, 79–304 (1911)

Morbeck, M.E.: Positional behaviour of *Colobus guereza*. A preliminary quantitative analysis. In: Kondo, S., Kawai, M., Ehara, A., Kawamura, S. (eds.) Symposia of the 5th Congress of the International Primatological Society, pp. 331–343. Tokyo Science Press, Tokyo (1975)

Morbeck, M.E.: Leaping, bounding and bipedalism in *Colobus guereza*: a spectrum of positional behavior. Yearb. Phys. Anthropol. **20**, 408–420 (1977)

Morbeck, M.E.: Forelimb use and posoitional adaptation in *Colobus guereza*. In: Morbeck, M.E., Preuschoft, H., Gomberg, N. (eds.) Environment, Behavior, and Morphology: Dynamical Interactions in Primates, pp. 95–117. G. Fischer-Verlag, New York (1979)

Morphisto: Evolution der schwimmenden Vierfüsser – Parallelentwicklungen für die Fortbewegung im Wasser. Ouerschnitte, Materialien für Unterrichtsvorbereitung und Selbststudium Jahrgang **3** (8) (2008)

Morris, S.C., Caron, J.-B.: A primitive fish from the Cambrian of North America. Nature. **512**, 419–422 (2014)

Morton, D.J.: Evolution of the human foot II. Am. J. Phys. Anthropol. **7**, 1–52 (1924)

Morton, D.J.: The Human Foot, its Evolution, Physiology and Functional Disorders, 2nd edn. Columbia University Press, New York (1948)

Moya-Sola, S., Köhler, M.: A *Dryopithecus* skeleton and the origin of great-ape-locomotion. Nature. **375**, 156–159 (1996)

Moya-Sola, S., Köhler, M.: The phylogenetic relationships of *Oreopithecus bambolii* Gervais 1872. Comptes Rendues, Academie des Sciences, Paris 324:141–148 (1997)

Moya-Sola, S., Köhler, M., Rook, L.: Evidence of hominid-like precision grip in the hand of the Miocene ape *Oreopithecus*. Proc. Natl. Acad. Sci. USA. **96**, 313–317 (1999)

Moya-Sola, S., Köhler, M., Alba, D.M., Casanovas-Vilar, I., Galindo, J.: *Pierolapithecus catalaunicus*, a new middle Miocene great ape from Spain. Science. **306**, 1339–1344 (2004)

Muller, M., van Leuwen, J.L.: The flow in front of the mouth of a prey-capturing fish. In: Duncker, H.R., Fleischer, G. (eds.) Functional Morphology in Vertebrates, Sonderband der Fortschr Zool. **30**, 222–227 (1983)

Muller, M., Osse, J.W.M.: Hydrodynamics of suction feeding in fish. Trans. Royal Soc. London. **37**, 51–135 (1984)

Nachemson, A.: Measurements of intradiscal pressure. Acta Orthop. Scand. **28**, 269–289 (1959)

Nachemson, A.: The effect of forward leaning on lumbal intradiscal pressure. Acta Orthop. Scand. **35**, 314–328 (1965)

Nachemson, A.: The load on lumbar discs in different positions of the body. Clin. Orthop. **45**, 107–122 (1966)

Nachemson, A., Morris, J.M.: In vivo measurements of intradiscal pressure. Discometry, a method for the determination of pressure in the lower lumbar discs. J. Bone Joint Surg. **46A**, 1077–1196 (1964)

Nachtigall, W.: Vogelflügel und Gleitflug. Einführung in die aerodynamische Betrachtungsweise des Flügels. J. Ornith. **116**, 1–38 (1975)

Nachtigall, W.: Zur Bedeutung der Reynoldszahl und der damit zusammenhängenden strömungsmechanischen Phänomene in der Schwimmphysiologie und Flugbiophysik. On the significance of Reynold's number anf the fluid mechanical phenomena connected to it in swimming physiology and flight biophysics. In: Nachtigall, W. (ed.) Physiology of Movement – Biomechanics Fortschritte der Zoologie, vol. 24, pp. 13–56. G. Fischer-Verlag, Stuttgart (1977)

Nachtigall, W.: Biomechanik. Vieweg, Braunschweig (2000)

Nakano, Y.: Footfall patterns in the early development of the quadrupedal walking of Japanese macaques. Folia Primatol. **66**, 113–125 (1996)

Nakano, Y.: The effects of substratum inclination on locomotor patterns in primates. Z. Morphol. Anthropol. **83**, 189–199 (2002)

Nakatsukasa, M., Hayama, S., Preuschoft, H.: Postcranial skeleton of a monkey trained for bipedal walking and implications for functional adaptation. Folia Primatol. **64**, 1–29 (1995)

Nakatsukasa, M., Kunimatsu, Y., Nakano, Y., Ishida, H.: A new skeleton of the large hominoid from Nachola, Northern Kenya. Am. J. Phys. Anthropol. **30**, 235 (2000)

Nakatsukasa, M., Kunimatsu, Y., Nakano, Y., Takano, T., Ishida, H.: Comparative and functional anatomy of phalanges in *Nacholapithecus kerioi*, a midlle Miocene hominoid from northern Kenya. Primates. **44**, 371–412 (2003a)

Nakatsukasa, M., Tsuikawa, H., Shimizu, D., Takano, T., Kunimatsu, Y., Nakano, Y., Ishida, H.: Definitive evidence for tail loss in *Nacholapithecus*, an East African Miocene hominoid. J. Human Evol. **45**, 179–1186 (2003b)

Napier, J.R.: Studies of the hand in living primates. Proc. Zool. Soc. London. **134**, 647 (1960)

Napier, J.R.: Prehensility and opposability in the hands of primates. Symp. Zool. Soc. London. **5**, 115 (1961)

Napier, J.R., Napier, P.H.: A Handbook of Living Primates. London/New York, Acad. Press (1967)

Napier, J.R., Walker, A.C.: Vertical clinging and leaping in living and fossil primates. In: Starck, D., Schneider, R., Kuhn, H.-J. (eds.) Neue Ergebnisse der Primatologie, Progress in Primatology, pp. 66–69. G.Fischer, Stuttgart (1967)

Newton, I.: Mathematical Principles of Natural Philosophy. Juffu Societatis Regis ac Typis Josephi Streater, London (1687)

Nickel, R., Schummer, A., Seiferle, E.: Anatomie der Haustiere, Bd. 1: Bewegungsapparat. Parey, Berlin (1968)

Niemitz, C.: Zur funktionellen Anatomie der Papillarleisten und ihrer Muster bei *Tarsius bancanus borneanus* Horsfield, 1821. Z.f.Säugetierkunde 42, (1977)

Niemitz, C.: Locomotion and posture of *Tarsius bancanus*. In: Niemitz, C. (ed.) Biology of Tarsiers, pp. 191–226. Stuttgart/New York, Fischer-Verlag (1984)

Niemitz, C.: The evolution of primate skin structures in relation to to gravity and locomotion. In: Jouffroy, F.K., Stack, H., Niemitz, C. (eds.) Gravity, Posture and Locomotion in Primates, pp. 129–156. Il Sedicesimo, Firenze (1990)

Niemitz, C.: A theory on the evolution of human bipedalism. The "Amphibische Generalistentheorie". In: Zeller, U. (ed.) Origin and Evolutionary Transformation of Mammals, pp. 46–48. Natural History Museum, Humboldt-University, Berlin (2000)

Niemitz, C.: Kinematics and ontogeny of locomotion in monkeys and human babies. Z. Morphol. Anthropol. **84**(2/3), 383–400 (2002a)

Niemitz, C.: A theory on the habitual orthograde human bipedalism – the "Amphibische Generalistentheorie". Anthropol. Anz. **60**, 3–66 (2002b)

Niemitz, C.: Das Geheimnis des aufrechten Gangs – unsere Evolution verlief anders. C.H. Beck, München (2004)

Niemitz, C.: The evolution of upright posture and gait – a review and a new synthesis. Naturwissenschaften. (2010). https://doi.org/10.1007/s00114-009-0637-3

Nieschalk, U.: Fortbewegung und Funktionsmorphologie kleiner quadrupeder Halbaffen in Anpassung an unterschiedliche Habitate. Doctoral Thesis, Biological Faculty, RU Bochum (1991)

Nieschalk, U., Demes, B.: Biomechanical determinants of reduction of the second ray in Lorisinae. In: Preuschoft, H., Chivers, D.J. (eds.) Hands of Primates, pp. 225–234. Springer, Wien/New York (1993)

Nieschalk, U., Klauer, G.J.: Functional morphology of the palmar pads in *Loris tardigradus* and *Galago senegalensis*. Fortschr. Zool. **35**, 605–607 (1989)

Nobiling, G.: Die Biomechanik des Kieferapparates beim Stierkopfhai, *Heterodontus portusjacksoni (= Heterodontus philippi)*. Adv. Anat. Embryol. Cell Biol. **52**(6), 1–52 (1977)

Norman, J.R., Fraser, F.C.: Riesenfische, Wale und Delphine. Parey-Verlag, Berlin (1963)

Nyakatura, J.A., Andrada, E.: A mechanical link model of two-toed sloths: no pendular mechanics during suspensory locomotion. Acta Theriol. (2012). https://doi.org/10.1007/s13364-012-0099-4

Nyakatura, J.A., Andrada, E.: A mechanical link model of two-toed sloths: no pendular mechanics during suspensory locomotion. Acta Theriol. https://doi.org/10.1007/s13364-012-0099-4

Nyakatura, J.A., Andrada, E., Curth, S., Fischer, M.S.: Bridging "Romer's Gap": limb mechanics of an extant belly-dragging lizard inform debate on tetrapod locomotion during the Early Carboniferous. Evol. Biol. (2013). https://doi.org/10.1007/s11692-013-9266-z, ISSN 0071–3260

Nyakatura, J.A., Allen, V.R., Lauströer, J., Andikfar, A., Danczak, M., Ullrich, H.-J., Hufenbach, W., Martens, T., Fischer, M.S.: A three-dimensional skeletal reconstruction of the stem Amniote *Orobates pabsti* (Diadectidae): analyses of body mass, centre of mass position, and joint mobility. PLoS One. **10**(9), e0137284 (2015). https://doi.org/10.1371/journal.pone.0137284

Oemichen, E.: Le vol des oiseau. In: Grassé, J.P. (ed.) Traité de Zoologie, pp. 131–170. Tome XV, Masson (1950a)

Oemichen, E.: Locomotion terrestre. Natation – Plonge. In: Grassé, P.P. (ed.) Traité de Zoologie, pp. 171–184. Tome XV, Masson (1950b)

Oemichen, E.: Musculature de l'oiseau. In: Grassé, P.P. (ed.) Traité de Zoologie, pp. 108–130. Tome XV, Masson, Paris (1950c)

Ogihara, N.: Towards understanding the origin and evolution of human bipedalism from mechanical engineering perspectives. J. Soc. Mech. Eng. **115**(1126), 659–663 (2012). https://doi.org/10.1299/jsmemag.115.1126_659

Ogihara, N., Hirasaki, E., Kumakura, H., Nakatsukasa, M.: Ground-reaction-force profiles of bipedal walking in bipedally trained Japanese macaques, reconstructed on an anatomical model-matching method. J. Human Evol. **58**, (2007). 2222522–261

Ogihara, N., Oku, T., Andrada, E., Blickhan, R., Nyakatura, J.A., Fischer, M.: Planar covariation of limb elevation angles during bipedal locomotion in common quails (*Coturnix coturnix*). J. Exp. Biol. **217**, 3968–3973 (2014). https://doi.org/10.1242/jeb.109355

Okada, M.: Primate bipedal walking: comparative kinematics. In: Kondo, S. (ed.) Primate Morphophysiology, Locomotor Analysis and Human Bipedalism, pp. 47–58. Univ. Tokyo Press, Tokyo (1985)

Omura, A., Anzai, W., Endo, H.: Functional and morphological variety in trunk muscles of Urodela. J. Vet. Med. Sci. (2013)

Osse, J.W.M., Muller, M., van Leuwen, J.L.: The analysis of suction feeding in fish. In: Duncker, H.R., Fleischer, G. (eds.) Functional Morphology in Verterbrates. Sonderband der Fortschritte der Zoologie **30**, 217–221 (1985)

Otto, F.: Das Konstruktionsprinzip Pneu. Institut für Leichte Flächentragwerke, Stuttgart. 9 IL 35 Vorauflage (1977)

Owen, R.: On the classification and geographical distribution of the Mammalia. Beade Lecture, Cambridge, U.K. Parker, London (1859)

Oxnard, C.E.: Locomotor adaptation in the primate forelimb. Symp. Zool. Soc. London. **10**, 165–182 (1963)

Oxnard, C.: Some methodological factors in studying the morphological-behavioral interface. In: Morbeck, M.E., Preuschoft, H., Gomberg, N. (eds.) Environment, Behavior, and Morphology: Dynamical Interactions in Primates, pp. 183–207. G. Fischer-Verlag, New York (1979a)

Oxnard, C.: The morphological-behavioral interface in extant primates: some implications for systematics and evolution. In: Morbeck, M.E., Preuschoft, H., Gomberg, N. (eds.) Environment, Behavior, and Morphology: Dynamical Interactions in Primates, pp. 209–227. G. Fischer-Verlag, New York (1979b)

Pajor, N., Perry, S. F.: Muscle attachment of ribs in extant and extinct species. Internat. Congr. of Vert. Morphology. Paris. Abstract (2007)

Padian, K., Horner, J.R.: The evolution of bizarre structures in dinosaurs: Biomechanics, sexual selection, social selection or species recognition? Journal of Zoology **283**, 3–17 (2010)

Pakusa, T.: Experimentelle Untersuchungen zu Morphologie und funktion der Hüftgelenksbänder. Doctoral Thesis Medizinische Fakultät RU Bochum (1998)

Panagiotopoulou, O., Pataky, T.C., Day, M., Hensman, M.C., Hensman, S., Hutchinson, J.R., Clemente, C.J.: Foot pressure distributions during walking in African elephants (*Loxodonta africana*). R. Soc. Open Sci. **3**, 160203 (2016)

Papenfuß, H.D.: Aerodynamik des Automobils, Institut f. Thermo- und Fluiddynamik. Ruhr-Universität, Bochum (2004)

Parrington, F.R.: The Vertebrae of Early Tetrapods. Problemes Actuels de Paléontologie, pp. 269–279. Colloques International de CNRS, Paris (1967)

Partridge, T.C.: Paleoecology of the Pliocene and lower Pleistocene hominids of Southern Africa: how good is the chronological and paleoenvironmental evidence? South J. Sci. **82**, 80–83 (1986)

Partridge, T.C., et al.: (1999)

Patel, B.A., Maiolino, S.A.: Morphological diversity in the digital rays of primate hands. In: Kivell, T.L., et al. (eds.) The Evolution of the Primate Hand, Developments in Primatology: Progress and Prospects, pp. 55–100 (2013). https://doi.org/10.1007/978-1-4939-3646-5_4

Pauwels, F.: Gesammelte Abhandlungen zur Biomechanik. Springer, Heidelberg/Berlin (1965)

Pauwels, F.: Biomechanics of the locomotor apparatus. Springer, Berlin (1980)

Payne, R.C., Crompton, R.H., Isler, K., Savage, R., Vereeke, E.E., Günther, M.M., Thorpe, S.K.S., D'Aout, K.: Morphological analysis of the hindlimb in apes and humans. I Muscle architecture. J. Anat. **208**, 709–724 (2006)

Pennycuik, C.J.: On the running of the gnu (*Connochaetes taurinus*) and other animals. J. Exp. Biol. **63**, 775–793 (1975)

Perret, K.: Environmantal enrichment – Einfluss verschiedener Beschäftigungs-massnahmen auf das Verhalten von Schimpansen (*Pan troglodytes*) im Zoo. Doctoral Thesis Fakultät für Biologie, RU Bochum (1997)

Perry, S.F.: Atmungsorgane. In: Westheide, W., Rieger, R. (eds.) Spezielle Zoologie 2: Wirbel- moder Schädeltiere, 2nd edn, pp. 127–141. Spectrum akademischer Verlag, Heidelberg (2010)

Perry, S.F., Breuer, T., Pajor, N.: Structure and function of the sauropod repiratory system. In: Klein, N., Remes, K., Gee, C.T., Sander, P.M. (eds.) Biology of the Sauropod Dinosaurs, pp. 83–93. Indiana Univ. Press, Bloomington (2011)

Peters, D.S.: Über den Zusammenhang von biologichem Artbegriff und phylogenetischer Systematik. In: Schäfer, W. (ed.) Aufsätze und Reden der Senckenbergischen Naturforschenden Ges, vol. 18, pp. 1–39. W. Kramer, Frankfurt a.M (1970)

Peters, D.S., Gutmann, W.F.: Über die Lesrichtung von Merkmals- und Konstruktionsreihen. J. Zool. Syst. Evol. Res. **9**, 237–263 (1971)

Peters, A., Preuschoft, H.: External biomechanics of leaping in Tarsius and its morphological and kinematic cosequences. In: Niemitz, C. (ed.) Biology of Tarsiers, pp. 227–255. G. Fischer-Verlag, Stuttgart/New York (1984)

Pfretschner, H.U.: Structural reinforcement and crack propagation in enamel. In: Russell, D.E., Santoro, J.P., Sigogneau-Russell, D.E. (eds.) Teeth Revisited. Proc.7th Internat Symp. Dental Morphology, Paris 1986, Mem. Mus. Histoire Naturelle © 53, pp. 133–143 (1988)

Pickford, M.: The geological and faunal context of late Miocene hominid remains from Lukeino, Kenya. Contexte géologique et faunique des restes d'hominidés du Miocène supérieur de Lukeino, Kenya. Comptes Rendus Académie des Sciences Paris, Série IIA Sciences de la Terre et des Planètes, **332**(2):145–152 (2001)

Popper, K.R.: The Logic of Scientific Discovery. Hutchinson, London (1968)

Porro, L., Rayfield, D. E., Clack, J., Herrel, A., Adriaens, D.: Mandibular mechanics and feeding across the water-land transition: Finite Element Analysis of the early Tetrapod lower jaw. Oral comm. SVP- meeting in Berlin, Abstract (2014)

Portmann, A.: Squelette. In: Grassé, J.P. (ed.) Traité de Zoologie, Tome XV, pp. 78–107. Masson, Paris (1950)

Pouydebat, E., Coppens, Y., Gorce, P.: Évolution de la préhension chez les primates humains et non humains: La précision et l'utilisation d'outils revisitées. Anthropologie. **110**, 687–697 (2006)

Pouydebat, E., Gorce, P., Bels, V.: Biomechanical study of grasping according to the volume of the object: human versus nonhuman primates. J. Biomech. **42**(3–9), 266–272 (2009)

Pouydebat, E., Reghem, E., Borel, A., Gorce, P.: Diversity of grip in adults and young humans and chimpanzees (*Pan troglodytes*). Behav. Brain Res. **218**, 21–28 (2011)

Prandtl, L.: Tragflügeltheorie. I. und II Mitt., Nachrichten d. Kgl. Ges. d. Wiss. Goettingen, Math.-Phys. Klasse, 451–477 und 107–137 (1918 und 1919)

Preuschoft, H.: Muskeln und Gelenke der Hinterextremität des Gorilla. Morpholog. Jb. **101**, 432–540 (1961)

Preuschoft, H.: Beitrag zur Funktion des Pongidenfusses. Z. Morphol. Anthropol. **53**, 19–28 (1963a)

Preuschoft, H.: Muskelgewichte bei Gorilla, Orang-Utan und Mench. Anthropol. Anz. **26**, 308–317 (1963b)

Preuschoft, H.: Muskeln und Gelenke der Vorderextremität des Gorilla. Morphol. Jb. **107**, 99–183 (1965)

Preuschoft, H.: Statische Untersuchungen am Fuss der Primaten. I Phalangen und Metatarsalia. Z. Anat. Entw. Gesch. **129**, 285–345 (1969a)

Preuschoft, H.: The mechanical basis of the morphological differences in the skeletons of apes and man. Proc. 2nd Int. Congr. Primat., Atlanta, Karger, Basel, vol. 2, pp. 160–170 (1969b)

Preuschoft, H.: Functional anatomy of the lower extremity. In: Bourne, G.H. (ed.) The Chimpanzee, vol. 3, pp. 221–294. Krager, Basel (1970a)

Preuschoft, H.: Statische Untersuchungen am Fuss der Primaten. II Statik des ganzen Fusses. Z. Anat. Entw, Gesch. **131**:156–192 (1970b)

Preuschoft, H.: Body posture and mode of locomotion in early Pleistocene hominids. Folia Primatol. **14**, 209–240 (1971)

Preuschoft, H.: Functional anatomy of the upper extremity. In: Bourne, G.H. (ed.) The Chimpanzee, vol. 6, pp. 34–120. Karger, Basel (1973a)

Preuschoft, H.: Body posture and locomotion in some East African dryopithecinae. In: Day, M.H. (ed.) Human Evolution, pp. 13–46. Taylor & Francis, London (1973b)

Preuschoft, H.: Body posture and mode of locomotion in fossil primates – Method and example: *Aegyptopithecus zeuxis*. In: Symp. 5th Congress Internat. Primat. Soc, pp. 345–359. Japan Science Press, Tokyo (1974)

Preuschoft, H.: Funktionelle Anpassung Evoluierender Systeme. Aufsätze und Reden der Senckenbergischen Naturforschenden Ges. Waldemar Kramer, Frankfurt, pp. 98–117 (1976)

Preuschoft, H.: On the quality and magnitude of mechanical stresses in the locomotor system during rapid movements. Z. Morphol. Anthropol. **75**, 245–262 (1985)

Preuschoft, H.: Funktionsweise der Sehnen, Beanspruchung und Form der Phalangen und Metacarpalia. Verh. Anat. Ges. **81**, 279–292 (1987)

Preuschoft, H.: Kleine Menschaffen oder Gibbons. In: Grzimeks Enzyklopädie, Kindler-Verlag, München, Vol. II, pp. 326–356 (1988)

Preuschoft, H.: The external forces and internal stresses occurring in the distal limb segments of dressage and jumping horses. Z. Säugetierkd. **54**, 172–190 (1989a)

Preuschoft, H.: Biomechanical approach to the evolution of the facial skeleton of hominoid primates. In: Splechtna, H., Hilgers, H. (eds.) Trends in Vertebrate Morphology, pp. 421–431. G. Fischer, Stuttgart/New York (1989b)

Preuschoft, H.: Quantitative approaches to primate morphology. Folia Primatol. **53**, 82–100 (1989c)

Preuschoft, H.: Physikalisches über die Zügelführung. Freizeit im Sattel, 1: -10, 2: 90–92, 5: 242–244, 10: 562–563 (1990a)

Preuschoft, H.: Gravitiy in primates and its relation to body shape and locomotion. In: Jouffroy, F. K., Stack, H., Niemitz, C. (eds.) Gravity, Posture and Locomotion in Primates, pp. 109–127. Il Sedicesimo, Firenze (1990b)

Preuschoft, H.: Überlegungen eines Biomechanikers zum "pferdegerechten" Reiten. In: B. Schürer (ed): Zwischen Langeweile und Überforderung. Edition Schürer, Kirchheim (1993)

Preuschoft, H.: Zur Evolution de menschliche Becken- u nd Rumpfform. In: Biedermann, H. (ed.) Manualtherapie bei Kindern. Enke, Stuttgart, pp. 77–88 (1999)

Preuschoft, H.: Die Entwicklung des aufrechten Gangs beim Menschen. Praxis der Naturwissenschaften. Biologie in der Schule **50(1)**, 24–42 (2001)

Preuschoft, H.: What does arboreal locomotion mean exactly and what are the relationships between "climbing", environment and morphology? Z. Morphol. Anthropol. **83**, 171–188 (2002)

Preuschoft, H.: Mechanisms for the acquisition of habitual bipedality: are there biomechanical reasons for the acquisition of upright bipedal posture? J. Anat. **204**, 363–384 (2004)

Preuschoft, H.: Die Biomechanik des aufrechten Ganges. In: Conard, N.J. (ed.) Woher kommt der Mensch? 2. Auflage edn, pp. 36–73. Tübingen, Attempto-Verlag (2006)

Preuschoft, H.: Biomechanical Reasons for Developing the Characteristic Shapes of Ribs and Ribcages in Cursorial Mammals. SVP-Meeting Berlin, Abstract (2014)

Preuschoft, H.: Power grip or precision handling? What determines hand morphology in primates, including hominidae? Biol. J. Linnaean Soc. London. **20**, 1–13 (2018)

Preuschoft, H., Christian, A.: Statik und Dynamik bei Tetrapoden. In: Ganslosser, U. (ed.) Spitzenleistungen. Unglaubliches aus dem Tierreich, pp. 89–130. Filander-Verlag, Fürth (1999)

Preuschoft, H., Demes, B.: Biomechanics of brachiation. In: Preuschoft, H., Chivers, D.J., Brockelmann, W., Creel, N. (eds.) The Lesser Apes. Evolutionary and Behavioural Biology, pp. 96–118. Edinburgh Univ. Press, Edinurgh (1984)

Preuschoft, H., Demes, B.: Influence of size and proportions on the biomechanics of brachiation. In: Jungers, W.L. (ed.) Size and Scaling in Primate Biology, pp. 383–399. Plenum Press, New York/London (1985)

Preuschoft, H., Franzen, J.L.: Locomotion and biomechanics in Eocene mammals from Messel. Palaeobiodiversity Palaeoenvironment. **92**(4), 459–476 (2012)

Preuschoft, H., Fritz, M.: Mechanische Beanspruchungen im Bewegungsapparat von Springpferden. In: Nachtigall, W. (ed.) Physiology of Movement – Biomechanics, Bewegungsphysiologie-Biomechanik. Fortschritte der Zoologie, vol. 24, pp. 75–98. G. Fischer-Verlag, Stuttgart (1977)

Preuschoft, H., Gottstein, B.: Die Wirkung von Fahrleinen auf Pferdemäuler. Reistsportmagazin 6/99: 34–35, 10/99: 28–29, 3/00: 36–37 (1999)

Preuschoft, H., Gudo, M.: Die Schultergürtel von Wirbeltieren: Biomechanische Überlegungen zu den Bauprinzipien des Wirbeltierkörpers und zur Fortbewegung von Tetrapoden. Zentralblatt für Geologie und Paläontologie, Teil II, pp. 339–361 (2006)

Preuschoft, H., Günther, M.M.: Biomechanics and body shape in primates compared with horses. Z. Morph. Anthrop. **80**, 149–165 (1994)

Preuschoft, H., Hüllen-Kluge, K.: Studien zu den Bewegung von Springpferden. Wissenschaftl. Publikationen 9, FN-Verlag, Warendorf, pp. 120–146 (1987)

Preuschoft, H., Klein, N.: Torsion and bending in the neck and tail of sauropod dinosaurs and the function of cervical ribs: insights from functional morphology and biomechanics. Plos One. **8**(10), e78574 (2013)

Preuschoft, H., Tardieu, C.: Biomechanical reasons for the divergent morphology of the knee joint and the distal epiphyseal suture in hominoids. Folia Primatol. **66**, 82–92 (1996)

Preuschoft, H., Witte, H.: Biomechanical reasons for the evolution of hominine body shape. In: Coppens, Y., Senut, B. (eds.) Origine(s) de la Bipedie chez les Hominidés, pp. 59–77. CNRS, Paris (1991)

Preuschoft, H., Witte, H.: Die Körpergestalt des Menschen als Ergebnis biomechanischer Erfordernisse. In: Voland, E. (ed.) Evolution und Anpassung, pp. 43–74. S Hirzel, Stuttgart (1993)

Preuschoft, H., Witzel, U.: Biomechanical investigations on the skulls of reptiles and mammals. Senckenb. Lethaea. **82**, 207–222 (2002). Sonderband

Preuschoft, H., Witzel, U.: A biomechanical approach to craniofacial shape in primates, using FESA. Ann. Anat. **186**, 397–404 (2004)

Preuschoft, H., Witzel, U.: The functional shape of the skull in vertebrates: Which forces determine morphology? With a comparison of lower primates and ancestral synapsids. – Anatomical Record, Special Issue: Finite element Analysis in Vertebrate Biomechanics. Ed. C. Ross. pp. 402–413 (2005)

Preuschoft, H., Witzel, U.: The functional structure of the facial skull in hominoidea. Folia primatologica, **75**, 202–252 (2004)

Preuschoft, H., Reif, W., Müller, W.H.: Funktionsanpassungen in Form und Struktur an Haifischzähnen. Z. Anat. Entw.-Gesch. **143**, 315–344 (1974)

Preuschoft, H., Fritz, M., Krämer, T.: Eziehungen zwischen mechanischer Beanspruchung und Gestalt bei Wirbellosen: Spannungsanalyse an Muschelklappen. Neues Jahrbuch für Geologie und Paläontologie, **150**, 161–181 (1975)

Preuschoft, H., Fritz, M., Niemitz, C.: Biomechanics of the trunk in primates and problems of leaping in *Tarsius*. In: Morbeck, M.E., Preuschoft, H., Gomberg, N. (eds.) Environment, Behavior, and Morphology: Dynamical Interactions in Primates, pp. 327–345. G. Fischer-Verlag, New York (1979)

Preuschoft, H., Demes, B., Meyer, M., Bär, H.F.: Die biomechanischen Prinzipien im Oberkiefer von langschnauzigen Wirbeltieren. Z. Morphol. Anthropol. **76**, 1–24 (1983)

Preuschoft, H., Demes, B., Meyer, M., Bär, H.F.: Les principes mécaniques realisé dans le machoire des vertebras à museau long.- the biomechanical principles realised in the upper jaw of long snouted vertebrates. In: Sakka, M. (ed.) Définition et Origines de l'Homme, Editions CNRS, pp. 177–202 (1986a)

Preuschoft, H., Demes, B., Meyer, M., Bär, H.F.: The biomechanical principles realised in the upper jaw of long-snouted primates. In: Else, J.G., Lee, P.C. (eds.) Primate Evolution, pp. 247–264. Cambridge Univ. Press, London (1986b)

Preuschoft, H., Knisel, M., Fritz, M.: Die Bewegungen von Pferden beim Springen. In: Preuschoft, H., Fritz, M., Hüllen-Kluge, K., Knisel, G., Streitlein, I. (eds.) Studien zu den Bewegungen von Sportpferden, pp. 60–119. Wissenschaftl. Publikationen 9, FN-Verlag, Warendorf (1987)

Preuschoft, H., Günther, M.M., Hayama, S.: Curvature of the lumbar spine as consequence of mechanical necessities in Japanese macaques trained for bipedalism. Folia Primatol. **50**, 42–58 (1989)

Preuschoft, H., Reif, W.T., Loitsch, C., Tepe, E.: The function of labyrinthodont teeth: big teeth in shallow jaws. In: Schmitt-Kittler, N., Vogel, K. (eds.) Constructional Morphology and Evolution, pp. 151–1171. Springer, Berlin/Heidelberg (1991)

Preuschoft, H., Witte, H., Demes, B.: Biomechanical factors that influence overall body shape of large apes and humans. In: Matano, S., Tuttle, R.H., Ishida, H., Goodman, M. (eds.) Topics in Primatology. 3 Evolutionary Biology, Reproductive Endocrinology, and Virology, pp. 259–289. Univ. Tokyo Press, Tokyo (1992)

Preuschoft, H., Godinot, M., Beard, C., Nieschalk, U., Jouffroy, F.K.: Biomechanical considerations to explain important morphological characters of primate hands. In: Preuschoft, H., Günther, M. M., Christian, A., 1998: Size dependence in prosimian locomotion and its implications for the distribution of body mass. Folia Primatol. **69**, 60–81 (1993)

Preuschoft, H., Witte, H., Christian A., Recknagel, S.: Körpergestalt und Lokomotion bei grossen Säugetieren. Verhandlunge der Deutschen Zoologischen Gesellschaft **87**, 147–163 (1994a)

Preuschoft, H., Lesch, C., Witte, H., Loitsch C.: Die biomechanischen Grundlagen der Gangarten, insgesondere des Galopps. In: Knezevic, F. (ed.) Orthopädie bei Huf- und Klauentieren. Schattauer-Verlag, Stuttgart (1994b)

Preuschoft, H., Falaturi, P., Lesch, C.: Was spürt das Pferd vom Reiter? In: B. Schürer (ed): Zusammen Rücken. Edition Schürer, Kirchheim (1994c)

Preuschoft, H., Günther, M., Christian, A.: Size dependences in prosimian locomotion and their implications for the distribution of body mass. Folia primatologica **69** (Suppl. 1), 60–81 (1998)

Preuschoft, H., Schmidt, M., Hayama, S., Okada, M.: The influence of three-dimensional movements of the forelimb on the shape of the thorax and its importance on erect body posture. In: Franzen, J.L. (ed.) Walking Upright, pp. 9–24. Courier Forschungsinstitut Senckenberg, 243 (2003)

Preuschoft, H., Hohn, B., Distler, C., Witzel, U., Sick, H.: Ribs and rib cages in terrestrial vertebrates. Their mechanical function and stressing, analysed with the aid of FESA. Journal of Vertebrate Paleontology **25**, Suppl. to Nr. 3, p.101. Abstract only. (2005)

Preuschoft, H., Witzel, U., Hohn, B., Schulte, D., Distler, C.: Biomechanics of locomotion and body structure in varanids with spezial emphasis on the forelimbs. In: Horn, H.G., Boehme, W., Krebs, U. (eds.) Advances in Monitor Research III, Mertensiella (Supplement to Salamandra), pp. 59–78. Chimaera Buchhandelsgesellschaft, Frankfurt a.M (2007)

Preuschoft, H., Hohn, B., Stoinski, S., Witzel, U.: Why so huge? Biomechanical reasons fort he acquisition of large size in sauropod and theropod dinosaurs. In: Klein, N., Remes, K., Gee, C. T., Sander, P.M. (eds.) Biology of the Sauropod Dinosaurs, pp. 197–218. Indiana Univ. Press, Bloomington (2011)

Preuschoft, H., Hohn, B., Witzel, U.: Biomechanics and functional preconditions for terrestrial lifestyle and their influence on the general tetrapod bauplan. Centennary meeting of the Paläontologische Gesellschaft, Berlin. Abstract (2012a)

Preuschoft, H., Falaturi, B., Busching, K., Recknagel, S.T.: Welche Bedingungen bestehen für einen optimalen Sitz des Reiteres? FFP-Spezialheft Pferde gesund erhalten durch gutes Reiten. Dülmen, pp. 92–119 (2012b)

Preuschoft, H., Schönwasser, K.-H., Witzel, U.: Selective value of characteristic size parameters in hylobatids. A biomechanical approach to small ape size and morphology. ? In: Reichard, U., Hirai, H., Barelli, C. (eds.) Evolution of Gibbon and Siamang, pp. 229–265. Springer, New York (2016)

Puff, A., Rosemeyer, B.: Das Verhalten des Fußlängsgewölbes beim normalen Gehakt. Röntgenkinematographische Untersuchung zur Gelenkmechanik. Morph. Jb. **105**, 274–291 (1970)

Rage, J.-C.: Amphibians and sqamates in the Eocene of Europe. Palaeobiodiversity Palaeoenvironment. **92**(2), 445–458 (2012)

Rak, Y.: The neanderthal: a new look at an old face. J. Human Evol. **15**, 151–164 (1986)

Rao, G.: Über die Rolle der Elastizität in der Lokomotion: Das Kollagen-Myosin-Titin-Modell (CMT-Modell). Doctoral Thesis Medical Faculty, Ruhr-Universität Bochum (1998)

Rao, J.G., Witte, H., Lesch, C.: Elastizität als morphogenetisches Prinzip. Verh. Anat. Ges. **89**, 98 (1994). Abstract

Rao, J.R., Witte, H., Recknagel, S., Deli, M., Lesch, C., Preuschoft, H.: Titin: Schlüssel zum Verständnis Elastischer Mechanismen in der Locomotion. Verh. Anat. Ges. **90**, 98–99 (1995)

Rao, J.G., Witte, H., Recknagel, S., Christian, A., Preuschoft, H.: The amount of elastic energy storage in the human locomotor system is a function of posture. Verh. Anat. Ges. **91**, 154 (1996). Abstract

Rayfield, E.J.: Cranial mechanics and feeding in tyrannosaurus rex. Proc. R. Soc. Lond. B. **271**, 1451–1459 (2004)

Rayfield, E.J.: Using finite element analysis to investigate suture morphology: a case study using large carnivorous dinosaurs. Anat. Rec. **283**(A), 349–365 (2005)

Rayfield, E.J.: Finite element analysis and understanding the biomechanics and evolution of living and fossil organisms. Annu. Rev. Earth Planet. Sci. **35**, 541–576 (2007)

Recknagel, S.: Untersuchung der mechanischen Schwingungsübertragung an der oberen und unteren Extremität des Menschen, Fortschritt – Berichte VDI Reihe 17: Biotechnik, No 123 (1995)

Recknagel, S., Witte, H.: Bewegungen des Beckens und der Wirbelsäule bei stossartiger Krafteinleitung über die Beine. Z. Orthop. **134**, 214–218 (1996)

Reichard, U., Preuschoft, H.: Why is the siamang larger than other hylobatids? In: Reichard, U., Hirai, H., Barelli, C. (eds.) Evolution of Gibbon and Siamang, pp. 169–183. Springer, New York (2016)

Reif, W. E.: Protective and hydrodynamic function of the dermal skeleton of elasmobranches. N. Jb. Geol. Paläont. SFB 53, Rep. 1976-1978, pp. 133–143 (1978)

Reif, W.E.: Oberflächenstrukturen und –skulpturen bei schnell schwimmenden Wirbeltieren. In: Reif, W.E. (ed.) Paläontologiesche Kursbücher, vol. 1, pp. 141–158. Frank, München (1981a)

Reif, W.E.: Biophysik und Morphologie schneller Schwimmer. In: Reif, W.E. (ed.) Paläontologische Kursbücher, vol. 1, pp. 233–259. Frank, München (1981b)

Reilly, S.M., Elias, J.A.: Locomotion in *Alligator mississippiensis*: kinematic effects of speed and posture and their relevance to the sprawling-to-erect paradigm. J. Exp. Biol. **201**, 2559–2574 (1998)

Reilly, S.M., Jorgensen, M.E.: The evolution of jumping in frogs: morphological evidence for the basal anuran locomotor condition and the radiation of locomotor systems in Crown Group Anurans. J. Morphol. **272**, 149–168 (2011)

Reilly, M., McElroy, E.J., Odum, R.A., Hornyak, V.A.: Tuataras and salamanders show that walking and running mechanisms are ancient features of tetrapod locomotion. Proc. Royal Soc. **273**, 1563–1568 (2006). https://doi.org/10.1098/rspb.2006.3489

Ren, L., Butler, M., Miller, C., Schwerda, D., Fischer, M., Hutchinson, J.R.: The movements of limb segments and joints during locomotion in African and Asian elephants. J. Exp. Biol. **211**, 2735–2751 (2008). https://doi.org/10.1242/jeb.018820

Reynaga, C.M., Astley, H.C., Azizi, E.: Morphological and kinematic specializations of walking frogs. J. Exp. Zool. **2018**, 1–12 (2018)

Richmond, B.G.: Origin of human bipedalism: the knuckle-walking hypothesis revisited. Yearb. Phys. Anthropol. **44**, 70–105 (2001)

Richmond, B.G., Strait, D.S.: Evidence that humans evolved from a knuckle-walking ancestor. Nature. **404**, 382–384 (2000)

Richmond, B.G., Wright, B.W., Grosse, J., Dechow, P.C., Ross, C.F., Spencer, M.A., Strait, D.S.: Finite Elements in Functional Morphology. The Anatomical Record, **283A**(2), 259–274

Ridgway, S.H.: Mammals of the Sea. Charles C. Thomas, Springfield (1972)

Ridgway, S.H., Harrison, R.J.: Handbook of Marine Mammals, vol. 1 and 2. Acad Press, London (1981)

Ritter, D.: Axial muscle function during lizard locomotion. J. Exp. Biol. **173**, 1–10 (1996)

Ritter, D., Nassar, P., Fife, M., Carrier, D.: Epaxial muscle function in trotting dogs. J. Exp. Biol. **204**, 3053–3064 (2001)

Roach, N.T., Venkadesan, M., Rainbow, M.J., Lieberman, D.E.: Elastic energy storage in the shoulder and the evolution of high-speed throwing in *Homo*. Nature. **498**, 483–486 (2013)

Robinson, J.T.: Early Hominid Posture and Locomotion. Chicago University Press, Chicago (1972)

Rodewald, A., Zankl, H.: Hautleistenfibel. G. Fischer, Stuttgart (1981)

Röhrs, M.: Bemerkungen zur Bedeutung der Bergmannschen Regel. In: Kurth, G (ed.) Evolution und Hominisation, 2. Aufl. G. Fischer Verlag, Stuttgart, 58–74 (1968)

Rollinson, J., Martin, R.D.: Comparative aspects of of primate locomotions with special reference to arboreal cercopithecines. Symp. Zool. Soc. Lond. **48**, 377–427 (1981)

Romer, A.S.: Vertebrate Paleontology. University of Chicago Press, Chicago (1945)

Romer, A.S.: The Vertebrate Body. Shorter Version of the Second Edition. W.B. Saunders Comp, Philadelphia (1956)

Romer, A.S.: Vergleichende Anatomie der Wirbeltiere. Übersetzt von H Frick. Parey-Verlag, Berlin (1966)

Rook, L., Bondioli, L., Köhler, M., Moya-Sola, S., Macchiarelli, R.: Oreopithecus was a bipedal ape after all: evidence from the iliac cancellous architecture. Proc. Natl. Acad. Sci. USA. **96**, 8795–8799

Roos, C.: Phylogeny and classification of gibbons (Hylobatidae). In: Reichard, U.H., Hirai, H., Barelli, C. (eds.) Evolution of Gibbons Ans Siamang, pp. 151–165. Springer, New York (2016)

Rose, M.D.: Quadrupedalism in new and old world monkeys. Doctoral Thesis, University of London (1973a)

Rose, M.: Quadrupedalism in primates. Primates. **14**, 337–357 (1973b)

Rose, M.D.: Positional behavior iof natural populations: some quantitative results of a field study of Colobus guereza and *Cercopithecus aethiops*. In: Morbeck, M.E., Preuschoft, H., Gomberg, N. (eds.) Environment, Behavior, and Morphology: Dynamical Interactions in Primates. G. Fischer-verlaga, New York (1979)

Rosenberger, A.L., Preuschoft, H.: Evolutionary morphology, cranial biomechanics and origins of tarsiers and anthropoids. Palaeobiodiversity Palaeoenvironments. **92**(4), 507–526 (2012)

Ross, C.: Oral communication at meeting in Greifswald (2004)

Ross, C.F., Patel, B.A., Slice, D., Strait, D.S., Dechow, P.C., Richmond, B.G., Spencer, M.: Modeling masticatory muscle forces in finite element analysis: sensitivity analysis using principal components analysis. Anat Rec A. **283**, 288–299 (2005)

Ruff, C.B., Walker, A.: Body size and body shape. In: Walker, A., Leakey, R. (eds.) The Nariokotome Homo Erectus Skeleton, pp. 234–265. Springer, Berlin/New York (1993)

Salisbury, S.W., Romilio, A., Herne, M.C., Tucker, R.T., Nair, J.P. (eds.): The dinosaurian ichnofauna of the lower Cretaceous Broome Sandstone of the Walmadany area (James Price Point), Dampier Peninsula, Western Australia. Society of Vertebrate Paleontology Memoir 16, J. Vertebr. Paleontol. **36**(Supplement to N0 6), 1–152 (2017)

Sander, M., Christian, A., Clauss, M., Fechner, R., Gee, C.T., Griebeler, E.-M., Gunga, H.-C., Hummel, J., Mallison, H., Perry, S.F., Preuschoft, H., Rauhut, O.W.M., Remes, K., Tütken, T., Wings, O., Witzel, U.: Biology of the sauropod dinosaurs: the evolution of gigantism. Biol. Rev. **86**, 117–155 (2010)

Sarrafian, S.K., Kazarian, L.E., Toponzian, L.K., Sarrafian, L.K., Siegelman, A.: Strain variation in the components of the extensor apparatus of the finger during flexion and extension. J. Bone Joint Surg. **52**, 980–990 (1970)

Schaller, G.B.: The Mountain Gorilla. University of Chicago Press, Chicago/London (1963)

Schaller, N.U.: Structural attributes contributing to locomotor performance in the ostrich. Doctoral thesis Faculty of Natural Sciences, Frankfurt a.M. (2008)

Schanz, T., Lins, Y., Viefhaus, H., Barciaga, T., Läbe, S., Preuschoft, H., Witzel, U., Sander, M.: Quantitative interpretation of tracks for determination of body mass. PLoS One. **8**, e77606 (2013)

Scherf, H.: Computed tomography in paleoanthropology — an overview. Archaeol. Anthropol. Sci. **5**, 205–214 (2013). https://doi.org/10.1007/s12520-013-0128-5

Scherf, H., Tilgner, R., Hublin, J.J.: Effects of locomotion – morphological differences in humeral cancellous bone of hominids and their relation to habitual loading conditions of the shoulder joint. AAPA Abstracts, p. 229 (2009)

Schilling, N., Carrier, D.: Function of the epaxial muscles during trotting. J. Exp. Biol. **212**, 1053–1063 (2009)

Schilling, D., Preuschoft, H.: Ererbt oder erlernt? Formspezifische merkmale der Gesänge von Gibbons. Verh. Dtsch. Zool. Ges. **77**, 220 (1984)

Schilling, D., Singer, D., Diller, H.: Säugetiere, 181 Arten Europas. BLV-Verlag München. Wien, Zürich (1983)

Schlunk, A.: Das Verhalten von Weidetieren (Pferdd, Rind, Schaf und Ziege) bei der Nahrungsaufnahme. Unpublished masters thesis. Faculty Biology and Biotechnology, RU Bochum (2010)

Schmid, P.: Eine Rekonstruktion des Skelettes von A.L. 288–1 (Hadar) und deren Konsequenzen [A reconstruction of the skeleton of A.L. 288–1 (Hadar) and its consequences]. Folia Primatol. (Basel). **40**, 283 (1983). https://doi.org/10.1159/000156111

Schmid, P.: The trunk of the australopithecines. In: Coppens, Y., Senut, B. (eds.) Origine(s) de la bipédie chez les hominidés, pp. 225–234. CNRS, Paris (1991)

Schmid, P., Piaget, A.: Three-dimensional kinematics of bipedal locomotion. Z. Morphol. Anthropol. **80**, 79–87 (1994)

Schmid, P., Stratil, Z.: Growth changes, variations and sexual dimorphism of the gorilly skull. In: Else, J.G., Lee, P.C. (eds.) Primate Evoluiton, pp. 239–247. Cambridge Univ. Press, Cambridge, UK (1986)

Schmid, P., Churchill, S.E., Nalla, S., Weissen, E., Carlson, K.J., de Ruiter, D.J., Berger, L.: Mosaic morphology in the thorax of Australopithecus sediba. Science. **320**, (2013).: 1234598-1-1234598-5

Schmidt, M.: Zur Evolution der Fortbewegung der Primaten. Dissertation Biologisch-Pharmazeutische Fakultät Jena (2001)

Schmidt, M.: Forelimb proportions and kinematics: how are small primates different from other small mammals? J. Exp. Biol. **211**, 3775–3789 (2008). https://doi.org/10.1242/jeb.019802

Schmidt, M., Fischer, M.: Cineradiographic study of foelimb movements during quadrupedal walking in the brown lemur (*Eulemur fulvus*, Primates, Lemuridae). Am. J. Phys. Anthropol. **111**, 245–262 (2000)

Schmidt, M., Voges, D., Fischer, M.: Shoulder movement during quadrupedal locomotion in arboreal primates. Z. Morphol. Anthropol. **83**, 235–242 (2002)

Schmidt-Kittler N., Vogel, K. (eds.): Constructional morphology and Evolution. Springer-Verlag, Berlin (1991)

Schmidt-Kittler, N., Vogel, K.: Introduction: a new look at morphology. In: Schmidt-Kittler, N., Vogel, K. (eds.) Constructional Morphpology and Evolution, pp. 1–6. Springer, Heidelberg (1991)

Schmitt, D.: Forelimb mechanics during arboreal and terrestrial quadrupedalism in Old World monkeys. In: Strasser, E., Fleagle, J., Rosenberger, A.L., McHenry, H. (eds.) Primate Locomotion, pp. 175–200. Plenum Press, New York/London (1998)

Schmitt, D.: Doctoral thesis Stony Brook on GRFs in primates (2003a)

Schmitt, D.: Mediolateral reaction forces and forelimb anatomy in quadrupedal primates: implications for interpreting locomotor behavior in extinct primates. J. Hum. Evol. **44**, 47–58 (2003b)

Schmittbuhl, M., Le Minor, J.M., Schaaf, A.: Relative position and extent of the nasal and orbital openings in *Gorilla, Pan* and the human species from the study of their areas, and centres of area. Folia Primatol. **67**, 182–192 (1996)

Schneider, E.: Hasenartige In Grzimeks Enzyklopädie Bd. **4**, 254–305 (1988)

Schreyer, C., Ramm, H., Wagner, W.: Praktische Baustatik. Teubner, Stuttgart (1964)

Schultz, A.H.: The skeleton of the trunk and limbs of higher primates. Hum. Biol. **2**, 303–438 (1930)

Schultz, A.H.: Die Körperproportionen der erwachsenen catarrhinen Primaten, mit spezieller Berücksichtigung der Menschenaffen. Anthropol. Anz. **10**, 154–185 (1933)

Schultz, A.H.: Characters common to higher primates and characters specific to man. Q. Rev. Biol. **11**, (1936). 259-283, 425–455

Schultz, A.H.: Postembryonic age changes. In: Hofer, H., Schultz, A.H., Starck, D. (eds.) Primatologia I, Karger-Verlag, Basel, pp. 887–1014 (1956)

Schultz, A.H.: The relative length of the foot skeleton and its main parts in primates. Symp. Zool. Soc. Lond. **10**, 119–206 (1963a)

Schultz, A.H.: Relations between the lengths of the main parts of the foot skeleton in Primates. Folia Primatol. **1**, 150–171 (1963b)

Schultz, A.H.: Form und Funktion der Primatenhände. In: Rensch, A. (ed.) Handgebrauch und Verständigung bei Affen und Frühmenschen, pp. 9–24. Huber, Bern (1968)

Schultz, A.H.: Die Primaten. Editions Rencontre, Lausanne (1971)

Schultze, H.-P.: *Gnathostomata, Kiefermäuler.* In: W. Westheide & R. Rieger: *Spezielle Zoologie Teil 2: Wirbel und Schädeltiere*, 1. Auflage, Spektrum Akademischer Verlag Heidelberg Berlin, ISBN 3-8274-0307-3 (2004)

Schulz, M.H.H.: 1995: Die biomechanische Bedeutung der Papillarleistenmuster auf den Fingerbeeren der Primaten. Z. Morphol. Anthropol. **80**, 223–235 Master thesis, Fakultät für Biologie, RU Bochum, (1993)

Schulze, H., Meier, B.: Behavior of captive Loris tardigradus nordicus: a qualitative description, including some information about morphological bases of behaviour. In: Alterman, L., et al. (eds.) Creatures of the Dark. Plenum Press, New York (1995)

Schwartz, J.H., Tattersall, J.: The human chin revisitet: what is it and who has it? J. Human Evol. **38**, 367–409 (2000)

Schwenk, K.: An introduction to tetrapod feeding. In: Schwenk, K. (ed.) Feeding: Form, Function and Evolution in Tetrapod Vertebrates, pp. 21–61. Academic Press, New York (2000)

Seilacher, A.: Constructional morphology of sand dollars. Palaeobiology. **3**(3), 191–221 (1979)

Seilacher, A.: Self-organizing mechanisms in morphogenesis and evolution. In: Schmidt-Kittler, N., Vogel, K. (eds.) Constructional Morphology and Evolution, pp. 251–271. Springer, Heidelberg/Berlin (1991)

Semaw, S., Renne, P., Harris, J.W.K., Feibel, C., Sbernor, R.L., Fessecha, N., Mowbray, K.: 2.5-million-years-old stone tools from Gona, Ethiopia. Nature. **385**, 333–336 (1997)

Sennwald, G.: Das Handgelenk. Springer, Berlin (1987)

Senut, B., Pickford, M., Gommery, D., Mein, P., Cheboi, K., Coppens, Y.: First hominid from the Miocene (Lukeino formation, Kenya). Comptes Rendus Académie des Sciences Paris, Série IIA Sciences de la Terre et des Planètes, 332, 137–144 (2001)

Shabel, A.: The paleobiology of the robust australopithecines (***Paranthropus***): a test of the durophage model with carbon isotope analysis. J. Vertebr. Paleontol. **25**(supplement to number 3), 114A (2005)

Shipley, L.A., Spalinger, D.E., Gross, J.E., Thompdson-Hobbs, N., Wunder, B.A.: The dynamics and scaling of foraging velocity and encounter rate in mammalian herbivores. Funct. Ecol. **10**, 234–244 (1996)

Shubin, N., Daeschler, E.B., Jenkins, F.A.: The pectoral fin of *Tiktaalik roseae* and the origin of the tetrapod limb. Nature. **440**, 764–771 (2006)

Simons, E.L.: Primate Evolution: an Introduction to Man's Place in Nature. Macmillan, New York (1972)

Simpson, G.G.: The principles of classification and a classification of mammals. Bull. Am. Mus. Nat. Hist. 88–350 (1945)

Simpson, G.G.: Horses. Natural History Library, Anchor Books. Doubleday & Company, Inc. Garden City, and the Am. Mus. Nat. History, New York (1951/1961)

Simpson, S.W., Levin, N.E., Quade, J., Rogers, M.J., Semaw, S.: Ardipithecus ramidus postcrania from the Gona project area, Afar Regional State, Ethiopia. J. Human Evol. **129**, 1–45 (2019)

Sinclair, A.G., McNeil, A.R.: Estimates of forces exerted in the jaw muscles of some reptiles. J. Zool. **213**, 107–115 (1987)

Slijper, E.J.: Die Cetacea, vergleichend – anatomisch und systematisch. Capita Zoologica **7**, 1–600 (1937). Doctoral thesis Utrecht (1936)

Slijper, E.J.: Comparative biologic-anatomical investigations on the vertebral column and spinal musculature of mammals. Verh. Koningl. Nederlandsche Acad. Wetenschappen, Afd. Natuurkd (2) **42**, 1–128 (1946)

Snellius, O.: Die Abhängigkeit des Hirngewichts von dem Körpergewicht und den geistigen Fähigkeiten. Arch. Psychiatrie. **23**, 436–446 (1891)

Snodgrass, I.M., Gilbert, P.W.: A shark-bite meter. In: Gilbert, P.W., Matthewson, R.F., Rall, D.P. (eds.) Sharks, Skates and Rays, pp. 331–340. Johns Hopkins Press, Baltimore (1967)

Sokoloff, S.T.: The muscular anatomy of the chimpanzee foot. Gegenbauers Morphol. Jahrb. **119**, 86–125 (1972)

Sparrow, W.A.: Creeping patterns of human adults and infants. Am. J. Phys. Anthropol. **78**, 387–401 (1989)

Spears, I.R., Crompton, R.H.: Finite element stress analysis as a possible tool for reconstruction of hominid dietary mechanics. Z. Morphol. Anthropol. **80**, 3–17 (1994)

Spinozzi, G., Truppa, V., Lagana, T.: Grasping behavior in tufted capuchin monkeys (*Cebus apella*): grip types and manual laterality for picking up a small food item. Am. J. Phys. Anthropol. **125**, 30–41 (2004)

Sponheimer, M., Lee-Thorpe, J.A.: Isotopic evidence for the diet of an early hominid. Science. **283**, 368–370 (1999)

Sponheimer, M., Lee-Thorp, J., de Ruiter, D., Codron, D., Codron, J., Baugh, A.T., Thackeray, F.: Hominins, sedges, and termites: new carbon isotope data from the Sterkfontein valley and Kruger National Park. J. Hum. Evol. **48**(3), 301–312 (2005)

Sponheimer, M., Alemseged, Z., Cerling, T.E., Grine, F.E., Kimbel, W.H., Leakey, M.G., Lee-Thorp, J.A., Manthi, F.K., Reed, K.E., Wood, B.A., Wynn, J.G.: Isotopic evidence of early hominin diets. Proc. Natl. Acad. Sci. USA. **110**, 10513–11051 (2013)

Starck, D.: Embryologie, 3rd edn. Thieme, Stuttgart (1975)

Starck, D.: Vergleichende Anatomie der Wirbeltiere auf evolutionsbiologischer Grundlage. Volumes 1–3, Springer-Verlag, Berlin (1978–1981)

Stefen, C., Ibe, P., Fischer, M.: Biplanar X-ray motion analysis of the lower jaw movement during incisor interaction and mastication in the beaver (*Castor fiber* L. 1758). Mamm. Biol. **76**, 534–539 (2011). https://doi.org/10.1016/j.mambio.2011.02.001

Stefen, C., Habersetzer, J., Witzel, U.: Biomechanical aspects of incisor action of beavers (*Castor fiber* L.). J. Mammal. **97**(2), 619–630 (2015)

Steiner, R., Rozak, M.: L'anatomie comparée du Dugong.: Memoire d'Université Louis Pasteur, Strasbourg, pp. 1–44 (2001). http://animaldiversity.ummz.umich.edu; http://hom.t-online.de/home/rothausc/dugong.htm

Steitz, E.: Die Evolution des Menschen. 3. aufl., Schweizerbartsche Verlagsbuchhandlung (Nägele und Obermiller), Stuttgart (1993)

Stephan, H.: Quantitative Vergleiche zur phylogenetischen Entwicklung des Gehirns der Primaten mit Hilfe von Progressionsindices. Mitteilungen der Max-Planck-Gesellschaft. **2**, 63–86 (1967)

Stephan, H.: Evolution of primate brains: a comparative anatomical investigation. In: Russell, R. (ed.) The Functional and Evolutionary Biology of Primates, pp. 145–174. Aldine Atherton, Chicago (1972)

Stern, J.: Essentials of Gross Anatomy. Davies Comp, Philadelphia (1988)

Stern, J.T., Susman, R.L.: The locomotor anatomy of *Australopithecus afarensis*. Am. J. Phys. Anthropol. **60**, 279–317 (1983)

Stern, J.T., Susman, R.L.: "Total morphological pattern" versus the "magic trait": conflicting approaches to the study of early hominid bipedalism. In: Coppens, Y., Senut, B. (eds.) Origine(s) de la Bipédie chez les Hominidés, pp. 99–112. CNRS, Paris (1991)

Stern, J.T., Wells, J.P., Jungers, W.L., Vangor, A.K., Fleagle, J.G.: An electrographic study of the pectoralis maior in Atelines and *Hylobates*, with special reference to the evolution of a pars clavicularis. Am. J. Phys. Anthropol. **52**, 13–25 (1980)

Stoessel, A., Fischer, M.S.: Comparative intra-limb coordination in avian bipedal locomotion. J. Exp. Biol. **215**, 4055–4069 (2012)

Storch, V., Welsch, U.: Systematische Zoologie. G. Fischer, Stuttgart/New York (1991)

Strasser, E.: Hindlimb proportions, allometry, and biomechanics in Old World Monkeys (primates, cercopithecidae). Am. J. Phys. Anthropol. **87**, 187–213 (1992)

Streitlein, I., Preuschoft, H.: Kinematic der Trabtempi von Reitpferden. In: Preuschoft, et al. (eds.) Studien zu den Bewegung von Sportpferden, pp. 20–65. Wissenschaftl. Publikationen 9, FN-Verlag, Warendorf (1987)

Strotdrees, E.: Zur Kausalen Morphogenese am Chondrocranium von *Monodelphis domestica*. Med. Diss. Bochum (1998/1997)

Sudhaus, W., Rehfeld, K.: Einführung in die Phylogenetik und Systematik. G. Fischer Verlag, Stuttgart (1992)

Susman, R.: Comparative and functional morphology of hominoid fingers. Am. J. Phys. Anthropol. **50**(2), 215–236 (1979)

Susman, R.: New hominid fossils from Swartkrans formation (1979-1986 excavations): postcranial specimens. Am. J. Phys. Anthropol. **79**, 451–474 (1989)

Susman, R., Stern, J.: Locomotor behavior of early hominids: epistemiology and fossil evidence. In: Coppens, Y., Senut, B. (eds.) Origine(s) de la Bipedie chez les Hominidés, pp. 121–132. CNRS, Paris (1991)

Sverdlova, N., Witzel, U.: Principles of determination and verification of muscle forces in the human musculoskeletal system: muscle forces to minimize bending stress. J. Biomech. **43**, 387–396 (2010)

Sweet, S.S.: Comparative ecology of two small arboreal monitors from Northern Australia. In: Horn, H.-G., Böhme, W., Krebs, U. (eds.) Mertensiella, Advances in Monitor Research Dt. Ges. F. Herpetologie und Terrarienkunde, Rheinbach. Chimaira Buchhandelsgesellschaft, Frankfurt a.M (2007)

Takahashi, Y., Matsumura, A., Kimura, K.: The insertions of the cruropedal muscles and implications for the locomotor evolution in primates. Z. Morphol. Anthropol. **83**, 291–303 (2002)

Taleisnik, J.: The ligaments of the wrist. J. Hand Surg. **1**, 110–118 (1976)

Taleisnik, J.: The Wrist. Churchill Livingstone Publishers, Edinburgh (1988)

Tan, H., Wilson, A.M.: Grip and limb force limits to turning performance in competition horses. Proc. Royal Soc., B. (2011). https://doi.org/10.1098/rspb.2010.2395

Tappen, N.C.: A functional analysis of the facial skeleton with split-line technique. Am. J. Phys. Anthropol. **11**, 503–532 (1958)

Tappen, N.C.: Some relationships between split-line patterns and underlying structure in primate skeletons. In: Starck, D., Schneider, R., Kuhn, H.J. (eds.) Neue Ergebnisse der Primatologie, Progress in Primatology, pp. 80–89. G. Fischer-Verlag, Stuttgart (1967)

Tardieu, C., Preuschoft, H.: Ontogeny of the knee-joint in humans, great apes and fossil hominids: Pelvi-femoral relationships during postnatal growth in humans. Folia Primatol. **66**, 68–81 (1996)

Tattersall, J.: Palaeoanthropology: the last half-century. Evol. Anthropol. **7**, 2–16 (2000)

Tencer, A.F., Viegas, S.F., Cantrell, J., Chang, M., et al.: Pressure distribution in the wrist joint. J. Orthop. Res. **6**, 509–517 (1988)

Thenius, E.: Stammesgeschichte der Säugetiere (einschließlich der Hominiden). In: J.-G. Helmcke, D. Starck, H. Wermuth (eds.) Handbuch der Zoologie, Bd. **8**, 1–722 Walter de Gruyter, Berlin (1969)

Thomason, J.J.: Functional Morphology in Vertebrate Paleontology. Cambridge University Press, Cambridge (1995)

Thomasson, J.J., Russell, A.P., Morgeli, M.: Forces of biting, body size and masticatory muscle tension in the opossum *Didelphis virginianus*. Can. J. Zool. **68**, 318–324 (1990)

Thompson, N.E., Demes, B., O'Neill, M.C., Holowka, N., Larson, S.G.: Surprising trunk rotational capabilities in chimpanzees and implications for bipedal walking proficiency in early hominins. Nat. Commun. **6**, 8416 (2015). https://doi.org/10.1038/ncomms9416, www.nature.com/naturecommunications

Thorpe, S.K., Crompton, R.H.: The locomotor ecology of wild orang-utans (*Pongo pygmaeus abeli*) in the Gunung Leuser ecosystem, Sumatra, Indonesia. Am. J. Phys. Anthropol. **127**(1), 58–78 (2005)

Tinker, S.W.: Whales of the World. Bess Press, Inc., Honolulu (1988)

Toldt, C.: Brauenwülste, Tori supraorbitales, and Brauenbögen, Arcuus superciliares, und ihre mechanische Bedeutung. Mitt Anthropol. Ges. Wien. **44**, 235–315 (1914)

Townsend, H.G.G., Leach, D.H., Fretz, P.B.: Kinematics of the equine thoracolumbar spine. Equine Vet. J. **15**, 117–122 (1983)

Trappmann, E.: EMG-Vorstudie zur biomechanischen Beanspruchung des Schädels. Unpublished Masters Thesis, Fakultät für Biologie, Ruhr-Universität Bochum (1990)

Troelsen, P.V., Wilkinson, D.M., Seddighi, M., Allanson, D.R., Falkingham, P.L.: Functional morphology and hydrodynamics of plesiosaur necks: does size matter? J. Vertebr. Paleontol. **39**(2), e1594850 (2019)

Troje, N.F., Frost, B.J.: Head-bobbing in pigeons: how stable is the hold phase? J. Exp. Biol. **203**, 935–940 (2000)

Tuttle, R.H.: Knuckle-walking and the evolution of the hominoid hands. Am. J. Phys. Anthropol. **26**, 171–206 (1967)

Tuttle, R.H.: Quantitative and functional studies on the hands of the Anthropoidea. The Hominoidea. J. Morphol. **128**, 309–364 (1969a)

Tuttle, R.H.: Knuckle-walking and the problem of human origins. Science. **166**, 953–961 (1969b)

Tuttle, R.H.: Postural, propulsive and prehensile capabilities in the cheiridia of chimpanzees and other great apes. In: Bourne, G. (ed.) The Chimpanzee, vol. 2, pp. 167–253. Karger, Basel (1970)

Tuttle, R.H.: Parallelism, brachiation and hominoid phylogeny. In: Luckett, W.P., Szalay, F.S. (eds.) Phylogeny of the Primates, pp. 447–480. Plenum Press, New York (1975)

Tuttle, R.H.: Kinesiological inferences and evolutionary implications from Laetoli bipedal traits G-1, G2/3 and A. In Leakey, M.D., Harris, J.M. (eds.) Laetoli: A Pliocene site in northern Tansania. Oxford: Clarendon Press, 503–523 (1987)

Tuttle, R.H., Webb, D., Tuttle, N.L.: Laetoli footprint trails and the evolution of bipedalism. In: Coppens, Y., Senut, B. (eds.) Origine(s) de la Bipedie chez les Hominidés, pp. 187–198. CNRS, Paris (1991)

Tuttle, R.H., Webb, D., Baksh, M.: Footprints and gaits of bipedal apes, bears and barefoot people: perspectives on Pliocene tracks. In: Matano, S., Tuttle, R.H., Ishida, H., Goodman, M. (eds.) Topics in Primatology. 3 Evolutionary Biology, Reproductive Endocrinology, and Virology, pp. 221–242. Univ. Tokyo Press, Tokyo (1992)

Ungerechts, B.: Über die Hydrodynamik schnell schwimmender Wirbeltiere. Doctoral Thesis, Biological Faculty, RU Bochum (1980)

Usherwood, J.R., Davies, Z.T.S.: Work minimization accounts for footfall phasing in slow quadrupedal gaits. eLife. **6**, 1–14 (2017). https://doi.org/10.7554/eLife.29495

Van Damme, J., Aerts, P.: Kinematics and functional morphology of aquatic feeding in Australian snake-necked turtles (Pleurodira, Chelodina). J. Morphol. **233**, 113–125 (1997)

van der Merwe, J., Thackeray, F., Lee-Thorp, J.A., Luyt, J.: The carbon isotope ecology and diet of *Australopithecus africanus* at Sterkfontein, South Africa. J. Hum. Evol. **44**(2003), 581–597 (2003)

Van der Sluijs, L., Gerken, M., Preuschoft, H.: Comparative analysis of walking gaits in South American Camelids. J. Zool. (2010). https://doi.org/10.1007/s10764-010-9399-1

van Leeuwen, J., Muller, M.: Optimum sucking techniques for predatory fish. Trans. Royal Soc. London. **37**, 137–169 (1984)

van Leuwen, J.L., Muller, M.: Prey capture in fish. In: Duncker, H.R., Fleischer, G., (eds.) Functional Morphologie in Vertebrates. Sonderband. Fortschr. Zool. **30**, 229–232 (1985)

Van Wassenbergh, S.: Transitions from water to land: terrestrial feeding in fishes. In: Sasisekharan, R., Lee, S.L., Rosenberg, A., Walker, L.A. (eds.) The Science and Regulations of Naturally Derived Complex Drugs. Springer (2019). https://doi.org/10.1007/978-3-030-13739-7_5

Van Wassenbergh, S., Aerts, P., Herrel, A.: Hydrodynamic modelling of aquatic suction performance and intra-oral pressures: limitations for comparative studies. J. R. Soc. Interface. **3**, 507–514 (2006a). https://doi.org/10.1098/rsif.2005.0110

Van Wassenbergh, S., et al.: A catfish that can strike its prey on land. Nature. **440**, 381 (2006b)

Vassilieva, A.B., Poyarkov, N.A., Iizuka, K.: Pecularities of bony skeleton development in Asian clawed salamanders (*Onychodactylus*, Hynobiidae) related to embryonisation. Biol. Bull. **40**(7), 589–599 (2013)

Venkataramana, V.V., Krafta, T.S., Dominya, N.J.: Tree climbing and human evolution. Proc. Natl Acad. Sci. **110**(4), 1237–1242

Vereecke, E., D'Août, K., De Clercq, D., Van Elsacker, L., Aerts, P.: Dynamic plantar pressure distribution during terrestrial locomotion of bonobos (*Pan paniscus*). Am. J. Phys. Anthropol. **120**(4), 373–383 (2003)

Vereecke, E., D'Août, K., Van Elsacker, L., De Clercq, D., Aerts, P.: Functional analysis of the gibbon foot during terrestrial bipedal walking: plantar pressure distributions and three-dimensional ground reaction forces. Am. J. Phys. Anthropol. **128**(3), 659–669 (2005)

Viegas, S.F., Tencer, A.F., Cantrell, J.: Load transfer characteristics of the wrist, Part 1 and 2. J. Hand Surg. **12**, 971–985 (1987)

Viegas, S.F., Patterson, R., Peterson, P.: The effects of various load paths and different loads on the load transfer characteristics of the wrist. J. Hand Surg. **14**, 458–465 (1989)

Vilenski, J.A.: Primate quadrupedalism: how and why does it differ from that of typical quadrupeds? Brain Behav. Evol. **34**, 357–364 (1989)

Vilenski, J.A., Moore, A.M.: Utilisation of lateral- and diagonal sequence gaits at identical speeds by individual vervet monkeys. In: Matano, S., Tuttle, R.H., Ishida, H., Goodman, M. (eds.) Topics in Primatology. 3 Evolutionary Biology, Reproductive Endocrinology, and Virology, pp. 129–137. Univ. Tokyo Press, Tokyo (1992)

Visalberghi, E., Trinca, L.: Tool use in capuchin monkeys: distinguishing between performing and understanding. Primates. **30**(4), 511–521 (1989)

Visalberghi, E., et al.: Performance in a tool-using task by common chimpanzees (*Pan troglodytes*), bonobos (*Pan paniscus*), an orangutan (*Pongo pygmaeus*), and capuchin monkeys (*Cebus apella*). J. Comp. Psychol. **109**, 52–60 (1995)

Voges, D.: Roentgenkinematographische Untersuchungewn der Lokomotion von Saguinus Oedipus (Linné 1758), Callithricidae, Primates. Masters Thesis Fakultät für Biologie, Universität, Jena (1999)

Voigt, S.: Die Tetrapodenichnofauna des kontinentalen Oberkarbon und Perm im Thüringer Wald - Ichnotaxonomie, Paläoökologie und Biostratigraphie. Cuvillier Verlag, Göttingen (2005). 305 pp

Voigt, S.: Tetrapodenfährten. Schriftenreihe der Deutschen Gesellschaft für Geowissenschaften. **61**, 92–106 (2012)

Voigt, S., Ganzelewski, M.: Towards the origin of amniotes: Diadectomorph and synapsid footprints in the early Carboniferous of Germany. Acta Palaeontol. Pol. **55**(1), 57–72 (2010). https://doi.org/10.4202/app.2009.0021

Voigt, S., Haubold, H.: Analyse zur Variabilität der Tetrapodenfährte *Ichniotherium cottae* aus dem Tambacher Sandstein (Rotliegend, Unterperm, Thüringen). Hallesches Jahrbuch für Geowissenschaften B. **22**, 17–58 (2000)

Voigt, C., Berman, D.S., Henrici, A.C.: First well-established track-trackmaker association of Palaeozoic tetrapods based on *Ichniotherium* trackways and diadectid skeletons from the lower Permian of Germany. J. Vertebr. Paleontol. **27**(3), 553–570 (2007)

Voigt, S., Niedźwiedzki, G., Raczyński, P., Mastalerz, K., Ptaszyński, T.: Early Permian tetrapod ichnofauna from the intra-Sudetic Basin, SW Poland. Palaeogeogr. Palaeoclimatol. Palaeoecol. **313-314**, 173–180 (2012)

von Koenigswald, W.: Schmelzmuster und Morphologie in den Molaren von Arvicolidae (Rodentia). Abhdl. Senckenberg. Naturforsch. Ges. **539**, 1–129 (1980)

von Koenigswald, W.: Enamel structure in molars of arvicolidae (Rodentia, Mammalia), a key to functional morphology and phylogeny. In: Kurtén, B. (ed.) Teeth: Form, Function and Evolution, pp. 109–122. Columbia Univ. Press, New York (1982)

von Koenigswald, W.: Enamel modification in enlarged front teeth among mammals and the various possible reinforcement of the enamel. In: Russel, D.E., Santoro, J.P., Sigogneau-Russell, D. (eds.) Teeth Revisited. Proc. 7 the Internat. Symp. Dental Morphol. Paris 1986 Mem. Musée National d`Histoire Naturelle ©, vol. 53, pp. 148–165 (1988) von Koenigswald, W.: Ein ungewöhnliches Schmelzmuster in den Schneidezähnen von *Marmota* (Rodentia, Mammalia). N. Jb. Geol. Paläont. **180**, 53–73 (1990)

von Koenigswald W., Martin T., Billet G.: Enamel microstructure and mastication in *Pyrotherium romeroi* (Pyrotheria, Mammalia) Paläontologische Zeitschrift, **89(3)**, 593–609 (2015). https://doi.org/10.1007/s12542-014-0241-5

von Koenigswald, W., Pfretzschmer, H.U.: Biomechanics in the enamel of mammalian teeth. In: Schmidt-Kittler, N., Vogel, K. (eds.) Constructional Morphology and Evolution, pp. 113–125. Springer, Heidelberg (1991)

von Koenigswald, W., Sander, M. (eds.): Tooth Enamel Microstructure - Balkema, Rotterdam (1997)

von Koenigswald, W., Habersetzer, J., Gingerich, P.D.: Morphology and evolution of the distal phalanges in primates. In: Lehmann, T., Schaal, S.F.K. (eds.) The World at the Time of Messel, Conference Volume, pp. 91–94 (Abstract) (2011)

von Koenigswald, W., Habersetzer, J., Gingerich, P.D.: Pedal distal phalanges of the Eocene adapoids *Europolemur* and *Darwinius* compared to phalanges of *Notharctus* and other primates. – Paleobiology and Paleoenvironment. **92**, 539–565. Special issue Messel and the terrestrial Eocene ed. Lehmann, T & Schaal, SFK. https://doi.org/10.1007/s12549-012-0096-2 (2012)

Von Praun, F.H.A.: Zur Biomechanik der Lokomotion von *Bufo marinus*. Unpublished Masters Thesis, Fakultät für Biologie, Ruhr-Universität Bochum (1997)

Wainwright, S.A., Biggs, W.D., Currey, J.D., Gosline, J.M.: Mechanical Design in Organisms. E. Arnold, London (1976)

Walker, A., Leakey, R. (eds.): The Nariokotome *Homo Erectus* Skeleton. Springer, Berlin/New York (1993)

Wang, W.J., Crompton, R.H.: The role of load-carrying in the evolution of modern body proportions. J. Anat. **204**(5), 417–430 (2004)

Warsinger, D.: A waving sail can explain Sauropod dinosaurs heat tolerance and long tails – *review of a yet unpublished article* (2018)

Waterman, P.G.: Food acquisition and processing as a function of plant chemistry. In: Chivers, D.J., Wood, B.A., Bilsborough, A. (eds.) Food Acquisition and Processing in Primates, pp. 177–214. Plenum Press, New York/London (1984)

Wayne, R.K.: Developmental constants on limb growth in domestic and some wild canids. J. Zool. **210**, 381–399 (1986a)

Wayne, R.K.: Limb morphology in domestic and wild canids.: The influence of development on morphological change. J. Morphol. **187**, 301–319 (1986b)

Webb, P.W.: Hydrodynamics and energetics of fish propulsion. Bull. Fish. Res. Board Canada. **190**, 1–159 (1975)

Weissengruber, G.E., Egger, G.F., Hutchinson, J.R., Roenewald, H.B., Elsasser, I., Famini, D., Forstenpointner, G.: The structure of the cushions in the feet of African elephants (*Loxodonta africana*). J. Anat. **209**, 781–792 (2006). https://doi.org/10.1111/j.1469-7580.2006.00648.x

Westheide, W., Rieger, R. (eds.): Spezielle Zoologie. Bd 2. Wirbel- oder Schädeltiere, p. 182. G. Fischer, Stuttgart. ISBN 3-437-20515 (2004)

Westneat, M.W.: Skull biomechanics and suction feeding in fishes. In: Shadwick and Lauder (eds.), Fish Physiology, Fish Biomechanics, 23: Chapter 2: Elsevier (2006)

Wheeler, P.E.: The evolution of bipedality and loss of functional body hair in hominids. J. Hum. Evol. **13**, 8–91 (1984)

Wheeler, P.E.: The loss of functional body hair in man: the influence of thermal environment, body form and bipedality. J. Hum. Evol. **14**, 8–23 (1985)

Wheeler, P.E.: Body hair reduction and tract orientation in man: hydrodynamics or thermoregulatory aerodynamics? In: Roede, M., Wind, J., Patrick, J., Reynolds, V. (eds.) The Aquatic Ape: Fact or Fiction? , pp. 221–236. Souvenir Press LTD, London (1991)

Whipple, I.L.: The ventral surface of the mammalian cheiridium, with special reference to the condition found in man. Zeitschrift für Morphologie und Anthroplogie. **7**, 261–368 (1904)

White, T.D., Suwa, G., Asfah, B.: *Australopithecus ramidus*, a new species of early hominid from Aramis, Ethiopia. Nature. **371**, 306–312 (1994)

White, T.D., Asfaw, B., Beyene, Y., Haile-Selassie, Y., Lovejoy, C.O., Suwa, G., WoldeGabriel, G.: *Ardipithecus ramidus* and the paleobiology of early hominids. Science. **326**(5949), 75–86 (2009). https://doi.org/10.1126/science.1175802. PMID 19810190

Williams, S.A., Ostrofsky, K.R., Frater, N., Churchill, S.E., Schmid, P., Berger, L.R.: The vertebral column of *Australopithecus sediba*. Science. **340**, (2013).: 1232996–1 through 1232996–5

Witte, H.: Beiträge zur funktionellen Anatomie du Biomechanik elastischer Elemente im Bewegungsapparat. Habil-Schrift, Med. Fakultät RU Bochum (1996)

Witte, H., Preuschoft, H., Recknagel, S.: Human body proportions on the basis of biomechanical principles. Z. Morphol. Anthropol. **78**, 407–423 (1991)

Witte, H., Lesch, C., Preuschoft, H., Loitsch, C.: Die Gangarten der Pferde: Sind Schwingungsmechanismen entscheidend? Teil I: Pendelschwingungen der Beine bestimmen den Schritt. Pferdeheilkunde. **11**, 199–206 (1995a)

Witte, H., Lesch, C., Preuschoft, H., Loitsch, C.: Die Gangarten der Pferde: Sind Schwingungsmechanismen entscheidend? Teil II: Federschwingungen bestimmen den Trab und den Galopp. Pferdeheilkunde. **11**, 265–272 (1995b)

Witte, H., Recknagel, S., Lesch, C., Rao, J.G., Preuschoft, H.: Is elastic storage of quantitative relevance for the functional morphology of the human locomotor apparatus? Acta Anat. **158**, 106–111 (1997)

Witte, H., Preuschoft, H., Fischer, M.S.: The importance of the evolutionary heritage of locomotion on flat ground in small mammals for the development of arboreality. Z. Morphol. Anthropol. **83** (2/3), 221–233 (2002)

Witte, H., Fischer, M., Schmidt, M., Gruber, S., Ludwig, O., Hackert, R., Schilling, N., Voges, D., Hoffmann, H., Preuschoft, H.: Human bipedality: mechanical preconditions and morphological adaptations. Courier Forsch.-Inst. Senckenberg. **243**, 25–33 (2003)

Witte, H., Fischer, M. S., Preuschoft, H., Voges, D., Schilling, C., Ijspeert, A. J.: Quadruped locomotion. Chapter 31 in "Living machines". Oxford University Press, Oxford, UK, pp. 289–303 (2017)

Witzel, U.: Über die Spannungsverteilung in der Caövaria aufgrund physioloi´gischer und traumatischer Bescheunigungen des Schädels. Osteologie 3 (Supplement 1): 50 (1994)

Witzel, U., Hofmann, P.: Über den Einflusss der durch Beschleunigungskräfte belasteten dura mater encephali auf die Spannungsverteilung in der Calvaria. Osteologie 2 (Supplement 1): 37 (1993)

Witzel, U.: Virtual synthesis of the skull in Neanderthals by FESS. In: Condemi, S., Weniger, G.-C. (eds.) Continuity and Discontinuity in the Peopling of Europe: One Hundred Fifty Years of Neanderthal Study, Vertebrate Paleobiology and Paleoanthropology, pp. 203–211. Springer Science+Business Media B.V, Berlin (2011)

Witzel, U., Preuschoft, H.: The bony roof of the nose in humans and other primates. Zool. Anz. **238**, 103–115 (1999)

Witzel, U., Preuschoft, H.: The functional shape of the human skull, as documented by three-dimensional FE-studies. Anthropol. Anz. **60**, 113–135 (2002)

Witzel, U., Preuschoft, H.: Finite-Element model construction for the virtual synthesis of the skulls in vertebrates: case study of *Diplodocus*. – Anatomical Record, Special Issue: Finite element Analysis in Vertebrate Biomechanics. Ed. C. Ross. pp. 391–401 (2005)

Witzel, U., Preuschoft, H.: The mechanical reasons of the internal structure of finger tips. – 22nd IPS-Kongress in Edinburgh, 2008. (Abstract 572; Poster) (2008)

Witzel, U., Jakowski, J., Jöhren, P.: Analyse der Spannungsverteilung im spongiösen Sinus-Augmentat nach sekundärer Implantation mit Hilfe der Methode der finiten Elemente (FE). Target. **1**, 10–11. Straumann, (1998)

Witzel, U., Preuschoft, H., Sick, H.: The role of the zygomatic arch in the statics of the skull and its adaptive shape. Folia Primatol. **75**, 202–218 (2004)

Witzel, U., Mannhardt, J., Goessling, R., de Micheli, P., Preuschoft, H.: Finite element analysis and virtual synthesis of biological structures and their application to sauropod skulls. In: Klein, N.,

Remes, K., Gee, C.T., Sander, P.M. (eds.) Biology of the Sauropod Dinosaurs, pp. 171–181. Indiana Univ. Press, Bloomington (2011)

Witzmann, F., Schoch, R. R.: Feeding behaviour in the Triassic plagiosaurid *Gerrothorax* – a case study of aquatic feeding in temnospondyls. Centennary Meeting of Paläontologische Gesellschaft, Berlin. Terra Nostra Geo-Union Alfred Wegener-Stiftung, Berlin, pp. 261–262 Abstract (2012)

Wolff, J.E.A.: Die Funktionelle Gestalt der Menschlichen Unterkiefersymphyse. Minerva Press, München (1982)

Wolff, J.E.A.: A theoretical approach to the chin problem. In: Chivers, D.J., Wood, B.A., Bilsborough, A. (eds.) Food Acquisition and Processing in Primates, pp. 391–405. Plenum Press, London (1984)

Wolff, J.E.A.: Mechanik des Unterkiefers mit beweglicher Symphyse. Z. Morphol. Anthropol. **76**, 25–35 (1985)

Wrangham, R., Holland Jones, J., Laden, G., Pilbeam, D., Conklyn-Brittain, N.L.: The raw and the stolen. Cooking and the ecology or human origins. Curr. Anthropol. **40**, 567–594 (1999)

Wroe, S., Huber, D.R., Lowry, M., McHenry, C., Moran, K., Clausen, P., Ferrara, T.L., Cunningham, E., Dean, M.N., Summers, A.P.: Three dimensional computer analysis of white shark jaw mechanics: how hard can a great white bite? J. Zool. **276**, 336–342 (2008)

Yamasaki, N.: Dynamical analysis of bipedal walking by computer simulation. Jpn. J. Ergonomics. **11**, 105–110 (1975). (in Japanese)

Yamasaki, N.: Primate bipedal walking: computer simulation. In: Kondo, S. (ed.) Primate Morphophysiology, Locomotor Analyses and Human Bipedalism, pp. 105–130. Univ. Tokyo Press, Tokyo (1985)

Yamasaki, N.: The effects of gravity on the interrelationship between body proportions and brachiation in the gibbon. Hum. Evol. **5–6**, 543 (1990)

Yamasaki, N.: Biomechanical interrelationship among body proportion, posture and bipedal walking. In: Matano, S., Tuttle, R.H., Ishida, H., Goodman, M. (eds.) Topics in Primatology, vol. 3, pp. 243–257. Evolutionary Biology, Reproductive Endocrinology, and Virology. Univ. Tokyo Press, Tokyo (1992)

Yamasaki, N., Ishida, H., Kimura, T., Okada, M.: Biomechanical analysis of primate bipedal walking by computer simulation. J. Hum. Evol. **8**, 337–349 (1979). https://doi.org/10.1016/00472484(79)90057-5

Yamasaki, N., Ishida, H., Okada, M., Kondo, S.: Biomechanical evaluation and evolutionary models for prehabitual bipedalism. Ann. Sciences Nat. Zool. **13**, 159–168 (1983)

Youlatos, D., Gasc, J.-P.: Critical foraging patterns: head—first vertical descent in the Red Howler Monkey (*Alouatta seniculus*). Z. Morphol. Anthropol. **80**, 65–77 (1994)

Young R. W.: Human Origins and Evolution. R.W. Young, Los Angeles, CA ISBN- 13:978-1494367152 (2013)

Young, I.S., Alexander, R.M.N., Woakes, A.J., Butler, P.J., Anderson, L.: The synchronisation of ventilation and locomotion in horses (*Equus caballus*). J. Exp. Biol. **166**, 19–31 (1992)

Zachos, F.E. Title page, Mammalian Biology/Zeitschrift für Säugetierkunde **77**(6), ISSN 1616-5047 (2012)

Zhu, M., Yu, X., Ahlberg, P.E., Choo, B., Lu, J., Quiao, T., Qu, Q., Zhao, W., Jia, L., Blom, H., Zhu, Y.: A Silurian placoderm with osteichthyan-like marginal jaw bones. Nature. **502**, 188–193 (2013)

Index

H. Preuschoft, *Understanding Body Shapes of Animals*,
https://doi.org/10.1007/978-3-030-27668-3

B

C

D

M